HANDBOOK OF STANDARDS AND GUIDELINES IN HUMAN FACTORS AND ERGONOMICS

Human Factors and Ergonomics
Series Editors

Waldemar Karwowski
University of Central Florida, Orlando, USA

Pamela McCauley
University of Central Florida, Orlando, USA

HANDBOOK OF STANDARDS AND GUIDELINES IN HUMAN FACTORS AND ERGONOMICS

Second Edition

Edited by
Waldemar Karwowski, Anna Szopa,
and Marcelo M. Soares

CRC Press
Taylor & Francis Group
Boca Raton London New York

CRC Press is an imprint of the
Taylor & Francis Group, an **informa** business

2nd edition published 2021
by CRC Press
6000 Broken Sound Parkway NW, Suite 300, Boca Raton, FL 33487-2742

and by CRC Press
2 Park Square, Milton Park, Abingdon, Oxon, OX14 4RN

© 2021 Taylor & Francis Group, LLC

First edition published by CRC Press 2005

CRC Press is an imprint of Taylor & Francis Group, LLC

Library of Congress Cataloging-in-Publication Data
Names: Karwowski, Waldemar, 1953- editor. | Szopa, Anna, 1982- editor. |
Soares, Marcelo Marcio, editor.
Title: Handbook of standards and guidelines in human factors and ergonomics/edited by Waldemar
Karwowski, Anna Szopa & Marcelo Soares.
Description: Second edition. | Boca Raton : CRC Press, 2021. | Series: Human factors and ergonomics |
Includes bibliographical references and index.
Identifiers: LCCN 2020056179 (print) | LCCN 2020056180 (ebook) | ISBN 9781466594524 (hardback) |
ISBN 9780429169243 (ebook)
Subjects: LCSH: Human engineering.
Classification: LCC TA166 .H2775 2021 (print) | LCC TA166 (ebook) | DDC 620.8/20218--dc23
LC record available at https://lccn.loc.gov/2020056179
LC ebook record available at https://lccn.loc.gov/2020056180

ISBN: 978-1-4665-9452-4 (hbk)
ISBN: 978-1-032-01010-6 (pbk)
ISBN: 978-0-429-16924-3 (ebk)

Typeset in Times
by MPS Limited, Dehradun

Contents

SECTION I Standardization Efforts in Human Factors and Ergonomics

SECTION II Nature of HF/E Standards and Guidelines

SECTION III   Standards for Evaluation of Working Postures

SECTION IV   Standards for Manual Material Handling Tasks

SECTION V Standards for Human-Computer Interaction

SECTION VI Management of Occupational Safety and Health

SECTION VII Safety and Legal Protection Standards

SECTION VIII   Military Human Factors Standards

SECTION IX   Sources of Human Factors and Ergonomics Standards

Preface

The past decade has seen an increase in the application of human factors and ergonomics (HFE) and safety management; therefore, standards and guidelines in this area have been playing a more important role more than ever before. Recent developments facilitate the design of optimal working conditions regarding human safety, health, and system performance. Standardization efforts help recognize the significance of human factors and ergonomics discipline by focusing on the requirements that need to be considered during the design, development, testing, and evaluation of workplaces and systems. Standards and guidelines are developed through a process that aims to ensure that all interested and potentially affected parties—including representatives of industrial and commercial entities, government agencies, professional and consumer associations, and the public—have an opportunity to represent their interest and participate in the development process.

This handbook of standards and guidelines offers guidance on the design of work systems in relation to human capacities and limitations and represents the best available knowledge and practices—both in their explicit and tacit forms that can be used in system design, testing, and evaluation processes. Although the application of relevant HFE standards and guidelines by itself cannot always guarantee optimal workplace design, it can provide clear and well-defined requirements for the ergonomics design process.

Handbook on Standards in Ergonomics, Human Factors and Safety Management, second edition has been revised, updated and new chapters were included to cover new areas. The handbook offers a comprehensive review of and knowledge about the selected international and national standards and guidelines in ergonomics and human factors. The handbook consists of 43 chapters and is divided into nine sections:

1. Standardization Efforts in Human Factors and Ergonomics
2. Nature of HF/E Standards and Guidelines
3. Standards for Evaluation Working Postures
4. Standards for Manual Material Handling Task
5. Standards for Human-Computer Interaction
6. Management of Occupational Safety and Health
7. Safety and Legal Protection Standards
8. Military Human Factors Standards
9. Sources of Human Factors and Ergonomics Standards

The presented standards and guidelines—for a complex approach in assessing and designing the interaction between people and the systems with which they relate—are a source of knowledge for academics, professionals, consultants, and practitioners from a great variety of interrelated fields, including engineering, safety, health, manufacturing, product design, quality, and environment.

Editors
Waldemar Karwowski
Anna Szopa
Marcelo Soares

Editor Biographies

Waldemar Karwowski is a Pegasus Professor and Chairman of the Department of Industrial Engineering and Management Systems at the University of Central Florida, USA. He holds an M.S. in Production Engineering and Management from the University of Technology Wroclaw, Poland, and a Ph.D. in Industrial Engineering from Texas Tech University, USA. He was awarded D.Sc. in Management Science by the Institute for Organization and Management in Industry, Warsaw, and received the National Professorship title from the President of Poland (2012). Three Central European universities also awarded him Doctor Honoris Causa degrees. Dr. Karwowski served on the Board on Human Systems Integration, National Research Council, USA (2007–2011). He is currently a Co-Editor-in-Chief of the *Theoretical Issues in Ergonomics Science* Journal, the Editor-in-Chief of the *Human-Intelligent Systems Integration* Journal, and the Field Chief Editor of the *Frontiers in Neuroergonomics* Journal. Dr. Karwowski has over 550 research publications, including over 200 journal papers focused on ergonomics and safety, human performance, neuro-fuzzy systems, non-linear dynamics, human-centered AI, and neuroergonomics. He is a Fellow of the Ergonomics Society (currently the Chartered Institute of Ergonomics and Human Factors (UK)), Human Factors and Ergonomics Society (HFES), Institute of Industrial and Systems Engineers (IISE), and the International Ergonomics Association (IEA), and served as President of both HFES (2007) and IEA (2000–2003). He received the William Floyd Award from the Chartered Institute of Ergonomics & Human Factors, United Kingdom in 2017, and the David F. Baker Distinguished Research Award, Institute of Industrial and Systems Engineers in Atlanta, USA in 2020. He was also inducted into the Academy of Science, Engineering & Medicine of Florida (ASEMFL) at its Inaugural Annual Meeting in November 2020.

Anna Szopa got her M.S. in quality and technology management from the Silesian University of Technology in Poland in 2006. In 2012, she completed her Ph.D. in Management at Jagiellonian University in Krakow, Poland, where she later served as an assistant professor and an associate professor. She has a broad range of teaching experience—from extensive tutoring and individual mentoring to designing and teaching business, economics, entrepreneurship, innovation, and technology transfer courses. She served as a visiting scholar at the University of Maryland (USA), University of Central Florida (USA), University of Haifa (Israel), and Technical University of Prague (Czech Republic). Her primary research interests include university-industry relationships and university spin-off management. Her current research centers on testing and transferring accessible technologies to public organizations in the United States. She has been applying basic principles to the design, development, and management of information to meet diverse users' needs. She can assess the impact of existing or emerging technologies, employ state-of-the-art tools and techniques to create, manage, and analyze information, and demonstrate an understanding of critical issues—including the privacy, authenticity, and integrity of information. Dr. Szopa is a co-editor of books about entrepreneurship, innovations, and technology transfer. She is a member of several international scientific committees. She also regularly conducts peer reviews for journals and conferences, including *Theoretical Issues in Ergonomics Science*, *Review of Managerial Science*, International Conference on Applied Human Factors and Ergonomics, and Asia-Pacific UIIN Conference on University-Industry Engagement.

Marcelo Soares is currently a Full Professor at the School of Design, Hunan University, China, selected for this post under the 1000 Talents Plan of the Chinese Government. He is also a licensed Full Professor of the Department of Design at the Federal University of Pernambuco, Brazil. He holds an M.S. (1990) in Industrial Engineering from the Federal University of Rio de Janeiro, Brazil, and a Ph.D. from Loughborough University, England. He was a post-doctoral fellow at the Industrial Engineering and Management System Department, University of Central Florida. He served as an invited lecturer at the University of Guadalajara, Mexico, University of Central Florida, USA, and the Technical University of Lisbon, Portugal. Dr. Soares is Professional

Certified Ergonomist from the Brazilian Ergonomics Association, which he was president for seven years. He has provided leadership in ergonomics in Latin American and in the world as a member of the Executive Committee of the International Ergonomics Association (IEA). Dr. Soares served as Chairman of IEA 2012 (the Triennial Congresses of the International Ergonomics Association), held in Brazil. Professor Soares is currently a member of the editorial boards of Theoretical Issues in Ergonomics Science, Human Factors and Ergonomics in Manufacturing, and several journal publications in Brazil. He has about 50 papers published in journals, over 190 conference proceedings papers, and 110 books and book chapters. He has undertaken research and consultancy work for several companies in Brazil. Prof. Soares is co-editor of the "Handbook of Human Factors and Ergonomics in Consumer Product Design" and the "Handbook of Usability and User-Interfaces (UX)" published by CRC Press. His research, teaching, and consulting activities focus on manufacturing ergonomics, usability engineering, consumer product design, information ergonomics, and applications of virtual reality and neuroscience in products and systems. He also studies user emotions when using products and techniques in real and virtual environments based on biofeedback (electroencephalography and infrared thermography).

Contributors

Thomas J. Albin
Auburn Engineers Inc.
Auburn, AL, USA

Heiner Bubb
Department of Ergonomics Technical University of Munich
Boltzmannstrasse, Garching, Germany

Charles A. Cacha
Ergonix, Inc.
USA

Gustav Caffier
Federal Institute for Occupational Safety and Health
Berlin, Germany

Gerald Chaikin
US Army Hunan Engineering Laboratory Detachment Redstone Arsenal

Denis A. Coelho
Department of Electromechanical Engineering University of Beira Interior
Calçada Fonte do Lameiro, Covilhã, Portugal

Daniela Colombini
Research Unit "Ergonomics of Posture and Movement" (EPM)
Don Gnocchi Foundation
Milano, Italy

Nico J. Delleman
TNO Human Factors
Paris Descartes University
Paris, France

Karine Borges de Oliveira
Production Department, Engineering School of Guaratinguetá
Sao Paulo State University
Sao Paulo, Brazil

Henk J. de Vries
Erasmus University Rotterdam
Rotterdam, The Netherlands

Carlos Mauricio Duque dos Santos
DCA Ergonomics and Design
Brazil

Eduardo Ferro dos Santos
Department of Basic and Environmental Sciences, Engineering School of Lorena
University of Sao Paulo
Sao Paulo, Brazil

Jan Dul
Erasmus University Rotterdam
Rotterdam, The Netherlands

Anne Ferguson
British Standards Institution
London, United Kingdom

João N. O. Filipe
Department of Electromechanical Engineering
University of Beira Interior
Calçada Fonte do Lameiro, Covilhã, Portugal

Kaj Frick
National Institute for Working Life
Stockholm, Sweden

Anna Gembalska-Kwiecień
Silesian University of Technology

David A. Graeber
The Boeing Company
USA

Sandra Haydeé Mejías Herrera
Department of Industrial Engineering
Central University "Marta Abreu" of Las Villas
Santa Clara, Cuba

Sadao Horino
Kanagawa University
Yokohama, Japan

Joanna Jablonska
Faculty of Architecture
Wroclaw University of Science and Technology
Wroclaw, Poland

Alexis Jacoby
University of Antwerp, Belgium

Daniel P. Jenkins
DCA Design International
Warwick, United Kingdom

Waldemar Karwowski
University of Central Florida
Orlando, FL, USA

Robert S. Kennedy
RSK Assessments, Inc.
Orlando, FL, USA

Omar Kheir
University of Antwerp, Belgium

Kevin Leyva
University of Central Florida
Orlando, FL, USA

Falk Liebers
Federal Institute for Occupational Safety and
    Health
Berlin, Germany

Enrique Herrera Lugo
University of Guadalajara
Guadalajara (UDG), Mexico

João C. O. Matias
Department of Electromechanical Engineering
University of Beira Interior
Calçada Fonte do Lameiro, Covilhã, Portugal

Joe W. McDaniel
Air Force Research Lab
Ohio, United States

Atsuo Murata
Department of Intelligent Mechanical Systems
Okayama University
Tsushimanaka, Okayama, Japan

Anjali Nag
Former Scientist
National Institute of Occupational Health (Indian
    Council of Medical Research), Ahmedabad

India School of Environment & Disaster
    Management Ramakrishna Mission
    Vivekananda University
Narendrapur, Kolkata, India

P. K. Nag
School of Environment & Disaster Management
    Ramakrishna Mission Vivekananda University
Narendrapur, Kolkata, India

Masatoshi Nomura
NEC Corporation
Tokyo, Japan

Enrico Occhipinti
Research Unit "Ergonomics of Posture and
    Movement" (EPM)
Don Gnocchi Foundation
Milano, Italy

James M. Oglesby
University of Central Florida
Orlando, FL, USA

Akira Okada
Osaka City University
Osaka, Japan

Daniel Podgórski
Central Institute for Labour Protection
National Research Institute
Poland

Lilia R. Prado-León
University of Guadalajara
Guadalajara (UDG), Mexico

Robert W. Proctor
Department of Psychological Sciences
Purdue University
West Lafayette, USA

Pei-Luen Patrick Rau
Department of Industrial Engineering
Tsinghua University
Beijing, China

Anke Rissiek
Avalution GmbH
Europaallee, Kaiserslautern, Germany

Sohsuke Saitoh
Human Factor Co., Ltd.
Japan

Eduardo Salas
Department of Psychology, and Institute for
Simulation and Training
University of Central Florida
Orlando, FL, USA

Andréa Ferreira Santos
DCA Ergonomics and Design
Brazil

Rosangela Ferreira Santos
DCA Ergonomics and Design
Brazil

Peter Schaefer
München University of Technology
Munich, Germany

Karlheinz G. Schaub
Darmstadt University of Technology
Darmstadt, Germany

Andreas Seid
Human Solutions GmbH,
Kaiserslautern, Germany

Thomas J. Smith
Research Associate, Human Factors
School of Kinesiology University of
Minnesota Minneapolis, MN, USA

Kay M. Stanney
University of Central Florida
Orlando, FL, USA

Ulf Steinberg
Federal Institute for Occupational Safety and
Health
Berlin, Germany

Tom Stewart
System Concepts, Ltd.
London, United Kingdom

Kimberly Stowers
University of Central Florida
Orlando, FL, USA

Carol Stuart-Buttle
Stuart-Buttle Ergonomics
Philadelphia, PA, USA

Marcelo M. Soares
School of Design, Hunan University, China and
Department of Design
Federal University of Pernambuco
Recife, Brazil

Anna Szopa
Jagiellonian University
University of Central Florida

Rainer Trieb
Human Solutions GmbH
Kaiserslautern, Germany.

Elzbieta Trocka-Leszczynska
Faculty of Architecture
Wroclaw University of Science and Technology
Ul. B. Prusa, Wroclaw, Poland

Pete Underwood
DCA Design International
Warwick, United Kingdom

Henk F. van der Molen
Arbouw Foundation
Amsterdam, the Netherlands

Stijn Verwulgen
University of Antwerp

Kim-Phuong L. Vu
Department of Psychology
California State University Long Beach
Bellflower Blvd, Long Beach, USA

H. Willemse
Erasmus University Rotterdam
Rotterdam, The Netherlands

Radosław Wolniak
Silesian University of Technology

Wei Xu
Center for IT Human Factors Engineering
Intel Corporation
USA

Toshiki Yamaoka
Wakayama University
Wakayama, Japan

Kazuhiko Yamazaki
IBM Japan Ltd.
Tokyo, Japan

Koji Yanagida
SANYO Design Center Co., Ltd.
Japan

Section I

Standardization Efforts in Human Factors and Ergonomics

1 An Overview of International Standardization Efforts in Human Factors and Ergonomics

Anna Szopa and Waldemar Karwowski

CONTENTS

INTRODUCTION

This chapter provides an overview of the international and U.S.-based standards and guidelines in human factors and ergonomics (HFE). In general, standardization allows society to gather and disseminate technical information (Spivak & Brenner, 2001). Standards provide quality control and support legislation and regulations to ensure equal opportunity and operating international markets. One of the main purposes of standardization is to assure uniformity and interchangeability in each area of application. For example, standards may limit the diversity of sizes, shapes, or component designs and prevent a generation of unneeded variation of products that do not provide unique service. Harmonization of international standards reduces trade barriers; promotes safety; allows interoperability of products, systems, and services; and promotes common technical understanding (Wettig, 2002).

Standard is defined as a documented agreement containing technical specifications or other precise criteria to be used consistently as rules, guidelines, or definitions of characteristics to ensure that materials, products, processes, and services are fit for the purpose served by those making reference to the standard (International Organization for Standardization [ISO], 2004). The standardization process can be performed at the national, regional, and international levels (see Figure 1.1). The basis for worldwide standardization in all areas is provided mainly by the following organizations: the ISO, the International Electrotechnical Commission (IEC), and the International Telecommunications Union (ITU). Standards related to human factors and ergonomics are mainly developed by the ISO and the European Committee for Standardization (CEN).

In Europe, aside from the CEN, there are two other standardization organizations—the European Committee for Electrotechnical Standardization (CENELEC) and the European Telecommunications Standards Institute (ETSI). Their mission is to develop a coherent set of

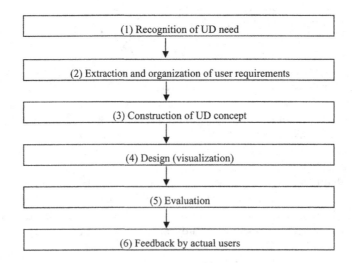

FIGURE 1.1   Hierarchy of standards levels.

voluntary standards that can serve as a basis for a Single European Market/European Economic Area (Wettig, 2002). At the national level, almost every nation has its own body for standards development. Examples of the national standardization organizations include the American National Standards Institute (ANSI), British Standards Institution (BSI), the Deutsches Institut für Normung (DIN), and the Association Française de Normalisation (AFNOR). In general, standards can also be prepared by technical societies, labor organizations, consumer organizations, trade associations, and governmental agencies.

International, regional, and national standards are distinguished by documented development procedures. These procedures have been designed to ensure that all interested parties that can be affected by a particular standard will have an opportunity to represent their interest and participate in the standards development process. For example, ISO standards are developed by technical committees consisting of experts from the industrial, technical, and business sectors who need the standards. Many national members of the ISO apply public review procedures to consult draft standards with the interested parties, including representatives of government agencies, industrial and commercial organizations, professional and consumer associations, and the general public (ISO, 2004). The ISO national bodies are expected to consider any feedback they receive and present a consensus position to the appropriate technical committees.

ISO STANDARDS FOR ERGONOMICS

The ISO is a worldwide federation of national standardization bodies from 146 countries. It is a nongovernmental organization that considers the diverse interests of users, producers, consumers, governments, and the scientific community. The mission of ISO is to promote the development of standardization and related activities in the world to facilitate the international exchange of goods and services and enhance cooperation in the areas of intellectual, scientific, technological, and economic activity (ISO, 2004).

In 1975, ISO formed the technical committee (TC) 159 to develop standards in the field of ergonomics (Parsons, Shackel, & Metz, 1995). The scope of the ISO/TC 159 activity has been described as the standardization in the field of ergonomics—including terminology, methodology, and human factors data. The ISO/TC 159 promotes the adaptation of working and living conditions to the anatomical, psychological, and physiological characteristics of man in relation to the physical, sociological, and technological environment. The main objectives of such standardization efforts are safety, health, well-being, and effectiveness (Parsons, 1995c).

All ISO standards can be obtained at the ISO website (http://www.iso.org). At present, the ISO/TC 159 organizational structure is administrated by the German Standards Association (DIN). The ergonomics standardization group consists of four subcommittees: SCI, SC3, SC4, and SC5. The areas of work of subcommittees and their organizational structure are presented in Table 1.1.

ISO STANDARDS FOR ERGONOMICS GUIDING PRINCIPLES

The TC159/SC1 subcommittee focuses on the standards related to the basic principles of ergonomics (see Table 1.2). The ISO 6385:2016 standard specifies the objectives for ergonomics system design and provides the definition of basic terms and concepts in ergonomics. This standard establishes ergonomic principles of the work system design as basic guidelines that should be applied in designing optimal working conditions with regard to human well-being, safety and health, and due consideration of technological and economic efficiency (Parsons, 1995a).

The ISO 10075 standard that deals with mental workload comprises three parts. The first part, General Terms and Definitions (1991), presents terminologies and main concepts. Part 2, Design Principles (1996), covers guidelines on the design of work systems—including tasks, equipment, workspace, and work conditions in reference to the mental workload. Part 3, Measurement and Assessment of Mental Workload, provides guidelines on measurement and assessment of mental

TABLE 1.1

Organizational Structure of ISO TC 159 "Ergonomics"

Committee	Title
TC 159/SC 1: Ergonomic guiding principles	
TC 159/SC 1/WG 1	Principles of the design of work systems
TC 159/SC 1/WG 2	Ergonomic principles related to mental work
TC 159/SC 1/WG 4	Usability of everyday products
TC 159/SC 3: Anthropometry and biomechanics	
TC 159/SC 3/WG 1	Anthropometry
TC 159/SC 3/WG 2	Evaluation of working postures
TC 159/SC 3/WG 4	Human physical strength: manual handling and force limits
TC 159/SC 3/WG 5	Ergonomic procedures for applying anthropometry and biomechanics standards
TC 159/SC 4: Ergonomics of human-system interaction	
TC 159/SC 4/WG 1	Fundamentals of controls and signaling methods
TC 159/SC 4/WG 2	Visual display requirements
TC 159/SC 4/WG 3	Control, workplace, and environmental requirements
TC 159/SC 4/WG 5	Software ergonomics and human-computer dialogues
TC 159/SC 4/WG 6	Human-centered design processes for interactive systems
TC 159/SC 4/WG 8	Ergonomic design of control centers
TC 159/SC 5: Ergonomics of the physical environment	
TC 159/SC 5/WG 1	Thermal environments
TC 159/SC 5/WG 2	Lighting environments
TC 159/SC 5/WG 3	Danger signals and communication in noisy environments

TABLE 1.2

ISO Standards for Ergonomics Guiding Principles

Reference Number	Title
ISO 6385:2016	Ergonomic principles in the design of work systems
ISO 10075:1991	Ergonomic principles related to mental workload—General terms and definitions
ISO 10075-2:1996	Ergonomic principles related to mental workload—Part 2: Design principles
ISO/FDIS 10075-3	Ergonomic principles related to mental workload—Part 3: Principles and requirements concerning methods for measuring and assessing mental workload
ISO/CD 20282-1	Ease of operation of everyday products—Part 1: Context of use and user characteristics
ISO/CD TS 20282-2	Ease of operation of everyday products—Part 2: Test method

workload and is currently at the stage of a Final Draft International Standard. The third part specifies the requirements for the measurement instruments to be met at different levels of precision in measuring mental workload. Because any human activity includes mental workload, the described standards on it are relevant to all kinds of work design (Nachreiner, 1995).

ISO STANDARDS FOR ANTHROPOMETRY AND BIOMECHANICS

The TC159/SC3 subcommittee focuses on the standards related to anthropometry and biomechanics. This subcommittee consists of four working groups: WG1, "Anthropometry"; WG2, "Evaluation of Working Postures"; WG3, "Human Physical Strength"; and WG4, "Manual

TABLE 1.3

Published ISO Standards and Standards Under Development for Anthropometry and Biomechanics

Reference Number	Title
ISO 7250:1996	Basic human body measurements for technological design
ISO 11226:2000	Ergonomics—Evaluation of static working postures
ISO 11228-1:2003	Ergonomics—Manual handling—Part 1: Lifting and carrying
ISO 14738:2002	Safety of machinery—Anthropometric requirements for the design of workstations at machinery
ISO 14738:2002/Cor 1:2003	
ISO 15534-1:2000	Ergonomic design for the safety of machinery—Part 1: Principles for determining the dimensions required for openings for whole-body access into machinery
ISO 15534-2:2000	Ergonomic design for the safety of machinery—Part 2: Principles for determining the dimensions required for access openings
ISO 15534-3:2000	Ergonomic design for the safety of machinery—Part 3: Anthropometric data
ISO 15535:2012	General requirements for establishing anthropometric databases
ISO/TS 20646-1:2014	Ergonomic procedures for the improvement of local muscular workloads—Part 1: Guidelines for reducing local muscular workloads
ISO/CD 11228-2	Ergonomics—Manual handling—Part 2: Pushing and pulling
ISO/CD 11228-3	Ergonomics—Manual handling—Part 3: Handling of low loads at high frequency
ISO/DIS 15536-1	Ergonomics—Computer manikins and body templates—Part 1: General requirements
ISO/CD 15536-2	Ergonomics—Computer manikins, body templates—Part 2: Structures and dimensions
ISO/FDIS 15537	Principles for selecting and using test persons for testing anthropometric aspects of industrial products and designs
ISO/DIS 20685	3D scanning methodologies for internationally compatible anthropometric databases

Handling and Heavy Weights." A list of published ISO standards and initiated standards in this area is presented in Table 1.3. The description of anthropometric measurements—which can be used as a basis for the definition and comparison of population groups—is provided by the ISO 7250:1996 standard. In addition to the lists of the basic anthropometric measurements, this document contains definitions and measuring conditions.

The three-part standard for the Safety of Machinery (ISO 15534) provides guidelines in determining the dimensions required for openings for human access to machinery. The first part of this standard (ISO 15534-1:2000) presents principles in determining the dimensions for opening the whole-body access to a machinery; the second part (ISO 15534-2:2000) specifies dimensions for the access openings. The third part of the safety of machinery standard (ISO 15534-3:2000) provides the requirements for the human body measurements (anthropometric data) that are needed for the calculations of access-opening dimensions in machinery specified in the previous parts of these standards (Parsons, 1995c). The anthropometric data are based on the static measurements of nude people and are representative of the European population of males and females.

The ISO 14738:2002 standard describes principles for deriving dimensions from anthropometric measurements and applying them to the design of workstations at nonmobile machinery. This standard also specifies the body space requirements for equipment during normal operation in sitting and standing positions. The ISO 15535:2012 standard specifies general requirements for anthropometric databases and their associated reports that contain measurements taken in accordance with ISO 7250. This standard presents information—such as characteristics of the user population, sampling methods, and measurement items and statistics—to make international comparison possible among various population segments.

The ISO 11228-1:2003 standard describes limits for manual lifting and carrying with consideration, respectively, of the intensity, frequency, and duration of the task. The recommended limits can be used in the assessment of several task variables and the health risks evaluation for the working population (Dickinson, 1995). This standard does not include the holding of objects (without walking), pushing or pulling of objects, lifting with one hand, manual handling while seated, and lifting by two or more people. Holding, pushing, and pulling objects are included in other parts of the ISO 11228 standard, which are currently at the stage of committee drafts. The ISO/TS 20646-1:2004 standard presents guidelines for the application of various ergonomics standards related to local muscular workload (LMWL) and specifies activities to reduce the level of LMWL.

ISO STANDARDS FOR ERGONOMICS OF HUMAN-SYSTEM INTERACTION

The TC159/SC4 subcommittee develops standards related to the ergonomics of human-system interaction. The subcommittee is divided into six working groups that consider the following topics: controls and signaling methods; visual display requirements; control, workplace, and environmental requirements; software ergonomics and human-computer dialog; human-centered design processes for interactive systems; and ergonomics design of control centers.

ISO Standards for Controls and Signaling Methods

The ISO 9355 standard on ergonomic requirements for the design of displays and control actuators provides guidelines for the design of displays and control actuators on work equipment, especially machines (see Table 1.4). A list of the parts of the ISO 9355 standard is presented in Table 1.4. Part 1 describes the general principles of human interactions with display and controls. The other two parts provide recommendations for the selection, design, and location of information displays (Part 2) and control actuators (Part 3). Part 4 covers general principles for the location and arrangement of display and actuators.

ISO Standards for Visual Display Requirements

The ISO 9241 Ergonomic Requirements for Office Work with Visual Display Terminals (VDTs) standard is one of the most important standards for ergonomic design (Eibl, 2005; Stewart, 1995). This standard presents general guidance and specific principles that need to be considered in the design of equipment, software, and tasks for office work with VDTs (see Table 1.5).

Part 1 of the ISO 9241 standard describes the underlying principles of the user performance approach. Part 2 describes how task requirements may be identified and specified in

---

TABLE 1.4

ISO Standards for Controls and Signaling Methods

Reference Number	Title
ISO 9355-1:1999	Ergonomic requirements for the design of displays and control actuators—Part 1: Human interactions with displays and control actuators
ISO 9355-2:1999	Ergonomic requirements for the design of displays and control actuators—Part 2: Displays
ISO/DIS 9355-3	Safety of machinery—Ergonomic requirements for the design of signals and control actuators—Part 3: Control actuators
ISO/DIS 9355-4	Safety of machinery—Ergonomic requirements for the design of displays and control actuators—Part 4: Location and arrangement of displays and control actuators

TABLE 1.5

ISO 9241 Ergonomic Requirements for Office Work with Visual Display Terminals (VDTs)

Reference Number	Title
ISO 9241-1:1997	Part 1: General introduction
ISO 9241-2:1992	Part 2: Guidance on task requirements
ISO 9241-3:1992	Part 3: Visual display requirements
ISO 9241-4:1998	Part 4: Keyboard requirements
ISO 9241-5:1998	Part 5: Workstation layout and postural requirements
ISO 9241-6:1999	Part 6: Guidance on the work environment
ISO 9241-7:1998	Part 7: Requirements for display with reflections
ISO 9241-8:1997	Part 8: Requirements for displayed colors
ISO 9241-9:2000	Part 9: Requirements for non-keyboard-input devices
ISO 9241-10:1996	Part 10: Dialogue principles
ISO 9241-11:2018	Part 11: Usability: Definitions and concepts
ISO 9241`-12:1998	Part 12: Presentation of information
ISO 9241-13:1998	Part 13: User guidance
ISO 9241-14:1997	Part 14: Menu dialogues
ISO 9241-15:1997	Part 15: Command dialogues
ISO 9241-16:1999	Part 16: Direct manipulation dialogues
ISO 9241-17:1998	Part 17: Form-filling dialogues

organizations and how task requirements can be incorporated into the system design and implementation process. Parts 3 through 9 provide assistance in the procurement and specification of the hardware and environmental components. Three parts present image quality requirements (performance specification) for different types of displays—black and white displays (Part 3), color displays (Part 8), and display with reflections (Part 7). Part 4 provides criteria for the keyboard; and Part 9 is for non-keyboard-input devices. Parts 5 and 6 are concerned with ergonomic principles of the design and procurement of computer workstations, workstation equipment, and work environment for office work with VDTs. Those two parts include issues such as technical design of furniture and equipment for the workplace, space organization, and workplace layout and physical characteristics of office work environment such as lighting, noise, and vibrations. Part 10 presents the core ergonomics principles that should be applied to the design of dialogs between humans and information systems. These principles are intended for the specifications, design, and evaluation of dialogs for office work with VDTs. Part 11 defines usability and specifies the usability evaluation in terms of user performance and satisfaction measures. Part 12 provides recommendations for information presented on the text-based displays and graphical user interfaces. Part 13 presents recommendations for different types of user guidance attributes of software interfaces such as feedback, status, help, and error handling. Parts 14–17 deal with different types of dialog styles: menus, commands, direct manipulation, and form filling.

The ISO 13406 standard provides additional recommendations to ISO 9241 with respect to visual displays based on flat panels. Two parts of this standard cover image quality requirements for the ergonomic design and evaluation of flat panel displays. ISO 14915 provides additional recommendations to ISO 9241 concerning multimedia presentations.

TABLE 1.6

ISO Standards for Software Ergonomics

Reference Number	Title
ISO 14915-1:2002	Software ergonomics for multimedia user interfaces—Part 1: Design principles and framework
ISO 14915-2:2003	Software ergonomics for multimedia user interfaces—Part 2: Multimedia navigation and control
ISO 14915-3:2002	Software ergonomics for multimedia user interfaces—Part 3: Media selection and combination
ISO/CD 23973	Software ergonomics for WWW-user interfaces

TABLE 1.7

ISO 11064 Ergonomic Design of Control Centers

Reference Number	Title
ISO 11064-1:2000	Part 1: Principles for the design of control centers
ISO 11064-2:2000	Part 2: Principles for the arrangement of control suites
ISO 11064-3:1999	Part 3: Control room layout
ISO 11064-4:2013	Part 4: Layout and dimensions of workstations
ISO/DIS 11064-6	Part 5: Environmental requirements for control centers
ISO/CD 11064-7	Part 6: Principles for the evaluation of control centers

ISO Standards for Software Ergonomics

ISO 14915: Software Ergonomics for Multimedia User Interfaces specifies recommendations and principles for the design of interactive multimedia user interfaces that integrate different media—such as static text, graphics, and images—and dynamic media—such as audio, animation, and video. This standard focuses on issues related to the integration of different media, whereas hardware issues and multimodal input are not considered. The standard consists of three parts (see Table 1.6) that address the general design principles (Part 1), multimedia navigation and control (Part 2), and media selection and combination (Part 3). The committee draft ISO/CD 23973 also considers ergonomics design principles for World Wide Web user interfaces.

ISO Standards for Ergonomic Design of Control Centers

ISO 11064: Ergonomic Design of Control Centers specifies requirements and presents principles for the ergonomic design of control centers (see Table 1.7). This standard is concerned with the following issues: principles for the design of control centers, control suite arrangements, control room layout, workstation layout and dimensions, displays and controls, environmental requirements, evaluation of control rooms, and ergonomic requirements for specific applications.

ISO Standards for Human-System Interaction

Two ISO standards focus on accessibility issues in the design of usable systems (see Table 1.8). ISO/AWI 16071 provides guidance on accessibility in reference to the software, whereas ISO/TS 16071:2003 addresses accessibility in reference to the human-computer interface. The guidelines on the human-centered design process throughout the life cycle of the computer-based interactive systems are described in the ISO 13407:1999 and ISO/TR 18529:2000n standards.

Usability methods supporting the human-centered design are described in the ISO/TR 16982:2002 standard. Further standards concerned with the human-system interaction address

TABLE 1.8

Published ISO Standards and Standards in Development for Human–System Interaction

Reference Number	Title
ISO 13407:1999	Human-centered design processes for interactive systems
ISO 1503:2008	Spatial orientation and direction of movement—Ergonomics requirements
ISO/TS 16071:2003	Ergonomics of human-system interaction—Guidance on accessibility for human-computer interfaces
ISO/TR 16982:2002	Ergonomics of human-system interaction—Usability methods supporting human-centered design
ISO/PAS 18152:2003	Ergonomics of human-system interaction—Specification for the process assessment of human-system issues
ISO/TR 18529:2000	Ergonomics—Ergonomics of human-system interaction—Human-centered lifecycle process descriptions
ISO 13406-1:1999	Ergonomic requirements for work with visual displays based on flat panels—Part 1: Introduction
ISO 13406-2:2001	Ergonomic requirements for work with visual displays based on flat panels—Part 2: Ergonomic requirements for flat panel displays
ISO/CD 9241-301	Ergonomic requirements and measurement techniques for electronic visual displays—Part 301: Introduction
ISO/CD 9241-302	Ergonomic requirements and measurement techniques for electronic visual displays—Part 302: Terminology
ISO/CD 9241-303	Ergonomic requirements and measurement techniques for electronic visual displays—Part 303: Ergonomic requirements
ISO/AWI 9241-304	Ergonomic requirements and measurement techniques for electronic visual displays—Part 304: User performance test method
ISO/CD 9241-305	Ergonomic requirements and measurement techniques for electronic visual displays—Part 305: Optical laboratory test methods
ISO/CD 9241-306	Ergonomic requirements and measurement techniques for electronic visual displays—Part 306: Field assessment methods
ISO/CD 9241-307	Ergonomic requirements and measurement techniques for electronic visual displays—Part 307: Analysis and compliance test methods
ISO/CD 9241-110	Ergonomics of human-system interaction—Part 110: Dialogue principles
ISO/CD 9241-400	Physical input devices—Ergonomic principles
ISO/AWI 9241-410	Physical input devices—Design criteria for products
ISO/AWI 9241-420	Physical input devices—Part 420: Ergonomic selection procedures
ISO/IEC 11581-1:2000	Information technology—User system interfaces and symbols—Icon symbols and functions Part 1: Icons—General
ISO/IEC 11581-2:2000	Information technology—User system interfaces and symbols—Icon symbols and functions Part 2: Object icons
ISO/IEC 11581-3:2000	Information technology—User system interfaces and symbols—Icon symbols and functions Part 3: Pointer icons
ISO/IEC 11581-5:2004	Information technology—User system interfaces and symbols—Icon symbols and functions Part 5: Tool icons
ISO/IEC 11581-6:1999	Information technology—User system interfaces and symbols—Icon symbols and functions Part 6: Action icons
ISO/IEC TR 9126-1:2001	Software engineering—Product quality—Part 1: Quality model
ISO/IEC TR 9126-2:2003	Software engineering—Product quality—Part 2: External metrics
ISO/IEC TR 9126-3:2003	Software engineering—Product quality—Part 3: Internal metrics
ISO/IEC TR 9126-4:2004	Software engineering—Product quality—Part 4: Quality in use metrics

issues such as development and design of icons (ISO/IEC 11581), design of typical controls for multimedia functions (ISO 18035), icons for typical WWW-browsers (ISO 18036), and definitions and metrics concerning software quality (ISO 9126).

ISO STANDARDS FOR ERGONOMICS OF THE PHYSICAL ENVIRONMENT

The ISO TC159 SC5 document describes an international standard in the area of ergonomics of the physical environment. The subcommittee in charge of this standard development is divided into three working groups (WGs): thermal environments (WG1), lighting (WG2), and danger signals and communication in noisy environments (WGS) (Table 1.9).

STANDARDS ON ERGONOMICS OF THERMAL ENVIRONMENT

The standards on ergonomics of thermal environments are concerned with the issues of heat stress, cold stress, and thermal comfort, as well as with the thermal properties of clothing and metabolic heat production due to work activity (Olesen, 1995). Physiological measures such as skin reaction due to contact with hot, moderate, and cold surfaces, and thermal comfort requirements for people with special requirements are also considered (see Tables 1.9 and 1.10). The comfort standard ISO 7730 provides a method for predicting human thermal sensations and the degree of discomfort, which can also be used to specify acceptable environmental conditions for comfort. This method is based on the predicted mean vote (PMV) and the predicted percentage of dissatisfied (PPD) thermal comfort indices (Olesen & Parsons, 2002). It also provides

---

TABLE 1.9
Published ISO Standards on the Ergonomics of the Thermal Environmental

Reference Number	Title
ISO 7243:2017	Ergonomics of the thermal environment—Assessment of heat stress using the WBGT (wet bulb globe temperature) index
ISO 7726:1998	Ergonomics of the thermal environment—Instruments for measuring physical quantities
ISO 7730:2005	Ergonomics of the thermal environment—Analytical determination and interpretation of thermal comfort using calculation of the PMV and PPD indices and local thermal comfort criteria
ISO 7933:2004	Ergonomics of the thermal environment—Analytical determination and interpretation of heat stress using calculation of the predicted heat strain
ISO 8996:2004	Ergonomics of the thermal environment—Determination of metabolic rate
ISO 9886:2004	Ergonomics—Evaluation of thermal strain by physiological measurements
ISO 9920:2007	Ergonomics of the thermal environment—Estimation of thermal insulation and water vapor resistance of a clothing ensemble
ISO 10551:2019	Ergonomics of the physical environment—Subjective judgment scales for assessing physical environments
ISO 11079:2007	Ergonomics of the thermal environment—Determination and interpretation of cold stress when using required clothing insulation (IREC) and local cooling effects.
ISO 11399:1995	Ergonomics of the thermal environment—Principles and application of relevant International Standards
ISO 12894:2001	Ergonomics of the thermal environment—Medical supervision of individuals exposed to extreme hot or cold environments
ISO 13731:2001	Ergonomics of the thermal environment—Vocabulary and symbols
ISO/TS 13732-2:2001	Ergonomics of the thermal environment—Methods for the assessment of human responses to contact with surfaces—Part 2: Human contact with surfaces at moderate temperature

methods for the assessment of local discomfort caused by drafts, asymmetric radiation, and temperature gradients. Other thermal environment standards address issues such as thermal comfort for people with special requirements (ISO TS 14415), responses on contact with surfaces at moderate temperatures (ISO 13732, Part 2), and thermal comfort in vehicles (ISO 14505, Parts 1–4). Standards concerned with thermal comfort assessment specify measuring instruments (ISO 7726), methods for estimation of metabolic heat production (ISO 8996), estimation of clothing properties (ISO 9920), and subjective assessment methods (ISO 10551). ISO 11399:1995 provides the information needed for a correct and effective application of international standards concerned with ergonomics of the thermal environment (Table 1.10).

STANDARDS ON COMMUNICATION IN NOISY ENVIRONMENTS

This set of standards considers communication in noisy environments—including warning, danger signals, and speech (see Table 1.11). The ISO 7731:2005 document specifies the requirements and test methods for auditory danger signals and provides guidelines for the design of the signals in public and workspaces. This document also provides definitions that guide the use of the standards concerned with noisy environments. The criteria for perception of the visual danger signals are

TABLE 1.10

ISO Drafts and Standards in Development on the Ergonomics of the Thermal Environment

Reference Number	Title
ISO/DIS 7730	Ergonomics of the thermal environment—Analytical determination and interpretation of thermal comfort using calculation of the PMV and PPD indices and local thermal comfort
ISO/FDIS 7933	Ergonomics of the thermal environment—Analytical determination and interpretation of heat stress using calculation of the predicted heat strain
ISO/FDIS 8996	Ergonomics of the thermal environment—Determination of metabolic rate
ISO/CD 9920	Ergonomics of the thermal environment—Estimation of the thermal insulation and evaporative resistance of a clothing ensemble
ISO/CD 11079	Ergonomics of the thermal environment—Determination and interpretation of cold stress when using required clothing insulation (IREQ) and local cooling effects
ISO/DIS 13732-1	Ergonomics of the thermal environment—Methods for the assessment of human responses to contact with surfaces—Part 1: Hot surfaces
ISO/DIS 13732-3	Ergonomics of the thermal environment—Touching of cold surfaces—Part 3: Ergonomics data and guidance for application
ISO/CD TS 14415	Ergonomics of the thermal environment—Application of International Standards to the disabled, the aged, and other handicapped persons
ISO/DIS 14505-1	Ergonomics of the thermal environment—Evaluation of thermal environment in vehicles—Part 1: Principles and methods for assessment of thermal stress
ISO/DIS 14505-2	Ergonomics of the thermal environment—Evaluation of thermal environment in vehicles—Part 2: Determination of equivalent temperature
ISO/CD 14505-3	Ergonomics of the thermal environment—Thermal environments in vehicles—Part 3: Evaluation of thermal comfort using human subjects
ISO 15265	Ergonomics of the thermal environment—Risk assessment strategy for the prevention of stress or discomfort in thermal working conditions
ISO/CD 15743	Ergonomics of the thermal environment—Working practices in cold: Strategy for risk assessment and management

TABLE 1.11

ISO Standards for Danger Signals and Communication in Noisy Environments

Reference Number	Title
ISO 7731:2003	Ergonomics—Danger signals for public and work areas—Auditory danger signals
ISO 11428:1996	Ergonomics—Visual danger signals—General requirements, design, and testing
ISO 11429:1996	Ergonomics—System of auditory and visual danger and information signals
ISO 9921-1:1996	Ergonomic assessment of speech communication—Part 1: Speech interference level (SIL) and communication distances for persons with normal hearing capacity in direct communication (SIL method)
ISO/TR 19358:2002	Ergonomics—Construction and application of tests for speech technology

provided in ISO 11428:1996, which addresses safety and ergonomic requirements and corresponding physical measurements.

ISO 11429:1996 specifies a system of danger and information signals in reference to different degrees of urgency. This standard applies to all danger signals that must be clearly perceived and differentiated — from extreme urgency to "all clear." Guidance on detectability is provided in terms of luminance, illuminance, and contrast, considering both surface and point sources. ISO 9921-1:1996 describes a method for the prediction of the effectiveness of speech communication in the presence of noise generated by machinery, as well as in other noisy environments. The following parameters are considered in this standard: an ambient noise at the speaker's position and ambient noise at the listener's position, a distance between the communication partners, and a variety of physical and personal conditions. The ISO/TR 19358:2002 standard deals with testing and assessment of speech-related products and services.

Standards on Lighting of Indoor Work Systems

ISO 8995 (1989): Principles of Visual Ergonomics – The Lighting of Indoor Work systems was developed by the ISO 159 SC5 WG2 "Lighting" group in collaboration with the International Commission on Illumination (CIE; Parsons, 1995b). This standard describes principles of visual ergonomics, identifies factors that influence human visual performance, and presents criteria for acceptable visual environments.

CEN STANDARDS FOR ERGONOMICS

The main aim of European standardization efforts is the development of a coherent set of voluntary standards that can provide a basis for a Single European Market/European Economic Area. The work of European standardization organizations is carried out in cooperation with worldwide bodies and the national standards bodies in Europe (Wettig, 2002). Members of the European Union (EU) and the European Fair-Trade Association (EFTA) have agreed to implement CEN standards in their national system and withdraw any conflicting national standards.

In 1987, the CEN established the CEN TCI22 "Ergonomics" as a body responsible for the development of European standards in ergonomics (Dul et al. 1996). The scope of the CEN TCI 22 "Ergonomics" has been defined as standardization in the field of ergonomics principles and requirements for the design of work systems and work environments—including machinery and personal protective equipment—to promote the health, safety, and well-being of the human operator and the effectiveness of the work (CEN, 2004). The organizational structure of the CEN TCI22 "Ergonomics" is presented in Table 1.12.

TABLE 1.12

Organizational Structure of CEN/TC 122

Working Group	Title
CEN/TC 122/WG 1	Anthropometry
CEN/TC 122/WG 2	Ergonomic design principles
CEN/TC 122/WG 3	Surface temperatures
CEN/TC 122/WG 4	Biomechanics
CEN/TC 122/WG 5	Ergonomics of human-computer interaction
CEN/TC 122/WG 6	Signals and controls
CEN/TC 122/WG 8	Danger signals and speech communication in noisy environments
CEN/TC 122/WG 9	Ergonomics of personal protective equipment (PPE)
CEN/TC 122/WG 10	Ergonomic design principles for the operability of mobile machinery
CEN/TC 122/WG 11	Ergonomics of the thermal environment
CEN/TC 122/WG 12	Integrating ergonomic principles for machinery design

The ISO and CEN have signed a formal agreement on technical collaboration (The Vienna Agreement) that established a close cooperation between these two standardization bodies. ISO and CEN decided to harmonize the development of their standards and to cooperate regarding the exchange of information and standards drafting. According to this agreement, the ISO standards can be adopted by CEN and vice versa. Table 1.13 presents published CEN ergonomics standards. The CEN ergonomic standards in development are shown in Appendix 1.1.

TABLE 1.13

Published CEN Standards for Ergonomics

CEN Reference	Title	ISO Standard
Ergonomics principles		
EN ISO 10075-1:2017	Ergonomic principles related to mental workload—Part 1: General issues and concepts, terms, and definitions	ISO 10075:1991
EN ISO 10075-2:2000	Ergonomic principles related to mental workload—Part 2: Design principles	ISO 10075-2:1996
ENV 6385:1990	Ergonomic principles of the design of work systems	ISO 6385:2016
EN ISO 6385:2016	Ergonomic principles in the design of work systems	ISO 6385:2016
Anthropometries and biomechanics		
EN 1005-1:2001	Safety of machinery—Human physical performance—Part 1: Terms and definitions	
EN 1005-2:2003	Safety of machinery—Human physical performance—Part 2: Manual handling of machinery and component parts of machinery	
EN 1005-3:2009	Safety of machinery—Human physical performance—Part 3: Recommended force limits for machinery operation	
EN 13861:2011	Safety of machinery—Guidance for the application of ergonomics standards in the design of machinery	
EN 547-1:2009	Safety of machinery—Human body measurements—Part 1: Principles for determining the dimensions required for openings for whole-body access into machinery	

(Continued)

TABLE 1.13 (Continued)

CEN Reference	Title	ISO Standard
EN 547-2:1996	Safety of machinery—Human body measurements—Part 2: Principles for determining the dimensions required for access openings	
EN 547-3:1996	Safety of machinery—Human body measurements—Part 3: Anthropometric data	
EN 614-1:2006	Safety of machinery—Ergonomic design principles—Part 1: Terminology and general principles	
EN 614-2:2000	Safety of machinery—Ergonomic design principles—Part 2: Interactions between the design of machinery and work tasks	
EN ISO 7250:1997	Basic human body measurements for technological design	ISO 7250:1996
EN ISO 14738:2008	Safety of machinery—Anthropometric requirements for the design of workstations at machinery	ISO 14738:2002
EN ISO 15535:2012	General requirements for establishing anthropometric databases	ISO 15535:2003
Ergonomics design of control centers		
EN ISO 11064-1:2000	Ergonomic design of control centers—Part 1: Principles for the design of control centers	ISO 11064-1:2000
EN ISO 11064-2:2000	Ergonomic design of control centers—Part 2: Principles for the arrangement of control suites	ISO 11064-2:2000
EN ISO 11064-3:1999	Ergonomic design of control centers—Part 3: Control room layout	ISO 11064-3:1999
EN ISO 11064-3:1999/AC:2002	Ergonomic design of control centers—Part 3: Control room layout	ISO 11064-3:1999/Cor. 1:2002
Human-system interaction		
EN ISO 13406-1:1999	Ergonomic requirements for work with visual display based on flat panels—Part 1: Introduction	ISO 13406-1:1999
EN ISO 13406-2:2001	Ergonomic requirements for work with visual displays based on flat panels—Part 2: Ergonomic requirements for flat panel displays	ISO 13406-2:2001
EN ISO 13407:1999	Human-centered design processes for interactive systems	ISO 13407:1999
EN ISO 13731:2001	Ergonomics of the thermal environment—Vocabulary and symbols	ISO 13731:2001
EN ISO 14915-1:2002	Software ergonomics for multimedia user interfaces—Part 1: Design principles and framework	ISO 14915-1:2002
EN ISO 14915-2:2003	Software ergonomics for multimedia user interfaces—Part 2: Multimedia navigation and control	ISO 14915-2:2003
EN ISO 14915-3:2002	Software ergonomics for multimedia user interfaces—Part 3: Media selection and combination	ISO 14915-3:2002
EN ISO 9921:2003	Ergonomics—Assessment of speech communication	ISO 9921:2003
Danger signals		
EN 457:1992	Safety of machinery—Auditory danger signals—General requirements, design, and testing	ISO 7731:1986, modified
EN 842:1996	Safety of machinery—Visual danger signals—General requirements, design, and testing	
EN 981:1996	Safety of machinery—System of auditory and visual danger and information signals	
Thermal environments		
EN 12515:1997	Hot environments—Analytical determination and interpretation of thermal stress using calculation of required sweat rate	ISO 7933:1989 modified
EN 27243:1993	Hot environments—Estimation of the heat stress on working man, based on the WBGT-index (wet bulb globe temperature)	ISO 7243:1989
EN 28996:1993	Ergonomics—Determination of metabolic heat production	ISO 8996:1990

TABLE 1.13 (Continued)

CEN Reference	Title	ISO Standard
EN ISO 10551:2019	Ergonomics of the physical environment—Subjective judgment scales for assessing physical environment	ISO 10551:1995
EN ISO 11399:2000	Ergonomics of the thermal environment—Principles and application of relevant International Standards	ISO 11399:1995
EN ISO 12894:2001	Ergonomics of the thermal environment—Medical supervision of individuals exposed to extreme hot or cold environments	ISO 12894:2001
EN ISO 7726:2001	Ergonomics of the thermal environment—Instruments for measuring physical quantities	ISO 7726:1998
EN ISO 7730:2005	Ergonomics of the thermal environment—Analytical determination and interpretation of thermal comfort using calculation of the PMV and PPD indices and local thermal comfort criteria	ISO 7730:1994
EN ISO 9886:2001	Evaluation of thermal strain by physiological measurements	ISO 9886:1992
EN ISO 9886:2004	Ergonomics—Evaluation of thermal strain by physiological measurements	ISO 9886:2004
EN ISO 9920:2009	Ergonomics of the thermal environment—Estimation of the thermal insulation and water vapor resistance of a clothing ensemble	ISO 9920:1995
ENV ISO 11079:1998	Evaluation of cold environments—Determination of required clothing insulation (REQ)	ISO/TR 11079:1993
EN 13202:2000	Ergonomics of the thermal environment—Temperatures of touchable hot surfaces—Guidance for establishing surface temperature limit values in production standards with the aid of EN 563	
EN 563:1994	Safety of machinery—Temperatures of touchable surfaces—Ergonomics data to establish temperature limit values for hot surfaces	
EN 563:1994/A 1:1999	Safety of machinery—Temperatures of touchable surfaces—Ergonomics data to establish temperature limit values for hot surfaces	
EN 563:1994/A 1:1999/AC:2000	Safety of machinery—Temperatures of touchable surfaces—Ergonomics data to establish temperature limit values for hot surfaces	
EN 563:1994/ AC:1994	Safety of machinery—Temperatures of touchable surfaces—Ergonomics data to establish temperature limit values for hot surfaces	
Displays and control actuators		
EN 894-1:1997	Safety of machinery—Ergonomics requirements for the design of displays and control actuators—Part 1: General principles for human interactions with displays and control actuators	
EN 894-2:1997	Safety of machinery—Ergonomics requirements for the design of displays and control actuators—Part 2: Displays	
EN 894-3:2000	Safety of machinery—Ergonomics requirements for the design of displays and control actuators—Part 3: Control actuators	

OTHER INTERNATIONAL STANDARDS RELATED TO ERGONOMICS

Because of historical and organizational factors, many ISO and CEN standards in the field of ergonomics have not been developed by technical committees ISO TC159 and CEN TC122. Some ergonomics areas covered by other ISO and CEN technical committees are presented in Table 1.14. A list of published ISO standards related to ergonomics area, but developed by other than the TC159 committee, is provided in Appendix 1.1.

TABLE 1.14

Ergonomic Areas Covered in Standards Developed by the other ISO and CEN Technical Committees (Dul, de Vlaming, & Munnik, 1996)

Topic	Technical ISO	Committee CEN
Safety of machines	TC 199	TC 114
Vibration and shock	TC 108	TC 211
Noise and Acoustics	TC 43	TC 211
Lighting		TC 169
Respiratory protective devices		TC 79
Eye protection		TC 85
Head protection		TC 158
Hearing protection		TC 159
Protection against falls	TC 94	TC 160
Foot & leg protection		TC 161
Protective clothing		TC 162
Radiation protection	TC 85	
Air quality	TC 146	
Assessment and workplace exposure		TC 137
Office machines	TC 95	
Information procession	TC 97	
Road vehicles	TC 22	
Safety color and signs	TC 80	
Graphical symbols	TC 145	

ILO GUIDELINES FOR OCCUPATIONAL SAFETY AND HEALTH (OSH) MANAGEMENT

Implementation of the ISO-standardized approach to management systems led to the view that this type of approach can also improve management of occupational safety and health. Following this idea, the International Labor Organization (ILO) developed voluntary guidelines on OSH management systems that reflect ILO values and ensure the protection of the workers' safety and health (ILO-OSH, 2001). The main objective of ILO is the promotion of social justice and internationally recognized human and labor rights (ILO, 2004). ILO represents the interests of the three parties equally: employers, employee organizations, and government parties.

The ILO-OSH (2001) provides recommendations concerning the design and implementation of OSH management systems (MS), in a way allowing the integration of OSH with the general enterprise management system. The ILO guidelines state that these recommendations are addressed to all who are responsible for occupational safety and health management. These guidelines are nonmandatory and are not intended to replace the national laws and regulations. The ILO-OSH document distinguishes two levels of the guideline's application: national and organizational. At the national level, ILO-OSH provides recommendations for the establishment of a national framework for OSH-MS. The guidelines suggest that this process should be supported by the provision of the relevant national laws and regulations. The establishment of the national framework for OSH-MS (Figure 1.2) includes the following actions (ILO-OSH):

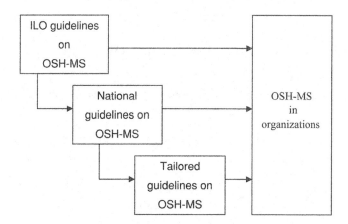

FIGURE 1.2 Establishment of national framework for the OSH-MS (ILO-OSH, 2001).

1. Nomination of competent institution(s) for OSH-MS
2. Formulation of a coherent national policy
3. Development of the national and tailored guidelines

At the organizational level, the ILO-OSH (2001) guidelines establish the responsibilities of the employers regarding occupational safety and health management and emphasize the importance of compliance with the applicable national laws and regulations. The ILO-OSH suggests that OSH-MS elements should be integrated into an overall organizational policy and management strategy actions. The OSH-MS in the organization consists of five main sections: policy, organization, planning and implementation, evaluation, and action for improvement. These elements correspond to Deming's cycle of a Plan-Do-Check-Act, internationally accepted as the basis for the "system" approach to management. The OSH-MS main sections and their elements are listed in Table 1.15.

The ILO-OSH (2001) guidelines require the establishment of the OSH policy by the employer in consultation with workers and their representatives and defines the content of such policy. These guidelines also indicate the importance of OSH policy integration and compatibility with other management systems in the organization. The guidelines also emphasize the necessity of worker participation in the OSH management system in the organization. Workers should be consulted regarding OSH activities and should be encouraged to participate in OSH-MS, including a safety and health committee. The organizing section of the guidelines underlines the need for allocation of responsibility and accountability for the implementation and performance of the OSH management system to the senior management. This section also includes requirements related to competence and training in the field of OSH and defines the needed documentation and communication activities.

The planning and implementation section of the ILO-OSH (2001) guidelines includes the elements of initial review, system planning, development and implementation, OSH objectives, and hazard prevention. The initial review identifies the actual states of the organization in regard to the OSH and creates the baseline for the OSH policy implementation. The evaluation section consists of performance monitoring and measurement, investigation of work-related diseases and incidents, audit, and management review. The guidelines require carrying out internal audits of the OSH-MS according to the established policies. Action for improvement includes the elements of preventive and corrective action and continual improvement. The last section underlines the need for continual improvement of OSH performance through the development of policies, systems, and techniques to prevent and control work-related injuries and diseases.

TABLE 1.15

ILO-OSH-MS Main Sections and Their Elements (ILO-OSH, 2001)

Section	Elements
Policy	3.1. Occupational safety and health policy
	3.2. Worker participation
Organizing	3.3. Responsibility and accountability
	3.4. Competence and training
	3.5. OSH management system documentation
	3.5. Communication
Planning and implementation	3.6. Initial review
	3.7. System planning and implementation
	3.8. Occupational safety and health objectives
	3.9. Hazard prevention
Evaluation	3.10. Performance monitoring and measurement
	3.11. Investigation of work-related incidents and their impact on BHP
	3.12. Audit
	3.13. Management review
Action for improvement	3.15. Preventive and corrective action
	3.16. Continual improvement

HUMAN FACTORS AND ERGONOMICS/STANDARDS IN THE UNITED STATES

U.S. Government Standards for Human Factors and Ergonomics

Among the human factors and ergonomics (HFE) U.S. government standards, two documents are usually mentioned as basics: a military standard providing human engineering design criteria (MIL-STD-1472) and man–system integration standards (NASA-STD-300; Chapanis, 1996; McDaniel, 1996). In addition, there are more specific standards that have been developed by U.S. government agencies such as the Department of Defense, Department of Transportation, Department of Energy, and the U.S. Nuclear Regulatory Commission. Additionally, many handbooks that contain more detailed and descriptive information concerning human factors and ergonomics guidelines, preferred practices, methodology, and reference data needed during the design of equipment and systems have also been developed. These handbooks help in the use and application of relevant government standards during the design process. A list of human factors standards used by the U.S. government agencies and a description of their scope can be found in the Index of Government Standards (Human Factors Standardization SubTAG [HFS SubTAG], 2004).

Human Factors Military Standards (Department of Defense)

A set of consensus-type military standards was developed by human factors engineers from the U.S. military's three services (Army, Navy, and Air Force), industry, and technical societies. As a result of the standardization reform of the late 1990s, most of the single-service standards were canceled and were integrated into a few Department of Defense (DoD) standards and handbooks. However, the distinction between the two main categories of human factors military standards—general (MIL-STD-1472 and related handbooks) and aircraft (JSSG 2010 and related handbooks)—remains unchanged, which reflects the criticality of aircraft design. A list of the main military standards and handbooks is presented in Table 1.16.

TABLE 1.16

Military Standards and Handbooks for Human Factors and Ergonomics (Based on HFS SubTAG, 2004)

Document Number	Title	Date	Source
Standards			
MIL-STD-882D	Standard Practice for System Safety	2000	http://assist.daps.dla.mil/docimages/0001/ 95/78/std882d.pd8
MIL-STD-1472F	Human Engineering	1999	http://assist.daps.dla.mil/docimages/0001/ 87/3 1/milstdl4.pdl
MIL-STD-1474D	Noise Limits	1997	http://assist.daps.dla.mil/docimages/0000/ 3 1/59/1474d.pdl
MIL-STD-1477C	Symbols for Army Systems Displays	1996	http://assist.daps.dla.mil/docimages/0000/ 42/03/69268.pd9
MIL-STD-1787C	Aircraft Display Symbology	2001	This is a controlled distribution document
Handbooks			
DOD-HDBK-743A	Anthropometry of U.S. Military Personnel	1991	http://assist.daps.dla.mil/docimages/0000/ 40/29/54083.pd0
MIL-HDBK-759C	Human Engineering Design Guidelines	1995	http://assist.daps.dla.mil/docimages/0000/ 40/04/mh759c.pd8
MIL-HDB K-7 67	Design Guidance for Interior Noise Reduction in Light-Armored Tracked Vehicles	1993	http://assist.daps.dla.mil/docimages/0000/ 13/24/767.pd1
MIL-HDB K-1473A	Color and Marking of Army Materiel	1997	http://assist.daps.dla.mil/docimages/0000/ 85/40/hdbkl473.pd6
MIL-HDB K-1908B	Definitions of Human Factors Terms	1999	http://assist.daps.dla.mil/docimages/0001/ 81/33/1908hdbk.pd9
Mil-HDBK-46855	Human Engineering Requirements for Military Systems Equipment and Facilities		

The basic human engineering principles, design criteria, and practices required for the integration of humans with systems and facilities are established in the MIL-STD-1472 F Human Engineering Design Criteria for Military Systems, Equipment, and Facilities. This standard document can be applied not only to the military but also to the commercial design of all systems, subsystems, equipment, and facilities. The MIL-STD-1472 F includes requirements for displays, controls, control-display integration, anthropometry, groundwork space design, environment, design for maintainability, design of equipment for remote handling, small systems and equipment, operational and maintenance ground and shipboard vehicles, hazards and safety, aerospace vehicle compartment design requirements, and man-computer interface. The MIL-STD-1472 also includes a nongovernmental standard ANSI/HFS 100 on VDT workstations. After standardization reform, the design data and information part of MIL-STD-1472F were removed and inserted into the MIL-HDBK-759.

Another important military standardization document is the MIL-HDBK-46855 A Human Engineering Requirements for Military Systems Equipment and Facilities. This handbook presents human engineering program tasks, procedures, and preferred practices. MIL-HDBK-46855 covers topics such as analysis functions, including human performance parameters, equipment capabilities, and task environments design; test and evaluation; and workload analysis, dynamic simulation, and data requirements. This handbook also adopted materials from the DoD-HDBK-763 Human Engineering Procedures Guide concerned with human engineering methods and tools, which remained stable over time. The newest rapidly evolving automated human engineering tools are not described in the MIL-HDBK-46855 but can be found at the Directory of Design Support Methods (DSSM) on the MATRIS website (http://dtica.dtic.mil/ddsm/).

Other military standards cover topics such as standard practice for conducting system safety (MIL-STD-882D); acoustical noise limits, testing requirements, and measurement techniques (MIL-STD-1474D); physical characteristics of symbols for army systems displays (MIL-STD-1477C); and symbology requirements for aircraft displays (MIL-STD-1787C). The definitions for all human factors standard documents are provided in the MIL-HDBK-1908B Department of Defense Handbook: Definitions of Human Factors Terms.

Other U.S. Governmental Standards: NASA and FAA

A list of other relevant government standards is provided in Table 1.17. The NASA-STD-3000 provides generic requirements for space facilities and related equipment, important for proper human-system integration. This document is integrated in the website that also offers video images from space missions and illustrates the human factors design issues. This standard document is not limited to any specific NASA, military, or commercial program and can be applied to almost all kinds of equipment. The NASA-STD-3000 consists of two volumes: Volume I – Man-Systems Integration Standards presents all of the design standards and requirements and Volume II – Appendices contains the background information related to the standards. The NASA-STD-3000 covers the following areas of human factors: anthropometry and biomechanics, human performance capabilities, natural and induced environments, health management, workstations, activity centers, hardware and equipment, design for maintainability, and facility management.

The standards of the Federal Aviation Administration (FAA) are concerned with the following topics: human factors design criteria oriented to the FAA mission and systems (HF-STD-001), design and evaluation of air traffic control systems (DOT-VNTSC-FAA-95-3), elements of the human engineering program (FAA-HF-001), evaluation of human factor criteria conformance of equipment that have an interface with both operator (FAA-HF-002) and maintainer (FAA-HF-003).

The Department of Energy (DOE) in their standard DOE-HDBK-1140-2001 provides the system maintainability design criteria for DOE systems, equipment, and facilities. The Federal Highway Administration (FHA) establishes standards concerning the development and operation of traffic management centers (FHWA-JPO-99-042). The FHA also describes human factor guidelines and recommendations for the design of Advanced Traveler Information Systems (ATIS), Commercial Vehicle Operations (CVO), and concerning accommodation of the older drivers and pedestrians. The Nuclear Regulatory Commission provides guidelines for the HFE conformance evaluation of the interface design of the nuclear power plant systems (NUREG-0700 & NUREG-0711). For federal and federally-funded facilities, FED-STD-795 has been developed, which establishes the standards for facility accessibility by physically handicapped persons.

Relevant OSHA Standards

The development of occupational safety and health standards in the United States is mandated by the general duty clause—Section 5(a)(1) of the Occupational Safety and Health Act of 1970—which states that "each employer shall furnish to each of his employees, employment and a place of employment which is free from recognized hazards that are causing or are likely to cause death or serious harm to his employees." In general, penalties related to deficient and unsafe working conditions have been issued under this general duty clause. The general duty clause has also been supplemented by the Americans with Disabilities Act (Public Law 101–336, 1990). The disabilities act has an important bearing on the ergonomics design of workplaces. The ADA prohibits disability-based discrimination in hiring practices and requires that all employers make reasonable accommodations in working conditions to allow qualified disabled workers to perform their job functions.

TABLE 1.17

USA Governmental Human Factors/Ergonomics Standards (Based on HFS SubTAG, 2004)

Document Number	Title	Date	Source
National Aeronautics and Space Administration			
NASA-STD-3000B	Man–Systems Integration Standards	1995	http://msis.jsc.nasa.gov
Department of Transportation, Federal Aviation Administration			
HF-STD-001	Human Factors Design Standard	2003	http://www.hf.faa.gov/docs/508/ docs/wjhtc/hfds.zip
DOT-VNTSC-FAA- 95-3	Human Factors in the Design and Evaluation of Air Traffic Control Systems	1995	http://www.hf.faa.gov/docs/ volpehndk.zip
FAA-HF-001	Human Engineering Program Plan	1999	http://www.hf.faa.gov/docs/did_ 001.htm
FAA-HF-002	Human Engineering Design Approach Document—Operator	1999	http://www.hf.faa.gov/docs/did_ 002.htm
FAA-HF-003	Human Engineering Design Approach Document—Maintainer	1999	http://www.hf.faa.gov/docs/did_ 003.htm
FAA-HF-004	Critical Task Analysis Report	2000	http://hfetag.dtic.mil/docs-hfs/faa- hf-004_critical_task_analysis_ report.doc
FAA-HF-005	Human Engineering Simulation Concept	2000	http://hfetag.dtic.mil/docs-hfs/faa- hf-005_human-engineering_ simulation.doc
Department of Transportation, Federal Highway Agency			
FHWA-JPO-99-042	Preliminary Human Factors Guidelines for Traffic Management Centers	1999	http://plan2op.fhwa.dot.gov/pdfs/ pdf2/edll0303.pdf
FHWA-RD-98-057	Human Factors Design Guidelines for Advanced Traveler Information Systems (ATIS) and Commercial Vehicle Operations (CVO)	1998	http://www.fhwa.dot.gov/tfhrc/ safety/pubs/atis/index.html
FHWA-RD-01-051	Guidelines and Recommendations to Accommodate Older Drivers and Pedestrians	2001	http://www.tfhrc.gov/humanfac/ 01105/cover.htm
FHWA-RD-01-103	Highway Design Handbook for Older Drivers and Pedestrians	2001	http://www.tfhrc.gov/humanfac/ 01103/coverfront.htm
Department of Energy			
DOE-HDBK-1140-2001	Human Factors/Ergonomics Handbook for the Design for Ease of Maintenance	2001	http://tis.eh.doe.gov/techstds/ standard/hdbkl140/ hdbkl1402001_partl.pdf
Multiple Departments			
FED-STD-795	Uniform Federal Accessibility Standards	1988	http://assist.daps.dla.mil/docimages/ 0000/46/05/53835.pd5

In 1990, OSHA issued a set of voluntary guidelines entitled "Ergonomics Program Management Guidelines for Meatpacking Plants" (OSHA 3123) that have been successfully used by many types of industries—including those from outside the food production business. In 2000, the U.S. federal government proposed the Ergonomics Program Rule (Occupational Safety and Health Administration [OSHA], 2000). The main elements of the standard included (a) training in basic ergonomics awareness, (b) providing medical management of work-related musculoskeletal disorders, (c) implementing a quick fix or going to a full program, and (d) implementing a full ergonomic program when indicated, including elements such as management leadership, employee participation, job hazard analysis, hazard reduction and control, training, and program evaluation. However, the regulation was repealed in March 2001.

Recently, OSHA has published three voluntary guidelines to assist employers of the specific types of industries in recognizing and controlling hazards. These guidelines are the:

- Nursing Home Guideline (issued on March 13, 2003)
- Draft Guideline for Poultry Processing (issued on June 3, 2003)
- Guideline for The Retail Grocery Industry (issued on May 28, 2004)

In addition, OSHA plans to develop additional voluntary guidelines with the use of a standard protocol (OSHA, 2004). The objective of this standard protocol is to establish a fair and transparent process for developing industry- and task-specific guidelines that will assist employers and employees in recognizing and controlling potential problems. By using this protocol, each set of guidelines will address a particular industry or task. It is intended that the industry- and task-specific guidelines will generally be presented in three major parts: (a) program management recommendations for management practices addressing ergonomic hazards in the industry or task; (b) worksite analysis recommendations for worksite and workstation analysis techniques geared to the specific operations that are present in the industry or task; and (c) hazard control recommendations that contain descriptions of specific jobs and that detail the hazards associated with the operation, possible approaches to controlling the hazard, and the effectiveness of each control approach. Because there are many different types of work-related hazards, injuries, and controls that vary from industry to industry and task to task, OSHA expects that the scope and content of the guidelines will vary.

OTHER RELEVANT STANDARDS FOR OCCUPATIONAL SAFETY AND HEALTH

In 2000, the National Safety Council (NSC), acting on behalf of the Accredited Standards Committee (ASC), issued a draft document (known as Z-365) entitled Management of Work-Related Musculoskeletal Disorders (MSD). The draft defines the following areas of importance in preventing work-related injuries: (a) management responsibility, (b) employee involvement, (c) training, (d) surveillance, (e) evaluation and management of work-related MSD cases, (f) job analysis and design, and (g) follow-up.

Independently in 2001, another ANSI committee—ASC Z-10 Occupational Health Safety Systems—was formed under the auspices of the American Industrial Hygiene Association (AIHA). The main objective of ASC Z-10 is to develop a standard of management principles and systems to improve the occupational safety and health in companies.

ANSI STANDARDS

A list of the HFE-relevant standards developed by the ANSI is presented in Table 1.18.

TABLE 1.18

List of ANSI Standards for Human Factor and Ergonomics

Reference Number	Title
ANSI/HFS 100-2007	American National Standard for Human Factors Engineering of Computer Workstations
BSR/HFES 100 DS	Human Factors Engineering of Computer Workstations
HFES 200-2008	Human Factor Engineering of Software User Interfaces
ASC Z-365	Management of Work-Related Musculoskeletal Disorders
ASC Z-10	Occupational Health Safety Systems
ANSI B11	Ergonomic Guidelines for the Design, Installation and Use of Machine Tools

AMERICAN NATIONAL STANDARD FOR HUMAN FACTORS ENGINEERING OF VISUAL DISPLAY TERMINALS—ANSI/IIFS 100-1988

Human Factors Engineering of Computer Workstations—BSR/HFES 100 Draft Standard

TheBSR/HFES 100 Human Factors Engineering of Computer Workstations (HFES 100) is a specification of the recommended human factors and ergonomic principles related to the design of the computer workstation (Albin, 2005). This standard is organized into four major chapters: (1) Installed Systems, (2) Input Devices, (3) Visual Displays, and (4) Furniture. The major topics described in each of these chapters are listed in Table 1.19.

Ergonomic Requirements for Software User Interfaces (HFES 200—1998)

The HFES/HCI200 Committee that operates under the auspices of the Human Factors and Ergonomics Society's Technical Standards Committee has been working on the development of a proposed American national standard for software user interfaces. This standard will provide requirements and recommendations for software interfaces with a primary focus on business and personal computing applications. The standard is related to the ISO 9241 series of user interface standards. The topics described in each section of the HFES 200 standard are listed in Table 1.20. These topics are (a) accessibility, (b) color, and (c) voice and telephony.

ANSI B11 Ergonomic Guidelines for the Design, Installation, and Use of Machine Tools

ANSI Bll Technical Report: Ergonomic Guidelines for the Design, Installation, and Use of Machine Tools is a consensual ergonomic guideline developed by the Machine Tool Safety Standards Committee (Bll) of the American National Standards Institute. The subcommittee responsible for the preparation of these guidelines consisted of representatives from manufacturing, higher education, safety, design, and ergonomics. The document specifies the ergonomic guidelines to assist the design, installation, and use of individual and integrated machine tools and auxiliary components in manufacturing systems.

The guidelines underline the importance of three basic ideas to achieve effective and safe design, installation, and use of the machine tools: (a) communication among all people that are involved with the machine tools (users, installers, manufacturers, and designers), (b) dissemination of the knowledge concerning ergonomic concepts and principles among all individuals, and (c) ability to apply ergonomic concepts and principle knowledge effectively to machine tools and auxiliary components. The document states that the provision of worker safety, work efficiency, and optimization of the whole production system requires consideration of the following ergonomics issues:

TABLE 1.19

Main Chapters and Described Topics of the Human Factors Engineering of Computer Workstations—BSR/HFES 100 Draft Standard

Chapter	Issues
Installed systems	Hardware components, noise, thermal comfort, and lighting
Input devices	Keyboards, mouse and puck devices, trackballs, joysticks, styli and light pens, tablets and overlays, touch-sensitive panels
Visual displays	Monochrome and color CRT and flat panel displays (viewing characteristics, contrast, legibility, etc.)
Furniture	Specifications for workstation components (chairs, desks, etc.); postures (reference postures, reclined sitting, upright sitting, declined sitting, and standing); anthropometry

TABLE 1.20

Topics Addressed in the Ergonomic Requirements for Software User Interfaces (HFES 200—1998)

Chapter	Issues
Accessibility	Keyboard input; multiple keystrokes
	Customization, repeat rates, acceptance delays
	Pointer alternative; accelerators; remapping, navigation
	Display fonts: size, legibility, styles, colors
	Audio output: volume and frequencies, customization, content, and alerts; graphic
	Color: palettes, background–foreground, customization, coding
	Errors and persistence; online documentation and help
	Customization: cursor, button presses, click interval, pointer speed, chording
	Window appearance and behavior: navigation and location, window focus, titles
	Input focus: navigation, behavior, order, location
Color	Color selection: chromostereopsis, blending and depth effects, use of blue and red, identification and contrast
	Color assignments: conventions, uniqueness and reuse, naming, cultural assignments
	General use consideration: number of colors, highlighting, positioning, and separation
	Special uses: warnings, coding, state indications, pointers, area identification
Voice and telephony	Speech recognition (input): commands, vocabularies, prompts, consistency, feedback, error handling, dictation
	Speech output: vocabularies message format, speech characteristics, dialog techniques, physical properties, alerting tones, stereophonic presentation
	Nonspeech auditory output: consistency, tone format, critical messages, frequency, amplitude interactive voice response
Technical sections	Presentation of information, user guidance, menu dialogs, command dialogs, direct manipulation, dialogue boxes, and form-filling dialogs windows

1. The variation of the employee's physiological and psychological characteristics such as strength and capacity.
2. Incorporation of ergonomic concepts and principles into new project, tool, machine, and work process at the beginning.
3. Routine tasks that are done precisely, rapidly, and continuously—especially tasks in hazardous environments that should be performed by machines.
4. Tasks that require judgment and integration of information—that is, tasks that humans do the best should be assigned to workers.
5. A system that does not consider human limits such as information handling, perception, reach, clearance, posture, or strength exertion can predispose an accident or injury.

The aforementioned documents also recommend matching the design of the tools and processes with the physical characteristics and capability of workers to ensure accommodation, compatibility, operability, and maintainability of the machine tools and auxiliary components.

ISO 9000:2000—QUALITY MANAGEMENT STANDARDS

Quality standards can also play an important role in assuring safety and health at the workplace. ISO stipulates that if a quality management system is appropriately implemented using the eight

Quality Management Principles (see following) and in accordance with ISO 9004, all an organization's interested parties should benefit. For instance, people in the organization will benefit from the (a) improved working conditions, (b) increased job satisfaction, (c) improved health and safety, (d) improved morale, and (e) improved stability of employment, whereas the society at large will benefit from the (a) fulfillment of legal and regulatory requirements, (b) improved health and safety, (c) reduced environmental impact, and (d) increased security.

As discussed by Hoyle (2001), the term ISO 9000 refers to a set of quality management standards. ISO 9000 currently includes three quality standards: ISO 9000:2000, ISO 9001:2000, and ISO 9004:2000. ISO 9001:2000 presents requirements, whereas ISO 9000:2000 and ISO 9004:2000 present guidelines. ISO first published its quality standards in 1987, revised them in 1994, and then republished an updated version in 2000. These new standards are referred to as the "ISO 9000:2000 Standards."

It is recommended that the ISO 9001:2000 standard be used if an organization is seeking to create a management system that provides confidence in the conformance of its products to established requirements. The standard recognizes that the word "product" applies to services, processed material, and hardware and software intended for, or required, by the customer (Hoyle, 2001).

The ISO 9000:2000 Standards apply to all kinds of organizations such as manufacturing, service, government, and education. The standards are based on eight quality management principles:

- Principle 1: Customer focus
- Principle 2: Leadership
- Principle 3: Involvement of people
- Principle 4: Process approach
- Principle 5: System approach to management
- Principle 6: Continual improvement
- Principle 7: Factual approach to decision making
- Principle 8: Mutually beneficial supplier relationships

There are five sections in the standard that specify activities that need to be considered when implemented in the quality management system. According to Hoyle (2001), following a description of the activities that are used to supply products, the organization may exclude parts of the product realization section that are not applicable to its operations. The requirements in the other four sections—such as quality management system, management responsibility, resource management, and measurement analysis and improvement—apply to all organizations and they need to demonstrate how it is applied to the organization's quality manual or other documentation. These five sections of ISO 9001:2000 define what the organization should consistently do to provide products that meet customer and applicable statutory or regulatory requirements and enhance customer satisfaction by improving its quality management system. ISO 9004:2000 can be used to extend the benefits obtained from ISO 9001:2000 to the employees, owners, suppliers, and society, in general.

ISO 9001:2000 and ISO 9004:2000 is harmonized in structure and terminology to assist the organization to move smoothly from one to the other. Both standards apply a process approach. Processes are recognized as consisting of one or more linked activities that require resources and must be managed to achieve predetermined output. The output of one process may directly form the input to the next process and the final product is often the result of a network or system of processes. The eight Quality Management Principles stated in ISO 9000:2000 and ISO 9004:2000 provide the basis for the performance improvement outlined in ISO 9004:2000. The ISO 9000 standards cluster also includes other 10000 series standards. Table 1.21 shows a list of the relevant standards and their purposes.

TABLE 1.21

The ISO 9000 Quality Management Standards and Guidelines

Standards and Guidelines	Purpose
ISO 9000:2015, Quality management systems—Fundamentals and vocabulary	Establishes a starting point for understanding the standards and defines the fundamental terms and definitions used in the ISO 9000 family to avoid misunderstandings in their use.
ISO 9001:2015, Quality management systems—Requirements	This is the required standard to be used to assess the organization's ability to meet customer and applicable regulatory requirements and thereby address customer satisfaction.
	It is now the only standard in the ISO 9000 family against which third-party certification can be carried.
ISO 9004:2018, Quality management—Quality if an organization—Guidelines to achieve sustained success	This guideline standard provides guidance for continual improvement of the organization's quality management system to benefit all parties through sustained customer satisfaction.
ISO 19011:2018, Guidelines for Auditing Management Systems	Provides the organization with guidelines for verifying the system's ability to achieve defined quality objectives. You can use this standard internally or for auditing your suppliers.
ISO 10005:2018, Quality management—Guidelines for quality plans	Provides guidelines to assist in the preparation, review, acceptance, and revision of quality plans.
ISO 10006:2017, Quality management—Guidelines for quality management in project	Guidelines to help the organization ensure the quality of both the project processes and the project products.
ISO 10007:2017, Quality management—Guidelines for configuration management	Gives the organization the guidelines to ensure that a complex product continues to function when components are changed individually.
ISO/DIS 10012, Quality assurance requirements for measuring equipment—Part 1: Metrological confirmation system for measuring equipment	Gives the organization guidelines on the main features of a calibration system to ensure that measurements are made with the intended accuracy.
ISO 10012-2:1997, Quality assurance for measuring equipment—Part 2: Guidelines for control of measurement of processes	Provides supplementary guidance on the application of statistical process control when this is appropriate for achieving the objectives of Part 1
ISO 10013:1995, Guidelines for developing quality manuals	Provides guidelines for the development and maintenance of quality manuals, tailored to your specific needs.
ISO/TR 10014:1998, Guidelines for managing the economics of quality	Provides guidance on how to achieve economic benefits from the application of quality management.
ISO 10015:2019, Quality management—Guidelines for competence management and people development	Provides guidance on the development, implementation, maintenance, and improvement of strategies and systems for training that affects the quality of products.
ISO/TS 16949:2009, Quality Management Systems—Particular requirements for the application of ISO 9001:2008 for Automotive production and relevant service part organization	Sector-specific guidance to the application of ISO 9001 in the automotive industry.

As discussed by Hoyle (2001), ISO requires that the organization determine what it needs to do to satisfy its customers; establish a system to accomplish its objectives; and measure, review, and continually improve its performance. More specifically, the ISO 9001 and 9004 requirements stipulate that an organization shall:

1. Determine the needs and expectations of customers and other interested parties
2. Establish policies, objectives, and a work environment necessary to motivate the organization to satisfy these needs
3. Design, resource, and manage a system of interconnected processes necessary to implement the policy and attain the objectives
4. Measure and analyze the adequacy, efficiency, and effectiveness of each process in fulfilling its purpose and objectives
5. Pursue the continual improvement of the system from an objective evaluation of its performance

ISO identified several potential benefits of using the Quality Management Standards. These benefits may include the connection of quality management systems to organizational processes, encouragement of a natural progression toward improved organizational performance, and consideration of the needs of all interested parties.

EMPIRICAL STUDIES ON SELECTED HF/E STANDARDS

The existing body of literature shows that not many empirical studies were conducted on human factors and ergonomics standards. This section provides a review of empirical studies on the human factors and ergonomics standards and their evaluation. Following the emergence of any standard, it is beneficial to investigate it with respect to (a) evaluation of the success of the standard and (b) estimation of the extent of the standard's usage. There are a few basic researches in the existing body of literature that primarily focused on the standard's validity, reliability, usability, and scope of application. Ideally, a standard should be evaluated before it formally becomes available to the practitioners.

SOFTWARE ENGINEERING AND HUMAN-CENTERED DESIGN

Though numerous standards currently exist for software engineering, it is argued that there is a poor adoption of the standards (Fenton, Littlewood, & Page, 1993) and that the extent to which software engineering standards are actually used is unknown (El Emam & Garro, 2000). El Emam and Garro conducted their study on ISO/IEC 15504, which is an international standard on software process assessment. El Emam and Jung (2001) also empirically evaluated the ISO/IEC 15504 assessment model.

As reported in El Emam and Jung (2001), ISO/IEC 15504 consists of nine parts: (a) concepts and introductory guide, (b) a reference model for processes and process capability, (c) performing an assessment, (d) guide to performing assessments, (e) an assessment model and indicator guidance, (f) guide to qualification of assessors, (g) guide for use in process improvement, (h) guide for use in determining supplier process capability, and (i) vocabulary. The architecture of the standard is two-dimensional. One dimension is the "process" dimension that consists of the five categories of processes. The other dimension is the "capability" dimension that consists of five levels and nine attributes. A 4-point achievement scale constitutes a rating scheme that rates each of the attributes. In this study, a questionnaire method was used to obtain data on the (a) use of the assessment model, (b) usefulness and ease of use, (c) meaningfulness of rating aggregation scheme, (d) usability of the rating scale, (e) usefulness of the indicators, and (f) understanding of the process and capability dimensions.

A total of 70 worldwide assessments were collected from Asia Pacific, Europe, and the United States. About 33% of the assessments were performed in organizations such as software production, other IT products, and services. The remaining assessments were done by distribution and logistics, business, and defense. Most of the assessors were either managers or senior technical personnel. Approximately 98% of the assessors have received at least one training on software process assessment. Most assessors positively responded to the usefulness and ease of use of the standard model. However, they indicated problems while rating the process attributes on the process dimension. They also had difficulty in rating achievement. In general, the study found that most assessors positively responded to the meaningfulness of the aggregated rating scheme, usability of the rating scale, the usefulness of the indicators, and comprehension of the process and capability dimensions.

In their study, Earthy, Jones, and Bevan (2001) reviewed ISO 13407 and associated ISO TR 18529, which are the standards for human-centered design processes for interactive systems. Earthy et al. argued that Human Factors have processes that can be managed and integrated with existing project processes and that this internationally agreed set of human-centered design processes provides a definition of the capability that an organization must possess to implement user-centered design effectively. Earth et al. further speculated that the standard can also be used to assess the extent to which a particular development project employs user-centered design.

Usability and Human-Computer Interaction

The body of literature indicates that the most research on ergonomics-related standards was done on different parts of the ISO 9241 standard series, which can be termed as usability or ergonomics of human-computer interaction standard. As reported in Bastien, Scapin, and Leulier (1999), user-interface design guidelines appear in different types such as design guides, style guides, guidelines compilations, principles, and general guides. For years, guidelines compilations have been given considerable emphasis in writing design guides. In this context, the most cited work is the compilation of guidelines by Smith and Mosier (1984), which contains 944 guidelines for the design of user interfaces in single sentence forms with illustrated examples.

Subsequently, Smith and Mosier (1986) conducted an empirical survey on the utilization of compiled guidelines. In their analysis of questionnaire responses, it was found that most respondents considered the compilation useful. However, the respondents also reported problems in its use. For instance, users had difficulty in (a) locating relevant guidelines within the compilation, (b) choosing which guidelines to use, and (c) translating general guidelines into specific design rules.

Bastien, Scapin, and Leulier (1999) conducted a pilot evaluation of the ISO/DIS 9241-10 "dialogue principles" that compared the part with other guidelines (termed Ergonomic Criteria) and a control group that did not use any guidelines. Each group evaluated a musical database application. The study found that the median time spent evaluating the application was lowest for the control group followed by the ISO and Ergonomic Criteria groups. Multiple comparisons between groups showed that only the control and the Ergonomic Criteria groups differed significantly. The study also found that the highest numbers of usability problems were identified by the Ergonomic Criteria group followed by the ISO and control groups. As in the first case, multiple comparisons between groups showed that only the control and the Ergonomic Criteria groups differed significantly. The study concluded that the Ergonomic Criteria group spent significantly more time evaluating the application than did the Control group. The Ergonomic Criteria group also uncovered significantly more usability problems, though no significant differences were found between the Control and the ISO groups as well as between the ISO and the Ergonomic Criteria groups.

Dzida and Freitag (1998) investigated the use of scenarios to validate analysis and design according to ISO 9241-11 "Guidance on Usability." To validate scenarios and requirements, Dzida and Freitag argued that "the process of requirement construction can be conceived of a task domain (represented in terms of a context scenario) into the initial model of an artifact (represented in terms of a use scenario or prototype)." In the study, the semantic structure of scenarios consisted of three parts: (a) context scenario, (b) concept of use, and (c) use scenario. A subsequent case study resulted in the development of a new context scenario that suggests a number of requirements that may induce an additional iteration step in the prototyping process. Dzida and Freitag concluded that by consensus, the analyst, the designer, and the user can decide not to consider the obsolete prototype version as a valid solution of the problem.

Several studies evaluated computer input devices with respect to ISO 9241-9 standard on "Requirements for Non-Keyboard Input Devices" by utilizing the Fitt's law (e.g., Douglas, Kirkpatrick, & MacKenzie, 1999; Isokoski & Raisamo, 2002; Keates et al., 2002; MacKenzie & Jusoh, 2001; MacKenzie, Kauppinen, & Silfverberg, 2001; Oh & Stuerzlinger, 2002; Poupyrev et al., 2004; Silfverberg, MacKenzie, & Kauppinen, 2001; Sohn & Lee, 2004; Zhai, 2004). These studies utilized different types of input devices (such as different types of mice, touchpads, trackballs, and joysticks). Soukoreff and MacKenzie reviewed several studies and compared the results of those that only utilized Fitt's law with those that utilized Fitt's law with the ISO 9241-9 standard. Soukoreff and MacKenzie throughput values lower for those studies that utilized both Fitt's law and ISO 9241-9. Furthermore, Soukoreff and MacKenzie provided seven specific recommendations that were argued to support and, in some instances, supplement the ISO 9241-9 standard on the evaluation of pointing devices.

In a recent study, Sohn and Lee (2004) introduced an ultrasonic pointing device system that was developed for an interactive TV system. The study evaluated the pointing device and compared it to other conventional pointing devices using ISO 9241-9. The results showed that the throughput of the new pointing device was slightly lower than the touchpad and trackball. The average movement time of the new pointing device was higher than that of the trackball and the touchpad. Further, the average error rate of the new pointing device was higher than that of the touchpad and the trackball.

Besuijen and Spenkelink (1998) reviewed ISO 9241-3 "Visual Display Requirements." According to the standard (ISO 9241-3), visual display quality was defined as sufficient if a total of 25 requirements were met. The requirements were based on physical measurements in luminance and spatial domains. Besuijen and Spenkelink (1998) identified several problems with the standard, such as (a) modeling problem—concerns with the inability of the standard to estimate a meaningful overall quality index and (b) compliance to the individual physical parameters for image quality (because there are many characteristics that have interaction effects). Following the review of the standard, Besuijen and Spenkelink (1998) proposed comparative user performance test methods that included both qualitative and quantitative performance tests, visual measurement, subjective evaluation of the image quality, and utilization of the Display Evaluation Scale (DES), which is already established as a valid and reliable measurement.

Becker (1998) reviewed ISO/DIS 13406-2 "Ergonomic Requirements for Flat Panel Displays" and ISO 9241-7 "Requirements for Display with Reflections" and stated that the rating and classification of the ergonomic performance of visual display terminals in the presence of ambient light sources requires assessment of the reflective properties of the display unit. In the study, Becker introduced a novel measurement of the two-dimensional "Bidirectional Reflectance Distribution Function" (BRDF). By utilizing this new approach and instrument, the study evaluated several properties of visual display performance (such as haze, gloss, and distinctness of image).

Another study by Umezu et al. (1998) investigated the measurement of specular and diffuse reflection (ISO 9241-7) of two types of video display units such as CRT (Cathode Ray Tube) and LCD (Liquid Crystal Display). The study utilized typical commercially available photometers to verify the certainty of specular reflection coefficient measurement from an ergonomic standard. The results showed that the measured value difference among the three different photometers was 20%, which the study concluded to be high. Umezu et al. (1998) argued that because the specular reflection coefficient itself was very small, a slight difference would create a high error ratio. The study also utilized inspectors to change working distance and site. The results showed that such changes did not significantly affect the ISO 9241-7 requirements.

Lindfors (1998) examined the accuracy and repeatability of ISO 9241-7 for testing CRT-type video display units. The display units were tested in various laboratory settings of Europe and Japan. One of the two hypotheses tested in the study was to investigate whether "the defined test methods are repeatable and reliable." The results showed that with respect to specular and diffuse reflection and interlaboratory variation, the hypothesis was found to be false. However, Lindfors (1998) argued that the problems associated with the rejection of the hypothesis could be covered and rectified either by interlaboratory agreement or by a future minor update of ISO 9241-7.

In a recent study, Marmaras and Papadopoulus (2003) investigated the extent to which ergonomic requirements for work on computers are met in Greek office workstations. The study assessed 593 office workstations using an assessment tool consisting of 70 assessment points. The study results showed that the ergonomic requirements that are independent of the specific characteristics of individual workspaces and environments, such as design standards for seats, monitors, and input devices, are adequately met. Ergonomic requirements that should take into consideration the specific characteristics and constraints of individual work content, workspaces, and environments (e.g., requirements dealing with workplace layout, environmental conditions, software, and work organization) are inadequately met. Based on these results, the study recommended the (a) enhancement of efforts for the application of ergonomic principles in the enterprises, (b) involvement of ergonomists in decision making, design, and selection of computerized office components, (c) spreading of ergonomics awareness and knowledge through training, and (d) development of office workplace design methods and tools and facilitation of the application of ergonomics principles.

Physical Environment

In their study, Griefahn and Brode (1999) investigated the sensitivity of lateral motions relative to vertical motions. The sensitivity was then compared to predictions provided by ISO 2631. Two experiments were conducted where lateral and vertical motions were applied consecutively or simultaneously and where the magnitude of a single- or dual-axis test signal was adjusted until it was judged as equivalent to a preceding single-axis reference motion of the same frequency. From the results of those two experiments, it was substantiated that ISO 2631 was qualitatively valid.

Another study by Griefahn (2000) examined the validity of the cold stress model of ISO/TR 11079. The study investigated the possible limitations of the applicability of the IREQ model and the necessities and possibilities to improve the model. In the study, 16 female and 59 male workers (16–56 years) were monitored during their work. According to their cold stress at the workplace, they were allocated to three groups. The first group of 33 participants were exposed to constant temperatures of more than 103°C, the second group of 32 participants were exposed to less than 103°C, and a third group of ten participants experienced frequent temperature changes of 133°C. The study also manipulated the degree of physiological workload. The first group of eight participants worked at metabolic rates of less than 100 W/m^2, the second group of 50 participants worked between 101 and 164 W/m^2, and the third group of 17 worked at more than 165 W/m^2. The results revealed that the IREQ model applies for air temperatures up to 15°C and for temperature changes of 13°C (at least) but needs to be improved with respect to gender. The study also found that the IREQ model did not apply sufficiently for high and largely varying workloads (165 W/m^2 and more). However, the study concluded that these situations were beyond currently available possibilities to protect workers adequately with conventional clothing material.

A recent theoretical study by Olesen and Parsons (2002) discussed the existing ISO standards related to thermal comfort. The existing thermal comfort standard (ISO 7730), metabolic rate (ISO 8996), and clothing (ISO 9920) were critically appraised with respect to validity, reliability, usability, and scope for practical application.

Ishitake et al. (2002) investigated the effects of exposure to whole-body vibration (WBV) and the ISO 2631/1-1997 frequency weighting on gastric motility. The gastric motility was measured by electrogastrography (EGG) in nine healthy volunteers. In the study, sinusoidal vertical vibration at a frequency of 4, 6.3, 8, 12, 16, 31.5, or 63 Hz was given to the subjects for ten minutes. The magnitude of exposure at 4 Hz was $1.0 \, \text{m/s}^2$ (RMS). The magnitudes of the other frequencies gave the same frequency-weighted acceleration according to ISO 2631-1. The pattern of the dominant frequency histogram (DFH) was changed to a broad distribution pattern by vibration exposure. The results showed that vibration exposure had the effect of significantly reducing the percentage of time that the dominant component had a normal rhythm and increasing the percentage of time there was tachygastria ($p < 0.05$). It was also found out that vibration exposure generally reduced the mean percentage of time with the dominant frequency in the normal rhythm component. There was a significant difference between the condition of no vibration and exposure to 4 and 6.3 Hz of vibration frequency ($p < 05$). From the findings, it was concluded that the frequency weighting curve given in ISO 2631/1-1997 was not adequate in evaluating the physiological effects of WBV exposure on gastric motility.

ENTERPRISE ENGINEERING AND INTEGRATION

Kosanke and Nell (1999), in their study, reviewed the existing ISO standards for enterprise engineering and integration. In their review, Kosanke and Nell (1999) provided a road map for standardization in enterprise engineering and integration. This road map essentially incorporated the existing ISO 15704 "Requirements for Enterprise Reference Architecture" with ISO 14258, 9000, 14000, and ENV 12204. Kosanke and Nell also advocated for future research needs with respect to process representation, human role representation, infrastructure integration, facilitation of terminology, and standards landscaping.

CONCLUSIONS

Although human factors and ergonomics standards cannot guarantee appropriate workplace design, they can provide clear and well-defined requirements and guidelines and, therefore, are the basis for good ergonomics design. Standards for workstation design and work environment can ensure the safety and comfort of working people by establishing requirements for the optimal working condition. By providing consistency of the human-system interface and improving ergonomics quality of the interface components, ergonomics standards can also contribute to the enhanced systems usability and overall system performance. This benefit is based on the general requirement of harmonization across different tools and systems to support user performance and avoid unnecessary human errors.

One of the most important benefits from standardization efforts is a formal recognition of the significance of ergonomics requirements and guidelines for system design on the national and international levels (Harker, 1995). The consensus procedure applied to standards development demands consultation with a wide range of commercial, professional, and industrial organizations. Therefore, the decision to develop standards and consensus of diverse organizations concerning the need for standards reflects the formal recognition that there are important human factors and ergonomic issues that need to be taken into account during the design and development of workplaces and systems.

Standards represent the essence of the best available knowledge and practice extracted from a variety of academic sources and presented in a way that is easy to use by professional designers and to implement this knowledge into the design process. The consensus procedure makes the standards under development known and available to the interested parties and the general public. Such a procedure also facilitates the dissemination and promotion of human factors and ergonomics knowledge across non-experts.

REFERENCES

Albin, T. J. (2005). Board of Standards Review/Human Factor and Ergonomics Society 100—Human Factors Engineering of Computer Workstations—Draft Standard for Trial Use. In Karwowski, W. (Ed.), *2005 Handbook of human factors and ergonomics standards and guidelines*, Lawrence Erlbaum Publishers.

Bastien, J. M. C., Scapin, D. L., & Leulier, C. (1999). The ergonomic criteria and the ISO/DIS 9241-10 dialogue principles: A pilot comparison in an evaluation task. *Interacting with Computers*, 11(3), 299–322.

Becker, M. (1998). Evaluation and characterization of display reflectance, *Displays*, 19, 35–54.

Besuijen, K., & Spenkelink, G. P. J. (1998). Standardizing visual display quality, *Displays*, 19, 67–76.

CEN. (2004). European Standardization Committee Web site. http://www.cenorm.be/cenorm/index.htm

Chapanis, A. (1996). *Human factors in systems engineering*. New York: Wiley.

Dickinson, C. E. (1995). Proposed manual handling international and European Standards. *Applied Ergonomics*, 26(4), 265–270.

Department of Industrial Relation. (2004). *California Department of Industrial Relation homepage*. http://www.dir.ca.gov/

Douglas, S. A., Kirkpatrick, A. E., & MacKenzie, I. S. (1999). Testing pointing device performance and user assessment with the ISO 9241, Part 9 standard. In *Proceedings of the ACM Conference on Human Factors in Computing Systems—CHI'99* (pp. 215–222). ACM: New York.

Dul, J., de Vlaming, P. M., & Munnik, M. J. (1996). A review of ISO and CEN standards on ergonomics. *International Journal of Industrial Ergonomics*, 17(3), 291–297.

Dzida, W., & Freitag, R. (1998), Making use of scenarios for validating analysis and design, *IEEE Transactions on Software Engineering*, 24(12), 1182–1196.

Earthy, J., Jones, B. S., & Bevan, N. (2001). The improvement of human-centered processes—Facing the challenge and reaping the benefits of ISO 13407. *International Journal of Human-Computer Studies*, 55, 553–585.

Eibl, M. (2005). International Standards of Interface Design. In W. Karwowski (Ed.), *Handbook of human factors and ergonomics standards and guidelines*. Lawrence Erlbaum Associates.

El Emam, K. & Garro, I. (2000). Estimating the extent of standards use: The case of ISO/IEC 15504. *The Journal of Systems and Software*, 53, 137–143.

El Emam, K., & Jung, H.-W. (2001). An empirical evaluation of the ISO/IEC 15504 assessment model. *The Journal of Systems and Software*, 59, 23–41.

Fenton, N., Littlewood, B., & Page, S. (1993). Evaluating software engineering standards and methods. In R. Thayer & R. McGettrick (Eds.), *Software engineering: A European perspective*, IEEE Computer Society Press: Silver Spring, MD, (pp. 463–470).

Griefahn, B. (2000). Limits of and possibilities to improve the IREQ cold stress model (ISO/TR 11079): A validation study in the field. *Applied Ergonomics*, 31, 423–431.

Griefahn, B., & Brode, P. (1999). The significance of lateral whole-body vibrations related to separately and simultaneously applied vertical motions: A validation study of ISO 2631. *Applied Ergonomics*, 30, 505–513.

Harker, S. (1995). The development of ergonomics standards for software. *Applied Ergonomics*, 26(4), 275–279.

Hoyle, D. (2001). *ISO 9000: Quality systems handbook*. Oxford: Butterworth Heinemann.

Human Factors Standardization SubTAG (HFS SubTAG). (2004). Index of Government Standards on Human Engineering Design Criteria, Processes, and Procedures. Version 1 (Draft), March 15, 2004. *Department of Defense, Human Factors Engineering Technical Advisory Group*. Web site: http://hfetag.dtic.mil/docs/index_govt_std.doc

ILO (2004). *International Labor Organization Web site*, http://www.ilo.org/public/english/index.htm

ILO-OSH (2001). Guidelines on occupational safety and health management systems, 1LO-OSH 2001. Geneva, International Labour Office, http://www.ilo.org/public/english/protection/safework/managmnt/guide.htm

Ishitake, T., Miyazaki, Y., Noguchi, R., Ando, H., & Matoba, T. (2002). Evaluation of frequency weighting (ISO 2631-1) for acute effects of whole-body vibration on gastric motility, *Journal of Sound and Vibration*, 253(1), 31–36).

Isokoski, P., & Raisamo, R. (2002). Speed-accuracy measures in a population of six mice. In *Proceedings of the Fifth Asia Pacific Conference on Human-Computer Interaction—APCHI 2002* (pp. 765–777). Beijing, China: Science Press.

ISO (2004). *International Standardization Organization Web site.* http://www.iso.org/iso/en/ISOOnline. openerpage

Keates, S., Hwang, F., Langdon, P., Clarkson, P. J., & Robinson, P. (2002). Cursor measures for motion impaired computer users. In *Proceedings of the Fifth ACM Conference on Assistive Technology—ASSETS 2002.* (pp. 135–142). New York: ACM.

Kosanke, K., & Nell, J. G. (1999). Standardization in ISO for enterprise engineering and integration, *Computers in Industry*, 40, 311–319.

Lindfors, M. (1998). Accuracy and repeatability of the ISO 9241-7 test method. *Displays*, 19, 3–16.

MacKenzie, I. S., & Jusoh, S. (2001). An evalution of two input devices for remote pointing. In *Proceedings of the Eighth IFIP Working Conference on Engineering for Human-Computer Interaction—EHCI2001.* (pp. 235–249). Springer-Verlag, Heidelberg, Germany.

MacKenzie, I. S., Kauppinen, T., & Silfverberg, M. (2001). Accuracy measures for evaluating computer pointing devices. In *Proceedings of the Human Factors in Computing Systems, CHI Letters*, 3(1), 9–16.

Marmaras, N., & Papadopoulus, S. (2003). A study of computerized offices in Greece: Are ergonomic design requirements met? *International Journal of Human-Computer Interaction*, 16(2), 261–281.

McDaniel, J. W. (1996). The demise of military standards may affect ergonomics. *International Journal of Industrial Ergonomics*, 18(5–6), 339–348.

Nachreiner, F. (1995). Standards for ergonomics principles relating to the design of work systems and to mental workload. *Applied Ergonomics*, 26(4), 259–263.

Oh, J.-Y., & Stuerzlinger, W. (2002). Laser pointers as collaborate pointing devices. In *Proceedings of Graphics Interface* (pp. 141–149). AK Peters and CHCCS.

Olesen, B. W. (1995). International standards and the ergonomics of the thermal environment. *Applied Ergonomics*, 26(4), 293–302.

Olesen, B. W., & Parsons, K. C. (2002). Introduction to thermal comfort standards and to the proposed new version of EN ISO 7730. *Energy and Buildings*, 34(6), 537–548.

Occupational Safety and Health Administration (OSHA) (2000). Ergonomics Program Rule. *Federal Register*, 65, 220.

OSHA (2004). *Occupational Safety and Health Administration Web site.* http://www.osha-sic.gov

Parsons, K. C. (1995a). Ergonomics and international standards. *Applied Ergonomics*, 26(4), 237–238.

Parsons, K. C. (1995b). Ergonomics of the physical environment: International ergonomics standards concerning speech communication, danger signals, lighting, vibration and surface temperatures. *Applied Ergonomics*, 26(4), 281–292.

Parsons, K. C. (1995c). Ergonomics and international standards: Introduction, brief review of standards for anthropometry and control room design and useful information. *Applied Ergonomics*, 26(4), 239–247.

Parsons, K. C., Shackel, B., and Metz, B. (1995). Ergonomics and international standards: History, organizational structure and method of development. *Applied Ergonomics*, 26(4), 249–258.

Poupyrev, I., Okabe, M., & Maruyama, S. (2004). Haptic feedback for pen computing: Directions and strategies. In *CHI'04 Extended Abstracts on Human Factors in Computing Systems* (pp. 1309–1312). Vienna Austria.

Public Law 101–336 (1990). Americans with Disabilities Act. *Public Law 336 of the 101st Congress, enacted July* 26, 1990.

Poupyrev, I., Okabe, M., & Maruyama, S., (2004). Haptic feedback for pen computing: Directions and strategies. *Extended Abstracts of the ACM Conference on Human Factors in Computing Systems—CHI 2004* (pp. 1309–1312). New York: ACM.

Silfverberg, M., MacKenzie, I. S., & Kauppinen, T. (2001). An isometric joystick as a pointing device for handheld information terminals. In *Proceedings of Graphics Interface of Canadian Information Processing Society* (pp. 119–126). Toronto, Canada.

Smith, S. L., & Mosier, J. N. (1984). The user interface to computer-based information systems: A survey of current software design practice. *Behaviour & Information Technology*, 3, 195–203.

Smith, S. L., & Mosier, J. N. (1986). Guidelines for designing user interface software. Retrieved from http://citeseerx.ist.psu.edu/viewdoc/download?doi=10.1.1.84.8930&rep=rep1&type=pdf2.

Sohn, M., & Lee, G. (2004). SonarPen: An ultrasonic pointing device for and interactive TV. *IEEE Transactions on Consumer Electronics*, 50(2), 413–419.

Spivak, S. M., & Brenner, F. C. (2001). *Standardization essentials: Principles and practice.* New York: Dekker.

Stewart, T. (1995). Ergonomics standards concerning human-system interaction: Visual displays, controls and environmental requirements. *Applied Ergonomics*, 26(4), 271–274.

Umezu, N., Nakano, Y., Sakai, T., Yoshitake, R., Herlitschkc, W., & Kubota, S. (1998). Specular and diffuse reflection measurement feasibility study of ISO 9241 Part 7 method. *Displays*, 19, 17–25.

Wettig, J. (2002). New developments in standardization in the past 15 years—Product versus process related standards. *Safety Science*, 40 (1-4), 51–56.

Zhai, S. (2004). Characterizing computer input with Fitt's law parameters—The information and non-information aspects of pointing. *International Journal of Human-Computer Studies*, 61, 791–809.

APPENDIX 1.1 HFE STANDARDS PUBLISHED BY OTHER THAN TC 159 ISO TECHNICAL COMMITTEES

CIE—International Commission on Illumination

ISO/CIE 8995:2002   Lighting of indoor workplaces

JTC 1/SC 6—Telecommunications and information exchange between systems

ISO/EEC 10021-2:2003   Information technology—Message Handling Systems (MHS): Overall architecture

JTC 1/SC 7—Software and system engineering

ISO/IEC TR 9126-4:2004   Software engineering—Product quality—Part 4: Quality in use metrics

ISO/IEC 12119:1994   Information technology—Software packages—Quality requirements and testing

ISO/IEC 12207:2008   Systems and software engineering—Software life-cycle processes

ISO/IEC 14598-1:1999   Information technology—Software product evaluation—Part 1: General overview

ISO/IEC 14598-4:1999   Software engineering—Product evaluation—Part 4: Process for acquirers

ISO/IEC 14598-6:2001   Software engineering—Product evaluation—Part 6: Documentation of evaluation modules

ISO/IEC 15288:2008   Systems and software engineering—System life-cycle processes

ISO/IEC TR 15504-5:1999   Information technology—Software Process Assessment—Part 5: An assessment model and indicator guidance

ISO/IEC 15910:1999   Information technology—Software user documentation process

ISO/IEC 18019:2004   Software and system engineering—Guidelines for the design and preparation of user documentation for application software

ISO/IEC TR 19760:2003   Systems engineering—A guide for the application of ISO/IEC 15288 (System life-cycle processes)

ISO/IEC 20926:2009   Software and systems engineering—Software measurement—IFPUG functional size measurement Method 2009

ISO/IEC 20968:2002   Software engineering—Mk II Function Point Analysis—Counting Practices Manual

JTC 1/SC 22—Programming languages, their environments, and system software interfaces

ISO/IEC TR 11017:1998   Information technology—Framework for internationalization

ISO/IEC TR 14252:1996   Information technology—Guide to the POSIX Open System Environment

ISO/IEC TR 15942:2000   Information technology—Programming languages—Guide for the use of the Ada programming language in high-integrity systems

JTC 1/SC 27—IT Security techniques

ISO/IEC TR 13335-4:2000   Information technology—Guidelines for the management of IT Security—Part 4: Selection of safeguards

ISO/IEC 21827:2008   Information technology—Security Techniques—Systems Security Engineering—Capability Maturity Model® (SSE-CMM®)

JTC 1/SC 35—User interfaces

ISO/IEC 15411:1999   Information technology—Segmented keyboard layouts

ISO/IEC 18035:2003   Information technology—Icon symbols and functions for controlling multimedia software applications

TC 8/SC 5—Ships' bridge layout

ISO 8468:2007   Ships and marine technology—Ship's bridge layout and associated—Requirements and guidelines

ISO 14612:2004   Ships and marine technology—Ship's bridge layout and associated equipment—Additional requirements and guidelines for centralized and integrated bridge functions

TC 8/SC 6—Navigation

ISO 16273:2003   Ships and marine technology—Night vision equipment for high-speed craft—Operational and performance requirements, methods of testing, and required test results

TC 20—Aircraft and space vehicles

ISO/TR 10201:2001   Aerospace—Standards for electronic instruments and systems

TC 20/SC 1—Aerospace electrical requirements

ISO 6858:2007   Aircraft—Ground support electrical supplies—General requirements

TC 20/SC 14—Space systems and operations

ISO 16091:2018   Space systems—Integrated logistic support

ISO 17399:2003   Space systems—Man-systems integration

TC 21/SC 3—Fire detection and alarm systems

ISO 12239:2010   Smoke alarms using scattered light, transmitted light or ionization

TC 22/SC 3—Electrical and electronic equipment

ISO 11748-2:2001   Road vehicles—Technical documentation of electrical and electronic systems—Part 2: Documentation agreement

ISO/TR 15497:2000   Road vehicles—Development guidelines for vehicle based software

TC 22/SC 13—Ergonomics applicable to road vehicles

ISO 2575:2010   Road vehicles—Symbols for controls, indicators, and tell-tales

ISO 3958:1996   Passenger cars—Driver hand-control reach

ISO 4040:2009   Road vehicles—Location of hand controls, indicators, and tell-tales in motor vehicles

ISO 6549:1999   Road vehicles—Procedure for H- and R-point determination

ISO/TR 9511:1991   Road vehicles—Driver hand-control reach—In-vehicle checking procedure

ISO/TS 12104:2003   Road vehicles—Gearshift patterns—Manual transmissions with power-assisted gear change and automatic transmissions with manual-gearshift mode

ISO 12214:2018   Road vehicles—Direction-of-motion stereotypes for automotive hand controls

ISO 15005:2002   Road vehicles—Ergonomic aspects of transport information and control systems—Dialogue management principles and compliance procedures

ISO 15007-1:2014   Road vehicles—Measurement of driver visual behavior with respect to transport information and control systems—Part 1: Definitions and parameters

ISO/TS 15007-2:2014   Road vehicles—Measurement of driver visual behavior with respect to transport information and control systems—Part 2: Equipment and procedures

ISO 15008:2017   Road vehicles—Ergonomic aspects of transport information and control systems—Specifications and compliance procedures for in-vehicle visual presentation

ISO/TS 16951:2004   Road vehicles—Ergonomic aspects of transport information and control systems (TICS)—Procedures for determining priority of on-board messages presented to drivers

ISO 17287:2003   Road vehicles—Ergonomic aspects of transport information and control systems—Procedure for assessing suitability for use while driving

TC 22/SC 17—Visibility

ISO 7397-1:1993   Passenger cars—Verification of driver's direct field of view—Part 1: Vehicle positioning for static measurement

ISO 7397-2:1993   Passenger cars—Verification of driver's direct field of view—Part 2: Test method

TC 23/SC 3—Safety and comfort of the operator

ISO 4254-1:2013   Agricultural machinery—Safety—Part 1: General Requirements

ISO/TS 15077:2002   Tractors and self-propelled machinery for agriculture and forestry—Operator controls—Actuating forces, displacement, location and method of operation

TC 23/SC 4—Tractors

ISO 4253:1993   Agricultural tractors—Operator's seating accommodation—Dimensions

ISO 5721-1:2013   Agricultural Tractors—Requirements, test procedures and acceptance criteria for the operators' field of Vision—Part 1: Field of vision to the front

TC 23/SC 7—Equipment for harvesting and conservation

ISO 8210:1989	Equipment for harvesting—Combine harvesters—Test procedure

TC 23/SC 14—Operator controls, operator symbols and other displays, operator manuals

ISO 3767-1:2016	Tractors, machinery for agriculture and forestry, powered lawn and garden equipment—Symbols for operator controls and other displays—Part 1: Common symbols
ISO 3767-2:2016	Tractors, machinery for agriculture and forestry, powered lawn and garden equipment—Symbols for operator controls and other displays—Part 2: Symbols for agricultural tractors and machinery
ISO 3767-3:2016	Tractors, machinery for agriculture and forestry, powered lawn and garden equipment—Symbols for operator controls and other displays—Part 3: Symbols for powered lawn and garden equipment
ISO 3767-5:2016	Tractors, machinery for agriculture and forestry, powered lawn and garden equipment—Symbols for operator controls and other displays—Part 5: Symbols for manual portable forestry machinery

TC 23/SC 15—Machinery for forestry

ISO 11850:2011	Machinery for forestry—General Safety Requirements

TC 23/SC 17—Manually portable forest machinery

ISO 8334:2007	Forestry machinery—Portable chain-saws—Determination of balance and maximum holding moment
ISO 11680-1:2011	Machinery for forestry—Safety requirements and testing for pole-mounted powered pruners—Part 1: Machines fitted with an integral combustion engine
ISO 11680-2:2000	Machinery for forestry—Safety requirements and testing for pole-mounted powered pruners—Part 2: Machines for use with a back-pack power source
ISO 11681-1:2011	Machinery for forestry—Portable chain-saw safety requirements and testing—Part 1: Chainsaws for forest service
ISO 11681-2:2011	Machinery for forestry—Portable chain-saws safety requirements and testing—Part 2: Chainsaws for tree service
ISO 11806-1:2011	Agricultural and forestry machinery—Safety requirements and testing for portable hand-held, powered brush cutters and grass trimmers—Part 1: Machines fitted with integral combustion engine
ISO 11806-2:2011	Agricultural and forestry machinery—Safety requirements and testing for portable hand-held, powered brush cutters and grass trimmers—Part 2: Machines for use with back-pack power unit
ISO 14740:1998	Forest machinery—Backpack power units for brush-cutters, grass-trimmers, pole-cutters and similar appliances—Safety requirements and testing

TC 23/SC 18—Irrigation and drainage equipment and systems

ISO/TR 8059:1986	Irrigation equipment—Automatic irrigation systems—Hydraulic control

TC 38—Textiles

ISO 15831:2004	Clothing—Physiological effects—Measurement of thermal insulation by means of a thermal manikin

TC 43/SC 1—Noise

ISO 11690-1:2020	Acoustics—Recommended practice for the design of low-noise workplaces containing machinery—Part 1: Noise control strategies
ISO 15667:2000	Acoustics—Guidelines for noise control by enclosures and cabins

TC 46—Information and documentation

ISO 7220:1996	Information and documentation—Presentation of catalogues of standards

TC 59/SC 3—Functional/user requirements and performance in building construction

ISO 6242-1:1992	Building construction—Expression of users' requirements—Part 1: Thermal requirements
ISO 6242-2:1992	Building construction—Expression of users' requirements—Part 2: Air purity requirements
ISO 6242-3:1992	Building construction—Expression of users' requirements—Part 3: Acoustical requirements

TC 67—Materials, equipment and offshore structures for petroleum, petrochemical, and natural gas industries

ISO 13879:1999	Petroleum and natural gas industries—Content and drafting of a functional specification
ISO 13880:1999	Petroleum and natural gas industries—Content and drafting of a technical specification

TC 67/SC 6—Processing equipment and systems

ISO 13702:2015	Petroleum and natural gas industries—Control and mitigation of fires and explosions on offshore production installations—Requirements and guidelines
ISO 15544:2000	Petroleum and natural gas industries—Offshore production installations—Requirements and guidelines for emergency response
ISO 17776:2016	Petroleum and natural gas industries—Offshore production installations—Major accidents hazard management during of new installations

TC 69—Applications of statistical methods

ISO 10725:2000	Acceptance sampling plans and procedures for the inspection of bulk materials

TC 72/SC 5—Industrial laundry and dry-cleaning machinery and accessories

ISO 8230-1:2008	Safety requirements for dry-cleaning machines Part 1: Common safety requirements
ISO 8230-2:2008	Safety requirements for dry-cleaning machines Part 2: Machines using Perchloroethylene
ISO 8230-3:2008	Safety requirements for dry-cleaning machines Part 3: Machines using combustible solvents
ISO 10472-1:1997	Safety requirements for industrial laundry machinery—Part 1: Common requirements

TC 72/SC 8—Safety requirements for textile machinery

ISO 11111:1995	Safety requirements for textile machinery

TC 85/SC 2—Radiation protection

ISO 17874-1:2010	Remote handling devices for radioactive materials—Part 1: General requirements

TC 92/SC 3—Fire threat to people and environment

ISO/TS 13571:2012	Life-threatening components of fire—Guidelines for the estimation of time to compromised tenability in fires

TC 92/SC 4—Fire safety engineering

ISO/TR 13387-1:1999	Fire safety engineering—Part 1: Application of fire performance concepts to design objectives

TC 94/SC 4—Personal equipment for protection against falls

ISO 10333-6:2004	Personal fall-arrest systems—Part 6: System performance tests

TC 94/SC 13—Protective clothing

ISO 11393-4:2018	Protective clothing for users of hand-held chainsaws—Part 4: Performance requirements and test methods for protective gloves
ISO 13688:2013	Protective clothing—General requirements Clothing for protection against contact
ISO 16603:2004	with blood and body fluids—Determination of the resistance of protective clothing materials to penetration by blood and body fluids—Test method using synthetic blood
ISO 16604:2004	Clothing for protection against contact with blood and body fluids—Determination of resistance of protective clothing materials to penetration by blood-borne pathogens—Test method using Phi-X 174 bacteriophage

TC 101—Continuous mechanical handling equipment

ISO/TR 5045:1979	Continuous mechanical handling equipment—Safety code for belt conveyors—Examples for guarding of nip points

TC 108/SC 2—Measurement and evaluation of mechanical vibration and shock as applied to machines, vehicles, and structures

ISO 14964:2000	Mechanical vibration and shock—Vibration of stationary structures—Specific requirements for quality management in measurement and evaluation of vibration

TC 108/SC 4—Human exposure to mechanical vibration and shock

ISO 2631-1:1997	Mechanical vibration and shock—Evaluation of human exposure to whole-body vibration—Part 1: General requirements
ISO 2631-2:2003	Mechanical vibration and shock—Evaluation of human exposure to whole-body vibration—Part 2: Vibration in buildings (1 Hz to 80 Hz)
ISO 2631-4:2001	Mechanical vibration and shock—Evaluation of human exposure to whole-body vibration—Part 4: Guidelines for the evaluation of the effects of vibration and rotational motion on passenger and crew comfort in fixed-guideway transport systems
ISO 2631-5:2018	Mechanical vibration and shock—Evaluation of human exposure to whole-body vibration—Part 5: Method for evaluation of vibration containing multiple shocks
ISO 5349-1:2001	Mechanical vibration—Measurement and evaluation of human exposure to hand-transmitted vibration—Part 1: General requirements

TC 108/SC 4—Human exposure to mechanical vibration and shock

ISO 5982:2019	Mechanical vibration and shock—Range of idealized values to characterize human biodynamics response under vertical vibration
ISO 6897:1984	Guidelines for the evaluation of the response of occupants of fixed structures, especially buildings and off-shore structures, to low-frequency horizontal motion (0.063 to 1 Hz)
ISO 8727:1997	Mechanical vibration and shock—Human exposure—Biodynamic coordinate systems
ISO 9996:1996	Mechanical vibration and shock—Disturbance to human activity and performance—Classification
ISO 10068:2012	Mechanical vibration and shock—Mechanical impedance of the human hand-arm system at the driving point
ISO 13090-1:1998	Mechanical vibration and shock—Guidance on safety aspects of tests and experiments with people—Part 1: Exposure to whole-body mechanical vibration and repeated shock
ISO 13091-1:2001	Mechanical vibration—Vibrotactile perception thresholds for the assessment of nerve dysfunction—Part 1: Methods of measurement at the fingertips
ISO 13091-2:2003	Mechanical vibration—Vibrotactile perception thresholds for the assessment of nerve dysfunction—Part 2: Analysis and interpretation of measurements at the fingertips

TC 121/SC 1—Breathing attachments and anaesthetic machines

ISO 7767:1997 Oxygen monitors for monitoring patient breathing mixtures—Safety requirements

TC 121/SC 3—Lung ventilators and related equipment

ISO 8185:2007	Respiratory tract Humidifiers for medical use—Particular requirements for respiratory humidification systems
IEC 60601-1-8:2006	Medical electrical equipment—Part 1–8: General requirements for safety and essential performance—Collateral standard: General requirements, tests, and guidance for alarm systems in medical electrical equipment and medical electrical systems-amendment 2
IEC 60601-2-12:2001	Medical electrical equipment—Part 2–12: Particular requirements for the safety of lung ventilators—Critical care ventilators

TC 123/SC 5—Quality analysis and assurance

ISO 12307-1:1994   Plain bearings—Wrapped bushes—Part 1: Checking the outside diameter

TC 127/SC 1—Test methods relating to machine performance

ISO 8813:1992	Earth-moving machinery—Lift capacity of pipelayers and wheeled tractors or loaders equipped with side boom

TC 127/SC 2—Safety requirements and human factors

ISO 2860:1992	Earth-moving machinery—Minimum access dimensions
ISO 2867:2011	Earth-moving machinery—Access systems
ISO 3164:2013	Earth-moving machinery—Laboratory evaluations of protective structures—Specifications for deflection-limiting volume

ISO 3411:2007	Earth-moving machinery—Physical dimensions of operators and minimum operator space envelope
ISO 3449:2005	Earth-moving machinery—Falling-object protective structures—Laboratory tests and performance requirements
ISO 3450:2011	Earth-moving machinery—Wheeled or high speed rubber-tracked machines—Performance requirements and test procedures for brake systems
ISO 3457:2003	Earth-moving machinery—Guards—Definitions and requirements
ISO 3471:2008	Earth-moving machinery—Roll-over protective structures—Laboratory tests and performance requirements
ISO 5006:2017	Earth-moving machinery—Operator's field of view—Test method and performance criteria
ISO 5010:2019	Earth-moving machinery—Wheeled machines—Steering requirements
ISO 5353:1995	Earth-moving machinery, and tractors and machinery for agriculture and forestry—Seat index point
ISO 6682:1986	Earth-moving machinery—Zones of comfort and reach for controls
ISO 7096:2020	Earth-moving machinery—Laboratory evaluation of operator seat vibration
ISO 8643:2017	Earth-moving machinery—Hydraulic excavator and backhoe loader lowering control device—Requirements and tests
ISO 9244:2008	Earth-moving machinery—Machine safety labels—General principles
ISO/TR 9953:1996	Earth-moving machinery—Warning devices for slow-moving machines—Ultrasonic and other systems
ISO 10262:1998	Earth-moving machinery—Hydraulic excavators—Laboratory tests and performance requirements for operator protective guards
ISO 10263-1:2009	Earth-moving machinery—Operator enclosure environment—Part 1: Terms and definition
ISO 10263-2:2009	Earth-moving machinery—Operator enclosure environment—Part 2: Air filter element test method
ISO 10263-3:2009	Earth-moving machinery—Operator enclosure environment—Part 3: Pressurization test method
ISO 10263-4:2009	Earth-moving machinery—Operator enclosure environment—Part 4: Heating ventilating and conditioning (HVAC) test method and performance
ISO 10263-5:2009	Earth-moving machinery—Operator enclosure environment—Part 5: Windscreen defrosting system test method
ISO 10263-6:2009	Earth-moving machinery—Operator enclosure environment—Part 6: Determination of effect of solar heating
ISO 10533:1993	Earth-moving machinery—Lift-arm support devices
ISO 10567:2007	Earth-moving machinery—Hydraulic excavators—Lift capacity
ISO 10570:2004	Earth-moving machinery—Articulated frame lock—Performance requirements
ISO 10968:2020	Earth-moving machinery—Operator's controls
ISO 11112:1995	Earth-moving machinery—Operator's seat—Dimensions and requirements
ISO 12117:2008	Earth-moving machinery—Laboratory test and performance requirements for protective structures of excavators—Part 2: Roll-over protective structures (ROPS) for excavators of over 6 T
ISO 12508:1994	Earth-moving machinery—Operator station and maintenance areas—Bluntness of edges
ISO 13333:1994	Earth-moving machinery—Dumper body support and operator's cab tilt support devices
ISO 13459:2012	Earth-moving machinery—Trainer seat-Deflection limiting volume, space envelope and performance requirements
ISO 17063:2003	Earth-moving machinery—Braking systems of pedestrian-controlled machines—Performance requirements and test procedures

TC 130—Graphic technology

ISO 12648:2006	Graphic technology—Safety requirements for printing press systems
ISO 12649:2004	Graphic technology—Safety requirements for binding and finishing systems and equipment

TC 131/SC 9—Installations and systems

ISO 4413:2010   Hydraulic fluid power—General rules and safety requirements for systems and their components

ISO 4414:2010   Pneumatic fluid power—General rules and safety requirements for systems and their components

TC 136—Furniture

ISO 5970:1979   Furniture—Chairs and tables for educational institutions—Functional sizes

TC 163/SC 2—Calculation methods

ISO 13790:2008   Energy performance of buildings—Calculation of energy use for space heating and cooling

TC 171/SC 2—Application issues

ISO/TR 14105:2011   Document management—Change management for successful Electronic document management system (EDMS) implementation

TC 172/SC 9—Electro-optical systems

ISO 11553-1:2020   Safety of machinery—Laser processing machines—Part 1: Laser Safety requirements

ISO 11553-2:2007   Safety of machinery—Laser processing machines—Part 2: Safety requirements for hand-held laser processing devices

ISO 11553-3:2013   Safety of machinery—Laser processing machines—Part 3: Noise reduction and noise measurement requirements for hand-held laser Methods for laser processing machines and hand-held processing devices and associated auxiliary equipment (accuracy grade 2)

TC 173—Assistive products for persons with disability

ISO 11199-1:1999   Walking aids manipulated by both arms—Requirements and test methods—Part 1: Walking frames

ISO 11199-2:2005   Walking aids manipulated by both arms—Requirements and test methods—Part 2: Rollators

ISO 11334-1:2007   Assistive products for walking manipulated by one arm—Requirements and test methods—Part 1: Elbow crutches

ISO 11334-4:1999   Walking aids manipulated by one arm—Requirements and test methods—Part 4: Walking sticks with three or more legs

TC 173/SC 3—Aids for ostomy and incontinence

ISO 15621:2017   Absorbent incontinence aids for urine and/or faeces—General guidance on evaluation

TC 173/SC 6—Hoists for transfer of persons

ISO 10535:2006   Hoists for the transfer of disabled persons—Requirements and test methods

TC 176/SC 1—Concepts and terminology

ISO 9000:2015   Quality management systems—Fundamentals and vocabulary

TC 176/SC 2—Quality systems

ISO 9004:2018   Quality management—Quality of an organization-Guidelines to achieve sustained success

TC 178—Lifts, escalators

ISO/TS 14798:2009   Lifts (elevators), escalators, and moving walks—Risk assessment and reduction methodology

TC 184—Industrial automation systems and integration

ISO 11161:2007   Safety of Machinery-Integrated manufacturing systems—Basic requirements

TC 184/SC 4—Industrial data

ISO 10303-214:2010   Industrial automation systems and integration—Product data representation and exchange—Part 214: Application protocol: Core data for automotive mechanical design processes

TC 184/SC 5—Architecture, communications, and integration frameworks

ISO 15704:2019   Enterprise modelling and architecture—Requirements for enterprise-referencing architectures and methodologies

ISO 16100-1:2009   Industrial automation systems and integration—Manufacturing software capability profiling for interoperability—Part 1: Framework

TC 188—Small craft

ISO 15027-3:2012   Immersion suits—Part 3: Test methods

TC 199—Safety of machinery

ISO 12100-2:2003   Safety of machinery—Basic concepts, general principles for design—Part 2: Technical principles

ISO 13849-1:2015   Safety of machinery—Safety-related parts of control systems—Part 1: General principles for design

ISO 13851:2019   Safety of machinery—Two-hand control devices—Principles for design and selection

ISO 13856-1:2013   Safety of machinery—Pressure-sensitive protective devices—Part 1: General principles for design and testing of pressure-sensitive mats and pressure-sensitive floors

ISO 14121-1:2007   Safety of machinery—Risk assessment—Part 1: Principles

ISO 14123-2:2015   Safety of machinery—Reduction of risks to health resulting from hazardous substances emitted by machinery—Part 2: Methodology leading to verification procedures

ISO/TR 18569:2004   Safety of machinery—Guidelines for the understanding and use of safety of machinery standards

TC 204—Intelligent transport systems

ISO 15623:2013   Intelligent transport systems—Forward vehicle collision warning systems—Performance requirements and test procedures

TC 210—Quality management and corresponding general aspects for medical devices

ISO 14969:1999   Quality systems—Medical devices—Guidance on the application of ISO 13485 and ISO 13488

ISO 14971:2019   Medical devices—Application of risk management to medical devices

TC 212—Clinical laboratory testing and in vitro diagnostic test systems

ISO 15190:2020   Medical laboratories—Requirements for safety

ISO 15197:2013   In vitro diagnostic test systems—Requirements for blood-glucose monitoring systems for self-testing in managing diabetes mellitus

TMB—Technical Management Board

IWA 1:2005   Quality management systems—Guidelines for process improvements in health service organizations

ISO/IEC Guide 50:2014   Safety aspects—Guidelines for child safety in standards and other specification

ISO/IEC Guide 71:2014   Guidelines for standards developers to address the needs of older persons and persons with disabilities

CASCO—Committee on conformity assessment

ISO/IEC 17025:2017   General requirements for the competence of testing and calibration laboratories

2 Positioning Ergonomics Standards and Standardization

Jan Dul and Henk J. de Vries

CONTENTS

INTRODUCTION

Ergonomics is just one of the fields where standards have been developed. In comparison to other fields, ergonomics standardization is relatively young, only some decades old, whereas technical fields like telecommunication, electrotechnology, mechanical engineering, and civil engineering have a standardization tradition of more than a century, and the origins of these standards can be found in many centuries before Christ in China, Mesopotamia, and Egypt. In those times, it was often the emperor who took the initiative. For example, the Chinese emperor Qing-Shihuang set compulsory standards on measurement, seeds, cloth sizes, and weaponry (Wen, 2004).

Is ergonomics standardization an area with exceptional characteristics, or is it just one of the many areas of standardization? In the first case, ergonomic standardization can learn lessons from the experiences in other fields. The same applies to the second case—but we should reckon the typical differences.

In this chapter, we distinguish between standards (the result of standardization, usually a document) and standardization (the process of making standards). We draw some general lines on standards and standardization, describe ergonomics standards and standardization, and subsequently examine to which extent the situation in ergonomics differs from the situation of standards and standardization in general.

In Part I, we start with a general description of standards and standardization. In Part II, we describe ergonomics standards and standardization. We conclude that the goals of ergonomic standards and standardization differ from those that generally apply and discuss the possible lessons.

PART I: GENERAL STANDARDS AND STANDARDIZATION

Standards

Standards are very common in everyday life. What the A4 series of paper sizes, specifications of credit cards, ISO 9000 requirements for quality management systems, the SI system of units (SI = Système International d'Unités), McDonald's product and service specifications, and the specifications of the GSM mobile phone system have in common is that they are used repeatedly by many people and, therefore, are laid down by standards. Standards can be considered as a lubricant for the modern industrial society.

Definition

The definition of a standard used by formal organizations is *a document, established by consensus and approved by a recognized body, that provides, for common and repeated use, rules, guidelines, or characteristics for activities or their results, aimed at the achievement of the optimum degree of order in a given context* (ISO/IEC, 1996). However, not all the aforementioned examples fit this definition—not all standards are consensus-based or approved by a recognized body, and standards may have another format than a document, like a software. Therefore, De Vries (1999) developed another definition: *A standard is an approved specification of a limited set of solutions to actual or potential matching problems, prepared for the benefits of the party or parties involved, balancing their needs, and intended and expected to be used repeatedly or continuously, during a certain period, by a substantial number of the parties for whom they are meant.*

Users

Standards can be used by companies but also by other groups in society, such as governments and testing organizations. We distinguish between direct users—parties that read the standard to apply it—and indirect users—parties that have a stake in the application of the standard. A party can be a person, a group of people, or an organization.

Direct Users

In general, the category of direct users includes the following:

- Designers: parties that design products, services, and processes to make a design that meets the criteria laid down in a standard
- Testers: parties that test products, services, or processes against the requirements in a standard or that use a test method specified in a standard
- Advisors: parties that provide advice concerning designing or testing according to a standard
- Regulators: parties that develop regulations, other requirements, standards, rules, or laws based on a standard

The activities of these direct users of standards are carried out by, or for, organizations like companies, testing laboratories, certification bodies, consultancy firms, and governmental agencies. Standardization can be important for these organizations; standards may apply to, or can be developed for, the products or services delivered or purchased by these organizations or the business processes in these organizations.

Indirect Users

Indirect users can be end-users of a standard: consumers, workers, and the public (e.g., safety and environmental issues). These groups normally do not read standards, but the results of applying them in the product and processes are important for these groups. Other indirect users can be special interest groups, such as branches of business organizations (for the common interests of their members), trade unions (for the interests of workers), and environmental pressure groups. In the next sections, we discuss the interests of three main direct and indirect users of standards: companies, governments, and consumers.

Companies

Companies can be both direct and indirect users of a standard. In company practice, the main aim of using standards is to contribute to business results and to the effectiveness and efficiency of the organization. Standards can reduce the costs of products and services. Meeting or not meeting certain standards can be the difference between success and failure in the market. The general aims of standardization include (after Sanders, 1972):

- Reducing the growing variety of products and procedures
- Enabling communication
- Contributing to the functioning of the overall economy
- Contributing to safety, health, and protection of life
- Protecting the consumer and community interests
- Eliminating trade barriers—although standards at the national or regional level can also create trade barriers, according to Hesser and Inklaar (1997).
  More specifically, the following issues illustrate the importance of standards for companies (Schaap & De Vries, 2004):

Within the company:

- Not re-inventing the wheel but using existing solutions (laid down in standards) that have already been well thought out.
- Bringing procedures into line with what is normal elsewhere, so that cooperation is simpler, and purchase is cheaper.
- Efficient working by repeatedly using the same solution.
- Using recognized requirements in the field of, for example, quality and safety.
- Cheaper purchasing due to economies of scale and more transparency in the market that results in an increase in price competition.
- Less cost of stock and logistics by making use of standard solutions.
- More possibilities for outsourcing.
- Fewer problems in the field of occupational health, safety, and environmental issues.

Outside the company:

- Being able to demonstrate the quality of products and services (using test methods laid down in standards).
- Giving clients confidence: the product (the production method) meets accepted requirements.

- Being allowed to bring products onto the market because conformity to standards may be a means to demonstrate conformity to legal requirements.
- Being able to bring products onto the market because they meet requirements (laid down in standards) that are important for customers.
- Being successful with products because they meet customers' wishes and are compatible with other products.

Standards may also have disadvantages for companies.
Within the company:

- Too much formalization may cause too much routine for employees, resulting in dissatisfaction.
- Standards for processes may hinder process improvement.
- Standards for products or services may hinder product innovation.

Outside the company:

- In case the company is not able to meet the requirements in a standard, it will lose market share.
- In case a competitor is better able to meet the requirements, the company faces a competitive disadvantage.
- Standards may make the market more transparent and, in this way, cause an increase in price competition at the cost of profit margins.

De Vries (1999) suggested that the importance of standards is growing because:

1. Companies, in general, can no longer be regarded as isolated organizations, not only in trade transactions but also in their technical operations. Especially in the area of information and communication technology (ICT), companies are connected to other companies. Also, in other areas, technical specifications chosen by the company must fit the specifications of the company's environment. For example, ICT without standardization is impossible, and the chemical composition of petrol should not differ per country.
2. The tendency to concentrate on core business and to contract other activities makes it necessary to agree with suppliers on, for instance, product specifications, product data, communication protocols, and the quality of the production and delivery processes. Because the company usually has several suppliers, each with several customers, the most profitable way to solve these matching problems is by using widely accepted standards.
3. There is a tendency to pay more attention to quality and environmental issues in a systematic way. This has increased the need for management systems standards, such as the ISO 9000 series of standards for quality management and the ISO 14000 series for environmental management. It has also increased the need for other standards because management systems cause companies to perform activities in a structured way; standards for products, production means, and information systems contribute to the structure needed.
4. There is an increasing need to provide confidence to customers and other stakeholders, which can be partly achieved through certification. Certification is a procedure by which a third party gives written assurance that a product, process, or service conforms to specified requirements (ISO/IEC, 1996). In general, these requirements are laid down in standards.

Governments

Most of the benefits and risks of standards for companies also apply to governments when governmental agencies use them to improve business processes and performance. Additionally, the following issues seem to be important for governments (De Vries, 1999):

- Governments may support standardization as a part of the government's general role in stimulating business performance and international trade.
- Governments may supplement, simplify, or improve their legal system with standardization by making references to standards in laws.
- By standardization, the market may govern itself with less regulation from the government.
- International standards may help to remove trade barriers.

Also, for governments, the importance of standards seems to be growing, because (De Vries, 1999):

1. Globalization of trade increases the need for international standardization.
2. Within the European Union and the European Free Trade Association, the choice for one single market without barriers to trade causes the replacement of different national standards by European ones. This makes export to several countries easier. There is no longer a need to produce different variants of products to meet different standards in different countries. However, companies that mainly serve national markets, especially in smaller countries, have been confronted with a substantial increase in the number of standards that are used.
3. Both at the European and national level, there is a general tendency to link standards to legislation in a way that provides detailed requirements that correspond to global requirements laid down in laws. This causes an increase in the number of standards and in obligations to use them.

Consumers. For consumers, standards may contribute to (ANEC, 2003):

- Accessibility/design for all
- Adaptability
- Consistent user interfaces
- Ease of use
- Functionality of solutions
- Service quality
- System reliability and durability
- Health and safety
- Environmental issues
- Information (product information, directions for use, and system status information)
- Reliability of information
- Privacy and security of information
- Interoperability and compatibility
- Multicultural and multilingual aspects
- More transparency in the market, enabling consumers to better compare products and services from different suppliers on price, quality, and other characteristics

Additionally, one can argue that standards may also lead to lower prices due to economies of scale and increased price competition.

Classification

De Vries (1998) provided several classifications of standards related to the contents of the standards, the intended users, or the process of developing the standards. For the purpose of this book, we concentrate on content-related classifications.

Entities

The content of a standard is concerned with entities or relations between entities. Thus, standards can be classified according to these. An entity may be a person or group of persons; a "thing" such as an object, an event, an idea, or a process. "Things" include plants and animals. Gaillard (1933, p. 33) provides a rather complete list of possible entities; a combination of the first two types of entities (e.g., a car with a driver).

Matching problems can concern:

- Matching thing–thing (e.g., bolts and nuts)
- Matching man–thing (e.g., safety requirements)
- Matching man–man (e.g., procedures and management systems).

Horizontal, Vertical

Standardization organizations often distinguish between horizontal and vertical standards. Horizontal standards set general requirements for a collection of different entities—for instance, biocompatibility criteria for medical devices. Vertical standards set several requirements for one kind of entity—for instance, medical gloves.

Basic, Requiring, and Measurement

Distinctions can be made among basic standards, requiring standards, and measurement standards (De Vries, 1998).

- Basic standards provide structured descriptions of (aspects of) interrelated entities to facilitate human communication about these entities or to be used in other standards. Examples are terminology standards, standards providing quantities and units, standards providing classifications or codes, and standards providing systematic data or reference models.
- Requiring standards set requirements for entities or relations between entities. These can include specifications of the extent to which deviations from the basic requirements are allowed. There are two subcategories: performance standards and standards that describe solutions.
  - Performance standards set performance criteria for the solution of matching problems. They do not prescribe solutions. Performance standards can include specifications of the extent to which deviations from the basic requirements are permissible.
  - Solution-describing standards describe solutions for matching problems.

In general, companies and other stakeholders in standardization prefer performance standards rather than standards that describe solutions (Le Lourd, 1992, p. 14), but most developing countries prefer descriptive standards with a large number of technical details (Hesser & Inklaar, 1997, p. 38). The percentage of performance standards is growing, at the expense of standards that prescribe certain solutions.

- *Measurement standards* provide methods to check whether requiring standards criteria have been met.

Interference, Compatibility, and Quality. Simons (1994) distinguishes among interference standards, compatibility standards, and quality standards:

- Interference standards set requirements concerning the influence of an entity on other entities. Examples are standards on safety, health, environmental issues, and EMC (electromagnetic compatibility with respect to electrical disturbances). Companies often must use interference standards because of governmental requirements. They, therefore, have no choice. They must use them.
- Compatibility standards concern the fitting of interrelated entities with one another to enable them to function together—like the specifications for films, cameras, and cell phones. Choices regarding compatibility standards are often commercial decisions that have a direct impact on market share. So, although the choices are up to the company, it is often the market situation that strongly influences these choices.
- Quality standards set requirements for entity properties to assure a certain level of quality. Examples of quality standards include the ISO 9000 quality management standards, film sensitivity standards (to enable standard film processing), measurement standards, and standards for company procedures. quality standards are often related to the company's business operations.

STANDARDIZATION

Definition

The activity of making standards may be called standardization. Standardization is concerned with establishing and recording a limited set of solutions for actual or potential matching problems directed at benefits for the party or parties involved and intending and expecting that these solutions will be repeatedly or continuously used during a certain period by a substantial number of the parties for whom they are meant. A matching problem should be understood as a problem of interrelated entities that do not harmonize with each other; solving it means determining one or more features of the entities that harmonize with one other or one or more features of an entity and its relationship with other entities (De Vries, 1997).

Developers

The main stakeholders in most standardization projects are the manufacturers of products and services, the professionals that support their development, and the customers that buy them. Other stakeholders include the organizations that represent these stakeholders (e.g., branch of business organizations, professional societies, and consumer organizations), governmental agencies, organizations for testing and certification, consultancy firms, research institutes, universities, and special interest groups, such as trade unions and environmental pressure groups. Many times, the manufacturers dominate the standardization.

Standards can be developed by companies, industrial consortia, branch of business organizations, professional associations, governmental agencies, or—last but not the least—formal standardization organizations. In the remaining part of this chapter, we concentrate on formal standardization. Three levels of formal standardization can be distinguished: national, regional, and global.

National Standardization

More than 150 countries have a national standardization organization (NSO). Well-known examples are the Association Française de Normalisation (AFNOR), American National Standards Institute (ANSI), British Standards Institution (BSI), Deutsches Institut für Normung, Germany (DIN), Japanese Industrial Standards Committee (JISC), Standardization Administration of China (SAC), and Standards Australia. In Northwest and Central Europe and Italy, Australia, the United States, and Japan, such organizations were founded in the first decades of the 20th Century. Before and shortly after the Second World War, India, South Africa, and several South American and other European countries followed. More than 100 other NSOs are less than 30 years old

(Toth, 1997). Most of the first and second generation of NSOs were founded by people from the industry with the aim to establish national standards. The NSOs provided the platform for national stakeholders to agree on standards and publish them.

NSOs usually distinguish the following steps for the development of national standards: (a) request, (b) assignment to a committee, (c) drafting, (d) public comment, (e) review of comments, (f) approval, (g) publishing, (h) publicity, (i) implementation, and (j) evaluation. The following description is mainly based on a Dutch publication (NEN, 2004a, pp. 12–13), with additions from the British Standard 0-2 (BSI, 1997a, pp. 16–25). Terms have been taken from BS 0-2; other NSOs may use different terms. Procedures for steps 1–8 may slightly differ per NSO but are similar to NEN and BSI procedures. Steps 9 and 10 are up to the standards users, though NSO procedures require the evaluation of standards after a certain period.

1. *Request*. Any company, organization, person, or the NSO itself can indicate a need for new standards or improvements in existing standards. Most NSOs have Sector Boards that are responsible for dealing with such requests. They may ask for advice from a standardization committee or have a feasibility study carried out. To decide on the proposal for a new work item, they need justification, including:
   * Reasons for standardization: What advantages can be expected for whom?
   * The topic of standardization: Is it (technically) convenient for standardization?
   * The amount of support in the market, including willingness to finance the project.
   * Reasons, if any, to standardize at the national level instead of at the—preferred—international or regional level.
     The title and scope of the standard need to be clear; when necessary, further specifications may be added, as well as a schedule for development.

2. *Assignment to a committee*. The Sector Board will decide on the proposal. When positive, the new work item will be assigned to a technical committee (TC). When no TC exists, a new one may be set up. A TC may establish a subcommittee (SC) or working group (WG) to handle the topic. A WG exists only for the time necessary to draft one or more standards. When the standards are ready, the WG is discontinued. An SC is a more permanent committee, responsible for a field of activities. This, however, does not exclude the SC from being disbanded. Interested parties are invited to get involved. Additionally, the new work item and the establishment of a new TC, SC, or WG—if any—is publicly announced so that representatives of all interested parties have the opportunity to join the committees. Often, in the first meeting, its composition is discussed, and organizations that were overlooked still get an invitation to participate. In general, the committee one step higher in the hierarchy must agree on this.

3. *Drafting*. The responsible TC, SC, or WG prepares a first draft based on professional expertise, deliberations, and consensus. Sometimes, research is carried out to obtain data to be used in the standard. Often, the committee initiates the discussion by using an initial document brought by one of the participants. The standard may require testing, to be carried out either by participants or—as part of a special program set up for the project—by a testing organization. When the TC has delegated the work to an SC or WG, their approval is followed by the TC's approval. After this, NSO employees check the draft in conformity with standards in adjacent areas, and with the NSOs rules for the drafting and presentation of standards. Some NSOs have laid down these rules in a national "standards for standards." Examples include the American national standard SES 1 (SES, 1995), British standard BS 0-3 (BSI, 1997b), the French standard X 00-001 (AFNOR, 1993), and the German standard DIN 820-2 (DIN, 1996).

4. *Public comment*. The draft standard is published for comments. It is announced in the NSO's regular—such as its monthly magazine and website—and selected media—such as

the specialists' journals. Sometimes, copies are sent to experts and to interested parties not represented in the committees. Other NSOs are notified and get the opportunity to comment. The party that requested the standard—if not a committee member—also gets a copy. Parties get a certain amount of time, mostly a few months, to give their comments, if any.

5. *Review of comments.* The TC, or its SC or WG, discusses the comments and uses them to improve the standard. Sometimes, major contributors are invited to a meeting to discuss their comments. All contributors are informed about the committee's decision on their comments.

6. *Approval.* The TC decides on the proposed standard. NSO officers check it again for conformity to standards in adjacent areas and to rules for drafting and presentation of standards. If it is decided that the standard is still appropriate, but the content or structure of the document is changed significantly because of the comments received, a second draft for public comment may be issued.

7. *Publishing.* The NSO does the final editing and publishes the standard.

8. *Publicity.* The NSO uses its own media—such as its journal or website—to announce the new standard. Additionally, press releases are sent, media events are organized, seminars or courses are held, and so forth.

9. *Implementation.* Of course, it is up to the companies and other organizations to implement the standard. Some NSOs provide support for this by means of written guidelines, courses, or advice.

10. *Evaluation.* Most NSOs review their standards after five years. The responsible committee then decides to withdraw, revise, or maintain them unchanged. Of course, market developments may be a cause for earlier revision. The evaluation of standards is a problem when the responsible TC has been disbanded. Then the NSO Bureau must decide whether the standards should remain in the collection.

Nowadays, the role of NSOs has shifted to the preparation of the national standpoint in international standardization. In the field of electrotechnology, there has been international standardization from the outset, since 1906. Often, international standards are adopted as national standards, whether translated into the national language or not. The NSO provides the national selling point for such standards. In addition to their role in standards development and standards selling, many NSOs serve their national market by providing information related to standards in the form of informative books or courses. More than 50% of NSOs provide activities in the field of certification. In many developing countries, the institutes also play a role in accreditation, testing, and metrology (De Vries, 1999).

International Standardization: Global

Formal standardization at the global level comprises the International Telecommunication Union (ITU), the International Electrotechnical Commission (IEC), and the International Organization for Standardization (ISO). ISO and IEC have been established by NSOs and, in turn, recognize only one NSO per country. This is the main reason countries with lots of sectoral standardization organizations, such as Japan, Norway, and the United States, also have an NSO. The NSO must be an intermediary between the national stakeholders and the global and regional standardization arena. Therefore, NSOs tend to form national TCs that reflect the committee structure of ISO/IEC.

ISO and IEC have a hierarchy of committees for standards development and approval. The "lowest" level includes working groups (WG), where experts develop standards. A WG develops one or a few standards and can then be disbanded. Decision making on standards is done in Technical Committees (TCs). Many TCs delegated a part of their activities to an in-between level: subcommittees (SCs). Participants in TCs and SCs are national delegates, whereas NSO committee members represent national stakeholder groups. Delegates are expected to speak and act on behalf of their country, not of their company or stakeholder group. NSO committees formulate the national standpoints for voting or for input by participation in committees at the international level. In national standardization, decisions are based on consensus. ISO and IEC use the consensus principle within committees that draft standards but voting also takes place at several crucial stages of development. NSOs vote on behalf of their country. The main stages in developing an ISO standard can be found in ISO (2004).

International Standardization: Regional

In many regions of the world, special standardization organizations have been established (see Table 2.1).

Most of these organizations do not develop their own standards but provide a forum to discuss standards matters of common interest. And the list is not complete—other, more informal, forms of cooperation exist. For example, among the Baltic countries, in Turkey and its neighboring countries, and China, Japan, and South Korea. In one region, there is a huge amount of standards development: Europe. CEN, CENELEC (electrotechnical standardization), and ETSI (standardization in the field of telecommunication) have developed thousands of standards. The standards development processes of CEN are almost identical to those of ISO. A main difference is that in ISO, each member country has one vote, no matter its size, whereas CEN operates a system of weighted voting. A second difference is that once a European standard has been adopted, CEN members must implement it by making it a national standard—either by publication of an identical text or by endorsement—and by withdrawing any conflicting preexisting national standards, whether they agree to the contents of this European standard or not. In the case of ISO standards, they are free to implement a standard or not. Though in the case of European standards, it is obligatory to adopt them as national standards, this does not imply that the use is obligatory as well. In principle, ISO—as well as CEN standards and national standards—are all voluntary standards. Only through reference in legislation or through market forces they may become more or less compulsory.

TABLE 2.1
Regional Formal Standardization Organizations

Region	Organization
Africa	African Regional Organization for Standardization (ARSO)
	Common Market for Eastern and Southern Africa (COMESA)
Arab countries	Arab Industrial Development and Mining Organization (AIDMO)
	Standardization and metrology organization for the Gulf Cooperation Council (GSMO)
Europe	Comité Européen de Normalisation (CEN)
	Comité Européen de Normalisation Electrotechnique (CENELEC)
	European Telecommunication Standards Institute (ETSI)
Most countries of the former Soviet Union	Euro-Asian Council for Standardization, Metrology, and Certification (EASC)
South Asia	South Asian Association for Regional Cooperation (SAARC)
Southeast Asia	ASEAN Consultative Committee for Standards and Quality (ACCSQ)
Asia/Pacific	Pacific Area Standards Congress (PASC)
North America	NAFTA Committee on Standards-Related Measures (CSRM)
Caribbean	Caribbean Common Market Standards Council (CCMSC)
South America	Comisión Panamerican de Normas Técnicas (COPANT)
	Comité MERCOSUR de Normalización (CMN)

Combined Development of European and International Standards

The political wish to have one common European market without barriers to trade has caused an enormous growth in European standardization. Most European standards are identical to the international ones, but they are developed for the simple reason that they must be implemented into the national standards systems of all EU and EFTA member countries—this does not apply for ISO standards. In fact, the same standards are developed within two committees—one at the global and one at the European level. Of course, this is not efficient, so CEN and ISO agreed to establish the possibility of a tight cooperation in standards development. In this "Vienna Agreement" (IEC and CENELEC have a comparable agreement called "Dresden Agreement"), CEN and ISO agreed on four main methods of achieving common standards (ISO/CEN, 1991; Smith, 1995):

1. Adoption of existing ISO standards by CEN, using a short procedure within CEN. If it seems that an ISO standard will need to be revised to meet Europe's requirements, the ISO committee will be given the opportunity to consider whether it is prepared to revise the standard.
2. Submission of work items developed by ISO to parallel approval procedures in ISO and CEN.
3. Submission of work items developed by CEN to parallel approval procedures in CEN and ISO.
4. Submission of European standards for adoption by ISO using a short procedure within ISO.

One might argue that such cooperation strengthens Europe's position within ISO. Because of the need to arrive at common European standards, European countries might join forces with ISO. Moreover, they have the majority in more than 50% of all ISO committees. In this way, Europe could impose its standards to the rest of the world. In some exceptional cases, this has occurred indeed, but almost always it appears that the Europeans do not agree with one another and that the controversies in international standardization are not between Europe and the rest of the world.

European Legislation

In some cases, there is a close relation between standards and legislation. At the European level, this applies, especially to the so-called New Approach. In the "Old Approach," member states imposed their own technical specifications and conformity controls for manufactured products. Any technical harmonization across the European Union relied on agreeing directives for individual products. The requirements in these directives then had to be implemented into the legal systems of the member states. European directives set detailed requirements mainly on product safety. It took many years to agree on the safety requirements and, often, the directive was already outdated once it was ready. Therefore, it was decided to replace this with a New Approach where the directives set only the essential requirements on safety, health, or environment, and these requirements are formulated globally. Linked to these directives are developed European standards where detailed requirements or test methods are laid down. A company that meets the relevant standards is assumed to meet the general requirements set in the directives. Thus, implementing the standards is an efficient way to meet the legal requirements. The company, however, can meet these requirements in another way. Though principally voluntary, in practice, these standards are almost obligatory. Conformity to requirements in the directives is indicated by means of the CE mark (CE = Conformité Européenne', see Leibrock, 2002, and Huigen, Inklaar, & Paterson, 1997).

Stakeholder Participation

An important issue in standardization is the involvement of stakeholders in the process of developing standards. In practice, not all relevant stakeholders are involved in standardization, and there can even be one-sided representation (De Vries, Verheul, & Willemse, 2003). In most TCs, the industry is overrepresented. As a result, the developed standard may not be acceptable for other relevant potential users. De Vries, Feilzer, and Verheul (2004) have listed 27 barriers to the participation of stakeholders in formal standardization.

To avoid the underrepresentation of a relevant stakeholder, a model has been developed to identify and select relevant stakeholders to be involved in standardization (Willemse, 2003). This model has been applied to ergonomics standardization (Dul, Willemse, & De Vries, 2003). In Chapter 20 of this book, the model and the application to ergonomics standardization are presented (Willemse, De Vries, & Dul, 2006).

PART II. ERGONOMICS STANDARDS AND STANDARDIZATION

During two million years of human history, implicit knowledge has been used to fit natural tools and artifacts to the needs of the individual human being. After the Industrial Revolution, which introduced mass production, a need arose for explicit knowledge about the interactions between the human and the complex man-made environment (Dul, 2003a). From this need, the discipline of ergonomics emerged. Product ergonomics developed as a field that can fit mass products to the individual consumer needs and characteristics. Production ergonomics developed as a field that can fit complex technical-organizational systems to the needs and characteristics of workers. In 1949, the first-ever national ergonomics society was established in the United Kingdom, where the name "ergonomics" was coined to name the field. This was followed by the creation of the International Ergonomics Association (IEA) in 1961, which nowadays represents some 19,000 ergonomics scientists and practitioners worldwide. In the early 1970s, the IEA decided to initiate the development of ergonomic standards (Parsons & Shackel, 1995) and requested the ISO to start ergonomics standardization. The aim was to design products and (production) processes to optimize the well-being and performance of consumers and workers interact with the products and production systems. Since that time, a large number of ergonomics standards have been developed for several types of products, production systems, and work environments.

ERGONOMICS STANDARDS

Definition

Ergonomics is a broad scientific and professional field, which is described by the IEA as follows: Ergonomics (or human factors) is the scientific discipline concerned with the understanding of interactions among humans and other elements of a system, and the profession that applies theory, principles, data, and methods to design to optimize human well-being and overall system performance. In this chapter, we limit ourselves to two types of ergonomics standards: global standards from ISO and regional (European) standards from CEN. The reason is that it appears that most ergonomics standards are developed by these organizations. Figure 2.1 shows the growth of the number of ISO and CEN standards over the years.

Ergonomics standards can also be developed by companies, consortia of companies, or industry branch organizations but these standards, if available, are not considered here. We estimate that in other fields of standards (e.g., electrotechnology), the number of standards from companies or consortia is considerably larger than that in the ergonomics field. We also estimate that several standards from companies and branches of industry organizations are, at least partly, based on formal standards from ISO and CEN.

Within ISO and CEN, we consider only standards that have been developed by the technical committees ISO TC 159 "Ergonomics," and CEN TC 122 "Ergonomics," although ergonomics standards have also been developed by other technical committees (Dul, De Vlaming, & Munnik, 1996).

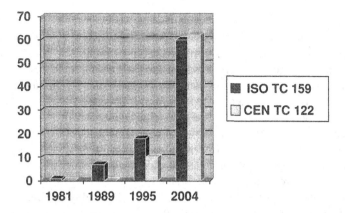

FIGURE 2.1   The number of ergonomics standards developed by ISO TC159 "Ergonomics" and CEN TC122 "Ergonomics."
Source: Dul et al., 2004; Dul, De Vlaming, & Munnik, 1996; Metz, 1991

Hence, in the remainder of this chapter, the definition of ergonomics standards is formal standards developed by ISO TC 159 or CEN TC 122.

Users

In some cases, potential users of ergonomics standards can be identified by looking at the field of application mentioned in these standards. In the "mother standard" of all ergonomics standards, many potential users are mentioned. The "mother standard" is the first ergonomics standard. It was published in 1981 as ISO 6385 "Ergonomic Principles in the Design of Work Systems," and revised in 2003 (Eveleens, 2003). This standard describes the design steps for a (work) system and the required ergonomics inputs to design an ergonomic system:

1. Formulation of goals of the system
2. Analysis and allocation of functions between human and technology
3. Design of the concept of the system
4. Detailed design (the term work is used in a broad sense)
   - Design of work organization
   - Design of work tasks
   - Design of jobs
   - Design of work environment
   - Design of work equipment, hardware, and software
   - Design of workspace and workstation

5. Realization, implementation, and validation

This standard mentions the following users: "users of this standard will include managers, workers (or representatives), professionals such as ergonomists, project managers and designers who are involved in the design of work systems."

Direct Users

A party (person, group of people, organization) that reads an ergonomics standard and uses it for professional activities is considered a direct user of the ergonomics standard. In particular, product and production system designers, testers, and advisors are potential direct users of ergonomics standards.

Designers

The objective of ergonomics is to contribute to the design of the technical and organizational environment. Hence, product and process designers are important direct users. Many ergonomics standards are formulated in general terms because the specific requirements depend on the organizational context in which the standard is used. For example, the requirements for the maximum mass for safe manual lifting of a load also depend on the lifting frequency and duration, and other organizational factors. Some ergonomics scientists emphasize that the body of scientific knowledge in certain areas is not yet sufficient to formulate specific standards to limit the exposure to risk factors. For example, Fallentin et al. (2001) evaluated standards that limit the physical workload and concluded that "the scientific coherency of specific quantitative criteria was limited, whereas general process-type standards were more favorable."

However, such generally formulated standards cannot be readily applied by designers (Dul, et al., 2004). Wulff, Westgaard, and Rasmussen (1999) found, for instance, that designers may not understand the general ergonomic recommendations, do not know how to make them concrete in specific situations, or do not consider them important enough if they conflict with other design requirements. Then, ergonomic advisors can assist the designers to apply ergonomics standards in specific situations. Most standards other than ergonomics standards, are product standards with specific requirements on technical issues, and these can be more simply implemented by designers.

Testers

Several parties use ergonomic standards to test whether products or processes meet ergonomics requirements. For example, testers can judge whether certain products (e.g., hand tools, VDU-screens, and chairs) meet the ergonomics requirements formulated in a standard. Also, product standards can be used by consumer organizations to judge the ergonomic quality of a consumer product. Occupational Health Services or the Labor Inspectorate can use ergonomics standards (e.g., on physical and mental workload) to evaluate production systems and work environments for performing a risk assessment of the work situation.

Recently, the International Ergonomics Association has taken an initiative to stimulate testing of the ergonomics quality of products and processes. The aims of this so-called Ergonomics Quality In Design (EQUID) initiative are:

- To define process criteria and requirements for the ergonomic design of products, work systems, and services
- To define a system for accrediting certifying bodies that will assess the ergonomics quality in design, using the relevant criteria and requirements
- To design, implement, and manage a system for regularly assessing and updating the process requirements for the ergonomic design of products, work systems, and services
- To design, implement, and manage a system for regularly evaluating and improving the accreditation program.

In this initiative, the product or process itself is not tested, but whether the ergonomic input during the development process of the product was appropriate. This test plan in ergonomics deviates from the testing practice in most other fields, where many standards provide methods for testing products, services, or processes.

Advisors

The initiative to develop ergonomic standards came from the International Ergonomics Association and ergonomics experts have been heavily involved in their development. Hence, in contrast to general standardization that is dominated by manufacturers of products and services, ergonomics standards seem to be dominated by ergonomics experts and may primarily reflect the professional insights about what can be considered as "good ergonomics." Putting ergonomics knowledge into standards may have supported the transfer of ergonomics knowledge and may have given status to the ergonomics discipline. Ergonomics advisors are probably the most important direct users of ergonomics standards. Ergonomists can use the standards when contributing to the design or testing of products and (production) processes within companies and other organizations. A review among certified European ergonomists (Breedveld, 2005) showed that 69% of these ergonomists regularly or always use international ergonomics standards in their work for the past year, whereas only 3% never used such standards last year. Most of the ergonomists were internal or external consultants, giving ergonomics advice to companies.

Regulators

Certain organizations may use ergonomics standards for regulatory purposes. Industrial branch organizations may adopt or adapt ergonomics standards for their branch of industry. Trade unions can use ergonomics standards to develop their own guidelines to be used for the interest of the workers. Governments can be direct users of ergonomics standards when using the standard in health and safety regulation. In Europe, several ergonomics CEN standards are related to European legislation to stimulate free trade within Europe, by setting the same minimum health and safety requirements for machinery produced in European countries. In particular, the Safety of Machinery Directive of 1998 (European Union, 2003) formulates general requirements for health and safety related to machinery and refers to CEN standards for specific requirements (see List of CEN Standards Related to Legislation section). Other European directives that have a relationship to ergonomics are the Use of Work Equipment Directive (European Union, 1989), Manual Handling Directive (European Union, 1990a), and the Work with Display Screen Equipment Directive (European Union, 1990b). These directives are part of the European social policy on worker protection, but no reference is made to specific CEN standards.

Indirect Users

In ergonomics standards, by definition, end-users of products (e.g., consumers) and end-users of (production) processes (e.g., workers) are essential entities; one of the objectives of ergonomics is to contribute to the well-being of the end-users (see previous definition of ergonomics). End-users normally do not read standards, but the result of applying these standards is important for them. In many design teams, the ergonomist can be considered as the "voice" of the end-user.

Other indirect users of ergonomics standards can be a branch of business organizations (for the common interests of their members) and trade unions (for the interests of workers), when they do not use standards in a direct way as described earlier.

In the next sections, we discuss the interests of the main direct and indirect users of ergonomics standards: companies, governments, and workers and consumers.

Companies

Companies can have an interest in applying ergonomics because the integration of ergonomics in the design of products and processes can contribute to social and economic business goals and strengthen the company's competitive advantage (Dul, 2003a, 2003b). When ergonomics standards are used to integrate ergonomics in the design of (production) processes, worker satisfaction and worker motivation may improve, sick leaves may reduce, and the worker performance can increase in terms of reduction of human error (quality) and increase of human output (productivity; see, for example, Breedveld, 2005, and Dul et al., 2004).

Ergonomics standards can also contribute to the development of user-friendly products, which better meet customer demands and improve customer satisfaction. Other reasons for companies to apply ergonomics standards is that it can help prevent occupational health and safety problems or are a useful means to obey Occupational Health and Safety regulations. Also, companies may want to meet ergonomics standards to demonstrate to customers and other stakeholders "good social practice" and social responsibility, both in the products that are sold as well as in the (production) processes.

In contrast to most other standards where economic goals are the major driver, ergonomics standards may be primarily used for the social dimension in terms of worker's satisfaction, health, and safety, without disregarding the economic dimension (Dul et al., 2004).

Governments Other Than Regulator

Governments can be users of ergonomics standards in a role other than a regulator (see List of CEN Standards Related to Legislation section). Similar to companies, governments also have the role of employer and can use ergonomics standards to improve employee satisfaction and improve service quality (Cook et al., 2003). Also, government agencies deliver products or services that may benefit from the use of ergonomics standards to improve the satisfaction of citizens. Governments can use other instruments than legislation to stimulate the development and application of ergonomics standards as a contribution to better health and safety of the population. For example, in Germany, the government supports the Commission for Occupational Health and Safety and Standardization (KAN, 2004) that observes the standardization process and ensures that standards makers devote sufficient attention to Occupational Health and Safety. Members of the KAN are employer and employee organizations, the State, the statutory accident insurance institutions, and the German national standardization institute DIN.

Within a government, ministries of social affairs, health, or labor may be interested to stimulate the development and use of ergonomics standards. However, in the case of standards in general, ministries of economic affairs may be interested to stimulate standardization.

Workers and Consumers

When ergonomics standards are applied to production systems, workers can expect that working conditions and occupational health and safety criteria are considered. Unions and work councils within organizations can represent workers when working conditions are evaluated with ergonomics standards.

When products are developed according to ergonomics standards, customers can expect that besides functionality, health and safety and user-friendliness are also assured. Consumer organizations represent consumers when testing products based on ergonomics standards.

With ergonomics standards, both workers and consumers benefit from its use in production systems and products. Other standards are primarily beneficial for suppliers and their customers. The latter may be consumers or professional users.

List of ISO Standards

Since the first ergonomics ISO standard was published in 1981, the number of ISO ergonomics standards has increased rapidly (see Figure 2.1 and Table 2.2). Tables 2.3 and 2.4 show ergonomics ISO standards that have been published or are in preparation by ISO TC 159 (Dul et al., 2004). Certain standards show up both as a published standard and as a standard in preparation. In

TABLE 2.2

ISO Standards on Ergonomics

	1989 (Metz, 1991)	1995 (Dul, De Vlaming, & Munnik, 1996)	2004 (Dul et al., 2004)
ISO published	7	18	60
ISO in preparation	14	31	25
Total	21	49	85

TABLE 2.3
Published Standards from ISO TC 159

ISO 6385:2004 Ergonomic principles in the design of work systems

ISO 7243:1989 Hot environments—Estimation of the heat stress on working man, based on the WBGT index (wet bulb globe temperature)

ISO 7250:1996 Basic human body measurements for technological design

ISO 7726:1998 Ergonomics of the thermal environment—Instruments for measuring physical quantities

ISO 7730:1994 Moderate thermal environments—Determination of the PMV and PPD indices and specification of the conditions for thermal comfort

ISO 7731:2003 Danger signals for workplaces—Auditory danger signals

ISO 7933:1989 Hot environments—Analytical determination and interpretation of thermal stress using calculation of required sweat rate

ISO 8996:1990 Ergonomics—Determination of metabolic heat production

ISO 9241-1:1997 Ergonomic requirements for office work with VDTs—Part 1: General introduction

ISO 9241-2:1992 Ergonomic requirements for office work with VDTs—Part 2: Guidance on task requirements

ISO 9241-3:1992 Ergonomic requirements for office work with VDTs—Part 3: Visual display requirements

ISO 9241-4:1998 Ergonomic requirements for office work with VDTs—Part 4: Keyboard requirements

ISO 9241-5:1998 Ergonomic requirements for office work with VDTs—Part 5: Workstation layout and postural requirements

ISO 9241-6:1999 Ergonomic requirements for office work with VDTs—Part 6: Guidance on the work environment

ISO 9241-7:1998 Ergonomic requirements for office work with VDTs—Part 7: Requirements for display with reflections

ISO 9241-8:1997 Ergonomic requirements for office work with VDTs—Part 8: Requirements for displayed colors

ISO 9241-9:2000 Ergonomic requirements for office work with VDTs—Part 9: Requirements for non-keyboard- input devices

ISO 9241-10:1996 Ergonomic requirements for office work with VDTs—Part 10: Dialogue principles ISO 9241-11:1998 Ergonomic requirements for office work with VDTs—Part 11: Guidance on usability

ISO 9241-12:1998 Ergonomic requirements for office work with VDTs—Part 12: Presentation of information

ISO 9241-13:1998 Ergonomic requirements for office work with VDTs—Part 13: User guidance

ISO 9241-14:1997 Ergonomic requirements for office work with VDTs—Part 14: Menu dialogues

ISO 9241-15:1997 Ergonomic requirements for office work with VDTs—Part 15: Command dialogues ISO 9241-16:1999 Ergonomic requirements for office work with VDTs—Part 16: Direct manipulation dialogues

ISO 9241-17:1998 Ergonomic requirements for office work with VDTs—Part 17: Form-filling dialogues ISO 9355-1:1999 Ergonomic requirements for the design of displays and control actuators—Part 1: Human interactions with displays and control actuators

ISO 9355-2:1999 Ergonomic requirements for the design of displays and control actuators—Part 2: Displays

ISO 9886:2000 Evaluation of thermal strain by physiological measurements

ISO 9920:1995 Ergonomics of the thermal environment—Estimation of the thermal insulation and evaporative resistance of a clothing ensemble

ISO 9921:2003 Ergonomic assessment of speech communication—Part 1: Speech interference level and communication distance for persons with normal hearing capacity in direct communication (SIL method)

ISO 10075:1991 Ergonomic principles related to mental workload—General terms and definitions

ISO 10075-2:1996 Ergonomic principles related to mental workload—Part 2: Design principles

ISO 10551:1995 Ergonomics of the thermal environment—Assessment of the influence of the thermal environment using subjective judgment scales

ISO 11064-1:2000 Ergonomic design of control centers—Part 1: Principles for the design of control centers

ISO 11064-2:2000 Ergonomic design of control centers—Part 2: Principles for the arrangement of control suites

ISO 11064-3:1999 Ergonomic design of control centers—Part 3: Control room layout

ISO/TR 11079:1993 Evaluation of cold environments—Determination of requisite clothing insulation (IREC)

(Continued)

TABLE 2.3 (Continued)

ISO 11226:2000 Ergonomics—Evaluation of static working postures

ISO 11228-1:2003 Ergonomics—Manual Handling—Part 1: Lifting and carrying

ISO 11399:1995 Ergonomics of the thermal environment—Principles and application of relevant International Standards

ISO 11428:1996 Ergonomics—Visual danger signals—General requirements, design, and testing

ISO 11429:1996 Ergonomics—System of auditory and visual danger and information signals

ISO 12894:2001 Ergonomics of the thermal environment—Medical supervision of individuals exposed to extreme hot or cold environments

ISO 13406-1:1999 Ergonomic requirements for work with visual displays based on flat panels—Part 1: Introduction

ISO 13406-2:2001 Ergonomic requirements for work with visual displays based on flat panels—Part 2: Ergonomic requirements for flat panel displays

ISO 13407:1999 Human-centered design processes for interactive systems

ISO 13731:2001 Ergonomics of the thermal environment—Vocabulary and symbols

ISO/TS 13732-2:2001 Ergonomics of the thermal environment—Methods for the assessment of human responses to contact with surfaces—Part 2: Human contact with surfaces at moderate temperature

ISO 14738:2002 Safety of machinery—Anthropometric requirements for the design of workstations at machinery

ISO 14915-1:2003 Software ergonomics for multimedia user interfaces—Part 1: Design principles and framework

ISO 14915-2:2003 Software ergonomics for multimedia user interfaces—Part 2: Multimedia navigation and control

ISO 14915-3:2003 Software ergonomics for multimedia user interfaces—Part 3: Media selection and combination

ISO 15534-1:2000 Ergonomic design for the safety of machinery—Part 1: Principles for determining the dimensions required for openings for whole-body access into machinery

ISO 15534-2:2000 Ergonomic design for the safety of machinery—Part 2: Principles for determining the dimensions required for access openings

ISO 15534-3:2000 Ergonomic design for the safety of machinery—Part 3: Anthropometric data

ISO 15535:2003 General requirement for establishing anthropometric databases

ISO/TS 16071:2003 Ergonomics of human-system interaction—Guidance on accessibility for human-computer interfaces

ISO/TR 16982:2002 Ergonomics of human-system interaction—Usability methods supporting human-centered design

ISO/TR 18529:2000 Ergonomics—Ergonomics of human-system interaction—Human-centered life-cycle process descriptions

ISO/TR 19358:2002 Ergonomics—Construction and application tests for speech technology

TR, Technical Report; TS, Technical Specification.

that case, the standard in preparation is a revision of the published standard; after publication, it will replace the existing standard. In Table 2.5, the ISO standards are organized according to ergonomics topics. This table shows that most standards are developed for the topics "Physical Environment" and "Visual information, VDTs, and software."

List of CEN Standards

Since the first ergonomics CEN standard was published in 1990 as ENV 26385, which was an adoption of the above ISO 6385, the number of CEN standards increased considerably (see Figure 2.1 and Table 2.6). Tables 2.7 and 2.8 show ergonomic CEN standards that have been published or are in preparation by CEN TC 122 (Dul et al., 2004). It turns out that several CEN and ISO standards are identical. This is a result of a policy of CEN and ISO to harmonize the development of their standards, according to the Agreement on technical cooperation between ISO and CEN (Vienna Agreement) of 1991 (CEN, 2004; ISO/CEN, 1991). The topics covered in CEN standards are presented in Table 2.9. Most standards are developed for the topics "Safety of Machinery," "Workplace and Equipment Design," "Physical Environment," and "Visual Information, VDTs, and Software."

TABLE 2.4
ISO Standards from ISO TC 159, in Preparation

ISO/DIS 7730 Ergonomics of the thermal environment—analytical determination and interpretation of thermal comfort using calculation of the PMV and PPD indices and local thermal comfort

ISO/FDIS 7933 Ergonomics of the thermal environment—analytical determination and interpretation of heat stress using calculation of the predicted heat strain

ISO/FDIS 8996 Ergonomics of the thermal environment—Determination of metabolic heat rate

ISO/CD 9241-10 Ergonomics of human-system interaction—Part 10: Dialogue principles

ISO/9920 Ergonomics of the thermal environment—Estimation of the thermal insulation and evaporative resistance of a clothing ensemble

ISO/CD 10075-3 Ergonomic principles related to mental workload—Part 3: Measurement and assessment of mental workload

ISO/CD 11064-4 Ergonomic design of control centers—Part 4: Layout and dimensions of workstations

ISO/CD 11064-6 Ergonomic design of control centers—Part 6: Environmental requirements

ISO/CD 11064-7 Ergonomic design of control centers—Part 7: Principles for the evaluation of control centers

ISO/CD 11079 Evaluation of the thermal environment—Determination and interpretation of cold stress when using required clothing insulation (IREQ) and local cooling effects

ISO/CD 11228-2 Ergonomics—Manual handling—Part 2: Pushing and pulling

ISO/CD 11228-3 Ergonomics—Manual handling—Part 3: Handling of low loads at high frequency

ISO/DIS 13732-3 Ergonomics of the thermal environment—Touching of cold surfaces—Part 3: Ergonomics data and guidance for application

ISO/CD 14505-1 Ergonomics of the thermal environment: Thermal environment in vehicles—Part 1: Principles and method for assessment for thermal stress

ISO/CD 14505-2 Ergonomics of the thermal environment: Thermal environment in vehicles—Part 2: Determination of equivalent temperature

ISO/CD 14505-3 Ergonomics of the thermal environment: Thermal environment in vehicles—Part 3: Evaluation of thermal comfort using human subjects

ISO/FDIS 15265 Ergonomics of the thermal environment—Risk assessment strategy for the prevention of stress or discomfort in thermal working conditions

ISO/DIS 15536-1 Ergonomics—Computer manikins and body templates—Part 1: General requirements

ISO/DIS 15536-2 Ergonomics—Computer manikins and body templates—Part 2: Structures and dimensions

ISO/DIS 15537 Principles for selecting and using test persons for testing anthropometric aspects of industrial products and designs

ISO/CD 15743 Ergonomics of the thermal environment—Working practices in cold: Strategy for risk assessment and management

ISO/CD 20282-1 Ease of operations of everyday products—Part 1: Context of use and user characteristics

ISO/CD 20282-2 Ease of operations of everyday products—Part 2: Test method

ISO/PRF TS 20646 Ergonomic procedures for the improvement of local muscular loads

ISO/CD 20685 3D scanning methodologies for internationally compatible anthropometric databases

CD, Committee Draft, registered draft standard; DIS, Draft International Standard, registered draft standard; FDIS, Final Draft International Standard, registered for formal approval.

List of CEN Standards Related to Legislation

As stated earlier, certain CEN standards have a legal status. In the so-called New Approach, several CEN standards are related to legislation formulated in European Directives. For the ergonomics field, certain standards are related to the Machinery Directive 98/37/EC (European Union, 2003). This Directive puts generally formulated essential requirements on safety and health

TABLE 2.5

ISO Standards from Tables 2.3 and 2.4, Organized According to Ergonomics Topics[a]

1. General design principles

ISO 6385, ISO 13407

2. Safety of machinery

ISO/FDIS 14738, ISO 15534-1, ISO 15534-2, ISO 15534-3

3. Physical environment

Noise/speech: ISO 7731, ISO 9921, ISO 11428, ISO 11429 ISO/TR 19358

Climate: ISO 7243 ISO 7726, ISO 7730, ISO 7933, ISO 8996, ISO 9241-6, ISO 9886, ISO 9920, ISO 10551, ISO/TR 11079, ISO 11399, ISO 12894, ISO 13731, ISO/TS 13732-2, ISO/DIS 13732-3

4. Physical workload

ISO 11226, ISO 11228-1

5. Mental workload

ISO 9241-2, ISO 10075, ISO 10075-2, ISO/CD 10075-3

6. Workplace and equipment design

General: ISO 9241-5, ISO 9241-6, ISO 11064-1, ISO 11064-2, ISO 11064-3

Anthropometry: ISO 7250, ISO 14738, ISO 15534-1, ISO 15534-2, ISO 15534-3, ISO 15535, ISO/DIS 15536-1, ISO/DIS 15537

7. Visual information, VDTs, and software

General: ISO 9241-1, ISO 9241-2, ISO 9241-3, ISO 9241-4, ISO 9241-5, ISO 9241-6, ISO 9241-7, ISO 9241-8, ISO 9241-9, ISO 13406-1, ISO 13406-2, ISO 16071

Software: ISO 9241-10 ISO 9241-11 ISO 9241-12 ISO 9241-13 ISO 9241-14 ISO 9241-15 ISO 9241-16, ISO 9241-17, ISO 13407, ISO 14915-1, ISO 14915-2, ISO 14915-3, ISO/TR 16982, ISO/TR 18529

8. Displays and controls

ISO 9241-4, ISO 9355-1, ISO 9355-2, ISO 11428, ISO 11429

9. Personal protection equipment

Notes

a   Standards can be listed under more than one topic.

TABLE 2.6

CEN Standards on Ergonomics

	1989 (Metz, 1991)	1995 (Dul, De Vlaming, & Munnik, 1996)	2004 (Dul et al., 2004)
CEN published	0	10	62
CEN in preparation	0	38	27
Total	0	48	89

when using machinery. Linked to this directive, the CEN standards shown in Table 2.10 give detailed requirements. A company that meets these standards is assumed to meet the general requirements set in the Directive. Apart from ergonomics standards that are related to legislation, ergonomics standards can also be useful for governments to stimulate voluntary actions in the field of Occupational Health and Safety. Because ergonomics in general and ergonomics standards, in particular, have (by definition, see subsection Definition) both a social AND an economic goal (Dul, 2003b), governments have two reasons to stimulate the use of ergonomics standards. By emphasizing this dual goal, ergonomics standards may be a positive incentive for companies that design both healthy and efficient production systems.

TABLE 2.7

Published Standards from CEN TC 122

EN 457:1992 Safety of machinery—Auditory danger signals—General requirements, design, and testing (ISO 7731:1986 modified)

EN 547-1:1996 Safety of machinery—Human body measurements—Part 1: Principles for determining the dimensions required for openings for whole-body access into machinery

EN 547-2:1996 Safety of machinery—Human body measurements—Part 2: Principles for determining the dimensions required for access openings

EN 547-3:1996 Safety of machinery—Human body measurements—Part 3: Anthropometric data

EN-563:1994 Safety of machinery—Temperature of touchable surfaces—Ergonomics data to establish temperature limit values for hot surfaces

EN 614-1:1995 Safety of machinery—Ergonomic design principles—Part 1: Terminology and general principles

EN 614-2:2000 Safety of machinery—Ergonomic design principles—Part 2: Interactions between the design of machinery and work tasks

EN 842:1996 Safety of machinery—Visual danger signals—General design requirements, design, and testing

EN 894-1:1997 Safety of machinery—Ergonomics requirements for the design of displays and control actuators—Part 1: General principles for human interactions with displays and control actuators

EN 894-2:1997 Safety of machinery—Ergonomics requirements for the design of displays and control actuators—Part 2: Displays

EN 894-3:2000 Safety of machinery—Ergonomics requirements for the design of displays and control actuators—Part 3: Control actuators

EN 981:1997 Safety of machinery—System of auditory and visual danger and information signals

EN 1005-1:2001 Safety of machinery—Human physical performance—Part 1: Terms and definitions

EN 1005-2 Safety of machinery—Human physical performance—Part 2: Manual handling of machinery and component parts of machinery

EN 1005-3:2002 Safety of machinery—Human physical performance—Part 3: Recommended force limits for machinery operation

EN ISO 7250:1997 Basic human body measurements for technological design (ISO 7250:1996)

EN ISO 7726:2001 Ergonomics of the thermal environment—Instruments for measuring physical quantities (ISO 7726:1998)

EN ISO 7730:1995 Moderate thermal environments—Determination of the PMV and PPE indices and specification of the conditions for thermal comfort (ISO 7730: 1994)

EN ISO 9241-1:1997 Ergonomic requirements for office work with VDTs—Part 1: General introduction (ISO 9241-1:1997)

EN ISO 9241-2:1993 Ergonomic requirements for office work with VDTs—Part 2: Guidance on task requirements (ISO 9241-2:1992)

EN ISO 9241-3:1993 Ergonomic requirements for office work with VDTs—Part 3: Visual display requirements (ISO 9241-3:1992)

EN ISO 9241-4:1998 Ergonomic requirements for office work with VDTs—Part 4: Keyboard requirements (ISO 9241-4:1998)

EN ISO 9241-5:1999 Ergonomic requirements for office work with VDTs—Part 5: Workstation layout and postural requirements (ISO 9241-5:1998)

EN ISO 9241-6:1999 Ergonomic requirements for office work with VDTs—Part 6: Guidance on the work environment (ISO 9241-6:1999)

EN ISO 9241-7:1998 Ergonomic requirements for office work with VDTs—Part 7: Requirements for display with reflections (ISO 9241-7:1998)

EN ISO 9241-8:1997 Ergonomic requirements for office work with VDTs—Part 8: Requirements for displayed colors (ISO 9241-8:1997)

(Continued)

TABLE 2.7 (Continued)

EN ISO 9241-9:2000 Ergonomic requirements for office work with VDTs—Part 9: Requirements for non-keyboard-input devices (ISO 9241-9:2000)

EN ISO 9241-10:1996 Ergonomic requirements for office work with VDTs—Part 10: Dialogue principles (ISO 9241-10:1996)

EN ISO 9241-11:1998 Ergonomic requirements for office work with VDTs—Part 11: Guidance on usability (ISO 9241-11:1998)

EN ISO 9241-12:1998 Ergonomic requirements for office work with VDTs—Part 12: Presentation of information (ISO 9241-12:1998)

EN ISO 9241-13:1998 Ergonomic requirements for office work with VDTs—Part 13: User guidance (ISO 9241-13:1998)

EN ISO 9241-14:1999 Ergonomic requirements for office work with VDTs—Part 14: Menu dialogues (ISO 9241-14:1995)

EN ISO 9241-15:1997 Ergonomic requirements for office work with VDTs—Part 15: Command dialogues (ISO 9241-15:1997)

EN ISO 9241-16:1999 Ergonomic requirements for office work with VDTs—Part 16: Direct manipulation dialogues (ISO 9241-16:1999)

EN ISO 9241-17:1998 Ergonomic requirements for office work with VDTs—Part 17: Form filling dialogues (ISO 9241-17:1998)

EN ISO 9886:2001 Evaluation of thermal strain by physiological measurements (ISO 9886:1992)

EN ISO 9920 Ergonomics of the thermal environment—Estimation of the thermal insulation and evaporative resistance of a clothing ensemble (ISO 9920:1995)

EN ISO 9921:2003 Ergonomics—Assessment of speech communication (ISO 9921:2003)

EN ISO 10075-1:2000 Ergonomic principles related to mental workload—Part 1: General terms and definitions (ISO 10075:1991)

EN ISO 10075-2:2000 Ergonomic principles related to mental workload—Part 2: Design principles (ISO 10075-2:1996)

EN ISO 10551:2001 Ergonomics of the thermal environment—Assessment of the influence of the thermal environment using subjective judgment scales (ISO 10551:1995)

EN ISO 11064-1:2000 Ergonomic design of control centers—Part 1: Principles for the design of control centers (ISO 11064-1:2000)

EN ISO 11064-2:2000 Ergonomic design of control centers—Part 2: Principles for the arrangement of control suites (ISO 11064-2:2000)

EN ISO 11064-3:1999 Ergonomic design of control centers—Part 3: Control room layout (ISO 11064-3:1999)

ENV ISO 11079:1998 Evaluation of cold environments—Determination of required clothing insulation (IREC) (ISO/TR 11079:1993)

EN ISO 11399:2000 Ergonomics of the thermal environment—Principles and application of relevant International Standards (ISO 11399:1995)

EN 12515:1997 Hot environments—Analytical determination and interpretation of thermal stress using calculation of required sweat rate (ISO 7933:1989 modified)

EN ISO 12894:2001 Ergonomics of the thermal environment—Medical supervision of individuals exposed to extreme hot or cold environments (ISO 12894:2001)

EN 13202:2000 Ergonomics of the thermal environment—Temperatures of touchable hot surfaces—Guidance for establishing surface temperature limit values in production standards with the aid of EN 563

EN ISO 13406-1:1999 Ergonomic requirements for work with visual display based on flat panels—Part 1: Introduction (ISO 13406-1:1999)

EN ISO 13406-2:2001 Ergonomic requirements for work with visual displays based on flat panels—Part 2: Ergonomic requirements for flat panel displays (ISO 13406-2:2001)

EN ISO 13407:1999 Human-centered design processes for interactive systems (ISO 13407:1999)

EN ISO 13731:2001 Ergonomics of the thermal environment—Vocabulary and symbols (ISO 13731:2001)

EN 13861:2002 Safety of machinery—Guidance for the application of ergonomics standards in the design of machinery

(Continued)

TABLE 2.7 (Continued)

EN ISO 14738:2002 Safety of machinery—Anthropometric requirements for the design of workstations at machinery (ISO 14738:2002)

EN ISO 14915-1:2002 Software ergonomics for multimedia user interfaces—Part 1: Design principles and framework (ISO 14915-1:2002)

EN ISO 14915-2:2003 Software ergonomics for multimedia user interfaces—Part 2: Multimedia control and navigation (ISO 14915-2:2003)

EN ISO 14915-3:2002 Software ergonomics for multimedia user interfaces—Part 3: Media selection and combination (ISO 14915-3:2002)

EN ISO 15535:2003 General requirements for establishing an anthropometric database (ISO 15535:2003)

ENV 26385:1990 Ergonomic principles of the design of work systems (ISO 6385: 1981)

EN ISO 27243:1993 Hot environments—Estimation of the heat stress on working man, based on the WBGT-index (wet bulb globe temperature) (ISO 7243:1989)

EN 28996:1993 Ergonomics—Determination of metabolic heat production (ISO 8996:1990)

ENV, Preliminary European Standard.

The obligatory implementation of European standards may also lead to problems. An example is the European standard EN 1335–1 "Office furniture—Office work chair—Part 1: Dimensions—determination of dimensions" (CEN, 2000). Although this standard was not developed by CEN TC 122 and is therefore not listed in one of the tables presented here, the standard includes some important ergonomics issues, such as specification for the adjustment of the seat height. Because Dutch people are, on average, the tallest people in the world but the population also includes small people due to its cultural diversity, the range of motion for height adjustments specified in the European standard is too small for the Dutch population. The Dutch TC on office furniture developed a deviating Dutch guideline for a Dutch work chair that provides guidance on how to tackle this situation (NEN, 2004b).

Classification

Entity

Because ergonomics deals with the interaction between a person and the environment, a major characteristic of all ergonomics standards is that the person is always one of the entities concerned. Some standards only deal with the entity person (e.g., human body measurements), other standards deal with both the person and the technical or organizational environment, in particular when matching a person and a "thing." In most other standards, the matching entities concern "things"—for example, a product or a part thereof.

Horizontal, Vertical

The current set of ergonomics standards contains both "horizontal" (general) and "vertical" (specific) standards. ISO 6385 on ergonomic design principles is an example of a horizontal standard. Examples of "vertical standards" are ISO 9241 on ergonomic requirements for visual display terminals and ISO 11064-3 on control room layout. Also, in the set of other standards, both vertical and horizontal standards can be found.

In the environmental field, aside from (horizontal) environmental standards (the ISO 14000 series of standards for environmental management and hundreds of standards that specify methods to measure pollution), initiatives have been taken to get more attention for environmental aspects in (vertical) product standards (CEN, 1998; Commission of the European Communities, 2004; ISO, 1997). In this way, a specific field attempts to get more impact in other fields. In ergonomics, we have not observed a similar strategy.

TABLE 2.8

Standards from CEN TC 122, in Preparation

prEN 614-lrev Safety of machinery—Ergonomic design principles—Part 1: Terminology and general principles. Under approval

prEN 894-4 Safety of machinery—Ergonomics requirements for the design of displays and control actuators—Part 4: Location and arrangement of displays and control actuators. Under development

prEN 1005-4 Safety of machinery—Human physical performance—Part 4: Evaluation of working postures and movements in relation to machinery. Under approval

prEN 1005-5 Safety of machinery—Human physical performance—Part 5: Risk assessment for repetitive handling at high frequency. Under approval

prEN ISO 6385 rev Ergonomic principles in the design of work systems (ISO/FDIS 6385:2003). Ratified[a]

prEN ISO 7730 rev Ergonomics of the thermal environment—Analytical determination and interpretation of thermal comfort using calculation of the PMV and PPD indices and local thermal comfort (ISO/DIS 7730:2003). Under approval

prEN ISO 7933 Ergonomics of the thermal environment—Analytical determination and interpretation of heat stress using calculation of the predicted heat strain (ISO/DIS 7933:2003). Under approval

prEN ISO 8996 rev Ergonomics—Determination of metabolic heat production (ISO/DIS 8996:2003). Under approval

prEN ISO 9886 rev Ergonomics—Evaluation of thermal strain by physiological measurements (ISO FDIS 9886:2003). Under approval

prEN ISO 9920 rev Ergonomics of the thermal environment—Estimation of the thermal insulation and evaporative resistance of a clothing ensemble. Under development

prEN ISO 10075-3 Ergonomic principles related to mental workload—Part 3: Measurement and assessment of mental workload (ISO/DIS 10075-3:2002). Under approval

prEN ISO 11064-4 Ergonomic design of control centers—Part 4: Layout and dimensions of workstations (ISO/DIS 11064-4:2002). Under development

prEN ISO 11064-6 Ergonomic design of control centers—Part 6: Environmental requirements for control centers (ISO/DIS 11064-6:2003). Under approval

prEN ISO 11079 Evaluation of cold environments—Determination of required clothing insulation (IREQ) (will replace ENV ISO 11079:1998). Under development

prEN ISO 13732-1 Ergonomics of the thermal environment—Methods for the assessment of human responses to contact with surfaces—Part 1: Hot surfaces (ISO/DIS 13732-1:2003). Under approval

prEN ISO 13732-3 Ergonomics of the thermal environment—Touching of cold surfaces—Part 3: Ergonomics data and guidance for application (ISO/DIS 13732-3:2002). Under approval

prEN 13921-1 Personal protective equipment—Ergonomic principles—Part 1: General requirements for the design and the specification. Under approval

prEN 13921-3 Personal protective equipment—Ergonomic principles—Part 3: Biomechanical characteristics. Under approval

prEN 13921-4 Personal protective equipment—Ergonomic principles—Part 4: Thermal characteristics. Under approval

prEN 13921-6 Personal protective equipment—Ergonomic principles—Part 6: Sensory factors. Under approval

prEN 14386 Safety of machinery—Ergonomic design principles for the operability of mobile machinery. Under approval

prEN ISO 14505-1 Ergonomics of the thermal environment: Thermal environment in vehicles—Part 1: Principles and method for assessment for thermal stress. Under development

prEN ISO 14505-2 Ergonomics of the thermal environment: Thermal environment in vehicles—Part 2: Determination of equivalent temperature. Under development

prEN ISO 15536-1 Ergonomics—Computer manikins and body templates—Part 1: General requirements (ISO/DIS 15536-1:2002). Under approval

prEN ISO 15537 Principles for selecting and using test persons for testing anthropometric aspects of industrial products and designs (ISO/DIS 15537:2002). Under approval.

prEN ISO 20685 3D scanning methodologies for internationally compatible anthropometric databases. Under development

prEN ISO 23973 Software ergonomics for World Wide Web user interfaces. Under development

Under development, active work item which has not yet reached the stage of inquiry; Under approval, active work item at a stage between the beginning of the inquiry and the end of formal vote; Ratified, work item at a stage between ratification and publication; Rev, standard under revision.

Basic, Requiring, Measurement. "Basic," "requiring," and "measurement" standards show up in the tables presented here. Almost all ergonomics standards belong to the category of "Requiring standards." For most of these standards, it is difficult to make the difference between performance standards and solution-describing standards. Some of the standards are basic standards, such as ISO 10075, which includes terms and definitions on mental workload, and data standards such as ISO 15534-3, with data on human body dimensions.

Some standards are "Measurement standards." Examples are ISO 7726 on methods for measuring physical quantities of the thermal environment and ISO/TS 13732-2 on methods for measuring human responses to contact with cold or hot surfaces.

Compared with other fields of standardization, the number of measurement standards in ergonomics is relatively small.

Interference, Compatibility, Quality. Because ergonomics deals with the interaction between a person and the environment, most ergonomics standards are interference standards, setting requirements on the influence of the entity "environment" on the entity "person." Examples are standards on human-system interaction or standards on manual handling of loads.

In most other areas of standardization, compatibility standards (e.g., in information technology) or quality standards (e.g., in management systems) are dominant.

Ergonomics Standardization

Definition

Ergonomics standardization is the activity of making ergonomics standards. As mentioned before, in this chapter, we concentrate on ergonomics standards; hence, ergonomics standardization from ISO and CEN because—in contrast with other fields of standardization—most ergonomics standards are developed by ISO and CEN and not by companies or consortia of companies. Therefore, our definition of ergonomics standardization is ergonomics standards-making by ISO and CEN.

Developers

National Standardization. It is common practice that national standardization organizations adopt ISO and—in Europe—CEN standards as national standards. This is an obligation for CEN standards, but it also occurs with many ISO standards. For example, the "mother ergonomics standards" ISO 6385 "ergonomic principles in the design of work systems" has been adopted by national standardization organizations in many countries.

Additionally, national standardization organizations develop their own national standards on ergonomics. In Chapter 20 of this book, Stuart-Buttle (2006) gives an overview of national ergonomics standards in the United States, Australia, and Japan.

International Standardization: Global

In the early 1970s, the International Ergonomics Association decided to initiate the development of ergonomic standards (Parsons & Shackel, 1995) and requested the ISO to set up a technical committee on ergonomics. In 1974, the ISO established TC 159. The scope of this TC is standardization in the field of ergonomics, including terminology, methodology, and human factors data. Since then until 2004, 60 standards have been developed and published (Dul et al., 2004), and new standards are being developed or revised within four subgroups on the following topics:

- Ergonomic guiding principles
- Anthropometry and biomechanics
- Ergonomics of human-system interaction
- Ergonomics of the physical environment

TABLE 2.9

CEN Standards from Tables 2.7 and 2.8, Organized According to Ergonomics Topics[a]

1. General design principles

EN 614-1, prEN 614-1, EN 614-2, prEN ISO 6385, EN 13407, ENV 26385

2. Safety of machinery

EN 457, EN 547-1, EN 547-2, EN 547-3, EN 563, EN 574 EN 614-1, prEN 614-1, EN 641-2, EN 842,

EN 894-1, EN 894-2, EN 894-3, EN 894-4 EN 981, EN 1005-1, EN 1005-2, EN 1005-3, prEN 1005-4,

prEN 1005-5, EN 13861, prEN 14386, prEN ISO14738

3. Physical environment

Noise/speech: EN 457, EN 981, EN ISO 9921

Climate: EN 563, EN ISO 7243, EN ISO 7726. EN ISO 7730, prEN ISO 7730, prEN ISO 7933, EN ISO 8996,

prEN ISO 8996, EN ISO 9241-6, EN ISO 9886, prEN ISO 9886, EN ISO 9920, prEN ISO 9920, EN ISO 10551,

ENV ISO 11079, prEN ISO 11079, EN ISO 11399, EN 12515, EN ISO 12894, EN 13202, EN ISO 13731,

prEN ISO 13732-1, prEN ISO 13732-3, prEN ISO 14505-1, prEN ISO 14505-2, EN ISO 27243, EN 28996

4. Physical workload

EN 1005-1, EN 1005-2, EN 1005-3, prEN 1005-4, prEN 1005-5

5. Mental workload

EN 614-2, EN 9241-2, EN ISO 10075-1, EN ISO 10075-2, prEN ISO 10075-3

6. Workplace and equipment design

General: EN ISO 9241-5, EN ISO 9241-6, EN ISO 11064-1, EN ISO 11064-2, EN ISO 11064-3,

prEN ISO 11064-4, prEN ISO 11064-6, prEN 14386

Anthropometry: EN 547-1, EN 547-2, EN 547-3, EN ISO 7250, EN 14738, EN ISO 15535, prEN ISO 15536-1,

prEN ISO 15537, prEN ISO 20685

7. Visual information, VDTs, and software

General: EN ISO 9241-1, EN ISO 9241-2, EN ISO 9241-3, EN ISO 9241-4, EN ISO 9241-5, EN ISO 9241-6,

EN ISO 9241-7, EN ISO 9241-8, EN ISO 9241-9, EN ISO 13406-1, EN ISO 13406-2

Software: EN ISO 9241-10, EN ISO 9241-11, EN ISO 9241-12, EN ISO 9241-13, EN ISO 9241-14,

EN ISO 9241-15, EN ISO 9241-16, EN ISO 9241-17, EN ISO 13407, EN ISO 14915-1, EN ISO 14915-2,

EN ISO 14915-3, prEN ISO 23973

8. Displays and controls

EN ISO 9241-4

9. Personal protection equipment

prEN 13921-1, prEN 13921-3, prEN 13921-4, prEN 13921-6

Notes

a   Standards can be listed under more than one topic.

In 2004, the secretariat of TC 159 is with the German Standardization Institute DIN. ISO TC 159 has 24 countries that participate in developing the standards and 29 observing countries.

International Standardization: Regional

Since the 1980s, Europe has developed towards a free internal market without barriers to trade. For this common market, common standards were desire for the safety requirements of machinery. In 1989, CEN established the Technical Committee TC 122 "Ergonomics" to address ergonomics requirements in relation to the safety of machinery. The first European ergonomics standard was published in 1990 as ENV 26385, which was an adoption of the above ISO 6385. Afterward, CEN has published ergonomics standards on the safety of machinery and other ergonomics issues. The scope of CEN TC 122 is standardization in the field of ergonomics principles and requirements for the design of work systems and work environments, including machinery and personal protective

TABLE 2.10

Ergonomics CEN Standards Linked to the European Machinery Directive

EN 457:1992 Safety of machinery—Auditory danger signals—General requirements, design, and testing (ISO 7731:1986 modified)

EN 547-1:1996 Safety of machinery—Human body measurements—Part 1: Principles for determining the dimensions required for openings for whole-body access into machinery

EN 547-2:1996 Safety of machinery—Human body measurements—Part 2: Principles for determining the dimensions required for access openings

EN 547-3:1996 Safety of machinery—Human body measurements—Part 3: Anthropometric data

EN-563:1994 Safety of machinery—Temperature of touchable surfaces—Ergonomics data to establish temperature limit values for hot surfaces

EN 614-1:1995 Safety of machinery—Ergonomic design principles—Part 1: Terminology and general principles

EN 614-2:2000 Safety of machinery—Ergonomic design principles—Part 2: Interactions between the design of machinery and work tasks

EN 842:1996 Safety of machinery—Visual danger signals—General design requirements, design and testing

EN 894-1:1997 Safety of machinery—Ergonomics requirements for the design of displays and control actuators—Part 1: General principles for human interactions with displays and control actuators

EN 894-2:1997 Safety of machinery—Ergonomics requirements for the design of displays and control actuators—Part 2: Displays

EN 894-3:2000 Safety of machinery—Ergonomics requirements for the design of displays and control actuators—Part 3: Control actuators

EN 981:1997 Safety of machinery—System of auditory and visual danger and information signals

EN 1005-1:2001 Safety of machinery—Human physical performance—Part 1: Terms and definitions

EN 1005-2 Safety of machinery—Human physical performance—Part 2: Manual handling of machinery and component parts of machinery

EN 1005-3:2002 Safety of machinery—Human physical performance—Part 3: Recommended force limits for machinery operation

EN ISO 7250:1997 Basic human body measurements for technological design (ISO 7250:1996)

EN ISO 14738:2002 Safety of machinery—Anthropometric requirements for the design of workstations at machinery (ISO 14738:2002)

equipment to promote the health, safety, and well-being of the human operator and the effectiveness of the work systems. Until 2004, CEN TC 122 has developed and published 62 standards (Dul et al., 2004). New standards are being developed or revised within nine working groups on the following topics:

- Anthropometry
- Ergonomic design principles
- Surface temperatures
- Biomechanics
- Ergonomics of human-computer interaction
- Signals and controls
- Danger signals and speech communication in noisy environments
- Ergonomics of personal protective equipment
- Ergonomics of the thermal environment

In 2004, the secretariat of CEN TC 122 is with the German Standardization Institute DIN.

Cooperation ISO-CEN

Many ergonomics standards have been developed in cooperation between ISO and CEN, within the framework of the Vienna Agreement. Since 1989, CEN has adopted many ergonomics ISO standards developed by ISO TC 159. Also, ISO and CEN have used parallel approval procedures for work items developed by either of them. Common standards are listed as "EN ISO" standards in Tables 2.7 and 2.8.

Stakeholder Participation

A model that has been developed to identify and select relevant stakeholders to be involved in standardization (Willemse, 2003) has been applied to ergonomics standardization (Dul, Willemse, & De Vries, 2003). The model and the application to ergonomics standardization are presented in Chapter 6 of this book (Willemse, De Vries, & Dul, 2006). One conclusion is that ergonomics experts seem to be major participants in ergonomics standardization and that there is an under-representation of other relevant stakeholders, for example, designers.

CONCLUSIONS

ERGONOMICS STANDARDS

During the past 30 years, ISO and CEN have published—or have prepared—more than 150 ergonomics standards. Because of this high production of new standards in the last three decades, it seems that some duplication of work, inconsistencies, and contradictions have occurred (Nachreiner, 1995). In future standards development, a reduction of overlap and more clear relationships among the standards are desirable. Feedback on the existing collection of standards—answers to questions like "are the standards known?" "are the standards used?" and "are the standards considered to be useful?"—could help further the development of ergonomics standardization as well. Also, the principle of "user participation" advocated by most ergonomists could be better applied to the ergonomics standardization; it is suggested to strengthen the involvement of relevant stakeholders in future ergonomics standardization, especially designers of products and production processes.

ERGONOMICS STANDARDS VERSUS OTHER STANDARDS

The large number of ergonomics standards is only a small fraction of all available ISO and CEN standards. In the introduction we raised the question: "Is ergonomics standardization an area with exceptional characteristics, or is it just one of the many areas of standardization?" By comparing ergonomics standards and standardization with other standards and standardization, we conclude that there are major differences (Table 2.11). When considering Table 2.11, we stress that many similarities also exist between ergonomics standards and other standards and that this table is only a rough estimate of the differences that we observed, and we existing nuances were not considered; hence, the table is open for discussion. Nevertheless, with respect to standardization (making of standards), we observed the following differences:

- Ergonomics standardization started only recently, whereas standardization in most other fields is about a century old, with roots from thousands of years ago.
- Ergonomics experts are the major initiators and developers of ergonomics standards, in contrast to manufacturers in other fields of standardization.
- Ergonomics standards are mainly formal standards developed by formal standardization organizations, whereas other standards are also developed by companies and consortia of companies.

TABLE 2.11

Summary of Major Differences between the Majority of Ergonomics Standards and the Majority of Other Standards

Aspect	Other Standards	Ergonomics Standards
Standardization		
Age	Thousands of years	30 years
Initiator/developer	Manufacturer	Ergonomics expert
Organization of the standardization	Standardization organization, company, consortium	Standardization organization
Standards		
Aim	Economic	Social
Government interest	Economic affairs	Social affairs
User groups	Designer, Tester	Ergonomics expert
End users	Consumer	Worker
Application	Specific, technical	General
Object/Entity	Product, test method	Human, process

With respect to standards (the results of standardization) we observed the following major differences:

- The main aim of ergonomics standards is social (well-being of the workers or the consumers), whereas the major aims of other standards are economic.
- Ministries of social affairs, health, or labor may have governmental interest in ergonomic standards—for example, support of regulation on occupational health and safety. In other fields, ministries of economic affairs may support standardization in general to stimulate the economy.
- The major user group of ergonomics standards are ergonomics experts. For other standards, designers and testers are the major user groups.
- The end-users of ergonomics standards are primarily workers, whereas the end-users of other standards are primarily consumers.
- Most ergonomics standards apply to technical and organizational aspects of (work) situations. Other standards usually apply to specific technical aspects of products.
- Ergonomics standards usually concern humans or the interaction of humans with (production or work) processes, whereas other standards concern primarily products or test methods.

Lessons

What can we learn from these differences? We feel that the following questions can be raised to contribute to the discussion on the future of ergonomics standards:

- Should we attempt to integrate more economic goals ("total system performance") in ergonomics standards?
- Should we, then, look for governmental support from ministries of economics affairs, apart from existing support from ministries of social affairs, health, and labor?
- Should we involve the designers, manufacturers, and consumers more in the development or in ergonomics standards, according to the ergonomic principle of user participation?

- Should we stimulate that more ergonomics standardization is organized by companies and consortia of companies?
- Should we develop more (technical) standards that can be applied to specific situations?
- Should we develop more standards for products and test methods?

It is not easy to give answers to these questions. But even if answers can be given, one main question remains: "Is it better to develop ergonomic standards as a separate field (as was done during the past 30 years), or is it better (similar to the standardization policy in the environmental field) to integrate ergonomics in standardization activities of other fields?" If so, the ergonomics field may gain more impact but may also lose identity.

REFERENCES

AFNOR. (1993). *X 00-001 Normes françaises—Règles pour la rédaction et la présentation—Conseils pratiques*. Paris: Association Française de Normalisation.

ANEC. (2003). *Consumer requirements in standardisation relating to the information society*. Brussels: European Association for the Co-ordination of Consumer Representation in Standardisation.

Baraton, P., & Hutzler, B. (1985). *Magnetically induced currents in the human body. Technology Trend Assessment No. 1*. Geneva: International Electrotechnical Commission.

Breedveld, P. (2005). *Factors influencing perceived acceptance and success of ergonomists within European organizations. MSc thesis, Rotterdam School of Management, Erasmus University Rotterdam, The Netherlands*.

BSI. (1997a). *BS-01A standard for standards—Part1: Recommendations for committee procedures*. London: British Standards Institution

BSI. (1997b). *BS-03 A standard for standards—Part 3: Specification for structure, drafting and presentation*. London: British Standards Institution.

CEN. (1998). *CEN Guide 4 Guide for the inclusion of environmental aspects in product standards*. Brussels: Comité Européen de Normalisation.

CEN. (2000). *EN 1335-1 Office furniture—Office work chair—Part 1: Dimensions—Determination of dimensions*. Brussels: Comité Européen de Normalisation.

CEN. (2004). *European standardization in a global context*. Brussels: European Committee for Standardization (CEN).

Commission of the European Communities. (2004). *Integration of environmental aspects into European standardisation. COM(2004)130 final*. Brussels: Commission of the European Communities.

Cook, L. S., Bowen, D. E., Chase, R. B., Dasu, S., Stewart, D. M., & Tansik, D. A. I. (2003). Human issues in service design. *Journal of Operations Management* 20(2), 159–174.

De Vries, H. J. (1997). Standardization—What's in a name? *Terminology—International Journal of Theoretical and Applied Issues in Specialized Communication* 4 (1), (rectification in 55–83).

De Vries, H. J. (1998). The classification of Standards. *Knowledge Organization*, 25(3), 79–89.

De Vries, H. J. (1999). *Standardization—A business approach to the role of National Standardization Organizations*. Boston/Dordrecht/London: Kluwer Academic Publishers.

De Vries, H. J., Verheul, H., & Willemse, H. (2003). Stakeholder identification in IT standardization processes. In J. L. King & K. Lyytinen (Eds.), *Proceedings of the Workshop on Standard Making: A Critical Research Frontier for Information Systems* (pp. 92–107). Seattle, WA, December 12–14, 2003.

De Vries, H. J., Feilzer, A., & Verheul, H. (2004). Removing barriers for participation in formal standardization. In F. Bousquet, Y. Buntzly, H. Coenen, & K. Jakobs (Eds.), *EURAS Proceedings 2004. Aachener Beiträge zur Informatik, Band 36, Wissenschaftsverlag Mainz in Aachen, Aachen* (pp. 171–176).

DIN. (1996). *DIN 820 Teil 2 Normungsarbeit—Teil 2: Gestaltung von Normen*. Berlin: Beuth Verlag GmbH.

Dul, J., De Vlaming, P. M., & Munnik, M. J. (1996). A review of ISO and CEN standards on ergonomics. *International Journal of Industrial Ergonomics*, 17(3), 291–291.

Dul, J. (2003a). *De mens is de maat van alle dingen—Over mensgericht ontwerpen van producten en processen*. (Man is the measure of all things—On human-centered design of products and processes). *Inaugural address. Erasmus Research Institute of Management*, Erasmus University Rotterdam.

Dul, J. (2003b). The strategic value of ergonomics for companies. In H. Luczak & K. J. Zink. (Eds.), *Human factors in organizational design and management VII*. (pp. 765–769). Santa Monica: IEA Press.

Dul, J., Willemse, H., & De Vries, H. J. (2003). Ergonomics standards: Identifying stakeholders and encouraging participation. *ISO-Bulletin (September)*, 19–23.

Dul, J., De Vries, H., Verschoof, S., Eveleens, W., & Feilzer, A. (2004). Combining economic and social goals in the design of production systems by using ergonomics standards. *Computers & Industrial Engineering* 47(2–3), 207–222.

European Union. (2003). Commission communication in the framework of the implementation of Directive 98/37/EC of the European Parliament and of the Council of 22 June 1998 in relation to machinery amended by directive 98/79/EC, *Official Journal of the European Union, Cl92*, 2–29.

European Union. (1989). Council Directive 89/655/EEC of 30 November 1989 concerning the minimum safety and health requirements for the use of work equipment by workers at work (second individual Directive within the meaning of Article 16 (1) of Directive 89/391/EEC, *Official Journal of the European Union* L 393, 30.12.1989, p. 13.

European Union. (1990a). Council Directive 90/269/EEC of 29 May 1990 on the minimum health and safety requirements for the manual handling of loads where there is a risk particularly of back injury to workers (fourth individual Directive within the meaning of Article 16 (1) of Directive 89/391/EEC), *Official Journal of the European Union* L 156, 21.06.1990, p. 9.

European Union. (1990b). Council Directive 90/270/EEC of 29 May 1990 on the minimum safety and health requirements for work with display screen equipment (fifth individual Directive within the meaning of Article 16 (1) of Directive 89/391/EEC), *Official Journal of the European Union* L 156 of 21.06.1990, p. 14.

Eveleens, W. (2003). A basic ergonomic standard. How to provide optimal working conditions for personnel. *ISO-Bulletin (June)*, 3–6.

Fallentin, N., Viikari-Juntura, E., Waersted, M., & Kilbom, A. (2001). Evaluation of physical workload standards and guidelines from a Nordic perspective. *Scandinavian Journal of Work Environment & Health*, 27(Suppl. 2), 1–52.

Gaillard, J. (1933). *A study of the fundamentals of industrial standardization and its practical application, especially in the mechanical field*. NV W.D. Meinema: Delft.

Hesser, W., & Inklaar, A. (1997) Aims and functions of standardization. In W. Hesser & A. Inklaar (Eds.), *An introduction to standards and standardization* (pp. 39–45). DIN Normungskunde Band 36. Berlin/Vienna/Zürich: Beuth Verlag, pp. 33–45.

Huigen, H. W., Inklaar, A., & Paterson E. (1997). Standardization and certification in Europe. In W. Hesser & A. Inklaar (Eds.), *An introduction to standards and standardization* (pp. 230–251). Berlin: Beuth Verlag.

ISO. (1997). *ISO Guide 64: Guide for the inclusion of environmental aspects in product standards*. Geneva: International Organization for Standardization.

ISO. (2004). Stages of the development of International Standards. International Organization for Standardisation. http://www.iso.org/iso/en/stdsdevelopment/whowhenhow/proc/proc.html

ISO/CEN. (1991). *Agreement on technical cooperation between ISO and CEN (Vienna Agreement) First Revision*. Geneva: ISO.

ISO/IEC. (1996). *ISO/IEC Guide 2 Standardization and related activities—General vocabulary* (7th edition). Geneva: International Organization for Standardization/International Electrotechnical Commission.

KAN. (2004). *Kommission Arbeitsschutz und Normung*. http://www.kan.de.

Leibrock, G. (2002). *Methods of referencing standards in legislation with an emphasis on European legislation. Enterprise Guides*. Brussels: European Commission, Enterprise Directorate-General, Standardisation Unit.

Le Lourd, Ph. (1992) *La normalisation et l'Europe—Secteurs de l' agro-alimentaire, du bois, et de l'eau*. Paris: Editions Romillat/AFNOR.

Metz, B. G. (1991). Outcomes of international standardization work in the field of ergonomics between 1995 and 1989. In Y. Quéinnec & F. Daniellou (Eds.), *Designing for everyone* (pp. 980–987). London: Taylor and Francis.

NEN. (2004a). *Handleiding Commissieleden* (3rd edition). Delft: NEN.

NEN. (2004b). De Nederlandse Werkstoel. *NormalisatieNieuws*, 13(2), 1.

Nachreiner, F. (1995). Standards for ergonomics principles relating to design of work systems and to mental work load. *Applied Ergonomics*, 26(4), 259–263.

Parsons, K. C., & Shackel, B. (1995). Ergonomics and international standards: History, organisational structure and method of development. *Applied Ergonomics*, 26(4), 249–258.

Sanders, T. B. R. (Ed.), (1972). *The aims and principles of standardization*. Geneva: International Organization for Standardization.

Schaap, A., & De Vries, H. J. (2004). *Evaluatie van normalisatie-investeringen—Hoe MKB-bedrijven kunnen profiteren van deelname aan normalisatie.* Zoetermeer: FME-CWM.

SES (1995). *SES 1, Recommended practice for standards designation and organization—An American National Standard.* Dayton, OH: Standards Engineering Society.

Simons, C. A. J. (1994). *Kiezen tussen verscheidenheid en uniformiteit. Inaugural address,* Rotterdam: Erasmus University Rotterdam.

Smith, M. A. (1995). *Vienna Agreement of Technical Cooperation between ISO and CEN. ISO/IEC Directives seminar 1995,* Geneva: ISO.

Stuart-Buttle, C. (2006). Overview of national and international standards and guidelines. In W. Karwowski (Ed.), *Handbook on standards and guidelines in ergonomics and human factors* (pp. 133–147). Mahwah, NJ: Lawrence Erlbaum Associates.

Toth, R. B. (Ed.). (1997). *Profiles of national standards-related activities.* NIST Special Publication 912, Gaithersburg, MD: National Institute of Standards and Technology.

Wen, Z. (2004). Reform and change—An introduction to China standardization. In *Presentation at the 11th International Conference of Standards Users IFAN 2004, 2004-11-11-12,* Amsterdam/Delft: NEN.

Willemse, H. (2003). *Management van normalisatie. Een model voor het managen van normalisatie (In Dutch: Management of standardization. A model for managing standardization).* MSc thesis, Rotterdam School of Management, Erasmus University Rotterdam.

Willemse, H., De Vries, H. J., Dul, J. (2006). Balancing stakeholder representation: An example of stakeholder involvement in ergonomics standardization. In W. Karwowski (Ed.), *Handbook on standards and guidelines in ergonomics and human factors* (pp. 149–156). Mahwah, NJ: Lawrence Erlbaum Associates.

Wulff, I. A., Westgaard, R. H., and Rasmussen, B. (1999). Ergonomic criteria in large-scale engineering design—II Evaluating and applying requirements in the real work of design. *Applied Ergonomics,* 30 (3), 207–221.

3 Ergonomic Performance Standards and Regulations
Their Scientific and Operational Basis

Thomas J. Smith

CONTENTS

Full Citation
  Smith, T. J. (2019). Ergonomic performance standards and regulations—their scientific and operational basis. In W. Karwowski (Ed.), *Handbook of standards and guidelines in human factors and ergonomics* (2nd Edition, Chap. 4). Mahwah, NJ: Lawrence Erlbaum Associates. November.

INTRODUCTION AND BACKGROUND

This chapter addresses the use of performance standards in occupational health and safety (OHS) regulations, with particular reference to ergonomic performance standards. Sections below deal with: (1) some relevant definitions and a historical perspective on OHS performance standards; (2) how performance standards differ from prescriptive standards; (3) an introduction to ergonomic performance standards and an analysis of their scientific and operational basis plus evidence regarding their criterion validity; and (4) recommendations regarding the application of ergonomic performance standards.

DEFINITIONS

No consensus has emerged regarding the descriptive terminology applied to OHS standards and regulations, particularly in ergonomic standards. Because there are instances of uncertainty, redundancy, and ambiguity in the terminology employed by different authors in this area, Table 3.1 is provided to clarify the meanings of various terms in this chapter. It is not assumed that the definitions in Table 3.1 necessarily apply to other chapters in this handbook.

It is first important to clarify the distinctions between a law or regulation, a standard, and a guideline or recommendation. The terms *law* and *regulation*—considered synonymous here—both apply to a statement that requires legal compliance. In the OHS field, the meaning associated with the term *standard* is ambiguous. For example, usage by the U.S. Department of Labor Occupational Safety and Health Administration (OSHA) indicates that a regulation and a standard are synonymous (OSHA, 1980). However, Webster's dictionary definition—"anything recognized as correct by common consent, by approved custom, or by those most competent to decide"—does not imbue the term with legal status. This is the meaning adopted in this chapter, based on the definition provided (Table 3.1) by the American Academy of Pediatrics, American Public Health Association, & National Resource Center for Health and Safety in Child Care University of Colorado Health Sciences Center at Fitzsimons (2002). Because it typically emanates from an entity with professional standing and credentials, a standard is a statement that carries a strong, but not regulatory, incentive for compliance. A *guideline* and *recommendation*—considered synonymous here—are statements that carry less incentive for compliance than a standard.

The basic objective of an OHS/ergonomic regulation, standard, guideline, or recommendation is to mitigate a hazard that poses a risk to the health and safety of the worker. There are essentially two distinct ways of formulating a requirement of this kind—as a performance standard or as a prescriptive standard. This chapter focuses on the former, but it is important to define both approaches to OHS/ergonomic standard writing.

Typically, the term *performance regulation* rather than *performance standard* is used by authors concerned with a performance-based approach to formulation of OHS regulations (Bryce, 1983, 1985a, 1985b). The term *performance standard* is ambiguous because it has been used in the OHS and ergonomic literature to refer to two different concepts. As noted by Fallentin et al. (2001) and Westgaard and Winkel (1996), physical workload standards (Table 3.1) first were systematically delineated about 100 years ago as performance criteria concerned with the mitigation of OHS hazards associated with work demands that give rise to decrements in behavioral performance (Definition 2 in Table 3.1). Extending the earlier work of Lavoisier in the 18th century, Taylor, Gilbreth, and others developed performance guidelines early in the 20th century for the physical workload capacity of workers that were directed at optimizing production efficiency (Fallentin, et al., 2001).

On the other hand, as outlined in the next subsection, performance regulations and standards that specify who is responsible for meeting a specified requirement have been known since ancient times. As applied to OHS, this sense of the term "performance standard" refers to a formulation that explicitly identifies the person (or persons) whose performance is responsible for meeting an OHS hazard mitigation objective specified in the standard (Table 3.1, Definition 1). In the remainder of this chapter, the use of the terms "performance standard" or "performance regulation" is understood to refer to Definition 1 in Table 3.1.

There is also some uncertainty regarding the terminology applied to an OHS standard/regulation that is not performance-based. In this case, the formulation for the standard/regulation refers to what the OHS requirement is, but not who is responsible for meeting the requirement. In this type of formulation, the terms *prescriptive*, *specification*, or *physical standard/regulation* (among others) have been applied (Table 3.1) (Bryce, 1983, 1985a, 1985b; Fallentin, et al., 2001; Smith, 1973; Westgaard and Winkel, 1996).

TABLE 3.1
Definitions of Terms Relevant to OHS/Ergonomic Standards

Term	Definition	References[1]
Behavioral hazard	A manifestation of worker or organizational system behavior that creates an OHS hazard. An unsafe act.	2–4
Code	An entire system of rules or laws.	Webster
Engineering hazard	An engineering design factor that creates an OHS hazard. An unsafe condition.	2–4
Environmental Hazard	An environmental design factor that creates an OHS hazard. An unsafe condition.	2–4
Ergonomic Standard/ Regulation	A standard/regulation directed at OHS hazard mitigation through ergonomic intervention targeting correction of a specified workplace design factor or condition. Ergonomic standards address a broad range of workplace design issues, such as physical workload, workplace, equipment, workstation, job, or warning design factors and conditions.	10–13
Generic Performance Standard/Regulation	A performance standard/regulation that specifies the person or persons whose performance is responsible for meeting an OHS hazard mitigation objective but does not specify how that objective should be achieved.	7–9
Guideline	A statement of advice or instruction pertaining to practice. Like a recommendation, it originates in an organization with acknowledged professional standing. Although it may be unsolicited, a guideline is developed in response to a stated request or perceived need for such advice or instruction.	5
Hazard	An engineering, behavioral, or operational factor that elevates the risk of detrimental performance by one or more workers (employees or managers) or by an organizational system.	6
Law	That which is laid down or fixed. A rule laid down or established, whether by custom or as an expression of the will of a person or power able to enforce its demands.	Webster
Operational Hazard	An OHS hazard arising because of interaction between behavioral factors and physical design factors.	4, 6
Performance Standard/ Regulation	A standard/regulation that explicitly identifies the person or persons whose performance is responsible for meeting an OHS hazard mitigation objective specified in the standard (Definition 1). A standard/regulation directed towards the mitigation of an OHS hazard associated with work demands that give rise to decrements in behavioral performance (Definition 2).	4, 7–9 10–12
Physical Hazard	An engineering or environmental design factor that creates an OHS hazard. An unsafe condition.	2–4
Physical Standard	A standard directed towards the mitigation of a physical OHS hazard.	2–4
Physical Workload Standard	An ergonomic standard directed towards the mitigation of work performance and/or musculoskeletal health hazards arising out of worker exposure to work demands related to physical exertion and/or biomechanical forces.	10–12
Prescriptive Standard/ Regulation	A standard/regulation that prescribes how a specific OHS hazard is to be mitigated, without any indication as to whose performance is	4, 7–9

(*Continued*)

TABLE 3.1 (Continued)

Term	Definition	References[1]
	responsible for mitigating the hazard. Typically directed towards the mitigation of physical OHS hazards.	
Process Standard/ Regulation	A qualitative performance or prescriptive standard/regulation that defines a program approach for purposes of mitigating one or more specified OHS hazards.	11
Qualitative Standard/ Regulation	A process standard/regulation that lacks precise, quantitative, numerical compliance criteria.	10–12
Quantitative Standard/ Regulation	A performance or prescriptive standard/regulation that specifies precise, quantitative, numerical compliance criteria.	10–12
Recommendation	A statement of practice aimed at providing an OHS benefit to the population served, usually initiated by an organization or a group of individuals with expertise or broad experience in the subject matter. A recommendation is not binding on the practitioner; that is, there is no obligation to carry it out. A statement may be issued as a recommendation because it addresses a fairly new topic or issue, because scientific supporting evidence may not yet exist, or because the practice may not yet enjoy widespread acceptance by the members of the organization or by the intended audience for the recommendation.	5
Regulation	A regulation originates in an agency with either governmental or official authority, has the power of law, and usually accompanied by an enforcement activity. A regulation often imbues a previous recommendation or guideline with legal authority. The components of a regulation vary by topic addressed and by area and level of jurisdiction. Because a regulation prescribes a practice that every agency or program must comply with, it usually is the minimum or the floor below which no agency or program should operate.	5
Rule	(1) A regular established method of procedure or action; or (2) a code.	Webster
Specification Standard/ Regulation	A prescriptive standard/regulation.	4, 7–9
Standard	A statement that defines a goal of practice. It differs from a recommendation or a guideline in that it carries a great incentive for universal compliance. It differs from a regulation in that compliance is not necessarily required for legal operation. It usually is legitimized or validated based on scientific or epidemiological data, or—when this evidence is lacking—it represents the widely agreed upon, state-of-the-art, high-quality level of practice. An entity that does not meet a standard may incur disapproval or sanctions from within or outside the organization. Thus, a standard is the strongest criteria for practice set by an organization or association.	5

[1]Numbers refer to references cited in the following footnotes.
[2,3]Heinrich (1931, 1959); [4]Smith (1973); [5]American Academy of Pediatrics, American Public Health Association, and National Resource Center for Health and Safety in American Academy of Pediatrics, American Public Health Association, & National Resource Center for Health and Safety in Child Care University of Colorado Health Sciences Center at Fitzsimons (2002); [6]Smith (2002a); [7–9]Bryce (1983, 1985a, 1985b); [10]Fallentin et al. (2001); [11]Westgaard and Winkel (1996); [12]Viikari-Juntura (1997); [13]Dul, de Vlaming, and Munnik (1996).

It is assumed here that the meanings of each of these terms are synonymous. However, in the opinion of these authors, none of these terms are entirely satisfactory. After all, an OHS requirement formulated as a performance standard/regulation obviously offers a prescription or specification about mitigation of a specified hazard. However, in line with customary usage, the terms "prescriptive standard" or "prescriptive regulation" will be used in this chapter to refer to a formulation for an OHS/ergonomic standard/regulation that specifies what the requirement is, without any specification as to responsibility for compliance.

Let us close this section by addressing the distinction between a *process* or *qualitative standard*, and a *quantitative standard* (Table 3.1). Whereas a quantitative standard specifies precise, quantitative, numerical compliance criteria for meeting an OHS/ergonomic requirement, a qualitative standard lacks such criteria and instead specifies a process or program approach to mitigate one or more OHS hazards (Fallentin et al., 2001; Viikari-Juntura, 1997; Westgaard & Winkel, 1996). It is assumed here that the terms "process standard" and "qualitative standard" have synonymous meanings. It also should be emphasized that qualitative and quantitative standards each can be formulated as either prescriptive or performance standards, although quantitative standards are typically formulated as prescriptive standards.

HISTORY OF PERFORMANCE STANDARDS

The formulation of regulations, standards, and guidelines that explicitly state whose performance is responsible for meeting specified compliance requirements have an ancient history. Table 3.2 provides some examples relevant to health and safety, dating back to ancient Mesopotamia. The most widely known of these, likely familiar to hundreds of millions of people worldwide, is the 6th commandment. Its literal formulation (scanning right to left) is "no murder," but this is inaccurate. The first letter of the second Hebrew word—ה—is the imperfect inflection of "to murder" meaning "you will murder" (Harrison, 1955, p. 69)—hence, the translation.

Six of the remaining ten commandments—numbers 2, 3, and 7-10—are similarly formulated as performance standards. Thus, well over 3,000 years ago, ancient Jews understood that God's fundamental words, passed on to them through Moses, not only prescribed basic rules of behavior but also carried the connotation of personal responsibility for performance in complying with these rules—the essence of a performance standard formulation.

It also is noteworthy that the term "law"—with an Old Norse etymology dating back to development by the Vikings of a code of common law first written down in the Icelandic Jonsbok Manuscript in the 12th century (www.mnh.si.edu/vikings/, Room 6)—refers to a rule laid down or established (Table 3.1). A rule, in turn, is defined as a regularly established method of procedure or action. Thus, there are connotations of both prescription and performance in our modern understanding of what a law means.

RECENT TRENDS IN PERFORMANCE STANDARDS AND REGULATIONS

Despite their ancient origins, performance standards/regulations have received little analytical attention from those in the HF/E or legal communities concerned with the formulation of OHS or ergonomic rules. In their recent reviews of ergonomic standards, no reference to the sense of performance standards of Definition 1 (Table 3.1) is made by Dul, de Vlaming, & Munnik (1996), Fallentin et al. (2001), Viikari-Juntura (1997), or Westgaard and Winkel (1996). An 11/19 search of all issues of three major journals published by the Human Factors and Ergonomics Society (HFES) (*Human Factors, Proceedings of the HFES, Ergonomics in Design*) found no publications dealing with performance standards/regulations, in Table 3.1, Definition 1 sense of the term. A Google search for all combinations of "ergonomic performance" followed by "standards"/"standard" or "regulations"/"regulation" found only one site

TABLE 3.2

Some Ancient Performance Standards Relevant to Health and Safety

Standard		Origin	Approximate Date[1]	Reference
Original Text	**English Translation**			
Not Available	When a lion kills [animals in an enclosed yard], his keeper shall pay all damages, and the owner of the yard shall receive the killed animals.	Ancient Mesopotamia (Akkadian through Neo-Babylonian periods)	2250–550 BCE	www.fordham.edu/ halsall/ancient/ 2550mesolaws.html
Not Available	If anyone opens his ditches to water his crop but is careless, and the water flood the field of his neighbor, then he shall pay his neighbor for his loss.	Code of Hammurabi, Article 55	1955–1913 BCE	www.yale.edu/ lawweb/avalon/ medieval/ hammint.htm
Not Available	When the builder has built a house for a man and his work is not strong, and if the house he has built falls in and kills a householder, that builder shall be slain.	Code of Hammurabi, Article 229	1955–1913 BCE	Foliente (2000, p. 13)
לֹא תִּרְצָח	You shall not murder.	6th Commandment	1300 BCE	Plaut (1981, p. 554)
Not Available	No person shall hold meetings in the City at night.	The Twelve Tables (original Roman Code of Law)	451–450 BCE	www.csun.edu/ ~hcfl1004/ 12tables.html

Notes

1 BCE = before the current era

containing the author's treatment of the topic (Smith, 2002b; http://www.doli.state.mn.us/pdf/ overview.pdf).

Nevertheless, over the past three decades, a number of jurisdictions have adopted performance regulations as part of their OHS regulatory code. In Canada, the OHS Acts of every province and territory feature routine and extensive use of performance regulations, prompted in no small part by a 1981 report ("Reforming Regulations") from the Economic Council of Canada encouraging Canadian governments to establish a national set of OHS regulatory provisions that would provide uniform and consistent OHS legislation for all Canadian workers directed at reducing occupational injuries and diseases (Bryce, 1983). This report referenced the developing use of performance regulations and pointed out that a performance-based approach allowed employers to make cost-efficient choices in complying with OHS regulatory provisions.

In a similar vein, the OHS Acts for every state and territory in Australia also make extensive use of performance regulations. Website links to the OHS Acts for governmental jurisdictions in both Canada and Australia may be found at http://www.eu-ccohs.org and http://www.nohsc-eu.gov.au/ respectively.

There also are U.S. examples of deployment of OHS performance regulations. Thus, although they are built largely around prescriptive regulations, the Occupational Safety and Health Standards (OSHA, 1980) feature 218 performance-based "employer shall," and 90 "employee shall" provisions (http://www.osha-slc.gov/pls/oshaweb/owastand.display_standard_group?p_ toc_level=1&p_part_number=1910&p_text_version=FALSE). In Minnesota, the AWAIR Act

makes extensive use of performance regulations to target workplace accident and injury reduction (Minnesota Department of Labor and Industry, 1993).

Another health and safety domain in which the performance-based approach to standard
setting is receiving growing emphasis is that of building performance (Foliente, 2000;
International Council for Building Research Studies and Documentation [CIB], 1982; Prior and
Szigeti, 2003). The application of performance-based building regulations dates to the code of
Hammurabi, where Article 229 mandated the death penalty for any builder whose building
collapsed and killed an occupant (Table 3.2). Prior and Szigeti (IBID) summarized a record of
growing application of performance-based building codes and regulations in the U.S., Europe,
and Australasia over the past five decades. More broadly, the World Trade Organization, recognizing that prescriptive codes and standards represent major non-tariff barriers to trade,
adopted the following clause in its Agreement on Technical Barriers to Trade in 1997
(Foliente, 2000):

> "Wherever appropriate, members shall specify technical regulations based on product requirements in
> terms of performance rather than design or descriptive characteristics."

The basic philosophy underlying this approach is summarized by CIB (1982): "the performance
approach is, first and foremost, the practice of thinking and working in terms of ends rather than
means." Foliente (2000) notes the advantage of this approach for building construction:

> "The most serious problem with the prescriptive approach is that it serves as a barrier to innovation.
> Improved and/or cheaper products may be developed, yet their use might not be allowed if construction
> is governed by prescriptive codes and standards."

A prominent example of the application of performance standards outside the realm of
OHS is the ISO 9001 quality management and ISO 14000 environmental management
standards, to which many thousands of companies worldwide are certified, and which
are formulated exclusively as performance standards (International Organization for
Standardization, 1994).

The perspective on OHS/ergonomic performance regulations offered in this chapter is informed
by two primary background sources. The first is the seminal chapter by K.U. Smith (1973), the first
publication (to our knowledge) to provide a systematic human factors rationale for a performance-
based approach to the formulation of OHS regulations. The publications of Bryce (1983, 1985a,
1985b) are the second source, generated in the course of his work with the Province of Alberta,
Canada on systematic incorporation of performance regulations into their occupational health and
safety act. Examples in the following sections of how performance regulations are used in OHS
legislation are drawn from the OHS Regulation for the Province of British Columbia (B.C.),
Canada, which features extensive use of performance regulations (Workers' Compensation Board
of British Columbia, 1999c).

PERFORMANCE VERSUS PRESCRIPTIVE REGULATIONS IN OCCUPATIONAL HEALTH AND SAFETY LEGISLATION

A performance standard may contain both performance and prescriptive criteria. However, a
generic performance standard or regulation (Table 3.1) describes what the OHS objective is
(the prescriptive criterion) and who is responsible to meet it (the performance criterion) but does
not prescribe explicitly how the objective should be achieved (Bryce, 1983). This section compares
the performance versus the prescriptive approach to the formulation of regulations promulgated in
actual OHS legislation.

To illustrate some key differences between these two approaches, consider the performance-based regulation dealing with general requirements for machine guarding adopted by B.C. versus the prescriptive regulation adopted by OSHA.

Example 1 The B.C. regulation reads as follows (Workers' Compensation Board of British Columbia, 1999a, Section 12.2, Safeguarding requirement, p. 12-1):

'12.2 Safeguarding requirement

The employer must ensure that machinery and equipment is fitted with adequate safeguards that

- **a.** protect a worker from contact with hazardous power transmission parts,
- **b.** ensure that a worker cannot access a hazardous point of operations, and
- **c.** safely contain any material ejected by the work process that could be hazardous to a worker.'

In contrast, the comparable OSHA regulation (OSHA, 1980, 1910 Subpart O—Machinery and Machine Guarding, 1910.212, General requirements for all machines) reads:

'1910.212 General requirements for all machines 1910.212(a)

Machine guarding.

1910.212(a)(1)

Types of guarding. One or more methods of machine guarding shall be provided to protect the operator and other employees in the machine area from hazards such as those created by point of operation, ingoing nip points, rotating parts, flying chips, and sparks. Examples of guarding methods are-barrier guards, two-hand tripping devices, electronic safety devices, etc.

1910.212(a)(2)

General requirements for machine guards. Guards shall be affixed to the machine where possible and secured elsewhere if, for any reason, attachment to the machine is not possible. The guard shall be such that it does not offer an accident hazard in itself.

1910.212(a)(3)

Point of operation guarding.

1910.212(a)(3)(i)

Point of operation is the area on a machine where work is actually performed upon the material being processed.

1910.212(a)(3)(ii)

The point of operation of machines whose operation exposes an employee to injury shall be guarded. The guarding device shall be in conformity with any appropriate standards therefore, or, in the absence of applicable specific standards, shall be so designed and constructed as to prevent the operator from having any part of his body in the danger zone during the operating cycle.

1910.212(a)(3)(iii)

Special hand tools for placing and removing material shall be such as to permit easy handling of material without the operator placing a hand in the danger zone. Such tools shall not be in lieu of other guarding required by this section but can only be used to supplement the protection provided.

1910.212(a)(3)(iv)

The following are some of the machines which usually require point of operation guarding:

1910.212(a)(3)(iv)(a)

Guillotine cutters.

1910.212(a)(3)(iv)(b)

Shears.

1910.212(a)(3)(iv)(c)

Alligator shears.

1910.212(a)(3)(iv)(d)

Power presses.

1910.212(a)(3)(iv)(e)

Milling machines.

1910.212(a)(3)(iv)(f)

Power saws.

1910.212(a)(3)(iv)(g)

Jointers.

1910.212(a)(3)(iv)(h)

Portable power tools.

1910.212(a)(3)(iv)(i)

Forming rolls and calendars.

1910.212(a)(4)

Barrels, containers, and drums. Revolving drums, barrels, and containers shall be guarded by an enclosure that is interlocked with the drive mechanism so that the barrel, drum, or container cannot revolve unless the guard enclosure is in place.

1910.212(a)(5)

Exposure of blades. When the periphery of the blades of a fan is less than seven (7) feet above the floor or working level, the blades shall be guarded. The guard shall have openings no larger than one-half (1/2) inch.

1910.212(b)

Anchoring fixed machinery. Machines designed for a fixed location shall be securely anchored to prevent walking or moving.'

END Example 1

Example 1 provides a framework for addressing in the following subsections: (1) generic versus narrowly focused OHS performance regulations; (2) formulation of performance and prescriptive OHS regulations; (3) advantages of performance relative to prescriptive OHS regulations; and (4) disadvantages of performance relative to prescriptive OHS regulations.

Generic Versus Narrowly Focused OHS Performance Regulations

There are two general types of OHS performance regulations, *broadly focused or generic* (Example 1) and *narrowly focused* (Bryce, 1983, 1985a). An OHS generic performance regulation embodies the following major elements:

- It identifies a particular class of participant in an OHS system whose performance is responsible for meeting a specified OHS objective;
- It describes an OHS objective but does not explicitly dictate or prescribe how the objective should be achieved;
- It represents a single, complete regulatory statement whose meaning and intent are clearly stated without being accompanied by numerous qualifying provisions;
- It is conceptually broader in its application than a prescriptive regulation, and as such, its regulatory scope encompasses a broader range of hazards and hazard abatement possibilities; and
- It is neither hazard-specific nor industry-specific in its formulation and, therefore, is potentially applicable to any workplace.

Referring to Example 1 above, the B.C. generic performance regulation dealing with general requirements for machine guarding differs from the parallel OSHA prescriptive regulation with respect to all these major elements. Noteworthy differences that merit emphasis are those relative to the B.C. regulation and formulation of the OSHA regulation: (1) is longer and more complicated; (2) contains a number of narrow qualifying provisions; and (3) is both hazard- and industry-specific.

In contrast to a generic OHS performance regulation, a narrowly focused OHS performance regulation is hazard-specific and may also be industry-specific. Consequently, relative to the former, the formulation of the latter more closely resembles that of a prescriptive regulation.

However, as the following example illustrates, a narrowly focused OHS performance regulation still exhibits key differences from a prescriptive regulation. Example 2 contrasts a narrowly focused OHS performance regulation promulgated by the Province of British Columbia, Canada, for control of toxic and hazardous substances with a prescriptive regulation promulgated by OSHA for the same purpose.

Example 2 The B.C. regulation (Workers' Compensation Board of British Columbia, 1999b, Sections 5.48 & 5.55, pp. 5–10 to 5–12) reads:

'5.48 Exposure limits

The employer must ensure that a worker's exposure to a substance does not exceed the exposure limits listed in Table 5-4.

5.49 through 5.54

[not included here]

5.55 Type of controls

1. If there is a risk to a worker from exposure to a hazardous substance by any route of exposure, the employer must eliminate the exposure, or otherwise control it below harmful levels and below the applicable exposure limit listed in Table 5-4 by
   a. substitution,
   b. engineering control,
   c. administrative control, or
   d. personal protective equipment.
2. When selecting a suitable substitute, the employer must ensure that the hazards of the substitute are known and that the risk to workers is reduced by its use.
3. The use of personal protective equipment as the primary means to control exposure is permitted only when
   a. substitution, or engineering or administrative controls are not practicable,
   b. additional protection is required because engineering or administrative controls are insufficient to reduce exposure below the applicable exposure limits, or
   c. the exposure results from temporary or emergency conditions only.'

The comparable OSHA regulation (OSHA, 1980, 1910 Subpart Z—Toxic and Hazardous Substances, 1910.1000, Air contaminants) reads:

'1910.1000 Air contaminants

An employee's exposure to any substance listed in Tables Z-1, Z-2, or Z-3 of this section shall be limited in accordance with the requirements of the following paragraphs of this section.

1910.1000(a) through **1910.1000(d)**

[not included here]

1910.1000(e)

To achieve compliance with paragraphs (a) through (d) of this section, administrative or engineering controls must first be determined and implemented whenever feasible. When such controls are not feasible to achieve full compliance, protective equipment or any other protective measures shall be used to keep the exposure of employees to air contaminants within the limits prescribed in this section. Any equipment and/or technical measures used for this purpose must be approved for each particular use by a competent industrial hygienist or other technically qualified person. Whenever respirators are used, their use shall comply with 1910.134.'

END Example 2

Table 5-4 referenced by the B.C. regulation and Tables Z-1 through Z-3 referenced by the OSHA regulation each contain prescriptive exposure limits for a long list of chemicals that are identical for the two regulations.

Example 2 illustrates the point that even though a narrowly focused performance regulation may contain prescriptive elements, it still may differ in important ways from a purely prescriptive regulation. Thus, even though both the B.C. and the OSHA regulations establish specific, prescriptive exposure limits for each chemical listed, only the B.C. regulation:

- explicitly specifies that it is the employer's responsibility to ensure that a worker's exposure to a listed chemical does not exceed the exposure limits;
- allocates to the employer the responsibility for determining exactly how the prescribed exposure limit for each chemical will be achieved;
- does not explicitly specify the particular means the employer must use to satisfy the objective of controlling worker exposure to any of the listed chemicals.

Formulation of Performance and Prescriptive OHS Regulations

This section compares and contrasts basic considerations underlying the formulation, or drafting, of performance and prescriptive OHS regulations (Bryce, 1985a). A well-drafted prescriptive OHS regulation typically exhibits the following features, and/or satisfies the following criteria, in its formulation.

1. It clearly states the specific means that must be applied or satisfied in compliance with the objective stated in the regulation. That is, it provides explicit instructions about what to do and how to do it, for purposes of regulatory compliance. Little or no latitude is provided for pursuing alternative strategies for compliance.
2. The class of personnel with performance responsibility for compliance with the stated regulatory objective is not explicitly identified in the regulation. However, virtually without exception, the implication is that the employer has responsibility for compliance.
3. A prescriptive regulation is most often promulgated when:
   - the nature of the hazard, whose mitigation is targeted by the regulation, is such that at least one specific hazard mitigation strategy can be identified;
   - one defined method, approach, or design feature prescribed in the regulation will definitely accomplish the hazard mitigation objective of the regulation (a prescriptive provision may lay out a series of alternatives for hazard mitigation, but each alternative typically is formulated as a prescriptive option);
   - a particular, defined safety standard needs to be adopted and applied in a regulatory framework; and/or
   - certification of regulatory compliance is required by a certifying body or individual.

4. It may be difficult to ascertain the underlying rationale for the regulatory prescription (the "why" behind the stated objective) from the stated prescriptive provisions.

In contrast, a well-drafted OHS performance regulation typically satisfies the following criteria, and/or exhibits the following features, in its formulation.

1. It should clearly specify the condition(s) under which the provision will apply.
2. It should identify a class of personnel whose work performance is responsible for meeting the stated regulatory objective.
3. It should clearly indicate the regulatory objective to be achieved using appropriate descriptive language and/or definitions pertaining to the mitigation of a specified hazard or hazards.
4. Criteria for gauging whether compliance with the stated regulatory objective or not has been achieved should be readily apparent from the formulation of the regulation, without need for recourse to administrative or legal interpretation—anyone referencing the regulation able to readily determine that compliance with its regulatory objective has been satisfied represents a key test of a meaningful performance regulation.
5. The performance objective should be phrased to be potentially applicable to a range of possible contexts or situations involving the hazard addressed by the stated objective.
6. Use of general qualifying terms, such as "suitable" or "sufficient," should be avoided, unless such terms are accompanied by additional guidance in the regulation that clearly indicates what will be considered to be suitable, sufficient, etc.
7. It can be generic or narrowly focused (previous section).
8. The specified objectives can include or be accompanied by qualifying provisions. A regulation with such "qualified performance objectives" combines a performance objective with a description of provisions that need to be considered for purposes of compliance with the objective.
9. It can be combined with prescriptive provisions in a given section of OHS regulations (see previous section).
10. Its promulgation generally is advisable under the following circumstances: (1) when the hazard of concern is chronic (i.e. long-standing) in nature; and/or (2) when the primary deficiency in the regulatory code is lack of information for the employer about mitigation of a particular hazard—hazards whose mitigation may best be served by such "informational" regulations typically are of a low-risk nature.
11. The performance requirements specified in the regulation should be clearly compatible with the performance capabilities and limitations of the class of person specified in the regulation.

To illustrate the rationale for criterion 11, it would be reasonable to require an employer, but not an employee or even a supervisor, to establish OHS policies and programs that comply with a promulgated OHS code.

To illustrate some of the above criteria regarding performance regulation formulation, the following example contrasts a more loosely with a more carefully drafted performance regulation. The more loosely drafted regulation, the so-called OSHA "general duty" clause (OSHA, 1970, Section 5—Duties), reads as follows.

Example 3 '5. Duties

a. Each employer -
   1. shall furnish to each of his employees employment and a place of employment which are free from recognized hazards that are causing or are likely to cause death or serious physical harm to his employees;
   2. shall comply with occupational safety and health standards promulgated under this Act.

b. Each employee shall comply with occupational safety and health standards and all rules, regulations, and orders issued pursuant to this Act which are applicable to his own actions and conduct.'

In contrast, the B.C. general duty regulations for employers and employees (Workers' Compensation Board of British Columbia, 1999c, Sections 115–124, pp. xiii–xv) read as follows.

'Section 115 General duties of employers

1. Every employer must
   a. ensure the health and safety of
      i. all workers working for that employer, and
      ii. any other workers present at a workplace at which that employer's work is being carried out, and
   b. comply with this Part, the regulations, and any applicable orders.

2. Without limiting subsection (1), an employer must
   a. remedy any workplace conditions that are hazardous to the health or safety of the employer's workers,
   b. ensure that the employer's workers
      i. are made aware of all known or reasonably foreseeable health or safety hazards to which they are likely to be exposed by their work,
      ii. comply with this Part, the regulations, and any applicable orders, and
      iii. are made aware of their rights and duties under this Part and the regulations,
   c. establish occupational health and Board of British Columbia, 1999c, Sections 115–124, pp. xiii–xv) read as safety policies and programs in accordance with the regulations,
   d. provide and maintain protective equipment, devices, and clothing in good condition as required by regulation and ensure that these are used by the employer's workers,
   e. provide to the employer's workers the information, instruction, training, and supervision necessary to ensure the health and safety of those workers in carrying out their work and to ensure the health and safety of other workers at the workplace,
   f. make a copy of this Act and the regulations readily available for review by the employer's workers and, at each workplace where workers of the employer are regularly employed, post and keep posted a notice advising where the copy is available for review,
   g. consult and cooperate with the joint committees and worker health and safety representatives for workplaces of the employer, and
   h. cooperate with the board, officers of the board, and any other person carrying out a duty under this Part or the regulations.

Section 116 General duties of workers

1. Every worker must
   a. take reasonable care to protect the worker's health and safety and the health and safety of other persons who may be affected by the worker's acts or omissions at work, and
   b. comply with this Part, the regulations, and any applicable orders.

    2. Without limiting subsection (1), a worker must

      a. carry out his or her work in accordance with established safe work procedures as required by this Part and the regulations,

      b. use or wear protective equipment, devices, and clothing as required by the regulations,

      c. not engage in horseplay or similar conduct that may endanger the worker or any other persons,

      d. ensure that the worker's ability to work without risk to his or her health or safety, or the health or safety of any other person, is not impaired by alcohol, drugs, or other causes,

      e. report to the supervisor or employer

        i. any contravention of this Part, the regulations, or an applicable order of which the worker is aware, and

        ii. the absence of or defect in any protective equipment, device or clothing, or the existence of any other hazard, that the worker considers is likely to endanger the worker or any other person,

      f. cooperate with the joint committee or worker health and safety representative for the workplace, and

      g. cooperate with the board, officers of the board and any other person carrying out a duty under this Part or the regulation.'

END Example 3

The B.C. general duty regulation also goes on to specify general duties for supervisors, owners, suppliers, and directors and officers of a corporation.

Referring to criteria 1–11 listed above for a well-drafted performance regulation, the B.C. general duty regulation largely or entirely conforms to criteria 1–8 and 11 and can be considered to represent a more carefully drafted generic performance regulation. The OSHA general duty regulation largely or entirely conforms to criteria 1–3, 5, and 7, and can be considered to represent a more loosely drafted generic performance regulation. For example, in contrast to the qualifying provisions in the B.C. regulation (Part 2 in Sections 115 and 116), there are no qualifying provisions in the OSHA regulation that make it readily apparent whether compliance has been achieved or not (criteria 4 and 8).

Furthermore, in the OSHA regulation, the precise meanings of the terms "free of recognized hazards" and "serious physical harm" are unclear (criterion 6), in the sense that the regulation itself provides no guidance either as to what constitutes serious harm or as to how an employer should perform to recognize hazards and remove them from the workplace. The B.C. regulation also refers to health and safety hazards, but the mandate for employer performance is "to remedy any workplace conditions that are hazardous" (Section 115-(2)-a), the implication being that the nature of a particular hazardous workplace condition has become known to the employer through prior effect.

Finally, it is not clear from the OSHA general duty regulation for employees how a U.S. worker should go about fulfilling the performance requirement that mandates employee awareness and understanding of the full range and scope of "occupational safety and health standards and all rules, regulations, and orders issued pursuant to this Act which are applicable to his own actions and conduct." It turns out that sprinkled throughout the OSHA Act are information provision and training requirements aimed at assisting employees in meeting this performance requirement. Nevertheless, no guidance is provided in the section of the act itself as to how an employee should satisfy the compliance performance requirement (criterion 11).

In contrast, although the B.C. regulation has a similar performance provision mandating employee compliance with OHS regulations (Section 116-(1)-b), this is accompanied by explicit qualifying provisions that proscribe employee behaviors (such as horseplay and substance abuse) that may compromise compliance. Furthermore, there are two qualifying provisions for employer performance designed to assist with worker compliance, one dealing with training (Section 115-(2)-e) and the other with posting of information (Section 115-(2)-f). These explicit qualifying provisions in the B.C. general duty regulations for both employees and employers provide clear guidance for B.C. workers regarding the way their compliance performance requirement is to be achieved.

ADVANTAGES OF PERFORMANCE RELATIVE TO PRESCRIPTIVE OHS REGULATIONS

Arguments for building OHS legislation around performance rather than prescriptive regulations may be categorized into three classes—relating to regulatory efficiency, regulatory effectiveness, and the goal of regulatory intervention. The subsections below deal with each of these areas.

Regulatory Efficiency of OHS Performance Regulations

The term regulatory efficiency used here refers to the volume and detail of regulatory provisions that may be required to achieve a particular OHS regulatory objective. In contrast to prescriptive OHS regulations, generic OHS performance regulations typically are more efficient, in the sense that their formulation is designed to address a broad range of hazards in a manner that lacks excessive detail and complexity (Bryce, 1983). Example 1 dealing with machine guarding in the preceding section illustrates this point. That is, relative to the more efficient B.C. machine guarding generic performance regulation, the less efficient prescriptive OSHA machine guarding regulation: (1) is longer and more complicated; (2) contains a number of narrow qualifying provisions; and (3) is both hazard- and industry-specific.

When it comes to OHS regulation, is regulatory efficiency a desirable objective? An affirmative answer to this question is suggested by considerations that, relative to a longer, more detailed regulation, a regulation with moderate length, detail, and complexity: (1) is easier to understand; (2) is likely to more effectively support OHS information transmittal and training; (3) may be more readily interpretable by workers, employers, OHS specialists, and regulatory bodies; and (4) thereby may encourage a greater dedication to compliance. Because the appropriate research is lacking, these predictions remain speculative at this point.

Regulatory Effectiveness of OHS Performance Regulations

As used here, the term regulatory effectiveness refers to the degree to which the stated objective of an OHS regulation—mitigation of a specified hazard—is in fact realized. It is the apparent lack of regulatory effectiveness that represents one of the strongest arguments against the use of prescriptive OHS regulations. Documentary evidence for such a lack is derived largely from evaluations of the effectiveness of U.S. OSHA regulations, regulations that are largely prescriptive in nature. A rather consistent finding is that OSHA prescriptive regulations directed at mitigating specified hazards do not reliably and predictably result in accident and injury prevention, the putative goal of the regulations (Smith, 2002a).

For example, Jones in 1973 documented the limitations of dependence on prescriptive engineering control regulations in bringing down occupational injury rates to acceptable levels. Gill and Martin (1976) echo this finding with the claim that the prescriptive engineering approach to safety management is not sufficient to prevent all accidents, and that innovative hazard control strategies such as performance standards and worker participation should be considered to achieve further improvement in accident prevention. Smith et al. (1971) and Gottlieb and Coleman (1977) report that inspections carried out by OSHA are relatively ineffective in preventing accidents because most hazards cannot be identified by traditional workplace inspections that rely on prescriptive regulations. Results from these and other studies suggest that, in general, only 5-25% of accidents can be avoided by rigorous compliance with conventional safety standards (Ellis, 1975; Smith, 1979). OSHA itself has noted that "OSHA's own statistics appear to indicate that 70–80% of all deaths and injuries each year are not attributable to a violation of any OSHA specification standard" (Occupational Safety and Health Reporter, 1976, p. 684).

What is the basis for this apparent lack of effectiveness of prescriptive OHS regulations in preventing occupational accidents and injuries? The answer offered here is that a prescriptive regulation only addresses one part of what a hazard actually represents. In operational terms, a hazard manifests itself as a human factor's consequence of performance-design interaction, the fundamental concern of human factors/ergonomic (HF/E) science (T.J. Smith, 1993, 1994, 1998, 2002a; Smith, Henning, and Smith, 1994). That is, the realization of the effect of an OHS hazard (degradation in operational performance, accident, injury, or death) depends upon the interaction of worker behavior and performance with a physical workplace condition. The assumption is that because their formulation ignores this operational reality of the actual nature of OHS hazards, prescriptive OHS regulations are inherently limited in their impact on the consequences of such hazards.

It is from this perspective that Smith (1973) advocates the use of OHS performance standards as the most effective strategy in achieving the hazard mitigation and accident and injury prevention goals of OHS regulation. This author introduces and defines the term *operational hazards* (IBID; Smith, 1979, 1988, 1990) to refer to hazards that arise as a consequence of highly variable (relative to both within- and between-worker performance) and largely unpredictable behavioral interactions of individuals or groups with existing physical (microergonomic) or organizational (macroergonomic) workplace design conditions.

From these human factors systems perspective, therefore, most occupational hazards represent incidents or conditions that exist or arise through the variable, sometimes transitory, and typically synergistic convergence of organizational, behavioral, operational, and physical-environmental factors specific to particular tasks, job operations, and socio-technical systems. The complexity of this interaction mocks simplistic attempts to pigeonhole hazards into a limited number of categories (e.g., physical versus behavioral), defeat efforts to attribute occupational accidents, injuries, and safety problems to one type of hazard or another, and stymies abatement strategies based on dissective, prescriptive assumptions about the nature of hazards.

The basic assumption underlying this analysis is that, in today's workplace, operational hazards represent the most prevalent source of work-related performance decrements, accidents, and/or injuries. Typically, these are not addressed by prescriptive OHS regulations, are not detected by outside inspectors, and are not abated by remedial approaches prompted by standardized accident and injury investigations. Further, operational hazards are context-specific to particular workers, tasks, jobs, and operations, in a manner that mirrors the context specificity observed generally with human behavior and performance (Smith, 1998; Smith, Henning, and Smith, 1994). These considerations suggest that guided by the dictates of the Pareto principle (Juran, 1964, 1995), OHS regulations should focus on operational hazards as a major priority for hazard mitigation and accident and injury prevention.

The effectiveness argument, therefore, posits that in contrast to prescriptive regulations, OHS performance regulations are better designed to provide this focus because their formulation embodies the closed-loop relationship between worker behavior and specified workplace conditions characteristic of operational hazards. What evidence is available to support this prediction?

No systematic research is yet available to provide a conclusive answer to this question.

Negative evidence cited above regarding the limited effectiveness of prescriptive OHS regulations is suggestive. More persuasive, in a narrow sense, is the demonstrated effectiveness of B.C. ergonomics performance regulations in reducing musculoskeletal injuries (MSIs), presented in the section for Ergonomic Performance Regulations. Yet, a third line of evidence stems from research on the ability of workers to recognize and control workplace hazards.

The basic premise of this research—grounded in the concept of worker self-regulation of hazard management (Smith, 2002a)—is that compared with managers, safety professionals, or government regulators, workers often know as much or more about their jobs, about job-related hazards that they encounter daily, and about how to reduce or eliminate those. That is, whereas prescriptive OHS regulations tend to focus narrowly on physical-environmental hazards, workers tend to identify a broader range of hazards, often related to job performance, to dynamic operational circumstances, and/or intermittent or transitory work situations or conditions, for which no explicit prescriptive regulations or standard operating procedures exist.

Studies on hazard self-regulation by workers support the conclusion that workers represent a relatively untapped resource for hazard control. Thus, Swain (1973, 1974) refers to workers as subject-matter experts on their work situations, whose input to management can serve as a key contributor to safety performance. He notes that worker hazard data represents a rare but effective source of information about near-accidents and potentially unsafe situations. Hammer (1972, 1976) likewise presents evidence showing that workers have a high awareness of job- related hazards and that asking them directly represents the best way of obtaining this information. Both authors note marked discrepancies between worker-provided hazard data versus that provided by more conventional specialist reports and post-accident investigations.

That workers can recognize hazards and have knowledge and skill in self-controlling them to improve safety performance, likewise, has been tested with positive results in a series of cross-industry studies conducted through the University of Wisconsin Behavioral Cybernetics Laboratory (Cleveland, 1976; Coleman and Sauter, 1978; Coleman and Smith, 1976; Gottlieb, 1976; Kaplan and Coleman, 1976; Kaplan, Knutson and Coleman, 1976; Richardson, 1973; M.J. Smith, 1994). Findings from this research reveal that employees are one of the best sources of information about the day-to-day hazards of their jobs, and support the following major conclusions:

- Workers generally identify hazards that most directly affect their personal safety. Unsafe conditions and operating procedures often are specified, but it is not uncommon for possibly hazardous actions or behaviors to be candidly cited.
- Worker-identified hazards largely comprise those of immediate self-concern.
- Studies of six Wisconsin metal processing plants (Cleveland, 1976; Coleman and Sauter, 1978) found that employee-identified hazards bear a significantly closer resemblance to actual causes of accidents than those specified in prescriptive OHS regulations or found by state inspections.
- Workers typically can clearly delineate their actions in avoiding, reducing, or controlling hazards.
- Workers tend to identify health hazards in terms of their acute toxic effects rather than their long-term potential for causing illness or disease. There is substantial awareness of engineering defects in ventilation or personal protection that may cause an increased risk of toxic exposure.
- Employers provided with worker hazard data generally conclude that: (1) employees can provide more and better information about job hazards than other available sources; (2) workers often identify hazards that the management is totally unaware of; and (3) survey results are of value in establishing priorities for safety and HM procedures.

What do these findings regarding the ability of workers to understand job-related hazards and their mitigation have to do with the effectiveness of OHS performance regulations? First, the findings tend to validate performance regulations dealing with employee duties, featured in the OHS acts for B.C. and other Canadian provinces which call for the employee to "carry out his or her work in accordance with established safe work procedures as required by this Part and the regulations"

(Example 3). Second, unlike prescriptive OHS regulations, both workers and generic OHS performance regulations are more attuned to the nature and significance of operational hazards—the former in terms of awareness and understanding and the latter in terms of design and scope. The argument, therefore, is that the consequences of a regulation mandating performance that will accommodate mitigation of operational hazards are likely to mirror those of efforts to call upon worker expertise for the same purpose.

Performance Regulations and the Purpose of OHS Regulatory Intervention

This brings us to perhaps the most provocative argument regarding the putative benefits of OHS performance regulations, an argument that gets at the fundamental purpose of OHS regulatory intervention. The two basic questions are, is regulatory intervention necessary, and if so, how should regulatory provisions be formulated to best accomplish their objectives? The first question represents a matter of active debate regarding ergonomics regulations (Fallentin et al., 2001), and will not be considered here. The second question has been considered in depth by Bryce (1983), with the conclusion that there is an inherent contradiction between the avowed purpose of OHS regulatory intervention, and the use of prescriptive regulations to achieve this purpose.

To understand the logic of this argument, consider the OSHA regulations. They comprise many hundreds of prescriptive provisions—almost all narrowly focused, many of them task- and/or industry-specific (OSHA, 1980). The OSHA general duty clause (Example 3) mandates that each employer shall furnish "a place of employment... free from recognized hazards that are causing or are likely to cause death or serious physical harm to his employees." The presumption of this mandate is that it is the employer's responsibility to ensure that his/her workplace is free from hazards that otherwise might violate the regulation. The code then goes on to describe hundreds of hazards that the employer should consider for compliance with the mandate. Yet the prescriptive formulation for many of the regulations not only describes what the hazard is, it goes on to prescribe how the hazard should be mitigated. In other words, because of the way the regulations are drafted, it is not the employer but the regulatory body—the government—that has assumed major responsibility for ensuring that the workplace is free of dangerous hazards by dictating to the employer, under the force of law, both the need for action and the course of action.

Clearly, there are political, socioeconomic, and legal implications of OHS regulations—particularly on this perspective on the inherently contradictory nature of prescriptive OHS regulations—that go beyond the scope of this chapter. From a human factors perspective, however, it is probably fair to say that neither the employer nor the government find entirely satisfactory a situation, arising primarily as a consequence of the prescriptive regulatory approach, in which the former abrogates substantial responsibility to the latter for prescriptive guidance on how a workplace should be operated to protect worker health and safety.

As the analysis of Bryce (1983) suggests, there have been three primary influences behind the accretion of today's voluminous code of largely prescriptive OHS regulations in the U.S. The first is the influence and tradition of English common law, with its emphasis on a prescriptive approach to legislative drafting. A second, related influence is that an early, major impetus for OHS regulations was to control worker exposure to hazardous substances through the formulation of prescriptive exposure limits. A third influence is the lack of rationalization over the years of the OHS regulatory code, related in no small part to a lack of appreciation on the part of both legislators and drafting bodies that primarily informational regulations targeting (typically lower risk) hazards need not necessarily contain the detailed guidance that regulations targeting higher risk hazards might require.

Because they inform the employer of the existence and nature of a hazard that requires mitigation but leave it up to the employer to determine how the mitigation might be accomplished, generic OHS performance regulations avoid the inherently contradictory design of prescriptive OHS regulations as regards regulatory intervention. This may be deemed beneficial for both employers and regulatory bodies alike. Moreover, as noted earlier, if mitigation of a particular hazard (such as a hazardous substance) is best served by the prescriptive approach (e.g., exposure limits), then a narrowly focused performance regulation can be drafted that is accompanied by prescriptive provisions (Example 2).

DISADVANTAGES OF PERFORMANCE RELATIVE TO PRESCRIPTIVE OHS REGULATIONS

Some of the major advantages of performance regulations cited in the previous section also may be construed as major disadvantages in terms of translating the abstract aims of OHS regulatory intervention into practical reality. That is, relative to prescriptive regulations, the brevity, breadth, and prudent avoidance of detailed prescriptive guidance on the part of generic performance regulations creates the potential for a more efficient, effective, and flexible regulatory approach. Yet these same qualities conceivably can make the approach more difficult to implement.

One such difficulty is that reliance on generic performance regulations may make the OHS inspector's job more difficult. That is, when it comes to citing employers for violations of OHS regulations, government inspectors tend to find it easier to cite violations of a specific regulation, as opposed to citing violations of a generic performance regulation that may apply to a specific situation. Referring to Example 1 above, an OSHA inspector may find it relatively straightforward to conclude that a fan less than seven feet above the floor with a guard opening of one inch represents a clear violation of the prescriptive regulation (OSHA, 1980, 1910 Subpart O—Machinery and Machine Guarding, 1910.212, General requirements for all machines, Section 1910.212(a)(5)) mandating that, "...when the periphery of the blades of a fan is less than seven (7) feet above the floor or working level, the blades shall be guarded. The guard shall have openings no larger than one-half (1/2) inch." On the other hand, the inspector may find it more difficult to decide whether this condition represents a clear and unequivocal violation of a generic performance standard that states, "The employer must ensure that machinery and equipment is fitted with adequate safeguards which...ensure that a worker cannot access a hazardous point of operations" (Workers' Compensation Board of British Columbia, 1999a, Section 12.2, Safeguarding requirement, p. 12-1, Section 12.2) or not. In other words, relative to reliance on the prescriptive regulation, reliance on the generic performance regulation for purposes of regulatory intervention regarding machine guarding may call for a more judgmental, and therefore less definitive and predictable, decision on the part of the inspector.

A second, related disadvantage of generic performance regulations is that relative to prescriptive standards, the former may provide less explicit guidance to employers about how to control specific workplace hazards that may pose a risk for job-related injury or illness. This concern can be considered particularly applicable to small business operations that typically feature less formalized and less informed safety management systems.

Yet a third disadvantage is that generic performance regulations pose a greater challenge for worker safety training in that, again, such regulations provide less explicit guidance regarding the job-related injury or illness risk associated with specific workplace hazards.

Given these disadvantages, it is not surprising that OHS performance regulations may not enjoy the trust and support of the labor community (Hansen, 1985). The fundamental source of labor concern in this regard is that, by their very nature, performance regulations place an undeniable emphasis on employer self-reliance and self-regulation when it comes to the mitigation of OHS hazards. Labor may not be prepared to accept that employers will, in fact, follow through with satisfactory performance in self-responsibility for OHS hazard mitigation.

More surprisingly perhaps is that some employers may also view the installation of performance regulations with suspicion, for much the same reasons. For example, as Bryce (1983) points out, employers in industries that are regulated by relatively separate, well-defined codes of prescriptive OHS regulations (such as construction, mining, etc.) may have structured their OHS management programs around such codes, and therefore may feel more comfortable in maintaining the regulatory status quo.

ERGONOMIC PERFORMANCE REGULATIONS

Ergonomic regulations represent one type of OHS regulation. Foregoing sections have dealt with the latter because of limited and relatively recent experience with the former. However, available evidence suggests that observations and conclusions summarized above regarding performance regulations are specifically applicable to ergonomics regulations.

In their comprehensive review of ergonomics standards and regulations, Fallentin et al. (2001) summarized the major characteristics of a total of nine government-promulgated general ergonomics standards and regulations. All but one deal with the mitigation of hazards predisposing to work-related MSIs (upper limb or low back). Only four of these have established regulatory enforcement status, namely, general ergonomics rules promulgated by the states of California and Washington in the U.S. (State of California, 1997; Washington State Department of Labor and Industries, 2000), by the Province of British Columbia in Canada (Workers' Compensation Board of British Columbia, 1999d, Sections 4.46 to 4.53, pp. 4–7 to 4–8), and by Sweden (National Board of Occupational Safety and Health NBOSH, 1998). In 2004, the Washington ergonomic rule was rescinded as the result of a statewide referendum.

The nine government-promulgated general ergonomics standards and regulations reviewed by Fallentin et al. (2001) feature a mix of prescriptive, narrowly focused performance plus prescriptive and generic performance regulations. As examples of prescriptive ergonomics standards dealing with musculoskeletal load, Westgaard and Winkel (1996) cite the following early formulations along with more recent European Committee for Standardization (CEN) formulations, the latter possibly influenced by the former.

Barnes (1937):

"The two hands should begin as well as complete their motions at the same time." "Work should be arranged to permit an easy and natural rhythm wherever possible."

Grandjean (1969):

"Arm movements should be in opposition to each other or otherwise symmetrical."

CEN-standard prEN 614-1, December 1991 (Westgaard and Winkel, 1996): "Repetitive movement that causes illness or injury shall be avoided."

"Work equipment shall be designed to allow the body or parts of the body to move in accordance with their natural paths and rhythms of motion."

The ergonomics rules promulgated by Washington State (Washington State Department of Labor and Industries, 2000) and B.C. (Workers' Compensation Board of British Columbia, 1999d, Sections 4.46 to 4.53, pp. 4–7 to 4–8) offer an instructive contrast between reliance on primarily prescriptive versus exclusively generic performance regulations for purposes of mitigating MSI hazards. The Washington State regulation mandated that employers with "caution zone jobs" first must determine whether these jobs have work-related musculoskeletal disorder (WMSD) hazards and if so, must reduce WMSD hazards so identified. As summarized in Example 4 below, this approach features narrowly focused performance combined with detailed prescriptive regulations.

Example 4 Guidance as to what constitutes a "caution zone job" first is provided with a narrowly focused performance regulation (Washington State Department of Labor and Industries, 2000, Section WAC 296-62-05103):

'WAC 296-62-05105 What is a 'caution zone job'?

Employers having one or more 'caution zone jobs' must comply with Part 2 of this rule. 'Caution zone jobs' may not be hazardous but do require further evaluation.'

This clause is accompanied first by a series of qualifying provisions, and then by 12 detailed prescriptive provisions that define a 'caution zone job' in relation to postural, upper limb, and lifting risk factors.

Part 2 of the regulation (IBID, Section WAC 296-62-05130) offers employers what are termed general or specific performance options.

'WAC 296-62-05130 What options do employers have for analyzing and reducing WMSD hazards?

General Performance Approach

(1) The employer must analyze 'caution zone jobs' to identify those with WMSD hazards that must be reduced. A WMSD hazard is a physical risk factor that by itself or in combination with other physical risk factors has a sufficient level of intensity, duration, or frequency to cause a substantial risk of WMSDs. The employer must use hazard control levels as effective as the recommended levels in widely used methods such as... [eight different methods prescribed here].
(2)-(4) [qualifying performance provisions not included here]
(5) Employers must reduce WMSD hazards as described below... [design change, work practice, and personal protective equipment hazard control prescriptions specified here].'

This formulation actually represents a narrowly focused performance regulation with qualifying performance and prescriptive provisions.

The formulation for what the regulation terms the specific performance option for analyzing and reducing WMSD hazards (IBID, Section WAC 296-62-05130) reads:

'Specific Performance Approach

The employer must analyze 'caution zone jobs' to identify those with WMSD hazards that must be reduced. A WMSD hazard is a physical risk factor that exceeds the criteria in Appendix B of this rule.'

Appendix B that this clause refers to contains carefully defined prescriptive WMSD hazard exposure criteria—for awkward postures (shoulders, neck, back, and knees), high hand force (pinching or gripping), highly repetitive motion of the neck, shoulders, elbows, wrists, or hands, and repeated impact of the hands or knees—that, if exceeded, qualify a job as a caution zone job. This formulation represents a narrowly focused performance regulation with qualifying prescriptive provisions only.

The regulation also contains provisions pertaining to employee education and involvement, plus hazard mitigation evaluation.

END Example 4

In contrast, the B.C. regulatory approach targets the same objective as the Washington State approach—mitigation of WMSD hazards—with generic performance regulations only, as illustrated in Example 5.

Example 5 Relevant portions of the B.C. ergonomics rule (Workers' Compensation Board of British Columbia, 1999d, Sections 4.46 to 4.53, pp. 4–7 to 4–8) read:

'ERGONOMICS (MSI) REQUIREMENTS

Section 4.47 Risk identification

The employer must identify factors in the workplace that may expose workers to a risk of musculoskeletal injury (MSI).

Section 4.48 Risk assessment

When factors that may expose workers to a risk of MSI have been identified, the employer must ensure that the risk to workers is assessed.

Section 4.49 Risk factors

The following factors must be considered, where applicable, in the identification and assessment of the risk of MSI:

a. the physical demands of work activities, including: (i) force required, (ii) repetition, (iii) duration, (iv) work postures, and (v) local contact stress;
b. aspects of the layout and condition of the workplace or workstation, including: (i) working reaches, (ii) working heights, (iii) seating, and (iv) floor surfaces;
c. the characteristics of objects handled, including: (i) size and shape, (ii) load condition and weight distribution, and (iii) container, tool, and equipment handles;
d. the environmental conditions, including cold temperature; and
e. the following characteristics of the organization of work: (i) work-recovery cycles, (ii) task variability, (iii) work rate.

Section 4.50 Risk control

1. The employer must eliminate or, if that is not practicable, minimize the risk of MSI to workers.
2. Personal protective equipment may only be used as a substitute for engineering or administrative controls if it is used in circumstances where those controls are not practicable.
3. The employer must, without delay, implement interim control measures when the introduction of permanent control measures will be delayed.

Section 4.51 Education and training

Section 4.52 Evaluation

Section 4.53 Consultation

Not included here.' END Example 5

Although the Washington State and B.C. ergonomics rules share the same regulatory purpose, their respective approaches differ markedly. Both rules are quite similar in informing employers what the specific regulatory objectives are—hazard identification, analysis and control, education, employee involvement, and hazard mitigation evaluation. However, with its prescriptions regarding exposure criteria for hazard identification, hazard analysis methods, hazard reduction measures, and so forth, only the Washington State rule goes on to carefully define how the employer must go about meeting these objectives. On the other hand, except for one prescriptive provision pertaining to personal protective equipment (Section 4.50), the B.C. generic performance rule leaves it up to the employer to determine how the objectives will be met.

Arguably, there are advantages to both approaches outlined above. Nothing in the B.C. rule matches the useful prescriptive guidance offered to employers by the Washington State rule about what methods to consider for purposes of the MSI hazard analysis. Alternatively, the B.C. rule is more informative regarding the range of potential MSI risk factors that an employer should consider for an MSI hazard identification and analysis.

In the last analysis, regardless of the formulation approach, it is the effectiveness and the scientific legitimacy of an ergonomics rule that matters most. The B.C. rule, as outlined in the next two subsections, appears to satisfy the former criterion, and arguably is superior to the Washington State rule regarding the latter.

REGULATORY EFFECTIVENESS OF GENERIC ERGONOMIC PERFORMANCE REGULATION

In their review of ergonomic standards and regulations, Fallentin et al. (2001, Table 12) report that no data are yet available from any jurisdiction regarding the jurisdiction-wide effectiveness of an ergonomics rule in reducing adverse health effects. However, information provided to the first author by the Senior Ergonomist for B.C. (K. Behiel, 2002, personal communication) indicates that the proposal and ultimate promulgation of the B.C. ergonomics rule has been associated with a steady decline in the province-wide MSI incidence rate between 1994 and 2001.

Data illustrating this conclusion are presented in Figure 3.1, which shows that the MSI incidence rate in B.C. decreased more or less steadily by about one-third from 1.83 to 1.26 injuries per full-time-equivalent employees per year, between 1994 and 2001. The B.C. ergonomics rule (Workers' Compensation Board of British Columbia, 1999d, Sections 4.46 to 4.53, pp. 4–7 to 4–8) was first proposed to the provincial legislature in 1994, and subsequently promulgated in 1998.

The significance of the observations in Figure 3.1 should be treated cautiously. The data are from only one jurisdiction—whether a comparable impact of ergonomics rules will occur in other jurisdictions remains to be established. Moreover, it is possible that the observed decline in MSI incidence rate would have occurred even if the ergonomics rule had not been introduced.

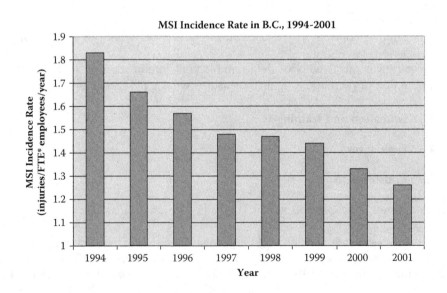

FIGURE 3.1   MSI incidence rate in B.C., 1994–2001 (*FTE=full time equivalent).

Nevertheless, the association documented in Figure 3.1 between consideration and promulgation of an ergonomics rule in B.C. and reduction of the province-wide MSI incidence rate is suggestive and points to two plausible conclusions. The first is that regulatory attention to ergonomics may have a positive impact on MSI injury reduction—even in the absence of actual regulatory intervention (note decline in Fig. 3.1 in incidence rate between 1994 and 1998). One interpretation is that the introduction of the rule in 1994 prompted greater attention to ergonomics and to MSI reduction among provincial employers after 1994, even before the rule was promulgated in 1998.

The second and more important conclusion is that an ergonomics rule built around generic performance regulations can be effective in achieving a positive health impact by reducing MSIs. This, in turn implies, that the possible advantages of formulating OHS regulations using the generic performance approach may outweigh possible disadvantages (previous section). The B.C. experience thus joins those documented for dozens of other non-governmental operations in demonstrating the effectiveness of ergonomic intervention in reducing MSIs (Karsh, Moro, and Smith, 2001; Westgaard and Winkel, 1997).

There is case evidence from B.C. supporting this second conclusion. Prompted by the impetus of the B.C. ergonomics rule, an Industrial Musculoskeletal Injury Reduction Program was introduced into the B.C. sawmill industry in 1998 (McHugh, 2002). Industry-wide since that time, MSI claims for B.C. sawmills have dropped to 44%, and the cost per MSI claim is estimated to have declined over 20%.

SCIENTIFIC LEGITIMACY OF GENERIC ERGONOMIC PERFORMANCE REGULATIONS

Fallentin et al. (2001) evaluate what they term the scientific coherency of the ergonomic standards and regulations addressed in their review. Two of the three North American jurisdictions with promulgated ergonomics rules—B.C. and Washington State—both are judged to have reasonable scientific legitimacy, in relation to sufficiency of the factual basis of the regulations (both rules) and to the adequacy of the program elements (B.C. rule). The scientific legitimacy of the California rule is not similarly evaluated by these authors.

These authors, as well as Radwin et al. (2002), Viikari-Juntura (1997), and Westgaard and Winkel (1996), also cite extensive epidemiological and biomechanical evidence pointing to the conclusion that excessive work-related external physical loading is related to an elevated risk for both back pain and upper limb pain, discomfort, and/or impairment or disability. Thus, to mitigate work-related MSIs, the scientific justification of establishing an ergonomics standard or regulation aimed at controlling worker exposure to excessive physical loading of the low back and/or upper limb seems unequivocal (Bernard, 1997; National Institute of Occupational Safety and Health [NIOSH], 2000).

Nevertheless, all the authors cited above call attention to a significant gap in this body of evidence—the lack of empirical support for definitive exposure-dose and dose-response relationships between exposure to different levels of external physical loading and consequent proportional decremental effects on musculoskeletal function of the low back or upper limb.

Without firm evidence for such relationships, the scientific legitimacy of establishing prescriptive ergonomics rules that mandate regulatory intervention when external physical loads exceed explicit, fixed levels remains open to question. For example, Westgaard and Winkel (1996) concluded their review of occupational musculoskeletal load guidelines as a basis for intervention by noting, "at present, guidelines to prevent musculoskeletal disorders can only give directions, not absolute limits." Yet both the Washington State ergonomics rule (2000) and the repealed U.S. federal OSHA ergonomics rule (OSHA, 2000) prescribe a series of exact external physical loading limits related to occupational exposure durations for different upper limb and low back postures and movements, above which ergonomic intervention for at-risk jobs is mandated.

Fallentin et al. (2001), Radwin et al. (2002), and Westgaard and Winkel (1996) all present conceptual models that depict the roles and influences of multiple factors in the development of musculoskeletal disorders of both the low back and upper limb. The models underscore the multiplicity of both biological (person) factors and workplace design factors whose interaction, in relation to exposure magnitude, duration, and frequency, may influence the etiology of MSIs. The general tenor of analysis in these reviews is that it is insufficient scientific knowledge and understanding of the complexity of this interaction that has forestalled clear delineation of the relationships between relative dose of external physical loading, and relative early and late musculoskeletal response. Thus, Fallentin et al. (2001) note:

'Insufficient knowledge of the mechanisms involved implies that there are no criteria available to define whether the responses studied are relevant intermediate variables between an assumed target tissue dose and disease or disorder. In a risk analysis approach this lack of knowledge regarding exposure-dose relationship and dose-response relationship introduces a large element of doubt.'

In the remainder of this section, I call attention to an additional source of variability influencing the musculoskeletal loading-response relationship—not considered in the models referenced above—that raises the distinct possibility that delineating a well-defined exposure-dose and dose-response calibration for this relationship represents a scientifically unobtainable goal (Smith, 2002b). Given that definitive research on this problem area currently is lacking (above), this prediction obviously remains speculative at this point. The major purpose in raising this possibility is to suggest that putative problems with scientific legitimacy inherent to the prescriptive approach to ergonomics regulations may be avoided by adopting generic performance regulations for purposes of ergonomics rule-making.

The additional source of variability alluded to above arises because of the basic properties of motor behavioral control—that is, we are concerned with kinesiological variability in musculoskeletal performance. Scientific justification for prescriptive limits of movement patterns associated with MSI risk must deal with three kinesiological sources of variability inherent to movement behavior: those related to the degrees-of-freedom (DOF) problem, the context specificity problem, and the imperfect control system problem.

The degrees-of-freedom problem pertains to the fact that the number of degrees of freedom (DOF) in a three-dimensional (3D) space inherent to movements of either the upper limb or the trunk is greater than the number of degrees of freedom of a target with which these movements may interact. Specifically, relative to three DOF of a fixed target in space, reaching and grasping a target by the upper limb features seven DOF (mediated by the shoulder, elbow, and wrist), and lifting of a target by the trunk and knees features four DOF. As a consequence, there are an infinite number of different 3D trajectories that an individual may deploy to move the upper limb to reach and grasp a target, or to move the trunk to lift a target (Rasmussen, 1991).

There are two operational behavioral consequences of the DOF problem. One is that movement behavior is highly context-specific—most of the variability observed in movement performance is attributable to the design of the environment in which performance occurs (Smith, 1998; Smith, Henning, and Smith, 1994). Context specificity arises because during development and maturation, to deal with the DOF problem, an individual develops a constellation of neural models of movement patterns that enable the individual to move effectors in a relatively reliable and re-producible manner (but see below) to interact with targets. The term commonly applied to this process is motor learning.

However, because everyone interacts with performance environments of distinctive designs during development and maturation, context specificity in movement behavior is different from individual to individual. In other words, confronted with common tasks with the same design (such as eviscerating a turkey, or assembling a small part), different workers are likely to deploy different patterns of movement behavior to interact with the design. This is because, during development and maturation, each individual has established a distinctive repertoire of neural models to guide movement behavior.

A third source of kinesiological variability in movement behavior is related to the fact that humans are imperfect control systems (Schmidt & Lee, 1999). Empirical evidence shows that when interacting with a target in a repetitive fashion, successive movement patterns executed by an individual may be remarkably similar but never identical. This is a feature of all effector movements. Such disparity in fidelity of movement patterning is exacerbated under conditions of fatigue or emotional stress. In other words, confronted with common tasks with the same design, the movement patterns of any given worker in interacting with the design may vary over time.

These considerations suggest that when it comes to delineating possible sources of between- and within-worker response variability that may underlie differential susceptibility, and lack of consistent exposure-dose and dose-response relationships observed in the etiology of MSIs, kinesiological variability in movement behavior is superimposed upon variability attributable to other factors specified in conceptual models referenced above (Fallentin et al., 2001; Radwin et al., 2002; Westgaard and Winkel, 1996). Indeed, given that within- and between-subject timing and accuracy of motor behavioral responses can vary over a two- to five-fold range as a consequence of context-specific influences (Smith, Henning, and Smith, 1994), it is possible that kinesiological factors represent the most prominent source of variance in on-the-job musculoskeletal performance.

The foregoing analysis suggests that the goal of developing reliable and reproducible exposure-dose and dose-response relationships relating musculoskeletal effects of exposure to the design attributes of a given on-the-job task may be scientifically unrealistic. Within- and between-worker variability intrinsic to behavioral guidance of movements, outlined above, means that the degree of musculoskeletal damage (the response) engendered by a given 'dose' of task effort (such as lifting or upper limb repetition at a particular frequency) is likely to be inherently unpredictable for a given work over time, and from worker to worker under identical task demand conditions.

If further research supports this prediction, the implication is that there may be fundamental questions of scientific legitimacy associated with prescriptive ergonomic regulations that reference MSI risk to explicit duration limits for postures or movements of the low back and/or upper limb. The further implication is that relying on generic performance regulations in the formulation of ergonomic rules for mitigating MSI hazards avoids this problem.

CONCLUSIONS

HF/E science is concerned with designing performance environments to effectively accommodate the capabilities and limitations of performers in those environments. This chapter deals with the question of how best to design OHS/ergonomic regulations and standards to accommodate the capabilities and limitations of those who must enforce and those who must abide by such codes.

The first thing to be said about this question is how little attention it has received on the part of both the HF/E and legal communities. Two to three decades ago, while advocating for a performance-based approach to formulating OHS regulations, K.U. Smith (1973) and Bryce (1983, 1985a, 1985b) defined a series of basic HF/E and regulatory issues and concerns associated with designing regulations for mitigating OHS hazards. Since then, however, there has been no follow-up to this work and no systematic evaluation of HF/E design of OHS regulatory code.

The analyses of these authors, coupled with subsequent regulatory experience in the U.S. and Canada, points to several distinct advantages associated with the application of performance-based OHS regulations based on considerations of efficiency, effectiveness, and the purpose of regulatory intervention. Because they focus on assigning performance responsibilities in meeting specified regulatory objectives, without detailing exactly how the objectives are to be achieved, performance regulations are undeniably more efficient than prescriptive regulations. If regulatory attention to specific hazards and/or specific industries is required, narrowly focused performance regulations combined—if needs be—with qualifying prescriptive provisions can be applied. Examples 1 and 2 in Section Performance Versus Prescriptive Regulations in Occupational Health and Safety Legislation illustrate these points.

Adoption of an ergonomics rule in B.C. based on generic performance regulations (Example 5) associated with a decline in the MSI incidence rate in the province (Fig. 3.1) provides putative, case-based evidence for the effectiveness of OHS performance regulations. However, a broader, default argument for performance regulations along these lines may be mounted on evidence regarding the relative lack of effectiveness of prescriptive regulations. The U.S. OSHA code is built largely around prescriptive regulations. However, various authors and OSHA itself cite evidence indicating that strict adherence on the part of U.S. employers to OSHA regulations would prevent only a fraction of occupational accidents and injuries that occur (T.J. Smith, 2002a). This evidence implies that the design of prescriptive regulations is deeply flawed, in the sense that the OSHA code lacks criterion validity—the regulatory outcome of the prescriptive approach does not meet the regulatory intention.

K.U. Smith (1973, 1979, 1988, 1990) calls attention to the likely basis of this design defect by pointing out that most job-related hazards are operational hazards, involving the interaction of a system between the behavioral performance of a worker and some design attribute of the work environment. From this perspective, the low criterion validity of OHS prescriptive regulations may be attributed to the fact that they are concerned only with the latter component of the interaction. In contrast, because they address how performance is supposed to interact with a specified design objective, performance regulations arguably are better designed to deal with the nature and pervasiveness of operational hazards.

The purported intention of OHS regulatory intervention represents a third argument favoring performance regulations. In specifying both the regulatory objective and how the objective is to be achieved, the prescriptive approach arguably fosters a babysitting or handholding relationship between the regulatory body and those being regulated. It strains credulity to believe that, in a general sense, either party finds such a relationship entirely satisfactory. Performance regulations avoid this shortcoming by specifying the objective and the party responsible, but not the method, for hazard mitigation.

When it comes to the specific case of ergonomic rules aimed at mitigation of MSI hazards, the prescriptive approach confronts the fundamental question of scientific legitimacy. The lack of empirical dose-response evidence linking hazard exposure levels and musculoskeletal performance decrements is attributed here to inherent within- and between-worker variability in the control of movement behavior. This variability makes it unlikely that it will ever be possible to document any consistent, graded relationships between exposure to job-related movement demands and consequent musculoskeletal effects. Yet, prescriptive ergonomic rules aimed at MSI hazard mitigation assume that such relationships exist (Example 4)—a scientific conundrum that a performance-based approach avoids.

The foregoing arguments favoring OHS performance regulations are not meant to suggest that this regulatory approach represents a panacea for dealing with OHS hazards. Major potential disadvantages of the performance-based approach include: (1) the lack of explicit guidance they provide regarding the need for regulatory intervention; (2) the consequent difficulty they pose for the regulatory process in terms of unequivocally establishing whether compliance or a violation has occurred; and (3) their reliance on self-reliance and self-responsibility on the part of the employer for purposes of hazard mitigation implies a trust in employer performance that may not always be justified.

Beyond putative advantages and disadvantages associated with the performance-based approach to OHS regulations, a broader difficulty is the lack of systematic, multi-jurisdictional, comparative research on operational experience with performance versus prescriptive OHS regulations. Consequently, the analysis presented here must be viewed as partially conceptual in nature. In the context of this handbook, the underlying message of this chapter is that those concerned with promulgating HF/E and OHS standards and regulations should pay as much attention to the design of their formulation as to their actual content. Those favoring prescriptive regulations must confront shortcomings in efficiency, effectiveness, and intervention philosophy inherent to this approach. Those favoring performance regulations must confront potential difficulties with application inherent to this approach. It is assumed here that ensuring the ultimate success of a given OHS standard or regulation must start with settling upon an appropriate HF/E design for its formulation. It is our judgment that, relative to the prescriptive approach, the performance-based approach to the latter is more likely to give rise to the former, particularly as regards to ergonomics regulations.

REFERENCES

American Academy of Pediatrics, American Public Health Association, & National Resource Center for Health and Safety in Child Care (University of Colorado Health Sciences Center at Fitzsimons) (2002). *Caring for our children. Cational health and safety performance standards: Guidelines for out-of-home child care* (2nd edition). Elk Grove Village, IL: American Academy of Pediatrics.

Barnes, R. M. (1937). *Motion and time study*. New York: Wiley.

Bernard, B. P. (Ed.). (1997). *Workplace factors. A critical review of epidemiologic evidence for work-related musculoskeletal disorders of the neck, upper extremity, and low back*. Cincinnati, OH: National Institute for Occupational Safety and Health.

Bryce, G. K. (1983). Some comments and observations on the application of generic performance regulations in occupational health and safety legislation. In *Paper presented to the 42nd Annual Meeting of the Canadian Association of Administrators of Labour Legislation*, September 28, 1983.

Bryce, G. K. (1985a). Performance regulations and beyond. *British Columbia Workers' Health Newsletter*, 11, 1–6.

Bryce, G. K. (1985b). *The concept and implications of performance regulations in occupational health and safety*. In *Presentation to British Columbia Worklife Forum*, October 30, 1985.

Cleveland, R. J. (1976). *Behavioral safety codes in select industries*. Madison, WI: Wisconsin Department of Industry, Labor and Human Relations.

Coleman, P. J., & Sauter, S. L. (1978). The worker as a key control component in accident prevention systems. In *Presentation to the 1978 Convention of the American Psychological Association*, Toronto, Ontario, Canada.

Coleman, P. J., & Smith, K. U. (1976). *Hazard management: Preventive approaches to industrial injuries and illnesses*. Madison, WI: Wisconsin Department of Industry, Labor and Human Relations.

Dul, J., de Vlaming, P. M., & Munnik, M. J. (1996). Guidelines. A review of ISO and CEN standards on ergonomics. *International Journal of Industrial Ergonomics*, 17, 291–297.

Ellis, L. (1975). A review of research on efforts to promote occupational safety. *Journal of Safety Research*, 7(4), 180–189.

Fallentin, N., Viikari-Juntura, E., Waersted, M., & Kilbom, Å. (2001). Evaluation of physical workload standards and guidelines from a Nordic perspective. *Scandinavian Journal of Work Environment & Health*, 27(Suppl. 2), 1–52.

Foliente, G. C. (2000). Developments in performance-based building codes and standards. *Forest Products Journal*, 50, 12–21.

Gill, J., & Martin, K. (1976). Safety management: reconciling rules with reality. *Personnel Management*, 8(6), 36–39.

Gottlieb, M. S. (1976). *Worker's awareness of industrial hazards: An analysis of hazard survey results from the paper mill industry*. Madison, WI: Wisconsin Department of Industry, Labor and Human Relations.

Gottlieb, M. S., & Coleman, P. J. (1977). *Inspection impact on injury and illness totals*. Madison, WI: Wisconsin Department of Industry, Labor and Human Relations.

Grandjean, E. (1969). *Fitting the task to the man*. London: Taylor & Francis.

Hammer, W. (1972). *Handbook of systems and product safety*. Englewood Cliffs, NJ: Prentice-Hall.

Hammer, W. (1976). *Occupational safety management and engineering*. Englewood Cliffs, NJ: Prentice-Hall.

Hansen, K. (1985). Performance regulations – A naive hope for workplaces without conflict. *British Columbia Workers' Health Newsletter*, 10, 1–6.

Harrison, R. K. (1955). *Biblical Hebrew*. Bungay, Suffolk, England: Hodder & Stoughton.

Heinrich, H. W. (1931). *Industrial accident prevention. A scientific approach* (1st edition). New York: McGraw-Hill.

Heinrich, H. W. (1959). *Industrial accident prevention. A scientific approach* (4th edition). New York: McGraw-Hill.

International Council for Building Research Studies and Documentation (1982). Working with the performance approach in building (CIB report, publication 64). Rotterdam, Netherlands: International Council for Building Research Studies and Documentation.

International Organization for Standardization (1994). *International Standard ISO 9001. Quality systems—model for quality assurance in design, development, production, installation and servicing.* Geneva, Switzerland: International Organization for Standardization.

Jones, D. F. (1973). *Occupational safety programs—Are they worth it?* Toronto, Ontario: Labour Safety Council of Ontario, Ontario Ministry of Labour.

Juran, J. M. (1964). *Managerial breakthrough. A new concept of the manager's job.* New York: McGraw-Hill.

Juran, J. M. (1995). *Managerial breakthrough. The classic book on improving management performance* (revised ed). New York: McGraw-Hill.

Kaplan, M. C., & Coleman, P. J. (1976). *County highway department hazards: A comparative analysis of inspection and worker detected hazards.* Madison, WI: Wisconsin Department of Industry, Labor and Human Relations.

Kaplan, M. C., Knutson, S., & Coleman, P. J. (1976). *A new approach to hazard management in a highway department.* Madison, WI: Wisconsin Department of Industry, Labor and Human Relations.

Karsh, B.-T., Moro, F. B. P., & Smith, M. J. (2001). The efficacy of workplace ergonomic interventions to control musculoskeletal disorders: a critical analysis of the peer-reviewed literature. *Theoretical Issues in Ergonomics Science,* 2(1), 23–96.

McHugh, A.-R. (2002). Ergonomic tool kits give sawmills bang for their buck. *Worksafe. The WCB Prevention Magazine on Occupational Health and Safety Issues,* 3(6, November–December), 10–11.

Minnesota Department of Labor and Industry (1993). *An employer's guide to developing a workplace accident and injury reduction (AWAIR) program.* Saint Paul, MN: Minnesota Department of Labor and Industry, Occupational Safety and Health Division.

National Board of Occupational Safety and Health (NBOSH) (1998). *Ergonomics for the prevention of musculoskeletal disorders (AFS 1998:1).* Stockholm, Sweden: NBOSH.

National Institute of Occupational Safety and Health (2000). *NIOSH testimony to OSHA. Comments on the proposed ergonomics program (29 CFR Part 1910, Docket No. S-777).* Cincinnati, OH: National Institute for Occupational Safety and Health.

Occupational Safety and Health Reporter (1976). Washington, DC: Bureau of National Affairs.

OSHA (1970). *Occupational safety and health act of 1970 (Public Law 91-596, 91st Congress, S.2193, December 29, 1970).* Des Plaines, IL: U.S. Department of Labor, Occupational Safety and Health Administration, Office of Training and Education.

OSHA (1980). *Code of federal regulations.* 29, Labor *(Part 1910, Occupational Safety and Health Standards).* Washington, DC: Office of the Federal Register, National Archives and Records Service, General Services Administration.

OSHA (2000). *Ergonomics program standard: final rule. Federal register 2000 (65(220), 68262-870).* Washington, DC: Office of the Federal Register, National Archives and Records Service, General Services Administration.

Plaut, W. G. (1981). *The Torah. A modern commentary.* New York: Union of American Hebrew Congregations.

Prior, J. J., & Szigeti, F. (2003). *Why all the fuss about performance based building.* http://www.auspebbu.com/files/Why%20all%20the%20fuss%20-%20PBB.pdf.

Radwin, R. G., Marras, W. S., & Lavender, S. A. (2002). Biomechanical aspects of work- related musculoskeletal disorders. *Theoretical Issues in Ergonomics Science,* 2(2), 153–217.

Rasmussen, D. A. (1991). *Human motor control.* San Diego, CA: Academic Press.

Richardson, V. L. (1973). *Hazard surveys at select employers.* Madison, WI: Wisconsin Department of Industry, Labor and Human Relations.

Schmidt, R. A., & Lee, T. D. (1999). *Motor control and learning. A behavioral emphasis* (3rd Ed). Champaign, IL: Human Kinetics.

Smith, K. U. (1973). Performance safety codes and standards for industry: The cybernetic basis of the systems approach to accident prevention. In J. T. Widner (Ed.), *Selected readings in safety* (pp. 356–370). Macon, GA: Academy Press.

Smith, K. U. (1979). *Human-factors and systems principles for occupational safety and health.* Cincinnati, OH: NIOSH, Division of Training and Manpower Development.

Smith, K. U. (1988). Human factors in hazard control. In P. Rentos (Ed.), *Evaluation and control of the occupational environment* (pp. 1–7). Cincinnati, OH: NIOSH, Division of Training and Manpower Development.

Smith, K. U. (1990). Hazard management: principles, applications and evaluation. In *Proceedings of the Human Factors and Ergonomics Society 38th Annual Meeting* (pp. 1020–1024). Santa Monica, CA: Human Factors and Ergonomics Society.

Smith, M. J. (1994). Employee participation and preventing occupational diseases caused by new technologies. In G. E. Bradley & H. W. Hendrick (Eds.), *Human factors in organizational design and management – IV* (pp. 719–724). Amsterdam: North-Holland.

Smith, M. J., Bauman, R. D., Kaplan, R. P., Cleveland, R., Derks, S., Sydow, M., & Coleman, P. J. (1971). *Inspection effectiveness.* Washington, DC: OSHA.

Smith, T. J. (1993). The scientific basis of human factors—A behavioral cybernetic perspective. In *Proceedings of the Human Factors and Ergonomics Society 37th annual meeting* (pp. 534–538). Santa Monica, CA: Human Factors and Ergonomics Society.

Smith, T. J. (1994). Core principles of human factors science. In *Proceedings of the Human Factors and Ergonomics Society 38th Annual Meeting* (pp. 536–540). Santa Monica, CA: Human Factors and Ergonomics Society.

Smith, T. J. (1998). Context specificity in performance – The defining problem for human factors/ergonomics. In *Proceedings of the Human Factors/Ergonomics Society 42nd annual meeting* (pp. 692–696). Santa Monica, CA: Human Factors and Ergonomics Society.

Smith, T. J. (2002a). Macroergonomics of hazard management. In Hendrick, H. W. and Kleiner, B. (Eds.), *Macroergonomics* (pp. 199–221). Mahwah, NJ: Lawrence Erlbaum.

Smith, T. J. (2002b). Ergonomics task-force recommendations. (Ed.), In *The state of ergonomics in Minnesota. A summary of the Minnesota Department of Labor and Industry's Ergonomics Task-Force activities and recommendations* (pp. 12–17). Saint Paul, MN: Minnesota Department of Labor and Industry.

Smith, T. J., Henning, R. H., & Smith, K. U. (1994). Sources of performance variability. In G. Salvendy & W. Karwowski (Eds.), *Design of work and development of personnel in advanced manufacturing* (Chap. 11, pp. 273–330). New York: Wiley.

State of California (1997). *Repetitive motion injuries. Cal/OSHA standards, California code of regulations (Title 8, Division 1, Department of Industrial Relations, Chapter 4, Division of Industrial Safety, Subchapter 7, General Industry Safety Orders, Group 15, Noise and Ergonomics, Article 106, Ergonomics, Section 5110).* Sacramento, CA: Cal/OSHA.

Swain, A. D. (1973). An error-cause removal program for industry. *Human Factors, 15*(3), 207–221.

Swain, A. D. (1974). *The human element in systems safety: A guide for modern management.* London: Industrial and Commercial Techniques.

Viikari-Juntura, E. R. A. (1997). The scientific basis for making guidelines and standards to prevent work-related musculoskeletal disorders. *Ergonomics, 40*(10), 1097–1117.

Washington State Department of Labor and Industries (2000). *Ergonomics. Washington Industrial Safety and Health Act (Section WAC 296-62-051).* Olympia, WA: Washington State Department of Labor and Industries.

Westgaard, R. H., & Winkel, J. (1996). Guidelines for occupational musculoskeletal load as a basis for intervention: a critical review. *Applied Ergonomics, 27*(2), 79–88.

Westgaard, R. H., & Winkel, J. (1997). Ergonomic intervention research for improved musculoskeletal health: a critical review. *International Journal of Industrial Ergonomics, 20*, 463–500.

Workers' Compensation Board of British Columbia (1999a). *General requirements. Occupational Health & Safety Regulation. B.C. Regulation 296/97, as amended by B.C. Regulation 185/99. Book 2, General Hazard Requirements, Parts 5-19. Part 12 – Tools, Machinery and Equipment.* Vancouver, BC: Workers' Compensation Board of British Columbia.

Workers' Compensation Board of British Columbia (1999b). *Controlling exposure. Occupational Health & Safety Regulation. B.C. Regulation 296/97, as amended by B.C. Regulation 185/99. Book 2, General Hazard Requirements, Parts 5-19. Part 5 – Chemical and Biological Substances.* Vancouver, BC: Workers' Compensation Board of British Columbia.

Workers' Compensation Board of British Columbia (1999c). *Division 3 – General duties of employers, workers and others. Occupational Health & Safety Regulation. B.C. Regulation 196/97, as amended by*

B.C. Regulation 185/99. Book 1, Core Requirements, Parts 1-4. Part 3 – Occupational Health and Safety. Vancouver, BC: Workers' Compensation Board of British Columbia.

Workers' Compensation Board of British Columbia (1999d). *Ergonomics (MSI) requirements. Occupational Health & Safety Regulation. B.C. Regulation 196/97, as amended by B.C. Regulation 185/99. Book 1, Core Requirements, Parts 1-4. Part 4: General Conditions*. Vancouver, BC: Workers' Compensation Board of British Columbia.

Section II

Nature of HF/E Standards and Guidelines

4 Standardization Efforts in the Fields of Human Factors and Ergonomics

James M. Oglesby, Kimberly Stowers, Kevin Leyva, and Eduardo Salas

CONTENTS

INTRODUCTION

As new technology and systems are introduced to aid human performance and living, it is well documented that improper design and application of these tools can negatively impact performance in different ways, such as increasing stress and compromising safety (Driskell & Salas, 1996; Sharit,

2006). To meet the challenges and risks of improper system development, standardization relied upon to establish criteria for products and processes in a variety of contexts. This does not only apply to elements in workplaces, but also to products used in our daily routine such as motor vehicles, appliances, and the construction of buildings. To help guide the design and development of manufactured goods, the utilization of standards helps in evaluating their functionality and effectiveness.

Standards are meant to provide an economic and social benefit for workers and consumers in the manufacturing of products; laws are created to promote safety and streamline various functions. Priest, Wilson, and Salas (2005) indicate several impacts that standards have had on the safety of personnel and financial benefits, such as the number of lives saved in intensive care in the medical field, workplace injury prevention, and decrease in vehicle accident fatalities over the last several decades despite increasing traffic.

Standards are used to obtain a manner of conformity in the design and development of products, processes, formats, and procedures to optimize economic activity efficiency (Hemphill, 2009). Many uses of standards are relevant in the field of human factors and ergonomics and are developed by national and international institutions, government, research societies, and private organizations. Standards can guide researchers and practitioners to investigate and determine the current operations of industries, as well as identify the needs and expectations for organizations to optimize safety and productivity.

As technology changes over time, standards must constantly be developed and to consider new developments. For example, the nuclear power domain is currently undergoing a significant change where power plant control room systems are expected to transition from analog to digital interfaces (O'Hara et al., 2008). Additionally, with the increasing access to mobile technology, including smartphones and tablets, many people utilize these platforms to conduct tasks as an alternative to desktop computers (e.g., Hemphill, 2009). New standards and guidelines are needed to accommodate the development, testing, and maintenance of newly introduced products throughout the system life cycle.

The purpose of this chapter is to update the Priest, Wilson, and Salas (2005) standards guide based on developments that have occurred over the past decade by standard development organizations (SDOs). The format of this chapter will follow the version in the previous edition of the *Handbook of human factors and ergonomics standards* chapter and will include progress of standards associated with human factors and ergonomics developed in the past decade. An overview of the different types of standards developed and their purposes is provided, as well as a discussion regarding the historical importance of standardization. Following this, we will discuss recent developments regarding the process of federal institutions, SDOs, and liaison organizations involved in creating and managing standards. The chapter will also provide information on accessible databases available to human factors professionals in finding and reviewing guidelines in multiple areas. Due to the length limitations for chapters in this handbook, our intent is not to provide an exhaustive list of institutions and databases that capture all human factors related guidelines but to identify selections considered to be strongly related to human factors and ergonomics. This information is meant to help professionals in the design and development of technology, rules, training, and workplace processes.

DEFINITION OF STANDARDS

The National Standards Policy Advisory Committee's (1978, p. 6) defines a *standard* as "a prescribed set of rules, conditions, or requirements concerning definitions of terms; classification of components; specification of materials, performance, or operations; delineation of procedures; or measurement of quantity and quality in describing materials, products, systems, services, or practices." In a simpler manner, Pember indicates that a standard could be considered a published document that provides specifications and procedures utilized to ensure that a product is fit for its intended purpose. Standards are meant to guide designers to ensure that products made meet the established criteria for safety and performance. Additionally, they provide a common language to communicate design requirements and facilitate goods between the buyer and seller, as well as protect the welfare of the public (ASTM, n.d.).

Standards are developed by a committee made of multiple stakeholders that have some sort of investment on the subject that the standards are meant to address. A demand for standardization can come from various agencies, such as governments, trade associations, industries, and consumers. The committees responsible for developing standards are experts in the field that attempt to achieve a consensus of what criteria should be in place, and the exact process for committee functions can vary between SDOs (ASTM, n.d.).

ASTM International (ASTM, n.d.) identifies five types of standards based on the degree of consensus needed for their development and use. A *company standard* is based on the consensus of employees within an organization, such as protocol for company policies and workplace procedures. A *consortium standard* refers to a standard that is agreed upon by a group of organizations formed to undertake an activity beyond the means of a lone organization. An *industry standard* is based on a consensus among many companies within an association or a professional society. *Government standards* are those that are approved of by a government agency, although these could be developed by either an agency or be adopted from already developed standards from the private sector. Finally, *voluntary consensus standards* are developed and approved by representatives of all sectors that have an interest in how the standard is used. The stakeholders that achieve consensus for these types of standards can include manufacturers, consumers, and representatives and are primarily utilized for commercial and regulatory purposes.

Standards can take a variety of forms, ranging from criteria for the characteristics of products being developed (e.g., functions, features, measurements, etc.) to the process for how products are manufactured and tested, and human activities and management specifications. For example, Veldman and Widmann (2000) constructed a framework that identifies three types of standards based on what is being relegated in the entire process: (1) product standards, (2) technical process standards, and (3) management process standards. *Product standards* refer to defining particular functions, measurements, features, and potential testing criteria for a manufactured product. *Technical process standards* involve criteria that specify the manufacturing and testing of a product. *Management process standards* are associated with personnel activity, which the authors consider to be more ambiguous based on the complexity of the manufacturing process. The International Organization of Standards (ISO) and the International Electrotechnical Commission (IEC) provide a taxonomy based on the most common purposes of standards development (ISO/IEC, 2004). Table 4.1 displays how standards can be formed to meet the needs and requirements of various aspects involved in operational functioning and product development.

HISTORY OF STANDARDS

The development of standards can be traced back to as early as 7000 BC in Egypt, where cylindrical stones were used as standard units of weight (Breitenberg, 1987). Aside from Egypt, other areas around the world—China, India, Arabia, and Mesopotamia—have been identified to have utilized standards for over thousands of years. One of the earliest documented attempts of establishing standards in the Western world was in the year 1120 AD. King Henry I of England ordered that the ell—the ancient yard—was to be the length of his forearm and established as the standard unit of length throughout his kingdom (Breitenberg, 2009).

One of the first instances of standardization during the colonization of the United States was developed in Boston in 1689 when a law was passed, requiring all manufactured bricks to be made by the dimensions of 9×4×4 inches. This standardization was in reaction to a fire that destroyed the city, and civil servants standardized building materials to ensure that the rebuilding process was undertaken in a quick and cost-effective approach (Breitenberg, 2009). Significant impacts for the United States continued in the late 1700s when Eli Whitney—who many consider "the Father of Standardization"—implemented the concept of mass production. Vice-president Thomas Jefferson

TABLE 4.1

The ISO/IEC (2004) Taxonomy for Different Types of Standards-Based on Their Purpose

Standards Term	Definition
Basic standard	Wide range coverage, general provisions for a particular field
Terminology standard	Standards of terms and definitions used in the process
Testing standard	Standards of testing methods, including sampling method, statistical analysis, and data collection
Product standard	Requirements to be met by one or a group of products to fit its purpose
Service standard	Requirements to be met by a service to establish its purpose
Interface standard	Requirements pertaining to the compatibility of products at their points of interconnection
Standard on data to be provided	Identification of what values or other data to be shared for specifying a product, process, or service

contracted Whitney to produce 10,000 muskets. Whitney developed the idea of streamlining the production of the muskets by making each component identical to one another, therefore being interchangeable and providing an opportunity for more efficient manufacturing. Whitney standardized the process for producing the muskets, from setting the measurements and characteristics for each component to training workgroups on each step in the manufacturing process. Whitney demonstrated the benefits of standardization to the U.S. Congress by assembling ten working muskets from randomly selected components he brought and convinced the congress that standardization can increase manufacturing output by allowing mass production (Breitenberg, 2009).

Aside from production, other historical events led to the recognition of standards being used to promote safety and accident prevention. A large fire that occurred in the year 1904 in Baltimore demonstrates how standardization could have mitigated catastrophic consequences. Fire engines from neighboring areas—and as far as New York—rushed to douse the flames, only to find out that the hose attachments for fire hydrants in their local cities did not fit the Baltimore hydrants. This lack of standardization led to most of the fire engines being useless and left the 30-hour long blaze to destroy over 1500 buildings, along with the city's electric, telephone, telegraph, and power facilities. With this valuable lesson learned, the standardization of hydrant and hose attachments saved the town of Fall River, Massachusetts, from a similar fate when a large fire broke out 23 years later, requiring the help of 20 neighboring city fire departments to help control the blaze (Breitenberg, 1987, 2009).

Some of the most significant examples of standardization in the United States involved (1) the standards of the railroad track gauge and (2) the standards associated with traffic signals. In 1886, a U.S. standard compelled all railroad gauges to be of uniform size, allowing the interchangeability of railroad cars to travel on all tracks between location and companies. This allowed for cross-country transportation and cargo transfer while still providing current benefits. For road transportation, the early 20th century witnessed a variety of designs in traffic signals from state-to-state. These differences included different colors of lights that were used (e.g., purple, blue, yellow), as well as the symbolic representation of what these lights meant (e.g., red meant "stop" in one state, but "go" in another). A national standard for colors was created by the American Association of State Highway Officials (AASHTO), the National Bureau of Standards (now known as the National Institute of Standards and Technology), and the National Safety Council to limit confusion among drivers that traveled across state lines. Due to the amount of traffic currently present on today's roads and highways, this standardization is perceived to have tremendous benefits for the U.S. transportation system (Breitenberg, 2009).

One of the first approaches of standardization at the international level occurred during World War II due to a mismatch of the specification of materials and tools between nations in the Allied Forces. The discrepancy in the size of screws used to construct military vehicles limited the ability to repair and replace parts of the tanks between American and English forces. These problems were eventually alleviated in 1948 by the development of an international standard screw thread and paved the way to meet the needs of reducing inventory and increasing compatibility at a large scale. Since then, standards have continually increased at the national and international levels to unify the manufacture and process of various products (Breitenberg, 2009).

Today, standards are involved in nearly every part of our daily lives. Breitenberg (2009) indicates that approximately 80% of all the world's manufactured goods are affected by standards and regulations incorporated by standards. Standards are not only important for protecting consumers, but industries must also attend to them to remain competitive and efficient in today's economy. A large degree of effort is put into standardization, and there are many individuals and institutions that are involved in establishing the criteria in regulating manufacture and processes.

The United States' standards development system itself is composed of a variety of institutions. Based on the roles and responsibilities of these organizations, there exist three different types of institutions: (1) liaison, (2) standards development organizations (SDOs), and (3) standards databases. Liaison organizations are institutions that communicate between agencies within the United States and abroad regarding standards development and implementation. SDOs are groups that facilitate the development of standards that are typically involved in product development, processes, and patenting. In the United States, there are approximately 130 scientific professional societies, 300 trade associations, and 40 private-sector organizations that are identified as SDOs (Hemphill, 2005). Many institutions provide access to published standardization documents for investigating guidelines by topic, date, and agency. The following will provide examples of each of these types of organizations relevant in the field of human factors, as well as recent developments associated with standardization.

STANDARDS INSTITUTIONS

LIAISON ORGANIZATIONS

American National Standardization Institute (ANSI)

ANSI (ansi.org) is considered the peak organization that promotes the development of standards to enhance the competitiveness of the United States' economy and quality of life (Hemphill, 2009). ANSI was established in 1918 and is the largest archive of standards in the nation. It houses standards from many accredited member organizations, covering topics ranging from technological, nuclear, and robot safety to many others. ANSI currently houses over 10,000 American National Standards (ANS) developed by hundreds of SDOs, which themselves are split into smaller expert groups who all conform to ANSI Essential Requirements. The U.S. government often uses many of the standards developed by ANSI, creating impact beyond member organizations. Additionally, ANSI is the United States' representative in the ISO, allowing the U.S. government to participate in the development of international standards.

ANSI has been involved in multiple efforts in recent years. In 2008, ANSI launched an online tool that allows the users to see changes to current standards, as well as modify the standards by adding links, notes, images, and other content to help documents be more thorough and help users collaborate on improvements to the resources (ANSI, 2008). In 2009, ANSI collaborated with Citation Technologies to launch a web database to house standards, giving the user the ability to search through standards using keywords (ANSI, 2009). Additionally, this database allows users to annotate and connect standards to their own company policy, providing customized experiences (ANSI, 2008). As a leader in standard development, ANSI offers accreditation programs for organizations who want to stand out in the competition by assuring the customer that they meet the well-established standards of a trusted institute consisting of over 125,000 companies.

Panels and collaboratives were created in ANSI to address the needs and requirements for specific topics in each domain. While each domain has different concerns, all members share in the need for standardization and the desire to stay informed on what is happening in their industry and how to get involved to push needed changes. Many of these panels and collaboratives seek knowledge, initiatives, and standards of other organizations in that domain and determine if there are common gaps in standards that have yet to be addressed or have recently been agreed upon.

There are currently seven active standards panels and collaboratives. Standard panels work toward creating standards, roadmaps for the domain, and coordination between different parties within the domain. An example of this is the Electric Vehicles Standards Panel (EVSP), which seeks to provide standards in three domains: vehicles, infrastructure, and support services. Each of these domains has its own set of challenges that will need to be overcome to have a successful implementation of electric vehicles. On the other hand, collaboratives exist to provide a place to identify existing standards for that domain and assist other parties in creating new standards. For example, the Homeland Defense and Security Standardization Collaborative (HDSSC) is an example of a collaborative that actively identifies standardization needs and solutions. Table 4.2 provides a description of the seven active ANSI panels and collaboratives, along with a description of their focus.

National Institute of Standards and Technology (NIST)

NIST was created to perform innovative research to advance science in industry through measurement and standards. NIST was established in 1901 within the U.S. Department of Commerce, and the influence of their standard has saved companies and individuals millions of dollars in topics such as encryption and computer security, stock exchange, and many others. Due

---

TABLE 4.2
ANSI Panels and Collaboratives with Descriptions of Their Focus

ANSI Panels and Collaboratives	Description
ANSI Network on Smart and Sustainable Cities (ANNSC)	Coordinating on voluntary standards and related activities for sustainable cities in the United States and abroad.
ANSI Energy Efficiency Standardization Coordination Collaborative (EESCC)	Assessing of energy-efficient activity standardization landscape.
ANSI Homeland Defense and Security Standardization Collaborative (HDSSC)	Assisting government agencies and other sectors in accelerating development and adoption of standards critical to homeland security and defense.
ANSI Nanotechnology Standards Panel (ANSI-NSP)	Coordinating the development of standards associated with nanotechnology, such as nomenclature/terminology, materials properties, and testing.
Nuclear Energy Standards Coordination Collaborative (NESCC)	The joint initiative of ANSI and NIST in responding to the current needs of the nuclear industry.
Electric Vehicles Standards Panel (EVSP)	Collaborating between public and private sector stakeholders involved in the safe mass deployment of electric vehicles in the United States.
ANSI Network on Chemical Regulation	Enabling manufacturers and stakeholders to gain consensus regarding domestic and global chemical regulations.

to the extensive financial support from government appropriations, fees, and funding from outside agencies, the NIST was able to accomplish many objectives from the previous chapter as it continues to function today.

In 2010, NIST released a series of *Recommended Practice Guide* titled "How to Measure", which focused on how to measure correctly, what to measure, what strategy to use, and how to understand the results. Each guide was on a different topic, written by an expert(s) in that domain. Examples of measurement guides include Fractography of Ceramics and Glasses, X-Ray Topography, Particle Size Characteristics, and The Fundamentals of Neutron Powder Diffraction. As 3D printing becomes more common, the NIST is researching the use of additive manufacturing and developing standards to improve consistency and quality. One of the main objectives of the NIST is to provide measurements, standards, data, and tools—which is why in 2012, the NIST opened Boulder's Precision Measurement Laboratory to conduct high precise measurements in a highly controlled environment. Some of the projects conducted in this laboratory include magnetic resonance imaging calibration, microfabricated devices, material composition analysis, and quantum computing.

STANDARD DEVELOPMENT ORGANIZATIONS (SDOs)

American Society of Civil Engineers (ASCE)

The American Society of Civil Engineers (ASCE) is the oldest engineering society in the United States (www.asce.org). Containing over 145,000 members from 174 countries, this organization plays a leading role in the engineering field by providing codes, standards, professional conferences, and continuing education for its members. In the area of standards development, ASCE provides guidance on large structures that affect the public, including the development, implementation, use, and safety risks of systems such as automated transportation. ASCE also has a forum, *The Civil Engineering Blog & News Network*, which shares ideas to help engineers work more effectively. Topics in this forum include stress, worker productivity, and management efficiency.

American Society of Mechanical Engineers (ASME)

The American Society of Mechanical Engineers (ASME) considers itself the leading developer of standards and codes used internationally to guide mechanical engineering (www.asme.org). Firms from more than 100 countries currently use these codes, which cover many topics—including engineering design, standardization, safety, and performance. The goal of ASME is to promote the practice of multidisciplinary engineering internationally. Currently, ASME offers 48 performance test codes that assess the accuracy and efficiency of various mechanical systems, as well as suggest performance measures specific to the type of system being assessed. Additionally, ASME offers guidance relevant to safety in mechanical systems through its Board on Safety Codes and Standards (BSCS). The BSCS oversees safety standards, codes, accreditation, and certification for many systems that affect humans every day, including elevators and transit vehicles.

Association for the Advancement of Medical Instrumentation (AAMI)

The Association for the Advancement of Medical Instrumentation (www.aami.org) is a non-profit organization founded in 1967 that develops standards associated with the design of medical tools used in healthcare, as well as procedures associated with these devices (e.g., cleaning, handling, etc.). The AAMI standards program is composed of over 100 technical committees and working groups that develop standards and reports.

AAMI has been involved in the development of the *ANSI/AAMI HE75, 2009/(R)2013 Human factors engineering–Design of medical devices*, a comprehensive virtual encyclopedia that focuses on human factors and ergonomics considerations regarding the development of medical equipment. This guide provides resources in guiding researchers and practitioners on various topics such as user-centered design, error and risk management, usability testing, anthropometric considerations, alarm and display designs, and other areas of human factors interest that are important in maintaining effective performance and optimal safety. Access to the encyclopedia is available through the ANSI website along with other publications associated with standards and guidelines.

Federal Aviation Administration (FAA)

FAA—as the federal authority for aviation in the United States—oversees all civil aviation in this country (www.faa.gov). Its primary mission is to provide the safest and most efficient aerospace system possible. This is done through many efforts, including the research, licensure, certification, regulation, training, and testing of aircraft and pilots. The FAA also develops standards that aid in keeping airplanes, pilots, and passengers safe during all stages of flight.

The development of Technical Standards Orders (TSO) aids in aerospace safety by guiding manufacturers of aircraft in their creation of certain processes and materials. Specifically, a TSO is a standard for minimum performance to be met by specific materials for manufacturers to produce them. Receiving a TSO authorization only allows the development of the aircraft and not its use. As such, the development and implementation of a TSO offers only a small portion of the safety standards and guidelines that the FAA provides.

As some of the most safety-critical stages in flight is the launch and landing of the aircraft, the FAA addresses this concern in many ways, focusing primarily on the implementation of guidelines for airport safety. Specifically, the FAA manages airport and runway safety through federal regulation, directing the implementation of safety management systems, runway safety programs, and airport design and engineering standards. Airport engineering standards address every aspect of an airport—from the overall layout to radio control, vehicles, beacons, and operation safety.

Human Factors and Ergonomics Society (HFES)

HFES is the largest organization composed of human factors researchers and practitioners (www. hfes.org). HFES has been involved in the development of American national standards since the 1980s. HFES is an ANSI-accredited organization and has collaborated with many institutions to apply human factors knowledge into standards associated with safety, system interaction, and other important factors associated with human performance. The following provides recent work in standardization that HFES has taken part in for this purpose.

One of the significant developments within the past decade is multiple national standards that focus on the design of workplaces and products. In 2007, ANSI approved ANSI/HFES 100-2007—*Human Factors Engineering of Computer Workstations*. This document provides guidelines for various elements that involve the design and installation of computer workstations. Specific standards include the application of computer displays, input devices (e.g., keyboard, mouse, trackball, etc.), and furniture that compose the workstation. Overall, the purpose of the guide is to create workstations that optimize their usability by adjusting for users, maximizing comfort, and avoiding physical strain during work shifts. In addition to design specifications, HFES 100-2007 (which is acting as a replacement for the standards BSR/HFES 100-2002 and ANSI/HFS 100-1988) provides information on metrics for assessing certain characteristics of the computer workstation and user interaction.

ANSI/HFES 200—*Human Factors Engineering of Software User Interfaces*—is a set of standards developed by HFES and ANSI since 2006 and has been approved as an American National Standard in 2008. HFES 200 was developed to provide design requirements to increase the accessibility, learnability, and usability of software. The standards include guidelines for interface design that will primarily benefit the user interacting with the software; adopt materials from ISO 9241 (ISO standards associated with human-computer interaction) involved with interaction techniques; and cover recommendations for interactive voice response, presentation of content, and use of colors.

HFES has also been involved in developing HFES 300, *Guidelines for Using Anthropometric Data in Product Design*, which was published in 2004. This standard is designed to provide recommendations for developing products that can be used safely and efficiently, optimizing the comfort of the user by considering individual differences in the dimensions of the human body. HFES 300 is considered an initial effort to provide an overall approach in utilizing anthropometric means and percentiles in developing clothes, tools, workstations, and complex systems that promote usability for individuals or groups. The standard also provides methods for effectively applying anthropometric data, information on their utility, and a collection of resources to help practitioners in using body dimension data for application.

Institute of Electrical and Electronics Engineers Standards Association (IEEE-SA)

IEEE-SA (standards.ieee.org) focuses on developing standards through a process of gathering individuals and organizations of various backgrounds. The IEEE-SA elects a *Board of Governors* to facilitate the creation and management of *Standards Board* committees, who develop the standards. Members of both boards are elected by IEEE-SA members annually. Within the IEEE-SA, there are programs that focus on different goals of standard development and maintenance. Some of the programs, such as Industry Connections, help create new standards, products, and services by collaboration among organizations and individuals. As with many other standardization organizations, the IEEE-SA maintains international standards through their *Global Cooperation* program, allowing cooperation between industry, government, and civilian personnel around the world. Another important function of the IEEE-SA is in the applied setting. The *Arc Flash Research* project is an example of a response to accidents in the creation of a new project focusing on improving safety through the creation of standards for this specific domain. An important aspect of standard creation is the distribution and availability of such standards. *The IEEE Get Program* accomplishes this with public access to current standards free of charge.

To provide a robust standard, IEEE-SA maintains a multistep process for standard development. To begin, the Standard Development Organization is given by a sponsor. The SDO then needs to approve the request to develop a new standard and the sponsor needs to form the working group to discuss, review, and present information and come to an agreement. The group then creates a draft that is sent for sponsor balloting—which, if approved, is then sent to the Standards Board that decides if it will be published and distributed. The standard will later be revised or withdrawn and archived after—at most—ten years. This ensures that older standards are still relevant or replaced by an updated version. IEEE-SA currently has over 900 active standards and over 500 standards in development or modification. These standards are spread across 20 distinct domains and members represent more than 160 countries.

International Organization for Standardization (ISO)

ISO is the world's largest developer of voluntary international standards, composed of 163 member countries (www.iso.org). Established in 1947, ISO has published more than 19,500 international standards covering several industries, including technology, agriculture, and healthcare. The goal of ISO is to promote international coordination and trade, as well as unify standards internationally, providing equal support to all member countries.

ISO standards aim to promote both efficiency and safety, and these standards are reviewed every 3–5 years to ensure that they are up to date. One of ISO's most well-known standards, *ISO 9001*, is undergoing revisions from its 2008 version, including the placement of more emphasis on risk in quality management systems. This movement to emphasize risk can be applauded as a step forward in the field of human factors and safety.

National Fire Protection Association (NFPA)

NFPA (www.nfpa.org) was started to help in the development of standards that would improve safety and mitigate risks of fire, electrical, and other building safety issues. Every standard developed goes through a process of "voluntary consensus" by committees of around 7000 volunteers, adhering to values consistent with the American National Standards Institute (ANSI) of balance, openness, and fairness. Currently, the NFPA houses more than 300 codes and standards with a member base of over 75,000, from almost 100 countries, with connections to 80 national trade and professional organizations. All the standards developed by NFPA are available to view for free and copies are available to download for a fee.

In 2013, the NFPA launched the document authenticity program. This program helps ensure that the published standards the customers or members received are the correct and full document. The program places a seal on all purchased documents, which links to a website to verify authenticity. Additionally, numerous standards have been developed for enhancing the safety and avoidance of fire-related accidents. Table 4.3 provides a list of the most popularly accessed standards from the NFPA currently.

Occupational Health and Safety Administration (OSHA)

OSHA (www.osha.gov) is a standards development organization that is part of the U.S. Department of Labor that focuses on the development and regulation of requirements associated with workplace settings. OSHA was formed in 1970 in accordance with the passing of the *Occupational Safety and Health Act*. Since its establishment, and with the cooperation of employers, workers, and fellow stakeholders, work-related deaths and injuries have dropped by 65% (OSHA, 2014). While OSHA operates standards on a federal level, there are multiple state-level safety programs approved by OSHA for application. Table 4.4 provides a summary of the states that have their own OSHA-approved programs for the private level and public sectors.

OSHA covers a variety of standards in a range of domains related to workplace safety. Examples of standardizations developed include the processing and handling of dangerous and hazardous equipment and materials, the use of warning and labeling to indicate potential hazards, and the criteria for protective equipment and safeguards to minimize the potential of workplace injury. Table 4.5 provides a list of some of the most accessed generalized industry standards as indicated from the OSHA official website.

ASTM International

ASTM International (astm.org)—formally known as the American Society for Testing and Materials—is a non-profit voluntary standards development organization that was formed in 1898. Since the turn of the century, ASTM International's work has extended beyond the United States, with branch offices in Belgium, Canada, China, and Mexico. ASTM International has developed

---

TABLE 4.3

List of Some of the Most Popular NFPA Codes (According to NFPA Online Catalog)

Standard Code	Version Year	Name
NFPA 70E	2015	Standard for Electric Safety in the Workplace
NFPA 70	2014	National Electrical Code
NFPA 54	2015	National Fuel Gas Code
NFPA 25	2014	Standard for the Inspection, Testing, and Maintenance of Water-based Fire Protection Systems
NFPA 101	2015	Life Safety Code
NFPA 72	2013	National Fire Alarm and Signaling Code
NFPA 13	2013	Standard for the Installation of Sprinkler Systems
NFPA 99	2015	Health Care Facilities Code
NFPA 58	2014	Liquefied Petroleum Gas Code
NFPA 30	2015	Flammable and Combustible Liquids Code

TABLE 4.4
List of States with State Approved OSHA Safety Programs (OSHA, 2014)

OSHA Approved State Plans: Public and Private Sector		OSHA Approved State Plans: Public Sector Only		Federally Protected by OSHA	
• Alaska	• Minnesota	• Connecticut	• New Jersey	• Alabama	• Massachusetts
• Arizona	• Nevada	• Illinois	• New York	• Arkansas	• Mississippi
• California	• New Mexico			• Colorado	• Missouri
• Hawaii	• North Carolina			• Delaware	• Montana
• Indiana	• Tennessee			• Florida	• Nebraska
• Iowa	• Utah			• Georgia	• New Hampshire
• Kentucky	• Vermont			• Idaho	• North Dakota
• Maryland	• Virginia			• Kansas	• Texas
• Michigan				• Louisiana	
				• Maine	

TABLE 4.5
List of the Most Accessed OSHA General Industry Standards

Standards Code	Name	Description
1910.1030	Bloodborne Pathogens	Applies to occupational exposure to blood and potentially infectious materials
1910.1200	Hazard Communications	Ensures hazards of all chemicals produced and imported are transmitted to work personnel
1910.134	Respiratory Protection	Controls threats of occupational diseases caused by breathing contaminants such as smoke, dust, fumes, etc.
1910.95	Occupational Noise Exposure	Protects against the effects of excessive noise in work settings
1910.178	Powered Industrial Trucks	Safety requirements associated with the operation of specialized industrial trucks such as forklifts, hand trucks, tractors, etc.
1910.146	Permit-Required Confined Spaces	Safety requirements regarding procedures for employees working in confined spaces
1910.147	Lockout/Tagout	Protects operators conducting maintenance on machines and systems where the starting up or functioning of machine could harm the employee
1910.120	Hazardous Waste Operations and Emergency Response	Safety requirements involved in the operation and clean-up of hazardous waste materials
1910.23	Guarding Floor and Wall Openings and Holes	Protection requirements for floor and wall openings, including stairs, chutes, hatches, etc.
1910.132	Personal Protective Equipment	Requirements of clothing and equipment to protect the eyes, head, extremities, etc. for necessary work conditions

over 12,000 standards that are in use today, and the institution consists of 30,000 professionals who represent over 150 countries and focuses on many topics and areas that span across 143 technical committees. Many technical committees can be associated with human factors and ergonomics, including Occupational Health and Safety (E34), Pedestrian Walkway Safety and Footwear (F13), and Electrical Protective Equipment for Workers (F18).

Robotic Industries Association (RIA)

RIA is currently the only trade group in North America that exclusively serves the robotics industry (www.robotics.org). In existence since 1974, this institution has a wide range of members ranging from manufacturers, researchers, consultants, and users. RIA provides an extensive resource for all things robotics, drawing resources from recent updates, technical reports, case studies, and interaction with experts. RIA publishes its own safety standard, the *American National Standard for Industrial Robots and Robot Systems–Safety Requirements*, which emphasizes user safety and risk assessment in human-robot interaction. RIA has also developed its own robot integrator certification program, which includes safety training and testing of personnel to ensure integrators are following the industry's best practices.

Society of Automotive Engineers (SAE)

SAE is a standards development organization that focuses on the design of transportation vehicles for ground, air, and naval contexts (www.sae.org/standards). SAE's goals are to provide criteria and procedures that ensure that manufacturers conduct efficient processes and safe products associated with vehicles. The organization forms committees to develop guidelines and publish standards involving vehicles and components in the fields of aerospace, automotive, and commercial vehicles. Examples of topics that SAE standards include are associated with the characteristics of parts and equipment used during vehicle assembly, terminology of vehicle components, and testing methods in assessing the durability and purpose of vehicle components.

Recent efforts include the development of standards associated with the assembly and utilization of electric-powered vehicles. For example, SAE standard J2910 published in 2014 is involved in standardizing the design and testing of hybrid-electric trucks and buses concerning safety factors, as these larger vehicles use high voltage batters and motors that could introduce new hazards absent in smaller vehicles. In the aerospace sector, SAE ARP6467—*Human Factors Minimum Requirements and Recommendations for the Flight Deck Display of Data linked Notices to Airmen (NOTAMs)*—is associated with outlining the requirements for displaying information on portable electrical displays, as these devices are increasingly being used by personnel in the field. SAE attempts to develop safety and efficiency requirements for newly developed technologies in the area of transportation and will continue its efforts as the field advances continually.

United States Access Board

The U.S. Access Board (http://www.access-board.gov/the-board) aids in the regulation of accessibility across the United States by developing guidelines that inform standards for building structures. With the signing of the Architectural Barriers Act (ABA) of 1968, structures built with federal funds required full accessibility by all citizens of the population, including those with limited mobility (http://www.access-board.gov/the-board/laws/architectural-barriers-act-aba). This gave rise to the ABA Standards—a set of standards jointly established by the General Services Administration (GSA), the Department of Defense (DoD), the Department of Housing and Urban Development (HUD), and the U.S. Postal Service (USPS). In 1990, the Americans with Disabilities Act (ADA) was signed, expanding the breadth of accessibility requirements to state and local government facilities, as well as commercial facilities (www.ada.gov). Under this act, the ADA standards were developed and are currently enforced by the Department of Transportation (DOT) and the Department of Justice (DOJ). In addition to the standards mentioned thus far, the U.S. Access Board has guided and aided the development of many more standards. A full list of these standards and guidelines can be found on their website, offering a potential starting point for human factors and ergonomics professionals working with products being used by populations with limited access and mobility.

STANDARDS DATABASES

ASSIST Quick Search

The ASSIST Quick Search is the official database used by the Department of Defense (DoD), with over 111,000 technical documents provided (http://quicksearch.dla.mil). There are more than 33,000 active user accounts able to access documents from ASSIST; the online database provides individuals an opportunity to access standards documents by multiple approaches. Standards can be searched by keywords and terms, focus areas to narrow the search (i.e. filters), and a specified date range to search for documents specified at a particular time. Files in the database are free for access, although some documents can only be ordered if there is no online copy available. Additionally, some documents are controlled and cannot be accessed if the user does not provide the appropriate credentials.

Federal Vehicle Standards

The U.S. General Services Administration (GSA) provides a searchable database called the Federal Vehicles Standards (https://apps.fas.gsa.gov/vehiclestandards/) that gives the user options to choose the publication year, vehicle type, vehicle size, and vehicle weight to provide a standard minimum that the vehicle should comply with. Types of vehicles range from civilian vehicles (i.e. cars, trucks) to service vehicles (i.e. ambulances, wreckers). The minimum can include values for wheelbase, engine type, and fuel type, among many others.

Food and Drug Administration (FDA)

FDA has developed a database that includes recognized consensus standards that is accessible online (http://www.accessdata.fda.gov/scripts/cdrh/cfdocs/cfstandards/search.cfm), consisting of national and international medical device design guidelines. Users can search for guidelines by a number of parameters, including particular SDOs, type of standards, and publication date. This database allows access to the standards of 26 agencies, including the American Dental Association (ADA), ANSI, ASTM International, and ISO.

IEEE Xplore

IEEE Xplore is a web-based digital library that houses over three million documents in the fields of electrical engineering, computer science, and electronics. This coincides with the goals of the Institute of Electrical and Electronics Engineers (IEEE) who published it. The databases not only house technical standards, but also journals, proceedings, and eBooks from IEEE.

National Standards Systems Network (NSSN)

NSSN is an international database of standards made available to the public (www.nssn.org). Consisting of over 30,000 records, this database offers many search options, including the ability to search by developer and type of standard. The list of developers included exceeds 600 organizations and spans the breadth of the globe, making this an excellent international resource. Services offered by this database also include standards tracking and automated reporting, making it much easier for students and professionals in human factors and ergonomics to stay up to date on standards in organizations relevant to their work.

CONCLUSION

The information provided in this chapter is meant to be utilized as a guide to assist human factors and ergonomics professionals in identifying standards that can be used to design, maintain, and evaluate products and systems. This included discussing how standards can cover different aspects associated with design and testing and demonstrating through historical accounts how standardization addressed needs and positively impacted safety and productivity. Standards will continue to be created and changed as technology rapidly advances and increases in complexity. Therefore, the reliance on standards to maintain the production of safe and efficient products will be evident in future cases.

One consideration to know about researching standards for practice is that different institutions may have different standards for safety and development. For example, Grant and Hinze (2014) introduced how the OSHA Code for Fall Regulations (CFR) 29, Part 1926 and Title 8 California Code of Regulations have different thresholds for fall protection in the use of trusses in construction projects. Depending on the context of how human factors professionals are attempting to increase safety and productivity by utilizing standards, identifying what sets of standards apply, and meeting all of them (e.g., both federal and state-level standards) may be required to pass testing and evaluations.

It is the aim of this chapter to provide applicable information in finding standards for multiple purposes relevant to the field of human factors and ergonomics. This is meant to guide researchers and practitioners in identifying standards that can be used for designing and optimizing products and processes. Additionally, this chapter is meant to help human factors professionals that are interested in learning more about the development of national and international standards that currently exist and contribute to the growing knowledge involved in applying human factors knowledge to the real world.

REFERENCES

American National Standards Institute (ANSI). (n.d.). *ANSI—American National Standards Institute.* Retrieved 6 February 2015, from http://www.ansi.org/.

ANSI (2009). *Citation technologies and ANSI partner bring full collection of ISO standards to market on robust citation web-based platform.* Retrieved 5 February 2015, from http://www.ansi.org/news_publications/news_story.aspx?menuid=7&articleid=0fff6315-2e72-46fd-bfd9–49d63fb03d80.

ANSI (2008). *ANSI iPackages: New website helps organizations to share, annotate, and personalize ISO 14000 environmental standards.* Retrieved 5 February 2015, from http://ansi.org/news_publications/news_story.aspx?menuid=7&articleid=c22e7e07–2408-4f33-b92e-cdee3cebf566.

ASTM (n.d.). *The handbook of standardization.* Retrieved 10 February 2015, from http://www.astm.org/GLOBAL/images/Handbook_Eng.pdf.

Breitenberg, M.A. (2009). *The ABC's of standards activities* (NISTIR 7614). Gaithersburg, MD: National Institute of Standards and Technology.

Breitenberg, M.A. (1987). *The ABC's of standards-related activities in the United States* (NBSIR 87-3576). Gaithersburg, MD: U.S. Department of Commerce.

Driskell, J.E., & Salas, E. (1996). *Stress and human performance.* Mahwah, NJ: Lawrence Erlbaum Associates.

Grant, A., & Hinze, J. (2014). Construction worker fatalities related to trusses: An analysis of the OSHA fatality and catastrophic incident database. *Safety Science, 65,* 54–62.

Hemphill, T.A. (2009). Technology standards-setting in the US wireless telecommunications industry: A study of three generations of digital standards development. *Telematics and Informatics, 26,* 103–124.

Hemphill, T.A. (2005). Technology standards development, patent ambush, and US antitrust policy. *Technology in Society, 27,* 55–67.

ISO/IEC (2004). ISO/IEC Guide 2: Standardization and related activities – General vocabulary (8th edition). Retrieved 11 February 2015, from http://isotc.iso.org/livelink/livelink/fetch/2000/2122/4230450/8389141/ISO_IEC_Guide_2_2004_%28Multilingual%29_-_Standardization_and_related_activities_--_General_vocabulary.pdf?nodeid=8387841&vernum=-2.

National Standards Policy Advisory Committee (1978). *National policy on standards for the United States and a recommended implementation plan.* Washington, DC: National Standards Policy Advisory Committee.

O'Hara, J.M., Higgins, J.C., Brown, W.S., Fink, R. (2008). Human factors considerations with respect to emerging technology in nuclear power plants (BNL-79947-2008). Upton, NY: Brookhaven National Laboratory.

OSHA (2014). *All about OSHA.* Retrieved 16 February 2015, from https://www.osha.gov/Publications/all_about_OSHA.pdf.

Priest, H.A., Wilson, K.A., & Salas, E. (2005). National standardization efforts in ergonomics and human factors. In W. Karwowski (Ed.), *Handbook of human factors and ergonomics standards* (pp. 111–131). Mahwah, NJ: LEA, Inc.

Sharit, J. (2006). Human error. In G. Salvendy (Ed.), *Handbook of human factors and ergonomics* (3rd edition, pp. 708–760, New York: John Wiley & Sons.

Veldman, J.R., & Widmann, E.R. (2000). An ontology for standards. Retrieved 6 February 2015, from http://citeseerx.ist.psu.edu/viewdoc/download?doi=10.1.1.23.6732&rep=rep1&type=pdf.

5 Overview of National and International Standards and Guidelines

Carol Stuart-Buttle

CONTENTS

INTRODUCTION

The Human Factors and Ergonomics profession has multiple domains and specialty areas for which there are specific standards, and not all standards can be addressed in this handbook. So, the reader may have to go searching. Likewise, some nations may not have standards at the time of publication but the guidance in this section will assist the reader in finding standards and guidelines in the future.

Within a country or group of countries—such as the European Union (EU)—there are only a few standards that are legislated as mandatory. Typically, at the mandatory level, the standards are in general terms so that they do not go out-of-date. There are numerous non-mandatory ergonomics standards that are desirable to follow as they often provide substantive guidance to those that are mandatory. Standard-setting bodies mostly oversee the development of non-mandatory standards and guidelines and develop them by professional consensus or experts in the topic area. Such non-mandatory documents are guidelines, although they may be entitled to either standards or guidelines. On occasion, for example, in the United States (U.S.), a non-mandatory standard may be cited and enforced by legislation.

A specialty group or individuals from a specific practice domain may be involved in developing a standard that becomes widely adopted by users in other areas. Such standards can be hard to find by someone not affiliated with the standard developing group. An effective approach to seek specific standards is to find related societies or associations through which to make inquiries.

Many countries adopt international standards and European standards are enfolded into most European Union members' laws. There is a rising need for standardization as business becomes more global. Therefore, standards are always being developed or updated, so they should be periodically checked to ensure they are current.

SCOPE

This chapter provides general guidance to finding international, European, and national standards and guidelines on human factors/ergonomics. Also, specific guidance is given to the primary standards of the United States (except military standards that are discussed in another chapter of this handbook), Canada, the United Kingdom, Australia, New Zealand, and Japan. A chapter in this handbook discusses standards in China. Some of the more widely known ergonomics standards and guidelines are introduced and resources are shared for the main standard developing bodies. Details of many standards mentioned in this chapter are discussed in this handbook.

SEARCHING FOR STANDARDS

There are several groups that provide European and international standards. In addition, European countries, and many other countries, each have their own standards developing group. Individual country standardization companies can be found through central groups such as the European Committee for Standardization (CEN—www.cen.eu). The individual standardization committees or corporations of each country usually serve as the focal point of contact in obtaining international standards. The following groups are resources for obtaining European and international standards and, in some cases, standards of individual countries. There is a fee for most standards, especially those that are non-mandatory.

PERINORM (WWW.PERINORM.COM)

Perinorm provides a subscriber-based service that offers a database of international, European, and national standards. Standards of 23 countries are available—apart from European countries—including the United States, Japan, Australia, Turkey, and South Africa. The information offered is available in English, French, and German.

INTERNATIONAL ORGANIZATION FOR STANDARDIZATION (ISO) (WWW.ISO.ORG)

ISO has a membership of 164 national standards bodies. International, voluntary, consensus-based ISO standards are developed through the member bodies. The more than 22,878 ISO standards are organized into 97 general topic areas (international classification for standards (ICS)). There are also listings of standards by the Technical Committee (and some Product Committees), which totals to 325. The TC 159 is the Ergonomics Technical Committee that has four subcommittees: General Ergonomics Principles, Anthropometry and Biomechanics, Ergonomics of Human-System Interaction, and Ergonomics of the Physical Environment. Human factors and ergonomics ISO standards are discussed in detail in this handbook. ISO standards can be purchased from a national member body of ISO or from the ISO store. The national members are listed at www.iso.org, under standards.

INTERNATIONAL ERGONOMICS ASSOCIATION (IEA) (WWW.IEA.CC)

The IEA is another possible resource in finding standards for a country. The association is a coalition of 53 Federated Societies and some Affiliated Societies and organizations. Each listed Society has a point of contact in the member country. This can be another avenue in finding the standards and guidelines of that country.

OTHER INTERNATIONAL STANDARDS GROUPS

There are other international groups that have worked to standardize pertinent aspects of a group's area of interest. A few additional groups are listed below as resources.

International Telecommunication Union (ITU)	www.itu.int
International Civil Aviation Organization (ICAO)	www.icao.int
World Wide Web Consortium (W3C)	www.w3.org

Additional comprehensive resources for links to numerous standards are: standards.nasa.gov

National Aeronautics and Space Administration (NASA) Technical Standards Program (NASA requires the public to register to log on to the standards page).

www.faa.gov/regulations_policies/handbooks_manuals/

Federal Aviation Administration (FAA) website leads to many useful resources when human factors/ergonomics are entered in the site search.

EUROPEAN STANDARDS

European standards are addressed in detail in this handbook. The following websites are sources for European standards and related information.

European Union (EU) (www.europa.eu)

This website provides background on the European Union, which is an economic and political union of 28 countries. The EU has also evolved to include policy areas such as climate, environment, and health. Policy activity entails the development of directives that provide a minimum standard for health and safety at work. Directives are written at a general level and each country of the EU develops the details of how they choose to comply with the directive.

CEN (www.cen.eu)

The European Committee for Standardization (CEN) develops European standards. The Technical Board oversees the Business Operations Support Team (boss.cen.eu) that guides the development of standards in many areas with approximately 400 Technical Committees and Working Groups. The standards are often technical ones within an industry or product but may be of interest to some in the human factors ergonomics field. Through the CEN web page, there are links to each of the 34 member country's main standards-setting groups. These groups are not the legislative groups of the countries, but they are responsible for implementing the European Standards as national standards.

OSHA/EU Cooperation (www.useuosh.org)

The Occupational Safety and Health Administration (OSHA) of the United States and the European Union (EU) have a joint web page to facilitate communication and sharing of information. The site serves to access either European or U.S. agencies for safety and health at work and their respective legislation.

Other European Standards Organizations

Some additional European standards-setting organizations, other than the main ones related to each country, are:

www.cenelec.org
European Committee for Electrotechnical Standardization (CENELEC)
www.etsi.org
European Telecommunications Standards Organization (ETSI)

CANADIAN STANDARDS

BRITISH COLUMBIA (BC) (WWW.WORKSAFEBC.COM)

The Workers' Compensation Board of British Columbia (WorkSafeBC) issued the Occupational Health and Safety Regulation (OHSR) with legal requirements that must be met by all workplaces under the jurisdiction of WorkSafeBC. There are 33 parts to the OHSR; Ergonomics is listed under Part 4: General Conditions with its own subheading "Ergonomics Musculoskeletal Injury (MSI)

Requirements." Sections 4.46 to 4.53 require employers to identify factors that might expose workers to the risk of MSI, assess the identified risks, and eliminate or minimize these. Employees are to receive education and training and be consulted by the employers. The requirements also include evaluations of effectiveness.

There are other possibly relevant sections, such as 7: Noise, Vibration, Radiation and Temperature; 8: Personal Protective Clothing and Equipment; and 11: Fall Protection.

ONTARIO (ON) (WWW.LABOUR.GOV.ON.CA)

The Ministry of Labour, Training, and Skills Development under the Government of Ontario oversee the implementation of the Occupational Health and Safety Act (OHSA) of 1990. All employers must have a health and safety policy and program, and the Officers of corporations have a direct responsibility to take every precaution reasonable to protect workers from hazards. A joint labor and management Health and Safety Committee, which is responsible for health and safety in the workplace, is required in facilities with greater than 20 workers. The committee is to meet regularly to discuss health and safety concerns, review progress, and make recommendations. Workplaces of less than 20 are required to have a Health and Safety Representative. By 1995, employers had to certify that the members of their joint Health and Safety Committees were properly trained. Ergonomics is covered by the general duty clause of the OHSA. MSDs and material handling are required topics in the education regulation of the OHSA. Some of the industry-specific regulations—for example, for Farming, Film and Television, Fisheries, and Industrial Services—also call out specific hazards such as MSDs and material handling in the regulations.

CANADIAN STANDARDS ASSOCIATION GROUP (CSA GROUP) (WWW.CSAGROUP.ORG)

The CSA Group is the Canadian source for ISO standards. It also produces voluntary standards pertaining to many areas, one of which is in Occupational Health and Safety. There are two main voluntary standards relating specifically to human factors/ergonomics, one of which is CSA Z1004-12 *Workplace Ergonomics*—a management and implementation standard that "describes specific requirements and provides guidance for the systematic application of ergonomics principles to the development, design, use, management, and improvement of work systems." The other standard is CSA-Z412-17 *Office Ergonomics*—an application standard for workplace ergonomics. This standard complements the international standard CAN/CSA-ISO 9241 (with multiple parts) *Ergonomics of Human-System Interaction*.

UNITED STATES STANDARDS

OCCUPATIONAL SAFETY AND HEALTH (OSH) ACT (WWW.OSHA.GOV)

The primary mandatory standard of the U.S. is the Occupational Safety and Health Act of 1970. The section pertinent to ergonomics is the general duty clause, Section 5 (a) (1), that states:

> "Each employer shall furnish to each of his employees employment and a place of employment which is free from recognized hazards that are causing or are likely to cause death or serious harm to his employees."

Citations for ergonomics have been under the general duty clause. There are additional laws under the OSH Act that are relevant to safety and design, such as personal protective equipment, egress, working surfaces, noise, and machine guarding.

AMERICANS WITH DISABILITIES ACT (ADA) (WWW.ADA.GOV)

The mandatory act originated in 1991. The latest version includes several updates and amendments published in the Federal Register on December 2, 2016. It has some bearing on ergonomics. One part (ADA Title III) of the act addresses accessibility for the disabled and another part (ADA Title II) pertains to employment. Two main points of the act under the employment section are:

- The ADA prohibits disability-based discrimination in hiring practices and working conditions.
- Employers are obligated to make reasonable accommodations to qualified disabled applicants and workers unless doing so would impose an undue hardship on the employer. The accommodations should allow the employee to perform the essential functions of the job.

Often, modifications to a job to accommodate someone disabled benefit all workers. Defining the essential functions of a job may involve those responsible for ergonomics.

FEDERAL DRUG ADMINISTRATION (FDA) (WWW.FDA.GOV)

Medical devices have been regulated since 1976, with several amendments to the original law. However, in 1990, the Safe Medical Devices Act (SMDA) was passed to give the FDA authority to ensure proper device design. In 1996, the FDA revised *Good Manufacturing Practices (GMP) Act* to include a New Quality System Regulation that requires manufacturers to: verify device design; ensure that design requirements are met; be responsible for implementation; and ensure a device meets the needs of users and patients and ensure the design output meets design input requirements. The FDA describes the relevance of human factors in medical device design (https://www.fda.gov/medical-devices/human-factors-and-medical-devices/human-factors-implications-new-gmp-rule-overall-requirements-new-quality-system-regulation).

In 1998, the FDA issued a guidance document *Overview of FDA Modernization Act of 1997, Medical Device Provisions* (www.fda.gov/MedicalDevices/DeviceRegulationandGuidance/GuidanceDocuments/ucm094526.htm) that clarifies premarket testing of the safe use of a medical device and FDA approval. Since the law was promulgated, guidance on how to comply with the law has grown more explicit to include applying human factors and usability methodology to medical devices to optimize safety and effectiveness in design. The FDA website specifies the human factors involved in design input, design verification, and design validation. There are extensive resources with links to human factors related to medical devices and the importance of the role of the human factor. (www.fda.gov/MedicalDevices/DeviceRegulationandGuidance/HumanFactors/ucm124829.htm). The FDA has published several guides on the human factors data they expect on premarket studies and the administration offers many resources to help with compliance (www.fda.gov/MedicalDevices/DeviceRegulationandGuidance/default.htm).

ASSOCIATION FOR ADVANCEMENT IN MEDICAL INSTRUMENTATION (WWW.AAMI.ORG)

AAMI develops consensus voluntary standards and conformity assessment activities related to medical devices. The group works closely with other national and international standards-setting bodies. Since the FDA Modernization Act (noted above under FDA), AAMI has developed a section of their website specifically to AAMI Human Factors Standards because of the growing interest in the topic (www.aami.org/hfconnect/). The active AAMI Human Factors Committee worked to develop an American National Standards Institute (ANSI) standard, ANSI/AAMI HE75, 2009/(R)2013 *Human Factors Engineering–Design of Medical Devices* that complements the earlier document ANSI/AAMI/IEC 62366: 2007 *Medical Devices–Application of Usability*

Engineering to Medical Devices. The ANSI/AAMI HE75 is a comprehensive, practical document that is a useful resource to human factors/ergonomics professionals outside of medical device application. Additional practical documents are available at the AAMI website. AAMI also conducts training courses on human factor activities related to medical devices.

CALIFORNIA ERGONOMICS STANDARD (WWW.DIR.CA.GOV)

The California State Ergonomics Standard (title8/5110) is a mandatory state that became effective in 1997 and addressed formally diagnosed work-related repetitive motion injuries that have occurred to more than one employee. The employer must implement a program to minimize repetitive motion injuries through worksite evaluation, control of the exposures, and training. The repetitive motion injury law is often referred to as the ergonomics standard; however, Title 8 Worker Safety and Health Standards addresses many relevant safety standards for safety and design. The repetitive motion injury component of Title 8 is found under §5110, listed as Ergonomics (www.dir.ca.gov/title8/5110.html).

WASHINGTON STATE—FORMER ERGONOMICS STANDARD (WWW.LNI.WA.GOV)

In 2000, Washington State adopted a mandatory ergonomics rule that was repealed in November 2003. The rule required employers to look at their jobs to determine if there are specific risk factors that make a job a "caution zone job," as defined by the standard. All caution zone jobs must be analyzed; employees of those jobs are to participate and be educated, and the identified hazards must be reduced.

Although the rule is no longer in effect, much of the ergonomics information of the rule—such as evaluation tools—remains on the Washington State Labor and Industries website as resources under "Safety & Health," "Create a Safety Program," and "Sprains and Strains" (https://lni.wa.gov/safety-health/preventing-injuries-illnesses/sprains-strains/). Under Sprains and Strains, there are sections for Ergonomics Process, Evaluation Tools, and Solutions for Sprains and Strains. Several Ergonomics guidelines by industry are also under solutions—for example, agriculture, construction, food processing, healthcare, and a variety of manufacturing industries.

STATE LAWS ON SAFE PATIENT HANDLING (WWW.ASPHP.ORG)

The Association of Safe Patient Handling Professionals (ASPHP) provides legislative updates (https://www.asphp.org/wp-content/uploads/2011/05/SPH-US-Enacted-Legislation-02222015.pdf). There is no federal law on safe patient handling; however, from 2003–2015, 12 states have passed laws, regulations, or resolutions related to safe patient handling, with ten of those states requiring a comprehensive program in healthcare facilities. The states with safe patient handling legislation are California, Hawaii, Illinois, Maryland, Minnesota, Missouri, New York, Ohio, Rhode Island, Texas, and Washington.

OSHA STANDARDS (WWW.OSHA.GOV)

The Occupational Safety and Health Administration (OSHA) of the federal government released an Ergonomics Program rule in November 2000 that was repealed in March 2001 by a new administration. Federal citations for ergonomics are based on the General Duty Clause of The 1970 Occupational Safety and Health Act (see above).

OSHA is a resource for guidelines, e-tools, and fact sheets, as well as training.

The primary Guidelines are (www.osha.gov/SLTC/ergonomics/):

- Ergonomics Program Management Guidelines for Meatpacking Plants
- Solutions for the Prevention of Musculoskeletal Injuries in Foundries
- Guidelines for Nursing Homes: Ergonomics Prevention of Musculoskeletal Disorders
- Guidelines for Shipyards: Ergonomics Prevention of Musculoskeletal Disorders
- Guidelines for Retail Grocery Stores: Ergonomics Prevention of Musculoskeletal Disorders
- Prevention of Musculoskeletal Injuries in Poultry Processing

Other resources include e-tools and fact sheets, such as Baggage Handling (Airline Industry) eTool, Beverage Delivery eTool, and Computer Workstations eTool (www.osha.gov/SLTC/ergonomics/controlhazards.html). OSHA lists resources from the National Institute of Occupational Safety and Health (NIOSH) and Trade Associations, covering many types of work areas. There are links to help with the process of identifying problems in generating solutions, as well as providing training and seeking assistance.

ANSI Standards (www.ansi.org)

The American National Standards Institute (ANSI) is a voluntary consensus standard body that annually issues standards. The following are just a few of the main ergonomics standards that exist or are being developed; there are many others that relate to safety and ergonomics/human factors. Information about ANSI documents may be acquired through ANSI and many standards from other groups—including the International Standards Organization (ISO), the International Electrotechnical Commission (IEC), and the American Society for Testing and Materials (ASTM)—can be purchased from the ANSI website. Often, ANSI standards are purchased directly from the group (or publisher) responsible for developing the standard in coordination with ANSI. Although voluntary, at times, ANSI standards are cited in OSHA regulations.

There are laws—such as the National Technology Transfer and Advancement Act (Public Law 104-113) of 1996—that say all federal agencies and departments shall use technical standards published by voluntary consensus standards bodies. Exceptions to this are allowed but must be justified. There is also a movement in the government to stray from military standards and adopt private sector standards as much as possible.

The following ANSI standards are some of the general ergonomics ones that are less industry-specific.

    a. *Human Factors Engineering of Computer Workstations*–ANSI/HFES 100-2007 (www.hfes.org)

    The document provides information on ergonomics of computer workstations, including workstation design specifications for the anthropometric range of the 5th–95th percentile and four reference positions: reclined, upright, and declined sitting and standing. Several input devices are covered and there is a section on how to integrate all the workstation components for an effective system.

    b. *Occupational Health Safety Management Systems*—ANSI/AIHA/ASSE Z10-2012 (www.aiha.org, www.assp.org)

    The ANSI secretariat was shared between the American Industrial Hygiene Association (AIHA) and the American Society of Safety Engineers (ASSE) (now ASSP—American Society of Safety Professionals). The standard addresses management principles and systems to allow organizations to design and implement approaches that improve occupational safety and health. An *Occupational Safety and Health Design Package* includes both Z10-2012 and ANSI/ASSE Z590.3 -2011—the latter providing design guidelines for identifying and ameliorating occupational risks during the process of designing and redesigning.

    c. *Human Factors Engineering of Software User Interfaces*—ANSI/HFES 200-2008 (www.hfes.org)

This is a five-part standard developed by The Human Factors and Ergonomics Society (HFES) under the auspices of ANSI. It closely mirrors the ISO 9241 standard on visual display terminals, except for original parts that are on color, accessibility, and voice input/output.

d. Under the product design of the ANSI web page, there is a section on Ergonomics with links to many ergonomics-related standards (http://webstore.ansi.org/ergonomics/Default.aspx). Some of these ergonomics standards are on office, vehicle, machine control center, software, accessibility, handheld tools, and human-system interaction. Not all these standards are by ANSI; many of these links are from ISO or European standards.

ACGIH TLVs® (WWW.ACGIH.ORG)

The American Conference of Governmental Industrial Hygienists (ACGIH) develop and annually publish Threshold Limit Values (TLVs®) for chemical substances and physical agents. There are TLVs® for hand-arm vibration and whole-body vibration as well as for thermal stress. Two TLVs® added around 2001 to the annual TLV® document are for Hand Activity Level, which is intended for monotask jobs performed for four hours or more, and a Lifting TLV®. The lifting threshold is provided as three tables with weight limits based on the frequency and duration of lift and the horizontal distance and height at the start of the lift. Trunk twisting is not included in the tables.

NATIONAL INSTITUTE OF STANDARDS AND TECHNOLOGY (NIST) (WWW.NIST.GOV)

NIST helps develop standards that address measurement accuracy, documentary methods, conformity assessment and accreditation, and information technology standards. NIST has developed industry usability reporting guidelines that directly affect software ergonomics. The institute is also developing human factors guidelines to prevent disparities related to the adoption of electronic health records. NIST has many laboratories to develop technology and performance metrics, as well as standards to strengthen the U.S. economy. It hosts the Malcolm Baldrige National Quality Award, for which many organizations strive. The challenge is achieving excellence through the Baldrige management framework. The framework is broad and encompasses issues related to macroergonomics (www.nist.gov/baldrige/baldrige-award).

Standards.gov (www.nist.gov/standardsgov/about-standardsgov) is administered by the Standards Coordination Office at NIST. The purpose is to provide information about standards for conformity within the federal government; useful background and resources on standards for the industry and academia; and standards and conformity information to the public.

MISCELLANEOUS STANDARDS SETTING GROUPS

There are many other sources of standards that may be important to certain domains or specialties. A few of the other organizations that develop standards are:

American Society of Mechanical Engineers (ASME)	www.asme.org
American Society for Testing and Materials (ASTM)	www.astm.org
Institute of Electrical and Electronics Engineers (IEEE)	www.ieee.org
Society of Automotive Engineers (SAE)	www.sae.org

UNITED KINGDOM STANDARDS

MANDATORY REGULATIONS

Health and Safety Executive (www.hse.gov.uk)

The United Kingdom (U.K.) has a statutory Health and Safety at Work Act (1974) under which are regulations that are laws (www.hse.gov.uk/legislation/statinstruments.htm). These regulations are general but are interpreted in the "Approved Codes of Practice" documents developed by the Health and Safety Executive (HSE). Approved Codes of Practice have special legal status and can be used in prosecutions. Guidance documents interpret the law and provide further detail for compliance that can be especially useful, although they are not legally binding.

As with most member countries of the European Commission, some of the U.K. laws originate with the proposals from the EC. There are several useful regulations that are now readily available through a comprehensive list of topics that provide guidance and resources. The original 1992 set of regulations known as the "Six Pack" apply across all industries. They are Manual Handling Operations, Display Screen Equipment, Workplace (Health Safety & Welfare), Provision and Use of Work Equipment, Personal Protective Equipment at Work, and Management of Health & Safety at Work. These topics and updates can be found under the topics list (http://www.hse.gov.uk/guidance/topics.htm). The HSE has reorganized its information to include a section on Human Factors and Ergonomics (http://www.hse.gov.uk/humanfactors/index.htm) that guides the user to regulations and resources that is a more comprehensive view of the workplace—including topics such as fatigue and shiftwork, and organizational change.

NONMANDATORY STANDARDS

British Standards Institute (BSI) (www.bsigroup.com)

BSI is the primary standards-setting body in the U.K. and therefore, also works with the International Standards Organization (ISO). Some earlier British Standards have developed into ISO standards. ISO standards can be purchased through BSI. Two notable general publications at BSI are ISO 45001: *Occupational Health and Safety Management* and ISO 9001 *Quality Management*. Both standards are globally used. BSI is also a resource for many other standards with a search list by industry sectors. Training in the standards is also offered.

AUSTRALIAN STANDARDS

The Commonwealth Government has federal jurisdiction; in addition, there are six states and two mainland territories that each have their own laws. The approach overall is similar to the European model that relies on a general duty of care by employers and employees. The focus is risk management that is based on risk assessment and control. Most of the states and territories have their own general health and safety acts but they vary in the degree of development of their regulations and codes of practice. Since the 1980s, there has been an attempt to harmonize the laws in the jurisdictions, with limited success. In 2008, the Safe Work Australia Act established SafeWork Australia as an independent government statutory agency to lead the national policy development for work safety and health and workers' compensation with the intent to harmonize laws and regulations across jurisdictions.

SAFEWORK AUSTRALIA (WWW.SAFEWORKAUSTRALIA.GOV.AU)

The national agency SafeWork Australia (SWA) develops models of a Worker Health and Safety Act and accompanying regulations, as well as model Codes of Practice, working with the jurisdictions. Each jurisdiction either harmonizes with the models or produces a variation to enact as law. SafeWork Australia does not regulate and enforce work health and safety laws since the Commonwealth states and territories retain those responsibilities.

Since the federal restructure to SafeWork Australia, most jurisdictions have legislated the Worker Health and Safety Act based on the SafeWork Australia model. A few jurisdictions have passed some regulations based on the models. SafeWork Australia will only consider developing Model Regulations and Codes of Practice that are already widely adopted by jurisdictions. With the similarities of the laws through harmonization, SafeWork Australia is the central source for Australian Work Health and Safety laws—with links to each jurisdiction—and is a resource for the laws in each area.

A *Model Work Health and Safety Regulations* was updated in 2019 and expands on the Model Worker Health and Safety Act. There are also Model Guidance materials for the regulations (https://www.safeworkaustralia.gov.au/doc/model-work-health-and-safety-regulations). Hazardous manual tasks are included in the model regulation.

Codes of Practice go into further detail than regulations. One example is the Model Code of Practice for *Hazardous Manual Tasks* (2018) that provides definitions of terms, risk assessment approaches, managing risks of musculoskeletal disorders, and illustrations to identify risks and interventions (www.safeworkaustralia.gov.au/category/ergonomics). This particular Code of Practice can also be found under the section "Ergonomics." Searching on the term Human Factors brings up additional useful information, although not organized in a section of its own or with Ergonomics.

In Australia, each jurisdiction oversees and regulates the laws as well as local worker's compensation. Comcare (www.comcare.gov.au) oversees national, Commonwealth regulation and national workers' compensation for national employers.

JURISDICTIONS

Each jurisdiction clarifies the rules that are in effect as well as any updates. The jurisdictions can be accessed from WorkSafe Australia (www.safeworkaustralia.gov.au/). However, below is a list of each jurisdiction and their websites. Each website provides considerable guidance material for workers and businesses in their jurisdiction.

Australian Capital Territory: WorkSafe ACT	www.accesscanberra.act.gov.au/app/home/workhealthandsafety/worksafeact
New South Wales: SafeWork NSW	www.safework.nsw.gov.au
Northern Territory: NT WorkSafe	www.worksafe.nt.gov.au
Queensland: Workplace Health and Safety	www.worksafe.qld.gov.au
South Australia: SafeWork SA	www.safework.sa.gov.au
Tasmania: WorkSafe Tasmania	www.worksafe.tas.gov.au
Victoria: WorkSafe Victoria	www.worksafe.vic.gov.au
Western Australia: WorkSafe WA	www.commerce.wa.gov.au/worksafe

STANDARDS AUSTRALIA (WWW.STANDARDS.ORG.AU)

Standards Australia is an organization that is recognized as the main source for developing non-mandatory technical and business standards and the dissemination of Australian and international standards. Their principle is to adopt or closely align their standards with international standards whenever possible.

NEW ZEALAND

NEW ZEALAND—WORKSAFE (WWW.WORKSAFE.GOVT.NZ)

WorkSafe New Zealand is the interface between the government and business on safety and health. The structure of health and safety laws, regulations, and codes of practice are similar to those of Australia. In New Zealand, the Health and Safety at Work Act 2015 went into effect in April 2016. Regulations accompanying the Act include General Risk and Workplace Management; Worker Engagement, Participation, and Representation; Major Hazard Facilities; Asbestos; Adventure Activities; Mining Operations and Quarrying Operations; Petroleum Exploration and Extraction; Rates of Funding Levy; and Infringement Offences and Fees. The website is also a source of practical information with a whole section on Managing Health and Safety.

NEW ZEALAND—STANDARDS (WWW.STANDARDS.GOVT.NZ)

Standards New Zealand is a business unit within the government Ministry of Business, Innovation, and Employment. The entity develops standards as well as publishes and sells New Zealand and joint New Zealand-Australia standards. Similar to Standards Australia, they base standards on international standards as much as possible. Standards New Zealand is also the resource to buy international standards.

JAPANESE STANDARDS

Japan has a general Labour Standards Act, revised in 2010, that states employers should take measures to ensure reasonable working conditions and improve these. Additional laws supplement the general Labour Standards Act, including the Industrial Safety and Health Law. There is a national system that ensures the law is followed by giving guidance as well as thorough inspection on a regional basis.

MINISTRY OF HEALTH, LABOUR AND WELFARE (WWW.MHLW.GO.JP)

The national ministry of Health Labour and Welfare of Japan oversees the Industrial Safety and Health Act. In 2018, the ministry issued the 13th Occupational Safety and Health Program—which is similar to a five-year strategic plan to improve safety and health. The Labor Standards Bureau of the ministry supervises and guides industry to meet the Industrial Safety and Health Act and other related work laws, as well as manages the workers' compensation insurance. The offices of the Bureau are located in each of the 47 prefectures of Japan and there are at least 321 Labor Standard Inspection offices (as reported in 2013). An employer who hires more than ten people must have formal work rules of employment to prevent occupational disease and injury.

JAPAN INDUSTRIAL SAFETY AND HEALTH ASSOCIATION (JISHA) (WWW.JISHA.OR.JP)

JISHA provides services to industries to comply with the Industrial Safety and Health Law by helping them establish an occupational safety and health management system. They help with the promotion of physical and mental health and OSHA education. JISHA offers materials analytical services, on-site measurement services, and consultation. JISHA is also the source of the laws and ordinances used to be housed by the Japanese International Center for Occupational Safety and Health (JICOSH), which closed in 2008. Although the information can be accessed from JISHA, the information has not been updated since 2008.

National Institute for Occupational Safety and Health, Japan (JNIOSH) (www.jniosh.johas.go.jp)

JNIOSH was formed in 2006 from the merger of the National Institute of Industrial Health and the National Institute of Industrial Safety. In 2016, JNIOSH and the Japan Organization of Occupational Health and Safety (JOHAS) amalgamated. The focus of JNIOSH remains the same—conduct research in the areas of industrial safety and occupational health to ensure the health and safety of workers. The site hosts the center's journal, *Industrial Health,* and some other publications, as well as events.

There are many research centers and groups at JNIOSH, such as the Occupational Ergonomics Research Group, Occupational Stress Research Group, Research Center for Overwork-Related Disorders, and Center for Risk Management Research. The centers and groups are on the website, with a list of directors involved in each of the initiatives.

Japanese Standards Association Group (JSA) (www.jsa.or.jp)

The JSA Group is the main resource for purchasing voluntary standards. The group is involved in four main areas:

1. Developing national voluntary Japanese Industrial Standards (JIS) often in collaboration with related organizations. These JIS are numerous and are mostly very technical.
2. Management Techniques. This includes the application and practical aspects of quality engineering (Taguchi Methods), data analysis methods, and techniques addressing reliability and motivation.
3. Standardization. JSA conducts surveys and research into standardization, such as: graphical symbols; networking and software applications; management systems; and consumer protection. They are involved with ISO Technical Committees for standard development. JSA is the national source for ISO standards.
4. Conformity Assessment. JSA established national committees to deliberate on the technical issues related to draft international standards to ensure that Japan meets the obligations of a Trade Agreement.

Other offerings of the JSA Group include education and training and various certifications.

WEB PAGE REFERENCES

www.acgih.org	American Conference of Governmental Industrial Hygienists (ACGIH), U.S.
www.aiha.org	American Industrial Hygiene Association (AIHA), U.S.
www.ansi.org	American National Standards Institute (ANSI), U.S.
www.asme.org	American Society of Mechanical Engineers (ASME), U.S.
www.assp.org	American Society of Safety Professionals (formerly of Safety Engineers) (ASSP), U.S.
www.astm.org	American Society for Testing and Materials (ASTM), U.S.
www.ada.gov	Americans with Disabilities Act (ADA), U.S.
www.aami.org	Association for Advancement in Medical Instrumentation (AAMI), U.S.
www.asphp.org	Association of Safe Patient Handling Professionals (ASPHP) U.S.
www.accesscanberra.act.gov.au	Australian Capital Territory (ACT)
www.worksafebc.com	British Columbia (BC) Government, Canada
www.bsigroup.com	British Standards Institute (BSI), U.K.
www.dir.ca.gov	California Ergonomics Standard, U.S.
www.csagroup.ca	Canadian Standards Association, Canada

www.comcare.gov.au	Comcare, Australia
www.cenelec.org	European Committee for Electrotechnical Standardization (CENELEC)
www.cen.eu	European Committee for Standardization (CEN)
www.etsi.org	European Telecommunications Standards Organization (ETSI)
www.europa.eu	European Union (EU)
www.faa.gov	Federal Aviation Administration (FAA), U.S.
www.fda.gov	Federal Drug Administration (FDA), U.S.
www.hse.gov.uk	Health and Safety Executive (HSE), U.K.
www.hfes.org	Human Factors and Ergonomics Society (HFES), U.S.
www.icao.int	International Civil Aviation Organization (ICAO)
www.iea.cc	International Ergonomics Association (IEA)
www.iso.org	International Organization for Standardization (ISO)
www.itu.int	International Telecommunication Union (ITU)
www.ieee.org	Institute of Electrical and Electronics Engineers (IEEE), U.S.
www.jisha.or.jp	Japan Industrial Safety and Health Association (JISHA)
www.jsa.or.jp	Japanese Standards Association Group (JSA)
www.mhlw.go.jp	Ministry of Health, Labour and Welfare, Japan
standards.nasa.gov	National Aeronautics and Space Administration (NASA) Technical Standards Program, U.S.
www.jniosh.johas.go.jp	National Institute of Occupational Safety and Health (JNIOSH), Japan
www.nist.gov	National Institute of Standards and Technology (NIST), U.S.
www.safework.nsw.gov.au	New South Wales: SafeWork NSW, Australia
www.worksafe.govt.nz	New Zealand: WorkSafe New Zealand
www.worksafe.nt.gov.au	Northern Territory: NT WorkSafe, Australia
www.osha.gov	Occupational Safety and Health Administration (OSHA), U.S.
www.useuosh.org	Occupational Safety and Health Administration (U.S.) and European Union (EU) Cooperation
www.labour.gov.on.ca	Ontario Government, Canada
www.perinorm.com	Perinorm; private database of international, European, and national standards
www.worksafe.qld.gov.au	Queensland: Workplace Health and Safety, Australia
www.safeworkaustralia.gov.au	SafeWork Australia (SWA)
www.sae.org	Society of Automotive Engineers (SAE International), U.S.
www.safework.sa.gov.au	South Australia: SafeWork SA, Australia
www.standards.org.au	Standards Australia
www.standards.govt.nz	Standards New Zealand
www.worksafe.tas.gov.au	Tasmania: WorkSafe Tasmania, Australia
www.worksafe.vic.gov.au	Victoria: WorkSafe Victoria, Australia
www.lni.wa.gov	Washington State—Department of Labor and Industries, U.S.
www.commerce.wa.gov.au/ worksafe	Western Australia: WorkSafe WA
www.worksafebc.com	Workers' Compensation Board of British Columbia, Canada
www.w3.org	World Wide Web Consortium (W3C)

6 Balancing Stakeholder Representation

An Example of Stakeholder Involvement in Ergonomics Standardization

H. Willemse, Henk J. de Vries, and Jan Dul

CONTENTS

INTRODUCTION

During the last 50 years, ergonomics has developed rapidly as an independent science and profession with its own knowledge, methods, and networks. Scientists have published their expertise in increasing numbers of ergonomics journals. However, much of the ergonomic knowledge has not yet reached engineers and managers. Engineers could use the large number of ergonomics standards that have become available during the last two decades to integrate ergonomics knowledge in the design process. But somehow, most of these standards have not yet found their way there.

This chapter states that an important reason for the limited use of ergonomics standards is insufficient stakeholder involvement in standardization. Because one of the main principles in ergonomics is user involvement, intended users should be involved in the process of designing an environment to assure a fit between this environment and its user. In the development of ergonomics standards, however, this principle seems to be ignored. Ergonomics standardization is considered to be the concern of "professional ergonomists" only. Lack of involvement of users and other stakeholders may result in standards that fail to meet the users' needs.

CASE STUDY

The example of stakeholder participation in ergonomics standardization presented here is based on a case study performed at the Dutch national standardization body, NEN. The subject of the case study is the revision of the Dutch national standard, NEN 1824:1995, on the requirements for the space of office workplaces. The composition of the working group that revised the standard is analyzed to find out whether the working group was a well-balanced representation of all the stakeholders. An analysis will be made about which stakeholders have and have not been involved, and their reasons. This will offer the possibility to answer the question: "Was the established working group for the revision of NEN 1824:1995 a well-balanced representation of all of the stakeholders?"

In the following sections, this question is answered. First, a brief description of the development of the standard and its content is given. This is followed by a description of the methods that have been used to analyze the (im)balance of stakeholder representation.

The subsequent section describes the results of the stakeholder analysis: which stakeholders were actively involved, and who were not. This answers the research question of the case study. In addition, the situation concerning ergonomics standardization on Dutch national and international levels—in general—is considered to provide a first impulse to a discussion concerning stakeholder participation in ergonomics standardization.

HISTORY OF THE STANDARD

The standard presented in this case study is the Dutch national standard, NEN 1824, on the requirements for the space of office workplaces. The first version of this standard was published in 1990 and was developed at the request of the Dutch Department of Social Affairs and Employment as part of a policy to stimulate the improvement of working conditions in the Netherlands. The first version's title was NEN 1824:1990 "Ergonomics–Ergonomic Recommendations for the Dimensions of Office Rooms." It described the size of the floor area that was preferred from an ergonomic point of view.

In 1995, this standard was revised to reflect amendments in the Dutch Labor Law and other legislation. For this version, ergonomics was no longer used as the starting point—meaning, the standard would no longer classify as an ergonomics standard. The new version solely described the minimal requirements for the width, height, and amount of floor area of office workplaces. Its title was NEN 1824:1995 "Requirements for the Space and Height of Office Workplaces." The 1995 version confronted the users with difficulties concerning the interpretation of the standard for application, especially in innovative office concepts. The standard left too much room for disparity in interpretation, resulting in a lot of questions and even disputes between the Dutch Labor Inspectorate and facility managers on the appropriateness for, in most cases, call-center workplaces.

In 1999, it was decided that the standard should be revised again to consider not only the traditional office rooms but also other workstations like call-center workplaces.

At the beginning of the year 2000, the committee that coordinates the Dutch standardization activities on the field of ergonomics composed a working group. The committee selected a group of members that, in their opinion, had to be involved in the revision process. This working group commenced by bringing up additional working group members, and when it was decided that the working group contained sufficient representation, the revision of NEN 1824:1995 started. The total process, including the two startup meetings, took seven meetings within a period of 15 months.

In 2001, the revision was published as NEN 1824:2001 "Ergonomics–Ergonomic Requirements for the Space of Office Workplaces." This revised version consists of a total of ten pages, where the main content lists the factors that must be kept in mind when equipping an office. A calculation method is given for the minimum floor space, taking into account the employee, the desk, the drawers, and other factors. For the office desk, a distinction is made between a workplace with a flat monitor and one with a regular CRT monitor. In addition, minimum dimensions for the circulation space and dimensions for visual separation of workplaces are given. Furthermore, physical factors of workplaces—like the indoor climate, ventilation, and daylight—are discussed. In three appendices, explanations and examples of calculations for the minimum space of common workplaces are given.

The last revision is the subject of this case study. In the following sections, the composition of the working group will be compared with the stakeholders that need to be involved in standardization according to a theoretical stakeholder selection.

METHODS

This section describes the methods that were used to judge whether the representation of the working group that revised NEN 1824:1995 was well balanced. First, a stakeholder classification theory was used to define which stakeholders should be involved in ergonomics standardization in general. Second, the involvement of these stakeholders in the working group that revised NEN 1824:1995 was analyzed.

STAKEHOLDER CLASSIFICATION

The theory of Mitchell, Agle, and Wood (1997) on stakeholder classification for organizations was applied to the standardization process by Willemse, Verheul, and De Vries (2003).

The stakeholders are classified according to the following attributes: power, legitimacy, and urgency with respect to the standardization process. Power (P) entails the capability of an actor to bring about the outcomes one desires. A stakeholder possesses legitimacy (L) if the members of a working group, subcommittee, or committee recognize a certain party as a stakeholder. Urgency (U) is defined as the degree to which the stakeholder demands immediate attention for its interest. These three stakeholder attributes (P, L, and U) allow the classification of stakeholders in eight categories (Figure 6.1).

From this list of eight categories of stakeholders, four categories are selected to be relevant in the general standardization processes. These are the "definitive stakeholders" (P, U, L), the "dependent stakeholders" (U, L), the "dominant stakeholders" (P, L), and the "discretionary stakeholders" (L).

The "definitive stakeholders" have the resources to work on the development of the standard (P), consider the standard being developed as important (U), and their involvement is indisputable (L). These stakeholders can be the driving force of the standardization process. In this case example, ergonomics consultants and occupational health and safety services can be considered as "definitive stakeholders." In addition, large companies and the Dutch Department of Spatial Planning and the Environment—being large and important users of the standard—can be considered as "definitive stakeholders."

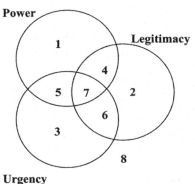

FIGURE 6.1 Stakeholder classification. These different stakeholder groups are labeled: 1, Dormant Stakeholder (P); 2, Discretionary Stakeholder (L); 3, Demanding Stakeholder (U); 4, Dominant Stakeholder (P, L); 5, Dangerous Stakeholder (P, U); 6, Dependent Stakeholder (U, L); 7, Definitive Stakeholder (P, U, L); 8, Nonstakeholder (–).
Source: Adapted from Mitchell, Agle, and Wood, 1997.

The "dependent stakeholders" are important for the general support of a standard (L) and see the importance of participating in the standardization process (U). These stakeholders, however, have limited (financial) resources to participate in the development of standards. Stakeholders for the revision of NEN 1824 are labor unions, architects, and office furnishers. Sector organizations that represent small- and medium-sized enterprises fall into this category as well. Through their representation of a large group of small- and medium-sized enterprises, the sector organizations become valuable working group members, whereas individual representatives from a small- or medium-sized enterprise are worthless.

The "dominant stakeholders" do not see immediate interest in joining the working group. In some way, however, they are of great importance for the standardization process. This can result from a certain position the stakeholder has—which, for example, makes the stakeholder important for the generation of support for the standard. This power base gives the stakeholder legitimacy to join the working group. For a standard for the space of office workplaces, the Dutch Department of Social Affairs and Employment is an important addition to the working group, as it is the legislative power in the field of working environment. The supervising power, the Dutch Labor Inspectorate, is also a party of importance.

The "discretionary stakeholders" are perceived to be important for the development of the standard (L), although they are not interested in participating themselves. A research institute is an example of a "discretionary stakeholder." Researchers can serve as an independent party to offer solutions to conflicts of interest between other stakeholders.

For the revision of NEN 1824:1995, the involvement of the following stakeholders is considered essential (Table 6.1).

REPRESENTATION

To answer the question: "Was the working group that was established for the revision of NEN 1824:1995 a well-balanced representation of all of the stakeholders?" the parties involved in the working group were compared to the list of stakeholders presented in Table 6.1.

The minutes of the meeting were analyzed to identify the parties involved in the revision of this standard, and those who were not. We assume that each of the working group members represented the line of business of the company that sent them. This was never explicitly discussed during working group meetings. Interviews were conducted with three working group members and the secretary to analyze the presence and role of the stakeholders in the working group.

TABLE 6.1
Essential Stakeholders of Ergonomics Standardization

Stakeholder Category	Stakeholder
"Definitive Stakeholder" (P, U, L)	• Ergonomics Consultants • Occupational Health and Safety Services • Large Employers • Department of Housing, Spatial Planning, and the Environment
"Dependent stakeholder" (U, L)	• Labor Unions • Small Companies (representation) • Architects • Office Furnishers
"Dominant Stakeholders" (P, L)	• Department of Social Affairs and Employment • Labor Inspectorate
"Discretionary stakeholder" (L)	• Research Institutes

Representatives of the stakeholders that were not represented were interviewed to find out if they were informed about the revision of the standard, and if they would have been interested in participating. In addition, the working group members were interviewed to learn why some stakeholders were not represented.

RESULTS

Table 6.2 shows the stakeholders that were represented in the working group and how many meetings the representatives attended. It turns out that the standard producing core of the group consisted of one large employer, one ergonomics consultant, and one representative from the Dutch Department of Housing, Spatial Planning, and the Environment; two representatives from the Institutes for Occupational Health and Safety Service; and one representative from a research institute.

Furthermore, the results show that six out of eleven essential stakeholders were not represented, and three stakeholders were represented by two representatives. In two of these cases, however, only one of the two representatives attended the meetings.

STAKEHOLDERS NOT INVOLVED

Three stakeholder groups—namely, the labor union, the office furnishers, and the architects or its sector organization—have not approached to participate in the working group, because it was thought that they would not be interested. However, interviews with some of these parties thought

TABLE 6.2
Representation and Attendance of Stakeholders of Ergonomics Standardization

| | | Attendance in Seven Meetings | |
| | | Representative | |
Stakeholder Category	Stakeholder	1	2
"Definitive stakeholder" (P, U, L)	• Ergonomics Consultants	• 0	• 5
	• Occupational Health and Safety Services	• 3	• 2
	• Large Employers		• 7
		0	(two times written feedback)
	• Department of Housing, Spatial Planning, and the Environment	• 6	
"Dependent stakeholder" (U, L)	• Labor unions	• Not Represented	
	• Small and Medium-sized Companies (representation)	• Not Represented	
	• Architects	• Not Represented	
	• Office Furnishers	• Not Represented	
"Dominant Stakeholders" (P, L)	• Department of Social Affairs and Employment	• Not Represented	
	• Labor Inspectorate	• Not Represented	
"Discretionary stakeholder" (L)	• Research Institutes	• 3	

that they use the standards in their operations and that they would have been interested in joining the workgroup. Most of the users interviewed indicate that they would have been able to offer some input in the working group from their practical experiences with the previous version of the standard.

The Department of Social Affairs and Employment has repeatedly been approached by the working group's secretary to join the group, but for the following reason, did not show any interest. The Dutch Occupational Health and Safety Act contains references to the standard NEN 1824:1995. This standard is seen as a possible elaboration of the legal framework for working conditions that are laid down in the act. The Department states that because it is a legislative institute, it cannot be involved in the development of a possible elaboration of the act because this might lead to a conflict of interest. This same story goes for the Dutch Labor Inspectorate, which is the supervising institute in the field of working conditions.

DISCUSSION

In this Dutch case, the ergonomics profession was overrepresented, and several other important stakeholders did not participate. The situation of overly represented professional ergonomists presented in our case study seems no exception in working groups and committees in the field of ergonomic standardization.

For example, the Dutch committee 302002 called "Physical Load" shadows the activities of ISO/TC 159/SC 3 "Anthropometry and Biomechanics" and CEN/TC 122 "Ergonomics." The Dutch committee gives feedback on standards being developed by these international committees. This committee consists of five representatives of research institutes and universities and two representatives from occupational health and safety institutes.

Another example is the Dutch committee 302015 called "Ergonomic Design Principles" which follows the activities of three international working groups. These are ISO/TC 159/SC1/WG1 "Ergonomic Guiding Principles," ISO/TC 159/SC 1/WG2 "Ergonomic Principles Related to Mental Workload," and CEN/TC 122/WG2 "Ergonomic Design Principles." This committee consists of representatives from two ergonomic consultancy companies, one research institute, and one occupational health and safety institute. In both examples, the working groups show an unbalanced stakeholder representation.

The problem of unbalanced stakeholder representation can also be considered at an international level. Most ergonomics standards are international and—as far as Western and Central Europe are concerned—European as well. The standardization bodies at these levels, ISO and CEN, operate the "country model," in which participants in technical committees represent their own country.

This raises the question of which people the ISO and CEN members should delegate. Let us take the example of the working group ISO/TC 159/SC3/WG2 "Working Postures." This working group consists of about two-thirds of ergonomic researchers, five standardization organization employees, a small number of ergonomic consultants, one government representative, and one industry representative.

Often, CEN and ISO members send employees of the national standards body. They are experienced in expressing the voice of their national standardization committee, but, in general, lack ergonomics expertise. The other option is to send ergonomics experts, but these may have the disadvantage of being less familiar with the standardization profession. The representation of countries by either an ergonomics expert or a standardization expert leads to an unbalanced stakeholder representation on the European and international level.

User involvement in standardization is thus even more neglected at a European and international level than in a Dutch national level. This seems inevitable, but it is—according to De Vries (1999, p. 68)—who provides a listing of advantages and disadvantages of the country model, at least partly compensated thanks to the country model. In this, participants in ISO are backed by a national standardization committee, to which "weak" stakeholders have easier access.

CONCLUSION

In this chapter, we present an example of stakeholder involvement in ergonomic standardization. This example has been selected to illustrate the situation of stakeholder involvement in ergonomics standardization.

It appears that many working groups responsible for developing and revising ergonomics standards are dominated by ergonomics professionals and that many other important stakeholders are not involved. We, therefore, state that the key to increasing the usage of ergonomic standards is to involve users, other than ergonomists, in the development of the standards. User involvement will allow the adoption of new points of view in ergonomic standardization, which will increase the usability and comprehensibility of ergonomic standards.

When ergonomics professionals desire for parties to adopt ergonomic standards, they should also adopt these parties as stakeholders in ergonomic standardization.

ACKNOWLEDGMENTS

A number of persons who contributed to the realization of the case study and this chapter deserve a special word of thanks.

Hugo Verheul—assistant to a professor at the Faculty of Technology, Policy, and Management at Delft University of Technology—was the project leader of a joint research of the Delft University of Technology and the Dutch standardization institute. For this research, case studies have been performed to improve the standardization process of the Dutch Standardization Institute. The improvements are aimed to increase stakeholder involvement in standardization processes and make standards more widely known.

Sandra Verschoof, standardization consultant at the Dutch standardization institute, has provided valuable information about standardization in the field of ergonomics.

Machiel van Dalen, head of the Taskforce Business Development of the Dutch standardization institute, has helped critically reviewing the standardization process and stakeholder involvement in the revision of the standard.

Michiel Kerstens has been very helpful in correcting the English and offering suggestions for textual improvements.

REFERENCES

De Vries, H. I. (1999). *Standardization—A business approach to the role of national standardization organizations*, Boston/Dordrecht/London: Kluwer Academic Publishers.

Mitchell, R. K., Agle, R. B., & Wood, D. J. (1997). Toward a theory of stakeholder identification and salience: Defining the principle of who and what really counts. *Academy of Management Review*, 22, 853–886.

Willemse, H., Verheul, H. H. M., & De Vries, H. J. (2003). *Stakeholderparticipatie bij de herziening van de norm NEN 1824*, Delft: C.P.S.

7 Ergonomics Standards in Ancient China

Pei-Luen Patrick Rau

CONTENTS

FULL NAME OF THE GUIDELINES

1. The height standard for the organization and remuneration of soldiers of the Song dynasty
2. The body size standard for soldier selection of Song dynasty
3. The weight standard for Song infantry armor

INTRODUCTION

The Song dynasty (960–1279) is famous for the development of military technology like permanent standing navy and gunpowder in the history of China. To defend against attacks from its northern enemies such as the Liao dynasty (916–1125), the Jin dynasty (1115–1234), and the Yuan dynasty (1206–1368), the Song dynasty established an army with millions of soldiers equipped with various kinds of weapons. The majority of the Song army was infantry because as an agricultural society, it could recruit many infantry soldiers quickly and easily. The Song's enemies were mainly nomadic people with strong cavalry. For example, the Jin dynasty—the ancestor of the Manchurians—wiped out the first Song central government (Northern Song) and forced them to move their capital to the south. The Manchurians that established the last dynasty in China, the Ching dynasty, were famous for their cavalry skill. Another of Song's enemies, the Mongolian descendants of Genghis Khan who were victorious in the 13th century on the Euro-Asian continent, was also famous for Mongolian cavalry.

To fight against cavalry, the Song had to design weapons, armor, and a strategy for their infantry troops. Long spears and heavy armor can help the infantry attack and repel cavalry. It was an important issue for them to design spears long enough for this purpose. Furthermore, if the armor to protect the soldiers was too heavy, it could hamper soldier mobility on the battlefield. Thus, selecting and training infantry soldiers to fight against cavalry was a significant issue. The Song expended great effort on their military system, strategy, and military technology during their 319-year reign. The Song selected, organized, and paid soldiers according to their height, body size, age, and fighting skills. Their central government built armories and developed national standards for manufacturing weapons and armor. The purpose of some of the national standards was to limit the weight of the equipment to match the physical capabilities of the soldiers. One national standard regulated the upper limit for the total weight for Song infantry armor and the weight of each part of the armor. This standard was issued by the 10th emperor (Gaozong, 1127–1162) of the Song dynasty.

OBJECTIVE AND SCOPE OF THE CHAPTER

The purpose of this chapter is to introduce three ergonomic standards of the Song dynasty: the height standard for the organization and remuneration of soldiers, the body size standard for soldier selection, and the weight standard for Song infantry armor. Since 960, all military volunteers had a height requirement for soldiers of the Song imperial army. In 1057, the remuneration system was linked to the anthropometric results and all recruited soldiers were grouped into five levels according to their height. The records in or after 1127 indicated that the height of the soldiers was used as the standard for organizing new troops. Based on the number of troops, the height distribution of soldiers that were adult Chinese males in the 12th century could be estimated. The Song dynasty measured the body size of soldiers for armor. It was recorded that soldiers who could not fit into the armor were retired in 1062. In 1134, the 10th emperor (Gao Zong) issued an order to regulate the upper limit for a Song infantry soldier's weight.

DISCUSSION OF THE SPECIFIC GUIDELINES

HEIGHT STANDARDS

The Song dynasty military was composed of six troop levels. The first level—the imperial army (Jin Jun)—was the standard army for fighting. The role of the imperial army was replaced by the central army (Tun Zhu Da Bing) and the new army (Xin Jun) in the Southern Song (Toktoghan, 1345; Zhou, Yang, and Wang, 1998; Wang & Yang, 2001). Height was one of the major criteria for the soldier selection in the Song imperial army, as recorded in the military section (Bin Zhi) of the official imperial history of the Song dynasty (Toktoghan, 1345). The unit of measurement was the ancient Chinese length unit (Chi); this gets larger as the years passed by. One Chi was officially recorded to equal about 29.6 cm in the Sui dynasty (581–618) and Tang dynasty (618–907) (Wei et al., 636; Qiu, 1994; Zheng, Tang, et al., 2000). But the excavated Song rulers were about 31.2 cm (Zheng, Tang, et al., 2000). The ancient Chinese inch (Cun) was one-tenth of one ancient Chinese meter.

The Song central government developed two approaches to measure the height of soldiers. The first approach used human models or soldier models (Bin Yang) (Toktoghan, 1345). The first Song emperor, Tai Zu (960–976), chose strong soldiers as references and sent them to recruiting locations in 960 (Toktoghan, 1345; in vol. 187, record (Zhi) 140, military (Bin) 1, imperial army (Jin Jun Shang), and vol. 193, record 146, military 7, recruitment (Zhao Mu Zhi Zhi)). The human model approach was replaced by the wood stick or sticks of equal length (Deng Zhang or Deng Chang Zhang). The wood sticks were used to measure the height of soldiers later on (Toktoghan, 1345; in vol. 187, record 140, military 1, imperial army, and vol. 193, record 146, military 7, recruitment).

Volunteers had to be tall enough to qualify as soldiers in the imperial army between 1008 and 1016 (the third Song emperor Zhen Zong). Short volunteers could only join the local army (Xiang Jun) at the second troop level. The local army was primarily responsible for local security and public service rather than fighting (Toktoghan, 1345; in vol. 193, record 146, military 7, recruitment). The volunteers who passed the lower height limit for the imperial army were distinguished by their height into five groups:

1. Under five Chi and five Cun (about 171.6 cm)
2. Between five Chi and five Cun to five Chi and six Cun (about 171.6 cm to 174.7cm)

3. Between five Chi and six Cun to five Chi and seven Cun (about 174.7 cm to 177cm).
4. Between five Chi and seven Cun to five Chi and eight Cun (about 177.8 cm to 181cm.)
5. Above five Chi and eight Cun (about 181cm)

The soldiers were assigned to different units or bases according to their height group (Toktoghan, 1345; in vol. 193, record 146, military 7, recruitment). The fourth Song dynasty emperor Ren Zong (1022–1063) established a complete recruiting and remuneration military system (Table 7.1). The standard had been changed several times in the Song dynasty:

1. In 1047, all imperial army volunteers had to be five Chi and seven Cun or taller (about 177.8 cm) (Toktoghan, 1345; in vol. 193, record 146, military 7, recruitment).
2. In 1051, the limit was reduced to five Chi and six Cun (about 174.7 cm) (Toktoghan, 1345; in vol. 194, record 147, military 8, selection criteria (Jian Xuan Zhi Zhi)).
3. The limit was again reduced as five Chi and three Cun (about 165.4 cm) in 1057 (Toktoghan, 1345; in vol. 194, record 147, military 8, selection criteria).

In 1057, the imperial army remuneration was linked to anthropometric data (Toktoghan, 1345; in vol. 193, record 146, military 7, recruitment).

- Soldiers taller than five Chi and seven Cun (about 177.8 cm) could be paid 1000 Song dollars (Qian).
- Soldiers who were from five Chi and five Cun to five Chi and seven Cun (about 171.6 cm to 177.8 cm) were paid 500 or 700 Song dollars.
- Soldiers who were about five Chi and five Cun (about 171.6 cm) were paid 300 or 400 Song dollars.
- Soldiers who were from five Chi and two Cun to five Chi and four Cun (about 162.2 cm to 168.5 cm) were paid 200 or 300 Song dollars.

The records in or after 1127 indicated that the height of the soldiers was used as the standard for organizing new troops (Toktoghan, 1345; in vol. 194, record 147, military 8, selection criteria).

- There were only enough soldiers taller than five Chi and eight Cun (about 181 cm) to form one troop (Table 7.2).

TABLE 7.1
The Height Standards in the Song Dynasty and Jin Dynasty

Year	Approach	Standard
960	Human model	No record
960–976	Sticks of equal length	No record
1008–1016	Sticks of equal length	About 171.6–181.0 cm (for assignment)
1047	Sticks of equal length	About 177.8 cm (lower limit)
1051	Sticks of equal length	About 174.7 cm (lower limit)
1057	Sticks of equal length	About 168.5–177.8 am (for remuneration)
1060	Sticks of equal length	About 165.4 cm (lower limit)
1081	Sticks of equal length	No record
1127 (?)	Sticks of equal length	About 165.4–181.0 cm (assignment)
1199	Sticks of equal length	About 1721.6–174.7 cm

TABLE 7.2

The Height Distribution of Soldiers in the Song Dynasty

Height (cm)	Number of Troops
181	1
177.8	4
176.3	6
174.7	6
171.6	33
170	17
168.5	19
166.9	14
165.4	21

- There were enough soldiers of about five Chi and seven Cun (about 177.8 cm) in height to form four troops.
- Soldiers who were about five Chi and six-and-half Cun (about 176.3 cm) formed six troops.
- Soldiers who were about five Chi and six Cun (about 174.7 cm) formed six troops.
- Soldiers who were about five Chi and five Cun (about 171.6 cm) formed 33 troops.
- Soldiers who were about five Chi and four-and-half Cun (about 170 cm) were assigned into 17 troops.
- Soldiers who were about five Chi and four Cun (about 168.5 cm) were assigned into 19 troops.
- Soldiers who were about five Chi and three-and-half Cun (about 166.9 cm) were assigned to 14 troops.
- There were 21 troops formed by soldiers who were about five Chi and three Cun (about 165.4 cm).

The number of soldiers in each troop may not be the same but are usually similar. The height distribution of soldiers that were adult Chinese males in the 12th century could be estimated by the number of troops. Most soldiers were about five Chi and five Cun (about 171.6 cm) or shorter. There were only a few taller than five Chi and seven Cun (about 177.8 cm) (Figure 7.1).

The Song dynasty was not the only dynasty that used height to select soldiers. One of their enemies—the Jin dynasty—also established height standards. The 5th emperor of the Jin dynasty, Zhang Zong (1189–1208), recruited volunteers from five Chi and five Cun to five Chi and six Cun (about 171.6 cm to 174.7 cm) as soldiers in the Jin imperial army in 1199 (Toktoghan, 1344; in vol. 4, record 25, military, imperial army).

Body Size Standards

Using height to select soldiers was not enough for the Song dynasty because height does not necessarily mean strength. The Song dynasty used armor to measure the body size of soldiers. It was recorded that soldiers who could not fit into the armor were retired in 1062 (Toktoghan, 1345; in vol. 193, record 146, military section 7, recruitment). Two years later (AD 1064), the same issue was brought up for the soldiers in the imperial army (Toktoghan, 1345; in vol. 194, record 147, military section 8, selection criteria). In 1068, the armor was officially considered a standard for selecting soldiers, and soldiers had to be strong enough to wear the armor to be recruited (Toktoghan, 1345; in vol. 194, record 147, military section 8, selection criteria) (Table 7.3).

1008-1016 A.D.   1047 A.D.   1051 A.D.   1057 A.D.   1127 A.D.

FIGURE 7.1 Height standards.

TABLE 7.3
The Body Size Standards in the Song Dynasty

Year	Approach	Standard
1062	The song infantry armor	Fitness
1064	The song infantry armor	Fitness
1068	The song infantry armor	Fitness

THE WEIGHT STANDARD FOR THE SONG INFANTRY ARMOR

Song armor had to be standardized before it could be used as a measurement tool. The standard for Song infantry armor was originally published in 1044 (Zeng & Ding, 1044; in vol. 13). The third emperor Ren Zong ordered two scholars to edit a military encyclopedia (Wu Jing Zong Yao). This encyclopedia covers famous campaigns, military organization, strategy, weaponry, etc. The design, manufacture, and function of weapons were described and standardized in this book. There are five parts of the Song infantry armor (Figure 7.2): helmet, neck covering, shoulder covering, chest covering, and belly and thigh covering.

In 1134, the 12th emperor (Gao Zong) issued an order to regulate the upper limit for the weight of the Song infantry armor and of each part of the armor (Toktoghan, 1345; in vol. 197, record 150, military section 11, weapons and armors). The armor was composed of 1825 pieces of steel; the

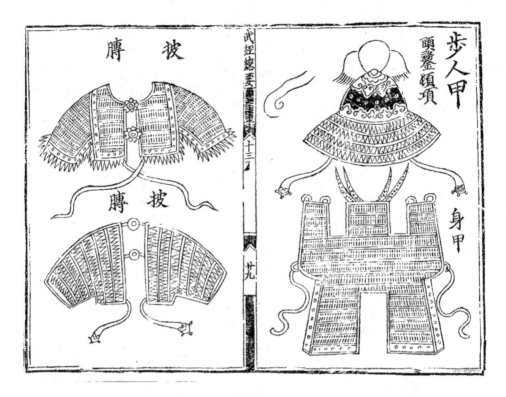

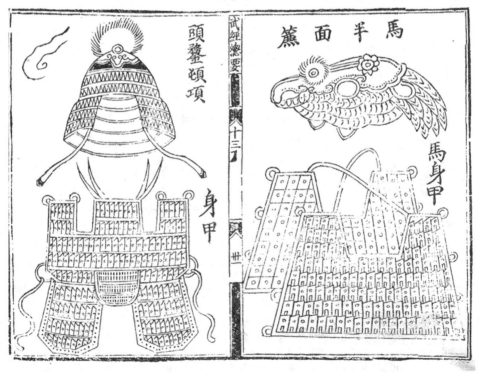

FIGURE 7.2 The Song infantry armor (Zeng & Ding, 1044, Reprinted by Zheng, 1988).

TABLE 7.4

The Weight Standards for the Song Infantry Armor

Component	Number of Piece of Steel	Weight for Each Piece (g)	Net Weight (g)
Helmet			637.5
Neck covering	311	9.4	2915.6
Shoulder covering	540	9.8	5265.0
Chest covering	332	17.6	5851.5
Belly and thigh covering	679	16.9	11,458.1
Leather string			3468.8
Total			29,596.5

total weight was from about 27 to about 30 kg. The upper limit was about 30 kg. The standards for each component are listed in the following (Table 7.4). It was stressed in the order that every piece of steel is weighted according to the standard. If the weight did not meet the requirement, that piece of steel would not be used.

- The helmet weighed about 637.5 g.
- The neck covering consisted of 311 pieces of steel. Each piece weighed about 9.4 g.
- The shoulder covering consisted of 540 pieces of steel. Each piece weighed about 9.8 g.
- The chest covering consisted of 332 pieces of steel. Each piece weighed about 17.6 g.
- The belly and chest covering consisted of 679 pieces of steel. Each piece weighed about 16.9 g.
- The leather string for fixing all of the pieces together weighed about 3468.8 g.

CONCLUSIONS

The ergonomics standards introduced in this chapter included the height standard for the organization and remuneration of soldiers, the body size standard for soldier selection, and the weight standard for the Song infantry armor. Some standards were issued from the central government and the emperor, indicating that human capabilities were considered an important factor from the 10th to 12th century in China. These standards indicated that the awareness of ergonomics could be traced to hundreds of years ago in human history.

REFERENCES

Qiu, G. (1994). *Length, volume, and weight measures in ancient China*. The Commercial Press: Taipei.

Toktoghan (1344). The official imperial histories of the Jin (Jin Shi). Reprinted by Chung Hua Book: Beijing, 1975.

Toktoghan (1345). *The official imperial histories of the Song (Song Shi)*. Reprinted by Chung Hua Book: Beijing, 1977.

Wang, X. and Yang L. (2001). *Military System in China (Hu Jun Jian Bang: Li Dai Bin Zhi)*, Wan Juan: Taipei.

Wei, Z., et al. (636). *The official imperial histories of the Sui (Shuishu)*. Reprinted by Chung Hua Book: Beijing, 2000.

Zeng, G., & Ding, D. (1044). *Military Encyclopedia (Wu Jing Zong Yao)*, cited in Ji (1783). The complete books in the four parts of the Imperial Library (Si Ku Chuan Shu).

Zheng, T., Tang, Q., et al. (2000). *Chinese history dictionary*. Shanghai Dictionary Publisher: Shanghai.

Zheng, Z. (1988). Chinese Ancient Print Series, Vol. 1. Shanghai Chinese Classics Publishing House: Shanghai.

Zhou, B, Yang, M., and Wang, Z. (1998). *The history of northern Song and southern Song*. Chung Hwa: Hong Kong.

8 Standard in Safety Management That Takes into Account Humans' Irrational Behavior and Decision-Making

Atsuo Murata

CONTENTS

INTRODUCTION

Human errors, crashes, collisions, or disasters are viewed from the cognitive, ergonomic (man-machine system), psychological, and organizational perspectives (Reason, 1990, 1997, 2016; Wiegman & Shappell, 2003). The direct causes or background factors (man, machine, media, and management factors) are identified from such perspectives. Based on such analysis, a lesson to be learned is extracted and made known to all people related so they can act accordingly.

Traditional safety management approaches tend to conclude that lessons from the investigation of the direct (superficial) causes can help us prevent other crashes, collisions, or disasters. However, despite such an approach, these cases continue to repeat worldwide (Dekker, 2006). Such an approach has been expressed as merely a ritual of disaster (Gladwell, 2008, 2009).

Reason (1990, 1997, 2008, 2013, 2016) regarded irrational judgmental heuristics and biases (Bazerman & Moore, 2001; Bazerman & Watkins, 2008; Bazerman & Tenbrunsel, 2012; Dobelli, 2013) as potential risk factors of human errors that lead to unfavorable or unexpected crashes, collisions, or disasters. However, Reason (1990) did not provide a systematic model of how such cognitive biases are related to distorted decision making and how they become a trigger for these occurrences. Dekker (2006) also pointed out a vicious cycle of repeated occurrences of unfavorable crashes, collisions, or disasters. He speculated that these are caused by hindsight or outcome bias. A preventive countermeasure without considering cognitive biases will not be effective in drastically disconnecting these crashes, collisions, or disasters.

The concept of normal accident (Perrow, 1999, 2011) is characterized by high complexity and the tight coupling of components in modern large-scale systems such as nuclear power plants, super express trains, or airplanes. High complexity induces invisibility or opaqueness, unpredictable interactions of components, and vulnerability to human errors. Tight coupling of components induces lower slack (for detail, see Mullainathan & Shafir, 2014) or fewer buffers of cognitive abilities, less time and flexibility, and unforgiveness (errors cannot be canceled). Tight coupling also induces a cascade of failures and amplifies the consequence of errors or failures. In a large-scale system, even seemingly unrelated components might be connected indirectly. The direct cause of the Challenger disaster was a malfunction of the O-ring. Gladwell (2008, 2009), referring to the concept of normal accident, stated that fixing the malfunction (direct cause) does not necessarily ensure the removal of all the infinite risks concerning the space shuttle launch. This indicates that the identification of only the direct causes is unhelpful in preventing another disaster. If the root causes behind the direct cause are not removed, the system cannot be safe.

It is difficult to rationally accept such a risk caused by the high complexity and tight coupling of a large-scale system, and we tend to get trapped in cognitive biases and optimistically ignore such a risk. We irrationally think that the identification of direct causes and the extraction of a lesson from it are effective in preventing other crashes, collisions, or disasters. Thus, our behavior or decision on safety is frequently distorted (cognitively biased). Here lies the limitation of the traditional standard of safety management. Therefore, we need to recognize in detail the factors contributing to the direct causes and develop a new standard of safety management that overcomes and compensates for the limitations of the traditional safety management approach.

We discuss how standards in safety management should be practiced in eradicating the occurrence of similar crashes, collisions, or disasters. First, the framework of traditional standards in safety management is discussed in detail to point out the limitations of such an approach. Based on the discussion on the limitations of traditional approaches, we discuss how a new standard of safety management should be developed to eradicate the vicious cycle. The discussion is based on the approach to think globally about a series of events without irrational (distorted) behavior or decision making. This chapter provides readers with a new standard of safety management from the viewpoint of humans' near-sighted and irrational behavior or decision making. In this manner, a new way of thinking or standard of safety management is advocated so that it contributes to cutting off the vicious circle of similar crashes, collisions, or disasters.

TRADITIONAL STANDARD IN SAFETY MANAGEMENT

Traditional approaches for managing human errors, unsafe behaviors, crashes, collisions, or disasters are summarized in Figure 8.1. These are the following seven direct causes of human errors: (1) improper design of man-machine system (interface), (2) insufficient consideration of the limitation of humans' cognitive information processing ability such as memory or perception, (3) insufficient consideration of humans' fatigue, stress, or workload, (4) lower motivation, (5) lack in skill, knowledge, or experience, (6) non-standardization of safety procedure, and (7) improper risk management. Figure 8.1 also shows that social systems—organizational management such as safety or organization culture—and regulation systems, as well as human errors, should be considered to prevent unsafe behaviors, crashes, collisions, or disasters. This indicates that safety management should be an integrated approach that considers human errors and organizational, managerial, and social factors that encompass humans.

Figure 8.2 shows the improvement (reduction) of human errors by pointing and naming of objects. The accuracy of response is improved when the object of instruction responds by pointing as compared otherwise. The response accuracy of the pointing and naming condition is higher than that of the condition with pointing and without naming. It must be noted that the response accuracy of the pointing and naming condition never becomes 100%, although the response accuracy is improved to a larger extent. The most important thing is that we should take appropriate recovery action from errors so that these do not lead to critical crashes, collisions, or disasters. A zero-human error system is actually impossible; the goal is to develop a mechanism that does not make human error lead to crashes, collisions, or disasters. Figure 8.3 shows progress in the mechanism for reducing operation errors. Persistent efforts to progress a mechanism to reduce human errors

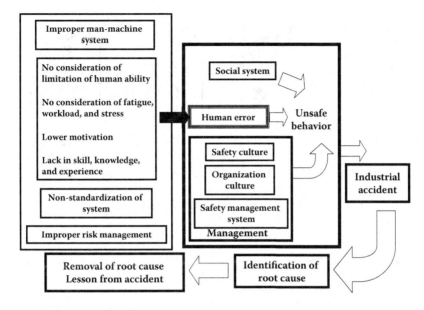

FIGURE 8.1   Traditional standard for managing human errors.

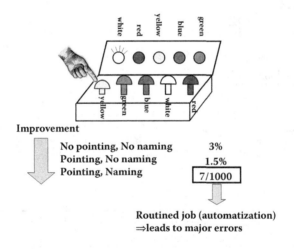

FIGURE 8.2   Reduction of human errors by pointing and naming.

must be continued to ensure safety. Figure 8.4 depicts four background factors behind human errors, crashes, collisions, or disasters. The four factors include man, machine, media, and management. Traditional safety management attempts to detect such background factors behind direct causes.

Figure 8.5 summarizes the process of aviation crash (Air Inter 148 crash) of Airbus A320 at Strasbourg, France in January 1990. The direct cause of this crash corresponded to a human error of the pilot. The TRK ■ FDA (Track ■ descending angle) mode had to be correctly chosen to ensure safe flight. Instead of the TRK ■ FDA mode, HDG ■ V/S (nose direction descending) mode was mistakenly chosen. The background factor behind this human error included man, machine, and management. Pilots were required to recognize instantly and accurately the different meanings of the display "33" in the Airbus A320. While the display "33" means the descending angle of

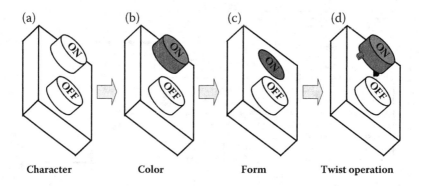

(a)
(b)
(c)
(d)
ON
OFF
Character
Color
Form
Twist operation

FIGURE 8.3 Advances in mechanisms for reducing operation errors.

Management
Media
Man
Hardware
Software
Machine
Management

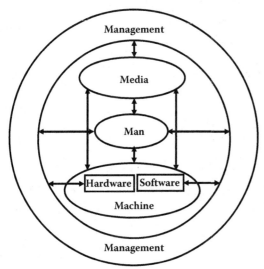

FIGURE 8.4 Four background factors of human errors, crashes, collisions, or disasters.

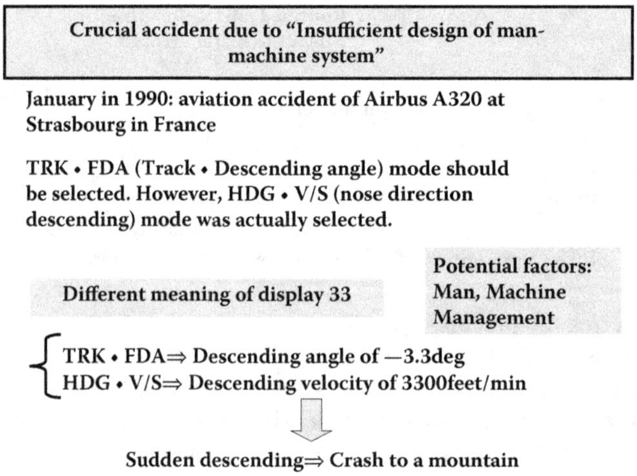

Crucial accident due to "Insufficient design of man-machine system"
January in 1990: aviation accident of Airbus A320 at Strasbourg in France
TRK • FDA (Track • Descending angle) mode should be selected. However, HDG • V/S (nose direction descending) mode was actually selected.
Different meaning of display 33
Potential factors: Man, Machine Management
TRK • FDA⇒ Descending angle of —3.3deg
HDG • V/S⇒ Descending velocity of 3300feet/min
Sudden descending⇒ Crash to a mountain

FIGURE 8.5 Flow chart of aviation crash of Airbus A320 at Strasbourg in France occurred in January 1990.

−3.3 degrees in the TRK ∎ FDA mode, this represents the descending velocity of 3300 ft./min in the HDG ∎ V/S mode. The direct cause is the mistake of the pilot to recognize the meaning of display "33." Although the pilot mistakenly chose the HDG ∎ V/S mode, he misunderstood that he correctly chose the TRK ∎ FDA mode. Contrary to the pilot's recognition that the airplane descended with an angle of −3.3 degrees, the airplane descended with a velocity of 3300 ft./min and crashed to the mountain.

Figures 8.6 and 8.7 show the difference in cockpit displays between Bowing B747 and Airbus A320. In Airbus A320, pilots are required to shift between the TRK ∎ FDA mode and the HDG ∎ V/S mode using only one switch. The situation of choice is shown on one cockpit display. The display of status is shown on the neighboring cockpit display. Therefore, pilots must be very careful in recognizing the airplane's status accurately. On the other hand, B747 is equipped with two separate switches (one for the TRK ∎ FDA mode and one for the HDG ∎ V/S mode). Generally

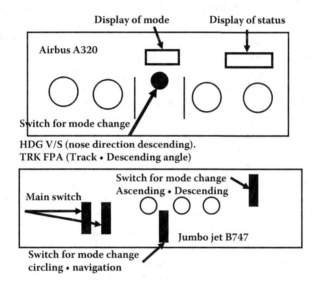

FIGURE 8.6   Differences of cockpit displays between Bowing Jumbo Jet B747 and Airbus A320.

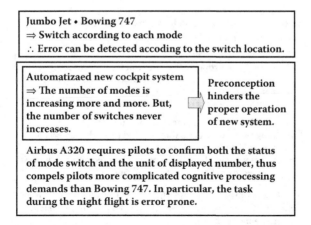

FIGURE 8.7   Summary of different cognitive information processing between Bowing Jumbo Jet B747 and Airbus A320.

speaking, errors can be more readily detectable in B747 (lower panel in Figure 8.6) than in A320 (upper panel in Figure 8.6), as there exist switches for each of the two modes (TRK ∎ FDA mode and HDG ∎ V/S mode). Insufficient design of a man-machine system is observable in A320. However, it must be noted that although the cockpit of B747 is less vulnerable to operation errors as compared with that of A320, the system of B747 is not perfectly safe. It is merely safer than A320 from the viewpoint of a man-machine system design. The ergonomically designed man-machine system is not an absolute one but a relatively better one. The reason regarding appropriate man-machine system design not being enough is mentioned in detail in the Limitation of Traditional Safety Management and How New Standard in Safety Management Should Be Practiced sections.

As shown in Figure 8.8, we cannot behave appropriately according to an operation manual, especially under emergency. Manuals are useless in such cases of emergency. We cannot use properly the function that is scarcely encountered as demonstrated in the Korean subway disaster in 2003, JR West Fukuchiyama line derailment crash in 2006, and the Fukushima Daiichi disaster in 2011. Figure 8.9 summarizes the China Airline 140 crash that occurred at Nagoya international airport in 1994. As shown in Figure 8.8, an appropriate operation under emergency is considerably difficult. The direct cause of this crash was a human error (a slip to turn on a turn-around mode). Although the pilots tried to turn the auto throttle system (ATS) off, he made a slip error and selected the ATS go-around mode. Although he firmly pushed the control level and attempted to cancel the "go-around" mode, he failed. As the pilots scarcely used the function and were not accustomed to the automated cockpit of Airbus A320, they panicked and eventually could not cancel the turn-around mode under emergency.

The automation of A300-600R was related to the crash. It is true that automation reduces human errors especially when non-skilled pilots are on board. The lack of management of technology (MOT) is also detectable as one of the background factors of the crash. The reason their automation technology induced such a crash must be analyzed in more detail. Airbus actually did not execute such an analysis. A similar trouble occurred in Finland in 1989. Airbus, however, did not actively receive a risk of crash and learn much from the Hiyari-Hatto incident that occurred in Finland—the organizational mindset to learn from that incident was apparently lacking. It is concluded that the organizational mindset of Airbus that did not actively learn from the Hiyari-Hatto incident must have led to a serious crash in Nagoya International Airport. Airbus did not attempt to further explore the root cause behind this crash.

Information processing under emergent situation
• cannot be conducted according to the manual.
• in many cases, manuals are useless.
• We cannot use properly the function that is scarcely used.

Korean subway fire accident in 2003
Nobody knew how to open the door manually.

JR West FukuchiyamaLine derailment accident
Although the cover glass was cracked to push a emergency
button, this did not operate.
⇒ In this vehicle, the power source changeover switch must
be changed from "Normal"to "Emergency".

In Fukushima Daiichi and Three Mile Island nuclear power
plants, the meltdown accident occurred due to the lack in
proper emergent actions

FIGURE 8.8 Examples of failures in information processing under emergent situation.

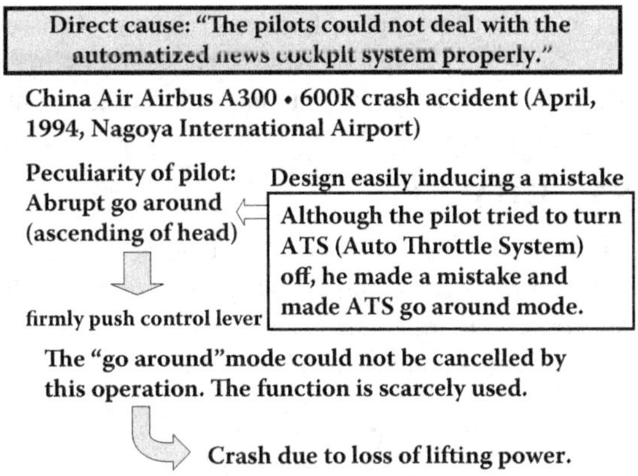

FIGURE 8.9 Summary of China Airline 140 crash occurred at Nagoya International Airport in 1994.

Traditional safety management does not further attempt to detect the common root cause of the two crashes (Air Inter 148 crash and China Air 140 crash). If the common root cause—that is, the disagreement between the engineer's design model and pilots' mental model—had been detected, this might be effectively used in preventing another crash. The traditional safety management approach apparently lacked a mindset or system to actively learn from failures and make effective use of it to prevent further failure (Syed, 2016).

LIMITATION OF TRADITIONAL SAFETY MANAGEMENT

Similar errors or accidents repeatedly occur despite obeying the procedure of the traditional safety management approach (Traditional Standard in Safety Management section) that identifies the direct cause and extracts a lesson. Limitations of standard in safety management lie here. To cut off the vicious cycle of crashes, collisions, or disasters, we need to make remarkable progress in the standards in safety management.

As mentioned in the aviation crash of Airbus A320 at Strasbourg, France in January 1990 (see Traditional Standard in Safety Management section and Figures 8.5–8.8), one of the direct causes was the insufficient design of man-machine system in A320. However, there was also a distorted thinking behind the direct cause. The engineers deviated from rational thinking that their automation design should be executed in consideration of the opinions of pilots who actually use the system as pointed out by Murata (2019). It is apparent that the engineers' decision to not adopt the opinion of pilots (users) was cognitively biased. The identification of only the direct cause is insufficient, and the root cause behind the direct cause must also be investigated and analyzed in detail to prevent another crash. We must notice that the gap of the mental model toward the automated cockpit system between engineers and users underlies the direct cause of the crash.

Murata and Moriwaka (2017) showed that it is frequently difficult to make safety compatible with economics or efficiency. Although it goes without saying that safety is more important than efficiency, our behavior or decision is frequently distorted, causing safety to be threatened. Consequently, this sometimes leads to crashes, collisions, or disasters. To avoid such an imbalanced behavior, the mechanism behind the cognitively biased behavior or decision—that is, the imbalance between safety and efficiency—should be discussed in more detail. Such an approach has not been attempted in the framework of traditional safety management.

The lack of detailed recognition of a root cause or background factor behind a direct cause without understanding human irrationality (cognitive biases, that is, the deviation of our behaviors from rational ones) leads to the unconscious repetition of errors, resulting in a vicious circle of similar crucial incidents, as pointed out by Murata, Nakamura, and Karwowski (2015). Therefore, understanding how cognitive biases distort decision making, prevent us from reaching a root cause or in-depth background factor, and lead to critical incidents is essential to avoid such situations of vicious circles. The next section mentions how our thinking or behavior is irrational and deviates from a rational one.

HUMAN'S IRRATIONAL BEHAVIOR

First, answer the following question: gin and tonic are mixed according to operations I and II in Figure 8.10. Which one is more after operation II, tonic in A or gin in B? An orthodox mathematical solution for this question is as follows.

The volumes of tonic and gin are W and Z in B, respectively, after operation II.

$$W = X\frac{X}{X+Y}, \ Z = X\frac{Y}{X+Y}$$

After moving Y of B to A, the volumes of tonic and gin in A are calculated as follows.

$$Y\frac{X}{X+Y}, \ Y\frac{Y}{X+Y}$$

The volumes of gin and tonic in A are as follows.

$$X - Y + Y\frac{X}{X+Y} + Y\frac{Y}{X+Y}$$
$$= \frac{(X-Y)(X+Y)+Y^2}{X+Y} + Y\frac{X}{X+Y}$$
$$= \frac{X^2}{X+Y} + \frac{XY}{X+Y}$$

Therefore, the volume of tonic in A is the same as that of gin in B.

However, the answer can be obtained without an orthodox mathematical solution as shown in Figure 8.11. In complicated systems, it is impossible to recognize the true nature of the system as a whole (forest) if we pay too much attention to individual subsystems (trees). Therefore, we cannot comprehend a phenomenon easily if we don't view the system as a whole. Unfortunately, we humans are not so good at reaching such an intuitive solution. Kahneman (2011) classified our thinking according to System1 (intuitive thinking) and System2 (deliberate thinking). The example here indicates that we cannot think accurately with System1, although we frequently rely on System1 to reach a solution. We don't afford to deliberate a situation using System2 and cannot

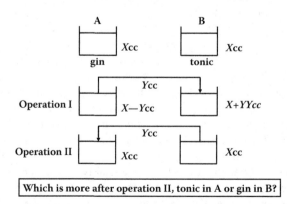

Which is more after operation II, tonic in A or gin in B?

FIGURE 8.10 Two safety approaches: safety approach based on hazard (risk of incident) detection and that based on safety confirmation.

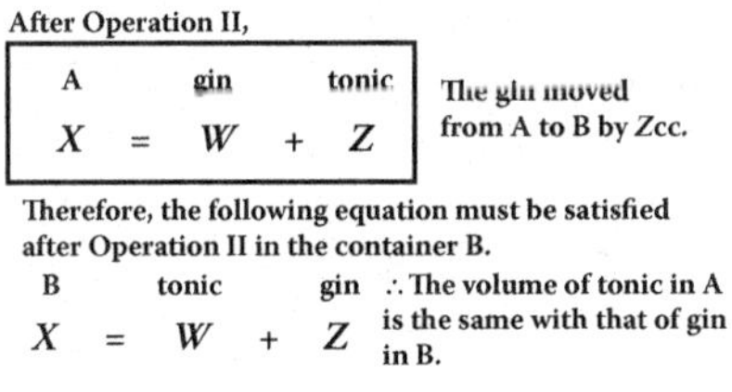

After Operation II,

A	gin	tonic	The gin moved
X =	W +	Z	from A to B by Zcc.

Therefore, the following equation must be satisfied
after Operation II in the container B.

B	tonic	gin	∴ The volume of tonic in A
X =	W +	Z	is the same with that of gin in B.

FIGURE 8.11   Global and quick (without mathematics) solution to the problem.

help relying on System1 under emergency. Therefore, our decision or behavior tends to be fallible and gets trapped into cognitive biases.

Next, we would like to solve the following question.

(Problem)

Hanako's car is falling apart. The car has a 10% probability of breaking down every mile. She is going to see her sick friend living ten miles away. She ponders the significance of driving a car with a 10% probability of breaking down each mile and finally figures that the probability of breaking down on the way to her friend's home cannot be much higher than 15%. She was shocked when her car broke down halfway. What was the actual probability of breaking down halfway?

The actual probability of failure is $1 - (1 - 0.1)^{10} = 0.651$ and is larger than our expectation. The behavior of Hanako was based on the under-evaluation of a failure risk. This under-evaluation is called the disjunction fallacy, and the probability of A1 OR A2 OR … OR A10 is under-evaluated. Even in such a rather simplified situation, we cannot evaluate a risk correctly.

A similar problem is given below.

(Problem)

Linda is 31 years old, single, outspoken, and bright. She majored in philosophy. As a student, she was deeply concerned with issues of social justice and discrimination and participated in anti-nuclear demonstrations. Which of the two alternatives is more probable?

A. She is a bank teller.
B. She is a bank teller (event A) and active in the feminist movement (event A1).

Probability Pr(A) is actually larger than Pr(B) (Pr(A) > Pr(A AND A1)). However, we tend to judge that Pr(A) < Pr(B) contrary to this rationality. We tend to over-evaluate the risk (probability) of Pr(A AND A1). This example also suggests that we cannot evaluate a risk correctly. This irrationality is called the conjunction fallacy.

The above-mentioned disjunction and conjunction fallacies belong to cognitive biases that lead to our irrational behavior. Cognitive biases cause our irrational behavior and decision, and it is possible that our distorted and irrational behavior becomes a trigger of critical disasters, crashes, or collisions (Murata, Nakamura, & Karwowski, 2015; Murata, 2017; Murata, 2019). We can usually spend our daily life or jobs without experiencing dangers, failures, or major incidents. If such states continue, we misunderstand that we can continue without such negative events. This leads to availability bias, confirmation bias, and normalcy bias (Dobelli, 2013).

Figure 8.12 shows the change of juvenile delinquency (number of murders) from 1980 to 2003 in Japan. This seems to indicate an increasing tendency of juvenile delinquency (number of murder). Figure 8.13 demonstrates that such thinking is apparently wrong. This corresponds to an availability bias by only using available data. The number of juvenile delinquency (murder) actually did not increase but remained at the same level from 1980 to 2003. It seems that we are not so good at viewing things from a holistic perspective.

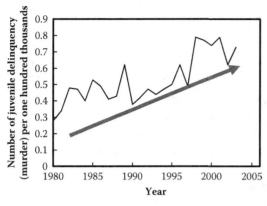

FIGURE 8.12 Change of number of juvenile delinquencies per one hundred thousand in Japan from 1980 to 2005. Does the number of juvenile delinquencies per one hundred thousand really tend to increase?

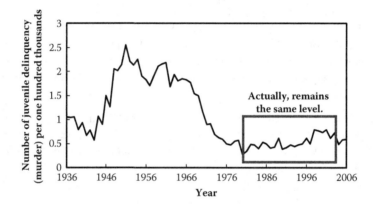

FIGURE 8.13 Change of number of juvenile delinquencies per one hundred thousand in Japan from 1936 to 2005. The number of juvenile delinquencies per one hundred thousand really from 1980 to 2005 remains the same level.

We tend to contribute to the increase of vicious crime to education (matter of mind). This is widespread throughout the whole society. Such an analysis of the cause is straightforward, does not require much time, and seems plausible. The cause of human errors is frequently attributed to individual mental problems (spiritualism). Such an analysis makes it impossible to comprehend the essence (or nature) and potential risk of social phenomenon or error. The promotion of attention or reinforcement of control (regulation) alone is not sufficient for the prevention of errors or accidents. We must appropriately analyze the background or risk factors of social phenomena or errors and accurately recognize the cause and effect. Can the awakening of attention and reinforcement of control activities or regulations reduce disasters, scandals, or violations? Figure 8.14 shows the correspondence of a top management to a few events that occurred in Japan. The root cause or source of the accident (why inattentively behaved) is not understood at all in such cases. Similar scandals or events are repeated endlessly if the root causes are not appropriately identified. This is the biased identification of cause and effect. It is by far easier for us to attribute the cause of mistakes or errors to the mentality or inattentiveness of the people related. However, such an identification of false cause and effect is never helpful in preventing another error, mistake, or violation. The recognition of humans' irrational thinking or decision is necessary to identify the root causes appropriately and reach a drastic solution. In other words, a drastic countermeasure is impossible without the identification of genuine cause and effect.

Figure 8.6 shows the necessity of learning the ability to overview many cases and identify factors observed in common to each case. Modeling and generalization of similar cases are

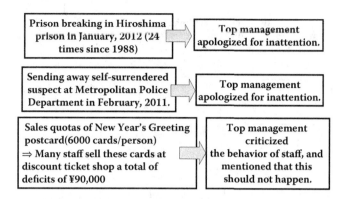

FIGURE 8.14 Factors observed in common behind three cases above.

necessary to cut off the vicious circle of similar crashes, collisions, or disasters. Such an approach is impossible if we get trapped in cognitive biases. Hindsight bias does not help at all in identifying the root cause behind the direct causes and generalizing the latter. An approach that can generalize the events without hindsight bias is essential for advanced safety management. Therefore, the understanding of humans' irrationality and the application of such characteristics to the advanced safety management strategy must be widespread.

HOW NEW STANDARD IN SAFETY MANAGEMENT SHOULD BE PRACTICED

The automation of operation in an airplane cockpit is conceptually acceptable because such automation basically aims to replace error-prone human operations with more accurate machines (automation). However, as mentioned in the correspondence of the designer's design model and the pilots' mental model in Murata (2019), Airbus does not listen to the opinion of pilots when they develop an automated cockpit operation system. Airbus should have made efforts to bridge the gap between automation and pilots. Without such an approach, it is actually difficult to develop an automated system that is truly usable and safe for pilots. With the progress of automation of operation in an airplane cockpit, the number of operation modes is increasing without considering the relationship between such a system and the pilots' ability of cognitive information processing. The number of switches in a cockpit panel does not increase accordingly. It is presumable that such preconception hinders the proper (error-free or less vulnerable to errors) operation of a new system. Airbus A320 requires pilots to confirm both the status of the mode switch and the unit of displayed number (see Figure 8.6) instantly and accurately. The units correspond to degrees and feet/minute in TRK ▪ FDA mode, and HDG ▪ V/S mode, respectively. Such requirements in A320 compel pilots to execute more complicated cognitive information processing demands than those in B747. The task is error-prone, particularly at night. Recently, the importance of management of technology (MOT) is emphasized increasingly. The background factor behind human error includes the lack in appropriate management of advanced automation technology. The management or organization should have had a mindset to consider the gap between engineers and pilots appropriately as pointed out in Murata (2019).

Similar problems were observed even in the China Air 140 crash. The mindset to not actively listen to and adopt the pilots' opinion in the design of automated cockpit (Murata, 2019) contributed as a root cause in this crash. Figure 8.7 summarizes the detailed root cause common in the Air Inter 148 crash and the China Air 140 crash. It is difficult to enhance safety further without competence to recognize irrationality (cognitive biases) at both individual and organizational levels and explore the root cause genuinely contributing to the crash (Figures 8.15 and 8.16).

Figure 8.8 shows the flow chart of an explosion at the U.S. power plant. The direct cause of the explosion was a human error—that is, the lapse of inserting a safety pin into a hand grenade. The

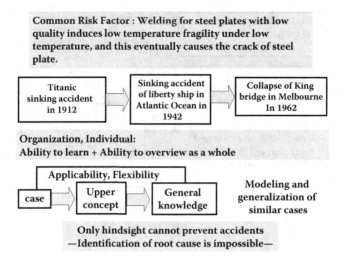

FIGURE 8.15 Necessity of learning ability to overview many cases and identify factors observed in common to each case.

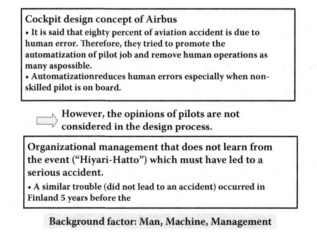

FIGURE 8.16 Detailed root cause in common to Air Inter 148 crash and China Air 140 crash.

identification of only the workers' error to forget to insert the safety pin into the hand grenade is insufficient to prevent another explosion. The background (man) factor behind this human error must be identified to cut off a vicious circle of errors. The increase of watchmen to monitor the lapse of safety pin insertion is not sufficient to prevent further explosion. It is necessary to recognize that we tend to be inevitably inattentive in monitoring tasks and cannot maintain a high arousal level. In short, it is actually impossible for us to execute monitoring tasks perfectly (without victims to human errors to omit an important task necessary for assuring safety). Therefore, with the full understanding and recognition of such disadvantages, a mechanism to overcome them must be practiced. In Figure 8.17, such a mechanism was put into practical use by developing a system to pick up a safety pin by a robot. This mechanism prevents the production line of the hand grenade to proceed to the next process if the safety pin is not inserted; the robot cannot pick up a hand grenade and proceed to the next manufacturing process if the safety pin is not inserted. In this manner, not the spiritualism but the mechanism to prevent human errors is necessary in safety management.

Figure 8.18 summarizes symptomatic treatments (see also Figure 8.8) that cannot get rid of root causes. It is impossible to eradicate the occurrence of similar crashes, collisions, or disasters if the

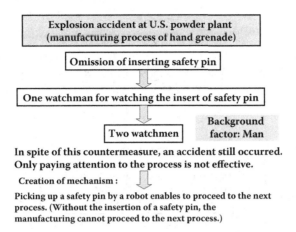

FIGURE 8.17 Flow chart of explosion at the U.S. powder plant.

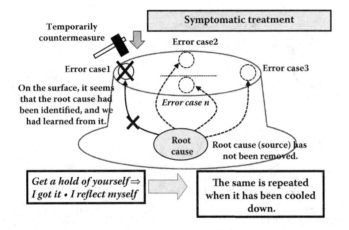

FIGURE 8.18 Symptomatic treatment that cannot get rid of root causes.

root cause is not removed completely. Figure 8.19 summarizes the concept to be free from symptomatic treatments in Figure 8.18 and the removal of a root cause. It is necessary for us to produce mechanisms or ways to remove the root causes of similar crashes, collisions, or disasters. Only the detailed investigation of the root cause behind direct causes certainly leads to the eradication of the risk of similar crashes, collisions, or disasters. Similar crashes, collisions, or disasters will continue to occur as long as the root cause is not removed completely. As mentioned in the Traditional Standard in the Safety Management section, the identification of the direct cause of crashes, collisions, or disaster is insufficient for the prevention of similar ones. The limitations of traditional standards in safety management lie in such an approach to identify direct causes, learn, and be cautious about these. It is impossible to eradicate a vicious circle of similar crashes, collisions, or disasters without the identification of a root cause behind direct causes. We should put emphasis on such an approach to identify the root cause behind direct causes in the practice of safety management.

Figure 8.20 is a sketch of Plant #1 in the Fukushima Daiichi nuclear power plant. The function loss of steam turbine generator and the loss of water circulation system were caused by a gigantic tsunami. An unexpectedly large earthquake also contributed to the loss of coolant. Figure 8.21 shows the potential direct causes of the Fukushima Daiichi disaster. There are a lot of direct causes

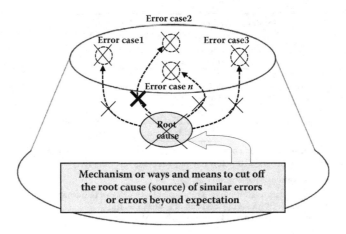

FIGURE 8.19 Freeing from symptomatic treatment and removal of root cause.

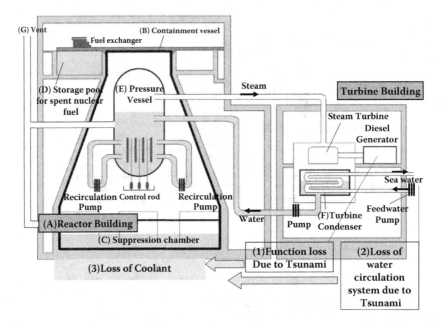

FIGURE 8.20 Direct causes of the Fukushima Daiichi disaster.

for the loss of cooling function in nuclear power plants. The conceivable direct causes include the following events: extra-large hurricane, landslip caused by torrential rain, hijacking of the central control room by terrorists, destruction of water circulation facility by terrorists, unexpected large-scale shock caused by a meteorite, destruction of laying pipes by sabotage, human errors like a direct cause of the Three Mile Island disaster, and an airplane crash at the pressure vessel. Figure 8.21 summarizes the hindsight bias in the Fukushima Daiichi disaster. The occurrence probability of earthquakes and Tsunami is higher after the Fukushima Daiichi disaster than before. Although there are a lot of causes that lead to the loss of cooling system and induce the meltdown of a nuclear reactor vessel, one tends to pay attention only to earthquakes and tsunamis and does not predict meltdown caused by other factors shown in Figure 8.21 if one gets trapped in hindsight bias. This means that we cannot address a root cause behind conceivable direct causes other than earthquakes and tsunami if we become victims of hindsight bias.

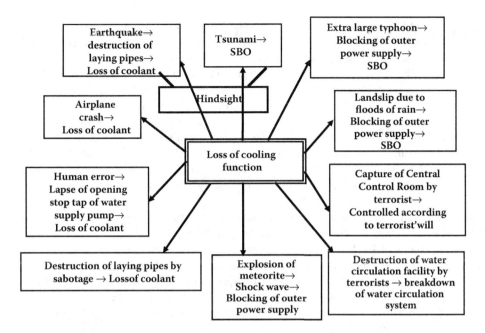

FIGURE 8.21 Hindsight bias in the Fukushima Daiichi disaster.

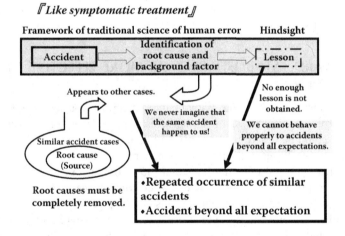

FIGURE 8.22 Necessity of getting out of symptomatic treatment and removal of root cause to disconnect a vicious circle of repetition of similar crashes, collisions, or disasters.

Figure 8.22 summarizes the necessity to get out of symptomatic treatment—to remove a root cause—and to disconnect a vicious circle of repetition of similar crashes, collisions, or disasters in addressing safety management. The framework of traditional safety management searches for direct causes of crashes, collisions, or disasters, extracts a lesson from the identified direct causes and pays attention to the direct causes so they do not contribute again. However, similar crashes, collisions, or disasters occur because of hindsight biases or the lack of pursuing a root cause behind direct causes. Therefore, the new standard in safety management should aim at exploring the influences of cognitive biases (such as hindsight bias) in behavior or decision making on safety practice. Standards without such considerations unconsciously and unexpectedly lead to critical unfavorable crashes, collisions, or disasters as mentioned in Murata, Nakamura, and Karwowski (2015). In conclusion, we must pay attention to irrational behaviors or decision making in safety management, be oriented to pursue a root cause behind direct causes, and make every effort to eradicate cognitive biases in critical decision making on safety management. It is difficult to learn from failures appropriately (Syed, 2016) if the root causes or in-depth background factors analysis of crashes, collisions, or disasters are not thoroughly conducted.

As mentioned in the Introduction, we are more vulnerable to cognitive biases and fallible in our behavior or decision. This tendency becomes more outstanding in systems with high complexity and tight coupling. To identify root causes and eradicate them completely, cognitive biases that occur at both the individual and organizational levels should be bore in mind so that distorted decisions or behavior would not contribute to crashes, collisions, or disasters. This would be effective in managing the unexpected and making safety management strategies progress more drastically.

If designers of man-machine systems or safety managers (workers) do not understand the fallibility of humans, the design of the systems or safety management could pose incompatibility issues between our actual behavior (decision) and rational one, which, in turn, will induce crucial errors or serious failures, especially in decision making. To avoid such incompatibility, we must focus on when, why, and how cognitive biases overpower our rationality, distort it, and lead us to make irrational and unsafe decisions. The proposed approach to safety management (restrain the influences of near-sighted viewpoints and cognitive biases as small as possible) must be the key perspective for a preventive approach to eradicate the root causes of crashes, collisions, or disasters. If we can identify the cognitive biases behind the direct causes and determine where they may be inherent or likely within specific systems or procedures, we can introduce appropriate interventions or countermeasures for safety management that will ensure that the recognized cognitive biases do not eventually manifest themselves as root causes of unfavorable critical crashes, collisions, or disasters.

SUMMARY: TOWARD ESTABLISHMENT OF MORE EFFECTIVE STANDARD IN SAFETY MANAGEMENT

The identification of direct causes alone without the extraction of a lesson never contribute to eradicating the unexpected occurrence of similar crashes, collisions, or disasters. Standards in safety management should be based on preventive thinking that pursues a root cause and restrain even the smallest cognitive biases in decision making. Preparedness for a risk or root cause means no worry; paying attention to all potential risks and root causes is essential for new standards in safety management. In other words, we should master competence to pay attention to the root cause or source of accidents or errors as a whole. This will help prevent the repeated occurrence of similar crashes, collisions, or disasters and those beyond one's expectation. Advanced safety management also should consider the cross-cultural factors that contribute to unsafe behaviors (Murata, 2017; Nisbett, 2004).

REFERENCES

Bazerman, M. H. and Moore, D. A. 2001. *Judgment in managerial decision making*. Harvard University Press: Cambridge, MA.
Bazerman, M. H. and Watkins, M. D. 2008. *Predictable surprises*. Harvard Business School Press: Cambridge, MA, USA.
Bazerman, M. H. and Tenbrunsel, A. E. 2012. *Blind spots: Why we fail to do what's right and what to do about it*. Princeton University Press: Princeton, NJ, USA.
Dekker, S. 2006. *The field guide to understanding human error*. Ashgate Publishing: Cornwall.
Dobelli, R. 2013. *The art of thinking clearly*. Harper: New York: NY.
Gladwell, M. 2008. The ethnic theory of plane crash, In *Outliers*, 206–261. Back Bay Books: New York.
Gladwell, M. 2009. Blowup, In *What the dog saw*, 345–358. Back Bay Books: New York.

Kahneman, D. 2011. *Thinking, fast and slow*. Penguin Books: New York.

Mullainathan, S. and Shafir, E. 2014. *Scarcity: The new science of having less and how it defines our lives*. Picador: New York, NY.

Murata, A., Nakamura, T., and Karwowski, W. 2015. Influence of cognitive biases in distorting decision making and leading to critical unfavorable incidents. *Safety* 1: 44–58.

Murata, A. 2017. Cultural influences on cognitive biases in judgment and decision making: On the need for new theory and models for accidents and safety, In *Modeling sociocultural influences on decision making – Understanding conflict, enabling stability*, ed. J.V. Cohn, S. Schatz, H.. Freeman, and D.J.Y. Combs, 103–109. CRC Press: Boca Raton, FL.

Murata, A. and Moriwaka, M. 2017. Anomaly in safety management: Is it constantly possible to make safety compatible with economy? In *Advances in safety management and human factors (Advances in intelligent systems and computing 604)* ed. P. Arezes, 45–54. Springer: London.

Murata, A. 2019. Safety engineering education truly helpful for human-centered engineering: Toward creation of mindset for bridging gap between engineers and users, In *Global advances in engineering creation*, ed. J. P. Mohsen, M. Y. Ismail, H. R. Parsaei, and W., Karwowski, 73–92. CRC Press: New York.

Nisbett, R. E. 2004. *The geography of thought: How Asians and Westerners think differently...and why*, Free Press: New York.

Perrow, C. 1999. *Normal accidents: Living with high-risk technologies*. Princeton University Press: Princeton, NJ.

Perrow, C. 2011. *The next catastrophe: Reducing our vulnerabilities to natural, industrial, and terrorist disasters*. Princeton University Press: Princeton, NJ.

Reason, J. 1990. *Human error*. Cambridge University Press: Cambridge.

Reason, J. 1997. *Managing the risks of organizational accidents revisited*. Ashgate Publishing: Surrey, UK.

Reason, J. 2008. *Human contribution—Unsafe acts, accidents, and heroic recoveries*. Ashgate Publishing: Surrey, UK.

Reason, J. 2013. *A life in error*. Ashgate Publishing: Surrey, UK.

Reason, J. 2016. *Organizational accidents revisited*. Ashgate Publishing: Surrey, UK.

Syed, M. 2016. *Black box thinking: Marginal gains and the secrets of high performance*. John Murray Publishers Ltd: New York, NY.

Wiegman, D. A. and Shappell, S. A. 2003. *A human error approach to aviation accident analysis—The human factors analysis and classification systems*. Ashgate Publishing: Burlington: VT.

9 The Brazilian Federal Standard Applied to Ergonomics

Marcelo M. Soares, Eduardo Ferro dos Santos, Carlos Mauricio Duque dos Santos, Rosangela Ferreira Santos, Andréa Ferreira Santos, and Karine Borges de Oliveira

CONTENTS

INTRODUCTION

The affirmation that the use of ergonomics leads to comfort, health, safety, and efficient performance in products and processes is well known and proven by several authors and practical actions (Hendrick, 1996; Parker, 1995; Rowan & Wright, 1994; Vallely, 2014). Books, guidelines, standards, and scientific and working papers on how to do so are available for use in several areas. They show that ergonomics adds economic (improved efficiency), social (safety and health), and environmental (comfort) benefits to them.

The benefits and information related to ergonomics in all areas must be known—what is the need to have governments draft punitive legislation for those who will not make use of ergonomics? Punishment is already intrinsic since no science is used to only benefit products and processes. The guidelines developed—which are based on state-of-the-art research—and actions published already serve as a guide to those who do not have information about ergonomics. The question does not seem to make sense, since ergonomics can give more profit, protection, and several other benefits. However, there are governments that need federal laws to ensure that the benefits of ergonomics are used, as with the case of Brazil.

Brazilian law applied to ergonomics has existed since 1978 (Brasil, MTE, 1978). Its initial approach was based on typing work, mainly of bank employees. There were changes to the law in 1990, and these addressed the conditions of furniture, typing tools, the handling and transportation of materials, environmental comfort, and the principles of the organization of work (Brasil. MTE, 1990). Currently, the government is proposing a reform of the law since it was last revised 30 years ago; the characteristic of work has also evolved during this time. Therefore, the issue for this discussion is regarding the standard.

THE EVOLUTION OF ERGONOMICS STANDARDS IN BRAZIL

Brazil has a system of federal laws on health and safety at work established by Ordinance 3214 of 1978. It was placed within the Consolidation of Labor Laws, called Regulatory Standards.

The regulatory standards were issued by the Ministry of Labor and Employment, and it is mandatory for public and private companies, public administration bodies, and the bodies of the Legislative, Executive, and Judiciary branches of government—which have employees governed by the Consolidation of the Laws of Work—to comply with them (Mendes & Campos, 2004). Currently, the Ministry of Labor and Employment has been extinguished at the behest of the current president; labor laws are now the responsibility of the Ministry of the Economy (Nazareth, Yered, & Bastos, 2019).

The Consolidation of Brazilian Labor Laws in Article 157, Item I states that companies are required to comply with the regulatory standards. In item II, it mentions that the company is obliged to instruct employees, by means of service orders, about precautions to take to avoid accidents at work or occupational diseases. In Article 158, it states that it is the employees' responsibility to observe the safety and occupational medical standards, and to collaborate with the company in applying what these states—it is wrong should the employee unjustifiably refuse or not comply with the instructions given by the employer (Meller Alievi et al., 2017). It is the obligation of employers and employees—the former must comply with the requirements of regulatory standards, and the latter must follow the employer's recommendations for them to display preventive behavior.

Before regulatory standards were drawn up, there was no standardization whatsoever. In reference to the Ergonomics Regulatory Standard, NR 17 follows the basis published in 1990. In 2007, two appendices were added—one specifically to analyze work in customer call centers and telemarketing, and the other for supermarket checkout operators (NR 17, 2007). In 2018, given the constant concerns with measurements of environmental comfort, it was suggested that measurement methods and minimum levels of illumination based on occupational hygiene standards be included in NR17.

It can be said that the NR 17 of 1990 is still the same as that of 1978, even though there were a few paragraphs that formed part of the review. In general, the use of NR 17 since 1978 represented a considerable advance in structuring workstations in Brazilian companies, mainly in the 90s. Before this standard was drawn, changes in the organization of work were made exclusively at the initiative of companies, and then it became mandatory. The current problem is that by virtue of federal laws, the standard has become a protocol—as it is regarded as a documentary means for conducting inspections—and ergonomics ends up masking the benefits that it brings.

In 2002, the Ministry of Labor published the Manual for Applying NR 17 (MTE. Ministério do Trabalho, 2002) and recommended that the work take part in solving existing problems in workstations. This manual started to be used in inspections, as the auditors felt they had operational difficulties in applying the requirements. Not only was it the fiscal auditors who had such difficulties, but this extended to all who sought to apply ergonomics in line with the requirements of the standard—given the complexity of the subject and the few pages of text about this. This manual that has more pages than the standard—since it contains the standard and commentaries on it—provides recommendations on what tools to use, such as the NIOSH (the US National Institute for Occupational Safety and Health) method for lifting loads, and detailed explanations in observing the organization of work. Based on this method, many of those involved improved the quality of their work. The only problem is that the mentality of Brazilian managers is that federal laws must be explicit. In the case of the manual, it is a study material without legal value, and thus, it is a publication that offers guidance.

PAST, PRESENT, AND FUTURE OF WORK

If Brazilian federal laws were created to ensure the benefits of ergonomics, they must accompany and contextualize the era in which we find ourselves, the current state of the art, and accompany the transformations that have occurred at work in the last 30 years and the emerging technologies. They must, therefore, be kept under constant review.

We are in an era where technologies emerge and dominate the means of communication and information, for which we are dependent on new work tools (smartphones, tablets, notebooks, wireless internet) and collaborative environments (cloud computing, big data, integration of systems) (Rüßmann et al., 2015). These technologies have become fundamental in our everyday lives, as they allow mobility, fast communication, interaction between people and processes, and the rapid acquisition and sharing of information and knowledge. Artificial intelligence, human-robot interaction, and physical cyber systems coming from the 4.0 generation we live in are having a profound impact on the systems of work (Walker, 2017).

This evolution has its positive side and a worrying side. When we look at it from the point of view of ergonomics, this development demands even more than Charles Chaplin's famous "modern times" film (Chaplin, 1936). What it showed was the future of work in Chaplin's perception at that time. What he was referring to was industrialization, mass production, the evolution concepts of the industrial era of industry 3.0 (Loughlin, 2019)—an age before the one we are now living in. In industry 3.0, the worker was subject to repetition and pressures coming from organizational factors. In the age of industry 4.0, what is required is the connection, the interaction between systems—where workers are no longer subject to repetition, but to intellectual development, the activities of human-machine interactions, the development of integrated systems, artificial intelligence in error prevention, autonomous data collection and interpretation, and big data. Industry 4.0 can lead to intense intellectual demands. The readiness, the attention needed to avoid mistakes, and the surfeit of information lead to a mental overload that has never been seen in previous ages (Harris, 2017).

There are companies that are not yet experiencing the Industry 4.0 that are stuck with concepts from previous eras and seek to blame humans for errors at work. Errors are not always made by humans and can also result from the interaction of systems (Myszewski 2012). System interactions are common today, as companies employ more machines than people. It is not a question of unemployment; technological development can increase job opportunities. It is a question of technological evolution—some time ago, human beings operated a machine with physical demands and, currently, supervise several systems, which place great mental demands on humans (Young et al., 2015).

In Industry 4.0, at any time and in any place, the worker responds to numerous demands that require ability and flexibility to perform various functions in addition to those related to their scope of work. Work, leisure, and family are mixed on the screen of a smartphone, and the workload is predominantly mental. Recent research has identified a series of new problems with consequences for workers' health—both physical and psychological—due to the new demands of work and the inappropriate use of technological innovations. The number of cases of mental disorders is increasing dramatically. In Brazil, mental illnesses have become the major cause of sick leave when, for example, stress and depression merge. New diseases emerge. Burnout Syndrome is recognized in Brazil as an Occupational Disease (Moura et al., 2019; Volino, 2019).

BRAZILIAN REGULATORY STANDARD FOR ERGONOMICS

Every evolution has its positive and negative side, it is necessary to control the negative side. This is done by enacting laws. Brazilian standards are currently being modified, which is an opportunity for the country to update its laws.

Brazil is one of the leading countries in terms of the number of published papers on ergonomics and has one of the largest ergonomics congresses promoted by any ergonomics association in the world. In addition, it is the largest economic power in Latin America. Therefore, if the current research developed by Brazilian researchers in this area and the current state of the art in ergonomics is used as a basis in updating NR 17, Brazil can have a standard of excellence. Therefore, there is a guiding question in analyzing this change: is the proposal for the new NR 17 adequate for the age we are living in?

The revision of the standards is necessary (Maas, Grillo, & Sandri, 2018). The news that the rules would change was announced from the early days of the new government. The changes have already started, and the government has set an implementation schedule for 2020. Improving the drafting of the standards and simplifying procedures is the government's focus, given Brazil's need for economic growth. NR 17 has been redefined and entered the arena of public consultation where society will set out its views (Governo Federal, 2019). A public hearing on the NR 17 changes took place on 11 September 2019.

The first paragraph of the new NR 17 presents etymological adjustments. The phrase "to set parameters" of the NR 17 of 1990—which, in fact, never set parameters—was changed to "set guidelines and requirements." The word "health" is also inserted as an objective of the standard; the phrase, "in order to provide maximum comfort, safety, and efficient performance" has become, "in order to provide maximum comfort, safety, health, and efficient performance" (Governo Federal, 2019).

During the hearing, other items of the proposal were presented. The importance of placing the phrase, "this standard applies to all work environments and situations" was highlighted. It is an unnecessary insertion for those who practice ergonomics as it would be more economical to use "work systems" (Governo Federal, 2019).

The text proposed in the new NR 17 emphasizes that the organization must conduct a preliminary survey of work situations that require adaptation to a workers' psychophysiological characteristics, thereby foreseeing specific action plans for use in the future regulatory standard of the Risk Management Program. This piece of text points out another regulatory standard—that of risk management—leaving the NR 17 more related to analysis and not to management. This can give the impression that the standard is a ritual of analysis (Governo Federal, 2019).

As for the need for the ergonomic analysis of work, the new NR 17 mentions that micro-enterprises or small companies are exempted from drawing up elaborating ergonomic work analysis. This fact has been generating a great deal of discussion in the ergonomics community.

This is absurd. The paragraph reinforces what is perceived in the practice of Brazilian ergonomics in companies—namely those that seek analysis conducted by protocols—only to satisfy the requirements of the law. NR 17 is perceived as an inspection tool and is not used to promote comfort, health, safety, and better performance. The improvement of processes is due to the practices of Lean, Six Sigma, UX, and other organizational practices that use theories present in ergonomics but, in the first place, they point out the best performance and are not held to account in any legislation whatsoever.

Aspects of work organization such as production standards, operating mode, time requirement, determination of the content of time, work rate, and task content are not explained in the new NR 17. All that has changed is in what order they appear in the standard; a new manual will be needed. The new wording also establishes control measures to prevent workers—when carrying out their activities—from being obliged to carry out tasks on a continuous and repetitive basis. The problem will remain subjective. After all, there are no applications of concepts of repeatability to all tasks.

The public hearing saw the participation of professionals and professional associations, such as occupational medicine, labor gymnastics, oil companies, the Public Ministry of Labor, and entities of Safety at Work technicians and of Physiotherapy of Work. Unfortunately, the Brazilian Ergonomics Association—ABERGO—was not present but later sent a note. We remind readers that, in Brazil, ergonomics is not a regulated profession but a science that is added to a profession. Some of those present expressed the importance of reinstating the issue of work intervals in the new NR 17, a subject that was withdrawn. Attention to the cognitive and psychosocial aspects of work was not also included. It has taken 30 years for the opportunity to adapt the standard to current demands; it is not known when we will have a new review. This would be the moment to include the cognitive and psychosocial aspects, which are important to the emerging technologies.

The text proposed for the new NR 17 was finalized and will be published officially in 2020. In comparison with the old NR17 text of 1990, the text being proposed has maintained the essence of looking at the drawing board, the tools, and physical risks. The cognitive, psychosocial aspects have been left out. The organizational ones have not provided better definitions. The environmental ones have been partially maintained. Management is for another regulatory standard. The theories of human-machine and human-computer interaction are not even part of it.

The standard would have a good application in the Chaplin scenario, in the age of Industry 3.0, well outside the state-of-the-art expected by researchers and practitioners of ergonomics—a step backward. The rates of mental illness that head the indicators of medical leave in Brazil will remain neglected. Cardiovascular diseases, heart attacks, and strokes will continue to be exclusively human factors, unrelated to work. All the evolution and leadership in research was not even consulted to improve the standard—another step backward.

eSocial: An Electronic Information System

eSocial is an electronic information system (big data) that has been implemented by the Brazilian government since 2014 (Governo Federal, 2014). It will soon become mandatory for all companies; it has been in the implementation phase since 2018. The electronic system unifies the way of sending labor, social security, tax, and tax information from all companies to the government. It is a system created to improve the accountability of Brazilian companies to centralize the sending of workers' labor, social security, tax, and land information. Its aim is to create a large and complete database to improve government control and oversight. This means that companies must enter information related to accidents at work, environmental risks, dangerous substances the worker comes in contact with, use of personal protective equipment, and ergonomic risks the employee is exposed to (Sekula & Michaloski, 2018).

In ergonomics, we are talking about a standardization of risks—a list of words that must be adopted and the simplification of human factors. To think that ergonomics is limited to a list of words is another backward step. Ergonomic risks must fit into a single field on the screen, which is insufficient if we were to put information about all kinds of work. It increases the speed at which companies fill in the information that the government wants to know since they just need to copy and paste. The only advantage is that the data are cross-checked, and the information is managed—which is of interest to the government. If there are differences, fines will be imposed. Ergonomics that must fit into a single field adapts the worker to the system; the opposite of the principle of ergonomics.

OTHER BRAZILIAN STANDARDS

In addition to the laws of the Brazilian government, there is also the consultation of technical standards by using ABNT—the Brazilian Association of Technical Standards (ABNT, 2019). They are not punitive; they orientate and are true guidelines. Because they do not have the power of law, they are not always used and may not even be well known.

ABNT has endeavored to disseminate the application of technical standards for ergonomics aimed at human-system interaction and furniture. Many of these standards were developed by the International Organization for Standardization (ISO) and have been translated into Portuguese. The knowledge contained in these documents guarantees good ergonomics in processes and products:

The ABNT/CEE-126–Special Study Committee on Ergonomics studies Human-System Interaction for the adoption of fundamental standards in Software Ergonomics (ABNT/CEE 2020). The EEC established the ABNT NBR ISO 9241 series related to human-system interaction and software ergonomics:

- ABNT NBR ISO 9241–110:2012–Ergonomics of human-system interaction, part 110, which establishes principles of human-machine dialogue with ergonomic design principles.
- ABNT NBR ISO/TR 9241–100:2012–Ergonomics of human-system interaction, part 100, which introduces standards related to software ergonomics.
- ABNT NBR ISO 9241–210:2011–Ergonomics of human-system interaction, part 210, focused on the human-centered project for interactive systems.
- ABNT NBR ISO 9241–151:2011–Ergonomics of human-system interaction, part 151, with guidelines for web user interfaces.
- ABNT NBR ISO 9241–11:2011–Ergonomic requirements for working with Visual Interaction devices, part 11, with guidelines on usability.
- ABNT NBR ISO 9241–12:2011–Ergonomic requirements for working with Visual Interaction devices, part 12, which guides the visual presentation of information.

ABNT/CB-15 stipulates the Brazilian Furniture Committee, which has also contributed decisively to the application of ergonomics in Brazil (ABNT/CB-015, n.d.). Office furniture standards focused on tables, chairs, cabinets, and workstations, which are widely accepted by the furniture industry. These guidelines are found in the set of technical standards of ABNT that deals with office furniture, such as:

- ABNT NBR 13962: 2006—Chairs
- ABNT NBR 13961: 2010—Cabinets
- ABNT NBR 13967: 2011—Workstations
- ABNT NBR 13962: 2006—Office furniture, especially chairs, with requirements and test methods
- ABNT NBR 13966: 2008—Office furniture, especially tables, which are classified and have physical characteristics

The Update Must Come From Us, Brazilian Ergonomists, and Not From the Government

Focusingonthe benefits of ergonomics—presenting everything that is good that it offers and not in complying with federal laws and imposing rules with the omission of what is important—is the path to good ergonomics. There are new challenges and uncertainties to be administered by using emerging technologies. We need to "understand work to transform it," more than anything else.

Ergonomists need to interact with users and stakeholders and leave the spreadsheets and tools aside. If there are only benefits from ergonomics as a science, there is a need for current guidelines and disclosure to companies—not government laws. Legislation should exist to regulate chaos, not what is good—in this case, the use of ergonomics.

Technology itself should offer greater adaptability, should use resources and ergonomics efficiently, as well as improve the exchange of information. To name a few examples, sensor-based technology can help automatically prevent accidents or health problems for workers in real-time. Devices connected to the Internet of Things make it possible to monitor and send safety information that includes an employee's biometrics. This will help companies reduce exposure to risks, thereby improving workplace safety more effectively.

If work changes, workers change too. If they are inserted in the development process by ensuring quality education and information, we will witness a cultural revolution within companies. More skills, participation, and well-being will drive this transformation. Therefore, the manual dimension will criss-cross with the intellectual dimension, and workers will be led to make independent decisions and intervene in solving problems. It is not standards laid down by the government that allow good ergonomics aligned with technological development to flourish. But if that is the function of the government, it should be at least pointed out that if standards are made in this way, we are not moving forward.

Is There a Framework That Supports Good Ergonomics?

This chapter presents three case studies of ergonomic analyses of the task performed in a textile factory in Brazil. They show the technological evolution in conducting the task of transporting and moving a trolley (or buggy) with cones of yarn, with reference to the NR-17 Regulatory Standard for Ergonomics, which has been an integral part of Brazilian Legislation on Health and Safety at Work since November 1990.

The ergonomic analyses of the task of transporting and moving trolleys/buggies with cones of textile yarn are shown at three different times—in 2013, 2016, and 2019—with different technologies. This demonstrates the evolution of the technology used in the transport process. In Case Study 1, what is characterized as the production model belonging to what we currently call Industry 2.0 (2nd generation industry—Manual Activities); Case Study 2 has characteristics of Industry 3.0 (3rd generation industry—Mechanized Activities); and Case Study 3 has characteristics of Industry 4.0 (4th Generation Industry—Automated and Robotized Activities).

It should be noted that this process of continuous improvement in tasks and at workstations is an integral part of the Participatory Ergonomics Management System that the company implemented in 2012 and has continued to use until now (December 2019).

It should be noted that the Ergonomics Management system is based on the Standard NR-17 for Ergonomics and that the objective of the system is to implement ergonomic improvements in workstations and production processes to satisfy and comply with Brazilian labor legislation.

A CASE STUDY WITH FOCUS ON APPLICATION OF THE BRAZILIAN REGULATORY STANDARD FOR ERGONOMICS

CONTEXTUALIZATION OF THE ERGONOMIC ANALYSIS

The demand for ergonomic analysis is due to the need to identify possible ergonomic inadequacies in the tasks and at workstations of a textile yarn factory while implementing improvements in working conditions and meeting the requirements of CLT—Consolidation of Labor Laws (Brazilian Labor Law) related to Occupational Health and Safety. Also, it is mandatory to comply with the NR-17 Regulatory Standard for Ergonomics of the Ministry of Labor and Employment of Brazil.

This study only considers the ergonomic analysis of a single task at three different periods of the process of continuous improvement of the Participatory Ergonomics Management system in the company.

We point out that the services performed were carried by a consultancy company hired by the textile factory—which entered a provision of services contract—and that when the activities began, the coverage and scope of the services to be performed was established.

We received general and specific information from the Industry's Health and Safety Service Unit (SESMT). The team consists of one occupational safety engineer, three safety technicians, and one occupational health physician. This information is related to issues concerning safety and health at work.

GENERAL OBJECTIVES OF THE ERGONOMIC ANALYSIS

The objective of the ergonomic analysis is to comply with the Regulatory Standard NR-17 for Ergonomics and diagnose the possible ergonomic inadequacies to which the plant's operators are exposed to correct, improve, and provide adequate working conditions for comfort, safety, and efficient performance.

The ergonomic analysis performed addresses the working conditions recommended in the Regulatory Standard NR-17 for Ergonomics, namely:

17.1 Lifting, Transport, and Individual Unloading of Materials;
17.2 Workstation Furniture;
17.3 Workstation Equipment;
17.4 Environmental Working Conditions; and
17.5 Organization of Work.

METHODOLOGY USED AT THE ERGONOMIC ANALYSIS

The work methodology described below was adopted to undertake the ergonomic analysis of the workstations and the tasks of operators, with an emphasis on the activities of the task and the organization of the work and their correlations with possible ergonomic inadequacies in working conditions. The ergonomic analysis was carried out in three stages, which we describe below. The methodology used in this current study is found in Santos (2011).

Survey of Data of the Current Situation

Visits "in loco" (at workstations and/or tasks) to conduct interviews with managers/operators to collect information relevant to the tasks; moreover, to make a photographic record and film the real work situation of the workstations and tasks analyzed.

Diagnosis of the Current Situation

Analysis and evaluation of the information collected in the interviews and general and specific observations of the tasks were performed. The references include "the perception of the operators" and ergonomic references for the operator to perform his/her tasks in conditions of comfort, safety, and efficiency without harming their physical, psychological, and cognitive health.

Technical Recommendations and Measures to Make Tasks Ergonomically Adequate

These consisted of a brief description of the technical recommendations and measures to make the following ergonomically adequate: the workstation, tasks, the production process, the environmental conditions, and the organization of work so that the company can implement these measures—according to criteria and priorities set by the company itself—with advice from the consulting firm.

The aim of these suggestions is to comply with the recommendations of the NR-17 Regulatory Standard for Ergonomics and other regulatory standards to which the NR-17 makes references, thereby aiming to neutralize or mitigate risk situations and/or ergonomic inadequacies related to the "psychophysiological" characteristics of the workers (operators), as explained in the standard referred to.

COVERAGE OF THE ERGONOMIC ANALYSIS

The ergonomic analysis carried out exclusively addressed the aspects recommended by the NR-17 for Ergonomics as requested by the company and which we consider relevant to the nature of the task, namely:

5.1 Characteristics of the Lifting, Transport, and Individual Unloading of Materials (item 17.1 of NR-17): in this item, the biomechanical aspects in the performance of the tasks were analyzed.

5.2 Characteristics of the Furniture Used (item 17.2 of NR-17): considering the comfort and usability ratio of the furniture that the operators use.

5.3 Characteristics of Workstation Equipment (item 17.3 of NR-17): considering the relationship of comfort and usability of the workstation equipment that the operators use.

5.4 Characteristics of Environmental Working Conditions (item 17.4 of NR-17): considering comfort conditions for noise level, lighting level, effective temperature index, relative humidity, and ventilation. Calibrated equipments for measurements of the environmental conditions were used as recommended by the Brazilian legislation on Occupational Health and Safety: luximeter, decibelimeter, TGD 100, and globe thermometer—all manufactured by Instrutherm.

5.5 Characteristics of the Organization of Work (item 17.5 of NR-17): considering the production standards, the operating mode, the time requirements, the pace of work, and the content of the tasks.

5.6 Technical Recommendations and Measures to make tasks ergonomically adequate: considering the measures of prevention and of the ergonomic adequacy of the workstations and tasks, as well as the suggestions that the employees made in the interviews that were conducted to comply with all the items of the NR-17 for Ergonomics.

CASES ANALYZED IN THIS STUDY

What follows are the ergonomic analyzes of the three cases, always addressing the five minimum items required by the regulatory standard NR-17 for Ergonomics.

CASE 1: TRANSPORTING CONES OF YARN WITH A MANUAL TROLLEY (INDUSTRY 2.0—2013)

Ergonomic Analysis of Working Conditions to Comply With NR-17 Standard for Ergonomics

Evaluation Parameter: Regulatory Standard NR-17 of the MTE—Ministry of Labor and Employment.

Task Analyzed: Transport of Cones of Yarn using a Manual Trolley with casters (wheels) (Figure 9.1).

Compliance With NR-17 Items

17.1 Lifting, Transport, and Handling of Materials;
17.2 Workstation Furniture;
17.3 Workstation Equipment;
17.4 Environmental Conditions (noise, lighting, temperature, relative humidity, and ventilation in the workplace);

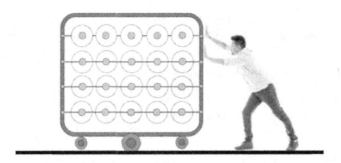

FIGURE 9.1   Operator manually pushing the trolley with cones of textile yarn on it.

17.5 Organization of Work (production rules, operating mode, time requirement, determination of the time content, and pace of work and task content).

In compliance with the requirements of the legislation (NR-17), we present a report of each item analyzed to identify the working conditions suggested in each item of NR-17 while considering the operators who perform the tasks. Oral interviews were carried out with the operators by the ergonomist and the occupational safety engineer. The latter also took measurements of the conditions of environmental comfort required by the regulatory standard NR-17.

The following is a brief description of each of the five items in the NR-17 Standard:

NR-17.1 Lifting, Transport, and Individual Handling of Materials

Operators carry out the tasks of placing the cones of textile yarn on the trolleys and manually transporting them with the cones from one sector to another within the factory. During this activity, the operators make a marked physical effort when handling the cones and in transporting the trolley with the cones, running the risk of developing skeletal muscle problems.

The operators carry out the activities while standing and are in constant movement throughout the sectors of the factory, transporting the trolleys with the cones of textile yarn. The operators take up postures with flexions and rotations of the spine when carrying out the tasks, especially when placing the cones on the trolley and removing them from it.

One cone has an average weight of 13 kg and one trolley holds an average of 40 cones (520 kg), while the trolley weighs 80 kg. Thus, the total weight transported is 600 kg.

Transport route of the cone trolley: 300 meters per trip (round trip), the average number of trips per working day with an eight-hour day: 120 trips, totaling an average of 3.6 kilometers while pushing the 600 kg manual trolley.

The operators do physical exercises in the workplace (they do stretch exercises for ten minutes at the start of the working day, three times a week). However, this is not enough to eliminate and/or reduce the physical fatigue that they mentioned they suffer from.

Under these conditions, operators find the working day exhausting. The ergonomist classifies it as a task of high ergonomic risk. This is at variance with what the NR-17 Standard for Ergonomics recommends, which requires the working conditions to be adapted to the psychophysiological characteristics of the operators.

NR-17.2 Workstation Furniture

No furniture is used to perform these tasks, nor is there a suitable bench for use when there are pauses in the task, as recommended by the NR-17 for Ergonomics; therefore, its specifications are not complied with.

NR-17.3 Equipment of the Workstation

In carrying out these tasks, the operators only use the cone trolley that is transported manually. This presents technical maintenance problems, which creates difficulties when moving it. The problem in the casters (difficulty of turning the wheel) creates discomfort and requires physical effort, which causes the operator to become fatigued.

Under these conditions, operators find the task exhausting and irritating, due to the excessive physical effort that needs to be made, which contributes to fatigue and stress (due to irritability). This disagrees with what the NR-17 Standard for Ergonomics recommends, which requires adapting the working conditions to the psychophysiological characteristics of the operators.

NR-17.4 Environmental Conditions of Work

The environmental conditions evaluated in the route and in the areas of moving and using the cone trolleys were collected during autumn (in Brazil) mornings and presented the following values:

- Lighting: minimum: 240 Lux/maximum: 320 Lux
- Noise: minimum: 78 dB/maximum: 84 dB
- Effective Temperature: 22.8°C
- Relative Air Humidity: 63%
- Air Speed: closed environment, with little ventilation

The environmental conditions partially comply with the NR-17 Standard for Ergonomics necessary to implement the adjustment of the sound pressure level, as the value is above what is recommended by the NR-17 standard. The other items (temperature, lighting, ventilation, and relative humidity) are within the recommended limits.

Operators use PPE—Personal Protective Equipment ("plug" type ear protector)—when carrying out tasks at the workstation. However, the reduction in the level of noise provided by the PPE is insufficient to provide the acoustic comfort recommended by NR-17.

Under these conditions, operators consider noise as an item of discomfort, and the safety engineer, after measuring the sound pressure level (dBA), confirms that it is above the level recommended by the NR-17 standard.

NR-17.5 Characteristics of the Organization of Work

Task Description (General)

1. Load the trolley with the cones, in sector 1;
2. Transport the manual trolley with cones from sector 1 to sector 2; and
3. Transport the empty cone trolley (without the cones) from sector 2 to sector 1.

The operators reported their tasks, and they were confirmed by the ergonomist "in loco."

Considerations Relating to the Organization of Work (Item 17.5)

a. Production Standards

The production and work safety standards established by the company are complied with by the operators. However, some operational procedures can be improved to avoid postural constraints, physical efforts, and the risk of accidents in the activities.

b. The Operating Mode

The operators carry out the activities prescribed by the company and partially establish their "modus operandi", as they must carry out the activities laid down in the operating procedure (a document of the quality control system) at the required time, depending on the type of production process.

c. Time Requirement

There is no requirement for operators to perform tasks in a minimum period. However, tasks are undertaken in line with production needs—which, in a way, is a time requirement—but the operators are not under time-pressure.

d. Determination of Content of Time

The operators carry out the activities of the task during the entire working day and it was found that they have neither downtime nor breaks during the working day (i.e. the occupation of time rate is practically 100%).

e. The Pace of Work

The pace of work is normal throughout the entire process according to what the operators reported. They partially manage their time —and, consequently, their pace of work—as this is dictated by production needs. In our perception, the pace of work at this post is 2 (moderate) for a Likert-type scale with three indicators, where 1 is low, 2 is moderate, and 3 is high.

As a result of the "moderate" pace of work and the high occupation of time rate, operators experience physical fatigue at the end of the working day, which can be a determining factor in triggering work-related occupational diseases (musculoskeletal problems and "stress").

f. The Content of the Tasks

Tasks are carried out in three shifts: from 06:00 to 14:00; from 14:00 to 22:00; and from 22:00 to 06:00, with a one-hour lunch break. The content of the tasks is compatible with the training and qualifications of the operators. They mentioned they did not have operational difficulties in performing the tasks/activities. However, they commented on the need to implement improvements in the maintenance of the cone trolleys because when they maneuver them, they need to make great physical effort, and this is exhausting.

Recommendations and Measures for Making Tasks Ergonomically Adequate

Item 17.1 of NR-17: Lifting, Transport, and Handling of Materials

- To improve the internal layout and the positioning of the trolleys on-site to facilitate handling them when they are maneuvered;
- To study the possibility of using a "tow-buggy" to transport the trolleys, thereby eliminating transporting the trolleys manually to avoid physical effort during the working day and experiencing muscle fatigue at the end of it; and
- To continue with preparatory physical exercises at work and create incentives to motivate operators to take part in these sessions.

Item 17.2 of NR-17: Workstation Furniture

- Provide benches to rest on to comply with NR-17 Ergonomics.

Item 17.3 of NR-17: Workstation Equipment

- Undertake maintenance of the cone trolleys more frequently (cleaning the wheels) or change the caster system (casters).

Item 17.4 of NR-17: Environmental Conditions at the Workstation

- Study the possibility of reducing the sound pressure level in the environment.

Item 17.5 of NR-17: Organization of Work

- Study the possibility of changing the process for transporting the cones [with a tow-buggy or even with an autonomous vehicle (robot)];
- Study the possibility of implementing breaks when performing tasks, which implies increasing the team of operators; and
- Study the possibility of implementing rotation of operators in sectors with different task to avoid fatigue and stress during the working day.

CASE 2: TRANSPORT OF CONE TROLLEYS USING A TOW-BUGGY (INDUSTRY 3.0—2016)

Ergonomic Analysis of Working Conditions to Comply With NR-17 Standard for Ergonomics

Evaluation Parameter: Regulatory Standard NR-17 of the MTE—Ministry of Labor and Employment.

Task Analyzed: Transport of Cones of Textile Yarn with a buggy that tows the Cone Trolleys (Figure 9.2).

Compliance With NR-17 Standard Items

- 17.1 Lifting, Transport, and Handling of Materials;
- 17.2 Workstation Furniture;
- 17.3 Workstation Equipment;
- 17.4 Environmental Conditions (noise, lighting, temperature, relative humidity, and ventilation in the workplace); and
- 17.5 Organization of Work (production rules, operating mode, time requirement, determination of the time content and pace of work, and content of tasks).

In compliance with the requirements of the legislation (NR-17 Standard), we present a report of each item analyzed to identify the working conditions suggested in each item of NR-17 Standard while considering the operators who perform the tasks. Oral interviews were carried out with the operators by the ergonomist and the occupational safety engineer. The latter also took measurements of the conditions of environmental comfort required by the regulatory standard NR-17 Standard.

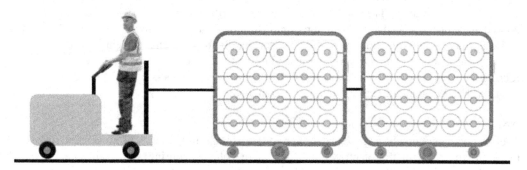

FIGURE 9.2　Operator using a tow-buggy with trolleys filled with cones of textile yarn.

The following is a brief description of each of the five items in NR-17:

NR-17.1 Lifting, Transport, and Individual Handling of Materials

Operators carry out the activities of the task that consists of placing the cones of yarn on the trolleys and transporting the trolleys with the cones from sector 1 to another sector 2 within the factory, using an electric tow-buggy. During the activity, operators make a marked physical effort when handling the cones.

However, they no longer make any physical effort in transporting the cone trolleys as before, which reduced physical fatigue at the end of the working day. The use of the tow-buggy visibly reduced the risk of developing skeletal muscle problems.

The operators stand to carry out their activities and move only when placing the cones on the trolleys and removing them.

The cone trolleys are transported by using an electric tow-buggy that the operator stands on and controls.

Operators take up postures with spinal flexions and rotations when placing the cones on the trolleys and removing them. One cone has an average weight of 13 kg and one cone trolley holds on average 40 cones (520 kg). The cone trolley is transported by the electric tow-buggy, thus eliminating the physical effort that the operator used to need to transport the cone trolley.

The transport route of the cone trolley: 300 meters per trip (round trip), the average of trips per working day with an eight-hour day: 120 trips, with an average of 3.6 kilometers, which is covered by the electric tow-buggy controlled by the operator. We point out that with the new tow-buggy, two trolleys are transported per trip—this doubles the number of cones transported per trip, reduces the pace of work, and doubles the capacity to transport cones of yarn.

The operators do physical exercises at work (they stretch for ten minutes at the start of the working day, three times a week). However, this is not enough to eliminate and/or reduce the physical fatigue mentioned by the operators.

In these conditions, the operators consider that the working day after operating the tow-buggy no longer leads them to have muscle fatigue. The ergonomist classifies it as a task of moderate ergonomic risk (due to the use of the tow-buggy, but they still handle the cones of yarn when loading and unloading the cone trolleys). This is partially in line with the NR-17 for Ergonomics recommendation, which requires adapting the working conditions to the operators' psychophysiological characteristics.

NR-17.2 Workstation Furniture

No furniture is used to perform these tasks, nor is there a suitable bench for use in the pauses of the task as recommended by the NR-17 Standard for Ergonomics. Therefore, this does not comply with the specifications of the NR-17 Standard for Ergonomics.

NR-17.3 Equipment of the Workstation

When the operators carry out the tasks, they use the electric tow-buggy and the cone trolleys that are pulled by the tow-buggy. The use of the electric tow-buggy contributed in eliminating the physical effort that the operator used to make in transporting the cone trolley from sector 1 to sector 2 manually.

Under these conditions, the operators consider that the task requires a light to moderate effort (only when handling cones of yarn when loading and unloading the cone trolleys), with breaks (during the working day, since this reduced the pace of work). This is partially in agreement with what the NR-17 Standard for Ergonomics recommends.

NR-17.4 Environmental Working Conditions

The environmental conditions evaluated along the route and in the areas of movement and use of the yarn trolleys with the electric tow-buggy were collected during spring (in Brazil), in the morning, and presented the following values:

- Lighting: minimum: 320 Lux/maximum: 420 Lux
- Noise: minimum: 78 dB/maximum: 92 dB
- Effective Temperature: 23.6°C
- Relative Air Humidity: 61%
- Air Speed: closed environment, with little ventilation.

The environmental conditions comply partially with the NR-17 for Ergonomics. The sound pressure level needs to be adjusted, as the value is above that recommended by the NR-17 standard. The other items (temperature, lighting, ventilation, and relative humidity) are within the recommended limits.

Operators use PPE—Personal Protective Equipment ("plug" type ear protector)—when carrying out tasks at the workstation. However, the reduction in noise level provided by PPE is insufficient to provide the acoustic comfort recommended by NR-17.

Under these conditions, operators consider noise as a discomfort item and the safety engineer, after measuring the sound pressure level (dBA), confirms that it is above the recommended by the NR-17 standard.

17.5 Characteristics of the Organization of Work

Description of the task of transporting the cone trolleys with a tow-buggy:

1. Implement the operational checklist of the tow-buggy (operational safety items);
2. Drive tow-buggy to the intermediate stock of sector 1;
3. Attach the 1st cone trolley to the hitch of the tow-buggy;
4. Attach the single hitch to the rear of the 1st cone trolley;
5. Attach the 2nd cone trolley to the single hitch of the 1st cone trolley;
6. Drive tow-buggy to sector 2 (final destination);
7. Uncouple the 2nd trolley from the single hitch coupled to the 1st trolley;
8. Uncouple the single hitch of the 1st trolley and place it on the support of the tow-buggy;
9. Uncouple the 1st Trolley from the tow-buggy;
10. Manually drive each tow-buggy to the respective box in sector 2;
11. Drive the tow-buggy to the box of the empty cone trolleys;
12. Repeat steps 3 through 9 for the empty cages.

The activities were reported by the operator and confirmed by the ergonomist "in loco" when monitoring the tasks and activities performed by the operator.

Considerations Relating to the Organization of Work (Item 17.5)

a. Production Standards

The production and work safety standards set by the company are complied with by the operators. However, some operational procedures can be improved to avoid postural constraints, physical efforts, and the risk of accidents in the activities (especially in loading the trolleys with the cones of yarn, which is done manually).

b. The Operating Mode

The operators carry out the activities prescribed by the company and partially establish their "modus operandi", as they must carry out the activities laid down in the operating procedure at the required time, depending on the type of production process.

c. Time Requirement

There is no requirement for operators to perform tasks in a minimum period. However, tasks are undertaken in line with production needs—which, in a way, is a time requirement—but the operators are not under time-pressure.

d. Determination of the Content of Time

The operators carry out the activities of the task during the entire working day and it is found that by using electric tow-buggies, they can pause during the shift (i.e. the occupation of time rate is practically 80% of the shift because according to the operators, it is possible to take breaks between trips).

e. The Pace of Work

The pace of work is normal throughout the process according to what the operators interviewed reported. They partially manage their time and, consequently, their pace of work, as this is dictated by production needs. In our perception, the pace of work at this post is 2 (moderate) for a Likert-type scale with three indicators, where 1 is low, 2 is moderate, and 3 is high.

As there is a "moderate" pace of work and the occupation of time rate is 80% of the working day, operators no longer experience physical fatigue at the end of their shift, which significantly reduced the likelihood of work-related occupational diseases (musculoskeletal problems and stress) being triggered.

f. The Content of the Tasks

Tasks are carried out in three shifts: from 06:00 to 14:00; from 14:00 to 22:00; and from 22:00 to 06:00, with a one-hour lunch break. The content of the tasks is compatible with the training and qualifications of the operators. They mentioned that they did not have operational difficulties in performing the tasks/activities and that since the electric tow-buggies started to be used, they no longer have muscle fatigue and tiredness at the end of their shift.

Recommendations and Measures for Making Activities Ergonomically Adequate

Item 1 of NR-17: Lifting, Transport, and Handling of Materials

- Study the possibility of facilitating the loading of trolleys with cones of yarn, which still requires physical effort and inadequate postures.
- Continue with preparatory physical exercises at work and create incentives to motivate operators to take part in these.

Item 2 of NR-17: Workstation Furniture

- Provide benches to rest on during breaks to comply with NR-17 for Ergonomics.

Item 3 of NR-17: Workstation Equipment

- Undertake the maintenance of the cone trolleys more frequently (cleaning the wheels) or change the caster system (casters).
- Study the possibility of "mechanizing" or "automating" the loading of trolleys with cones of yarn to eliminate the handling of cones, which requires physical effort and inadequate postures during this activity.

Item 4 of NR-17: Environmental Conditions at the Workstation

- Improve ventilation at the site (dissatisfaction with temperature during the spring).

Item 5 of NR-17: Organization of the Work

- Study the possibility of facilitating the loading of the trolleys with cones of yarn (mechanize/automate), which still requires physical effort and inadequate postures; and

- Study the possibility of implementing predetermined breaks when undertaking tasks, which implies increasing the team of operators.

CASE 3: TRANSPORT OF CONE OF YARN TROLLEYS USING AN AUTONOMOUS COLLABORATIVE ROBOT VEHICLE (COBOT) (INDUSTRY 4.0—2019)

Ergonomic Analysis of Working Conditions to Comply With NR-17 Standard for Ergonomics

Evaluation Parameter: Regulatory Standard NR-17 of the MTE—Ministry of Labor and Employment.

Task Analyzed: Transport of Cones of Yarn using an Autonomous Vehicle (Collaborative Robot—Cobot) (Figure 9.3).

(Collaborative Robot—Cobot)

Compliance With the Items of NR-17 for Ergonomics:

17.1 Lifting, Transport, and Handling of Materials;
17.2 Workstation Furniture;
17.3 Workstation Equipment;
17.4 Environmental Conditions (noise, lighting, temperature, relative humidity, and ventilation in the workplace); and
17.5 Organization of Work (production rules, operating mode, time requirement, determination of the content of time and place of work, and content of the tasks).

In compliance with the requirements of the legislation (NR-17), we present a report of each item analyzed to identify the working conditions suggested in each item of NR-17 while considering the operators who performed the task. Oral interviews were carried out with the operators by the ergonomist and the occupational safety engineer. The latter also took measurements of the conditions of the environmental comfort required by the regulatory standard NR-17.

There follows a brief description of each of the five items in NR-17:

NR-17.1 Lifting, Transport, and Individual Handling of Materials

The operators carry out the activities of the task which consists of placing the cones of textile yarn in the trolleys and transporting these trolleys with the cones from sector 1 to sector 2 inside the factory, using an autonomous vehicle (Cobot or Collaborative Robot). During the activity, operators make a marked physical effort when handling the cones.

However, they no longer make any physical effort when transporting the cones of yarn trolleys as before, which reduced the physical fatigue they feel at the end of the shift. The use of the collaborative robot visibly reduced the risk of developing skeletal muscle problems.

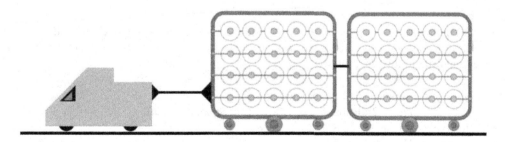

FIGURE 9.3 Autonomous vehicle for transporting trolleys packed with cones of textile yarn.

The operators stand when carrying out the activities and they only move when placing the cones in the trolleys and removing them.

The cones of yarn trolleys are transported by the autonomous vehicle—collaborative robot—and no longer by the operators.

The operators assume postures with flexions and spinal rotations when placing the cones in the trolleys. One cone has an average weight of 13 kg and one trolley holds an average of 40 cones (520 kg). The cones of yarn trolley are transported by the autonomous vehicle, eliminating the physical effort that the operator used to make to transport the trolley either manually or by using the tow-tuggy.

The transport route of the cones of yarn trolley, now pulled by Cobot: 300 meters per trip (round trip), an average of trips per working day with an eight-hour shift: 120 trips, totaling an average of 3.6 kilometers, which is covered by the autonomous vehicle. It is pointed out that with the new autonomous vehicle, two cones of yarn trolleys are transported per trip, which doubles the number of cones transported per trip, reduces the work rate, and doubles the capacity for transporting cones of yarn.

The operators do physical exercises in the workplace (there are stretching exercises for ten minutes at the start of a shift, three times a week), which helps minimize muscle fatigue and prevent the emergence of occupational diseases.

In these conditions, operators consider that their working day, since operating with the collaborative robot, no longer ends with muscle fatigue. The ergonomist classifies it as a low ergonomic risk task (due to the use of the collaborative robot, but the cones are still handled when loading the trolleys with cones of yarn). This partially complies with the recommendations of the NR-17 for Ergonomics, which requires the working conditions to be adapted to the operators' psychophysiological characteristics.

NR-17.2 Workstation Furniture

No furniture is used to perform these tasks, nor is there a suitable bench for operators to rest on when there are pauses in the task as recommended by the NR-17 for Ergonomics. Therefore, this does not comply with the specifications of this standard.

NR-17.3 Equipment of the Workstation

When the operators carry out the activities of the task, they use the autonomous vehicle. This contributed to eliminating the physical effort made by the operator in transporting the cones of yarn trolley from sector 1 to sector 2, for which the tow-buggy was previously used.

The fact that operators no longer carry out the task of transporting the cones of yarn trolley (which was done by operating the tow-buggy which pulled the cones of yarn trolley) by using the autonomous vehicle means that the operators no longer perform the transport activity.

Under these conditions, operators consider the task requires moderate physical effort (only when handling cones when the trolleys are loaded with these cones) and there are breaks during the shift; the pace of work has been reduced. This is totally in accordance with what the NR-17 for Ergonomics recommends.

NR-17.4 Environmental Working Conditions

The environmental conditions evaluated on the route and in the areas of movement and use of the cone trolleys with the collaborative robot were collected during winter (in Brazil), in the morning, and presented the following values:

- Lighting: minimum: 320 Lux/maximum: 420 Lux
- Noise: minimum: 72 dB/maximum: 78 dB (in the supply area)
- Effective Temperature: 20.8°C
- Relative Air Humidity: 60%
- Air Speed: closed environment, with little ventilation.

The environmental conditions partially comply with the NR-17 for Ergonomics, it being necessary to adjust the sound pressure level as it is above that recommended by the NR-17 standard. The other items (temperature, lighting, ventilation, and relative humidity) are within the recommended limits.

Operators use PPE—Personal Protective Equipment ("plug" type ear protector)—when performing tasks at the workplace. The new type of PPE used (plug type ear protector) allows a reduction of 15 dB (15 decibels). Therefore, the sound pressure level that is 78 dB is reduced to 63 dB, which is within the limit recommended by NR-17 for acoustic comfort (65 dB).

Under these conditions, operators consider the noise level to be comfortable.

NR-17.5. Characteristics of the Organization of Work

Description of the task of transporting cone trolleys with the autonomous vehicle—collaborative robot (Cobot):

1. Implement the operational checklist of the autonomous vehicle—collaborative robot (Cobot) (operational safety items);
2. Drive the collaborative robot (Cobot) to the intermediate stock of sector 1;
3. Attach the 1st cone trolley to the coupling of the Cobot;
4. Attach the single coupling to the rear of the 1st cone trolley;
5. Attach the 2nd cone trolley to the single coupling of the 1st cone trolley;
6. Drive the Cobot to sector 2 (final destination);
7. Uncouple the 2nd cone trolley from the single coupling attached to the 1st cone trolley;
8. Uncouple the loose coupling from the 1st cone trolley and place it on the Cobot support;
9. Uncouple the 1st cone trolley from the Cobot;
10. Manually take the cone trolley to the respective box in sector 2;
11. Drive the Cobot (just by pressing a button on it) to the empty box of cone trolleys; and
12. Repeat steps 3 through 9 for the empty cages.

The activities were reported by the operator and confirmed by the ergonomist "in loco" during the monitoring of the tasks and activities performed by the operator.

The detailed description of the tasks and activities is set out in detail in the Operational Procedure for the Task (an official document that describes the activities of the task of Transporting the Cone Trolley) with the autonomous vehicle—collaborative robot (Cobot).

Considerations Regarding the Organization of Work (Item 17.5)

a. Production Standards

The production and work safety standards established by the company are complied with by the operators. However, some operational procedures can be improved to avoid postural constraints, physical efforts, and the risk of accidents in the activities (especially in loading the trolleys with the cones, which is done manually).

b. The Operating Mode

The operators carry out the activities prescribed by the company and partially establish their "modus operandi," as they must carry out the activities prescribed in the operating procedure (which is a document of the quality control system) at the time required, depending on the type of production process.

c. Time Requirement

There is no requirement for operators to perform tasks in a minimum period. However, tasks are undertaken in line with production needs—which, in a way, is a time requirement—but the operators are not under time-pressure.

d. Determination of the Content of Time

The operators carry out the task activities during the entire working day and it appears that by using the autonomous vehicle, they get time for breaks during the working day (i.e. the occupation of time rate is practically 40% of the working day as the operators said breaks can be taken between trips and some tasks can even be performed outside the sector).

e. The Pace of Work

The pace of work is normal throughout the process according to what the operators interviewed reported. They partially manage their time and, consequently, their pace of work as this is dictated by production needs. In our perception, the work pace at this post is 1 (low) for a Likert-type scale with three indicators, where 1 is low, 2 is moderate, and 3 is high. Since the pace of work is "low" and the occupation of time rate is 40% of the working day, operators no longer experience physical fatigue at the end of their shift, which significantly reduces the likelihood of triggering work-related occupational diseases (musculoskeletal problems and stress).

f. The Content of the Tasks

Tasks are carried out in three shifts: from 06:00 to 14:00; from 14:00 to 22:00; and from 22:00 to 06:00, with a one-hour lunch break. The content of the tasks is compatible with the training and qualifications of the operators. The operators mentioned that they did not have operational difficulties in performing the tasks/activities and that due to the electric autonomous vehicle, they no longer have muscle fatigue and tiredness at the end of the working day.

Recommendations and Measures for Making Tasks Ergonomically Adequate

Item 1 of NR-17: Lifting, Transport, and Handling of Materials

- Study the possibility of facilitating and/or automating the loading of the trolleys with cones, which still requires physical effort and inadequate postures; and
- Continue with preparatory physical exercises at work and create incentives to motivate operators to take part in these.

Item 2 of NR-17: Workstation Furniture

- Provide benches to rest on during breaks to comply with NR-17 for Ergonomics.

Item 3 of NR-17: Workstation Equipment

- Carry out the maintenance of the cone trolleys more frequently (cleaning the wheels) or change the caster system (casters); and
- Study the possibility of "mechanizing" or "automating" the loading of the cone trolleys to eliminate the handling of cones, which requires physical effort and inadequate postures in the activity.

Item 4 of NR-17: Environmental Conditions at the Workstation

- Environmental conditions are adequate and recommendations are not necessary.

Item 5 of NR-17: Organization of Work

- To study the possibility of facilitating loading the trolleys with cones (to mechanize/to automate), which still requires physical effort and inadequate postures; and

- There is a possibility of reducing the team of operators by using the autonomous vehicle; there is no longer a need to drive a two buggy.

CONCLUSION

By carrying out the ergonomic analysis of the working conditions with the five minimum items required by the NR-17 Standard for Ergonomics of the Ministry of Labor Brazil as references, the inadequacies of the working conditions in the three cases analyzed could be identified. This also made it possible to make the recommendations of "ergonomic improvements" in the items evaluated, and implement these in a constant and evolutionary way, as verified in the three case studies presented.

As to the process of "continuous improvement" implemented in the task of transporting the cone trolleys from Sector 1 to Sector 2 provided by the Participatory Ergonomics Management System and the Zero Weight Program (with reference to the NR-17 standard for Ergonomics) implemented for six years (2013 to 2019), the process of ergonomic improvement shown in the three cases contributes to eliminating the main ergonomic risk of the task—"transporting" the cone trolleys manually (by pushing 600 kg for long periods of the working day).

The elimination of the ergonomic risk also contributes to eliminating the risk of developing occupational diseases, especially those arising from the inadequacies related to physical effort.

The use of the tow-buggy and, eventually, the autonomous vehicle Cobot, increased the sector's productivity since both the tow-buggy (Case 2) and the collaborative robot (Case 3) doubled the capacity of transporting the cone trolleys. The manual transport of cone trolleys was done one by one; the transport with the tow-buggy and with the collaborative robot is done with two cone trolleys at a time. In other words, 40 cones at a time are transported when the trolleys are maneuvered manually, while 80 cones are transported per trip when the tow-buggy or the collaborative robot is used.

Finally, the elimination of the tow-buggy operator reduced the number of operators for transporting cone trolleys. Operators could be reused in other sectors of the company, with less ergonomic risk in the task, and, consequently, better quality of life and well-being at work both from the point of view of physical demands and the cognitive demands of the task.

It should also be noted that the next step is to totally eliminate the ergonomic risks of the task. What is under study is the process of automating the loading of the cone trolleys (probably by using Cobots) to eliminate 100% of the physical effort made in this activity (manual loading the cone trolleys).

Note: we advise readers that at this time (December 2019), the Ergonomics Standard NR-17 for Occupational Health and Safety in Brazil is in the process of being "modernized/updated," and the same thing is happening with other regulatory standards related to Safety at Work. However, the current NR-17 Standard remains in force.

REFERENCES

ABNT. (2019). "ABNT—Associação Brasileira de Normas Técnicas." ABNT. http://www.abnt.org.br/.

ABNT/CB-015. (n.d.). "ABNT/CB-015—Comitê Brasileiro Do Mobiliário." Accessed February 3, 2020. http://www.abnt.org.br/cb-15.

ABNT/CEE-126. (n.d.). "ABNT/CEE-126—Comissão de Estudo Especial de Ergonomia Da Interação Humano-Sistema." Accessed February 3, 2020. http://www.abnt.org.br/cee-126.

Brasil. MTE. (1990). NR 17—Ergonomia. Brasil, MTE. http://trabalho.gov.br/images/Documentos/SST/NR/NR17.pdf.

Brasil, MTE. (1978). Portaria 3214 de 08 de Juho de 1978. http://trabalho.gov.br/participacao-social-mtps/participacao-social-do-trabalho/legislacao-seguranca-e-saude-no-trabalho/item/3679-portaria-3–214-1978.

Chaplin, C. (1936). Tempos Modernos (Modern Times)—FILME 968. Coleção Folha Charles Chaplin.

Governo Federal. (2014). "ESocial Governo Federal." 29/03/2017. http://portal.esocial.gov.br/institucional/conheca-o.

Governo Federal. (2019). "Consulta Pública NR 17." http://participa.br/secretaria-de-trabalho/secretaria-de-trabalho-norma-regulamentadora-17-nr-17.

Harris, D. (2017). *Engineering psychology and cognitive ergonomics*. London: Routledge https://doi.org/10.4324/9781315094465.

Hendrick, H. W. (1996). "The ergonomics of economics is the economics of ergonomics." *Proceedings of the Human Factors and Ergonomics Society Annual Meeting* 40 (1): 1–10. https://doi.org/10.1177/154193129604000101.

Loughlin, S. (2019). "Industry 3.0 to industry 4.0: Exploring the transition." *New Trends in Industrial Automation*. https://doi.org/10.5772/intechopen.80347.

Maas, L., Grillo, L. P., & Sandri, J. V. de A. (2018). "A saúde e a segurança do trabalhador sob competência de normas regulamentadoras frágeis." *Revista Brasileira de Tecnologias Sociais* 5 (1): 22. https://doi.org/10.14210/rbts.v5n1.p22-32.

Meller Alievi, A., Silveira, G. E., Merljak, L. V., & Libardoni, P. J. (2017). *A Nova Consolidação Das Leis Do Trabalho. A Nova Consolidação Das Leis Do Trabalho*. Editora CRV. https://doi.org/10.24824/978854441950.2.

Mendes, R., & Campos, A. C. C. (2004). "Saúde E Segurança No Trabalho Informal: Desafios E Oportunidades Para A Indústria Brasileira Saúde e Segurança No Trabalho Informal: Desafios e Oportunidades Para a Indústria." *Revista Brasileira de Medicina do Trabalho* 2, 209–223.

MTE. Ministério do Trabalho. (2002). *Manual de Aplicação Da Norma Regulamentadora N°17. Secretaria de Inspeção Do Trabalho – SIT*.

Moura, G., Brito, M., Pinho, L., Reis, V., Souza, L., & Magalhães, T. (2019). "Prevalência E Fatores Associados À Síndrome De Burnout Entre Universitários: Revisão De Literatura." *Psicologia, Saúde & Doença* 20(2): 300–318. https://doi.org/10.15309/19psd200203.

Myszewski, J. M. (2012). "Management responsibility for human errors." *TQM Journal* 24(4), 326–337. https://doi.org/10.1108/17542731211247355.

Nazareth, R. T., Yered, P. L., & Bastos, A. T. (2019). "A Simplificação e Desburocratização Da Segurança e Saúde No Trabalho Introduzidas Pela Norma Regulamentadora NR-01 | Nazareth | Anais Do Encontro Nacional de Pós Graduação." In *Anais Do Encontro Nacional de Pós-Graduação*. https://periodicos.unisanta.br/index.php/ENPG/article/view/2178/1676.

NR 17. (2007). "Norma Regulamentadora No 17—Ergonomia." *Brasil, MTE*, no. 17: 14. http://trabalho.gov.br/images/Documentos/SST/NR/NR17.pdf.

Parker, K. G. (1995). "Why ergonomics is good economics." *Industrial Engineering Norcross, Ga*. 27 (2), 41–47.

Rowan, M. P., & Wright, P. C. (1994). "Ergonomics is good for business." *Work Study* 43(8), 7–12. https://doi.org/10.1108/eum0000000004015.

Rüßmann, M., Lorenz, M., Gerbert, P., Waldner, M., Justus, J., Engel, P., & Harnisch, M. (2015). *Industry 4.0: The Future of Productivity and Growth in Manufacturing Industries*. Boston: The Boston Consulting Group.

Santos, C. M. D. (2011). *Ergodesign, legislation and standard: A contribution to the improvement in the poduct ergonomic quality, workstations and working conditions at production process* (In Portuguese). Ph.D. Thesis. Department of Production Engineering, UNIP-Paulista University, São Paulo, Brazil.

Sekula, E., & Michaloski, A. O. (2018). "The Events of Health & Safety in the Workplace Related to the E-Social." *Espacios* 39(32), 18–24.

Vallely, I. (2014). "Goog ergonomics is good economics." *Plant Engineer*. https://doi.org/10.1201/9781420006308.pt1.

Volino, W. (2019). "Síndrome De Burnout: Doença Ocupacional Presente Desde A Formação Até A Atuação Do Médico Especialista." In *Psicologia Da Saúde: Teoria e Intervenção*, 183–191. https://doi.org/10.22533/at.ed.70119120314.

Walker, A. M. (2017). "Tacit knowledge." *European Journal of Epidemiology*, *32*(4), 261–267. https://doi.org/10.1007/s10654-017-0256-9.

Young, M. S., Brookhuis, K. A., Wickens, C. D., & Hancock, P. A. (2015). "State of science: Mental workload in ergonomics." *Ergonomics*. https://doi.org/10.1080/00140139.2014.956151.

10 Design in the Absence of Standards

Daniel P. Jenkins and Pete Underwood

CONTENTS

INTRODUCTION

As is illustrated throughout this book, different industries or domains have their own specific standards to support the application of human factors. Ostensibly, at least, these standards serve three main purposes:

1. They describe the role of human factors in the design process—explaining what should happen at each stage of the design development process.
2. They provide guidance on the application of relevant tools and techniques—explaining which tools are available and how one might apply them.
3. They outline minimal acceptance criteria—for example, prescribing minimum text heights, maximum operation forces, etc.

As is evident from the other chapters in this book, the breadth and depth of this guidance differs between domains. Some industries are supported by detailed guidance covering most aspects of the design, whereas others provide high-level guidance. Ultimately, regardless of the domain, it is highly likely that design teams will be faced with situations where there is either no relevant standard available or the detail in the standard is ambiguous and further clarification is required. In these instances, there are two options available: (1) adopt a different standard or form of guidance or (2) derive guidance from first principles.

The first option—to adopt existing guidance—has a number of obvious advantages, most notably in terms of time and effort. The other clear advantage of adopting an existing standard or form of guidance is that it can simplify the process of gaining acceptance from regulators and reviewers—this is particularly relevant where several options are credible as it can provide an auditable justification for the chosen option. There will, however, be instances where the guidance is either unviable or inappropriate for the specific case in hand, necessitating derivation from first principles. The following two sections provide more detail on each of these cases.

ADOPTING STANDARDS AND GUIDANCE

As previously stated, adopting existing guidance is normally advantageous in terms of efficiency and gaining reviewer acceptance. While a lack of guidance is addressed in the next section, too much guidance can also pose a challenge—particularly when it is conflicting. The relative suitability of the available guidance will be context-dependent; however, as a rough heuristic, guidance materials can be ordered based upon their type. This hierarchy is structured in terms of the usefulness of the guidance and its ability to support a robust argument. This hierarchy can be broadly defined as follows, starting with the most suitable and ending with the least suitable:

- Applicable regulations
- Applicable domain-specific standards
- Generic international standards
- Generic regional standards
- Non-applicable regulations from parallel industries (e.g., train regulations applied to a tram or bus)
- Non-applicable domain-specific standards from parallel industries
- Non-applicable domain-specific standards from loosely connected industries (e.g., military standards applied to a consumer product)
- Domain-specific textbooks
- Generic textbooks
- Relevant journal papers
- Relevant conference papers

Applicable regulations naturally take the top position in the list as these are mandatory requirements. Applicable standards (e.g., from ISO, EN, BS, EEMUA), on the other hand, are, by definition, optional. However, if these are contractual requirements, they often must be treated as mandatory (although, unlike regulations, deviations may be negotiated). To manage expectations, applicable regulations and standards should be declared at the start of any project and agreed with all stakeholders (e.g., the design team, the client, reviewers, and regulators).

Following domain-specific standards, generic standards provide useful guidance on the more transferable aspects of human factors. For example, generic guidance on the design of control rooms is provided in a number of standards (e.g., EN11064, EEMUA 191, and EEMUA 201) regardless of whether the control room is for a power station, chemical plant, or control room for a piece of medical equipment. Other generic areas include human-computer interfaces, alarm management, lighting, thermal comfort, noise, and vibration. As a rule of thumb, those adopted as international standards are often more useful due to their wider acceptance.

The guidance provided across regulations and standards from different domains often has a common ancestry. As such, it is often highly compatible—after all, the humans at the center of these systems typically have consistent requirements and constraints. Where guidance is not available in applicable specific standards, domain-specific guidance from other industries may provide valuable information. An example of this would be the adoption of guidance on staircase design from building regulations for a vehicle. Wherever possible, it makes sense to seek guidance from similar domains that face similar constraints.

Textbooks can be used to provide additional guidance alongside standards. Guidance from key texts such as Pheasant and Haselgrave (2006, and its early editions) often form the basis of many standards relating to anthropometry. The obvious difference between textbooks and standards relates to the process of their creation. Standards are subject to a wider review and consultation process than would be typical for a textbook. Like standards and regulations, textbooks can be either generic or domain-specific. Domain-specific textbooks are often able to take greater account of the specific challenges of individual domains.

One challenge when sourcing information from multiple standards can be incompatibility between them. Where a conflict exists, it is advisable to record this conflict and clearly indicate which option is preferred along with some form of justification. Where possible, it is useful to define which standards will be used at the start of a document along with some indication of which standards will be used in what order. Working to a primary standard supported by others in a clear structure is often preferable to "cherry-picking" requirements from a wide range of standards to fit a design. Ultimately, a design becomes more justifiable if it is based on a smaller number of applicable standards.

Summary

In summary, the following points should be considered when adopting standards and guidance:

1. Focus on the most credible guidance first
2. Declare applicable guidance at the start of a project report
3. Nominate a lead guidance document
4. Avoid "cherry-picking" from between standards to meet a design case
5. Where conflict exists between standards, declare this along with the rationale for the chosen option

DERIVING GUIDANCE FROM FIRST PRINCIPLES

As previously discussed, deriving guidance from first principles can be a time-consuming process. However, it may be appropriate to do so when there is a lack of guidance available, or the guidance available is not suitable for the given context.

It may be the case that developing guidance to address all three key purposes outlined at the start of the chapter may not be necessary. For example, the role of human factors within the design project may already be defined and only a certain aspect of the project requires a novel combination of methods and/or acceptance criteria. However, for completeness, this chapter will describe how guidance can be developed to (1) describe the role of human factors in the design process, (2) provide direction on the application of relevant methods, and (3) outline minimal acceptance criteria.

Describing the Role of Human Factors in the Design Process

The type and amount of input a human factors team provides at the various stages of the design process will vary. This is likely to be influenced by the availability of staff and budgets and the novelty of the situation. Therefore, it is important to understand how this input will be combined with the activities of the wider design team. Fortunately, there is a considerable amount of guidance available on how to integrate human factors into the design process (e.g., ISO 9241-210; 62366). Such guidance often refers to a Human Factors Integration Plan (HFIP) as a systematic and documented means of defining various aspects of the project, for example:

- What the focus of the human factors activities will be (e.g., understanding and specifying the context of use, specifying user requirements, producing design solutions, and evaluating designs)
- When and where the human factors work will take place
- A description of the key inputs for the human factors activities
- The methods used to achieve the goals of the human factors activities
- A description of the outputs of the human factors activities
- Who the key stakeholders are (e.g., end-users, client organizations, internal stakeholders)
- Who will be delivering the human factors activity
- How the human factors work will be coordinated with the activities of the other design team members

Whether the human factors element of a project is written in a formalized HFIP document or captured elsewhere by the project team (in the form of a quotation or project plan), it is advisable to address and record the points listed above. By doing so, the role of human factors in the design process will be defined and can be communicated to project managers and the wider project team. On a practical level, any documentary evidence of human factors integration should be as clear and concise as possible to enhance its usability for the whole design team and maximize the scope for creative thinking (Wilson & Morrisroe, 2005). Some form of flow chart is commonly adopted for this purpose (see example in Figure 10.1). Color coding can highlight the activities of different roles. Furthermore, linking between activities can provide a clear indication of interdependencies.

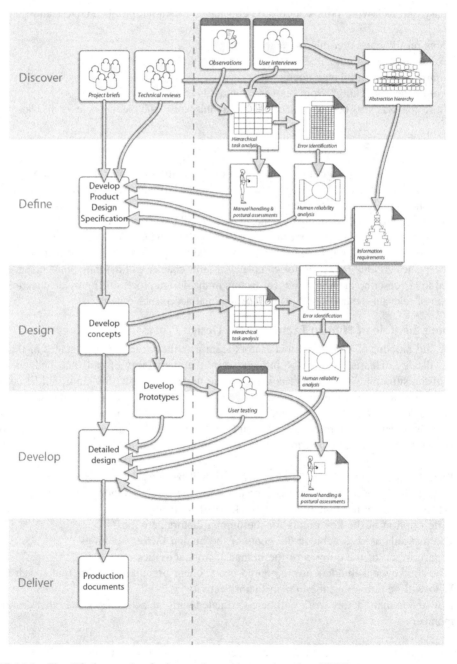

FIGURE 10.1   Simplified example of a human factors integration plan (HFIP).

Application of Relevant Tools and Techniques

The first step of the method selection process should be to determine what the aims of any data collection and analysis are. The aims of the activity of a human factor will vary between projects and across the different stages of the design process. At the start of the project, more exploratory tools should be selected to describe the constraints of a given domain and identify the needs of the users or stakeholders. Whereas, toward the end of a project, more objective measures are typically required that allow designs to be tested and assessed against acceptance criteria. Examples of methods that are best suited to a particular stage of the design process are provided in Figure 10.2.

To make a judgment about a method's suitability, it is important to understand what insights it can offer, its resource requirements, its benefits, and drawbacks, how it compares to alternative methods, and how it complements other methods. Various sources of information that provide these are available (e.g., Karwowski, Soares, & Stanton, 2011; Stanton et al., 2013; Wilson & Corlett, 2005).

Individual methods, clearly, each have their own strengths and weaknesses. Subjective methods, based on interviews, can provide rich detailed insights; however, they may not represent an accurate description of the activity as it may be observed. As such, it is normally advisable to combine a number of methods to balance these weaknesses out. An example of this may include a

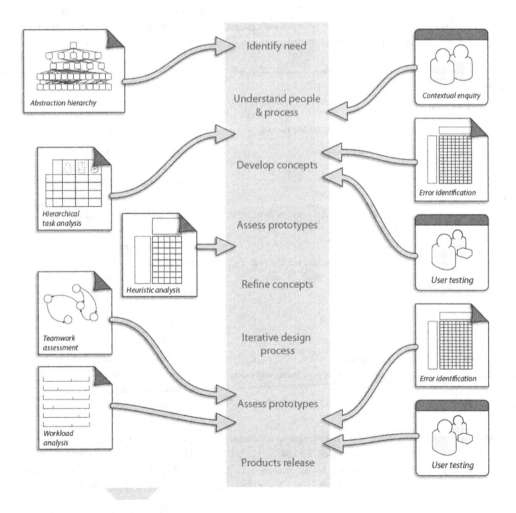

FIGURE 10.2 Application of human factors methods in the design process.
Source: Adapted from Stanton et al. (2013).

descriptive approach, such as Hierarchical Task Analysis, complemented with a more formative approach, such as Cognitive Work Analysis.

Outlining Minimal Acceptance Criteria

If minimal acceptance criteria need to be devised, it is necessary to consider what attribute of a product/ system they are assessing—for example, characteristics relating to safety, performance, or satisfaction. If no relevant guidance is available, then the acceptance criteria can be generated by consulting project stakeholders and/or subject matter experts. Alongside an appropriate range of end-users, these stakeholders/experts should include representatives from sales & marketing, maintenance, and management.

At the start of a project, a list of the factors that are important to a product or system can be established via interview. Simple techniques such as pair-wise comparisons can be used to establish a hierarchy of values.

Benchmarking of other products or systems can be a useful starting point for defining acceptance criteria. This may include legacy or competitor products. Likewise, they may include products from adjacent markets—for example, expectations for user experience may be set from other categories (e.g., car user interfaces defined by experiences from smart phones).

As a project nears its completion the importance of objective measurement of these values increases. It must be possible to evaluate a given characteristic to understand whether the relevant acceptance criteria have been met. Therefore, the acceptance criteria should refer to features of a product/system that can be directly or indirectly measured. Examples of these are captured in Table 10.1.

TABLE 10.1
Types of Metrics

Metric	What to Measure	Example Methods
Efficiency	Task completion time	Task analysis techniques
		Time and motion studies
		Process charting methods
		Link analysis
Efficacy	The key system metrics (these will be system-specific)	Team assessment methods
		Performance modeling
		System modeling tools
Safety	Error likelihood, manual handling risk	Safety analysis
		Human error identification
		Human error predictions
		Manual handling assessments
		Postural assessments
		Hazard identification
Intuitiveness and usability	User acceptance	User testing (interviews and observations)
	Compliance with accepted norms and expectations	Empathic modeling
		Heuristic checklists
		Workload assessments
		Interface analysis
Resilience	How many ways can the task be completed?	Task analysis techniques
	Who else can complete the task	FRAM
Inclusiveness	Key dimensions and forces, location and arrangement of components, color choices, labeling, and text characteristics	Desk-based assessments based on engineering drawings and anthropometric data sets
		Fitting trials using production products, prototypes, or mock-ups

Wherever possible, these acceptance criteria should be measurable and testable. The test method and the acceptance criteria should be unambiguously described (typically, in the HFIP).

Summary

Any guidance that is developed from first principles is likely to be questioned and scrutinized by reviewers. As such, it is important to clearly and concisely document the following:

1. The reason why guidance has not been adopted
2. A clear description of the process undertaken
3. A description of the rationale for the methods selected
4. A detailed description of the acceptance criteria along with an explicit link to real-world task performance.

CONCLUSIONS

Design in the absence of standards creates both opportunities and challenges. When working with regulators and reviewers, conformance to established guidance can expedite the review process by reducing the need to justify each decision. Accordingly, the adoption of standards and guidelines offers efficiencies both in the formulation of direction and in the review process.

In instances where the guidance available is unsuitable, bespoke guidance can be developed from first principles. This is particularly useful for very specific cases. Furthermore, it could be argued that innovative approaches to research and analysis are more likely to result from more innovative outputs.

For the vast majority of cases, some combination of prescribed and adopted guidance is expected along with elements of guidance derived from first principles.

REFERENCES

Karwowski, W., Soares, M. M., Stanton, N. A. eds. (2011). *Human factors and ergonomics in consumer product design: Methods and techniques*. CRC Press: Boca Raton, FL.

Pheasant, S. & Haselgrave, C. (2006). *Bodyspace: Anthropometry, ergonomics and the design of work*, 3rd edition. Taylor and Francis.

Stanton, N. A., Salmon, P. M., Rafferty, L. A., Walker, G. H., Baber, C. & Jenkins, D. P. (2013). *Human factors methods: A practical guide for engineering and design* (2nd edition). Ashgate: Aldershot.

Wilson, J. R. and Corlett, E. N. eds. (2005). *Evaluation of human work*, 3rd edition. Boca Raton: CRC Press, 2005.

Wilson, J. R. & Morrisroe, G. (2005). Systems analysis. In Wilson, J. R. and Corlett, E. N. (eds.) *Evaluation of human work*, 3rd and completely revised edition. Boca Raton: CRC Press.

11 Standards of Human Factors/Ergonomics in Cuba
Theoretic-Practical Considerations Developed

Sandra Haydeé Mejías Herrera

CONTENTS

INTRODUCTION

The main purpose of this chapter is to provide wide information about the development of ergonomic standards in Cuba, although some studies carried out through the years may be omitted since the information is dispersed. These standards are associated with the development of ergonomics in Cuba and the world, which makes it necessary to mention the diverse institutions, investigation groups, and universities that have supported this process in the country. Among these institutions are organs such as the Ministry of Work and Social Security, which have attended to the worries of ergonomists. Still, the country does not boast a society of ergonomics. The necessary competencies have not been defined for ergonomic action nor has a middle-term policy on ergonomic material been established. It is possible to recognize the line of work formed by the ministry mentioned earlier, which joins the technical committee for review, the update of the norms in this material, and the adoption of the norms of the International Organization for Standardization (ISO). The present chapter tends to initiate the spread of the work of ergonomists in the country where they have participated and supported—Latin American institutions such as the Latin American Union of Ergonomics (ULAERGO), the International Association of Ergonomics (IEA), and universities of the developed world as well as those of Latin America—which has allowed the existence of a group of specialists in diverse lines of work, hence promoting investigations.

EVOLUTION OF HUMAN FACTORS/ERGONOMICS IN CUBA

The evolution of ergonomics at the international level is characterized as passing from a *reactive design* to an approximately *proactive design* (Salvendy, 2006). Without a doubt, this step has marked important phases in this development; at the beginning, the focus was on the interactions of man-machine that has evolved to reach the interactions of man-technology. The present worries about technology show that ergonomists attend to a complete system that consists of persons, the organization, and its processes and equipment. All these happenings in the world have opened new doors to areas where ergonomists integrate with specialists of different fields to optimize the performance of productive systems, processes, services, workstations as well as increase the usage of products to a superior level.

Some considerations of Cuban specialists state that the concept of ergonomics was introduced at the beginning of the decade of the 70s. In this decade, the country was engaged in intensifying the development of its fragile economy inherited after the triumph of the revolution in 1959—characterized by small enterprises mainly along the line of artisanship with machines that were worn out in most cases and with a workforce that was almost not specialized.

It is daring to state the different moments that transpired in the introduction and development of ergonomics in Cuba right from the beginning to the present day. The author considers it worth a risk to establish the periods that mark the turning points of its adoption. Surely, other authors in the future will have the opportunity to enrich them and this chapter will mark the beginning for them. This refers to the continuation of the different development periods of ergonomics in Cuba.

First period (during the decade of the 70s): Existence of the need to install industries in all the new industrial regions and advanced technology for industrialization of the country.

Precisely in 1970, Engineer Angel Luis Portuondo—president some years later of the National Tribunal of Industrial Engineering for Doctorate defenses—introduced the concept of ergonomics in the post-graduate courses in Cuba. Universities of industrialized countries such as Canadian University Service Overseas (CUSO) started collaborating with the Industrial Engineering Faculty of José Antonio Echeverría Institute (ISPJAE) found in the capital of the country, and with the Industrial Engineering Faculty of Central University "Marta Abreu" of Las Villas (UCLV). These institutions received professors of high level from Magdeburg University in Germany.

In both universities of the country, they imparted courses on ergonomic topics, provided specialized bibliography, donated equipment to measure noise and illumination, and installed veloergometers in the laboratories of Industrial Engineering.

The training in this period led to Professor Angel Luis Portuondo obtaining a master's degree at Cranfield Institute of Technology in England and Professor Vicente López was the first Doctor in Ergonomics from the Institute of Pozna In Poland. The topic dealt with in these studies was "thermal comfort equation in subtropical climates."

They also developed other courses as part of an international project, where high-level professors of the University of Waterloo, Canada, and the University of Nottingham in England visited Cuba. In 1976, ergonomics was included as a subject in the curriculum of the course of Industrial Engineering in two universities: José Antonio Echeverría Institute and Central University "Marta Abreu" of Las Villas. Together, results of the investigations that had been carried out in those few years by ergonomists were introduced in classes.

The essential characteristic of this period was the international collaboration to train the first ergonomists who would impart classes in universities and start investigations in the country.

Second period (during the decade of the 80s): Professor Silvio Viña got the second doctorate in Poland with the topic of "ergonomic and physiological aspects of work in agriculture." The advance of this discipline in Poland and the contributions that these professors conveyed in their doctorates caused them to be considered as pioneers of ergonomics in Cuba. During 1980, deeper studies in ergonomic material were carried out. In 1988, the first doctorate of ergonomics was defended in Cuba by Professor Joaquín García of "Camilo Cienfuegos" University of Matanzas located in the west of the country. The topic of investigation was "a research in mental workload in the sugar industry."

Cane sugar was the principal product which the country used to export, and this industry had been promoted by the establishment of several sugar factories. Also, the number of workers associated with this period made it possible that the contributions made from the studies of doctorate thesis allowed to take important measures in the improvement of work conditions. This permitted ministerial organisms to set regulations to protect workers by guaranteeing adequate working hours, rest, and safety measures in the sugarcane fields particularly during the cutting activity and in the sugar factories during the production of cane sugar and its derivatives.

During these years, ergonomic laboratories with equipment designed by professors of ergonomics, Cuban technicians, and equipment received through the international collaboration with universities of industrialized countries were installed in the principal universities of the country.

Toward the end of this period, something important that distinguished the training in this discipline occurred—the incorporation of the study of ergonomics in the Institute of Industrial Design in Havana in 1988. This is the only institute until today that trains designers in different specialties. Including this discipline, this field of design opened opportunities for the development of systems and products with ergonomic attributes and made the activity of industrial design important in the growth period of the Cuban economy.

Other institutions have worked and are still working in the field of ergonomics. For example, the Institute of Work Studies and Investigations (IEIT), the Institute of Work Medicine, the Institute of Physical Culture, and the National Institute of Workers' Health (INSAT) are specialists of high prestige. They deal with particular areas of ergonomics and have specialized in important topics that accredit them to be consultants of enterprises at the national level and to be a decisive part of technical committees who work during the new century for the design and implementation of norms.

It is possible to declare that the role of these ergonomists in this period has been concentrated on the necessary investigations to carry out in the industries of the country, and post-graduate training is done on a small scale only under specific requests of institutions and universities.

Third period (during the decade of the 90s): The historic moment that was lived in the country at the beginning of the year 1990 was characterized by the beginning of the economic crisis owing to the occurrences that transpired in socialist countries. This makes some specialists consider that a decrease occurred in the activities related to ergonomics. However, despite the existing situation that affected the investigations, specialists of the Institute of Work Studies and Investigations continued working, as will be discussed in posterior epigraphs. In the Central University "Marta Abreu" of Las Villas, the software ERGOSIZES version 1.0 was made to create databases with anthropometric information, measure human body dimensions, and determine percentiles (CENDA, 1999).

The training did not cease despite the economic situation of the country and in this period, the program of post-graduate education in Human Resource Management commenced in the José Antonio Echeverría Institute. Professors and specialists of different universities and companies in the country participated and one of the lines of investigation was ergonomics, although the monographs developed about it were not numerous. The program allowed not only to elevate the academic level of professors and the investigative level of specialists of companies but also enabled those who were investigating in ergonomics to form work teams for ergonomic intervention.

This year marks the beginning of the education of professors in courses imparted by the MAPFRE Insurance of Spain through projects that competed with other participants from the whole of Latin America and Europe. The courses imparted by MAPFRE allowed the training of the youth and, in some cases, enabled them to continue this discipline as their line of investigation or confirm it.

Nevertheless, the investigations continued with an essential characteristic of being carried out in the form of isolated investigative groups, although the exchange and consultancy between the specialists of different regions of the country was always an essential and necessary compliment.

In this period, the ergonomists are faced with the essential challenge of evaluation and redesign of workplaces of diverse industrial activities in the country.

Fourth period (since 2000 until now): The interest in Human Resource Management and Management by Competences is renewed and intensified in the country and with it comes more interest in ergonomics and the present areas of interest. Professor Sandra Mejias of the Central University "Marta Abreu" of Las Villas defends her doctorate about the topic "macroergonomic tools for the improvement of work systems" (Mejías, 2003)—this was the second doctorate about this topic in the country, owing to the investigations of the ergonomics group of this university where monographs of undergraduate studies and post-graduate training were produced. The topics dealt with by this group included manufacturing companies (bio industries, shoe, and textile) and those of services (catering company), where the places and systems of work were evaluated by micro- and macroergonomics (Mejías, 2000, 2001, 2002). The Provincial Academic Prize was received for these results. In 2004, the software ERGOSIZES version 2.0 received the first national prize together with José Antonio Echeverría Institute in the National Concourse of Computation for students, being evaluated positively by the Ministry of Work and Social Security experts (CENDA, 2005).

This version of the software incorporates new dimensions, shows the image of the represented dimension of the human body in 3D, gives an explanation about where each dimension is located and calculates the percentiles that the client chooses to proceed with the design.

From that moment until 2012, there was intensified training in the system of Human Resources Management in the country. This was due to the need to improve these processes that were not at the level of international tendencies.

A consultancy company, together with the Ministry of Work and Social Security, created the Human Resource Network in 2007, which grouped specialists in human resources all over the country. For the first time, this company organized courses of ergonomics for specialists of companies. At the same time, bases for a management system—like the one mentioned before—are established; for the first time, such a system included among its norms indications related to ergonomics and the actions to be undertaken in the companies.

The ergonomics investigation group of Central University "Marta Abreu" of Las Villas worked with specialists from companies with the aim of training them in the processes of human resources. The topic of ergonomics in these trainings is dealt with when the system of human resources is imparted—though to a very limited extent, where the necessary physical and mental aspects for the selection of personnel, positioning of people in workplaces, and the ergonomic risks present among the detected risks during the elaboration of the profiles of workplaces are mentioned. Like this, the activity of the investigation group in ergonomic material decreases, and the results of the investigation are associated with particular processes of human resources.

The same happens in the José Antonio Echeverría Institute, where post-graduate training is dedicated to these processes of human resources. However, Professor Silvio Viñas leads the way for young investigators, as it is explained in the third epigraph of this chapter. At this moment, the number of universities that impart the course of Industrial Engineering has increased and, in 2010, approximately 16,000 Industrial Engineers graduated. These Engineers received courses in ergonomics but unfortunately, few of them apply it after in their occupational life.

New doctorates are defended in this period. In José Antonio Echeverría Institute, Professor Jordán Rodríguez evaluates the exposure to musculoskeletal risk factors (ERIN) (Rodríguez, 2011), and Professor Grethel Real of Matanzas University elaborates index for the ergonomic evaluation of waitresses in the tourism sector (Real, 2012).

Until now, training and investigations of professors, some specialists, and investigators in study centers continue with the desire to put in practice post-graduate programs (masters and doctorates) in ergonomics. With the help of the Latin American Union of Ergonomics and the International Association of Ergonomics, Cuba participates in regional and international conferences to divulge in the world investigations in this field. The first National Ergonomics Conference was held in 2012 supported by the Ministry of Work and Social Security and the Latin American Union of Ergonomics, where important and educative agreements in ergonomic material were reached. Its publication on the website of the Latin American Union of Ergonomics activated the region and revived the interest in ergonomics in Cuba. In the international conference of designers, a workroom was dedicated to ergonomics, and conferences of interest to the audience were imparted.

If this period were to be characterized, it would be important to note that the level of the investigations carried out as well as their quality has been the first challenge that has allowed to consolidate ergonomics in the country. The second is the level of training of Cuban ergonomists characterized by high rigor obtained from international programs of the first-world and Latin American programs of post-graduate that allows them to dedicate doctorate studies to professors of Latin American countries. Thirdly, the challenge of isolation among ergonomists in the country and uniting them in conferences has been overcome.

However, there are still important inadequacies that will be presented on continuation and show that today, the work that remains ahead is immense.

PRACTICAL APPLICATIONS OF INVESTIGATORS, PROFESSORS, AND WORK GROUPS FROM UNIVERSITIES AND INVESTIGATION CENTERS

The second epigraph shows how the investigations and results of studies in ergonomic material were developed attached to the different periods of transition of ergonomics in Cuba. Important centers and institutes have participated in these results. The Institute of Work Studies and Investigations already mentioned was inaugurated officially in 1974 and stopped being an institute in February 2013, assuming the functions of the Ministry of Work and Social Security. In various moments, the Institute of Work Protection that was united with the Institute of Work Studies and the Institute of Transport Investigations also had an important participation. Focusing on this, it can be mentioned that in the first period (decade of the 70s), important investigations and ergonomic applications were carried out in Cuba in topics such as agriculture machine design, analysis of manual work in agriculture, comfort thermic zones during work, determination of the physical work capacity of Cuban workers, and anthropometric measurements of the Cuban population. These results, as it was shown, were a product of the first trainings that were conducted in this material.

Precisely, the study of the Cuban population by the Department of Anthropology of Havana University in the first years of the 70s constituted an important result developed in the country for the first time, though less divulged and employed in the process of industrialization.

In the second period, important results related to the design of methodologies for the analysis of mental load in the sugar industry were obtained. As it was expressed before, sugar was the principal export product of our country and the study was directly related to the existing economic necessities. Nevertheless, the applications had a reduced extension and did not survive with time, although they enabled them to continue producing replications in other industries of the country. Other important results that cannot be left out are those related to the role played by the Institute of Work Studies and Investigations at the national level. For example, in 1981, this institute under the Ministry of Work and Social Security carried out an ergonomic study in the citric industry and in workplaces where packages of citric were manufactured for national and international consumption (IEIT, 1981). This industry has also been so important to the Cuban economy, but the design of corresponding matting, heights, and ranges had not been defined adequately. Two regions of the country are known for harvesting the highest quantity of this fruit and these studies which are conserved today as part of the patrimony of the work carried out, were directed toward these places. Although ergonomic thinking was still absent in entrepreneurs, it did not allow the applications to be maintained through the years and some of the actions put in practice were lost due to reparations and investments made.

In 1987, an analysis on the application of ergonomic requirements and work safety in new investments was done and the Institute of Work Studies and Investigations directed and applied these studies (IEIT, 1987). Also, the existent context demanded those studies in the country and this institute to carry out and respond to that necessity.

In the third period, exactly in 1993, three specialists of the Institute of Work Studies and Investigations participated in an ergonomic evaluation of distinct variants of design of a prototype of a Cuban bicycle (ISDI, 1993). At that moment, the reduction in the number of vehicles for transport in the country made it necessary to come up with variants related to the tradition of the people like the bicycle —not only used to pass time but also to travel to work. In the investigation, the physiological criteria of a prototype bicycle designed in the Institute of Industrial Design were discussed and recommendations for the design of a Cuban bicycle were formulated. More than one industry in the country started the manufacture of bicycles and this study was a step forward, where designers and ergonomists joined effort to produce a necessary solution for the country.

The designing of the first version of ERGOSIZES software in this period with the participation of undergraduate students and computer laboratory technicians marked the first step of the results obtained in the Industrial Engineering faculty of Central University "Marta Abreu" of Las Villas. It only remained in the reach of experts of the Institute of Work Studies and Investigations, entrepreneurs, and designers of the central region of the country who gave a positive opinion about it, but due to the magnitude of the economic cost implied, it was not possible to confirm databases of anthropometric data with it.

In 1995, in the bio-industry of the Institute of Biotechnology of Plants found in Central University "Marta Abreu" of Las Villas, a psychophysiological evaluation of the workplaces was done in the laminar flow cabins where workers were known to be affected by skeletal-muscular pains. The beginning of this study was not because of requests from the bio-industry but as an academic and investigative interest to study those workplaces. However, the results showed that the workers were not only affected in the hands and at the back, but they also had mental stress. As a result, measures were projected and included in an intervention program. The presence of students in this result was decisive because, together with the professor, they formed an ergonomic team acting as specialists and applying the techniques of diagnosis.

In 1996, another workgroup was formed in the Institute of Work Studies and Investigations to evaluate the work and efficiency conditions by ergonomic aspects in the review of injectants in the Institute of Serums and Vaccines "Carlos J. Finlay" (IEIT, 1999). This same institute put a consultancy on Safety, Health, and Work Conditions in the company CUBACEL, S.A. (IEIT, 1999). In the first case, the institute—under the scientific community—advanced by executing the processes in high quality, which maintained positive thinking and attitude toward the introduction of technologies and recommendations that led to the achievement of their objectives.

In 1996, the studies in the bio-industry under the Institute of Biotechnology of Plants continued and various students did course works and monographs of Industrial Engineering and a master's degree thesis with a strong component of ergonomics. The increase of investigative results is shown by the participation of more people in the process. Two programs of intervention were put in practice—with compromise from the administration—and could obtain an increase in productivity and work-life quality of the workers since the intervention program included formative, ergonomic, and organizational actions, recommendations on performance evaluation, and distribution of equipment at the workplace, among others. To the surprise of the top administration who was a member of the ergonomic team, there was an improvement in the work-life quality of the workers in the experimental group (Mejías, 2003).

In the fourth period, the Institute of Work Studies and Investigations established a consultancy on work conditions in the international center for traffic operation of the telecommunications company ETECSA (IEIT, 2000) and the evaluation of ergonomic conditions and safety in the cabin of a sugar cane harvester prototype Model 4000 (IEIT, 2001) belonging to a subproject to improve this machine used in the sugar industry.

In 2005, the Institute of Work Studies and Investigations proposed and verified experimentally the general and specific checklists for different types of risks (IEIT, 2005). In the verification, different institutions participated; present were the students of the Industrial Engineering course of Central University "Marta Abreu" of Las Villas who applied the checklists and collected a lot of important information that was handed over to the institute (IEIT, 2005). The study continued and a procedure for the identification of ergonomic risks in enterprises originating from a study of ergonomic risks associated with posture and effort was obtained as part of the work in this very institution (IEIT, 2009).

Although the investigation group of the Central University "Marta Abreu" of Las Villas had reduced its work when it started emphasizing investigations related to the human resource system as it was explained in epigraph two, it continued developing the macroergonomic tool—now called consultancy procedure or procedure of macroergonomic intervention. The development of this tool led to the reduction of its steps from seven to six through the fusion of steps 1 and 2, naming it "determination of the demands of different nature and definition of the ergonomic demands." Its application in a tourism agency (Mejías, 2000, 2002) and in ETECSA (Mejías, 2006) were presented together, with monographs defended for the degree of Master's in Human Resources Management.

Subsequently, an ergonomic evaluation of workplaces for the manufacture of twisted tobacco in an enterprise was carried out in 2010. Cuban tobacco (cigars and cigarettes) is a product of high demand both nationally and internationally and is an old industry with marked traditions where tables and seats have had particular characteristics that were not studied, which implies that ergonomic attributes were not incorporated (Mejías, 2010).

The investigation group led by Viña in José Antonio Echeverría Institute worked on various applications that had been divulged at the international level. Topics like diagnosis and design of workplaces in the tobacco industry (Viña, 2005), ergonomic redesigns in workplaces of packaging in the pharmaceutical industry (Rodríguez, Torres & Viña, 2007), and 3D model ergonomic redesigning were effected (Torres, Rodríguez, & Viña, 2008, 2010); methods to evaluate exposure to risk factors of muscular-skeletal disorders and their prevention were elaborated (Torres, Rodríguez, & Viña, 2011); and the manual manipulation of materials in the workplace of vaccines was evaluated and redesigned (Torres & Viña, 2012).

Maybe the result that has been most divulged and greatly motivated interest at a national level—and those who follow the results of the country in this material at the international level—is the individual risk assessment ergonomics tool (abbreviated as ERIN method in Spanish) for the evaluation of exposure to risk factors of muscular-skeletal disorders (MSDs). This method that can be used by specialists and people with minimum training allows to conduct a large-scale evaluation of these factors in dynamic and static tasks. This tool follows the guide of the International Ergonomics Association (IEA) and the World Health Organization (WHO). There is a worry about MSDs in the country just like in other countries, since they take second place for total permanent disability, exceeded only by cardiovascular diseases by 180 cases out of every 10,000 workers. During the year 2009, about half a million (426,233) medical certificates related to the muscular-skeletal system were presented by Cuban workers. These health problems caused the loss of more than six million workdays. Even though the proportion of the workplaces related to this ailment has not been determined, it is significant. Cuba is characterized by a health system that covers the necessities of the population and boasts more than 450 physiotherapy centers for the rehabilitation of injured people. It is considered that a great portion of the people attended to get the injuries at work where inadequate posture, static force, movements against the principles of the economy of movements are some of the causes of these injuries.

Studies done in the country by investigation groups and interventions carried out show that these risk factors can be reduced. However, proactive approximation would reduce losses and associated costs as a result of work-related diseases and rehabilitations.

In the present, investigations have not been stopped. Cuban ergonomists work in enterprises and universities, although the published results are mainly those of professionals associated with investigative groups of universities. Among the journals where there is a possibility of publishing these investigations are the Journal of Industrial Engineering at José Antonio Echeverría Institute of Technology and the Cuban Journal of Health and Work. The number of publications is not enough today and neither have they attended to the topics dealt with in the norms of ergonomic material.

The alliance between investigators of different universities and countries continues and characterizes the advance and practical applications that have been accomplished until today. The work developed from 2009 between the Central University "Marta Abreu" of Las Villas and the Business School of the University of Leeds (UK) improved the conception of the man-organization interphase and the facilities that it offers for the enhanced performance of management systems (Mejias, 2010). At the same time, the participation in the 8th International Conference on Manufacturing Research and the 18th congress of the International Ergonomics Association made it possible to demonstrate from the macroergonomic perspective the search for effectivity in collaborative supply chains (Mejias & Huaccho, 2010; 2012).

Much ahead, the publication of the work, "Macroergonomics Intervention Programmes: Recommendations for their design and implementation" in the Journal Human Factors and Ergonomics in Manufacturing and Services (Mejias & Huaccho, 2011) was an important precedent for ergonomists in the country. Publications in important journals at the international level that showed the points of view and recommendations of Cuba on this topic had never been achieved until that moment.

Other results are related to the macroergonomic diagnosis of the work system of the distribution and commercialization process of dispensed beer (Ramírez, Marrero, & Mejías, 2012; Marrero, Huaccho, & Mejias, 2013) and the determination of ergonomic demands in work systems of organic sugar stores (Marrero & Mejías, 2013).

As a major worry, the following question arises: how can the design of repetitive tasks with informative devices incorporated with those of control affect the response in emergency situations under increased industrialization and automation of processes in the country? (Mejías & Marrero, 2013). In the 27th Annual British Academy of Management Conference, some results obtained until this moment—owing to the alliance held for various years—are shown (Marrero, Huaccho, & Mejias, 2013).

As it can be observed, the number of investigations and applications carried out and have been extended to other sectors of the economy should be the first step for the elaboration, review, and adoption of norms related to ergonomics. As it will be shown, these steps have been taken—although not to the necessary level.

Various causes are responsible for the statements mentioned before: inadequate specialized equipment to carry out studies and measurements, the existence of few ergonomists in the country, and an insufficient number of graduates of post-graduate training programs to lead investigation groups in the entire country. Although these together paved way for the creation of the Cuban Ergonomics Association, they do not strengthen the bases to construct firm foundations that allow the emergence, development, and consolidation of a strong set of norms related to ergonomics.

DESIGNED AND ADOPTED STANDARDS ASSOCIATED TO ERGONOMICS

To achieve the current and future objectives of ergonomics, there is a need for support that may guide specialists and non-specialists in the designing of systems and products expressed in the guides and standards related to the topic. The comments made in the epigraphs before showed the aspects that distinguish the analyzed periods of the introduction and the advance of ergonomics in Cuba; some elements that observe in a primary form their interrelation with the norms in this material.

There are no doubts that the necessary bases have been created since training allowed gradual advances toward the increment of investigations directed at the search for positive solutions to current problems in the occupational activities of manufacturing and service enterprises and, on small scale, in the designing of goods for consumption in the country. On the other hand, there is work in projects for post-graduate training programs (Ph.D.) in ergonomics as a way of uniting and educating specialists and entrepreneurs in this discipline.

This will allow ergonomic thinking in designers and engineers to consolidate in the country, as well as in business managers who buy supplies such as products, equipment, and protective gear from abroad without considering the necessities of the Cuban population that had negative consequences in various situations of work.

In Cuba, as in other countries, standards have not been those that had the scoop to emerge. The way to the justification of their appearance has been long and has passed through important historic occurrences that supported their approval. In the first place, the triumph of the Cuban Revolution on the first of January 1959 marked a turning point for Cuban workers who worked in inhumane conditions—far from what the legal laws demanded for their protection to ensure good health and avoid accidents and work-related diseases. From this historic moment, the transformation into a more just society—one that cared for the well-being of man—opened doors for important steps after being subjected to several years of unjust governments, where the opportunities to enjoy workdays and adequate rest never existed as with the concern for the use of protective gear and respect of safety measures.

From then, all that was mentioned before would be a concern to institutions, enterprises, labor unions, and ministries in charge of practicing a set of measures that would ensure the quality of life of Cuban workers. In 1964, the Council of Ministers approved the General Bases of Work Protection and Hygiene and Article 49 of the Constitution of the Republic established that the State would guarantee the right to work protection, safety, and hygiene through the adoption of adequate measures for the prevention of accidents and work-related diseases. Also, another important measure was put in practice—law number 13 on the 28th of December 1977 dealt with Work Protection and Hygiene and Decree No. 101 on the 3rd of March 1982 complements the previous law and was in force until the first Code of Work was approved in the same period. This Law number 13 in Chapter II, Article 7 alluded to the necessity of study, investigation, and control of hygienic-sanitary aspects of work environment and psychophysiological behavior of man and the consequences he faced due to work influence, his organization, and environment. At the same time, the Ministry of Public Health oversaw the creation of integral rehabilitation programs (physical, psychological, and labor) for workers, specifying that specialized institutions be created as it was implemented in the following years. It established functions for other ministries, such as that of Education and that of Higher Education, to form principles and habits to work in a safe way. Like this, it assigned to labor organizations functions such as collaboration in scientific-technical investigations—which were carried out to improve work conditions. About a workday, this Law 13 dedicated spaces in the work routine and rest for the youth, pregnant women, and workers with reduced work capacity in different moments. This demonstrates that from the first period that marked the evolution of ergonomics, the standards already contained important aspects related to the performance of ergonomic material.

It is precisely on the 17th of June 2014 when a new Code of Work was established and, at the same time, the Resolution 283/14 appeared, considering 35 work-related diseases. This resolution—made by the Ministry of Work and Social Security and Ministry of Public Health in 1996—did not include work-related diseases linked to muscular-skeletal disorders and only incorporated 31 diseases. The list of these diseases now includes carpal tunnel syndrome, epicondylitis, and the chronic tenosynovitis of the hand and wrist; the studies carried out in the country demonstrated that they exist and are related to work activity.

The approval of these important standards recognized the ergonomic studies and advances during various years and was a sample of the level of knowledge acquired by specialists of the Ministry of Work and Social Security and Ministry of Public Health together with the work of labor organizations in the country.

After referring to these important standards, it is necessary to return to the stages that marked the evolution of ergonomics without deviating from the topic contained in this epigraph to continue making step by step reference to the norms that appear in the country.

The decade of the 70s—which constitutes the first period—showed a big concern by enacting Law 13 and the actions that it generated—involving various ministries, institutions, and labor organizations—during unfit historic conditions for ergonomic standards. As it was mentioned, the beginning of training and detailing general norms that included secluded ergonomic elements were predominant. However, two norms caused the rising concern for this discipline—the one that referred to ergonomic requirements of seats and the one that dealt with ergonomic requirements of a single individual and duo-individual beds. Table 11.1 shows those that were approved and put in practice during that period.

During the second period, the National Office of Normalization designed and approved a strong group of standards related to the safety considered. Most of them contained ergonomic elements in their epigraphs, although not in detail. Table 11.2 relates those cases to other standards where ergonomic focus can be identified. This makes this period distinguished by consisting of a major effort in design work and approval of standards related to ergonomics.

What is expressed before is combined with the publication of the Resolution 492 of 1980 on how to develop the investigation of work accidents, which opened the way for the organization and consideration of ergonomic aspects.

The search for bibliographic sources of these norms (OTN, 2014)—which, in some cases, were already subjected to revision or abolished—allows us to state that the 80s is characterized by a great emergence and formalization of standards—some of which were an obligation. The fundamental cause lies in the studies, investigations, and applications. Among the standards, exclusively the ergonomic ones, particular interest was in anthropometry, weight, and size of men and women under 19 years, ergonomic specifications for closets and cupboards in construction projects, as well as methods to evaluate physical work capacity. Even out of the safety area, a norm of Metrologic Assurance System related to methods and media of anthropometric verification was emitted. There is no doubt that most of these standards emerged because of the concern of specialists and technicians about the topic and the results obtained.

However, there is still no high correspondence between the standards formulated and the essential work developed in ergonomic investigations from universities and national institutes. Even though ergonomics exists as a line of investigation in some universities where the course of Industrial Engineering is imparted, some of the studies carried out are not fully registered. Therefore, not enough initiatives are taken to start creating other standards. Though there is continued reception of norms mainly from the area of safety and from specialists of this area, it should be recognized that they contain a strong ergonomic component in most cases.

As Soares (2006) states, in analyzing the difficulties in the development of ergonomics in four Latin American countries, a factor that contributes to the difficulty in the establishment of ergonomics and its standards has been not recognizing the benefits of implementing ergonomic principles within many occupational activities. In Cuba, it should be added that most entrepreneurs

---

TABLE 11.1
Standards in the Decade of the 1970s Which Included Ergonomic Elements

Code	Year	Title
Law 13	1977	Work Protection and Hygiene.
NC 19-04-01	1979	System of Norms of Work Protection and Hygiene Protection gear for workers. General requirements and classification.
NC 19-01-04	1979	System of Norms of Work Protection and Hygiene Dangerous and harmful factors of production. Classification.
NC 19-04-11	1979	System of Norms of Work Protection and Hygiene. Safety colors and signals.
NC 19-04-14	1979	System of Norms of Work Protection and Hygiene. Symbols of electric voltage. Form, dimensions, and technical requirements.
NC 53-17	1979	Seats. Ergonomic requirements.
NC 53-18	1979	Single individual and duo-individual beds. Ergonomic requirements.

TABLE 11.2

Standards Approved in the Decade of the 1980s

Code	Year	Title
NC 19-01-04	1980	System of norms of work protection and hygiene. Noise. General hygienic-sanitary requirements.
NC 19-01-05	1980	System of norms of work protection and hygiene. general vibration. General hygienic-sanitary requirements.
NC 19-00-03	1981	System of norms of work protection and hygiene. Load and work intensity. Evaluation criteria.
NC 19-04-15	1981	System of norms of work protection and hygiene. Screens of individual protection. Classification. General technical requirements.
NC 19-04-18	1981	System of norms of work protection and hygiene. Cranes. Signal colors for dangerous elements.
NC 19-01-08	1982	System of norms of work protection and hygiene. Manual machines. Acceptable levels of vibrations.
NC 19-02-21	1982	System of norms of work protection and hygiene. Sugarcane collection centers. General safety requirements.
NC 19-03-05	1982	System of norms of work protection and hygiene. Wrapping and packing. General safety requirements.
NC 19-03-10	1982	System of norms of work protection and hygiene. Manufacture of wood products. General safety requirements.
NC 19-04-10	1982	System of norms of work protection and hygiene. Rubber shoes for miners. General requirements.
NC 20-01	1982	Anthropometry. Weight and size of girls and boys under 19 years. Evaluation of an individual.
NC 53-43	1982	Elaboration of construction projects. Closets and cupboards. General ergonomic specifications.
NC 19-01-06	1983	System of norms of work protection and hygiene. Noise measurement in places where people are found.
NC 19-01-07	1983	System of norms of work protection and hygiene. Vibration. Methods of measurement. General requirements.
NC 19-01-10	1983	System of norms of work protection and hygiene. Noise. Determination of the sonorous potency. Method of orientation.
NC 19-01-12	1983	System of norms of work protection and hygiene. Determination of the levels of illumination in rooms and workplaces. Methods of measurement.
NC19-01-13	1983	System of norms of work protection and hygiene. Determination of the loss of audition. Method of measurement.
NC 19-01-21	1983	System of norms of work protection and hygiene. Manual machines. Methods of measurement of vibration parameters. General requirements.
NC 19-02-26	1983	System of norms of work protection and hygiene. Cranes. General safety requirements.
NC 19-02-27	1984	System of norms of work protection and hygiene. Cranes. Safety distances.
NC 19-02-35	1984	System of norms of work protection and hygiene. Cranes. Brakes. Safety requirements.
NC 19-03-02	1984	System of norms of work protection and hygiene. Production of crude sugar. General safety requirements.
NC 19-03-17	1984	System of norms of work protection and hygiene. Production of boards from bagasse. General safety requirements.
NC 19-03-23	1984	System of norms of work protection and hygiene. Operation of gable cranes. General safety requirements.
NC 19-02-38	1985	System of norms of work protection and hygiene. Control organs of work media. General safety requirements.
NC 19-02-42	1985	System of norms of work protection and hygiene. Cranes. Safety requirements for control cabins.
NC 19-02-46	1985	System of norms of work protection and hygiene. Cranes. Safety requirements for holding media.
NC 19-02-47	1985	System of norms of work protection and hygiene. Safety requirements for the manipulation of containers.

(Continued)

TABLE 11.2 (Continued)

Code	Year	Title
NC 19-03-22	1985	System of norms of work protection and hygiene. Cranes. Safety requirements for the simultaneous exploitation of various cranes.
NC 19-02-40	1986	System of norms of work protection and hygiene. Control organs of work media. Symbology.
NC 19-02-49	1986	System of norms of work protection and hygiene. Cranes. Symbology.
NC 19-03-30	1986	System of norms of work protection and hygiene. Process of cement production. General safety requirements.
NC 19-03-36	1986	System of norms of work protection and hygiene. Maintenance and repair of road vehicles and their components. General safety requirements.
NC 19-04-04	1986	System of norms of work protection and hygiene. Special clothes for protection. Classification. General requirements.
NC 19-00-07	1987	System of norms of work protection and hygiene. Methods for the evaluation of physical work capacity.
NC 19-02-44	1987	System of norms of work protection and hygiene. Cranes. Safety requirements for control systems.
NC 19-02-45	1987	System of norms of work protection and hygiene. Cranes. Safety mechanisms and devices. Terms and definitions.
NC 90-16-01	1987	Metrologic assessment. Sonometers. Methods and means of verification.
NC 19-00-08	1988	System of norms of work protection and hygiene. General technical and organizational techniques of the labor activity.
NC 19-01-19	1988	System of norms of work protection and hygiene. Emergency illumination. Classification and general requirements.
NC 19-02-15	1988	System of norms of work protection and hygiene. Tractors and self-propelled agricultural machines. Requirements for the workplace of the operator.
NC 19-02-32	1988	System of norms of work protection and hygiene. Seats for the media of work. Classification and general requirements.
NC 19-02-54	1988	System of norms of work protection and hygiene. Industrial sewing machines. Safety requirements.
NC 19-03-03	1988	System of norms of work protection and hygiene. Work of loading and unloading. General safety requirements.
NC 19-03-11	1988	System of norms of work protection and hygiene. Work of loading and unloading on the sea. Safety requirements.
NC 90-01-59	1988	System of norms of metrologic assurance. Anthropometers. Methods and means of verification.

hardly heard about the topic and would confuse the term in the question—"do you know what ergonomics is?"—with economics and would answer without knowing what it meant.

Here, it is important to clarify that there is no figure of an ergonomist as a profession. Neither does ergonomics identify as a specialty in ministries and institutes of investigation, which are responsible for the designing of solutions with a strong component of the topic.

The main cause from the author's point of view is that training of entrepreneurs had not been enough (it still isn't) and consultancy companies did not promote courses in this field. At the same time, not putting more emphasis on the work that deals with standards in force when this discipline is imparted during Industrial Engineering and post-graduate courses could have contributed to this.

The decade of the 90s affected the advance of ergonomics as explained in the epigraphs. The sources consulted in the search carried out in Territorial Normalization Offices have only showed that two standards with ergonomic criteria in their specifications were designed in this period. Table 11.3 shows them.

Besides this, the Technical Committee number 6 of Safety, Health, and Environment was created, which constituted the Subcommittee of Ergonomics in 1999 where university professors, specialists of companies, and investigators of various institutes participated.

In the fourth period (from 2000 to the present), it can be confirmed that work on standards was reinforced. Standards were not forgotten in the period before—only that, the concern was directed toward other problems in the economic and social sector that demanded special attention.

Resolution 37/2001, which establishes the procedures for the evaluation of risks, emerged during this period. A little later, it was replaced by Resolution 31/2002, which establishes a set of general procedures for the identification, evaluation, and control of risk factors in work, contributing to the good practices of work management and hygiene. Among these, emphasis is also put on ergonomic risks, although they are not known in companies in aspects such as identification and evaluation. In relation to this last resolution, another one is published—dedicated to work accidents (Resolution 19/2003)—which replaced Resolution 492/1980. Its appearance originated from the need to employ new procedures and focus on evaluate risks that determine accidents and incidents. Like this, new resolutions appear successively, showing top concern about safety and advancing toward the designing and implementation of Safety and Health Management Systems in work. Table 11.4 shows the most important resolutions that demonstrate the rise to superior conscience levels toward the safety required in processes of manufacture and services. Among these advances, ergonomics was seen to have benefited since it was divulged when the topic was dealt with.

There is also an increase in the number of standards published in relation to the topic of safety. However, the ones explained before are still maintained. On one hand, ergonomic questions that necessarily cannot be forgotten continue to be incorporated in many. On the other, new standards specifically related to ergonomics appear. Even so, the new ergonomic standards appear more

TABLE 11.3
Standards Formulated in the Third Period (Decade of the 1990s)

Code	Year	Title
NC ISO 13854	1999	Safety from machines. Minimum distances to avoid the squashing of parts of the human body.
NC 69-6	1999	Requirements of the contents and reach of technical services for touristic investments. Part 6: Requirements of technical documentation of interior designing.

TABLE 11.4
Resolutions Related to Safety

Code

Resolution 32/2001 creates the Registration and Approval Center of Personal Protection Equipment

Resolution 31/2002 General practical procedures for the identification, evaluation, and control of risk factors at work (replaces Resolution 37/2001)

Resolution 19/2003 Work accidents (replaces Resolution 492/1980)

Resolution 39/2007 General Bases of safety and health in work (replaces the bases of 1964)

Instruction 2/2008 Procedures for the implantation of the Safety and Health Management System at work

Resolution 51/2008 Work safety manual

frequently than before and those approved by the International Organization for Standardization (observe Table 11.5) were adopted. Without a doubt, work has been covered. This happened due to the increased concern to certify the Safety and Health Management Systems in the work of enterprises in the country. Another standard that has played an important role is the group of norms 3000 related to the implementation of the Integral Human Resource Management System. In both groups of norms—that is to say, those of the safety system and those of the human resource system—ergonomic designing and implementation of ergonomic programs are promoted.

TABLE 11.5
Standards Formulated in the Fourth Period (From 2000 to the Present)

Code	Year	Title
NC 107	2001	Safety and health in work and basic sanitation in areas of work. General safety requirements.
NC 116	2001	Safety and health in work. Basic ergonomic requirements to consider in workplaces and work activities.
NC 13852	2001	Machine safety. Safety distance to avoid reaching dangerous zones with superior limbs.
NC 124-1	2001	Safety and health in work. Machine safety. Basic concepts. General designing principles—Part 1: Basic terminology and methodology.
NC 124-2	2001	Safety and health in work. Machine safety. Basic concepts. General designing principles—Part 2. Technical principles and specifications.
NC ISO 14121	2002	Machine safety. Principles for the evaluation of risks.
NC ISO 3864-1	2003	Graphic symbols. Safety colors and signals—Part 2. Principal designs of safety signals in places of work and public areas.
NC ISO 8995	2003	Illumination in interior workplaces.
NC ISO 9241-5	2003	Ergonomic requirements for office work with data visualization screens. Part 5: Workplace lay out and postural requirements.
NC 341	2005	Work in confined spaces. General safety requirements.
NC 18001	2005	Safety and health in work. Safety and health management system in work. Requirements.
NC 18002	2005	Safety and health in work. Safety and health management system in work. Directions for the implantation of the NC 18001.
NC 18011	2005	Safety and health in work. General directions for the evaluation of safety and health management systems in work. Audit process.
NC 441	2006	Environmental health. Hygienic and sanitary requirements for safety.
NC 3000	2007	Integral human resource management system—Vocabulary.
NC 701	2009	Colors for the identification of pipes depending on the fluid being transported.
NC 704	2009	Buildings. Office spaces: arrangement of furniture and dimensions.
NC ISO 10075	2009	Ergonomic principles related to mental workload. Terms and general definitions.
NC ISO 10075-2	2009	Ergonomic principles related to mental workload. Part 2: principles of designing.
NC ISO 10075-3	2009	Ergonomic principles related to mental workload. Part 3: principles and requirements referring to the methods of measurement and evaluation of mental workload.
NC 869	2011	Safety and health in work. High-temperature environments. Estimation of thermic stress in work based on the WBGT (Wet Bulb and Globe Temperature) index.
NC 870	2011	Safety and health in work: Ergonomics-criteria of reference and physiological indicators for the evaluation of work intensity and physical workload.
NC 871	2011	Safety and health in work. Noise in the work environment. General hygienic and sanitary requirements.
NC ISO 13731	2011	Ergonomics in the thermal environment: Vocabulary and symbols.

Besides the advances discussed before, the body of auditors does not have enough specialists to carry out the assessment of these management systems. Even the inspection organs that belong to the Ministry of Work and Social Security found in the whole country have no specialists in ergonomics, which leads to the topic not being adequately addressed, although it may be present in the requirements to audit. The cause is focused on the lack of knowledge about the requirements of the norms related to ergonomics by auditors and inspectors.

RECOMMENDATIONS FOR THE DEVELOPMENT AND FORMULATION OF NEW NORMS FOR THE DESIGNING OF PRODUCTS, PROCESSES, AND SYSTEMS

Presently, there is a tendency to only adopt ISO norms in the country after approval by the Ergonomics Technical Subcommittee and consider existing necessities. It would be prudent to have these norms not only be the ones adopted since what has been discussed until here shows that studies and investigations have been consolidated. Besides, the necessity of more connection and coordination among these studies and the formulation of standards is evident. It should conduce to the valuation of more proposals such that this subcommittee decides which ones to approve.

The increase in the number of investigations should be maintained continuously and these should be more generalized in manufacturing companies and those in services. This will directly lead to the management knowing about the benefits of ergonomics and deciding to put it in practice. Through continuous training in this discipline, it is expected to significantly increase to higher levels with post-graduate programs that are already in place.

The lack of guides for ergonomic evaluations and the design of intervention programs is a shortcoming in the field of entrepreneurship. Only safety and health programs that do not involve ergonomic aspects in most cases are designed and approved in the country. Safety specialists manifest a lack of knowledge on how to incorporate these two elements which have become critical points today. Therefore, from the author's point of view, this is a primary problem to be solved under the increasing educational levels in progress.

Innumerable case studies where both questions have been put in practice are published in the bibliography—that is to say, evaluations and intervention programs. The proposed considerations include this problem from the point of view of ergonomists and designers, occasionally causing conflicts among their procedures. Their varying visions in the form of projecting this problem and its solutions have established a variety of methodologies that include different steps. Documentation of these methodologies is a good opportunity today to evaluate variants for the design of new standards.

For example, in the area of ergonomics, macroergonomics is a sub-discipline that has generated key tools in the process. Among them is Participative Ergonomics (PE). The reports in the bibliography revised confirm the advantages of making use of PE today (Noro & Imada, 1991; Haines et al., 2002; Kuorinka, 1997; de Jong & Vink, 2002; de Looze, van Rhijn, & Tuinzaad, 2003; van der Molen et al., 2005). With these events together, ergonomists focused on putting their competencies into practice so that ergonomic attributes are present in the design and development of consumer goods to ensure that they are safe, efficient, and reliable. Likewise, methodologies have emerged from designers' theoretic conceptions and these have accomplished important contributions to obtain superior quality products (Soares, 1998; Ahram, Karwowski, & Soares, 2011; Sala-Diakanda & Soares, 2011; Diniz & Soares, 2011).

However, some specialists recently stated that (Mejias & Soares, 2014): nevertheless, with the advances achieved in both disciplines (ergonomics and designing), a good integration of human factors in the work systems is not always achieved and neither does an optimal product suitability to the user and his context always exist. The first important problem to attend to would be the fractioning of the elements of each discipline during ergonomic intervention, where non-consideration of cultural factors of each region or social group is added in the present.

In a recent investigation, the need of transition towards the evaluation of systems and products involving physical, cognitive, social, organizational, environmental, and cultural aspects to improve working conditions and establish strategies of macroergonomics interventions from planning, evaluation of work systems, products, and environments to ensure that it is compatible with the needs, abilities, and limitations of the human factor was stated. They started from the hypothesis that better solution results can be obtained when there is an integration in the macroergonomics intervention tools—not only from the theoretic point of view of ergonomics and design but also from cultural ergonomics (CE).

This has supposed the necessity to continue investigating if a possibility exists to better optimize the human performance and suitability of a product to the user if the work system, product, and user are integrated in the macroergonomics intervention tools. At the same time, it is necessary to know if CE can be considered in this integration and contribute to better results. The author suggests that the proposal of a new standard for the evaluation of systems and products shows the knowledge of these results and applications that are carried out to propose the aspects that should be contemplated in ergonomic evaluations and how to implement them step by step. The standards existing at an international level should be consulted. However, their approval should not be made from the exact copy but should be adjusted to the existing economic conditions. It should foresee the knowledge level about the topic, specialists who will put it into practice, and the profile of functions that their work position demands.

A conceptual model was published last year as a result of these interrogations (See Figure 11.1). In it, the authors state conceptual formulations, knowledge, experience, and existent applications in specialized references and emphasized the necessity of a new interface in the historic development of Ergonomics: the Macroergonomics-Product interface also known as Man-Machine-Organization-Technology-Product Systems interface. The development of this interface in the present should allow overcoming the deficiencies that affect the work systems' performance and the products' usability that are studied with low levels of interconnection. The absence of evaluative causes of accidents in work systems provoked by an inadequate design of products or analysis of few work environment variables—the social, cultural, and organizational contexts—results in problem reports and the corresponding solutions proposed do not reach higher user performance levels, safety, efficiency, and satisfaction with the products that are used. The consideration of the interface from the model is a call to design procedures from its premises, characteristics, inputs, and outputs—which allow defining of ergonomic demands in a systemic and integral way in the first place. Secondly, the evaluation of work systems and products results in the characterization of a process and its workplaces—a direct and indirect users' profile that includes their quality of life and expectations among other elements of interest in the initial evaluation stage. The ergonomists should start constructing a Work System Profile and User Profile that allows studying what is lacking in a product that does not satisfy the necessities of a working system and how the actual work system conception makes difficult the performance with the product. What is mentioned above means passing to a superior stage of evaluation where the dysfunctions of the macroergonomics-product interface evaluate the existent compatibility level in a determined cultural context.

The outputs at the end of any procedure designed to evaluate should allow relying on a Work System Profile and Product Profile that contains the requirements needed to optimize the performance established by the work system and safe, efficient, and reliable products.

The second problem that must not be neglected is that regardless of how much the development of ergonomic intervention programs has advanced (Kleiner & Drury, 1999; Kleiner & Shewchuk, 2001; Dzissah et al., 2005; Kleiner, 2006; Villeneuve, Remijn, Lu, Hignett, & Duffy, 2007), it is still considered that many of them have some deficiencies that permit possible satisfactory improvements. This happens because the human factors study continues with an individual approach; still, the focus is on physical factors alone or on cognitive aspects, and sometimes, the essential worry is the anthropometric dimension to accomplish adequate design. Even, if the intervention tools detect a set of existent problems, it is observed that the associated intervention programs do not emerge as part of a systemic and integral analysis of ergonomic demands and it is reflected in a list of independent ergonomic actions to solve each particular problem found.

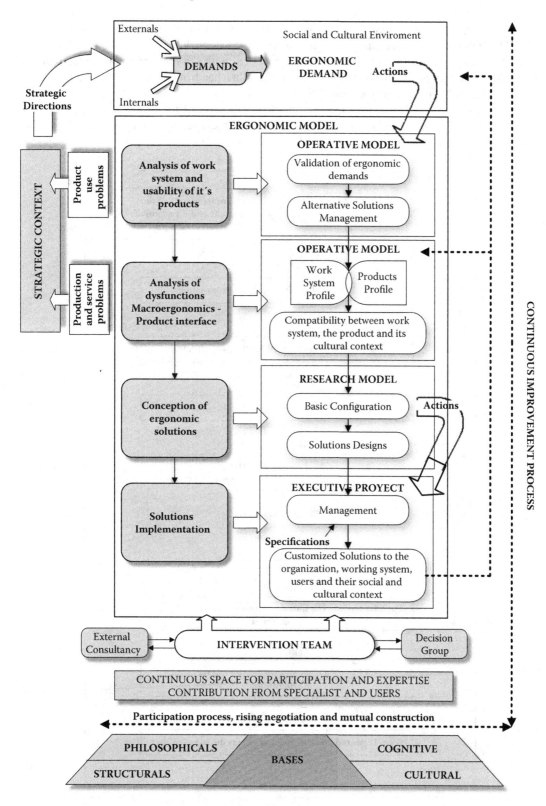

FIGURE 11.1 A conceptual model for the evaluation of work systems and products.
Source: Mejías & Soares, 2014. Adapted from Mejías, 2003.

Adopting the recommendations by Mejias and Huaccho (2011) (see Figure 11.2)—which are still being referenced by specialists at the international level—could constitute one of the variants to be analyzed for the formulation of standards that deal with the topic and allow specialists to fulfill their necessities in Cuban companies. On the other hand, it is precise that the standards formulated should consider key conceptual elements that these authors recommend on the definition of *"program,"* such that they indicate the "systematicity" of the process, its frequency, its preventive focus, as well as the proactive approach and objectives centered in achieving certain benefits within the framework of a given object of study.

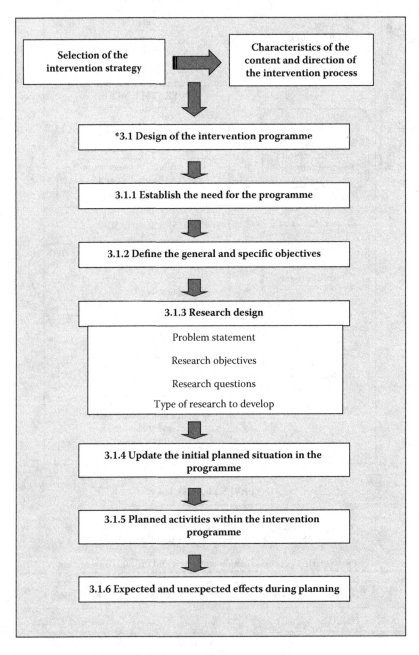

FIGURE 11.2   Steps for the design of the intervention program.
Source: Mejias & Huaccho, 2011.

Both problems that are dealt with in these epigraphs will not only allow the activity of ergonomists to attract more interest and concern from entrepreneurs but also contribute to the regulation of the actions related to ergonomics in audits and inspections by institutions that are responsible for the safety and health of Cuban workers.

REFERENCES

Ahram, T.; Karwowski, W., & Soares, M. M. (2011), Smarter products, user-centered systems engineering. In: Karwowski, W., Soares, M. M., & Stanton, N. *Human factors and ergonomics in consumer product design: Methods and techniques*, Boca Raton, CRC Press, (pp. 83–96).

Cavalcanti, J. (2003), Análise ergonómica da Sinalizacao de seguranca: um enfoque da ergonomia informacional e cultural. *Work, 41*, 3427–3432.

CENDA. (1999). Software Ergo-sizes. Register 04671-4671, Cuba.

CENDA. (2005). Software Ergo-sizes. V. 2.0. Register 1088-2005, Cuba.

de Jong, A. M., & Vink, P. (2002). Participatory ergonomics applied in installation work. *Applied Ergonomics, 33*, 439–448.

de Looze, M. P., van Rhijn, J. W., & Tuinzaad, B. (2003). A participatory and integrative approach to improve productivity and ergonomics in assembly. *Production Planning and Control, 14*, 174–181.

Diniz, R., & Soares, M. M. (2011). User-centered design method to attend users' needs during product design process: A case study in a public hospital in Brazil. In: Karwowski, W., Soares, M. M. and Stanton, N. *Human factors and ergonomics in consumer product design: Methods and techniques*. Boca Raton, CRC Press, (pp. 455–476).

Dzissah, J., Karwowski, W., Rieger, J., & Stewart, D. (2005). Measurement of management efforts with respect to integration of quality, safety, and ergonomics issues in manufacturing industry. *Human Factors and Ergonomics in Manufacturing, 15*, 213–232.

Haines, H., Wilson, J., Vink. P., & Koningsveld, E. (2002). Validating a framework for participatory ergonomics. *Ergonomics, 45*, 309–327.

Huaccho Huatuco, L., & Herrera, S. M. (2010). Towards effective supplier-customer collaboration: A macroergonomics perspective. In: Vitanof V. I., & Harrison D. (Eds.) Advances in Manufacturing Technology XXIV. Proceedings of the *8th International Conference on Manufacturing Research ICMR 2010*. 14th–16th September 2010. (pp 332–336), Durham University and Glasgow Caledonian University, ISBN: 9781905866519.

ISDI. (1993). Evaluación ergonómica en distintas variants de diseño de un prototipo de bicicleta cubana. *Informe de investigación terminada. Ministerio del Trabajo y Seguridad Social*, Cuba.

IEIT. (1981). Estudio ergonómico en puestos de trabajo de los envasaderos de cítricos. *Informe de investigación terminada. Ministerio del Trabajo y Seguridad Social*, Cuba.

IEIT. (1987). Diagnóstico sobre la aplicación de requisitos ergonómicos y de seguridad del trabajo en nuevas inversiones. *Informe de investigación terminada. Ministerio del Trabajo y Seguridad Social*, Cuba.

IEIT. (1999). Consultoría sobre seguridad, salud y condiciones de trabajo en la empresa CUBACEL S.A. *Informe de investigación terminada. Ministerio del Trabajo y Seguridad Social*, Cuba.

IEIT. (2000). Consultoría sobre condiciones de trabajo en el centro internacional de operación de tráfico de la empresa ETECSA. *Informe de investigación terminada. Ministerio del Trabajo y Seguridad Social*, Cuba.

IEIT. (2001). Evaluación de las condiciones ergonómicas y de seguridad en la cabina del prototipo de cosechadora cañera modelo 4000. *Informe de investigación terminada. Ministerio del Trabajo y Seguridad Social*, Cuba.

IEIT. (2005). Proyección y comprobación experimental de listas de chequeo para distintos tipos de riesgos. *Informe de investigación terminada. Ministerio del Trabajo y Seguridad Social*, Cuba.

IEIT. (2009). Riesgos ergonómicos asociados a la postura y el esfuerzo, un procedimiento para su identificación en la empresa. *Informe de investigación terminada. Ministerio del Trabajo y Seguridad Social*, Cuba.

Kleiner, B., & Drury, C. (1999). Large scale regional economic development: macroergonomics in theory and practice. *Human Factors and Ergonomics in Manufacturing, 9*, 151–163.

Kleiner, B., & Shewchuk, J. (2001). Participatory function allocation in manufacturing. *Human Factors and Ergonomics in Manufacturing, 11*, 195–212.

Kleiner, B. (2006). Macroergonomics: analysis and design of work systems. *Applied Ergonomics, 37*, 81–89.

Kuorinka, I. (1997). Tools and means of implementing participatory ergonomics. *International Journal of Industrial Ergonomics, 19*, 267–270.

Marrero, M., Huaccho, L., & Mejias, S. (2013). Macroergonomics evaluation of a beer distribution process. In Proceedings of the 27th Annual British Academy of Management Conference (BAM, 2013). In British Library's Management and Business Portal.Vol. 1, (pp. 1–20), Liverpool, Inglaterra: University of Liverpool Management School.

Marrero, M., & Mejías, S. (2013). La determinación de la(s) demanda(s) ergonómica(s): un paso clave para la elaboración de programas de intervención en los sistemas de trabajo de los almacenes de azúcar orgánico. *Revista Centro Azúcar, 40*(3), 40–45.

Mejías, S. (2000). Diagnóstico de la Gestión de Recursos Humanos y la calidad de vida laboral en el centro de elaboración SERVISA s.a. del grupo Cubanacán. *Informe de Investigación terminada. Centro de Información Científico-Técnica.* Universidad Central "Marta Abreu" de Las Villas. Cuba.

Mejías, S. (2001). Programa de intervención macroergonómica para la mejora de los sistemas de trabajo en la fábrica de calzado "Dinamo" de la empresa de Calzado VICALZA del grupo COMBELL. *Informe de Investigación terminada. Centro de Información Científico-Técnica.* Universidad Central "Marta Abreu" de Las Villas. Cuba.

Mejías, S. (2002). Programa de intervención macroergonómica para la mejora de la calidad de vida laboral y la calidad del servicio en el centro de elaboración SERVISA s.a. del grupo Cubanacán. *Informe de Investigación terminada. Centro de Información Científico-Técnica.* Universidad Central "Marta Abreu" de Las Villas. Cuba.

Mejías, S. (2003). Herramienta de intervención macroergonómica para el mejoramiento de los sistemas de trabajo. PhD Thesis, Universidad Central Marta Abreu de Las Villas, Cuba.

Mejías, S. (2006). Propuesta de un procedimiento de consultoria para el mejoramiento de los sistemas de trabajo en la Gerencia territorial de ETECSA, Villa Clara. *Informe de Investigación terminada. Centro de información Científico-Técnica.* Universidad Central "Marta Abreu" de Las Villas. Cuba.

Mejías, S. (2010). Evaluación ergonómica de los puestos de trabajo en la UEB Tabaco Torcido de Villa Clara. *Informe de investigación terminada. Centro de Información Científico-Técnica.* Universidad Central "Marta Abreu" de Las Villas. Cuba.

Mejias, S. (2010). Interfase Hombre – Organización: facilidades que brinda para el mejoramiento del desempeño de los Sistemas de Gestión. *XII Convención Internacional de las Industrias Metalúrgica, Metalmecánica y del Reciclaje (Metánica, 2010).* La Habana, Cuba.

Mejias, S., & Huaccho, L. (2011). Macroergonomics intervention programmes: recommendations for their design and implementation. *Human Factors and Ergonomics in Manufacturing and Services, 21,* 227–243.

Mejias, S., & Huaccho, L. (2012). Macroergonomics' contribution to the effectiveness of collaborative supply chain. *In Work: A Journal of Prevention, Assessment & Rehabilitation, 41*(1), pp. 2695–2700.

Mejías, S., & Marrero, M. (2013). Cómo el diseño de tareas repetitivas con dispositivos informativos y de control incorporados afecta la respuesta en situaciones de emergencia? *Encuentro Internacional de Diseño (FORMA, 2013). Simposio de Ergonomía en el Diseño.* Palacio de las Convenciones, La Habana.

Mejias, S., & Soares, M. (2014). The evaluation of work systems and products: Considerations from the cultural ergonomics. In *Advances in ergonomics in design, usability & special populations: Part I. Proceedings of the 2nd International Conference on Ergonomics In Design/Ergonomics Modeling, Usability & Special Populations.* (pp. 329–339) Kraków, Poland, Jagiellonian University.

Noro, K., & Imada, A. S. (1991). *Participatory ergonomics.* Taylor & Francis, London.

OTN. (2014). Catálogo de normas cubanas. *Oficina territorial de normalización,* Villa Clara, Cuba.

Ramírez, S., Marrero, M., & Mejías, S. (2012). Diagnóstico al sistema de trabajo del proceso de distribución y comercialización de cerveza dispensada en la empresa Cervecería Antonio Díaz Santana. In VII Conferencia de Ingeniería Industrial. 8va Conferencia Internacional de Ciencias Empresariales (CICE).

Real, G. L. (2012). Modelo y procedimientos para la intervención ergonómica en las camareras de piso del sector hotelero. Caso Varadero. PhD Thesis, Universidad de Matanzas, Cuba.

Rodríguez, Y., Torres, Y., & Viña, S. (2007). Rediseño ergonómico de puestos de Trabajo en líneas de envase de la industria farmacéutica. In: *Memorias del I Simposio Internacional de Ingeniería Industrial: Actualidades y Nuevas Tendencias;* Universidad de Carabobo: Venezuela.

Rodríguez Y. (2011). Erin: Método Práctico Para Evaluar La Exposición a Factores De Riesgo De Desórdenes Musculo-Esqueléticos. PhD Thesis. José Antonio Echeverría Institute of Technology.

Sala-Diakanda, S. N., & Soares, M. M. (2011). Model-based framework for influencing consumer products conceptual designs. In: Karwowski, W., Soares, M. M., & Stanton, N. *Human factors and ergonomics in consumer product design: Methods and techniques.* CRC Press: Boca Raton, (pp. 63–82).

Salvendy, G. (2006). *Handbook of human factors and ergonomics. Part I: The human factors function Tercera Edition.* John Wiley & Sons, Inc., 1–31.

Soares, M. M. (1998). Translating user needs into product design for disabled people: A study of wheelchairs. Ph.D. Thesis. Loughborough University.

Soares, M. (2006). Ergonomics in Latin America: Background, trends and challenges. *Applied Ergonomics, 37*(4), 555–561.

Taveira Alvaro, D., James, C., Karsh, B.-T., & Sainfort, F. (2003). *Quality management and the work environment: an empirical in a public sector organization.*

Torres, Y., Rodriguez, Y., & Viña, S. (2008). Cuban experience in modeling ergonomic redesigns of workplaces in pharmaceutical industry. In: Fifth International Cyberspace Conference on Ergonomics; Malaysia.

Torres, Y., Rodríguez, Y., & Viña, S. (2010). Ergonomics intervention using movement analysis system and 3D modeling techniques: A case from industrially developing country. In: Seventh International Conference on Prevention of Work-Related Musculoskeletal Disorders, PREMUS; August 29th to September 3rd; Angers, France.

Torres, Y., & Viña, S. (2012). Evaluation and redesign of manual material handling in a vaccine production centre's warehouse. *Work: A Journal of Prevention, Assessment and Rehabilitation, 41*, 2487–2491.

Torres, Y., Rodríguez, Y., & Viña, S., (2011). Preventing work-related musculoskeletal disorders in Cuba: An industrially developing country. *Work: A Journal of Prevention, Assessment and Rehabilitation, 38*, 301–306.

van der Molen, H. F., Sluiter, J. K., Hulshof, C. T. J., Vink, P., van Duivenbooden, C., Colman, R., & Frings-Dresen, M. H. W. (2005). Implementation of participatory ergonomics intervention in construction companies. *Scandinavian Journal Environment Health, 31*, 191–203.

Villeneuve, J., Remijn, S., Lu, J., Hignett, S., & Duffy, A. (2007). Ergonomics intervention in hospital architecture. *Meeting Diversity in Ergonomics*, 243–269.

Viña, S. (2005). Diagnosis and design of workplaces at Arca Factory. In Thatcher, A., James, J., & Todd, A. (eds). Fourth International Cyberspace Conference on Ergonomics. Johannesburg: International Ergonomics Association Press.

Section III

Standards for Evaluation of Working Postures

12 Standards in Anthropometry

*Andreas Seid, Rainer Trieb, Anke Rissiek,
and Heiner Bubb*

CONTENTS

WHAT IS ANTHROPOMETRY?

Anthropometric design is one of the most important disciplines within the area of ergonomic approaches. Products and working tools can be used safely, comfortably and in a healthy manner with the correct anthropometry layout. As in today's modern tailored production, an immediate and direct adaptation of individual dimensions and proportions cannot realistically be realized; data that can describe the diversity of human body dimensions and be used to adapt the general size of a technical device and that device's adaptability to human body dimensions is needed. The main objective here is to design these products in a way that can be used effectively and in a healthy and fatigue-free manner. In the case of highly developed products, a design that focuses on comfort can play a major role. The definition of anthropometry is derived from the statement that, "*Anthropometry* is the scientific measurement and statistical analysis of data about human physical characteristics and the application (engineering anthropometry) of this data in the design and evaluation of system, equipment, manufactured products, human-made environments, and facilities."

The most important objective of anthropometry is the *standardization* of data procurement and, consequently, the reproducibility of the results (Martin & Knussmann, 1988). Standardization is the means, whereby anthropometric data from different fact-finding and analysis sources can be compared and exchanged. Product sales on a global scale have now made the transferability of survey data from different parts of the world directly to the ergonomic design departments of companies significantly more important. This necessary requirement has been fulfilled by the international standardization of anthropometric data and processes. The extensive national standards were analyzed and summarized by scientific and industrial advisory boards and committees. The result is a collection of international standards where the measurement of human beings and the use of their body measurement data is described (see Table 12.1).

The most important objective of anthropometry is its application to product development. To apply it properly, however, we must first understand the creation of anthropometric tables. There are three relevant steps involved.

TABLE 12.1

ISO Standards for Anthropometry

ISO 7250	Basic human body measurements for technological design
ISO 14738	Safety of machinery, Anthropometric requirements for the design of workstations at machinery
ISO 15534-1	Ergonomic design for the safety of machinery, Part 1: Principles for determining the dimensions required for openings for whole-body access into machinery
ISO 15534-2	Ergonomic design for the safety of machinery, Part 2: Principles for determining the dimensions required for access openings
ISO 15534-3	Ergonomic design for the safety of machinery, Part 3: Anthropometric data
ISO 15535	General requirements for establishing anthropometric databases
ISO 15537	Principles for selecting and using test persons for testing anthropometric aspects of industrial products and designs

For the *measurement of anthropometric data*, intensive knowledge about the construction and biomechanical function of the human body is an absolute prerequisite. Since the 1920s, it has been accepted that the measurement of the human body must be carried out with the help of defined measuring points and distances, using the skeletal bone structure as a reference (Martin, 1914). Mechanical measurement tools were created because of this definition and these tools have hardly changed at all in the last 80 years. It is only in the last ten years that the development of contact-free body scanners has enabled the measurement of soft body parts to enhance and supplement the traditional procurement of data using the skeletal points method.

A further essential prerequisite for the quality of anthropometric data is the selection of a *random sample* that exactly matches the objective of the study or survey. A particular body measurement (like arm length, for instance) can only be correctly applied if the random sample also matches the future end-user of a product. It is logical that data from men and women should not be mixed, but the differentiability of the measurements can be extremely complex and extensive. Gender, age, region, and social environment are only a few of the many distinct factors. Dynamic changes in the population also play an important role. For over 100 years, the height of young people has been seen to be steadily increasing. This effect, described as *secular growth*, amounts to roughly one to two cm every ten years, depending on the region involved. Since the life cycle of high-quality products—like cars, for example—can amount to 25 years, this effect must be taken into consideration in the design of such products.

The measurement of anthropometric data takes place with the individual test person who provides a random sample. In contrast, the future end customer is taken into account when the data is used for product creation and design. This transition takes place through *statistical analysis* of the body measurement data. In the last few decades, the percentile has become the most important specific value. The declaration of a percentile x states that x percent of the surveyed basic population falls below the given measuring value. A developer can thus guarantee, relatively simple, that a defined number of customers can use the product. Here is an example: a vehicle designer must consider 95% of male American customers during the design of a vehicle interior. To do this, he takes the value of the 95th percentile of sitting height from the tables. This value amounts to 935 mm. If he now ensures that the distance between the driver's seat and the interior roof contour is at least 935 mm, he can guarantee that 95% of male Americans will be able to sit in this vehicle without their heads touching the roof.

When developing a product to match human requirements, the engineer or designer uses available anthropometric data. The application of anthropometric data is thus characterized, in that the dimensions of machines and workplaces are developed both with the help of statistical body measurement data and with knowledge about the morphology and the mechanics of the human body. The application typically takes place based on the percentile values, because the statistical data of the future user group is integrated by definition. Percentile length, width, breadth, and circumference are typically used, with percentile grasp and pedal space also availed for more complex tasks. The direct use of anthropometric data from literature is limited to problem areas where either individual body measurements or the envelope contour of a human being is sufficient to provide a representative answer. Direct applications can be resolved with sufficient accuracy by means of tables.

In the second case, statistical body measurements are assembled into synthetic models of the human being created. This can either be realized by means of templates or computer-supported manikins. Here, care must be exercised in assembling body data correctly because correlations of different strengths typically exist within data like this. For instance, when you add the 95th percentile leg length to the 95th percentile torso and 95th percentile head length, you will not get a 95th percentile body length; a body length that is much too long will be created instead. The explanation for this is that the 95th percentile body height is, on one hand, typically represented by relatively long legs, but by a more normal torso on the other. A correct assembly of individual anthropometric data can only be guaranteed if the relevant correlation tables are taken into account.

HOW TO PLAN AN ANTHROPOMETRIC SURVEY

CONCEPTUAL DESIGN OF A SURVEY

The Selection of the Test Persons

It would really be necessary to survey the whole population to create anthropometric tables. However, this only occurs very rarely; all potential astronauts, for instance, can be measured for the development of a space capsule. For the creation of generally valid anthropometric tables, a method must be sought by means of which the correct test can be selected from a very large population group. To create an anthropometric table of the American population, for example, a correct narrowing-down of the collective to be measured is necessary. It is well known that anthropometric data varies depending on population, human race, social group, specialized vocation, and so on. To get reliable values, certain demands must be made on the population of the sample. Data may never be taken from so-called "lump samples" as is the case, for instance, when university students are the main source of a survey. The sample should be representative. For example: in Europe, the population of the Southern countries is, on average, smaller than that of the Northern lands.

The correct selection of test persons takes place by a statistical narrowing-down of the population. In doing so, the population's physiological differentiation characteristics and its demography must be determined. This process is typically carried out based on earlier anthropometric surveys.

Screening Test

A screening test must be carried out (if none are available). Here, the test persons are chosen by genuine random selection, or a preselecting is made based on statistical information about the population. A screening test should comprise roughly 0.5% of the entire population if the results are to be analyzed in accordance with the traditional statistical process and partition/subdivision (Juergens, Aune, & Pieper 1989).

In Figure 12.1, you can see the distribution of the heights of 7,144 German military personnel (solid line) born in 1949 and taken in a random sample, compared to the heights of all German military personnel born in 1949 (356,000). The diagram shows that a good dovetailing of both surveys does indeed exist.

Physiological Differentiation Characteristics

In the second step following the screening test, the test person's collective must also be narrowed down in accordance with the application-relevant criteria (physiological differentiation characteristics). Just how important the correct limitation of the test person's collective can be is clearly shown in the following example. When manual workplaces in a car factory in Detroit must be set up, it makes no sense to include Japanese men or Indian women in the test collective—but when the driver's cockpit for that particular vehicle is being designed, the Japanese and Indians must be considered because the vehicle is going to be exported to those two countries.

As mentioned at the start, there are many physiological differentiation characteristics. The influence of these characteristics on the later application of the data must be very carefully checked during the selection of the random sample. The four most important characteristics are examined in the following sections.

Race and Birth Nationality

The influence that a test person's race has on anthropometric measurements is clearly apparent. Typically, Asians usually have a less robust physique than other races and are not as tall. In Table 12.2, the difference between German, Italian, and former Yugoslavian people living in Germany are shown. The 5th percentile body height differs by 45 mm (male) and 71 mm (female).

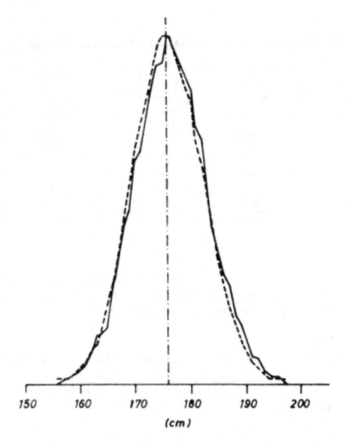

150   160   170   180   190   200

(cm)

FIGURE 12.1   Comparison of random samples and collectivity.

The same effect can also be seen within one country. In Europe, it is well known that in Italy, for instance, heights differ strongly; southern Italians are significantly smaller than those born in the north of the country. Juergens, Aune, and Pieper (1989) compared and categorized the results based on various international serial measurement surveys. Race-typical differentiation characteristics were established, and demographic distribution was also taken into account. Based on the statistical analyses of internationally valid survey results, Juergens suggests a distribution of the world population into 20 regions: North America, Latin America-Indo, Latin America European, Northern Europe, Central Europe, Eastern Europe, South-East Europe, France, Iberian Peninsula, North Africa, West Africa, Southwest Africa, Near East, Northern India, Southern India, Northern Asia, South China, Southeast Asia, Australia, and Japan.

Gender

Anthropometric studies are basically gender-differentiated because the dependency of the results on gender is very apparent. In literature, however, the anthropometric data of women are considered much less than are those of their male counterparts. This is because the vast majority of surveys were carried out in connection with military studies. Thanks to increasing female participation in the consumption of consumer goods, female data are playing an increasingly important role in product development, in particular.

TABLE 12.2

5th Percentile Body Height of German, Italian and Yugoslavian People (DIN 33402)

Nationality	Age	Gender	5th Percentil Body height (mm)
Germany	16-60	Male	1.629
Italy	16-60	Male	1.587
Former Jugoslavia	16-60	Male	1.632
Germany	16-60	Female	1.510
Italy	16-60	Female	1.439
Former Jugoslavia	16-60	Female	1.473

Age

The role a test person's age plays is also a parameter that obviously influences anthropometric data. Although young people can continue to grow until their 25th year, the bodies of those 45 years in age begin to change shape due to the process of involution.

Social Differentiation

As has been established by various surveys (e.g., Juergens, Habicht-Benthin, & Lengsfield, 1971 or Backwin & McLaughlin, 1964), there are also significant anthropometric data variances within nationality, race, and age group. Analyses show that social pedigree (school education and income of the test person as well as the school education and income of the parents) correlates to an increase in body height.

Overlapping Effect, Secular Growth

Secular growth is a very important phenomenon. In an observation carried out over an extended period, average anthropometric data (especially body height) was seen to increase depending on the year the data was acquired. This observation was carried out because the first anthropometric measurements were taken for the medical examination of soldiers (only for body height), as was effected in all national states by the introduction of conscription. Average body height can be seen to increase between 10 and 20 mm per decade. Until now, no indication of an end to this development has been found (Greil, 2001). Many theories have been put forward attempting to explain these phenomena. It is mostly assumed that the improvement of nutrition is the cause. However, despite the poor nutrition available for most people in Europe in the years during and after the Second World War, no collapse of the constant increase of average body height could be identified.

The data source and the field of application are, therefore, two factors that must be carefully considered in the application of data. For successful standardization, the validity and reliability of data must be guaranteed, but data acquisition is very costly in terms of time and money, so existing data is often quite old. Out-of-date data, for example, should never be used in the design of long-life industrial products (e.g., cars, aircraft, and railway carriages) for obvious reasons.

Overlapping Effect, Industrialization

A further overlapping effect is created by increasing internationalization and economically related population movement. It has been known for centuries that town and city dwellers are taller than their rural cousins. Worldwide industrialization is causing population relocation and people in the country are moving into the proximity of the industrial centers. Italy is an excellent example of this; the somewhat smaller southern Italians are emigrating to the industrial north of the country.

Although an increase in body height is to be expected from the effects of secular growth, statistical mean body height values are reduced, thanks to the mixing of large and small inhabitants. Similar effects can be seen in industrial countries that have a significant proportion of foreign residents. In Germany, for example, the integration of Italians and Turks (see DIN 33402) slow the secular growth effect. In the United States, too, the influx of Asian and Latin American immigrants causes statistical displacements.

The Selection of Anthropometric Measurements

In the last few decades, measurement performed on live persons (somatology) for anthropometric tables has become firmly established. To define anthropometric data, extensive measurements on skeletons (osteology) had to be carried out in the preliminary stages (Braeuer, 1988; Karoly, 1971). Juergens further developed these studies for ergonomic design in industrial anthropology (Juergens & Matzdorff, 1988). The studies describe the objective of industrial anthropology as, "...to design the environment of the human being in accordance with his physical needs, in the broadest sense. The acquisition of the morphology and biomechanics of the human beings is, thus, permanently oriented on a technical precondition, how they arise from the design of workplaces, utility objects, etc."

Although pure anthropometric measurements are derived only from the body and clearly identifiable measuring points, with industrial anthropological data, the demands of the products to be designed are also integrated. This means that for the definition and selection of measurements, it is not only body posture or movement that plays a key role but clothing as an influencing factor must also be considered. It may also be necessary to re-enact complex workplace situations for the test.

Axes and Reference Systems

To be able to describe and compare measuring points according to position and orientation, standardization of the axes and reference systems is necessary. In industrial anthropology, the reference to standing and seating areas has established itself. The prerequisite here is that these areas are flat and non-deformable. In various earlier studies (e.g., Lewin & Juergens, 1969), further reference points (e.g., heel point, contact point with the ischium protuberance) are defined, but these should be avoided because the positions involved are difficult to reproduce. As an alternative, contact points or contact areas (e.g., the pelvis or shoulder, see Figure 12.2) or functional, product-related reference points are used (e.g., seat reference point SRP).

The human head plays a special role due to its degree of mobility. As early as 1884, the *Frankfurt Horizontal* was established; it was named after the conference location. A test person must take up the following head posture: The connecting line between the lowest point of the bony eye socket and the highest point on the upper edge of the auditory canal must run horizontally,

Description: Projective maximum depth of the lower torso between the maximum anterior protrusion of the abdomen and the maximum posterior protrusion of the buttock. See figure 28.

Method: Subject sits fully erect with thighs fully supported and lower legs hanging freely, with rearmost point of the buttocks touching the surface of a vertical panel. Distance is measured from the vertical panel to the maximum anterior protrusion of the abdomen.

Instrument: Anthropometer.

FIGURE 12.2   Measurement with vertical panel (ISO 7250—4.2.17).

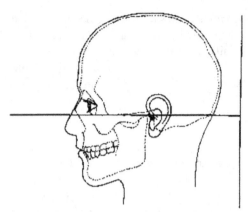

FIGURE 12.3 Frankfurt horizontal.

parallel to the standing area (see Figure 12.1). This definition is used for all measurements where the head is involved (Helbig, Juergens, & Reelfs, 1987).

The Influence of Postures

Since a human being typically takes up different postures while working or using products, very representative measurement approaches must be defined for anthropometric studies. Standing and sitting postures with various arm positions have established themselves as the most valid in all serial measurements.

Studies of standing postures are typically made of test persons standing ramrod-stiff. The feet are closed and parallel to one another and head posture is in accordance with the Frankfurt Horizontal. Nowadays, it is recognized that a stiff posture does not correspond to the physiological working posture and a more relaxed posture is preferred. Nevertheless, studies (Lewin & Juergens, 1969) have shown that reproducibility in the case of physiological postures is, at best, only inadequately attainable.

A similar effect was observed with sitting postures. A rotation of the pelvis occurs due to the curvature of the spinal column (kyphosis). This leads to deviations of up to 50 mm between stiff and relaxed sitting postures. These deviations are overlapped by two further influencing factors—the asymmetry of the human being and the increasing shortening of certain measurements as the day goes on.

Several studies (Gaupp, 1909; Ludwig, 1932) substantiate that the majority of human beings display *asymmetry*. Although around 75% of all adult right arms are longer than their left arms, the left leg is longer than the right in over 50% of the population. This asymmetry continues throughout the entire body; the different leg lengths lead to a tilt in the pelvis, which in turn leads to scoliosis (lateral curvature of the spine) and this, in turn, is followed by an asymmetry of the acromion (shoulder line). The average difference in asymmetry is between 10 and 20 mm in the case of the acromion (shoulders, right lower than the left), and 10 to 30 mm difference in arm lengths.

The cause of this asymmetry is probably to be found in the one-sided actions that we humans so often carry out. A close relationship can be seen between being right-handed (or left-handed) and the excess length of the more active right (or left) arm. The crossed asymmetry in the leg can be explained by the crossing of right-handed or left-handed dexterity and the anklebone. A right-handed person will typically jump with the left leg, while a left-handed person will move the right leg first.

The effect of the *daytime shortening* of torso and leg lengths can amount to a body height difference of between 20 and 40 mm. The reasons for this are the compression of the intervertebral discs of the spinal column, the articular cartilage of the legs, and the sinking of the foot arch.

SELECTION OF MEASUREMENT TOOLS

The Manual Measuring Tools

Martin (1914) is seen as the founder of a catalog of anthropometric measurement prescriptions and methods. He also developed the measurement setting that is named after him. Essentially, it consists of different calipers, specific vernier calipers, and a calibrated measuring tape that can take individual body dimensions (see Figure 12.4). The essential idea is to confine measuring (as far as possible) to measurements of bone-to-bone distances. In every case, bone areas are measured; these areas are immediately under the skin surface and can be clearly felt by hand, even in the case of corpulent subjects. In general, anthropometric data are taken from a subject who either stands on a table or sits on it with their legs dangling. The body posture is totally "stretched," and the head is kept in Frankfurt Horizontal. Measurements are generally taken from unclothed subjects (Bradtmiller et al., 2004).

Circumference measurements, on the other hand, show much more inner deviation. The actual use of the measuring tape is very important; it should be taut, with no loosely hanging section. The result depends on the "elastic" properties of the subject and the intention of the person carrying out the test.

Advantages and Disadvantages of Manual Measurement

Anthropometric measurement with manual measuring tools is a proven and standardized methodology. The true advantage of manual measurement is the determination of skeletal characteristics because they must be actually felt and measured.

One disadvantage of manual measurement can be seen in measurement accuracy and reproducibility. Kirchner et al. (1990) show that in manual measurement, inter-individual and intra-individual measurement value differences of up to 100 mm occur for body height or sitting height. A further disadvantage is that body shape and posture measurements of soft body parts can only be carried out with a great deal of effort. Many of these problems can be solved by means of the 3D body scanner.

Contact-Free Measurement with Body Scanning

For the past few years, the rapid development of laser and camera technologies has enabled the contact-free, accurate, and fast scanning of body surfaces by means of body scanners. A three-dimensional image of the test person can be created within just a few seconds.

Nowadays (Robinette et al., 2004) there are four different principles in use for three-dimensional measuring tools. *Contact devices* are biomechanical systems, by means of which a measuring point on the subject beings is contacted and the three-dimensional coordinates defined.

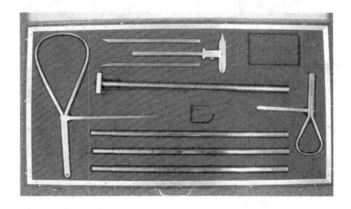

FIGURE 12.4   Measurement setting by Martin.

This genre includes mechanical measuring machines (Wenzel, Steinbichler, etc.) or mechanical measuring arms (e.g., Faro-Arm). *Tracking systems* follow markers attached to the subject with several detectors and calculate the corresponding three-dimensional coordinates of the markers by the triangulation method. This type includes tracking systems from a wide range of manufacturers (Vicon, Polhemus, Peak, Qualisys, etc.) based on the ultrasonic principle and systems based on optical and electromagnetic principles. *Stereophotogrammetry systems* employ several cameras that photograph the test person from different angles. The allocation of the skin points to be measured takes place by means of manual allocation or automatically, with the help of projected patterns.

3D body scanners (see Figure 12.5) detect the overall outer surface of the human being in a single pass. Using one or more projection units or lasers, lines or patterns are projected onto the subject. Various fixed or mobile CCD cameras detect the deformation of a line or pattern (triangulation method). Three-dimensional surface information is then calculated from the recorded deformation by means of conversion equations. A 3D "cloud" of points representing the skin surface of the test person is then created from the image information.

The body scanner approach has many advantages over traditional sizing surveys. Scanners are more precise and consistent than even trained human beings, requiring only 810 seconds per stance to capture the needed data. The entire point cloud also becomes available almost instantly, allowing it to be mined and analyzed along with thousands of other clouds, yielding information about the changing shape of our bodies and changes in body measurements. Since the scanning system captures shapes rather than measurements, the most important element of the body scanner

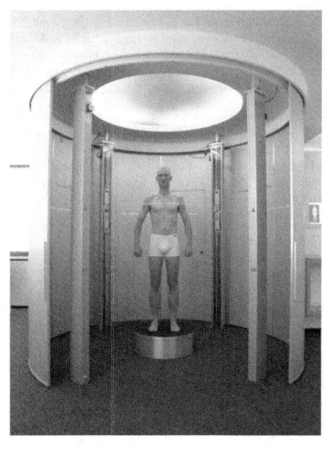

FIGURE 12.5   Body scanner.

is not in itself the method that is used to obtain the point cloud but the software that extracts the body measurements.

The beginning of the 21st century saw a new era of anthropometric surveys. CAESAR (Civilian American and European Surface Anthropometry Resource) was the first survey that used a 3D body scanner to acquire 3D scans of body surfaces as well as demographics data and traditional measurements (Blackwell et al., 2002; Robinette, Daanen, & Zehner, 2004). In 1999 and 2000, 11,000 subjects in Great Britain were scanned and manually measured (SizeUK). The "next step of evolution" was made in France in 2003–2005. 11,562 subjects were scanned using a new technology, so parallel manual measurements were no longer used (Godil & Ressler, 2009). The Swedish survey of 2004/2005 scanned more than 4,000 subjects, using the same technology and concept as the French survey. 2007 saw the initiation of several body scanning surveys in other countries like Spain and China. Carried out in Germany between September 2007 and February 2009, SizeGERMANY is the latest representative anthropometrical measurement survey of more than 13,000 men, women, and children. At more than 30 measurement locations distributed across the country, complete three-dimensional body shape information and more than 80 body measurements were acquired using 3D body scanners (Seidl, Trieb, & Wirsching, 2009).

The use of 3D body scanners poses a special challenge for the performance of the study or survey. On one hand—and for the first time—this technology enables the acquisition of reproducible and accurate skin surface and posture data. At the same time, however, comparability with manual studies and surveys must be ensured. This is why combined measurement methods have since become established. Geometrical markers are attached to relevant skeletal feature points (Robinette, Daanen, & Zehner, 2004). During the scan, both sets of information (the surface data of the subject and the marker positions) are detected by special software solutions, and body measurements are calculated in a fully automated process (see Figure 12.6). Integrated solutions like

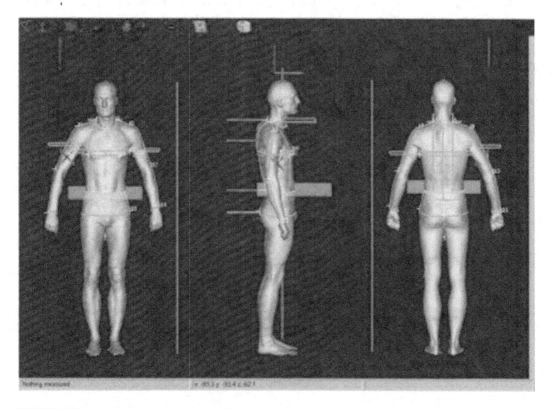

FIGURE 12.6 Automatic measurement of body dimensions.

this (i.e. body scanner, hardware, and matching software) are offered today by a few specialized companies (e.g., Human Solutions or TC2).

THE STATISTICAL ANALYSIS OF DATA

Fundamentals of Anthropometric Statistics

If we take only the average of the test data, the data received will not suffice. The discipline of statistics provides different ways of representing human measurements, especially in their variability—these are the range between the highest and the lowest observed measure, the so-called percentile (see the following), and the standard deviation. As it makes sense to only use the percentile for smaller and known populations, the application of the percentile is the norm in scientific anthropometry. The idea here is as follows: the persons surveyed are put into a sequence of small to large, going by the required measurement. The Xth percentile is defined by the measurements of the one subject who separates the X% lower part from the remainder (see Figure 12.7). In anthropometric tables, it is usually the 5th, the 50th, or the 95th that is taken (also the 1st and the 99th on occasion). If we take the measurement of the 5th percentile as an example, this means that 5% of the sample is lower than this. The 50th percentile divides the population of the sample into half smaller and half larger. This is not necessarily the mean value, however. This will only be the case if the distribution of the values is symmetric. Nevertheless, experience has shown that anthropometric distance measurements have a good chance of being normally distributed and, therefore, symmetric.

The percentile of a body measurement is derived from the distribution of a measurement. The pertinent distribution is determined from all the measurement values of a size. It is from this that the total frequency distribution is derived through integration. The associated percentile value can be taken from the total frequency distribution. At the ordinate (y-axis), the desired percentile value is selected (e.g., 5th percentile) and projected on to the cumulated frequency curve, while the associated measurement value is read from the abscissa (x-axis).

When the number of subjects in the sample is large enough and the sample's composition is representative of the required population, these values are good estimations of the true values of the population. The necessary number of subjects can be estimated by the rules of scientific statistics. Formula 1 gives the necessary number N of subjects in a sample, when the estimated value s of the standard deviation is known (e.g., from a pre-experiment). Under the assumption of the standard normal distributors, the value $z_{\acute{a}/2}$ can be determined by this tolerated error probability. n_P is the amount of the pre-experiment sample.

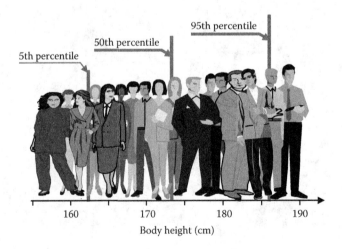

FIGURE 12.7 The percentile idea, illustrated by the example of body height.

Formula 1: Calculation of number of test subjects

$$N = \frac{4z_a^2/2^{s^{2n_p}}}{\Delta_{\text{crit}}^2 (n_p - 1)}.$$

For example: in a pre-experiment of $n_P = 40$ subjects for body height, a standard deviation of $s = 73.4$ mm has been observed. If the average of the body height of a certain population (e.g., the population of the people in the United States) is to be investigated, a representative composition (according to the human race, social situation, profession, age, and gender) of the people of the United States must be found. In this case, this value should be indicated with an accuracy of $\Delta_{\text{crit}} = 5$ mm and an error probability of $\alpha = 1\%$ (from statistic table $z_{0.5} = 2.576$)—according to formula 1 at least 5,885 subjects are necessary. If the demand can be reduced (e.g., acuity only 20 mm, error probability $\alpha \leq 10\%$), the necessary number of subjects would only be 150. Simplified formulas for the calculation of the number of subjects needed for a sample are printed in ISO 15535 (Annex A).

The Combinatorial Analysis of Body Measurements

In all international anthropometric studies and applications, it has become standard to use body height as a criterion. The input of a 5th, 50th, and 95th percentile type always originates solely from body height. During application, the error is very often made that this percentile value is also transferred to other individual or composite measurements. Here is an example of this: two-dimensional templates are still popular for vehicle interior design nowadays; if the 95th percentile type is now taken for the design of the roof height of a car (as mentioned earlier, the input relates only to body height), then a wrong result will be produced. For the design of the roof height, the sitting length of the torso (trunk length) is decisive, but this size has relatively little to do with body height (it correlates poorly). This effect is shown in Figure 12.8. Three persons have the same body height, but different proportions of leg length and torso height.

More modern anthropometric analyses take the correlation information on body measurements into account. The correlation coefficient specifies whether two measurements have a relatively minimally varying, proportional correlation. The correlation coefficient can take on a range of values from 0 (no correlation) to 1 (direct correlation).

Juergens (1992) shows that coefficients between 0 and 0.7 are so non-representative (from an ergonomic standpoint) that one measurement cannot be derived from the other. In Table 12.3, the correlations between a few other important measurements are itemized in an exemplary manner. You can see that body height still correlates well with crotch height (0.84), but when taken with sitting measurements, only a weaker correlation exists (0.74 with sitting height). In the case of shoulder breadth (0.37) and the circumference (0.13, 0.15), a significant correlation is no longer apparent. In contrast, you can see that the sitting dimensions (sitting height and elbow height) show a significantly higher correlation (0.53) than when taken against the body height. A similar effect can be observed in the case of circumferences—whereas correlation with body height hardly exists, in this case, a higher intercorrelation can indeed be seen (0.46).

For the purposes of practical application, it can be assumed that body height is only a sensible indicator for leg lengths and accessibility statements (Juergens, 1992). For measurements that occur in sitting positions, sitting height is a better control value than body height. In the case of circumferences, the same thing occurs—they display a correlation with one another and not with body height. This means that for the design of complex workplaces (e.g., vehicle interiors), we must revert to detailed task-specific tables or to a suitable correlation matrix. Since this is very complex, however, other means like computer-supported *manikins* (computer-supported models of humans) can be used.

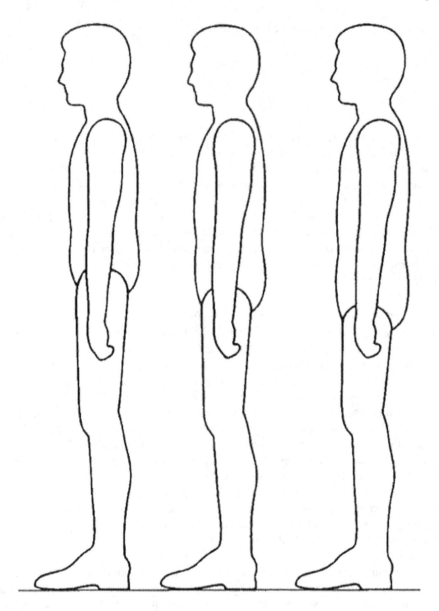

FIGURE 12.8　Different proportions with the same body height.

INTERNATIONAL STANDARDS FOR ANTHROPOMETRIC SURVEY DESIGN

ISO 15537, Principles for Selecting and Using Test Persons

For the correct selection of test persons for anthropometric tests, ISO has defined its own standard: ISO 15537. It was developed by the CEN/C 122 Work Group (Ergonomics) in cooperation with the SC3 Subcommittee of the ISO/TC 159 Work Group.

ISO 15537 was created for testing the anthropometric aspects of industrial products and designs that have direct contact with the human body. This standard is also for testing the safety aspects of products.

The procedure of test person selection is fully described in ISO 15537. Here, the selected procedure is primarily oriented on the development of machines and not as much on the creation of anthropometric tables.

TABLE 12.3

Correlation Matrix of Some Body Dimensions (Juergens, 1992)

	Body Height	Crotch Height	Sitting Height	Ellbow Height, Sitting	Shoulder Breadth	Chest Circumference	Waist Circumference
Body Height (4.1.2)		0.84	0.74	0.25	0.37	0.13	0.15
Crotch Height (4.1.7)			0.46	0.02	0.25	0.04	0.00
Sitting Height (4.2.1)				0.53	0.35	0.13	0.14
Ellbow Height, Sitting (4.2.5)					0.05	0.06	0.09
Shoulder Breadth (4.2.9)						0.21	0.30
Chest Circumference (4.4.9)							0.46
Waist Circumference (4.4.10)							

The following tasks are also described in ISO 15537: dimensions of products; critical anthropometry measurements; the combination of both; body types, screening, and detailed tests; selection of test persons for screening and detailed tests; and documentation with results. To help you understand this, a descriptive example of anthropometric tests for an elevator has been included in Annex A. Besides detailed, step-by-step instructions, the test procedure is also graphically summarized in a table.

ISO 15537 is supplemented by two tables. Table 12.1 of ISO 15537 contains 20 measurements of European persons (18 to 60 years old) with the 5th, 50th, and 95th percentile values. Table 12.2 of ISO 15537 shows 19 body measurements taken from worldwide anthropometric surveys. These have been analyzed according to physique types (small and large) and the percentile type has been entered. For the smaller type, the 5th, 50th, and 95th percentiles were calculated. The same procedure was used for the larger type, but the 95th percentile of the smaller type corresponds to the 5th percentile of the larger type (see Table 12.4).

ISO 7250—Basic Human Body Measurements for Technological Design

ISO 7250: 1996 provides a description of anthropometric measurements that can be used as a basis for comparing population groups. It describes the most important body measurements used in anthropometric surveys. As the authors remark, the body measurements should neither be a binding definition nor instructions for taking measurements—they should be used as a recommendation for tests, studies, and surveys. The ISO 7250 was also created by the SC3 Subcommittee of the ISO/TC 159.

ISO 7250: 1996 is currently one of the most important standards because it introduced an international harmonization of the body measurements to be acquired. 55 body measurements are defined based on detailed postures. Each of the body measurements has a clear-cut description (name and number), a sketch, a notation of the measurements, the recommended measuring method, and the measuring instrument. In Figure 12.9, the description of ISO 7250 is shown with an example of body height.

Further attributes of the ISO 7250 are given in a detailed definition of all technical terms as well as the measuring conditions (clothing of subject, support surfaces, body symmetry, and the measuring tools).

ISO 15535, General Requirements for Establishing Anthropometric Databases

ISO 15535 was defined in close conjunction with ISO 7250. The goal of the standard is to formulate a database structure and the associated reports. This standard was prepared by the same subcommittee as ISO 7250. Analog with the already discussed ISO standards, ISO 15535 defines terms, references, collection design, and data acquisition requirements like sample size, type of clothing, and accuracy of measuring instruments. A database format and the minimum content of the database are defined in a detailed description. The method for estimating the number of

TABLE 12.4

Nineteen Body Measurements of Worldwide Anthropometric Data in ISO 15537. Worldwide Human Body Measurements for Persons Between 25 and 45 Years of Age, Divided into Two Categories, "Smaller Type" and "Larger Type"

Human body measurement [a]	Smaller type [b]			Larger type [b]	
	Value, mm				
	P5	P50	P95/P5	P50	P95
Stature (body height)	1 390	1 520	1 650	1 780	1 910
Sitting height (erect)	740	800	870	935	1 000
Eye height, sitting	620	690	750	815	880
Forward reach (fingertips)	670	740	810	880	950
Shoulder (bideltoid) breadth	320	365	410	455	500
Shoulder (biacrominal) breadth	285	325	360	395	430
Hip breath, standing	260	300	335	375	410
Knee height	405	455	505	550	600
Lower leg length (popliteal height)	320	365	410	460	505
Elbow-grip length	270	305	340	375	410
Buttock-knee length	450	505	560	615	670
Buttock-heel length	830	920	1 010	1 100	1 190
Hip breadth, sitting	260	305	350	395	440
Hand length	140	155	170	185	200
Hand breadth at metacarpals	65	75	90	100	110
Foot length	200	225	250	275	300
Head circumference	475	505	540	570	600
Head length	160	175	185	195	205
Head breadth	120	135	145	160	170

NOTE   For children and elderly populations, separate data sets are sometimes are needed.

[a]   Source: Hans W. Jürgens, Ivar A. Aune, Ursula Pieper: *International Data on Anthropometry* ([12] in the Bibliography).

[b]   Both types should be considered when testing products designed for the whole world. "Smaller type" and "Larger type" categories are given if it is not possible to create a product design for the whole world. "Smaller type" data are based on females from "smaller type" populations. "Larger type" data are based on males from "larger type" populations.

subjects is described in Annex A; Annex B and C contain a printed definition and a printed example of an anthropometric datasheet. Annex D, E, and F focus on statistical analysis of data.

ISO 20685 (Draft), 3D Scanning Methodologies for Internationally Compatible Anthropometric Databases

To meet the demands that the strong growth of this technology has created, the ISO/TC 159 Technical Committee, Subcommittee SC3, is working on ISO Standard 20685. This is still at the draft stage at present since the topicality of the technology is relatively extensive. ISO 20685 goes into details about the technical characteristics of the hardware and software, the clothing of the test persons, and the methods for reducing errors in 3D scanning. Suitable postures for carrying out a study in body scanners are given special attention because the posture definitions of the ISO 7250 cannot be included in all systems for reasons of space. Here, the ISO 20685 suggests four postures: three standing and one sitting (see Figure 12.10).

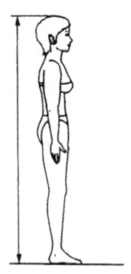

FIGURE 12.9 Description of body height (4.1.2) in ISO 7250.

Furthermore, ISO 20685 also discusses the different body measurement definitions in ISO 7250 and ISO 8559. Because ISO 7250 is primarily defined for ergonomic applications, ISO 8559 focuses on garment design users. Another important element in ISO 20685 is the functional properties of the software, by means of which the analysis of the scan data takes place. This also includes the interaction between 3D measurement and combined measurement with landmarks, plus features of manual and automated measurement and the selection, segmentation, and visualization of body parts and landmarks.

3D ANTHROPOMETRIC SIZE AND SHAPE SURVEY

INTRODUCTION

In recent years, a paradigm shift in anthropometric measurements of the population has begun. Classic manual measurement is increasingly being supplemented or even replaced by modern body scanning technology. In the U.S., the U.K., and the Netherlands, the classic manual measurement was combined with body scanning, while in France and Sweden, serial measurement surveys were carried out exclusively by body scanners. In 2006 and 2007, other large international serial measurement surveys were prepared and launched (e.g., Germany, Spain, Romania, Thailand, etc.) and in China, a preliminary survey of children preceded the serial measurement of approximately one million Chinese.

INTERNATIONAL SURVEYS WITH BODY SCANNING

Since 1998, a series of surveys where 3D body scanners were used as the core technology for data acquisition have been performed worldwide. Figure 12.11 shows the main surveys performed with 3D body scanning since 1998.

For the first time, the relatively young body scanning technology was used for the CAESAR project (Civilian American and European Surface Anthropometry Resource) (Robinette, Daanen, & Zehner 2004). The project was initiated by the Wright-Patterson Air Force Base, Ohio, United States, and TNO in the Netherlands. Beginning in the U.S. and Canada, followed by the Netherlands and Italy, 4,400 men and women between the ages of 18 and 65 were scanned and measured between April 1998 and early 2000. The results of the serial measurement survey in the form of 39 CDs with the original scan data—around 100 body measurements include 40 manual measurements and approximately 70 3D marker positions—and the socio-demographic survey are currently on sale via the SAE International.

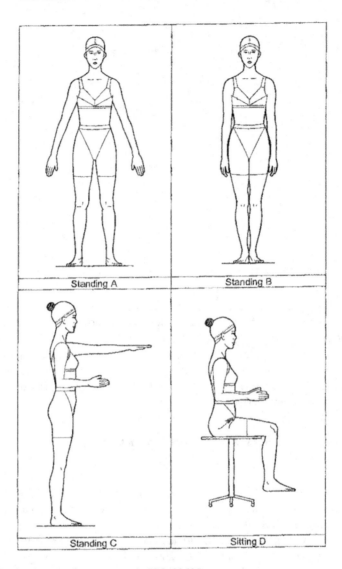

FIGURE 12.10   Defined measuring postures in ISO 20685.

In recent years, new measurement surveys have already taken place in some European countries, not least motivated by the ongoing efforts for the establishment of a harmonized European apparel sizes system. Taking place almost simultaneously with CAESAR and for the first time in Europe, the new technology was used in 1999/2000 on around 1,500 women during the target group-specific surveys for "underwear" to obtain current measurements for the design of athletic clothing in Europe. The "Women 60 Plus" survey conducted in Germany between 2001 and 2003 also showed significant deviations between current body measurements and the 1993/94 clothing tables for ladies' outer garments that were still valid during the years of the survey.

In the U.K. in 1999/2000, 11,000 men and women were measured, and these results deviated significantly from previous measurements. This resulted in 11,562 men, women, and children being measured in France between 2003 and 2005. These survey results showed that the French have become taller and considerably heavier. The results of this survey are now available for the apparel industry in the form of size tables that are also age-graded in special analyses.

2004/2005 saw 4,000 men, women, and children between the ages of 13 and 85 measured in Sweden. In Spain, a survey of more than 10,000 women of all ages was launched at 59

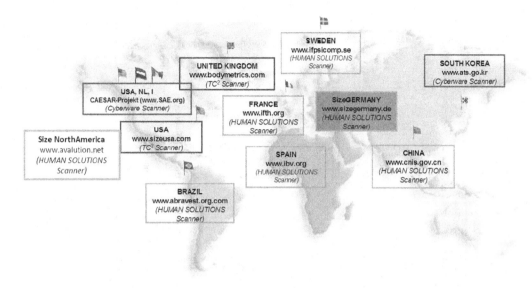

FIGURE 12.11   International surveys with body scanning (since 1998).

measurement locations in 2007. The key figures of the most important 3D body scanning serial measurement surveys can be seen in Table 12.5.

The SizeGERMANY survey is especially significant. In addition to four scanning postures, data acquisition included an extensive socio-demographic questionnaire in which market-specific issues for the participating sectors were addressed. What is special about this was that, for the first time, the results were posted on an internet portal, enabling usage-specific and targeting group-specific online analyses.

IMPLEMENTATION OF SIZEGERMANY SURVEY

The original objective of SizeGERMANY was the acquisition of representative data for use in the apparel industry and in technical ergonomics. The following features were defined as required for the concept of a "new generation" of serial measurement programs:

Measurement must take place solely with the help of body scanners in accordance with **ISO 20685.** Standard body shapes must also be analyzed as a supplement to body dimensions.

The analysis of the body dimensions must take place in compliance with the market standards for technical ergonomics in accordance with **ISO 7250** and for the apparel industry in accordance with **ISO 8559.**

The results are made available in the form of a web-based, interactive **data portal,** enabling companies to carry out product-specific analyses online. Human Solutions and the Hohensteiner Institutes were responsible for the content and the financial aspects of the project. The **guidelines of ISO 15535** were also incorporated.

The expected results were offered to the industry in various packages, in accordance with the scope of performance and market. More than 100 companies from the apparel industry (e.g., Adidas, BOSS, OTTO, H&M, C&A, Zara, QVC, Vaude, etc.) and the automotive industry (BMW, Opel, Porsche, VW, Peugeot, etc.) participated in the SizeGERMANY project.

Measurement of the test subjects began on the 1st of July 2007 and the final presentation took place on the 19th of February 2009. After a period of only 18 months, the following definite focal points were achieved:

• The measurement of the test subjects.

TABLE 12.5

Key Figures of the Most Important International 3D Surveys

Country	Germany	Spain	Sweden	France	England	Germany Underwear	Germany Women60+	CAESAR
Year	2007/2009	2007/2008	2004/2005	2003/2005	1999/2000	2001–2003	2000	1998–2000
Management	Human Solutions/BPI	IBV	IFP	IFTH	UCL	BPI	BPI	WPAB, TNO
Groups	Men, Women, Children	Women	Men, Women, Children	Men, Women, Children	Men, Women	Women	Women	Men, Women
Test Persons	12,000	10,000	4000	11,562	11,000	1500	1500	2400 USA/CA 2000 NL/I
Age Groups	6–910–1314–1718–2526–3535–454–6–55,56–65>65	18–65	13–85	5–6,7–89–10,11–1213–1415–17–18–2021–252–6–3536–4546–5556–70	16–2526–3536–4–546–5556–656–6–75, über 76	≤5051–6061–70ü–ber 70	≥2021–3031–404–1–5051–6061–70>70	18–65
Regions	4	6		4	3	5	2	4 Länder
Scanning Locations	30	59		37	8	5	2	
Postures	3 standing 1 sitting	2 standing 1 sitting	2 standing 1 sitting	1 standing 1 sitting	1 standing 1 sitting	1 standing 1 sitting	1 standing 2 sitting	2 standing 1 sitting
Markers	Partly	Yes	Yes	Yes (9)	No	No	No	Yes (72)
Body-Scanner	Human Solutions	Human Solutions	Human Solutions	Human Solutions	[TC]²	Human Solutions	Human Solutions	Cyberware/Vitronic
Results	Online-Portal	Charts & Report	Charts & Report	Charts & Report	Portal	Charts & Report	Charts & Report	Charts & Report on 39 CD's

- The analysis of the dimensions and proportions of men's, women's, and children's bodies.
- The derivation of all statistical characteristic measurements for the apparel and automotive industries.
- The development of new garment sizes and the creation of market share tables for men, women, and children.
- The analysis of the acquired data regarding body shapes and proportions and the development of 3D standard body shapes. The provision of an anthropometric typology for CAD-Tool RAMSIS (Seidl, 2004).
- The development of a secular growth simulation of the German population that will last until 2040.
- The provision of all data and analyses in a web-based online portal and the support of companies participating in the portal.

The Cornerstone—ISO 20685

In 2006, the ISO/TC 159 Technical Committee, Subcommittee SC3, announced the introduction of the ISO Standard 20685 (3D scanning methodologies for internationally compatible anthropometric databases). This standard goes into detail about the technical requirements of the body scanner and the characteristics of the hardware and software; the scan volume of a test person is defined, for example, and a calibration method is recommended. The clothing of the test persons and the methods for reducing errors in 3D scanning are also discussed. One further important element in ISO 20685 is the functional properties of the software, by means of which the analysis of the scan data is carried out. This also includes the interaction between 3D measurement and combined measurement with landmarks, plus features of manual and automated measurement and the selection, segmentation, and visualization of body parts and landmarks.

Suitable postures for carrying out a study with body scanners are specified (ISO 20685 suggests four postures, i.e. three standing and one sitting). ISO 20685 also discusses the different size definitions in ISO 7250 and ISO 8559 and recommends the data set.

Selection of Subjects

The primary requirement for a serial measurement program is to represent the entire population. In most of the serial measurements that were manually carried out, two-tenths of one percent of the population was selected as being representative (based on the statistics of the German Federal Office of Statistics) and subsequently measured. In some cases, only one individual target group (e.g., military personnel) was incorporated into the measurement statistics within the context of comparatively less intensive measurement procedures.

In addition to the incorporation of the ISO 15537, a new approach was used for the first time in SizeGERMANY: during the survey, current databases were constantly aligned with the statistically necessary data range (Seidl & Bubb, 2005). In this dynamic monitoring of test persons, each person was classified by gender, age, and regional distribution. Based on the accuracy requirements of the partners (10mm for body length and 95% confidence), this resulted in a sample size of at least 165 persons per group (Figure 12.12).

The number of test persons per class was aligned with the current data of the German Federal Office of Statistics. Due to another request to measure test persons of six years of age up to the maximum age, 18 age groups were specified. The first group began at six years of age, and the last age group was concluded by a group classed as open-end (see Figure 12.13). To achieve an exact regional distribution, 31 sites in four main regions (East, West, North, and South) where the surveys were carried out were selected. This resulted in a minimum requirement of 11,880 test persons (number of subjects = group size×number of age groups×number of regions).

During the survey of the test persons, their characteristics (age, region, and gender) were acquired and analyzed online daily. All the measurement teams had online access to this analysis

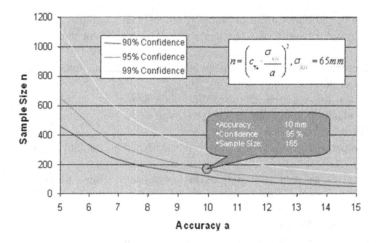

FIGURE 12.12   Relation of sample size and required accuracy.

through a web browser. If deviations occurred in the desired sample sizes, only more of the type of test persons that were missing in the classes were selectively measured (Figure 12.13).

Postures and Body Dimensions

In the selection of body dimensions, both ISO Standards 7250 and 8559 were rigorously adhered to. This is advantageous in that the results for an industrial user already correspond to the way he uses body dimensions for his product development. The results are also very easily compared with those of older surveys. In cooperation with the industry partners, 43 body dimensions were defined for the apparel industry (ISO 8559) and 37 body dimensions for technical ergonomics (ISO 7250). Based on the measurements, 15 special body dimensions were also defined (used in the creation of 3D human models, e.g., RAMSIS) (Robinette, Daanen, & Zehner 2004).

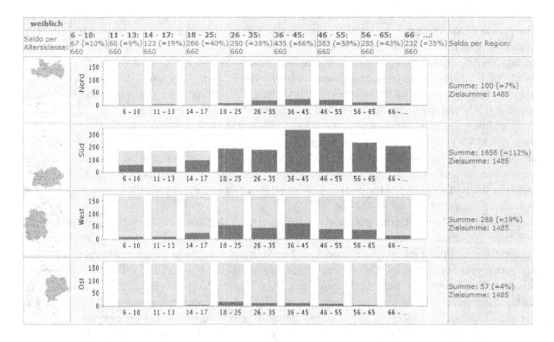

FIGURE 12.13   Web-based online analysis of sample sizes.

Four postures (three standing postures and one sitting posture) were necessary to analyze the 95 body dimensions. On the right in Figure 12.14, the standing posture is represented by a stretched and angled arm posture; this enables the measurement of functional dimensions in standing posture. The second posture from the right is a specific measurement posture for body circumferences. The third measurement posture from the right is a relaxed standing posture to determine the correct body lengths. The posture on the left in Figure 12.14 shows the sitting posture with bent arms to correctly acquire all necessary measurements in sitting posture.

The weight of the test persons was also electronically determined during scanning and more than 100 socio-demographic questions were asked by means of an electronic questionnaire.

Measurement Technology

VITUS Smart XXL body scanners with ANTHROSCAN software (www.human-solutions.com) were used to measure the test persons at the 31 sites. This device was specially developed for extensive serial measurement programs and it is the only one that complies with ISO 20685. These systems have proven themselves in numerous measurement surveys (Spain, France, Sweden, etc.).

The VITUS Smart XXL is a four-column, 3D body scanner that uses the optical triangulation measurement principle (laser technology and 100% safe for the eyes). The eight sensor heads have a measuring range of 2100 mm height, 1000 mm depth, and 1200 mm width (see ISO 20685). The level of accuracy of the system is extremely high: it has an average error (circumference) of less than 1 mm. Within just 12 seconds of scanning time, the sensors capture a density of 27 points/cm^2, meaning that approx. 550,000 3D points for each person are measured. Yet another advantage the system has is mobility: the weight of the measuring columns is less than 100 kg and the floor space required is only around 4.8 m².

Data Quality Management

A comprehensive and process-oriented data management process was used for SizeGERMANY to obtain exact body dimensions. This process is schematically displayed in Figure 12.15. Each

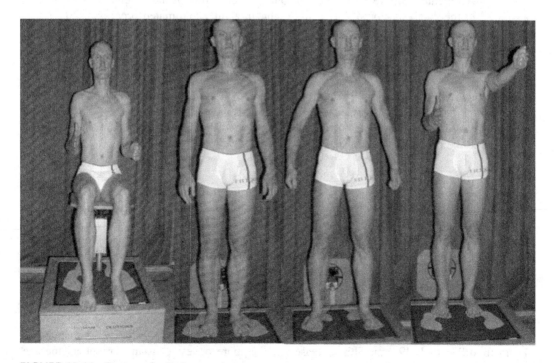

FIGURE 12.14 Postures for body scanning.

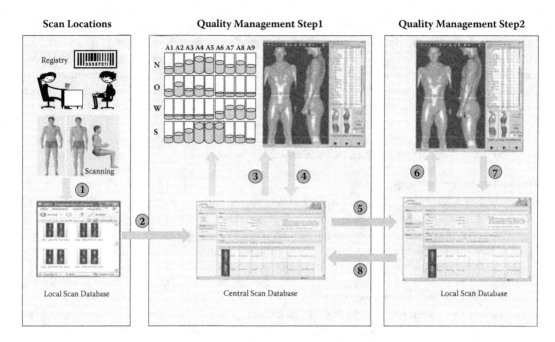

FIGURE 12.15 SizeGERMANY quality management.

test person was managed by a specialist during the entire measuring process. In a contact-free procedure, the test person was scanned in four different postures, wearing standardized apparel for measurement procedures. Then, the electronic questionnaire was completed. The entire timeframe involved totaled between 20 and 30 minutes per test person. The measurement data was first saved locally (Figure 12.15, Step 1) but after each day of measurement, the data was transferred to the Analysis Portal during the night, in a fully automated process (Step 2), and stored in a central database. Data integrity, scanner quality, and system calibration were checked during an additional quality control process, then the automatic analysis process is commenced. The initial analysis of the body dimensions took place automatically by software (Step 3 and Step 4).

In a second step, trained specialists used computers to check that the positions of the reference points were correct (Step 6). Every change in the dimensions (and the dimensions' history) were saved in a local database (Step 7). When the entire process was finished, the data was once again transferred to the central database (Step 8).

Results

To adequately gauge the change in the German population, it makes sense to compare the results with those of earlier serial measurements. The most up-to-date available reference, in this case, is the DIN 33402 of 1999–2002. This analysis is based on various summarized measurements taken in East and West Germany. The results (DIN 33402) were analyzed in four age groups: 18–25, 26–40, 41–60, and 61–65. To compare these results with those of SizeGERMANY, the SizeGERMANY results were again analyzed for the DIN 33402 age groups.

Table 12.6 shows the changes in dimensions for body height and chest circumference. The 50th percentile for body height increased by 42 mm from 1,750 mm to 1,792 mm. Height increase in the medium age groups amounts to more than 50 or 51 mm. The chest circumference of men shows a greater increase for the 50th percentile than for all the other age groups—namely, an increase of 74 mm from 975 to 1049 mm. This increase remains more or less constant throughout the age groups from 55 mm (18–25) to 93 mm (61–65).

TABLE 12.6

Difference SizeGERMANY to DIN33402

MEN	Age	DIN 33402 P50 [mm]	SizeGERMANY 2008 P50 [mm]	Difference P50 [mm]
Body Height	18-25	1790	1805	15
	26-40	1765	1815	50
	41-60	1735	1786	51
	61-65	1710	1746	36
	18-65	1750	1792	42
chest circumference	18-25	930	985	55
	26-40	970	1032	62
	41-60	995	1081	86
	61-65	990	1083	93
	18-65	975	1049	74

Fehler! Verweisquelle konnte nicht gefunden werden.

Table 12.7 displays the differences between DIN 33402 and SizeGERMANY for the 50th percentile for women. The body height of women increases throughout the age groups by 39 mm and waist circumference increases on average by 82 mm. Growth in the oldest age group (61–65), however, has increased by 123 mm from 815 to 938 mm. This marked increase could possibly be accounted for by the significantly greater number of SizeGERMANY subjects in this age group. A look at the sitting height shows that growth, in this case, is less than that of body height. This means that the increase in body height is mainly due to an increase in leg length.

TABLE 12.7

Difference SizeGERMANY to DIN33402

WOMEN	Age	DIN33402 P50 [mm]	SizeGERMANY P50 [mm]	Difference P50 [mm]
Körperhöhe	18-25	1660	1673	13
	26-40	1635	1683	48
	41-60	1615	1654	39
	61-65	1595	1619	24
	18-65	1625	1664	39
Waist Circumference	18-25	730	734	4
	26-40	780	775	-5
	41-60	820	859	39
	61-65	815	938	123
	18-65	730	812	82
Sitting Height	18-25	880	883	3
	26-40	870	886	16
	41-60	855	869	14
	61-65	840	848	8
	18-65	860	875	15

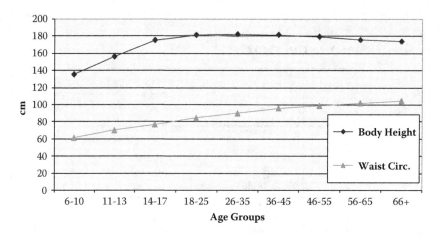

FIGURE 12.16   Relation of age groups to body dimensions.

Figure 12.16 shows the trend throughout the age groups for the most important characteristic measurements for men in SizeGERMANY. It is apparent that the average body height of the older age groups decreases. This decrease slows considerably in the case of the 26–35 and 36–45 age groups. This can be seen as an indication that the trend of secular growth in Germany is clearly slowing.

Yet another important trend can be seen in Table 12.8. Here, the 5th percentile and the 95th percentile of body height and chest circumference from SizeGERMANY are shown in comparison to DIN 33402. If you look at the population aged between 18 and 65 years, the 5th percentile (body height) increases by only 18 mm. In contrast, the 95th percentile (body height) shows a considerable increase of 85 mm (1,855 mm to 1,940 mm).

Chest circumference data behaves similarly. While the 5th percentile increases by 48 mm (from 870 to 918 mm), the 95th percentile (chest circumference) increases by 125 mm from 1,110 to 1,235 mm. In general, all body dimensions have one thing in common: the distribution function of the dimensions extends increasingly to the right.

TABLE 12.8
Change of 5th and 95th Percentiles

MEN	Age	DIN 33402		SizeGERMANY		Difference	
		P5 [mm]	P95 [mm]	P5 [mm]	P95 [mm]	P5 [mm]	P95 [mm]
Body Height	18-25	1685	1910	1683	1938	-2	28
	26-40	1665	1870	1690	1965	25	95
	41-60	1630	1835	1662	1923	32	88
	61-65	1605	1805	1636	1877	31	72
	18-65	1650	1855	1668	1940	18	85
Chest Circumference	18-25	845	1055	885	1133	40	78
	26-40	860	1090	917	1222	57	132
	41-60	880	1145	948	1249	68	104
	61-65	880	1125	951	1271	71	146
	18-65	870	1110	918	1235	48	125

In the meantime (2016 to 2019), a large-scale anthropometric measurement has been carried out in the U.S. and Canada under the name SizeNorthAmerica by Avalution, a company of the Human Solutions Group, following the same pattern used in SizeGERMANY. The SizeITALY project—with a total of 5,900 test persons—was carried out in the same way from 2012 to 2013. In contrast to the values collected in America, the data from the two European countries show that young men are significantly taller than old men. But in all three cases, it can be observed that the annual growth in length seems to come to a standstill, as the 18- to 25-year-old men are no further taller than the old men. Within SizeNorthAmerica (SizeNA), around 18,000 subjects have been measured altogether, among those—for the first time—are also children between six and 17 years. There are significant differences compared to the data that practically represent Europe through SizeGERMANY. Table 12.9 gives an overview of the differences in the 5th, 50th, and 95th percentile of the two-measurement series. Another difference is striking: the proportions (i.e. the ratio of stem length to body height) differ in the different regions. This is illustrated in Figure 12.17. Furthermore, the BMI from SizeNorthAmerica has the highest average values ever determined within a measurement survey. 30.8% of U.S. women and 22.4% of U.S. men show a normal weight (i.e. BMI between 18.5 to 24.9). This is a significantly smaller portion than in Germany; 53.9% German women and 35.2% German men. The pre-obesity status for women and men is comparable between the U.S. and Germany. However, 40% of U.S. adults show more than two times higher obesity rates compared to Germany. Due to a huge sociodemographic questionnaire, significant differences can be visualized between urban and rural populations, and between populations of different ethnic origins. As the U.S. is a typical immigration country, the population is correspondingly heterogeneous—which is why the SizeNorthAmerica data can also be filtered by different ethnic groups. The experience with country-specific measurement surveys has shown clear differences in body dimensions in different countries. Especially in a country like the U.S. where many people of different ethnic origins live, it is important to have access to a valid database for product development (Rissiek & Stöhr, 2019)

THE USE OF DATA

It is the objective of anthropometric data procurement to make high-quality data available for the development of ergonomic machines and products. Just as we have seen in the previous chapter, it is a relatively complex matter to select the right test persons and representative body measurements. With these results, easier problems—like the height of a door frame or the distance to an operating lever, for example—must be solved quickly and accurately. The development of more complex task problems is another matter entirely. Multifaceted questions must be answered during the development of a vehicle interior—for instance, where the seating position is, how the vision is looking out the windows, can the controls be reached and activated, how the pedal travel is, are

TABLE 12.9

Various Sizes 5th, 50th, and 95th Percentile Male According to SizeNA and Size Germany

	Men, SizeNorthAmerica			Men, SizeGERMANY		
	5th Perz.	50th Perz.	95th Perz.	5th Perz.	50th Perz.	95th Perz.
Height	163.7	175.7	187.7	165.9	178.5	193.2
Sitting height (erect)	84.5	91.4	97.8	85.4	92.2	99.4
Crotch height	73.8	82.2	90.9	73.2	81.3	90.9
Bust/chest girth	92.1	109.6	134.9	92.4	105.7	123.3
Waist girth	76.9	98.8	131.5	77.0	94.6	117.5
Weight [kg]	62.7	86.1	127.7	64.2	82.0	110.4

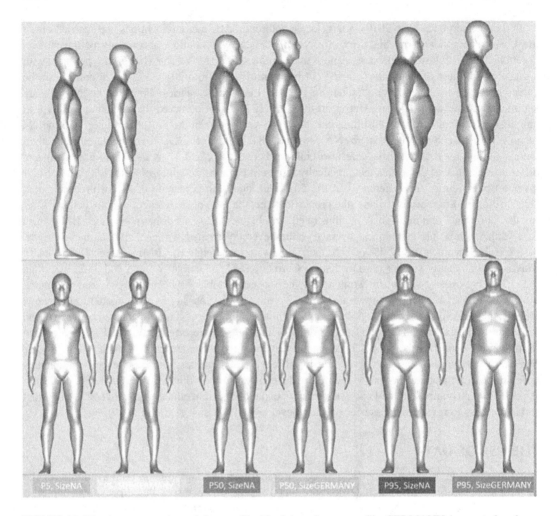

FIGURE 12.17   Average avatars, adult man, SizeNorthAmerica versus SizeGERMANY (generated on base of height, bust/chest, waist, and hip girth).

there collisions with door trims, how clearly can the mirror be seen, and so on. There is no guarantee that the answers to such problems can be given from tables; more powerful tools are needed for such complex matters.

An effect observed by Kirchner et al. (1990) must be included here as well, and that is the fact that different users produce different results from the same task. On one hand, this occurs due to the different test approaches used in different tables (social layers, age structures, and posture definitions) and, on the other, the transformation of the anthropometric data in the application is interpreted differently by designers and engineers, with the results being difficult to reproduce.

In the case of the application—as in the case of data procurement—standardized methods and working procedures must be adhered to. Juergens (1992) here differentiates between the *direct application* and the *indirect application* of anthropometric data.

DIRECT APPLICATION OF ANTHROPOMETRIC DATA

Use of Charts

In direct application, problem questions are answered in which the following are discussed.

Whether or not the body envelope (external contours) of the human being suffices.

Whether or not a single measurement for the design of the product to be developed is relevant.

The first question comes into play for all proximal (pertaining to close contact with the body) products. Clothing, shoes, and safety equipment can be derived directly from the tables. Nevertheless, the size of access openings for body parts must be directly dimensioned. In the case of the second question, the measurements are directly related to the technical object to be designed. A simple example of this is the standardized door height of buildings. To enable the majority of the population to go through the door without stooping, a corresponding measurement (e.g., 99th percentile) must be selected. Another example of this is the design of an office table. To enable everyone to reach the papers, telephone, computer, etc. on the table, its depth must be applied to the 5th percentile female (subject to functionality demands).

Very often, additional add-ons must be integrated into the design. A simple example of this kind of application would be the design of the roof height of the highest point in a vehicle cabin: the relevant measurement to consider, in this case, would be the maximum body height of the population. To get the measurement of the technical object, a safe distance of 10% to 25% must be added, depending on the type of application for which the vehicle cabin will be used.

In all these examples, only a limit value for the design of the products is used. With more complex designing tasks, a situation might arise where interior and exterior measurements have to be considered at the same time. The design of a vehicle cabin, for example, must be tackled in such a way that a tall man has enough room inside the cabin and short persons must also be able to operate the controls. Industrial products such as this should be designed in a manner that at least 90% of the population can use without any restrictions or hindrances. The values of the 5th percentile and the values of the 95th percentile are normally taken as the limit for this design. As anthropometric tables indicate the measurements for males and females separately, the demand for 90% usability can be fulfilled by taking the values of the 5th percentile for the female and the 95th percentile for the male.

Use of Data Portals

Printed tables, however, have a great disadvantage, because the only anthropometric data that can be used is data for which an analysis is available in tabular form. Any subsequent adjustment of the statistical analysis for the requirements of the ergonomist (such as adapting the age group) is very difficult. This is why the global development of online databases of anthropometric body dimension data has been launched in recent years.

WEAR (World Engineering Anthropometry Resource) is the name given to a group of international universities and research institutions that have joined forces to implement an international database for anthropometry. The database targets users from the fields of product development, apparel development, and the automotive and furniture industries (Gupta & Zakaria, 2014, WEAR, 2014).

In collaboration with the Institute Français du Textile et de l'Habillement (IFTH) and the Hohenstein Institute, Human Solutions developed iSize, an international body dimension portal. The anthropometric data of more than 100,000 men, women, and children from 13 nations are stored in the portal, and iSize is already used by more than 200 companies today.

The basis of this international system, however, was the development of the SizeGERMANY portal. The objective of this project was to process the results of the SizeGERMANY serial measurement survey for users in such a way that they can analyze the data themselves and adapt it to their own customers and products.

Human Solutions developed this portal as a web-based online portal. The challenge here was to develop a simple user interface and to provide a full range of analysis functions. Analyses for the apparel industry, however, require different functions than those used in ergonomics analyses.

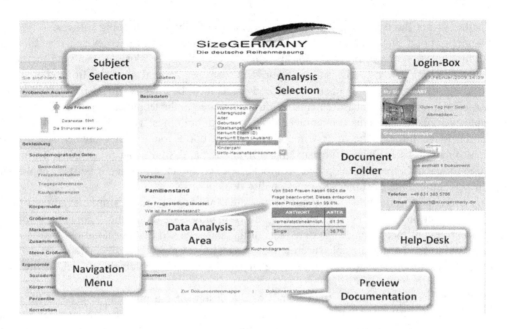

FIGURE 12.18 User interface portal.

Figure 12.17 shows the user interface of the SizeGERMANY portal. The left side contains the navigation throughout the entire portal, divided into subject selection and the navigation menu, which contains various menus for the apparel industry and technical ergonomics. In the center of the user interface is the area in which all the results of the analysis are displayed. On the right-hand side of the user interface are the login box, document folders, and the help desk.

The input window for subject selection is displayed in Figure 12.18. In the upper area here, the user can select whether all analyses should relate to men, women, boys, or girls. The region and the age group can also be selected here. In the central menu test, persons can be further filtered. Up to three socio-demographic entries can be input. In the lower menu area, the test subjects can be selected in accordance with any dimension area.

Based on this selection of test subjects, further analyses for ergonomics are made possible, namely

- socio-demographic data
- body dimensions
- percentiles
- correlations

The analysis on the portal can be accessed by standard web browsers which are available on the market. A desired analysis is calculated online on the central statistics server and made available to the user in the form of a preview and a detailed, automated datasheet.

Figure 12.18, for example, shows the first page of the datasheet using the example of the body height of women. All women between the ages of 18 and 55 from the region of South Germany were selected. Only married women or women living in a marital-like relationship were additionally selected. The portal automatically generates the entire analysis, including a description of the body dimensions, an output of the selection of the subjects, and a table with body dimensions and statistical characteristics. In this way, all socio-demographic data, body dimensions, and percentile calculations can be analyzed for each body dimension with any configuration of test subjects.

Figure 12.18 shows another CD analysis function: here, the correlation of every individual body dimension to any other body dimension for any group of test subjects can also be calculated. It also shows the body height to sitting height for all women. Each individual test subject is represented by a point in the diagram. The horizontal and vertical lines in the diagram show the 5th, 50th, and 95th percentiles of the corresponding body dimension.

The line which runs at an angle is the regression line. In this analysis sheet (which is generated online), both body dimensions of the correlation are described, together with the selection/number of test subjects. In the block on the right (the linear regression function), the percentile values and the correlation in accordance with Pearson are given numerically.

INDIRECT APPLICATION WITH TEMPLATES AND THREE-DIMENSIONAL MODELS

Although available tables can suffice for the estimation of the body measurements needed for simple workplace designs, sensible and meaningful design aids must be available for high-value and safety-relevant workplaces. One option that has proven its worth in the last few decades is the use of two-dimensional contour templates of percentile types (5th, 50th, and 95th) of body height. These are available in different sizes in either fixed or movable versions. Two-dimensional templates were developed to make the anthropometric data of the human being available to the designer (in direct association with the movement potentiality of the future user of the product design).

The development of a body contour template is a complex procedure, because not only must it portray the anthropometric data correctly, but it must also take body shape into account. A pure derivation from anthropometric tables is, therefore, not possible; additional body shape analyses must be carried out with the aid of image-providing processes—and in the case of movable templates, the correct anthropometric body part lengths must also be created for every posture position used. In templates like this, the very complex movements of human joints (e.g., combined rotation and displacement in the knee joint) must be mechanically simulated. Yet, another demand made on movable templates is the correct portrayal of the physiologically correct mobility areas. Here, linked (coupled) movements of several body elements must be simulated by means of intelligent mechanics. In Figure 12.19, this is represented by an example of the shoulder joint (Helbig & Juergens, 1977). The forward movement of the arm forces a displacement of the shoulder joint in an upward direction via the coupling with the scapula. If this movement is continued upwards, the clavicle (collarbone) will become increasingly involved in the movement.

In Figure 12.19, the "Kieler Puppe" is illustrated. This simulates the complex biomechanics of the most important human joints thanks to intelligent mechanical replacement designs. The back construction of three parts with coupled joints is worthy of special attention. It allows a very good simulation of the physiological movement of the spinal column (Helbig & Juergens, 1977).

In the last 40 years, a multitude of templates have been developed for various purposes. A few of these templates have found their way into various national and international standards: the *SAE template* is a 2D template developed especially used in the automotive industry. The data, design, and application are on record in the standard SAE J826a (Society of Automotive Engineers). This template has become extremely important in the automotive industry, being used not only for ergonomic design but also for the authorization of the vehicles themselves. The template consists only of the torso of the 50th percentile American and a 2D leg simulation of the 95th percentile American. In Figure 12.19, the SAE template (J826a) and the related SAE-Standards (J1100, J941, J1052, J1517, and J1516) are shown.

The *DIN 33408, body contour template for seats* was modeled on the SAE template. It is derived from the developments of the Kieler Puppe (Helbig & Juergens, 1977) and encompasses a total of six templates in three perspectives: front, side, and top view. Five body types are supported

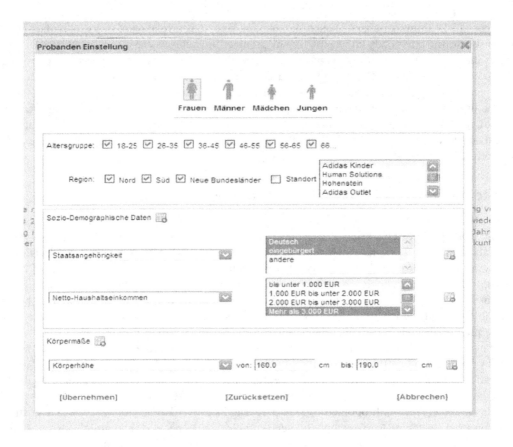

FIGURE 12.19 Input window for subject selection.

in total. For the male form, the 5th, 50th, and 95th percentiles are used, while for the female form, the 1st to the 5th and the 95th percentile are available.

There also exist various templates that offer a simplified representation of the human being in the form of body contour lines (e.g., DIN 33416 or Bosch template). These have never achieved the same breadth of distribution as the templates described earlier. It must be said, however, that the body contour templates are still extremely widespread in actual practice. This is due to the comparably economical purchase price and the manageability of the templates themselves.

The informational value of the templates is, however, very limited because neither the physique proportions regarding future product user population (age, race, living space, secular growth) can be sufficiently considered, nor can task-dependent postures or movements be simulated or evaluated from a comfort point of view.

Digital Human Models (DHM) for Ergonomic Design

The limitations of templates with regard to physiological movement simulation, anthropometric scalability, and the correct computation of anthropometric data correlations have in the last few years been, more or less, eliminated by modern computer technology. Since the 1980s, modern 3D CAD technologies have become established and this, in turn, leads to a need for 3D aids for the design of the human being. Although more than 150 ergonomic manikins are known to have been developed (Hickey, Pierrynowski, & Rothwell, 1985; Porter et al., 1993; Seidl & Speyer, 1992), only a few systems have been able to establish themselves successfully on a global industrial basis in the last few years. Thanks to the increasing impact of the DHM, more can now be ascertained about their potential capabilities.

Historical Development of the DHM

The development of computer models of human beings originated with NASA. In 1967, William A. Fetter, a Boeing employee, developed the optical computer model, First Man. Based on the anthropometric dimensions of a scalable 50th percentile man, it was used to check reach and accessibility in aircraft cockpits. Building on this development, there followed (between 1969 and 1977) the Second Man, Third Man, Fourth Man and Woman, BoeMan, and CAR models. Developments in the United States were further advanced with the Bubbleman, Tempus, and Jack models from Badler and the Chrysler Cyberman (Figure 12.20). In the mid-1980s, the Safework manikin was developed in Canada at the Ecole Politechnique in Montreal.

Various manikin approaches were also created and used in Europe. In 1984, in the Laboratoire d' Anthropologie Appliquee et d'Ecole Humaine in Paris, the Ergoman manikin was created. The mid-1980s in Germany saw the realization of the computer model Franky (within the framework of a BMFT program), the Heiner System (at the Technical University in Darmstadt), and the Anybody manikin (and of course its successor ANTHROPOS). Anybody was modeled on Oszkar, the development of a Hungarian management consultancy. In England, Sammie was created, the first manikin on the market for which the license was made industrially available to third parties.

Körpermaß

Körperhöhe

Definition nach ISO 7250, 4.1.02:
Vertikaler Abstand von der Standfläche zum höchsten Punkt des Kopfes (Vertex).

Die Auswertung beruht auf den Daten der Frauen
- aus den Altersgruppen 18-25, 26-35, 36-45 oder 46-55
- aus der Region Süd
- mit Familienstand: verheiratet/eheähnlich

Die Auswertung enthält 1081 Datensätze.

ALTER	MITTEL	STDABV	VAR	MIN	MAX	P1	P2.5	P5	P50	P95	P99
18-25	166.5	6.4	41.4	152.3	183.1	152.3	153.1	154.6	167.6	177.9	183.1
26-35	167.3	7.2	52.3	148.0	188.2	151.0	152.6	157.0	166.9	181.5	185.2
36-45	167.5	7.0	49.7	149.4	186.8	151.9	155.4	156.5	167.4	179.6	184.5
46-55	164.6	6.7	44.3	147.9	182.7	149.2	151.8	154.4	164.6	176.2	181.4
Gesamt	166.4	7.0	49.6	147.9	188.2	150.4	153.0	155.4	166.3	178.9	184.4

Maßeinheit ist 'cm'

FIGURE 12.20   Automated datasheet.

In the mid-1980s, the functional differentiation of the manikin came into being for the first time. While Sammie focused on the automotive industry, Ergomas and Champ DHM (later marketed as AnySim) were created for factory planning.

One development with a special rating is the RAMSIS manikin. It was developed for the ergonomic interior design of vehicles between 1987 and 1994 under contract to—and in cooperation with—the entire German automotive industry. This industrial project is arguably the most extensive ever for the development of a manikin.

The Five Basic Elements of a Manikin

In principle, manikins constitute the foundation for multifaceted usability in a wide range of different applications. The prerequisite here, however, is that the orientation of the model is tailored to the specific application area. The morphology and biomechanics of the human body are made available in the quality required for the specific problem. There are five basic elements that have to be taken into account in manikin development.

The *design of the manikin* considers the kinematics of the model—that is, the number and mobility of the joints and associated body elements as well as the accurate detail of the outer surface simulation that corresponds to the human skin. Here, there are significant characteristics for the design of the manikin (discussed in part in the chapter on 2D templates). The number of joints defines the kinematic flexibility of the manikin. The mathematical/biomechanical simulation of the joints influences the motion paths of all body elements that are coupled with a joint. The selection of *the* limit angle ranges of each joint ensures that even in the case of extreme postures, the behavior of the DHM will remain correct. Besides the physiological simulations of the skeleton, the realistic outer surface simulation of the human skin surface constitutes the second focal point in a manikin design. The surface of the body of the DHM must not only be correctly altered in an anthropometric and movement-dependent manner, but it must also fulfill beauty requirements because such systems are always used to present the analysis results.

The coupling of the DHM with suitable *anthropometric databases* is the basis of developing efficient products with the DHM. Conventional tables are integrated into most DHMs and most systems have interfaces for their own data input. Only a very few manikins, however, support further modern approaches like the calculation of anthropometric data via correlation coefficients (RAMSIS, Safework) (already discussed). The user can consider the correlation of individual body data and consequently compute physique models with selectable or relevant critical sizes, which represent the population much more realistically. Enhanced control parameters like age group, regional differentiation, or somatotypes are also selectable. Secular growth simulators (i.e. forecasting on the expected mean changes in size (growth in length)) of the population groups have also become available meanwhile (RAMSIS).

The third focal characteristic is *biomechanical posture and movement simulation.* In a sitting posture, for instance, a hip joint angle deviation of only 3° leads to a knee upper edge that will deviate more than 2 cm from the real body posture. In many models, postures are set using the "artistic feeling" of the user. Thanks to inverse kinematics, more modern systems now have solution processes available. These processes, leaning on robotics and simulating postures and movements by means of angle or energy minimization, consider the real movement behavior of the human being inadequately at best. Only a small number of systems have realistic calculation and modeling capabilities. The Ergomas/Ergoman system, for example, has a movement simulation that is based on real observations. RAMSIS has a posture simulator based on high-dimensional distribution functions that draw upon roughly 60,000 3D data values. Latest studies (Chaffin, 2004) indeed simulate movement sequences by taking into account the biomechanical characteristics of the body (maximum forces, mass moved, and distributions); but the necessary scientific data substructures are not yet fully available at the time of writing.

The fourth focal characteristic of a DHM is its *ability to analyze* the product to be developed. The user wants to obtain information about the ability of his design to match human requirements and perhaps some indications as to quality-increasing measures he could take. Whereas all ergonomic analysis and test procedures were integrated into DHM at the beginning of its development (e.g., OWAS, NIOSH, MtM), modern DHMs increasingly specialize in their own field of application. Although DHMs need a fast ergonomic appraisal with the aid of maximum force procedures (e.g., NIOSH), comfort, mirror view, and belt run all have a more major role to play in the designing of systems for, say, the automotive industry. Nevertheless, a few basic and analysis functionalities from the development of the DHM have meanwhile become established. In vision simulation, the view of the CAD design is reproduced using the manikin's eyes. Mathematical 3D hull (enveloping) surfaces for extremities are calculated in the simulation of spaces within reach (grasp spaces). The forecasting of maximum operating forces is also one of DHM's domains since the dependency of the maximum forces on the posture can be correctly taken into account.

The last focal characteristic of the DHM is *seamless integration into the design process*. A DHM can only be sensibly used if it is integrated into the design environment—in other words, if it is in the "virtual world" of the product. Today, there are two different solution approaches available. The *Standalone-System* is a computer program that loads product data via standardized CAD interfaces. The advantage of this solution is flexibility since it can work together with all available systems. One disadvantage to note is that the user must learn the ropes of a second computer program. With *CAD Integration*, the manikin—with all functionalities—is integrated into the CAD system of the user who doesn't have to leave his familiar surroundings—and he won't have to take the time to export and import product data. The disadvantage of this solution is the high price involved.

Systems Available in Industry

In the last few years, three ergonomic manikins have become established in professional applications, controlling more than 95% of the market. The main field of application of the manikin JACK (Figure 12.21) is animation and visualization in vehicle design and architecture. This model consists of 39 body elements; visualization takes place via area segment imagery with textures. The anthropometric database is based on the Human Solutions Library, on NASA data, and on in-house studies, surveys, and tests. Posture and movement simulation enables interactive movement of the entire manikin in real-time. Animation is based on robotics methods and realistic movement simulation is not stored. Numerous modules are available in the form of analysis tools for factory planning and vehicle development. The system is available either as a stand-alone version or integrated into SiemensNX. John Deere, British Aerospace, Caterpillar, Volvo, and General Dynamics as well as various universities use Jack.

The manikin *HUMAN BUILDER* (former Safework) was conceived for workplace design in factory planning and product design. The anthropometric database is based on U.S. Army data and encompasses an anthropometric body-type generator that can create statistical test samples. The posture and movement simulation enables the movement of short chains in the model's arms, legs, and torso by means of inverse kinematics. Vision simulation, fixed accessibility areas, a joint-dependent comfort evaluation, maximum force calculation, and a center of gravity analysis are available as analysis tools. Human Builder is available solely in the form of an integration unit with CATIA, ENOVIA, and DELMIA by Dassault Systems. It is used by Chrysler, Boeing, and various universities and academies.

The manikin *RAMSIS* (Figure 12.22) was developed for vehicle design. The anthropometric database encompasses a physique typology that takes body measurement correlation, desired percentile models, an international database with more than ten global regions, and a secular growth model into account. Posture and movement simulation is carried out by task-related

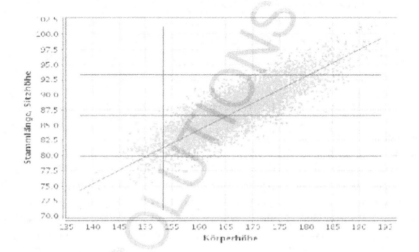

Korrelationen

Körperhöhe und Stammlänge, Sitzhöhe

Die Auswertung beruht auf den Daten aller Frauen und enthält 5946 Datensätze.

Maß 1:

Körperhöhe

Definition nach ISO 7250. 4.1.02:
Vertikaler Abstand von der Standfläche zum
höchsten Punkt des Kopfes (Vertex).

Maß2

Stammlänge, Sitzhöhe

Definition nach ISO 7250. 4.2.01:
Vertikaler Abstand von einer horizontalen
Sitzfläche zum höchsten Punkt des Kopfes (Vertex).

Von den 5946 gewählten Datensätzen fallen 5891
Datensätze in die Auswahl.

Maß2 = 0.44 * Maß1 + 13.33
Korrelation r = 0.85
Die Korrelation ist als hoch zu interpretieren.

Maß 1:		Maß 2:	
P5 =	153.4	P5 =	79.8
P50 =	164.9	P50 =	86.6
P95 =	179.0	P95 =	93.3

FIGURE 12.21 Correlation body height and sitting height.

animation, enabling the forecasting of the most probable posture. More than 80 functions for the analysis of vehicles and vehicle interiors are available (vision and mirror simulation, seat simulation, accessibility limits, comfort, belt analyses, etc.). The system is available as a standalone version for UNIX and Windows, as a CAD integration unit with CATIA, and as a programming library for independent applications. RAMSIS is used by more than 75% of all car manufacturers (Audi, BMW, Mercedes-Benz, General Motors, Ford, Porsche Volkswagen, Seat, Honda, Mazda, etc.), by aircraft manufacturers (Airbus), and manufacturers of heavy machinery (e.g., Bomag) (Figure 12.23, 12.24, 12.25, 12.26 and 12.27).

SUMMARY

Anthropometrics is the scientific measurement and statistic analysis of data about human physical characteristics and the application (engineering anthropometry) of this data in the design and evaluation of systems, equipment, manufactured products, and human-made environments and facilities.

Reliable serial measurement surveys form the basis for the application of anthropometric data. The planning and execution of surveys like this must be considered from different points of

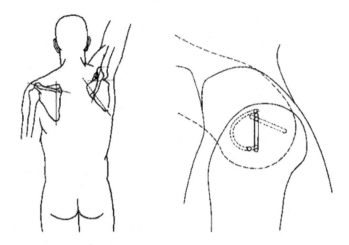

FIGURE 12.22   Mechanical simulation of the shoulder joint.

view. The population, dimensions to be measured, and proper measuring tools must be selected correctly from the outset. Comparability with surveys that have been already carried out must also be addressed, as with correct statistical analysis.

In the last ten years, 3D serial measurement surveys with body scanners have established themselves as state-of-the-art. A comparison shows the different conceptions of these surveys and the example of SizeGERMANY shows the entire process of a large serial measurement survey in detail.

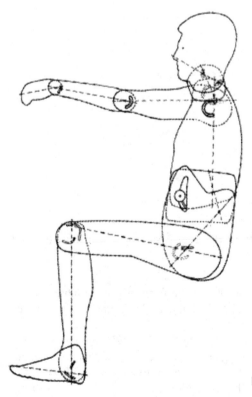

FIGURE 12.23   Template "Kieler Puppe."

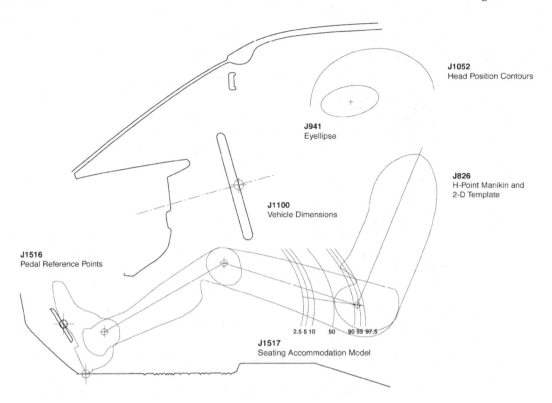

J1052
Head Position Contours

J941
Eyellipse

J826
H-Point Manikin and
2-D Template

J1100
Vehicle Dimensions

J1516
Pedal Reference Points

2.5 5 10   50   90 95 97.5

J1517
Seating Accommodation Model

FIGURE 12.24   SAE template.

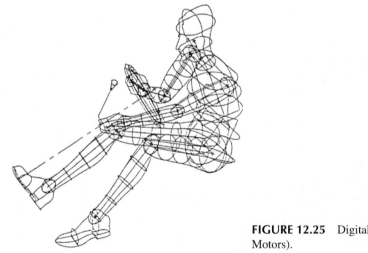

FIGURE 12.25   Digital Man Model Cyberman (General Motors).

The results of these surveys are typically made available in the form of printed tables. In the case of new series measurements, the results are also made available in the form of online portals, enabling the user to perform dynamic analysis.

Besides the application of tables, product development with the aid of 2D templates is also widespread. Here, the biomechanics of the human being can also be considered during product development. DHM represents the cutting-edge of today's technology. They enable the efficient use of anthropometric data in product design (thanks to biomechanical simulations of the

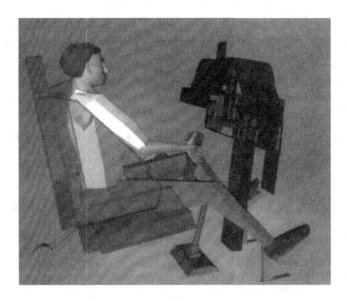

FIGURE 12.26   The DHM Jack.

FIGURE 12.27   The DHM RAMSIS.

human being), direct use of 3D data, the consideration of modern statistical correlation methods, and product-related functionalities.

REFERENCES

Backwin, H., & McLaughlin, S. D. (1964). Increase in Stature, Is the end in sight? *Lancet*.
Blackwell, S., Robinette, K., Daanen, H., Boehmer, M., Fleming, S., Kelly, et al. (2002). Civilian American and European surface anthropometry resource (CAESAR), *Final Report*, Volume 2.

Bradtmiller, B., Gordon, C. C., Kouchi, M. K., Juergens, H. W., & Lee, Y. (2004). Traditional anthropometry. In N. J. Delleman, C. M. Haslegrave, & D. B. Chaffin (Eds.), *Working postures and movements*, CRC Press LLC, (pp 18–29).

Braeuer, G. (1988). Osteometrie. In R. Martin, & R. Knussmann (Eds.), *Anthropologie: Handbuch der vergleichenden Biologie des Menschen*. Band 1. Fischer Verlag.

Chaffin, D. B. (2004). Digital human models for ergonomic design and engineering. InN. J. Delleman, C. M. Haslegrave, & D. B. Chaffin (Eds.), *Working postures and movements*, CRC Press LLC, (pp 18–29)

DIN 33402. (1978). Koerpermasse des Menschen. Part 1–3.

Gaupp, E. (1909). *Die normalen Asymetrien des menschlichen Koerpers*. Fischer Verlag, Jena.

Greil, H. (2001). *Wachstum und Variabilität im Körperbau und ihre Berücksichtigung bei industriellen Größensystemen. In Brandenburgische Umweltberichte* (pp. 62–76). Heft 10. Potsdam

Godil, A., & Ressler, S. (2009). Shape and size analysis and standards. In Duffy, G. V. (Eds.), *Handbook of digital human modeling. Research for applied ergonomics and human factors engineering*. CRC Press.

Gupta, D., & Zakaria, N. (2014). *Anthropometry, apparel sizing and design*. Google eBook: Elsevier, N. Z.

Helbig, K., & Juergens, H. W. (1977). Entwicklung einer praxisgerechten Körperumrissschablone des sitzenden Menschen. Forschungsbericht Nr. 187 der Bundesanstalt fuer Arbeitsschutz und Unfallforschung Dortmund.

Helbig, K., Juergens, H. W., & Reelfs, H. (1987). Augen-Kopf-Körper-Interaktion in der Vertikalebene am Beispiel des Mensch-Maschine-Systems. Ergonomische Studien Nr. 4. Bundesamt fuer Wehrtechnik und Beschaffung.

Hickey, D., Pierrynowski, M. R., & Rothwell, P. (1985). Man-modeling CAD programs for workspace evaluations. *Lecture Manuscript*, University of Toronto.

ISO 14738. (2002). Safety of machinery, Anthropometric requirements for the design of workstations at machinery.

ISO 15534-1. (2000). Ergonomic design for the safety of machinery, Part 1: Principles for determining the dimensions required for openings for whole-body access into machinery.

ISO 15534-2. (2000). Ergonomic design for the safety of machinery, Part 2: Principles for determining the dimensions required for access openings.

ISO 15534-3. (2000). Ergonomic design for the safety of machinery, Part 3: Anthropometric data.

ISO 15535. (2003). General requirements for establishing anthropometric databases.

ISO 15537. (2004). Principles for selecting and using test persons for testing anthropometric aspects of industrial products and designs.

ISO 20685. (2004) (Draft). 3D scanning methodologies for internationally compatible anthropometric databases.

ISO 3411. (1995). Earth-moving machinery, Human physical dimensions of operators and minimum operator space envelope.

ISO 7250. (1996). Basic human body measurements for technological design.

ISO 8559. (1989). Garment construction and anthropometric surveys, Body dimensions.

Juergens, H. W., Aune, I.A., & Pieper, U. (1989). Internationaler anthropometrischer Datenatlas. *Schriftenreihe der Bundesanstalt fuer Abreitsschutz*. Fb 587.

Juergens, H. W., & Matzdorff, I. (1988). Spezielle industrieanthropologische Methoden. In R. Martin, & R. Knussmann (Eds.), *Anthropologie: Handbuch der vergleichenden Biologie des Menschen*. Band 1. Fischer Verlag.

Juergens, H. W. (1992). Anwendung anthropometrischer Daten. Direkte und indirekte Anwendung. In H. Schmidtke (Ed.), *Handbuch der Ergonomie*. (A–3.3.1). Hanser Verlag.

Juergens, H. W., Habicht-Benthin, B., & Lengsfeld, W. (1971). Körpermasse 20jähriger Männer als Grundlage fuer die Gestaltung von Arbeitsgerät, Ausruestung und Arbeitsplatz. *Bundesministerium fuer Verteidigung*, Bonn.

Karoly, L. (1971). *Anthropometrie: Grundlagen der anthropologischen Methoden*. Fischer Verlag.

Kirchner, A., Kirchner, J.-H., Kliem, M., & Mueller, J. M. (1990). Räumlich-ergonomische Gestaltung. *Schriftenreihe der Bundesanstalt fuer Arbeitsschutz*. Verlag fuer neue Wissenschaft.

Lewin, T., & Juergens, H. W. (1969). *Ueber die Vergleichbarkeit von anthropometrischen Massen*. Zeitschrift Morph. Anthrop. 61.

Ludwig, W. (1932). *Das Rechts-Links-Problem im Tierreich und beim Menschen*. Springer Verlag, Berlin.

Martin, R. (1914). *Lehrbuch der Anthropologie*. Jena: Verlag Gustav Fischer.

Martin, R., & Knussmann, R. (1988). *Anthropologie. Handbuch der vergleichenden Biologie des Menschen*. Fischer Verlag.

Porter, J. M., Case, K., Freer, M. T., & Bonney, M. C. (1993). Computer-aided ergonomics design of automobiles. In B. Peacock, & W. Karwowski (Eds.), *Automotive ergonomics*, Taylor & Francis, (pp. 43–78).

Rissiek, A., & Stöhr, M. (2019). Body measurement and body shape knowledge for ergonomics—Methods and current trends, Presentation at the RAMSIS User Conference.

Robinette, K. M., Daanen, H. A. M., & Zehner, G. F. (2004). Three dimensional anthropometry. In N. J. Delleman, C. M. Haslegrave, & D. B. Chaffin (Eds.), *Working postures and movements*, CRC Press LLC, (pp 29–49)

Seidl, A. (1997). Computer man models in ergonomic design. In H. Schmidtke (Eds.), *Handbuch der Ergonomie* (Part A–3.3.3). Bundesamt für Wehrtechnik und Beschaffung. Hanser Verlag.

Seidl, A. (2004). The RAMSIS and ANTHROPOS human simulation tools. In N. J. Delleman, C. M. Haslegrave, & D. B. Chaffin (Eds.), *Working postures and movements*, CRC Press LLC, (pp. 445–453)

Seidl, A., & Bubb H. (2005). Standards in anthropometry. In Karwowski, W. *Handbook on standards and guidelines in ergonomics and human factors*, Lawrence Erlbaum Associates Publishers, (pp 169–196)

Seidl, A., Trieb, R., & Wirsching, H. J. (2009). SizeGERMANY – The new Anthropometric Survey. In *Proceeding* of *IEA Congress 2009. International Ergonomic Association.*

Seidl, A., & Speyer, H. (1992). RAMSIS: 3D-Menschmodell und integriertes Konzept zur Erhebung und konstruktiven Nutzung von Ergonomiedaten. In VDI-Bericht 948. *Das Mensch-Maschine-System im Verkehr*, VDI-Verlag Duesseldorf, (pp. 297–309)

WEAR. (2014). World Engineering Anthropometry Resource. www.bodysizeshape.com, www.ovrt.nist.gov.

13 Evaluation of Static Working Postures

Nico J. Delleman and Jan Dul

CONTENTS

INTRODUCTION

Pain, fatigue, and disorders of the musculoskeletal system may result from sustained inadequate working postures and repetitive movements that may be caused by poor work situations. Musculoskeletal pain and fatigue may themselves influence posture control, which can increase the risk of errors and may result in reduced quality of work or production and hazardous situations. Good ergonomic design and proper organization of work are basic requirements to avoid these adverse effects. ISO 11226 (2000) contains an approach to determine the acceptability of static working postures. The standard has been written by Working Group 2 "Evaluation of working postures" of Subcommittee 3 "Anthropometry and Biomechanics" of Technical Committee 159 "Ergonomics" of the International Organization for Standardization (ISO).

ISO 11226—SCOPE

ISO 11226 (ISO, 2000) establishes ergonomic recommendations for different work tasks. It provides information to those involved in the design, or redesign, of work, jobs, and products who are familiar with the basic concepts of general ergonomics, particularly in working postures. The standard specifies recommended limits for static working postures without any or with minimal external force exertion while considering body angles and time aspects. It is designed to provide guidance on the assessment of several task variables, allowing the health risks for the working population to be evaluated. The standard applies to the adult working population. The recommendations will give reasonable protection for nearly all healthy adults. The recommendations concerning health risks and protection are mainly based on experimental studies regarding the musculoskeletal load, discomfort/pain, and endurance/fatigue related to static working postures.

RECOMMENDATIONS

The standard starts with general recommendations. It is stated that work tasks and operations should provide sufficient physical and mental variation. This means a complete job, with sufficient variation of tasks (for instance, an adequate number of organizing tasks, an appropriate mix of short, medium, and long task cycles, and a balanced distribution of easy and difficult tasks), sufficient autonomy, opportunities for contact, information, and learning. Furthermore, the full range of workers possibly involved with the tasks and operations should be considered, especially body dimensions. With respect to working postures, the work should offer sufficient variation between and within sitting, standing, and walking. Awkward postures, such as kneeling and crouching, should be avoided whenever possible. It is stressed that measures meant to induce variations of posture should not lead to monotonous repetitive work.

EVALUATION PROCEDURE

The main part of the standard consists of specific recommendations to evaluate static working postures. The evaluation procedure considers various body segments and joints independently by one or two steps. The first step considers only the body angles (recommendations are mainly based upon risks for overloading passive body structures, such as ligaments, cartilage, and intervertebral disks). An evaluation may lead the result to be accepted and proceed to step 2, or not recommended. An evaluation result acceptable means that a working posture is acceptable only if variations of posture are also present (refer above). Furthermore, it is stated that every effort should be made to obtain a working posture closer to the neutral posture if this is not already the case. An evaluation result that goes to step 2 means that the duration of the working posture will also need to be considered (recommendations are based on endurance data). Extreme positions of joints should be evaluated as not recommended (Figures 13.1 and 13.2).

RELATIONSHIP TO OTHER STANDARDS

Evaluation procedures and recommendations concerning lifting and carrying, pushing and pulling, and handling of low loads at high frequency can be found in ISO 11228-1 (ISO, 2003), ISO/DIS 11228-2 (ISO, 2005a), and ISO/DIS 11228-3 (ISO, 2005b), respectively.

FIGURE 13.1   Typical working posture during sewing machine operation.

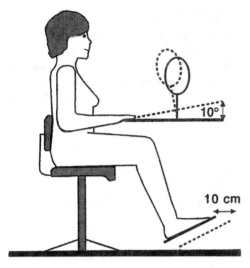

FIGURE 13.2 The four adjustments of the sewing machine workstation tested (2 table slopes × 2 pedal positions).

EXAMPLE OF APPLICATION—SEWING MACHINE OPERATION

The working posture of sewing machine operators is characterized by a forward inclined head and trunk and a flexed neck (Figure 13.1). In an effort to improve working conditions through the definition of recommendations for adjustment of the workstation, several operators worked for 45 minutes at each of the four workstation adjustments—that is, two-pedal positions × two table slopes (Figure 13.2). One pedal position was the average at the operators' own industrial workstations, whereas the other was 10 cm farther away from the front edge of the table. The latter condition was introduced to allow the operator to sit closer to the table and create a more upright trunk posture. One table slope was 0° (flat table), which is in accordance with the operators' own

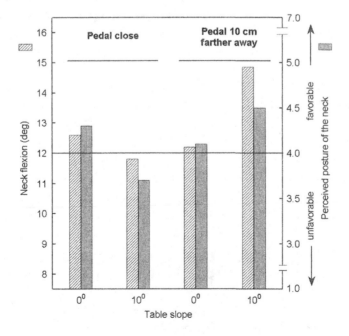

FIGURE 13.3 The four adjustments of the sewing machine workstation tested versus neck flexion and perceived posture of the neck (a score of 4 is exactly between favorable and unfavorable).

industrial workstations. The other table's slope was 10° (inclined toward the operator), which was introduced to create a more upright head posture. In the following, the focus is placed on the neck flexion. All neck flexions measured were within the range of 0–25°, which is acceptable according to ISO 11226 (Figure 13.3). However, the average ratings on neck posture provide data for establishing more detailed evaluation criteria (Figure 13.3). A neck flexion of about 15° turned out to be better than smaller neck flexions. These and other new scientific data are described by Delleman, Haslegrave, and Chaffin (2004).

REFERENCES

Delleman, N. J., Haslegrave, C. M., & Chaffin, D. B. (Eds.). (2004). *Working postures and movements—Tools for evaluation and engineering.* Boca Raton, FL: CRC Press.

ISO (2000). *ISO 11226 Ergonomics—Evaluation of static working postures.* Geneva, Switzerland: International Organization for Standardization.

ISO (2003). *ISO 11228-1 Ergonomics—Manual handling—Part 1: Lifting and carrying.* Geneva, Switzerland: International Organization for Standardization.

ISO (2005a). *ISO/DIS 11228-2 Ergonomics—Manual handling—Part 2: Pushing and pulling. Standard under preparation.* Geneva, Switzerland: International Organization for Standardization.

ISO (2005b). *ISO/DIS 11228-3 Ergonomics—Manual handling—Part 3: Handling of low loads at high frequency. Standard under preparation.* Geneva, Switzerland: International Organization for Standardization.

14 Evaluation of Working Postures and Movements in Relation to Machinery

Nico J. Delleman and Jan Dul

CONTENTS

INTRODUCTION

About one-third of the workers in the European Union (EU) are involved in painful or tiring positions for more than half of their time at work, and close to 50% of the workers are exposed to repetitive hand or arm movements (Paoli & Merllié, 2001). Pain and fatigue may lead to musculoskeletal diseases, reduced productivity, and deteriorated posture and movement control. The latter can increase the risk of errors and may result in reduced quality and hazardous situations.

EN 1005-4—SCOPE AND STATUS

The European Standard 1005-4 (CEN, 2005a) presents guidance when designing machinery or its components parts in assessing and controlling health risks due to machine-related postures and movements during assembly, installation, operation, adjustment, maintenance, cleaning, repair, transport, and dismantle. The standard specifies requirements for postures and movements without any or only minimal external force exertion. The requirements are intended to reduce the risks for nearly all healthy adults but could also have a positive effect on the quality, efficiency, and profitability of machine-related actions.

The standard has been written by Working Group 4 "Biomechanics" of Technical Committee 122 "Ergonomics" of the European Committee for Standardization (CEN), under a mandate given to CEN by the European Commission and the European Free Trade Association (EFTA). The standard applies to all machines traded within the EU. The technique and principles could, however, also be applied elsewhere and for other products.

RISK ASSESSMENT

The standard adopts a stepwise risk-assessment approach in assessing postures and movements as part of the machinery design process and provides guidance during the various design stages (as shown in Figure 14.1). The approach is based on the U-shaped model presented in Figure 14.2, which proposes that health risks increase when the task approaches either end of the curve—that is,

275

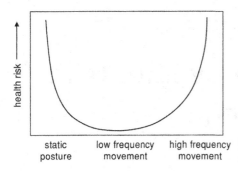

static posture   low frequency movement   high frequency movement

FIGURE 14.1   Flowchart illustrating the risk assessment approach.

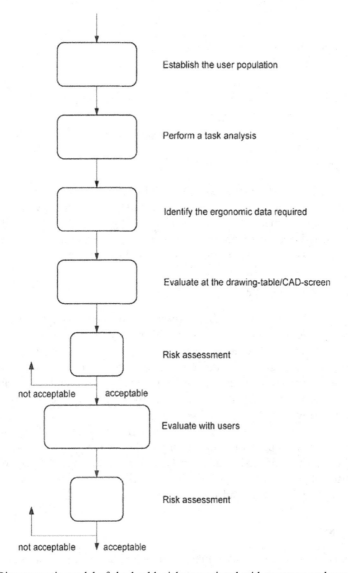

Establish the user population

Perform a task analysis

Identify the ergonomic data required

Evaluate at the drawing-table/CAD-screen

Risk assessment

not acceptable   acceptable

Evaluate with users

Risk assessment

not acceptable   acceptable

FIGURE 14.2   Diagrammatic model of the health risks associated with postures and movements.

if there is static posture (with little or no movement) or if movement frequencies are high. A distinction is made between:

- Evaluation without users—when there is no full-size model prototype of the machinery or its parts currently available.
- Evaluation with users—when a full-size model prototype of the machinery or its parts is available.

Postures and movements are evaluated according to the following scheme:

- Acceptable—The health risk is considered low or negligible for nearly all healthy adults. No action is needed.
- Conditionally acceptable—There exists an increased health risk for the whole or part of the user population. The risk shall be analyzed together with contributing risk factors, followed as soon as possible by a reduction of the risks (i.e. redesign); or, if that is not possible, other suitable measures shall be taken—for example, the provision of user guidelines to ensure that the use of the machine is acceptable.
- Not acceptable—The health risk cannot be accepted for any part of the user population. A redesign to improve the working posture is mandatory.

For postures or movements observed that are conditionally acceptable, a second step of the evaluation procedure is introduced. Acceptability may depend on the nature and duration of the posture and period of recovery, on the presence or absence of body support, or on movement frequency. Concerning static postures, the risk assessment is a simplified version of the procedure described in ISO 11226 (2000). Frequency-related risk assessment of movements is based on Kilbom (1994).

RELATIONSHIP TO OTHER STANDARDS

In the early phases of design (Figure 14.2), the standard refers to EN 614-1 (CEN, 1995) and EN-ISO 14738 (CEN/ISO, 2002). Concerning manual materials handling, force exertion, and repetitive work in relation to machinery, the designer is referred to Parts 2, 3, and 5 of EN 1005 (CEN, 2003, 2002, 2005b).

EXAMPLE OF APPLICATION—PRESS OPERATION

In the metal industry, presses are used to form objects and cut superfluous material. Usually, an operator must reach forward to place or take an object into or from the press (Figure 14.3). In an effort to improve working conditions through redesign of presses, several operators processed lightweight objects at four reach distances (70, 80, 90, and 100 cm). Each distance was tested for a period of 25 min. Movement frequencies of the trunk and the upper arm were around ten per minute. Table 10.1 provides the range of forward inclination of the trunk and the evaluation results by applying the standard, as well as the average operators' ratings. It is concluded that the evaluation results, according to the standard, are supported by the operator's ratings.

Tables 14.1 and 14.2 contain the range of forward elevation of the upper arms and the evaluation results by applying the standard, as well as the average operators' rating. According to the standard, upper arm elevations into the 20–60° zone are acceptable if frequencies are under ten per minute, and not acceptable otherwise. In the example, the movement frequency is around ten per minute, and elevations are in the upper part of the zone (reach distances 80–100 cm). In such a case, it is wise not to draw conclusions immediately. Here, the operators' ratings after a 25-min testing support a conclusion that reach distances up to 90 cm is acceptable. Nevertheless, longer testing is recommended, as the standard states: "Working periods of long duration and high movement frequencies are known to increase health

FIGURE 14.3   Forward reaching during press operation.

risks due to machine-related working postures and movements. Current knowledge only allows in part for quantitative evaluation of these risk factors." The standard also states: "Particularly if the machine may be used under the conditions mentioned above, it is strongly recommended that maximum improvement of the working posture be achieved, even if the working posture or movement is already assigned the outcome 'acceptable'." For this, it is recommended to do a detailed analysis using the newest scientific data concerning working postures and movements (Delleman, Haslegrave, & Chaffin, 2004).

TABLE 14.1

Forward Inclination of the Trunk, and Evaluation Results According to the Standard and the Operators for Various Reach Distances

Reach Distance	70 cm	80 cm	90 cm	100 cm
Trunk Inclination Forwards	0–10°	10–20°	20–30°	30–40°
Evaluation Result According to Standard	Acceptable	Acceptable	Not acceptable	Not acceptable
Evaluation Result According to Operators	Favorable	Favorable	Unfavorable	Unfavorable

TABLE 14.2

Forward Elevation of the Upper Arms, and Evaluation Results According to the Standard and the Operators for Various Reach Distances

Reach Distance	70 cm	80 cm	90 cm	100 cm
Upper Arm Elevation Forwards	30–40°	40–50°	40–50°	40–50°
Evaluation Result According to Standard	Acceptable–Not acceptable	Acceptable–Not acceptable	Acceptable–Not acceptable	Acceptable–Not acceptable
Evaluation Result According to Operators	Favorable	Favorable	Favorable	Favorable–Unfavorable

REFERENCES

CEN. (1995). *EN 614-1 Safety of machinery—Ergonomic design principles—Part 1: Terminology and general principles*. Brussels, Belgium: European Committee for Standardization.

CEN. (2002). *EN 1005-3 Safety of machinery—Human physical performance. Part 3: Recommended force limits for machinery operation*. Brussels, Belgium: European Committee for Standardization.

CEN. (2003). *EN 1005-2 Safety of machinery—Human physical performance—Part 2: Manual handling of machinery and component parts of machinery*. Brussels, Belgium: European Committee for Standardization.

CEN. (2005a). *EN 1005-4 Safety of machinery—Human physical performance. Part 4: Evaluation of working postures and movements in relation to machinery*. Brussels, Belgium: European Committee for Standardization.

CEN. (2005b). *prEN 1005-5 Safety of machinery—Human physical performance. Part 5: Risk assessment for repetitive handling at high frequency. Standard under preparation*. Brussels, Belgium: European Committee for Standardization.

CEN/ISO. (2002). *EN-ISO 14738 Safety of machinery—Anthropometric requirements for the design of workplaces at machinery*. Brussels, Belgium: European Committee for Standardization or Geneva, Switzerland: International Organization for Standardization.

Delleman, N. J., Haslegrave, C. M., & Chaffin, D. B. (Eds.) (2004). *Working postures and movements—Tools for evaluation and engineering*. Boca Raton, FL: CRC Press.

ISO. (2000). *ISO 11226 Ergonomics—Evaluation of static working postures*. Geneva, Switzerland: International Organization for Standardization.

Kilbom, Å. (1994). Repetitive work of the upper extremity: Part I—Guidelines for the practitioner, and Part II—The scientific basis (knowledge base) for the guide. *International Journal of Industrial Ergonomics, 14*, (51–57), 59–86.

Paoli, P., & Merllié, D. (2001). *Third European survey on working conditions 2000*. Luxembourg: Office for Official Publications of the European Communities.

15 Standards on Physical Work

Demands in the Construction Industry

Henk F. van der Molen and Nico J. Delleman

CONTENTS

INTRODUCTION

Building and construction is one of mankind's oldest activities. Looking at the pace of innovation in other sectors of industry, the construction industry should be characterized as conservative. Work still imposes physical demands, and work organization and methods remain traditional. In the Netherlands, however, the past decades have shown a growing emphasis on working conditions. In the construction industry, physical work demands are the most important cause of absence and disability. More than half of the cases of sick leave among Dutch construction site workers are the result of musculoskeletal complaints and disorders, mostly related to the lower back region (Arbouw, 1990–2002).

According to Hoonakker et al. (1992), in a group of about 1,000 Dutch construction workers, the following hazards were evident: repetitive work (61%), handling of heavy materials (59%), poor working posture (52%), and high-force exertion (49%). In addition, occupational profiles based on the results of periodic medical examinations show a high percentage of complaints related to the musculoskeletal system. To reduce or eliminate health risks relating to the back, neck, and limbs, it is necessary to know which tasks particularly overload construction workers. Consequently, guidelines and standards on physical workloads are mandatory. In 1997, Koningsveld and van der Molen (1997) described the history and future of ergonomics in building and construction at the first international ergonomics symposium as part of the 13th triennial congress of the IEA in Tampere. It was concluded that radical changes in working methods, work organization, and working conditions are a prerequisite for the future of companies and the well-being of their workers. Ergonomists can help in the improvement process by means of applying ergonomic standards, particularly with respect to physical work demands.

The section Guidelines on Physical Work Demands describes guidelines for assessing all aspects of the physical work demands of building tasks in the construction industry. The assessment of these demands based on an adequate task analysis is a particularly important step in reducing physical work demands. It is then possible to reduce physical work demands using technical, organizational, or individual measures. The following section (Applications of the Guidelines in Standards for Professions) describes the process of applying these instructions in specific standards for construction industry professions through the Arbouw documents (A-documents). The final section contains Dutch legislation relating to lifting on construction sites, based on the A-documents under Application of the Guidelines in Standards for Professions.

GUIDELINES ON PHYSICAL WORK DEMANDS

BACKGROUND

Most guidelines on physical work demands focus on the lifting of materials. The collective labor agreements for the construction industry in the Netherlands generally specify a maximum weight of 25 kg. In countries such as Australia, the United Kingdom, and the United States, it is common practice not to consider the object weight as the only risk factor in quantitative guidelines. Therefore, the so-called Arbouw guidelines follow this multifactor approach by also taking into consideration, for instance, handling frequencies, task duration, or whatever is known to most affect physical load in particular work activities. Quantitative guidelines concerning lifting and carrying, pushing and pulling static postures, and repetitive work have been developed at the request of the parties involved in negotiating the labor agreements. The first edition of the Arbouw guidelines was based on a literature review and discussions among experts on physical workload in 1992 and 1993. Subsequently, ten health and safety professionals evaluated the guidelines by actually working with them. In 2001, the guidelines were slightly revised to include the latest information from standards, reviews, and so forth. The health limits in the revised version of the Arbouw guidelines are based on Waters et al. (1993), Mital, Nicholson, and Ayoub (1993), Kilbom (1994a, 1994b), ISO/DIS 11226 (2000), NF X 35-106 (1985), and prEN 1005-4 (2002).

Evaluation Scheme

All construction guidelines are based on the following evaluation scheme:

Green zone: basic, non-increased health risk for ≥90% of men (P90 men).

Yellow zone: increased health risk. Action may be planned in stages; immediate action is to be pre-ferred. For the guidelines on external force exertion, such as those for lifting and car-rying, pushing and pulling, and repetitive arm work, this zone denotes that between 25-90% of men (P25 men–P90 men) can exert a certain force. Static postures and repetitive movements associated with an increased health risk are included in this zone in the case of a task duration of between one and four hours.

Red zone: strongly increased health risk. Immediate action is necessary. For the guidelines on ex-ternal force exertion (refer to previous), the zone denotes that ≤25% of men (P25 men) can exert a certain force. Static postures and repetitive movements associated with an increased health risk are included in this zone in the case of a task duration of over four hours.

The boundary between the green zone and the yellow zone is called the action limit (AL), whereas the boundary between the yellow zone and the red zone is called the maximum Arbouw limit (MAL). The AL–MAL concept has been used earlier for manual lifting (i.e. the National Institute for Occupational Safety and Health [NIOSH], 1981), whereas the green-yellow-red concept has also been described before for repetitive work (e.g., Hedén et al., 1993). The Arbouw guidelines use both concepts as one, with their own definitions and in a consistent way, for all kinds of physical work demands.

Criteria for AL and MAL

AL represents a health limit that should be the ultimate goal for action programs to get working conditions into the green zone. MAL was introduced for setting priorities within these programs. Strongly increased health risks (red) are to be tackled first, followed by those conditions associated with an increased health risk (yellow). Nonetheless, it should be realized that an essential element of priority setting is also to check whether possible actions are reasonably practicable in economic, organizational, and technical terms.

The MALs for external force exertion (lifting and carrying, whole-body pushing and pulling, and repetitive arm work) are set to a level at which a majority of the workers (75%) are not even able to exert that particular force. Employers do not desire such a situation in par-ticular as it excludes far too many workers from doing the jobs. So, there is a strong motivation for setting in motion a process of change, in addition to the obvious reasons for reducing health risks. Currently, most Dutch construction workers are male so the guidelines on maximum forces are based on male population data. It would be possible to formulate guidelines for females as well.

For static postures and repetitive movements, task duration was used to set AL and MAL. Primarily tasks lasting more than one hour a day are included in the risk assessment procedure, based on the results of a review by Kilbom (1994a, 1994b). For repetitive work, primary at-tention is on movement frequencies ≥2 and ≥10 per minute, respectively. Positions of body segments and joints are evaluated based on the European Committee for Standardization (CEN) and the International Organization for Standardization (ISO) standards. Those postures and movements observed that are associated with an increased health risk are classified in the yellow zone if the task lasts between one and four hours and classified in the red zone if it lasts more than four hours. As a matter of course, durations of tasks loading the same body region are to be added together. It is recommended that each evaluation should be interpreted in relation to the level of complaints for the associated body region for the group of workers involved. Although the four-hour limit—and, to a lesser extent, the one-hour limit—is arbitrary, the authors are of the opinion that the current procedure at least takes the risk factor task duration into con-sideration. If the application of the guidelines does not disclose an increased health risk, while an increased level of physical complaints for the group of workers involved exists, further analysis is considered necessary.

GUIDELINES FOR FIVE AREAS OF PHYSICAL WORK DEMANDS (LIFTING, PUSHING AND PULLING, CARRYING, STATIC LOAD, AND REPETITIVE WORK)

Guideline on Lifting (Mainly Based on NIOSH, 1981; Mital, Nicholson, & Ayoub, 1993; and NF X 35-106, 1985)

The following conditions are red:

- Weight more than 25 kg when in a standing position
- Weight more than 10 kg when in a sitting, squatting, or kneeling position
- Weight more than 17 kg when lifting one-handed
- Horizontal location (H) more than 63 cm
- Asymmetry angle (A) more than 135°
- Vertical location (V) more than 175 cm or less than 0 cm
- Frequency (F) more than 15 lifts/minute

The first two tables here distinguish between symmetric and asymmetric lifting, where a lifting index ≤1 is green, a lifting index of 1–3 is yellow, and a lifting index ≥3 is red. In the case of limited headroom, the lifting index found in one of these tables must be multiplied by a multiplier taken from the third table. An index greater than one means an increased health risk, concerning musculoskeletal complaints, and disorders in particular. A greater index means a higher risk. The fourth table presents the guidelines for lifting in a sitting, squatting, or kneeling position, or for one-handed lifting.

Lifting index for symmetric lifting (asymmetry angle [A] 0°), vertical location (V) 175 cm, vertical travel distance (D) 175 cm work, and lifting duration 8 hours and no recovery time. The figures in parentheses are for lifting under optimal symmetric conditions—vertical location (V) of 75 cm, vertical travel distance (D) of 0 cm, work and lifting duration of one hour, and recovery time for seven hours. A lifting index ≤1 is green, a lifting index 1–3 is yellow, and a lifting index ≥3 is red.

		Frequency		
Weight	Horizontal Location (*H*)	5–9 Lifts/Minute	1–5 Lifts/Minute	<1 Lift/Minute
	≤25 cm	*2.8 (0.5)*	*1.2 (0.3)*	0.5 (0.3) Green
0–5 kg	25–40 cm	4.2 (0.7)	*1.9 (0.5)*	0.9 (0.4)
	40–63 cm	7.1 (1.7)	*2.9 (0.8)*	*1.4 (0.6)* Yellow
	≤25 cm	5.9 (0.9)	*2.3 (0.6)*	*1.1 (0.5)*
5–10 kg	25–40 cm	8.3 (1.5)	3.7 (1.0)	*1.7 (0.8)*
	40–63 cm	14.2 (2.3)	5.9 (1.5)	*2.7 (1.3)*
	≤25 cm	8.3 (1.4)	3.5 (0.9)	*1.6 (0.8)*
10–15 kg	25–40 cm	12.5 (2.2)	5.6 (0.4)	*2.6 (1.2)*
	40–63 cm	21.4 (1.9)	8.8 (2.3)	4.1 (1.9) Red
	≤25 cm	11.1 (1.9)	4.7 (1.2)	2.2 (1.0)
15–20 kg	25–40 cm	16.6 (2.9)	7.4 (1.9)	3.5 (1.6)
	40–63 cm	28.5 (4.7)	11.7 (3.0)	5.4 (2.6)
	≤25 cm	13.8 (2.3)	5.8 (1.5)	2.7 (1.3)
20–25 kg	25–40 cm	20.8 (3.7)	9.3 (2.4)	4.3 (2.0)
	40–63 cm	35.7 (5.8)	14.7 (3.8)	6.8 (3.2)

The lifting index for asymmetric lifting (asymmetric angle [A] 135°), vertical location (V) 175 cm, vertical travel distance (D) 175 cm, work and lifting duration is hours, and no recovery time. The figures in parentheses are for lifting under optimal, asymmetric conditions—that is, vertical location (V) of 75 cm, vertical travel distance (D) of 0 cm, work and lifting duration for one hour, and recovery time of seven hours. A lifting index ≤1 is green, a lifting index of 1–3 is yellow, and a lifting index ≥3 is red. (See next page.)

| Weight | Horizontal Location (H) | Frequency (F) | | |
		5–9 Lifts/Minute	1–5 Lifts/Minute	<1 Lift/Minute
	≤25 cm	4.6 (0.8)	*2.0 (0.5)* Yellow	0.9 (0.5) Green
0–5 kg	25–40 cm	7.1 (1.3)	3.3 (0.8)	*1.5 (1.7)*
	40–63 cm	12.5 (2.0)	5.0 (1.3)	*2.4 (1.1)* Yellow
	≤25 cm	9.1 (1.6)	4.0 (1.1)	*1.9 (0.9)*
5–10 kg	25–40 cm	14.2 (2.6)	6.7 (1.7)	3.0 (1.4)
	40–63 cm	25.0 (4.0)	10.0 (2.6)	4.8 (2.3)
	≤25 cm	13.6 (2.5)	6.0 (1.6)	*2.8 (1.4)* Yellow
10–15 kg	25–40 cm	21.4 (3.9)	10.0 (2.5)	4.6 (2.1)
	40–63 cm	37.5 (6.0)	15.0 (4.0)	7.1 (3.4) Red
	≤25 cm	18.1 (3.3)	8.0 (2.1)	3.8 (1.8)
15–20 kg	25–40 cm	29.1 (5.1)	13.3 (3.4)	6.1 (2.9)
	40–63 cm	50.0 (8.0)	20.0 (5.3)	9.5 (4.5)
	≤25 cm	22.7 (4.1)	10.0 (2.7)	4.7 (2.3)
20–25 kg	25–40 cm	35.7 (6.4)	16.6 (4.2)	7.6 (3.6)
	40–63 cm	62.5 (10.0)	25.0 (6.6)	11.9 (5.7)

Limited Headroom Multiplier

	Multiplier
Fully Upright	1.00
95% Upright	1.67
90% Upright	2.50
85% Upright	2.60
80% Upright	2.78

Guidelines for lifting in a sitting, squatting, or kneeling position or for lifting one-handed infrequently.

| | Weight | |
	Green/Yellow Limit	Yellow/Red Limit
Sitting, Squatting, or Kneeling	4.5 kg	10.0 kg
One-Handed	7.5 kg	17.0 kg

For frequent lifting (≥2/minute) and longer durations (>1 hour/day), the guidelines on repetitive work apply.

Guideline on Pushing and Pulling (Mainly Based on Mital, Nicholson, & Ayoub, 1993, and NF X 35-106, 1985)

Pushing/Pulling With the Whole Body While Walking. The following tables distinguish among pushing to set an object in motion (initial force exertion), pulling to set an object in motion (initial force exertion), and pushing or pulling to keep an object in motion (sustained force exertion). The guidelines are valid for two-handed pushing and pulling for a whole working day (eight hours), and an optimal height of the hands during force exertion (95–130 cm). A detailed analysis is necessary if slipping is likely to occur (or actually occurs) or in the case of high movement speed, awkward posture, asymmetric force exertion (one-handed and course changes), or a bad view of the surroundings (surface, obstacles etc.). All cells of the tables contain a pushing and pulling index besides the green, yellow, or red evaluation result. An index greater than one means an increased health risk, concerning musculoskeletal complaints and disorders in particular. A greater index means a higher risk. Pushing to set an object in motion (initial force exertion)* is usually not possible because slipping is likely to occur.

Force	Frequency			
	2.5 × /Minute–1 × /Minute	1 × /Minute–1 × /5–Minutes	1 × /5 Minutes–1 × /8 Hours	≤1 × /8 Hours
0–25 kgf	1.0 Green	1.0 Green	0.9 Green	0.8 Green
25–30 kgf*	*1.2 Yellow*	*1.2 Yellow*	*1.1 Yellow*	1.0 Green
30–45 kgf*	*1.8 Yellow*	*1.7 Yellow*	*1.6 Yellow*	*1.5 Yellow*
45–50 kgf*	2.0 Red	*1.9 Yellow*	*1.8 Yellow*	*1.7 Yellow*
50–65 kgf*	2.6 Red	2.5 Red	2.3 Red	*2.2 Yellow*
>65 kgf*	>2.6 Red	>2.5 Red	>2.3 Red	>2.2 Red

Pulling to set an object in motion (initial force exertion)* is usually not possible because slipping is likely to occur.

Force	Frequency			
	2.5 × /Minute–1 × /Minute	1 × /Minute–1 × /5 Minutes	1 × /5 Minutes–1 × /8 Hours	≤1 × /8 Hours
0–20 kgf	1.0 Green	1.0 Green	1.0 Green	1.0 Green
20–40 kgf*	*2.0 Yellow*	*2.0 Yellow*	*2.0 Yellow*	*2.0 Yellow*
40–45 kgf*	2.3 Red	2.3 Red	*23 Yellow*	*23 Yellow*
45–50 kgf*	2.5 Red	2.5 Red	2.5 Red	*2.5 Yellow*
>50 kgf*	>2.5 Red	>2.5 Red	>2.5 Red	>2.5 Red

Pushing or pulling to keep an object in motion (sustained force exertion)* = usually not possible because slipping is likely to occur. — = combination of frequency and distance is not realistic. (See next page.)

Force	Distance (m)	Frequency			
		2.5/Minute–1/Minute	l/Minute–1/5 Minutes	1/5 Minutes–1/8 Hours	≤1/8 Hours
	2–15	0.8 Green	0.4 Green	0.4 Green	0.3 Green
0–5 kgf	15–30	—	0.8 Green	0.4 Green	0.3 Green
	30–60	—	—	0.6 Green	0.4 Green
	2–15	*1.5 Yellow*	0.9 Green	0.7 Green	0.6 Green
5–10 kgf	15–30	—	*1.5 Yellow*	0.8 Green	0.6 Green
	30–60	—	—	*1.2 Yellow*	0.9 Green
	2–15	2.3 Red	*1.3 Yellow*	*1.1 Yellow*	0.9 Green
10–15 kgf	15–30	—	2.3 Red	*1.3 Yellow*	0.9 Green
	30–60			*1.8 Yellow*	*1.3 Yellow*
	2–15	3.1 Red	1.7 Red	*1.5 Yellow*	*1.2 Yellow*
15–20 kgf	15–30		3.1 Red	*1.7 Yellow*	*1.2 Yellow*
	30–60	—		2.3 Red	*1.7 Yellow*
	2–15	3.8 Red	2.2 Red	*1.9 Yellow*	*1.5 Yellow*
20–25 kgf*	15–30		3.8 Red	*2.1 Yellow*	*1.5 Yellow*
	30–60			2.9 Red	2.2 Red
	2–15	4.6 Red	2.6 Red	2.2 Red	*1.8 Yellow*
25–30 kgf*	15–30		4.6 Red	2.5 Red	*1.8 Yellow*
	30–60	—	—	3.5 Red	2.6 Red
	2–15	>4.6 Red	>2.6 Red	>2.2 Red	>1.8 Red
>30 kgf*	15–30	—	>4.6 Red	>2.5 Red	>1.8 Red
	30–60			>3.5 Red	>2.6 Red

Pushing and Pulling With the Whole Body While Staying on the Spot. The guidelines for setting an object in motion (refer to the first two tables under Guidelines for Five Areas on Pushing and Pulling) also apply to pushing/pulling with the whole body while staying on the spot.

Pushing and Pulling With the Upper Limbs. The following guidelines are valid if the postures of the trunk and the upper limbs are evaluated as green (refer to the guideline on repetitive work), if the hands do not reach further forwards than three-fourths of the maximum reach distance (trunk upright), and if the hands are between the pelvis and shoulder height.

Guidelines for infrequent pushing and pulling with the upper limbs. A and B refer to the various types of force exertion distinguished in the guideline on repetitive work. (See page 216.)

Force		
	Green/Yellow Limit	Yellow/Red Limit
A	7.5 kgf	17.0 kgf
B	19.0 kgf	43.0 kgf

For frequent pushing/pulling (≥2/minute) and longer durations (>1 hour/day), the guidelines on repetitive work apply.

Guideline on Carrying (Mainly Based on Mital, Nicholson, & Ayoub, 1993)

The following condition is red:

- Weight of more than 25 kg
- Weight of more than 10.5 kg for infrequent one-handed carrying

The following guideline is valid for two-handed carrying for a whole working day (eight hours), with the hands at an optimal height (knuckle height and arms hanging down). All cells of the tables contain a carrying index besides the evaluation result of green, yellow, or red. An index greater than one means an increased health risk concerning musculoskeletal complaints and disorders, in particular. A greater index means a higher risk.

Weight	Distance (m)	Frequency			
		3/Minute–1/Minute	1/Minute–1/5 Minutes	1/5 Minutes–1/8 Hour	≤1/8 Hour
0–10 kg	≤2.0	0.5 Green	0.5 Green	0.4 Green	0.4 Green
	2–8.5	0.8 Green	0.6 Green	0.5 Green	0.4 Green
10–15 kg	≤2.0	0.8 Green	0.7 Green	0.7 Green	0.6 Green
	2–8.5	*1.2 Yellow*	0.9 Green	0.8 Green	0.6 Green
15–20 kg	≤2.0	*1.1 Yellow*	1.0 Green	0.9 Green	0.8 Green
	2–8.5	*1.5 Yellow*	*1.2 Yellow*	1.0 Green	0.8 Green
20–25 kg	≤2.0	*1.3 Yellow*	*12 Yellow*	*1.1 Yellow*	1.0 Green
	2–8.5	1.9 Red	*1.5 Yellow*	*1.3 Yellow*	1.0 Green
>25 kg	≤2.0	>1.3 Red	>1.2 Red	>1.1 Red	>1.0 Red
	2–8.5	>1.9 Red	>1.5 Red	>1.3 Red	>1.0 Red

Guideline for Infrequent One-Handed Carrying (Standing or Walking Up to 90 m).

	Weight	
	Green/Yellow Limit	Yellow/Red Limit
Infrequent One-Handed Carrying (Standing or Walking)	6 kg	10.5 kg

Guideline on Static Postures (Mainly Based on ISO/DIS 11226, 2000)

It is recommended to use the following guidelines to begin with tasks lasting longer than one hour (continuous or a total of distinct periods) per working day (N.B. shorter task durations cannot be considered safe in all cases). In the case that two or more tasks load the same body region (through static postures or repetitive work), the duration of these tasks are to be taken together. It is recommended to interpret the results of guideline application in relation to the complaints and disorders of the particular body region found for the group of employees involved. An evaluation result of yellow and red becomes yellow in the case of a task duration between one and four hours; it becomes red in the case of a task duration >4 hours.

The guidelines on the lower back, shoulder, and shoulder girdle, as well as on the neck and upper back, are based on a description of the actual posture (observed and measured) with respect to a reference posture—a sitting or standing posture with a non-rotated upright trunk, a non-kyphotic lumbar spine posture, and arms that hang freely while looking straight along the horizontal. Guidelines on sitting also include raised sitting

- Asymmetric trunk posture (axial rotation or lateral flexion): yellow/red
- For sitting: kyphotic lumbar spine posture (valid for trunk inclination between 0° and 20°): yellow/red
- * = with full trunk support: green; without full trunk support: consult an expert for evaluating holding time—recovery time regimes (see ISO/DIS 11226, 2000); if that is possible, the evaluation result is yellow in the case of a task duration between one and four hours and red in the case of a task duration >4 hours (expert guess of the authors); if two or more tasks load the same body region, the duration of these tasks are to be taken
- ** = with full trunk support: green
- Upper arm retroflection (i.e. elbow behind the trunk when viewed from the side of the trunk), upper-arm adduction (i.e. elbow not visible when viewed from behind the trunk), or extreme upper-arm external rotation: yellow/red
- Raised shoulder: yellow/red
- * = with full arm support: green; without full arm support: consult an expert for evaluating holding time— recovery time regimes (see ISO/DIS 11226, 2000); if that is not possible, the evaluation result is yellow in the case of a task duration between one and four hours and red in the case of a task duration >4 hours (expert guess of the authors); if two or more tasks load the same body region, the duration of these tasks are to be taken together.

Low Back			
	Trunk Inclination		
<0°	0°–20°	20°–60°	>60°
Yellow/Red**	Green	Green/Yellow/Red*	Yellow/Red

Shoulder and Shoulder Girdle		
Upper Arm Elevation		
0°–20°	20°–60°	>60°
Green	Green/Yellow/Red*	Yellow/Red

Neck and Upper Back

Green

- Head inclination 0–25° and neck flexion (i.e. head inclination minus trunk inclination) 0–25°.

Green/Yellow/Red

- Head inclination 25–85°
- With full trunk support, consult an expert for evaluating holding time—recovery time re-gimes (see ISO/DIS 11226, 2000); if that is not possible, the evaluation result is yellow in the case of a task duration between one and four hours and red in the case of a task duration >4 hours (expert guess of the authors); if two or more tasks load the same body region, the duration of these tasks are to be taken together
- Without full trunk support, the holding time for trunk inclination is critical and should be evaluated

Yellow/Red

- Head inclination <0°
- Changing to green in the case of full head support
- Head inclination >85°
- Neck flexion (head inclination minus trunk inclination) <0° (i.e. neck extension) or >25°
- Asymmetric neck posture—axial rotation or lateral flexion of the head versus the upper part of the trunk

Other Joints. The following conditions are Yellow/Red:

- Extreme joint positions
- For a kneeling position: no adequate knee protection® For standing (except when using a buttock rest): flexed knee
- For sitting: knee angle > 135° (180° = the upper leg in line with the lower leg), unless the trunk is inclined backward and fully supported (as in car driving)
- For sitting: knee angle <90° (180° = the upper leg in line with the lower leg)

Pedal Operation. The following conditions are yellow/red:

- Standing
- Sitting, leg-actuated
- Sitting, ankle-actuated, force exertion >5.5 kgf
- Sitting, ankle-actuated, pedal operating already by merely supporting foot weight

General. The following conditions are yellow:

- Standing continuously ≥1 hour/day
- Standing for a total of distinct periods ≥4 hours/day

Guideline on Repetitive Work (Mainly Based on NF X 35-106, 1985, and prEN 1005-4, 2002)

It is recommended to use the following guidelines to begin with tasks lasting longer than one hour (continuous or a total of distinct periods) per working day (N.B. shorter task durations cannot be considered safe in all cases). If two or more tasks load the same body region (through static postures or repetitive work), the duration of these tasks are to be taken together. It is recommended to interpret the results of guideline application in relation to the complaints and disorders of the particular body region found for the group of employees involved. An evaluation result of yellow and red becomes yellow in the case of a task duration between one and four hours and becomes red in the case of a task duration >4 hours.

- Asymmetric trunk posture (axial rotation and/or lateral flexion): yellow/red
  1. An increased health risk is present if ≥ 2/minute an evaluation result of yellow and red is found. (N.B. Lower frequencies cannot be considered safe in all cases; for example, with longer task duration.)

Shoulder and Shoulder Girdle		
Upper Arm Elevation		
0°–20°	20°–60°	>60°
Green	Yellow/Red 1	Yellow/Red 2

- Upper-arm retroflection (i.e. elbow behind the trunk when viewed from the side of the trunk), upper-arm adduction (i.e. elbow not visible when viewed from behind the trunk), or extreme upper-arm external rotation: yellow/red 2.

- Raised shoulder: yellow/red 2 neck flexion (i.e. head inclination minus trunk inclination) 0°–25°.
  1. An increased health risk is present if ≥10/minute an evaluation result of yellow and red is found. (N.B. lower frequencies cannot be considered safe in all cases, for example, with longer task duration.)
  2. An increased health risk is present if ≥2/minute an evaluation result of yellow and red is found. (N.B. lower frequencies cannot be considered safe in all cases, for example, with longer task duration.)

Low Back		
Trunk Inclination		
<0°	0°–20°	>20°
Yellow/Red 1	Green	Yellow/Red

Neck and Upper Back

Green

- Neck flexion (i.e. head inclination minus trunk inclination) 0–25°.

Yellow/red:

- Neck flexion (head inclination minus trunk inclination) <0° (i.e. neck extension) or >25° (1).
- Asymmetric neck posture, that is, lateral flexion or extreme axial rotation of the head versus the upper part of the trunk (1).
  1. An increased health risk is present if ≥ 2/minute an evaluation result of yellow and red is found. (N.B. lower frequencies cannot be considered safe in all cases.)

Other Joints. The following conditions are *yellow and red*:

- Extreme joint positions (1)
- For a kneeling position: no adequate knee protection
- For standing (except when using a buttock rest): flexed knee (1)
- For sitting: knee angle >135° (1) (180° = the upper leg in line with the lower leg), unless the trunk is inclined backward and fully supported
- For sitting: knee angle <90° (1) (180° = the upper leg in line with the lower leg)
   1. An increased health risk is present if ≥2/minute an evaluation result of yellow/red is found. (N.B. lower frequencies cannot be considered safe in all cases.)

Pedal Operation. The following conditions are *yellow and red*:

- Standing
- Sitting, leg-actuated

Force Exertion

A. General force limits, that is, one-handed or two-handed, sitting or standing almost all directions

A	Force	
Frequency	**Green and Yellow Limit**	**Yellow and Red Limit**
≥2 and <3/minute	3.0 kgf	6.5 kgf
≥3 and <4/minute	2.0 kgf	4.0 kgf
≥4 and <5/minute	1.5 kgf	3.0 kgf
≥5/minute	1.0 kgf	2.5 kgf

B. Specific force limits: fore and aft directions; upward and downward directions; force limits are also valid for combined fore and aft and upward and downward directions. For one-handed pinching and pedal operation, see NF X 35–106 (1985).

B	Force	
Frequency	**Green/Yellow Limit**	**Yellow/Red Limit**
≥2 and <3/minute	8.0 kgf	18.0 kgf
≥3 and <4/minute	5.0 kgf	11.5 kgf
≥4 and <5/minute	3.0 kgf	7.0 kgf
≥5/minute	2.0 kgf	4.5 kgf

APPLICATION OF THE GUIDELINES IN STANDARDS FOR PROFESSIONS

The Arbouw guidelines are intended for health and safety professionals. The guidelines show the user the most effective way to arrive at the green zone—or, in the short term, the yellow zone. In other words, they provide guidance as to whether actions are best directed toward the workplace (fixtures, transport, machines, and tools and objects), the work organization, and the workers. The results of analyses regarding a certain profession or job are discussed with employers and employees to arrive at a decision regarding actions that are reasonably easy to implement. Subsequently, a clear document is written for employers, employees, commissioners of work, architects, or manufacturers of equipment and tools. 15 A-documents (state-of-the-art documents) are available: lifting, paving, scaffolding, installing windowpanes, steel bending, finishing inclined roofs, finishing terrace roofs, carpentry window frames, carpentry roofs, joining bricks, bricklaying, cabling and piping, tile setting, laying natural stones, and ergonomics of crane cabins.

The recommendations in A-documents include only those solutions that are attainable in practice (http://www.arbouw.nl). Furthermore, the recommendations include improvements that have the greatest effect in the long term. Using the A-documents, employers and workers will be able to observe government regulations satisfactorily. NIOSH is now having some of these documents translated into English.

DUTCH LEGISLATION TAILORED ON CONSTRUCTION WORK

In the Dutch Working Conditions Act, the government lays down general rules regarding physical work demands. The main theme is that employers should prevent or at least limit the risks to the health and safety of workers from physically demanding work. The employer should devote special attention to these risks when drawing up a risk inventory and evaluation. The Working Conditions Act also gives specific rules for temporary and mobile worksites. According to these rules, a safety and health plan must be drawn up for work sites of any size; this plan must also include information about risks and measures to reduce the physical work demands.

The government developed specific policies to reduce physical work demands imposed by lifting and by manual work during construction work. In a policy guideline, the government announced specific legislation aimed at introducing standards for construction work, based mainly on the A-documents (see Application of the Guidelines in Standards for Professions). The full text of this legislation is given in the following.

POLICY GUIDELINES AMENDMENT TO WORKING CONDITIONS LEGISLATION

Amendment of policy guidelines decree with respect to working conditions legislation, in connection with the adoption of a policy guideline relating to lifting on worksites

Article 1

The policy guidelines with respect to working conditions legislation[1] are amended as follows:

Following policy guideline 5.2-2 Physical Demands in Day Nurseries, a policy guideline is inserted that reads as follows:

Policy Guideline 5.3 Lifting on Worksites

Basis: Article 5.3, paragraph 1, of the Working Conditions Decree.

The conditions of Article 5.3, paragraph 1, of the Working Conditions Decree, regarding lifting on work sites as referred to in Article 1.1, paragraph 2, under a, of the Working Conditions Decree, are met if the following guidelines are observed.

General

1. Manual lifting must be avoided or limited as far as reasonably possible.
2. The maximum allowable weight lifted manually by one person is 25 kg.
3. If there is sufficient space, the maximum allowable weight lifted manually by two persons jointly is 50 kg.
4. Until January 1, 2007, points 2 and 3 will not apply to work in the fitting and insulation sector, the furniture industry, the fitting-out sector, and the fitting of stairways in the carpentry sector, on condition that a covenant is concluded with these sectors by January 1, 2003 (December 31, 2003, for the fitting and insulation sector). This covenant must contain an agreement that before January 1, 2007, measures will be introduced by which the physical

work demands made on employees will be structurally reduced to a level comparable to the standards referred to in points 2 and 3.

5. Until July 1, 2003, points 2 and 3 will not apply to the fitting of exterior sun blinds, on condition that before January 1, 2003, a sector standardization will be developed by which the physical work demands made on employees will be structurally reduced to a level comparable to the standards referred to in points 2 and 3.

6. Until July 1, 2004, points 2 and 3 will not apply to the fitting, renovation, and maintenance of lifts, on condition that before January 1, 2004, a sector standardization will be developed by which the physical work demands made on employees will be structurally reduced to a level comparable to the standards referred to in points 2 and 3.

Specific

7. Roof rolls heavier than 25 kg must be transported mechanically. In situations where that is technically or organizationally impossible, roof rolls that are not heavier than 35 kg can be transported manually—contrary to point 2—to a maximum of five rolls per person per day.

8. Pavers heavier than 4 kg must not be processed manually.

9. Bricklaying and gluing may only take place without mechanical lifting aids if
   - The free workspace is at least 0.6 m;
   - The elements are lighter than 14 kg;
   - Elements weighing 4–14 kg must be lifted using both hands and processed to a maximum of 1.5 m above and not below standing level; and
   - Elements lighter than 4 kg must be processed to a maximum of 1.7 m and from at least 0.2 m above standing level.

10. With the exception of
    - Interior bricklaying or gluing directly under or from an upper floor, or
    - Bricklaying or gluing at ground level.

11. Concrete reinforcing steel and tools for processing heavier than
    - 17 kg must not be lifted with one hand.
    - 20 kg must not be lifted from less than 50 cm above the ground.

12. Scaffold elements heavier than 23 kg must not be lifted and transported manually by one person.

Article II. This decree will come into force on January 1, 2003.

This decree will be published in the Netherlands Government Gazette together with the explanation.

Explanation

Sphere of Application. This policy guideline relates to lifting on worksites and refers to sites as described in the Working Conditions Decree. In Article 1.1, paragraph 2, under "a," a worksite is described as a temporary or mobile site where civil engineering works or building works are carried out, a non-exhaustive description of which is included in Appendix I to the Council Directive 92/57/EEC of June 24, 1992 relating to the minimum safety conditions regarding health and safety at temporary and mobile worksites (PbEG L 245).

In the aforementioned appendix, the following civil engineering and building works are listed: (1) Excavation work; (2) Groundwork; (3) Construction; (4) Assembly and disassembly of prefabricated elements; (5) Fitting-out or equipping; (6) Refurbishment; (7) Renovation; (8) Repair; (9) Dismantling; (10) Demolition; (11) Upkeep; (12) Maintenance, painting, and cleaning work; (13) Redemption.

By way of explanation, it should be noted that window cleaning is not included in "12. Maintenance, painting, and cleaning work." Consequently, the policy guideline does not apply to the activities of window cleaners.

Objective, Starting Points. The policy guideline objective is to avoid or limit the dangers of physical work demands on the health and safety of employees as far as reasonably possible. In this context, physical work demands are understood to mean the work posture, movements, or exertions consisting of lifting, carrying, or moving, or supporting one or more loads in any other way (Article 1.1, paragraph 4, under "a," of the Working Conditions Decree). For the sake of brevity, the policy guidelines refer to lifting as lifting, carrying, moving, or supporting one or more loads. In this policy guideline, the A-documents of the Arbouw Foundation have been used as a starting point. These A-documents contain limit values and recommendations to reduce the physical work demands in the construction industry. In 1994, the "Lifting" A-document was published, which includes limit values and recommendations for the manual lifting of loads. In addition, specific A-documents have now been published for various professions and tasks in the construction industry. The following A-documents include passages relating to physical demands: Steel Bending, Paving, Fitting Windowpanes, Finishing Inclined Roofs, Cabling and Piping, Carpentry Roofs, Carpentry Window Frames, Finishing Terrace Roofs, Scaffolding, Lifting, Tile Setting. The A-documents series is still being extended. It is, therefore, possible that when new A-documents are published, supplementary conditions will be added to this policy guideline. In principle, the lifting assessment system in the specific A-documents is identical to that in the "lifting" A-document. Several striking deviations are incorporated in this policy guideline. The conditions of this policy guideline are met if the lifting and ergonomic recommendations described for the work in question in the A-documents are observed.

General. The "Lifting" A-document argues that in ideal situations, manual lifting must never take place. As a result, the policy guideline indicates that lifting must be avoided or limited as far as reasonably possible. Consequently, the starting point is that employers must prevent employees from being exposed to health and safety hazards by physical work demands.

Nevertheless, in certain cases that are provided for in Point 1 of the policy guideline, it is not possible to avoid this entirely. After all, there may be other pressing interests—for which the employer is also responsible—that may be damaged excessively as a result. In that case, the employer should weigh up the interests involved and take a decision according to reasonableness. When weighing up the options, the technical, operational, and economic feasibility of the proposed measures should be considered in relation to the possible social consequences for employees (illness and disability). If heavy or large loads must be lifted manually, the safety of the employee is paramount. This means that the following important factors must always be observed:

- Loads must be lifted with two hands.
- The load must be easy to grasp.
- The floor must be flat, stable, and not slippery.

In any case, the maximum liftable manual load for one person is 25 kg. The "Lifting" A-document contains even more requirements to obtain the optimal lifting situation. These requirements are derived from the NIOSH formula. In short, this boils down to the following: the maximum permissible weight may be lifted when the result of the NIOSH formula is optimal. The lower the result, and, therefore, the greater the deviation from the optimal situation, the lower the maximum manually lifted weight becomes. When circumstances are no longer optimal, the maximum liftable manual load (i.e. 25 kg.) will, in principle, be too heavy for manual lifting. The A-document, therefore, recommends applying the NIOSH formula when assessing the situation and using the result of this calculation as a guideline for the maximum liftable weight. Any loads exceeding the weight permitted by the formula pose a threat to health. In the construction industry, however, situations occur where the limit value according to the NIOSH formula is exceeded when there are no reasonable alternatives available. With these situations, the social partners have agreed that for the time being, feasible values will be used. For this purpose, the Maximum Arbouw Limit (MAL) is used in the "Lifting" A-document. This MAL value may increase to a maximum of three times the NIOSH limit. However, the upper limit of 25 kg is always maintained. For the time being, only the upper limit of 25 kg has been included in the policy guideline. However, that does not alter the fact that manual lifting should be avoided or limited as far as reasonably possible, where at least the MAL—and, preferably, the NIOSH limit—should be the goal.

In addition to lifting by one person, there are activities where it is usual, or at least possible, to carry out lifting activities with two persons. Examples include activities such as fitting frames, windowpanes, or (plastic) catch pits, cesspits, and street furniture (see the relevant A-documents). In these situations, a maximum jointly lifted weight of 50 kg is permitted. Here, too, the following important factors must always be observed:

- Loads must be lifted with two hands.
- The load must be easy to grasp.
- There must be satisfactory hand-load contact.
- The floor must be flat, stable, and not slippery.

Moreover, the surroundings must offer both persons sufficient space to choose an optimal position with respect to the load. In addition, the lifting task must be practically symmetric, enabling an even distribution of the load.

The recommendations for a further restriction of the lift weight if conditions are not optimal, as indicated previously for one person, also apply to lifting by two persons.

With regard to activities in the fitting and insulation sector, the furniture industry, and the fitting-out sector—including the fitting of wooden stairways in the carpentry sector, installing, renovating, maintaining lifts, and fitting exterior sun blinds—the provisions of points 2 and 3 of the policy guideline do not (yet) apply. However, the provisions of point 1 continue to apply, so in those cases, manual lifting must also be avoided or limited as far as reasonably possible. This refers to activities where, in accordance with the current state of science and technology, certain physical heavy work cannot yet be avoided.

As a result, this exception is subject to the condition that the fitting and insulation sector, the furniture industry, the fitting-out sector, and the carpentry sector will have reached agreements—within the framework of a covenant—to implement measures by January 1, 2007, to reduce the physical work demands on employees to the level included in this policy guideline. For the time being, the exception is valid until January 1, 2007, when the situation will be reconsidered.

The exception for lifts and exterior sun blinds is subject to the condition that by January 1, 2004 and January 1, 2003, respectively, these sectors will draw up standardizations to avoid and limit the physical demands involved in these activities. Agreements have been reached with these sectors to form a tripartite working group for that purpose. Subsequently, these sectors will be given extra six months to announce and implement these standardizations within the sector, before the policy guideline comes into force.

Regarding the exception for the fitting of stairways, it has been assumed that, in line with the "Stairway Fitting Work Instructions" of the Netherlands Carpenters' Association, a maximum of 12 flights of stairs are fitted daily by a two-person team (or a maximum of six flights of stairs per person) and that the working method referred to as "dompen" is used. "Dompen" is a method where the fitter does not need to lift more than half the weight of the stairway. The exception refers to the fitting of stairways and does not refer to carrying parts of the stairway to the place within the dwelling where it is to be fitted—that is, unloading it from a lorry or container. The exception is included because here, too, the current state of science and technology means that these activities cannot be avoided yet. It is intended to reach an agreement with the sector to develop the necessary research activities to reduce the physical demands associated with fitting stairways. When the time comes, the question of whether this research can be used as a basis for more specific conditions relating to the fitting of stairways can be included in the policy guideline.

As regards the exception of exterior sun blinds, it has been assumed that standard-sized sun blinds are fitted by a team of at least two persons. The exception refers to hanging and fitting sun blinds in brackets fitted earlier for that purpose. It is also been assumed that sun blinds intended for windows, and so forth, on the ground floor are fitted with the aid of ladders, with employees standing on separate ladders. A tripartite working group is mapping out the specific bottlenecks associated with these activities. This group will propose solutions before January 1, 2003 to structurally reduce the physical work demands on employees in this sector to a level comparable to the standards mentioned in points 2 and 3 of this policy guideline.

As a rule, the installation of lifts takes place during the late stage of construction, when a building is more or less completed. That means that there are sometimes only limited possibilities to move, lift, or position heavy materials mechanically. This applies even more during maintenance and renovation. This exception is intended for actions that cannot yet be carried out in accordance with points 2 and 3 of this policy guideline. A tripartite working group is mapping out the specific bottlenecks associated with these activities. This group will propose solutions before January 1, 2004 to structurally reduce the physical work demands on employees in this sector to a level comparable to the standards mentioned in points 2 and 3 of this policy guideline.

Specific. Practice has shown that a limit of 25 kg for rolls of bituminous roof covering (roof rolls) would impose unacceptable limits on the length of the roll or the thickness of the material. The "Terrace Roofs" A-document indicates that a weight of 35 kg is an acceptable compromise when the work involves only occasional manual lifting and carrying of roof rolls. That has been included in this policy guideline.

The limit values for lifting materials in other specific parts of this policy guideline have been copied from the "Paving," "Steel Bending," and "Scaffolding" A-documents. For the manual processing of pavers, a maximum permissible weight of 4 kg applies. This relates to the high frequency with which the pavers are exposed, which results in the weight making demands on the employee at an earlier stage.

The manual processing of glue blocks for bricklaying and building bricks of 14 kg or more is not permitted. This limit is also mentioned in collective labor agreements. For such elements, the use of technical lifting aids is, therefore, always necessary. Elements of 4–14 kg may only be processed manually if they are lifted with two hands and are processed at a maximum of 1.5 m above and not below standing level. Elements lighter than 4 kg may only be processed manually at a maximum of 1.7 m and from 0.2 m above standing level. Consequently, elements heavier than 4 kg may not be processed manually with one hand.

When laying the first and final layers of bricks internally between upper floors, it is not always possible to adhere to the aforementioned bricklaying heights. This is also the case when laying the first layers of bricks externally from ground level. This policy guideline contains an exception for such cases. Free workspace is understood to mean that the available depth for bricklayer and materials when laying is taking place from the scaffold or the minimum depth of the bricklaying console when laying is taking place from a console.

NOTE

1 Supplement to the Netherlands Government Gazette (2001, p. 239), last amended by decree of April 24, 2002 (Netherlands Government Gazette 84).

REFERENCES

Arbouw. (1990–2002). Surveys sick leave data in the Dutch construction industry.

Hedén, K., Andersen, V., Kemmlert, K., Samdahl-Høiden, L., Seppänen, H., & Wickström, G. (1993). Model for assessment of repetitive, monotonous work—RMW. In W. S. Marras, W. Karwowski, J. L. Smith, & L. Pacholski (Eds.), *The ergonomics of manual work*, London Washington, DC: Taylor & Francis, (pp. 315–317).

Hoonakker, P. L. T., Schreurs, P. I. G., Van der Molen, H. F., & Kummer, R. (1992). *Conclusions and evaluation of the research projects concerning the psychosocial workload of six professions in the construction industry* (in Dutch). Amsterdam: Arbouw.

ISO/DIS 11226. (2000). *Ergonomics—Evaluation of static working postures*. Geneva, Switzerland: ISO.

Kilbom, Å. (1994a). Repetitive work of the upper extremity: Part I—Guidelines for the practitioner. *International Journal of Industrial Ergonomics, 14*, 51–57.

Kilbom, Å. (1994b). Repetitive work of the upper extremity: Part II—The scientific basis (knowledge base) for the guide. *International Journal of Industrial Ergonomics, 14*, 59–86.

Koningsveld, E. A. P., & van der Molen, H. F. (1997). History and future of ergonomics in building construction. In *Abstracts from the first international symposium on ergonomics in building and construction*. Part of the 14th IEA congress. Tampere, Finland.

Mital, A., Nicholson, A. S., & Ayoub, M. M. (1993). *A guide to manual materials handling*. London/Washington, DC: Taylor & Francis.

NF X 35-106. (1985). *Ergonomie—Limites d'efforts recommandées pour le travail et la manutention au poste de travail*. Paris: AFNOR.

NIOSH. (1981). *Work practices guide for manual lifting* (NIOSH Technical Report No. 81-122). Cincinnati, OH: U.S. Department of Health and Human Services, National Institute for Occupational Safety and Health.

prEN 1005-4. (2002). *Safety of machinery—Human physical performance—Part 4: Evaluation of working postures and movements in relation to machinery*. Brussels, Belgium: CEN.

Waters, T. R., Putz-Anderson, V., Garg, A., & Fine, L. J. (1993). Revised NIOSH equation for the design and evaluation of manual lifting tasks. *Ergonomics, 36*, 749–776.

16 Safety of Machinery — Human Physical Performance

Manual Handling of Machinery and Component Parts of Machinery

Karlheinz G. Schaub

CONTENTS

INTRODUCTION

MANUAL MATERIALS HANDLING (MMH) DURING MACHINE OPERATION

In industrialized and developing countries, the machinery sector is an important part of the engineering industry and may be one of the industrial mainstays of the economy. However, social costs arising from accidents or sick leave may be directly caused by hazardous machinery operations. Manual handling of machinery or component parts of machinery can lead to a high risk of injury to the musculoskeletal system if the loads to be handled are too heavy, handled at high frequencies for long durations, or in awkward postures. Manually applied effort is often required by operators working with machines for their intended purpose. Risks exist if the design of the machinery is not in accordance with ergonomic design principles.

When designing and constructing machinery for manual handling, the designer should ensure a safe and ergonomically designed machine in the market. An ergonomic risk assessment during the early design stage of machinery—as well as appropriate measures to reduce the risk by redesign, if necessary—may help reach this aim. It also eliminates the need for cost and time-intensive machine alterations later at the shop floor level when health and safety inspectors discover hazardous situations during the intended machine operations.

ERGONOMIC EVALUATION OF MANUAL MATERIALS HANDLING TASKS

In a worldwide overview, many regulations and evaluation tools on manual materials handling exist. Though most of the regulations on manual materials handling are edited by national institutions of occupational health and safety, the regulations and associated evaluation tools differ from one country to another as it concerns their complexity, their level of protection, and their way of presentation. To provide a general global common approach, ISO 11228 (Ergonomics—Manual Handling, Parts 1 and 2) was created. Like national or regional regulations (e.g., EU-Directive [europa.eu.int] on manual materials handling (90/269/EEC, 1990) international standards address to employers and employees (Meyer et al., 1998)).

However, for designers and manufacturers of machinery that involve manual materials handling activities, no international regulations or guidelines exist at the moment. In the European Union, the Machinery Directive bridges that gap.

THE EU-MACHINERY DIRECTIVE

Whereas existing national health and safety provisions providing protection against the risks caused by machinery must be approximated in the EU to ensure free movement on the market of machinery without lowering existing justified levels of protection in the Member States, the provisions of the EU-Machinery Directive (98/37/EC) concerning the design and construction of machinery—essential for a safer working environment—shall be accompanied by specific provisions concerning the prevention of certain risks to which workers can be exposed at work, as well as those based on the organization of safety of workers in the working environment.

The EU-Machinery Directive addresses to designers and manufacturers of machinery solely (not to employers and employees) and demands an EC declaration of conformity—represented by the CE mark on a machine—to be carried out. An easy way to declare conformity is to pinpoint that machines had been designed in accordance with harmonized CEN standards mandated by the European Commission (see earlier chapters on ergonomically relevant EU Directives). EN 1005 ("Safety of Machinery—Human Physical Performance") highlights ergonomic requirements concerning the operator's physical effort during machine operation.

Part 2 of this standard deals with "Manual handling of machinery and component parts of machinery." The design criteria given in this standard can be used by the designer when making risk assessments.

EN 1005-2 provides relevant data for the aspects of working posture, load, frequency, and duration.

EN 1005-2 is of relevance for all designers and manufacturers of machinery that involve manual materials handling in any phase of a machine's "life cycle" (construction, transport and commissioning, and use and decommissioning) for the "intended user population" in its "intended use"—including a "foreseeable misuse."

EN 1005-2 helps ensure compliance with the EU Machinery Directive and reduces social costs arising from hazardous machinery operations (Ringelberg & Schaub, 1997; Schaub et al., 1996; Schaub, 1997; Schaub & Landau, 1998).

MANUAL HANDLING OF MACHINERY AND COMPONENT PARTS OF MACHINERY

Subsidiary Standards

Guidance to the work on EN 1005-2 was given by several type A standards mandated under the Machinery Directive:

- EN 292-1 "Safety of Machinery—Basic Concepts, General Principles for Design—Part 1: Basic Terminology, Methodology"
- EN 292-2 "Safety of Machinery—Basic Concepts, General Principles for Design—Part 2: Technical Principles and Specifications" give some rough information on general ergonomic aspects
- EN 614-1, Safety of Machinery—Ergonomic Design Principles—Part 1: Terminology and General Principles
- EN 1050, Safety of Machinery—Risk Assessment
- EN 1005-1, Safety of Machinery—Human Physical Performance—Terms and Definitions

For a better understanding and proper application of EN 1005-2, it is highly recommended to read the previously mentioned standards first.

Structure of the Standard

EN 1005-2 consists of four main chapters and five annexes. Most of them are required and predefined due to the drafting rules for CEN standards (e.g., chaps. 0–3 and Annex ZA: Introduction, Scope, Normative References, Definitions and Relationship between this European Standard and the EU Directive for Machinery, chapter 4 represents the core of the standard: "Recommendations for the Design of Machinery and Component Parts Where Objects are Lifted, Lowered and Carried"). See also ENV26385. Annexes A–D (population characteristics and system design, recommended thermal comfort requirements, risk assessment worksheets, and bibliography) offer supplementary information.

A foreword introduces the standard. It presents information on the document status and the Technical Committee that has prepared it (CEN/TC 122 "Ergonomics"). This standard was mandated by the European Commission and the European Free Trade Association and supports essential requirements of EU Directive(s).

Reference is given to the other parts of EN 1005 "Safety of Machinery—Human Physical Performance":

- Part 1: Terms and Definitions
- Part 2: Manual Handling of Machinery and Component Parts of Machinery
- Part 3: Recommended Force Limits for Machinery Operation
- Part 4: Evaluation of Working Postures in Relation to Machinery
- Part 5: Risk Assessment for Repetitive Handling at High Frequency

Part 5 is still under preparation. This part (Part 2) and Part 4 are currently submitted to CEN inquiry. For definitions in that standard, refer to EN 1005-1.

Introduction and Scope

The introduction describes general aspects of manual materials handling and the associated risk of injury for the musculoskeletal system. Parts of the text are taken over from EN 614-1, the corresponding type A standard. The standard requires machinery designers to adopt a three-stage approach to:

- Avoid manual handling activities wherever possible
- Utilize technical aids
- Further reduce the inherent level of risk by optimizing handling activities

The standard applies to the manual handling of objects of 3 kg or more (and for carrying of less than 2 m). Lower weights are dealt with by EN 1005-5. As addressed to designers and manufacturers, there is little focus on the carrying of objects. The designers should eliminate the need to carry objects during machinery operation. Whenever necessary, carrying should be limited to a distance of one or two steps (less than 2 m). As mandated within the Machinery Directive, the standard provides information on ergonomic design and risk assessments concerning lifting, lowering, and carrying during the whole "life cycle" of a machine—that is, construction, transport and commissioning, use and decommissioning, disposal, and dismantling.

The standard does not cover the holding and pushing or pulling of objects. Hand-held machines and manual materials handling while seated are excluded as well.

RECOMMENDATIONS FOR THE DESIGN OF MACHINERY AND COMPONENT PARTS

To minimize the health risks emerging from manual handling of machinery or component parts, the designer should establish whether hazards exist. If they do, a risk assessment should be carried out (chap. 4.3 of the Standard). In general, hazards should be removed by excluding the need for manual handling activities. Where this is not possible, technical aids for the handling of machinery and component parts should be provided and machinery and component parts should be (re)designed in accordance with ergonomic design principles. When machinery, component parts, or technical aids introduce awkward postures, reference should be made to EN 1005-4. When pushing and pulling are introduced, EN 1005-3 should be considered.

Population Characteristics

Table 16.1 applies to the general working population. This information is in accordance with measurements of maximum energetic capacity, subjective estimation of tolerability limits, and objective measurements of physical capabilities.

TABLE 16.1

Population Percentages in Relation to Measurement Criteria and the Object Mass

Options	Psychophysical Data Indicating Tolerability Capacity	Measurements of Forces Indicating Limits	Measurements on the Maximum Metabolic Ability Limits
10 kg	99% (F + M)	99% (F + M)	99% (F + M)
	99% F	99% F	99% F
	99.9% M	99.9% M	99.9% M
20 kg	95% (F + M)	95% (F + M)	95% (F + M)
	90% F	90% F	80–85% F
	99.9% M	99.9% M	99% M
25 kg	85% (F + M)	85% (F + M)	85% (F + M)
	75% F	75–75% F	70% F
	99.9% M	99.9% M	99% M

F, Female; M, Male.

System Design

For the system design, the following interrelated aspects from EN 1005-2 Annex A should be considered:

- **Constrained postures**. Thought should be given to the design and redesign of machines that lead to constrained working postures and monotonous work. In these circumstances, discomfort and fatigue increases rapidly and muscular efficiency falls. In addition, machinery should be designed to minimize static postures as far as possible.
- **Acceleration and movement accuracy**. Acceleration places higher force requirements and strain on the body. Movement accuracy increases the time needed for manual handling and increases muscle effort. Accuracy of precise positioning should be supplied by the design.
- **Discomfort, fatigue, and stress on the operator**. Research and experience in the industry have shown that preventing discomfort, fatigue, and stress during physically demanding work reduces ill-health and increases output. It is important to consider three factors:

    a. Physiological effort required
    b. The amount of work in constrained postures
    c. The large variation in individual susceptibility to fatigue

RISK ASSESSMENTS

General Information

The risk assessment model presented involves three methods. These methods have the same basis but differ in their complexity of appliance. The most efficient approach is to begin the risk assessment by applying Method 1 (the simplest procedure) and use Methods 2 or 3 only if the assumptions or operational situations identified in Method 1 are not met (see Figure 16.1). If Method 3 fails, Chapter 4.2 of the standard offers recommendations for the redesign of machinery. Each method requires three steps to be carried out as described in the chapter "Worksheets."

The risk assessment takes respect of the parameters shown in Table 16.2.

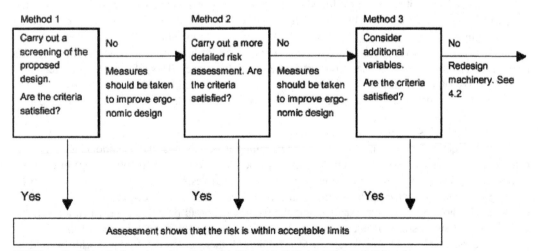

NOTE   Consider further steps to reduce risk factors to their lowest possible level

FIGURE 16.1 Flowchart identifying the stepwise approach to assessment.

TABLE 16.2

Hazards Taken into Account in the Risk-Assessment Model of EN 1005-2

Hazards[a]

Posture of the Trunk	Posture of the Arms
Frequency of Operation	Work Duration
One-handed Operations[b]	Manual Handling by Two Persons[b]
Coupling Conditions (Worker $\Longleftrightarrow$ Load)	Additional physically demanding tasks[b]

Notes

[a] Assumptions for the working conditions are moderate ambient thermal environment, unrestricted standing posture, good coupling between feet and floor, smooth lifting, objects to be lifted are not very cold, hot, or contaminated.

[b] Only available in risk-assessment method 3.

The risk-assessment model follows widely the revised National Institute of Occupational Safety and Health (NIOSH) (1991) approach (National Institute of Occupational Safety and Health [NIOSH], 1994; Andersson, 1999; Garg, 1989; Waters et al., 1999), supplemented by additional parameters from other international sources (Directorate of the Danish Labour National Inspection Service for Machinery, 1986; Federal German Institution of Occupational Safety and Health [FIOSH], 1994/1997) NF X 35-10.

Contrary to other national or international standards, EN 1005-2 (like all harmonized CEN standards realized for support of the Machinery Directive) addresses the designer and manufacturer of machinery only. Therefore, it does not focus on a "general working population" (or 90% of it) but aims toward the "intended user population" (see Table 16.1). On one hand, the target population of these standards could include elderly people and children as well as domestic applications of machinery; on the other, the target population could be a highly selected collective of young or middle-aged male workers on, for example, an oil-drilling platform in the arctic sea.

Therefore, a variable reference mass is introduced into the risk-assessment model. This might conflict, on one hand, with the struggle for equal chances of male and female workers when applying for jobs, which require considerable physical effort, as gender is—next to age—one of the primary influences on physical capabilities. On the other, highly protective ergonomic regulations could have negative effects on the foundation and development of small- and medium-sized companies. The reference mass plays a similar role as the mass constant in the NIOSH approach. This reference mass is regarded to be the maximum weight that should be manipulated manually under "ideal" conditions. With reference to the hazards described in Table 12.3, this reference mass is lowered by means of multipliers.

Risk-Assessment Methods

Method 1 is a quick screening method. First, the user must check whether the operational assumptions made (e.g., two-handed operation only, handling by one person only, and smooth lifting) in this risk assessment correspond with the work situation to be analyzed. If yes, he or she may choose one of three working examples characterized by a critical mass, a critical vertical mass displacement, or a critical frequency. If one of the working examples fits the work situation to be analyzed, the risk assessment was carried out successfully. If not all the operational assumptions made fit and none of the working examples correspond to the work situation to be analyzed, the user should continue with method 2. Method 2 is an easy-to-handle method as opposed to screening Method 1, which indicates risks.

In comparison with Method 1, some additional risk factors can be considered in Method 2. The user must calculate a risk index (RI) as a quotient of the actual mass to be handled and the recommended mass limit. The recommended mass limit is composed of a reference mass and several multipliers (e.g., horizontal multiplier, vertical multiplier, and distance multiplier), which may be selected from precalculated tables. If the risk index is ≤0.85, the risk may be regarded as tolerable (green). A risk index between 0.85 and 1.0 indicates that a significant risk exists (yellow). It is recommended to apply Method 3 to identify how the risk may be reduced, how to redesign the machinery, or to ensure that the risk is tolerable. An RI of ≥1.0 (red) means that redesign is necessary. The design can be improved by changing the situations that lead to low multipliers.

Method 3 is an extended assessment method that assesses risk in a thorough way and is supplemented by additional risk factors not presented in Methods 1 and 2 (e.g., one-handed lifting; lifting by two persons; and additional physical workload next to lifting, lowering, and carrying). The calculation procedure used in this method is similar to that used in Method 2.

WORKSHEETS

The risk-assessment procedures are described in detail in chapter 4.3 of the Standard. Worksheets for the three methods of the risk-assessment model are contained in Annex C and may be copied for application purposes (see Tables 12.4–12.8). The risk-assessment model consists of three methods increasing the level of complexity. The first method is a quick screening procedure to assess the task. Method 2 must be applied if the screening procedure indicates risks. This method tasks account for additional risk factors. It is advisable to begin the risk assessment by applying Method 1 (the simplest procedure) and use Method 2 or 3 only if the assumptions or operational situations identified in Method 1 are not met.

Each method requires three steps to be carried out:

Step 1: Consider the reference mass (see Table 16.3).

Step 2: Assess the risk factors according to the worksheet.

TABLE 16.3
Reference Mass (M_ref) Taking into Consideration the Intended User Population 6

Field of Application	M_{ref} [kg]	Percentage of: F and M	F_{emales}	M_{ales}	Population Group	
Domestic Use[a]	5	Data not available			Children and the Elderly	Total population
	10	99	99	99	General domestic population	
Professional Use (General)[b]	15	95	90	99	General working population	General working population
	25	85	70	90	Including the young and elderly	
					Adult working population	
Professional Use (Exceptional)[c]	30	Data not available			Special working population	Special working population
	35					
	40					

Notes

[a] When designing a machine for domestic use, 10 kg should be used as a general reference mass in the risk assessment. If children and the elderly are included in the intended user population, the reference mass should be lowered to 5 kg.

[b] When designing a machine for professional use, a reference mass of 25 kg should not be exceeded in general.

[c] Although every effort should be made to avoid manual handling activities or reduce the risks to the lowest possible level, there may be exceptional circumstances where the reference mass might exceed 25 kg (e.g., where technological developments or interventions are not sufficiently advanced). Under these special conditions, other measures have to be taken to control the risk according to EN 614 (e.g., technical aids, instructions, or special training for the intended operator group).

TABLE 16.4
Worksheet 1: Risk-Assessment for Method 1

Risk Assessment: Method 1—Screening by Means of Critical Values

EN 1005 Safety of Machinery—Human Physical Performance—Part 2: Manual Handling of Machinery and Component Parts of Machinery

This method provides a quick screening procedure to identify whether the handling operation represents a risk to the operators). Step 2 requires one of three critical operational situations (cases 1–3) to be selected. The limiting condition is that all assumptions for handling operations are fulfilled.

Step 1: Consider the reference mass.

Identify the intended user population and select the reference mass (M_{ref}) according to the intended user population (Table 16.3).

Step 2: Carry out the risk-assessment.

Please tick the following criteria for the handling operation, if met:
- Two-handed operation only
- Unrestricted standing posture and movements
- Handling by one person only
- Smooth lifting
- Good coupling between the hands and the objects handled
- Good coupling between the feet and floor
- Manual handling activities, other than lifting, are minimal
- The object to be lifted are not cold, hot, or contaminated
- Moderate ambient thermal environment

If one or more of these criteria are not met, refer to Method 2.

If all criteria are met, then select one of the following critical variables. These apply to a work shift of eight hours or less.

Case 1 *Critical mass*
- The load handled does not exceed 70% of the reference mass-selected from Table 16.3.
- Vertical displacement of the load is ≤25 cm and between hip and shoulder height.
- The trunk is upright and not rotated.
- The load is kept close to the body.
- The frequency of lifts is equal to or less than 0.00333 Hz (one lift every five min).

Case 2 *Critical vertical mass displacement*
- The load handled does not exceed 60% of the reference mass-selected from Table 16.3.
- Vertical displacement of the load is not above shoulder height or below knee height.
- The trunk is upright and not rotated.
- The load is kept close to the body.
- The frequency of lifts is ≤0.00333 Hz (one lift every five min).

Case 3 *Critical frequency*
- The load handled does not exceed 30% of the reference mass-selected from Table 16.3.
- Vertical displacement of the load is ≤25 cm and between hip and shoulder height.
- The frequency of lifts is equal to or less than 0.08 Hz (five lifts every min).
- The trunk is upright and not rotated.
- The load is kept close to the body.

Or
- The load handled does not exceed 50% of the reference mass-selected from Table 16.3.
- Vertical displacement of the load is ≤25 cm and between hip and shoulder height.
- The frequency of lifts is ≤ 0.04 btz (two and a half lifts every min).

(Continued)

TABLE 16.4 (Continued)

- The trunk is upright and not rotated.
- The load is kept close to the body.

Step 3: Select the action required.

If the design fits one of the operational situations (cases 1–3) described above, the risk assessment has been carried out successfully.

If none of the operational situations are satisfied, or any of the criteria specified in step 2 are not met, either

⟹ Consider modifying or redesigning the machinery, or

⟹ Use a more detailed risk-assessment procedure to identify critical risk factors (Methods 2).

TABLE 16.5
Worksheet 2a: Risk Assessment Worksheet for Method 2: Part a

Risk Assessment: Method 2—Estimation by tables

EN 1005 Safety of Machinery—Human Physical Performance—Part 2: Manual Handling of Machinery and Component Parts of Machinery

Step 1: Consider the reference mass.

Identify the intended user population and select the reference mass (M_{ref}) according to the intended user population (Table 16.3).

Step 2: Carry out the risk assessment.

Please indicate (tick), whether the handling operation meets the following criteria:

- Two-handed operation only
- Unrestricted standing posture and movements
- Handling by one person only
- Smooth lifting
- Good coupling between the feet and floor
- Manual handling activities, other than lifting, are minimal
- The objects to be handled are not cold, hot, or contaminated
- Moderate ambient thermal environment

If one or more of these criteria are not met, refer to Method 3.

If all criteria are met, then determine the level of risk by:

  1.. calculating the recommended mass limit ($R_{ML.2}$) using the multipliers provided in Table 16.3.

  2.. calculating the risk index (RI) as follows:

$$\text{risk index}\left(R_I = \frac{\text{actual mass}}{R_{ML}} = \frac{[kg]}{[kg]}\right)$$

Step 3: Select the action required.

- $R_I \le 0.85$ the risk may be regarded as tolerable.
- $0.85 < R_I < 1.0 \implies$ significant risk exists. It is recommended that:

⟹ Method 3 is applied to identify how the risk may be reduced,

⟹ The machinery be either redesigned, or

⟹ Ensure, that the risk is tolerable.

- $R_I \ge 1.0$ Redesign is necessary. The design can be improved by changing the situations that lead to low multipliers.

TABLE 16.6
Worksheet 2b: Risk Assessment Worksheet for Method 2: Part b

Risk Assessment: Method 2—Estimation by Tables
EN1005 Safety of Machinery—Human Physical Performance—Part 2: Manual Handling of Machinery and Component Parts of Machinery
Table C.2—Calculation of the recommended mass limit (R_{ML2})

Reference mass (M_{ref})　　　　　　　　　　　　　　　　　　　　　　　　　　　　　R_{ML2} =

Reference Mass [kg] (see Table 16.3)									M_{ref}
Vertical multiplier (V_M)								×	
Vertical Location (cm)	0	25	50	75	100	130	>175		V_M
Factor	0.78	0.85	0.93	1.00	0.93	0.84	0.00		
Distance Multiplier (D_M)								×	
Vertical Displacement (cm)	25	30	40	50	70	100	>175		D_M
Factor	1.00	0.97	0.93	0.91	0.88	0.87	0.00		
Horizontal Multiplier (H_M)								×	
Horizontal Location (cm)	25	30	40	50	55	60	>63		H_M
Factor	1.00	0.83	0.63	0.50	0.45	0.42	0.00		
Asymmetric Multiplier (A_M)								×	
Angle of Asymmetry (°)	0	30	60	90	120	135	>135		A_M
Factor	1.00	0.90	0.81	0.71	0.62	0.57	0.00		

Coupling Multiplier (C_M)　　　　　　　　　　　　　　　　　　　　　　　　　　　×

	Good	Fair	Poor	C_M
Quality of Grip Description	Load length ≤ 40 cm; Load height ≤ 30 cm; good handles or hand-held cutouts. Easy to handle loose parts and objects with wrap-around grasp and without excessive wrist deviation.	Load length ≤ 40 cm; load height ≤ 30 cm; and poor handles or hand-held cutouts or 90° finger flexion. Easy to handle loose parts and objects with 90° finger flexion and without excessive wrist deviation.	Load length > 40 cm; or load height > 30 cm; or difficult to handle parts, sagging objects or asymmetric center of mass, unstable contents or hard to grasp object, or use of gloves.	
Factor	1.00	0.95	0.90	

Frequency multiplier (F_M) dependent on work duration (d)　　　　　　　　　　　×

						Frequency			F_M
	Hz	0.0033	0.0166	0.0666	0.1000	0.1500	0.2000	>0.2500	F_M
	[lifts/min]	0.2	1	4	6	99	12	>15	
Work	$d \le 1$ h	1.00	0.94	0.84	0.75	0.52	0.37	0.00	
Duration (d)	1 h < d ≤ 2 h	0.95	0.88	0.72	0.50	0.30	0.00	0.00	
	2 h < d 8 h	0.85	0.75	0.45	0.27	0.00	0.00	0.00	

$R_{ML2} = M_{ref} \times V_M \times D_M \times H_M \times A_M \times C_M \times F_M$ 　　　　　　　　　= 　[kg]

TABLE 16.7
Worksheet 3a: Risk Assessment Worksheet for Method 3: Part a

Risk Assessment: Method 3—Calculation by Formula

EN 1005 Safety of Machinery—Human Physical Performance—Part 2: Manual Handling of Machinery and Component
　Parts of Machinery

Step 1: Consider the reference mass.

Identify the intended user population and select the reference mass (M_{ref}) according to the intended user population
　(Table 16.3).

Step 2: Carry out the risk assessments.

Please indicate (tick) whether the handling operation meets the following criteria:

(Continued)

TABLE 16.7 (Continued)

- Unrestricted standing posture and movements
- Smooth lifting
- Good coupling between the feet and floor
- The objects to be handled are not cold, hot, or contaminated
- Moderate ambient thermal environment

If one or more of the criteria are not met, consider ways of meeting each of the criteria. Refer to chapter 4 of this standard. If all criteria are met, calculate the recommended mass limit (R_{ML}).

Case 1   If the recommended mass limit (R_{ML2}) is already known (calculated during Method 2), then calculate the recommended mass limit (R_{ML}) as follows:

$$R_{ML} = R_{ML2} \times O_M \times P_M \times A_1 \text{ [kg]},$$

where

O_M one-handed operation	if true $O_M = 0.6$	otherwise $O_M = 1.0$
P_M two-person operation	if true $P_M = 0.85$	otherwise $P_M = 1.00$
A_T additional physically demanding tasks	if true $A_T = 0.8$	otherwise $A_T = 1.0$

In case the recommended mass limit (R_M) has not been calculated, then calculate the recommended mass limit (R_{ML}) as follows:

$$R_{ML} = M_{ref} \times V_M \times H_M \times A_M \times C_M \times F_M \times O_M \times P_M \times A_T$$

The following definitions apply:

$V_M = 1 - 0.003|V - 75|$	if $V < 0$ cm, $V_M = 0.78$	if $V > 175$ cm, $V_M = 0$
$D_M = 0.82 + 4.5/D$	if $D < 25$ cm, $D_M = 1$	if $D > 175$ cm, $D_M = 0$
$A_M = 1 - (0.0032A)$		if $A > 135°$, $A_M = 0$
$H_M = 25/H$	if $H < 25$ cm, $H_M = 1$	if $H > 63$ cm, $H_M = 0$

M the reference mass from
    Table 16.3 in kg

V vertical location of the load, in cm

D vertical displacement of the load,
    in cm

H horizontal location of the load,
    in cm

A angle of asymmetry, in degree

C_M coupling multiplier from
    Table 16.3

F_M frequency multiplier from
    Table C.2

O_M one-handed operation	if true $C_M = 0.6$	otherwise $O_M = 1.0$
P_M two-person operation	if true $P_M = 0.85$	otherwise $P_M = 1.0$
A_T additional physically demanding tasks	if true $A_T = 0.8$	otherwise $A_T = 1.0$

Calculate the risk index (R_1) as follows:

$$\text{Risk index}(R_1) = \frac{\text{actual mass}}{R_{ML}} = \frac{[kg]}{[kg]}$$

Step 3: Select the action required.

- $R_I \leq 0.85$ the risk may be regarded as tolerable.
- $0.85 < R, < 1.0$ Significant risk exists, it is recommended to:$\Rightarrow$ redesign the machinery or $\Rightarrow$ ensure, that the risk is tolerable.
- $R_I \geq 1.0$ Redesign is necessary. The design can be improved by changing the situations that lead to low multipliers.

TABLE 16.8
Worksheet 3b: Risk Assessment Worksheet for Method 3: Part b

Risk Assessment: Method 3—Calculation by Formula

EN 1005 Safety of Machinery—Human Physical Performance—Part 2: Manual Handling of Machinery and Component Parts of Machinery

Table C.3—Coupling Multiplier (CM)

Quality of Grip	*Good*	*Fair*	*Poor*
Description	Load length ≤ 40 cm; load height ≤ 30 cm; good handles or hand-hold cut-outs. Easy to handle loose parts and objects with wrap-around grasp and without excessive wrist deviation.	Load length ≤ 40 cm; load height ≤ 30 cm; **and** poor handles **or** hand-hold cut-outs or 90° finger flexion. Easy to handle loose parts and objects with 90° finger flexion and without excessive wrist deviation.	Load length > 40 cm; **or** load height > 30 cm; **or** difficult to handle parts, sagging objects, asymmetric center of mass or unstable contents, hard-to-grasp object, **or** use of gloves.
Factor	1.00	0.95	0.90

Table C.4—Frequency Multiplier (F_M)

Frequency		2 h < d ≤ 8h		Work Duration d 1 h < d ≤ 2 h		d ≤ 1 h	
[Hz]	[lifts/min]	V^a ≤ 75 cm	V ≥ 75 cm	V ≥ 75 cm	V ≥ 75 cm	V < 75 cm	V ≥15 cm
≤0.00333	≥0.2	0.85	0.85	0.95	9.95	1.00	1.00
0.00833	0.5	0.81	0.81	0.92	0.92	0.97	0.97
0.01666	1	0.75	0.75	0.88	0.88	0.94	0.94
0.03333	2	0.65	0.65	0.84	0.84	0.91	0.91
0.05000	3	0.55	0.55	0.79	0.79	0.88	0.88
0.06666	4	0.45	0.45	0.72	0.72	0.84	0.84
0.08333	5	0.35	0.35	0.60	0.60	0.80	0.80
0.10000	6	0.27	0.27	0.50	0.50	0.75	0.75
0.11666	7	0.22	0.22	0.42	0.42	0.70	0.70
0.13333	8	0.18	0.18	0.35	0.35	0.60	0.60
0.15000	9	0.00	0.15	0.30	0.30	0.52	0.52
0.16666	10	0.00	0.13	0.26	0.26	0.45	0.45
0.18333	11	0.00	0.00	0.00	0.23	0.41	0.41
0.20000	12	0.00	0.00	0.00	0.21	0.37	0.37
0.21666	13	0.00	0.00	0.00	0.00	0.00	0.34
0.23333	14	0.00	0.00	0.00	0.00	0.00	0.31
0.25000	15	0.00	0.00	0.00	0.00	0.00	0.28
>0.2500	>15	0.00	0.00	0.00	0.00	0.00	0.00

Notes

[a] V is the vertical location.

Step 3: Identify the action required (Tables 16.4–16.8):

- No action is necessary if the risk level is tolerable.
- Redesign if the risk level is not tolerable or check that the risk is tolerable.
- Use a more complex risk-assessment method.

ACKNOWLEDGMENTS

Dedicated to CEN/TC 122/WG 4, who developed this standard, especially to my colleagues from the writing group.

REFERENCES

90/269/EEC. (1990). *Council Directive of 29 May 1990 on the minimum health and safety requirements for the manual handling of loads where there is a risk particularly of back injury to workers (fourth individual Directive within the meaning of Article 16 (1) of Directive 89/391/EEC).*

Andersson, G. B. J. (1999). Point of view: Evaluation of the revised NIOSH lifting equation, a cross-sectional epidemiologic study. *Spine 24* (4), 395.

Directorate of the Danish National Labour Inspection Service for Machinery. (1986). *Heavy lifts "backaches" compendium 5)*, Copenhagen.

EN 292-1. *Safety of machinery—Basic concepts, general principles for design—Part 1: Basic terminology, methodology.*

EN 292-2. (1991). *Safety of machinery—Basic concepts, general principles for design—Part 2: Technical principles and specifications.*

EN 614-1. *Safety of machinery—Ergonomic design principles—Part 1: Terminology and general principles.*

EN 1005-1. *Safety of machinery—Human physical performance—Part 1: Terms and definitions.*

EN 1005-2. *Safety of machinery—Human physical performance—Part 2: Manual handling of machinery and component parts of machinery.*

EN 1005-3. *Safety of machinery—Human physical performance—Part 3: Recommended force limits for machinery operation.*

EN 1005-4. *Safety of machinery—Human physical performance—Part 4: Evaluation of working postures in relation to machinery.*

EN 1005-5. *Safety of machinery—Human physical performance—Part 5: Risk assessment for repetitive handling at high frequency.*

EN 1050. *Safety of machinery—Risk assessment.*

Federal German Institution of Occupational Safety and Health (FIOSH). (1994/1997). *Guideline on safety and health protection during manual handling.* Special edition 9 and 43 of the Series of the Federal German Institution of Occupational Safety and Medicine, Berlin.

Garg, A. (1989). An evaluation of the NIOSH Guidelines for Manual Lifting, with special reference to horizontal distance. *American Industrial Hygiene Association Journal 50*(3), 157–164.

ISO 11228-1. *Ergonomics–Manual handling–Part 1: Lifting and carrying.* ISO 11228-2, Ergonomics–Manual handling–Part 2: Pushing and pulling.

Meyer, J. R., Colombini, D., Heden, K., Ringelberg, A., Viikari-Juntura, E., Boocock, M., Lobato, J. R., & Schaub, Kh. (1998). European directive (90/269) for the prevention of risks in manual handling tasks. In Health Service Section—International Social Security Association (Ed.), *2nd Internationales Symposium, Low back pain in the health care profession—risk and prevention.* Hamburg, Germany, September 10–11, 1998.

National Institute of Occupational Safety and Health. (1991, May). *Scientific support documentation for the revised 1991 Lifting Equation.* (Technical contract reports). Cincinnati, OH: National Institute for Occupational Safety and Health. U.S. Department of Commerce, National Technical Information Service, Springfield, VA.

National Institute of Occupational Safety and Health (NIOSH). (1994, January). *Applications manual for the revised NIOSH Lifting Equation.* U.S. Department of Health and Human Services, Public Health Service, Centres for Disease Control and Prevention, National Institute for Occupational Safety and Health, Cincinnati, OH. 45226.

NF X 35-10. Acceptable limits of manual load carrying for one person.

Ringelberg, J. A., & Schaub, Kh. (1997). Background of prEN 1005, Part 2. Manual handling. In IEA'97 (Ed.), *From experience to innovation* (Vol. 3, p. 570). The 13th Triennial Congress of the International Ergonomics Association, Tampere, Finland, June 29–July 4, 1997.

Schaub, Kh., Boocock, M., Grevé, R., Kapitaniak, B., & Ringelberg, A. (1996). The implementation of risk assessment models for musculoskeletal disorders in CEN standards. In N. Fallentin, & G. Sjogaard (Eds.), *Proceedings of the Symposium Risk Assessment for Musculoskeletal Disorders* (pp. 73–74). Nordic Satellite Symposium under the auspices of ICOH'96, Copenhagen, Denmark September 13–14, 1996.

Schaub, Kh. (1997). Manual handling of machinery and component parts of machinery. In IEA'97 (Ed.), *From experience to innovation* (Vol. 3, pp. 574–577). The 13th Triennial Congress of the International Ergonomics Association, Tampere, Finland, June 29–July 4, 1997.

Schaub, Kh., & Landau, K. (1998). The EU machinery directive as a source for a new ergonomic tool box for preventive health care and ergonomic workplace and product design. In P. A. Scott, R. S. Bridger, & J. Charteris, (Eds.), Global Ergonomics, Proceedings of the Ergonomics Conference (p. 219–224). Cape Town, South Africa, September 9–11.

Waters, T. R., Baron, S. L., Piacitelli, L. A., Andersen, V. P., Skov, T. Haring-Sweeney, M. Wall, D. K., & Fine, L. J. (1999, February). Evaluation of the revised NIOSH lifting equation, *Spine 24*(4), 386–394. http://europa.eu.int/comm/enterprise/newapproach/standardization/harmstds/reflist/machines.html

Section IV

Standards for Manual Material Handling Tasks

17 Repetitive Actions and Movements of the Upper Limbs

Enrico Occhipinti and Daniela Colombini

CONTENTS

Full Name of the Standards
   EN 1005-5: Safety of machinery—Human physical performance—Part 5: Risk assessment for repetitive handling at high frequency.
   ISO 11228-3: Manual handling Part 3: Handling of low loads at high frequency.

INTRODUCTION

Working tasks that require manual repetitive actions at high frequency may cause the risk of fatigue, discomfort, and musculoskeletal disorders. A proper risk assessment and management should seek to minimize these health effects by considering a variety of risk factors related to the duration of exposure, the frequency of actions, the use of force, the postures and movements of the body segments, the lack of recovery periods, and other additional factors (Colombini et al., 2001).

   To this regard, two parallel standards have been produced in 2007, respectively, by CEN and ISO:

   • EN 1005-5: Safety of machinery—Human physical performance—Part 5: Risk assessment for repetitive handling at high frequency (CEN, 2007).
   • ISO 11228-3: Manual handling Part 3: Handling of low loads at high frequency (ISO 11228-3, 2007)

Though the two mentioned drafts are devoted to different targets, they are conceptually similar and can be presented in the same context.

SCOPE OF THE STANDARDS

EN 1005-5: Risk Assessment for Repetitive Handling at High Frequency

This European standard presents guidance to the designer of machinery or its component parts in controlling health risks due to machine-related repetitive handling at high frequency.

   The standard has been prepared to be a harmonized standard as defined by the U.E. "Machinery Directive" and associated EFTA regulations but, for procedural reasons, has been published to only be an informative voluntary standard.

   It applies mainly to designers and producers of new machinery and assembly lines for professional use operated by the healthy adult working population.

   The machinery designer has to specify reference data for the action frequency of the upper limbs during machinery operation. The standard presents a risk assessment method and gives guidance to the designer on how to reduce health risks for the operator.

ISO 11228-3: Handling of Low Loads at High Frequency

This International Standard establishes ergonomic recommendations for repetitive work tasks involving the handling of low loads at high frequency (by upper limbs).

   The standard provides information for all those involved in the design (or redesign) and risk assessment of manual repetitive tasks and jobs.

   It is designed to provide guidance on several task variables, evaluating the health risks for the working population. It applies to the adult working population; the recommendations will give reasonable protection for nearly all healthy adults.

MAIN DEFINITIONS

WORK TASK

An activity or activities required to achieve an intended outcome of the work system (e.g., stitching of cloth, the loading or unloading of pallets).

REPETITIVE TASK

Task characterized by repeated cycles.

WORK CYCLE

A sequence of (technical) actions that are always repeated the same way.

CYCLE TIME

The time elapsed—from the moment one operator begins a work cycle to the moment that the same work cycle is repeated (in seconds).

TECHNICAL ACTION (MECHANICAL)

Elementary manual actions required to complete the operations within the work cycle, such as holding, turning, pushing, cutting.

REPETITIVENESS

Quality of task when a person is continuously repeating the same cycle, technical actions, and movements in a significant part of a normal workday.

FREQUENCY

The number of technical actions per minute.

FORCE

The physical effort required of the operator to execute the task.

POSTURE AND MOVEMENTS

The positions and movements of the body segment(s) or joint(s) required to execute the task.

RECOVERY TIME

The period of rest following a period of activity that allows restoration of musculoskeletal function (in minutes).

ADDITIONAL RISK FACTORS

Other factors for which there is evidence of a causal or aggravating relationship with work-related musculoskeletal disorders of the upper limbs (e.g., vibration, local pressure, cold).

CONTENTS OF THE STANDARDS

GENERAL RECOMMENDATIONS

Manual repetitive tasks, if unavoidable, should be designed in a way such that they can be performed adequately with respect to the force required, the posture of the limbs, and the foreseeable presence of recovery periods. In addition, tasks and related machines should be designed to allow variation in movements. Additional factors (like vibration, cold, etc.) have to be considered.

Data from recent epidemiological studies on workers exposed to repetitive movements of upper limbs allow those involved in the design or redesign of workplaces, tasks, and jobs to forecast—from exposure indices—the occurrence of the consequent UL-WMSDs (Occhipinti & Colombini, 2004, 2007). The adequate situation occurs when the exposure index corresponds to a forecast of occurrence of WMSDs as observed in a reference working population not exposed to occupational risks for the upper limbs.

RISK ASSESSMENT

When manual repetitive tasks are present, then a risk assessment approach should be adopted. This should follow a four-step approach:

1. hazard identification;
2. risk estimation by simple methods;
3. risk evaluation by detailed methods (if necessary);
4. risk reduction.

The international literature reports that the "frequency of upper limbs action" has been connected to other risk factors like force (the more the force, the lower the frequency), posture (the more the joint excursion, the longer the time necessary to carry out an action), and recovery periods (if well distributed during the shift, they increase the recovery of muscles) (Colombini et al., 2001). The technical action is identified as the specific characteristic variable relevant to repetitive movements of the upper extremities. The technical action is factored in by its relative frequency during a given unit of time.

The hazard identification and simple risk estimation procedures are largely based on different experiences and proposals of the literature (Keyserling et al., 1993; Schneider, 1995; Silverstein, Fine, & Armstrong, 1987; Colombini, Occhipinti, & Grieco, 2002); the detailed risk evaluation procedures are substantially based on the OCRA Index method proposed by the authors (Colombini, Occhipinti, & Grieco, 2002; Occhipinti, 1998).

Due to different scopes and targets, the two mentioned standards have slight differences when presenting specific procedures for risk assessment—that aspect will be separately and synthetically detailed in the following paragraphs.

EN 1005-5

Hazard Identification

The first stage of the risk assessment is to identify whether hazards that may expose individuals to a risk of injury exist. If such hazards are present, then a more detailed risk assessment is necessary.

In *EN 1005-5*, the "no hazard" option (for the designer) is present when machinery and the related task imply no cycles or a cyclic task where perception of cognitive activities are clearly prevalent. For all the machinery/task combinations where cyclic manual activities are foreseen, risk estimation shall be applied. To this end, the designer shall identify and count the technical actions (for each upper limb) needed to carry out the task (NTC); define the foreseeable duration of the cycle time (FCT); consider the foreseeable duration of work and frequency of recovery periods (a general duration of 240–480 minutes of a task during one shift with, at least two breaks of ten minutes, plus the meal break are to be considered); and consider the possibility of rotation on different tasks when designing a machinery in the context of an assembly line.

Risk Estimation by Simple Methods (Method 1)

The presence of acceptable characteristics for all the considered risk factors is verified. When the characteristics described are fully and simultaneously present, it is possible to affirm that exposure to repetitive movements is acceptable. Where one or more of the listed characteristics for the different risk factors are not satisfied, the designer shall use a more detailed evaluation. The acceptable characteristics of the risk factors are listed in Table 17.1. It is to be underlined that the final acceptable frequency of action per minute was set to 40, given that the designer considers a reference organizational scenario (task duration of 240–480 minutes with at least two usual breaks of ten minutes plus meal break during the shift) and not the "best" scenario (almost one break of ten minutes every hour of repetitive work) that should lead to a higher acceptable frequency of actions per minute.

Risk Evaluation by Detailed Method (Method 2)

If one or more of the acceptable conditions reported in the previous step are not satisfied, the designer shall describe more analytically each risk factor that interferes with the frequency of actions. Since different risk factors can be present in different combinations and degrees, it is possible to expect many levels of risk.

The level of risk is assessed with reference to the OCRA method (Colombini, Occhipinti, & Grieco, 2002). The OCRA index, when assessing a single repetitive task in a shift (mono task job), is given by the ratio between the foreseeable frequency (**FF**) of technical actions needed to carry out the task, and the reference frequency (**RF**) of technical actions for each upper limb. This is a

---

TABLE 17.1

List of Acceptable Characteristics of the Risk Factors in EN 1005-5

Absence of force or use of force at the same conditions exposed in EN 1005-3

Absence of awkward postures and movements considering the same conditions exposed in EN 1005-4 as summarized below:

- The upper arm postures and movements are in the range between 0° and 20°.
- The articular movements of the elbow and wrist do not exceed 50% of the maximum articular range.
- The kinds of grasp are "power grip" or "pinch" lasting not more than 1/3 of the cycle time.

Low repetitiveness. This occurs when:

- The cycle time is more than 30 seconds.
- The same kinds of actions are not repeated for more than 50% of the cycle time.

Absence of additional factors (physical and mechanical factors). This occurs when:

- The task should not include hand/arm vibration, shock (such as hammering), localized compression on anatomical structures due to tools, exposure to cold, use of inadequate gloves for grasping, etc.

The frequency of upper limb actions (for each arm) is less than 40 actions/minute.

- To compute the frequency of actions/minute use the following formula:

FF = NTC × 60/FCT

where: FF is the foreseeable frequency of actions per minute; FCT is the foreseeable duration of the cycle time in seconds; NTC is the number of technical actions (for each upper limb) needed to carry out the task.

particular procedure for monotask jobs. For multitask jobs, one can refer to a specific annex (see also OCRA Index in ISO 11228-3).

In this context: **OCRA Index = FF/RF**

The foreseeable frequency (number per minute) of technical actions needed to carry out the task (**FF**) is given by the formula already reported in Table 17.1.

The following formula calculates the reference frequency (number per minute) of technical actions (**RF**) on a work cycle base:

$$RF = CF \times Po_M \times Re_M \times Ad_M \times Fo_M \times (Rc_M \times Du_M)$$

where:

CF = "constant of frequency" of technical actions per minute = 30

Po_M; Re_M; Ad_M; Fo_M = multipliers for the risk factors postures, repetitiveness, additional, force.

Rc_M = multiplier for the risk factor "lack of recovery period"

Du_M = multiplier for the overall duration of repetitive task(s) during a shift.

When designing a machinery-related task, evaluate the reference frequency of the technical actions within a work cycle that is representative of the task under examination. The analyses shall include the main risk factors that the designer can influence with the consequent choice of a specific multiplier for each risk factor. These multipliers will decrease from 1 to 0 as the risk level increases. The risk factors and the corresponding multiplier, influenced by the designer, are:

- Awkward or uncomfortable postures or movements (posture multiplier) (Po_M);
- High repetition of the same movements (repetitiveness multiplier) (Re_M);
- Presence of additional factors (additional multiplier) (Ad_M);
- Frequent or high force exertions (force multiplier) (Fo_M).

The other factors considered in the formula ($Rc_M \times Du_M$) are generally out of the direct influence of the designer and, consequently, they will be considered in this context as a constant, reflecting a common condition of repetitive task duration of 240–480 minutes/shift with two breaks of ten minutes each plus the meal break. If other "daily repetitive task duration" and/or "breaks or recovery periods" scenarios are foreseen (less duration; more recovery periods), reference action frequency can be higher: special tables are provided for this in an annex.

In practice, to determine the reference frequency (per minute) of technical actions (**RF**), proceed as follows:

- start from CF (30 actions/minute);
- CF (the frequency constant) must be weighted (by the respective multipliers) considering the presence and degree of the following risk factors: force (**Fo$_M$**), posture (**Po$_M$**), repetitiveness (**Re$_M$**), and additional (**Ad$_M$**);
- apply the constant that considers the multiplier for repetitive task duration (**Du$_M$**) and the multiplier for recovery periods (**Rc$_M$**);
- the value obtained represents the reference frequency (per minute) of technical actions (**RF**) for the examined task in the common condition of at least two breaks of ten minutes (plus the lunch break) each, in a shift maximum of 480 minutes.

Posture Multiplier (Po$_M$)

If the conditions described in method 1 for posture are present, the multiplier factor is 1. If those conditions are not present, use the indications in Table 17.2 to obtain the specific multiplier:

At the end of the analysis of awkward postures, choose the lowest multiplier Po_M (that corresponds to the worst condition) between the posture and movements of the elbow, wrist, and hand (type of grip).

The designer, at this step, shall also consider shoulder postures and movements.

To this end, the designer shall check that:

- the conditions in ISO EN 14738 and EN 1005-4 are satisfied.
- the arms are not held or moved at about shoulder level for more than 10% of cycle time (Punnett et al., 2000).

If one of those two conditions occurs, a risk of shoulder disorders exists and should be accurately considered. However, at the moment, the standard was prepared, there were no available data for identifying a specific Po_M for shoulders; consequently, Po_M for shoulders was not included in EN 1005-5. This problem has been partially solved when preparing the ISO TR 12295 (ISO TR 12295, 2014), and the Po_M for the shoulder was considered (Table 17.3).

TABLE 17.2
Multiplier for Awkward Postures (Po_M)

Awkward Posture	Portion of the Cycle Time			
	Less Than 1/3 From 1% to 24%	1/3 From 25% to 50%	2/3 From 51% to 80%	3/3 More Than 80%
ELBOW supination (≥ 60°)	1	0.7	0.6	0.5
WRIST extension (≥ 45 °) or flexion (≥ 45°)				
HAND pinch or hook grip or palmar grip (wide span)				
Elbow pronation (≥ 60°) or flexion/ extension (≥ 60°)	1	1	0.7	0.6
WRIST radio-ulnar deviation (≥ 20°)				
HAND power grip with narrow span (≤ 2 cm)				

TABLE 17.3
Multiplier for Awkward Postures (Po_M) of Shoulders

Shoulder flexion/abduction (upper arm elevation) more than 80°					
Percentage of the cycle time	10%	20%	30%	40%	≥50%
Posture multiplier (P_M)	0.7	0.6	0.5	0.33	0.07
Shoulder in mild elevation (flexion or abduction between 45° and 80° or extension >20°)					
Percentage of the cycle time	1/3 from 25% to 50%		2/3 from 51% to 80%		3/3 more than 80%
Posture multiplier (P_M)	0.7		0.6		0.5

From ISO TR 12295-2014.

Repetitiveness Multiplier (Re$_M$)

When the task requires the performance of the same technical actions of the upper limbs for at least 50% of the cycle time—or when the cycle time is shorter than 15 seconds—the corresponding multiplier factor (Re$_\mathbf{M}$) is 0.7. Otherwise, Re$_\mathbf{M}$ is equal to 1.

Additional Multiplier (Ad$_M$)

The main additional factors are (non-exhaustive list): the use of vibrating tools, gestures implying countershock (such as hammering), the requirement for absolute accuracy, the localized compression of anatomical structures, exposure to cold, use of gloves interfering with handling ability, and high pace completely determined by the machinery. If additional factors are absent for most of the task duration, the multiplier factor equals 1. Otherwise, the additional factor multiplier Ad$_\mathbf{M}$ equals:

- 1 if one or more additional factors are present for less than 25% of the cycle time.
- 0.95 if one or more additional factors are present for 1/3 (from 25% to 50%) of the cycle time.
- 0.90 if one or more additional factors are present for 2/3 (from 51% to 80%) of the cycle time.
- 0.80 if one or more additional factors are present for 3/3 (more than 80%) of the cycle time.

Force Multiplier (Fo$_M$)

If the criteria described in Method 1 are satisfied, the multiplier is 1. If these conditions are not met, use Table 17.4 to determine the force multiplier (Fo$_\mathbf{M}$) that applies to the average level of force as a function of time.

The force level (upper row) is given as a percentage of the Maximal Isometric Force (F_b) as determined in EN 1005-3 (step A). As an alternative, a value derived from the application of the CR-10 Borg-scale can be used (second row) (Borg, 1998). Use $Fo_\mathbf{M} = 0,01$ when the technical actions require "peaks" above 50% of F_b, or when a score of 5 (or more) in the CR-10 Borg scale for almost 10% of the cycle time is present. The values in the Table 17.4 can be interpolated if intermediate results are obtained.

Predetermined Value (Constant) for the Repetitive Task Duration Multiplier (Du$_M$) and the Multiplier for Recovery Periods (Rc$_M$)

Since the multipliers (Du$_\mathbf{M}$ and Rc$_\mathbf{M}$) considered in the formula are generally out of the direct influence of the designer, they are considered here as unique constants reflecting a common condition as:

Du$_\mathbf{M}$ = 1 (multiplier for overall repetitive task duration of 240–480 minutes)

Rc$_\mathbf{M}$ = 0.6 (for a foreseeable presence of two breaks of ten minutes and a lunch break in a repetitive task duration of 240–480 minutes per shift). Therefore: **(Rc$_\mathbf{M}$× Du$_\mathbf{M}$) = 0.6.**

TABLE 17.4

Multiplier Relative to the Different Use of Force (Fo$_M$)

Force Level in % of F_b	5	10	20	30	40	≥50
CR-10 Borg Score	0.5 very, very weak	1 very weak	2 weak	3 moderate	4 somewhat strong	≥5 strong/very strong
Force multiplier (Fo$_\mathbf{M}$)	1	0.85	0.65	0.35	0.2	0.01

Final Evaluation by Method 2 and Criteria for Risk Reduction

For jobs with a single repetitive task, the OCRA Index is obtained by comparing for each upper limb the foreseeable frequency (FF) of technical actions needed to carry out the repetitive task and the reference frequency (RF) of technical actions, as previously calculated.

Table 17.5 supplies the relevant values of the OCRA Index to assess the risk in relation to the three-zone rating system (green, yellow, red) and to decide for consequent actions to be taken.

The criteria of Table 17.5 were defined in relation to the available literature regarding the occurrence of upper limb WMSDs in working populations not exposed to repetitive movements of the upper limbs; both the association between OCRA Index and prevalence of persons are affected (PA) by (one or more) UL-WMSDs. Details about the procedure used for identifying the critical values of the OCRA Index are given in a specific annex of the standard. In synthesis based by the authors (Occhipinti & Colombini, 2004, 2007), the association between the OCRA Index (independent variable) and the prevalence of Persons Affected (PA) by one or more UL-WMSDs (dependent variable) can be summarized by the following simple regression linear equation:

$$PA = 2.39(\pm 0.14) \times OCRA.$$

On the other side, by using the PA variable in a not exposed population, reference limits were established starting from the 95th percentile (PA=4.8%) as the "driver value" for the so-called green limit, and from twice the 50th percentile (PA=7.4%) as the "driver value" for the so-called red limit. Those "driver" values of PA expected in a reference working population (not exposed) have been compared with the regression equation at the level corresponding to the 5th percentile. In such a way, by adopting a prudential criterion of assessment of not acceptable (yellow) or at risk (red) results, it was possible to find the OCRA values corresponding respectively to the green and red limits and discriminate the green, yellow, and red areas as reported in Table 17.5.

In practice:

- The green limit means that, just above that level, in the exposed working population area forecasted, the PA values higher than the 95th percentile (PA = 4.8%) expected in the reference (not exposed) population constitute almost 95% of the cases.
- The red limit means that, just above that level, in the exposed working population area forecasted, the PA values higher than twice the 50th percentile expected in the reference (not exposed) population constitute almost 95% of the cases,

Annexes

The EN 1005-5 standard is completed by the following annexes:

- Annex A (informative): Identification of technical action
- Annex B (informative): Posture and types of movements
- Annex C (informative): Force

TABLE 17.5

Classification of OCRA Index Results for Evaluation Purposes

OCRA Risk Index	Zone	Risk Evaluation
≤2.2	Green	Acceptable
2.3–3.5	Yellow	Conditionally acceptable
>3.5	Red	Not acceptable

- Annex D (informative): Association between the OCRA index and the occurrence of upper limbs WMSDs: criteria for the classification of results and forecast models
- Annex E (informative): Influence of recovery period and work time duration in determining the overall number of reference technical actions within a shift (RTA) and, consequently, the OCRA index.
- Annex F (informative): An application example of risk reduction in a mono-task analysis.
- Annex G (informative): Definition and quantification of additional risk factors.
- Annex H (informative): Risk assessment by Method 2 when designing "multitask" jobs.

ISO 11228-3

ISO 11228-3 is explicitly based on a consensus document regarding the exposure assessment of upper limb repetitive movements prepared by the IEA TC on MSD (Colombini et al., 2001)

Hazard Identification

In determining if a hazard is present, attention should be given to the following factors: repetition; posture and movement; force; duration and insufficient recovery; and additional risk factors (object characteristics, vibration, impact forces, environment, work organization, psychosocial factors). For each factor, a brief statement explains when it is to be considered as a hazard.

It is to be underlined that ISO TR 12295 (ISO TR 12295, 2014) has recently established so-called "Key Enters" to decide when it is advisable to use the standard of ISO 11228 series.

The key enter for ISO 11228-3 is reported in Table 17.6.

If the answer to the key enter is *YES*, it is suggested to use the standard, eventually performing a "quick assessment" along with the indications of the TR or directly using the simple risk assessment methods in ISO 11228-3.

Simple Risk Assessment

When performing a simple assessment of monotask jobs, the standards primarily use a specific checklist and evaluation model given in an annex.

There are four parts in the estimation procedure: Part A) preliminary information describing the job task; Part B) hazard identification and risk estimation checklist; Part C) overall evaluation of the risk; and Part D) remedial action to be taken.

The checklist adopts a six-step approach considering the four primary physical risk factors (repetition, high force, awkward posture and movements, insufficient recovery) as well as any other additional risk factors that may be present. Initial consideration is given to the prevalence of

---

TABLE 17.6
Key Enter for the Application of ISO 11228-3

Application of ISO 11228-3

Are there one or more repetitive tasks of the upper limbs with a total duration of one hour or more per shift?   NO   YES

Where the definition of "repetitive task" is:

task characterized by repeated work cycles

or

tasks during which the same working actions are repeated for more than 50% of the time.

If NO, then this standard is not relevant. Go to the next Key Question regarding the other standards.

If YES, then go to step 2 of Quick Assessment.

From ISO TR 12295.

work-related health complaints and/or work changes that may have been implemented by the operator/supervisor.

As an alternative, other simple risk assessment methods could be adopted and they are reported in a specific annex; among them, it is worth recalling the tools OCRA Checklist, QEC, PLIBEL, and others.

As a result of the overall classification of risk, the following action should be taken:

GREEN ZONE: No action required.
YELLOW ZONE: The risk shall be further estimated, analyzed together with contributing risk factors, and followed as soon as possible by redesign. Where redesign is not possible other measures to control the risk shall be taken.
RED ZONE: The work could be harmful. It is advisable to evaluate the task more accurately using Method 2. Action to lower the risk (e.g., redesign, work organization, worker instruction and training) is necessary.

Risk Evaluation by Detailed Method

If the risk assessed by Method 1 is in the "yellow" or "red" zone, it is recommended to perform a more detailed risk assessment also for a better choice and follow-up of the remedial measures to be taken. If the job is composed of two or more repetitive tasks (multitask job), it is recommended to use a detailed method.

For detailed risk assessment, OCRA (occupational repetitive action) is the preferred method. It is recommended for the specific purposes of this part of ISO 11228 because, given the knowledge at the time of publication, it considers all the relevant risk factors, is also applicable to "multitask jobs," and provides criteria—based on extensive epidemiological data—in forecasting the occurrence of UL-WMSD in exposed working populations.

Other detailed risk assessment methods can be used, as a secondary choice, for a detailed risk assessment—they are the methods of the STRAIN INDEX and of the HAL/ACGIH TLV for monotask handwork (both methods are shortly reported in a specific annex).

The risk evaluation performed by using the "preferred" OCRA method adopts just the same procedures presented for EN 1005-5, the only relevant difference is that here—as in the original OCRA method—the Actual number of Technical Actions carried out during the work shift (ATA) and the Reference number of Technical Actions (RTA) (for each upper limb) are directly computed by considering the multipliers for "daily duration of repetitive work" and "recovery periods." In practice, the OCRA Index is given by the formula:

$$\textbf{OCRA Index} = \frac{\text{number of technical actions actually carried out in the shift}(\textbf{ATA})}{\text{number of reference technical actions in the shift}(\textbf{RTA})}$$

The overall actual number of technical actions carried out within the shift (ATA) can be calculated by multiplying F_j for the net duration of each repetitive task/s analyzed and summing the results of each repetitive task.

$$\textbf{ATA} = \textbf{S}(\textbf{F}_j \times \textbf{D}_j)$$

Where:

D_j is the net duration (in minutes) of the task j.
F_j is the frequency of actions per minute of task j.

The following general formula calculates the overall number of **reference** technical actions within a shift (RTA):

$$_{J=1}^{n}\textbf{RTA} = \sum [\textbf{CF} \times (\textbf{Fo}_{Mj} \times \textbf{Po}_{Mj} \times \textbf{Re}_{Mj} \times \textbf{Ad}_{Mj}) \times \textbf{D}_j] \times (\textbf{Rc}_M \times \textbf{Du}_M)$$

Where:

n = number of repetitive task/s performed during the shift.

j = generic repetitive task.

CF = "constant of frequency" of technical actions per minute = 30.

Fo_{Mj}; Po_{Mj}; Re_{Mj}; Ad_{Mj} = multipliers for the risk factors force, postures, repetitiveness, additional in each j repetitive task.

Dj = net duration (in minutes) of the repetitive task j/

Rc_M = multiplier for the risk factor "lack of recovery period".

Du_M = multiplier according to the overall duration of all repetitive tasks during a shift.

All the multiplier factors are identical to those given in EN 1005-5—except for the Rc_M and Du_M multipliers that are determined by the criteria given in the followings and detailed in Tables 17.7 and 17.8.

Recovery Period Multiplier (RcM)

A recovery period is when one or more muscle-tendon groups are basically at rest.

The following can be considered as recovery periods:

1. breaks (official or non-official) including the lunch break;
2. visual control tasks;
3. periods within the cycle that leave muscle groups totally at rest consecutively for at least ten seconds almost every few minutes.

For repetitive tasks, the reference condition is represented by the presence for each hour of repetitive task of work breaks of at least 8–10 minutes, consecutively or—for working periods lasting less than one hour—in a ratio of 5:1 between work time and recovery time.

In relation to these reference criteria, it is possible to consider how many hours, during the work shift, do not have an adequate recovery period. It requires the observation, one by one, of the single hours that make up a working shift: for each hour, a check must be made if there are repetitive tasks and if there are adequate recovery periods. Based on the presence or absence of adequate

TABLE 17.7

Elements for the Determination of the Recovery Period Multiplier (Rc_M)

Number of hours without adequate recovery	0	1	2	3	4	5	6	7	8
Multiplier Rc_M	1	0.90	0.80	0.70	0.60	0.45	0.25	0.10	0

TABLE 17.8

Elements for the Determination of the Duration Multiplier (Du_M)

Time (in minutes) devoted to repetitive task(s) during a shift	<121	121–180	181–240	241–300	301–360	361–420	421–480	>480
Duration multiplier (Du_M)	2.0	1.7	1.5	1.3	1.2	1.1	1.0	0.5

recovery periods within every hour of repetitive work, the number of hours with "no recovery" is counted. Consequently, it is possible to determine the Rc_M multiplier according to Table 17.7.

Overall Duration of Manual Repetitive Tasks and Duration Multiplier (Du_M)

Within a working shift, the overall duration of manual repetitive tasks is important to determine the overall risk for upper limbs. When one or more repetitive manual tasks last for a relevant part (7–8 hours) of the shift, the Du_M is equal to 1. As the overall duration decreases (e.g., six hours), the duration multiplier (Du_M) increases, allowing more RTA. In other contexts, however, there may be differences with respect to these scenarios (e.g., regularly working over-time, part-time work, repetitive manual tasks for only a part of a shift); the multiplier (Du_M) considers these changes with respect to usual exposure conditions. Table 17.8 gives the values of Du_M in relation to the overall daily duration of manual repetitive tasks.

Risk Index Calculation and Risk Evaluation

The OCRA Index is obtained by comparing, for each upper limb, the actual number of technical actions carried out during the work shift (ATA) and the reference number of technical actions (RTA).

The risk classification criteria (green, yellow, red) are identical to the one given for EN 1005-5 and reported in Table 17.5.

Risk Reduction

A proper risk assessment is the basis for appropriate choices in risk reduction.

Risk reduction can be achieved by combining, in different ways, improvements in different risk factors and should consider, among other things:

- the avoidance and limitation of repetitive handling, especially for long daily durations without proper recovery periods or at high frequencies,
- proper design of the task, workplaces, and work organization using existing international standards and introducing adequate task variation,
- proper design of the objects, tools, and materials handled,
- proper design of the work environment, and
- individual workers' capacities and level of skill for the specific task.

A specific annex gives more detailed information about risk reduction options.

Annexes

The ISO 11228-3 standard is completed by the following rich annexes:

- Annex A (informative): Risk assessment—General framework and information on available methods
- Annex B (informative): Method 1—Simple risk assessment checklist
- Annex C (informative): Method 2—OCRA method for detailed risk assessment (also includes suggestions for multiple repetitive tasks analysis)
- Annex D (informative): Other methods for detailed risk assessment
- Annex E (informative): Risk reduction

REFERENCES

Borg G. A. V. (1998). *Borg's perceived exertion and pain scales*, Human Kinetics Europe.
CEN. (2007). *EN 1005-5: Safety of machinery—Human physical performance—Part 5: Risk assessment for repetitive handling at high frequency.*

Colombini D., Occhipinti E., Delleman N., Fallentin N., Kilbom A., & Grieco A. (2001) Exposure assessment of upper limb repetitive movements: a Consensus Document. *In International encyclopaedia of ergonomics and human factors* (pp 52–66), Ed. W. Karwowski, Taylor & Francis, London and New York.

Colombini D., Occhipinti E., & Grieco A. (2002). *Risk assessment and management of repetitive movements and exertions of upper limbs: Job analysis, OCRA risk index, prevention strategies and design principles.* Ergonomics book series. Elsevier. Vol. 2.

Keyserling W.M., Stetson D.S., Silverstein B., & Brower M.L. (1993). A check list for evaluating ergonomic risk factors associated with upper extremity cumulative trauma disorders. *Ergonomics 36*, 807–831.

ISO 11228-3 (2007). Manual handling. Part 3: Handling of low loads at high frequency.

ISO TR 12295 (2014). *Ergonomics—Application document for ISO standards on manual handling (ISO 11228-1, ISO 11228-2 and ISO 11228-3) and evaluation of static working postures (ISO 11226).*

Occhipinti E. (1998). OCRA, a concise index for the assessment of exposure to repetitive movements of the upper limbs. Ergonomics, *41* (9), 1290–1331.

Occhipinti E., & Colombini D. (2004). Metodo OCRA: Aggiornamento dei valori di riferimento e dei modelli di previsione dell'occorrenza di UL-WMSDs nelle popolazioni lavorative esposte a movimenti e sforzi ripetuti degli arti superiori. La *Medicina del Lavoro, 95–4*, 305–319.

Occhipinti, E., & Colombini, D. (2007). Updating reference values and predictive models of the OCRA method in the risk assessment of work-related musculoskeletal disorders of the upper limbs. *Ergonomics, 50* (11), 1727–1739.

Punnett L., Fine L. J., Keyserling W. M., & Chaffin D. B. (2000). Shoulder disorders and postural stress in automobile assembly work. Scandinavian Journal of Work, Environment and Health, *26*(4), 283–291.

Schneider S. (1995). OSHA's Draft standard for prevention of work-related Musculoskeletal Disorders. *Applied Occupational and Environmental Hygiene, 10*(8), 665–674.

Silverstein B. A., Fine L. J., Armstrong T. J. (1987). Occupational factors and carpal tunnel syndrome. *American Journal of Industrial Medicine, 11*, 343–358.

18 Ergonomics of Manual Handling—Part 1
Lifting and Carrying

Karlheinz G. Schaub

CONTENTS

INTRODUCTION

ERGONOMIC AND ECONOMIC ASPECTS OF MANUAL MATERIALS HANDLING

Disorders of the musculoskeletal system are common worldwide. They play an important role in occupational health and are one of the most frequent disorders. In a world characterized by the globalization of national economies, it is desirable to specify recommended limits for manual materials handling activities on an internationally accepted basis. This helps improve and maintain healthy and safe working conditions that enhance quality and productivity and reduces sick leave and personal suffering from work-related muscular disorders. Harmonized international regulations also help reduce advantages in economic competition based on poor and hazardous ergonomic design.

    Due to the progress in ergonomic knowledge, a comprehensive risk assessment should be used to derive recommended limits for manual materials handling, and attention should be paid to the mass of objects in combination with working postures, frequency, and duration of manual handling that persons may be reasonably expected to exert when carrying out activities associated to manual handling. An ergonomic approach has a significant impact on reducing the risks of lifting and carrying. Of particular relevance is a good design of the work, especially the tasks and the workplace, which may include the use of appropriate aids.

RELATION TO OTHER STANDARDS, REGULATIONS, AND GUIDELINES

Manual materials handling is one of the classical topics in the ergonomics of physical workload. Many national, regional (e.g., European http://europe.eu.int/eur-lex/), and international (e.g., ILO) regulations and guidelines exist in that field.

In the 1970s, the first multifactorial methods for the calculation of load limits for the manual handling of materials were created in Germany (Schaub & Landau, 1997).

In the United States, the NIOSH work practices guide and the corresponding lifting equation was published in 1981 and revised in 1991. Currently, the NIOSH lifting-equation is being used in many other countries to derive limitations for manual materials handling tasks.

In Europe, manual materials handling is regulated by the "Council Directive 90/269/EEC of 29 May 1990 on the minimum health and safety requirements for the manual handling of loads where there is a risk particularly of back injury to workers." As this directive offers a good approach for a risk assessment of manual-handling activities, it served as a basis for the development of ISO 11228-1 and is described in detail later.

Next to this directive, manual materials handling is addressed in Europe by the European Union (EU) Machinery Directive (98/37/EC) that addresses designers and manufacturers of machinery, namely, EN 1005, a set of harmonized CEN standards that supports this directive.

LIFTING AND CARRYING

Part 1 of ISO 11778 provides a step-by-step approach to estimating the health risks of manual lifting and carrying; at each step, recommended limits are proposed. In addition, practical guidance for manual handling is given in the annexes.

The risk-assessment model presented estimates the risk associated with a manual material handling task. It also takes into consideration the hazards (unfavorable conditions) related to the manual lifting and the time spent with manual handling activities. Unfavorable conditions could be high masses to be manipulated or awkward postures required during the lifting process, such as twisted or bent trunks or a far reach. This Standard provides information on both repetitive and nonrepetitive lifting.

SUBSIDIARY STANDARDS

Guidance to the work on ISO 11228-1 was given by several ISO and CEN Standards mentioned in the following and listed as normative references:

- ISO 7730, Moderate Thermal Environments—Determination of the PMV (predicted mean vote) and PPD (predicted percentage of dissatisfied) Indices and Specification of the Conditions for Thermal Comfort
- ISO 11226, Ergonomics—Evaluation of Working Postures
- ISO 14121, Safety of Machinery—Principles of Risk Assessment
- ISO/IEC Guide 51, Safety Aspects—Guidelines for Their Inclusion in Standards
- EN 614-1, Safety of Machinery—Ergonomic Design Principles—Part 1: Terminology and General Principles
- EN 614-2, Safety of Machinery—Ergonomic Design Principles—Part 2: Interactions Between the Design of Machinery and Work Tasks

- EN 1005-2, Safety of Machinery—Human Physical Performance—Part 2: Manual Handling of Machinery and Component Parts of Machinery
- EN-ISO 7250, Basic List of Definitions of Human Body Measurements for Technical Design

STRUCTURE

ISO1 1228-1 consists—next to the Foreword and Introduction—of four main chapters and four annexes (including bibliography). Most of them are required and predefined in drafting rules.

Chapter 3 offers definitions that apply within this Standard.

Chapter 4, "Recommendations," represents the core of the Standard and contains the following subheadings:

- Ergonomic Approach
- Risk Assessment
- Risk Estimation and Risk Evaluation
- Manual Lifting
- Cumulative Mass of Manual Lifting and Carrying
- Risk Reduction
- Additional Considerations

Annexes A–C offer detailed supplementary information as the main part is kept relatively short.

ISO 11228 "Ergonomics—Manual handling" consists of the following parts:

- Part 1: Lifting and Carrying
- Part 2: Pushing and Pulling
- Part 3: Handling of Low Loads at High Frequency

INTRODUCTION AND SCOPE

The introduction describes general aspects of manual materials handling and the associated risk of injury for the musculoskeletal system. An ergonomic approach that includes guidelines for a good design of work is implemented in this standard and aims to have a substantial impact on reducing the risk of lifting and carrying. Recommended limits in the standard consider the intensity, the frequency, and the duration of tasks and are derived from four major research approaches: epidemiological, biomechanical, physiological, and psychophysical. The Standard provides information for designers, employers, employees, and other persons engaged in work, job, and product design.

The Standard applies to the manual handling of objects of 3 kg or more in vocational and nonoccupational activities. Lower weights are dealt with in Part 3 of this Standard. The Standards provide information for designers, employers, employees, and others involved in work, job, and product design. The Standard applies to an eight-hour working day and to moderate walking speed on horizontal level surfaces.

The Standard does not cover the holding and pushing or pulling of objects. One-handed lifting and lifting by two or more people as well as hand-held machines and manual materials handling while seated are excluded as well.

ERGONOMIC APPROACH

To minimize the health risks emerging from manual handling, it should be avoided whenever possible. Where this is not feasible, a risk assessment should ensure that health risks are on an acceptable low level. Guidance for the ergonomic design of manual materials handling tasks is given in Annex A of this standard. Other considerations for the ergonomic design of manual materials handling tasks are mentioned in Chapter 4, "The EU Manual Handling Directive."

RISK ASSESSMENT, RISK ESTIMATION, AND RISK EVALUATION

In general, a risk assessment consists of five stages: peril recognition, hazard identification, risk estimation, risk evaluation, and risk reduction. For further information, reference is made to ISO 14121, EN 1050, and ISO/IEC Guide 51. (For information about hazard identification, refer to Annex A of the Standard.) The risk assessment presented in the following takes into consideration the mass, grip, and position of the object, as well as the frequency and duration of the handling task.

The risk assessment consists of a step-by-step approach for the evaluation of manual lifting and carrying. When recommended limits are exceeded, the tasks analyzed should not be carried out manually, but adopted in a way that the risk assessment model will be satisfied (see Figure 18.1).

Providing information and training for employees should not be used as the only measure to ensure a safe way of manual handling.

Risk estimation and evaluation is presented as a step-by-step model, consisting of five steps. When the questions asked in each step have to be answered with "no," adaptation of the manual handling task is necessary. (For appropriate actions refer to Annex A of the Standard.)

Steps 1–3 apply to the limits while lifting; Steps 4 and 5 consider the carrying and are described in detail in Figure 18.2.

Step 1

As an initial screening for nonrepetitive lifting tasks under ideal conditions (see Table 18.1), the mass of the object to be manipulated should be considered in relation to the reference mass as described in Table 18.2. This table is taken from EN 1005-2. If the mass to be manipulated exceeds the reference mass, adaptation is necessary.

To estimate the influence of an unfavorable posture, use the risk assessment model equation in step 3 with a frequency multiplier of "1." The horizontal multiplier will indicate the severity of a possible far reach. Vertical distance and asymmetry multipliers will show the negative influence of a twisted or bent trunk.

Step 2

In any case—but especially for repetitive lifting tasks—the mass of the object must be considered in combination with the lifting frequency as described in Figure 18.3. This figure offers limitations for mass—frequency combinations and lifting durations of less than one hour per day or 1–2 hours per day, respectively. The absolute maximal frequency for lifting is 15 lifts per minute. In this case, the total duration of liftings shall not exceed one hour per day, and the object mass shall not exceed 7 kg.

For repetitive manual lifting under ideal conditions, lifting durations of ≤2 hours per day, the risk assessment for lifting has been finished successfully. Otherwise, step 3 must be carried out. In case of the absence of carrying, the risk assessment may stop here (Table 18.3).

Step 3

The risk assessment to be carried out in this step is presented in Table 18.4 (see also ISO 11228-1, Annex A.6.2). The risk assessment described follows widely the NIOSH '91 approach, except that the NIOSH "mass constant" is replaced by the variable "reference mass" described in Table 18.2. It also considers other methods (Schaub & Landau, 1997).

The recommended limits are derived from a risk assessment model with the following assumptions listed in Table 18.4.

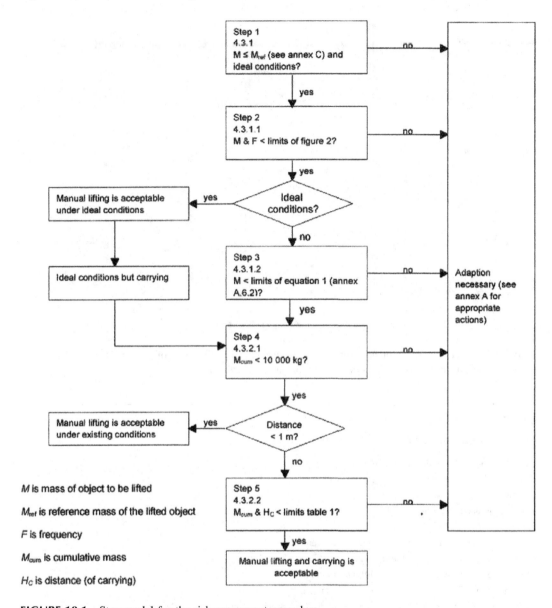

FIGURE 18.1 Step model for the risk assessment procedure.

The primary task variables include the following data (see Figure 18.3):

- Object mass (M) in kg
- Horizontal distance (H) in meters (m) measured from the midpoint of the line joining the ankles
- To the midpoint at which the hands grasp the object while in the lifting position
- Vertical location (V) in meters (m) determined by measuring the distance from the floor to the point at which the hands grasp the object
- Vertical travel displacement (D) in meters (m) from origin to the destination of lift
- Frequency of lifting (F) is the average number of lifts per minute (min)
- Duration of manual lifting in hours (h)
- Angle of asymmetry (A) in degrees (in Figure 18.4 as α)
- Quality of gripping (C)

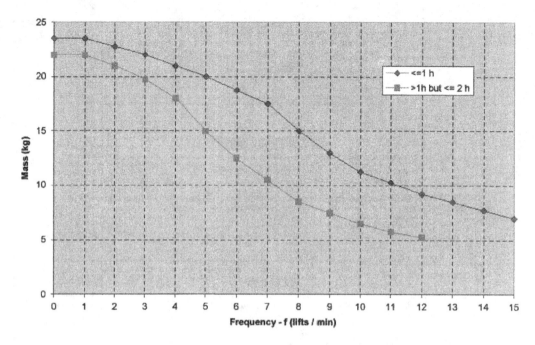

FIGURE 18.2 Maximum frequency for manual lifting related to the mass of the object under ideal conditions for two different lifting durations, in correspondence with Table 18.3.

TABLE 18.1
"Ideal" Condition for Manual Handling

Moderate ambient thermal environment

Two-handed operation only

Unrestricted standing posture

Handling by one person only

Smooth lifting

Good coupling between the hands and the objects handled

Good coupling between the feet and the floor

Manual handling activities, other than lifting, are minimal

The objects to be lifted are not cold, hot, or contaminated

Vertical displacement of the load is ≤0.25 m and does not occur below or above the knuckle

Shoulder height

Trunk upright and not rotated

Load kept close to the body

The limit for the mass of the object is derived using the following equation:

$$M \leq M_{ref} \times H_M \times V_M \times D_M \times A_M \times F_M \times C_M, \tag{18.1}$$

where M_{ref}[1] is the reference mass, H_M is the horizontal distance multiplier, derived from Equation (18.2); V_M is the vertical location multiplier, derived from Equation (18.3); D_M is the vertical displacement multiplier, derived from Equation (18.4); A_M is the asymmetry multiplier, derived

TABLE 18.2

Reference Mass (M_{ref}) Taking into Consideration Different Populations

Field of Application	M_{ref} (kg)	F and M	Females	Males	Population Group	
		Percentage of				
Domestic Use[a]	5	Data not available			Children and the elderly	Total population
	10	99	99	99	General domestic population	
Professional Use (General)[b]	15	95	90	99	General working population including the young and old	General working population
	25	85	70	90	Adult working population	
Professional Use (Exceptional)[c]	30	Data not available			Special working population	Special working population
	35					
	40					

Notes

[a] When designing a machine for domestic use, 10 kg should be used as a general reference mass in the risk assessment. If children and the elderly are included in the intended user population, the reference mass should be lowered to 5 kg.

[b] When designing a machine for professional use, a reference mass of 25 kg should not be exceeded in general.

[c] Although every effort should be made to avoid manual handling activities or reduce the risks to the lowest possible level, there may be exceptional circumstances where the reference mass might exceed 25 kg (e.g., where technological developments or interventions are not sufficiently advanced). Under these special conditions, other measures have to be taken to control the risk according to EN 614 (e.g., technical aids, instructions, and special training for the intended operator group).

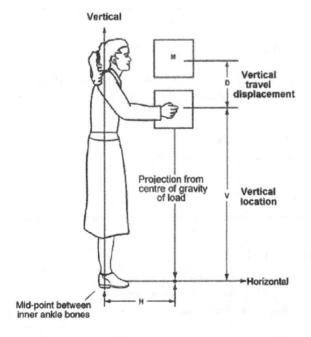

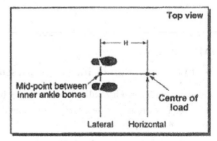

FIGURE 18.3 Task variables.

TABLE 18.3

Frequency Multiplier (F_M) of Equation (18.1)[a]

Frequency (Lifts/min)	Continuous, Repetitive Lifting Task Duration					
	≤1 hour		1 but <2 hours		>2 but <8 hours	
	$V < 0.75$ m	$V \geq 0.75$ m	$V < 0.75$ m	$V \geq 0.75$ m	$V < 0.75$ m	$V \geq 0.75$ m
≥0.2	1.00	1.00	0.95	0.95	0.85	0.85
0.5	0.97	0.97	0.92	0.92	0.81	0.81
1	0.94	0.94	0.88	0.88	0.75	0.75
2	0.91	0.91	0.84	0.84	0.65	0.65
3	0.88	0.88	0.79	0.79	0.55	0.55
4	0.84	0.84	0.72	0.72	0.45	0.45
5	0.80	0.80	0.60	0.60	0.35	0.35
6	0.75	0.75	0.50	0.50	0.27	0.27
7	0.70	0.70	0.42	0.42	0.22	0.22
8	0.60	0.60	0.35	0.35	0.18	0.18
9	0.52	0.52	0.30	0.30	0.00	0.15
10	0.45	0.45	0.26	0.26	0.00	0.13
11	0.41	0.41	0.00	0.23	0.00	0.00
12	0.37	0.37	0.00	0.21	0.00	0.00
13	0.00	0.34	0.00	0.00	0.00	0.00
14	0.00	0.31	0.00	0.00	0.00	0.00
15	0.00	0.28	0.00	0.00	0.00	0.00
>15	0.00	0.00	0.00	0.00	0.00	0.00

[a] If $V < 0.75$ m, $F_M = 0.00$.

from Equation (18.5); E_M is the frequency multiplier (see Table 18.3); and C_M is the coupling multiplier for quality of gripping.

The multipliers for Equation (18.1)(18.1) are obtained from Equations (18.2) to (18.5) and Table 18.3 to Table 18.6. If such a multiplier exceeds a value of 1, its value should be taken as 1.

$$H_M = 0.25/H \quad \text{If H} \leq 0.25, \text{ then } H_M = 1$$
$$\text{If H} > 0.63, \text{ then } H_M = 0 \tag{18.2}$$

TABLE 18.4

Assumptions Made in the Risk-Assessment Model

Are only valid for two-handed, smooth lifting with no sudden acceleration effects (i.e. jerking)

Cannot be used for tasks where the worker is partly supported (e.g., one foot not on the floor)

Width of the object 0.75 m or less for populations with smaller statues (body height)

Are only valid for unrestricted lifting postures

Are only valid when good coupling exists (i.e. handholds are secure and shoe/floor slip potential is low)

Are only valid under favorable conditions

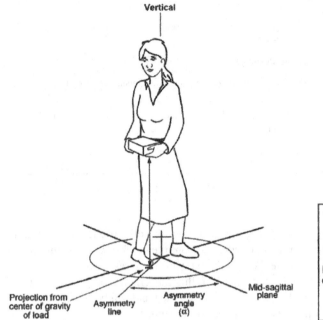

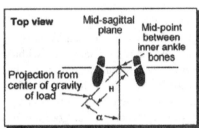

FIGURE 18.4 Angle of asymmetry.

$$V_M = 1 - 0.3 \times 10.75 - V| \quad \text{If } V > 1.75, \text{ then } V_M = 0$$
$$\text{If } V < 0, \text{ then } \quad V_M = 0 \tag{18.3}$$

$$D_M = 0.82 + 0.045/D \quad \text{If } D > 1.75, \text{ then } D_M = 0$$
$$\text{If } D < 0.25, \text{ then } D_M = 1 \tag{18.4}$$

$$A_M = 1 - 0.0032 \times A \quad \text{If } A > 135°, \text{ then } A_M = 0 \tag{18.5}$$

The equation has to be calculated for both the start and endpoint of each task. Endpoint calculations may only be of importance if there is a definite precision placement involved. If the item is thrown into place without undue stress on the body in the extended position, then calculating the endpoint value is not necessary.

The appropriate frequency multiplier, F_M, is determined by first considering the continuous duration of the repetitive lifting task and then the duration of the rest period that immediately follows the repetitive lifting task.

The categories of continuous, repetitive lifting tasks, their durations, and the required duration of the rest period that is to immediately follow the lifting task are provided in Table 18.5.

It is critical to note that the combination of the work period and the rest period must be jointly considered to be a work-rest cycle, wherein the rest period provides sufficient opportunity for the worker to recover following a continuous period of lifting-related work. Accordingly, if two successive work periods are separated by a rest period of inadequate duration, then the worker cannot adequately recover, and the entire period—the two work periods plus the rest period—must be treated as if it were a single, continuous work period.

The determination of F_M is then accomplished via entry into Table 18.3.

TABLE 18.5

Continuous Lifting Tasks and Their Required Rest Periods

Categories	Definitions	Required Resting Period
Short Duration	≤1 hour	≥120% of the duration of the continuous, repetitive lifting task
Medium Duration	>1 hour but <2 hours	≥30% of the duration of the continuous, repetitive lifting task
Long Duration	≥2 hours but ≤8 hours	No amount is specified; normal morning, afternoon, and lunch breaks are presumed

Note: For respective frequency coefficients, see Table 18.3.

The quality of gripping is defined as:

Good: if the object can be grasped by wrapping the hand comfortably around the handles or hand-hold cutouts of the object without significant deviations from the neutral wrist posture or the object itself without causing excessive wrist deviations or awkward postures

Fair: if the object has handles or cutouts that do not fulfill the criteria of good quality of gripping or if the object itself can be grasped with a grip in which the hand can be flexed about 90°

Poor: if the criteria of good or fair quality of gripping are not fulfilled

Step 4

Under ideal conditions, the cumulative mass (product of mass and frequency of carrying, which may have been limited in steps 1 and 2) manipulated per shift should not exceed 10,000 kg.

For carrying, the reference mass should never exceed 25 kg, and the frequency is limited to a maximum of 15 times per minute.

If the carrying distance is less than 1 m, manual handling is acceptable under the existing conditions. For longer distances (i.e. 20 m), it is substantially lower (6,000 kg/eight hours). For nonideal conditions and carrying distances greater than 1 m, step 5 has to be applied.

This step considers data from NF × 35–106, N × 35–109, Grieco et al. (1997); Garg, Chaffin, and Herrin 1978; Genaidy & Ashfour 1987.

Step 5

Maximum cumulative masses for several distances and frequencies are contained in Table 18.7. If the cumulative masses described in Table 18.1 are not exceeded, the entire risk assessment has been carried out successfully. Otherwise, adaptation is necessary.

Table 18.7 provides the limits in kilograms per minute that should protect against the excess of local load—in kilograms per hour—which should protect against excess of general load—and in kilograms per eight hours—and limit the long-term risk. The limits are not the simple

TABLE 18.6

Coupling Multiplier for the Quality of Gripping (C_M)

Quality of Gripping	Height < 0.75 m	Height ≥ 0.75 m
Good	1.00	1.00
Fair	0.95	1.00
Poor	0.90	0.90

multiplications, because the risks for short-term, medium-term, and long-term are qualitatively different. The last column of the table takes the examples of different combinations of mass and frequency. These examples show that the limits in kilograms per minute cannot be always applied because of the limits of maximal mass and frequency (5 kg × 15/min = 75 kg/min even for a distance of 1 m, and 25 kg cannot be lifted more than 1/min—see Figure 18.2).

In the practical application of the Standard, the limits of maximal mass and frequency have the priority; when those limits are respected, the limits to carrying have to be applied.

Under unfavorable environmental conditions—or when lifting from and to low levels, for example, below knee height or when the arms are lifted above the shoulders—the recommended limits for cumulative mass for carrying in Table 18.7 should be substantially reduced (at least by one-third).

Risk Reduction

Risk reduction can be achieved by reducing the need for manual handling activities or by minimizing or excluding hazards resulting from the task, the object, the workplace, the work organization, or the environmental conditions. Examples are given in Annexes A.2–A.5 of the Standard.

Annexes

Although the main part of the standard is relatively compact and describes the risk assessment mainly in a five-step approach, Annex A offers detailed background information on the following:

- Avoidance of manual handling (A1)
- Design of task, workplace, and work organization (A2)
- Design of the object (A3)
- Design of the handling of live objects (A4)
- Design of the work environment (A5)
- (for "ideal conditions for manual materials," see Table 18.1)
- Assessment method for recommended limits for mass, frequency, and object position for repetitive and nonrepetitive lifting tasks (A6). Risk assessments are based on the assumptions listed in Table 18.4.

TABLE 18.7

Recommended Limits for Cumulative Mass Related to Carrying Distance (for General Working Population)

| Distance (m) | kg/min | Maximum Cumulative Mass | | Example of M*f |
		kg/hour	kg/8 hour	
20	15	750	6,000	5 kg × 3/min15 kg × 1/min25 kg × 0.5/min
10	30	1,500	10,000	5 kg × 6/min15 kg × 2/min25 kg × 1/min
4	60	3,000	10,000	5 kg × 12/min15 kg × 4/min25 kg × 1/min
2	75	4,500	10,000	5 kg × 15/min15 kg × 5/min25 kg × 1/min
1	120	7,200	10,000	5 kg × 15/min15 kg × 8/min25 kg × 1/min

Note 1. In the calculation of the cumulative mass, a reference mass of 15 kg is used for the general working population and frequency of carrying—15 times/min.

Note 2. The total cumulative mass of lifting and manual carrying should never exceed 10,000 kg/day, whichever is the daily duration of work.

Note 3. 23 kg is included in the 25 kg.

- Individual considerations (A7)
- Information and training (A8)

Annex B offers two examples of an assessment and ergonomic approach of manual handling of (live) objects. Annex C offers the table for the reference mass (see Table 18.2).

WORKSHEETS

Unfortunately, neither the main part nor the annexes contain any worksheets that may be copied for application purposes.

THE EU-MANUAL HANDLING DIRECTIVE

This Directive, which is the 4th individual Directive within the meaning of Article 16(1) of Directive 89/391/EEC, lays down minimum health and safety requirements for the manual handling of loads where there is a risk—particularly of back injury to workers. It obliges the employer to take appropriate organizational measures, or use the appropriate means—in particular, mechanical equipment—to avoid the need for the manual handling of loads by workers. Where this cannot be avoided, the employer shall take the appropriate organizational measures, use the appropriate means—or provide workers with such means—to reduce the risk involved in the manual handling of such loads, having regard to Annex I of this directive. Wherever the need for manual handling of loads by workers cannot be avoided, the employer shall organize workstations in a way that makes such handling as safe and healthy as possible and:

- Assess, in advance if possible, the health and safety conditions of the type of work involved and, in particular, examine the characteristics of loads, taking account of Annex I
- Take care to avoid or reduce the risk particularly of back injury to workers, by taking appropriate measures, considering, in particular, the characteristics of the working environment and the requirements of the activity, taking account of Annex I

Workers or their representatives shall be informed of all measures to be implemented, pursuant to this Directive, with regard to the protection of safety and of health.

Employers must ensure that workers or their representatives receive general indications and, where possible, precise information on:

- The weight of a load
- The center of gravity of the heaviest side when a package is eccentrically loaded

Employers must ensure that workers receive proper training and information on how to handle loads correctly and the risks they might be open to, particularly if these tasks are not performed correctly, having regard to Annexes I and II.

Annex I describes the characteristics of the load.

The manual handling of a load may present a risk particularly of back injury if it is:

- Too heavy or too large
- Unwieldy or difficult to grasp
- Unstable or has contents likely to shift
- Positioned in a manner requiring it to be held or manipulated at a distance from the trunk, or with a bending or twisting of the trunk
- Likely, because of its contours or consistency, to result in injury to workers, particularly in the event of a collision

A physical effort may present a risk particularly of back injury if it is:
- Too strenuous
- Only achieved by a twisting movement of the trunk
- Likely to result in a sudden movement of the load
- Made with the body in an unstable posture

The characteristics of the work environment may increase a risk particularly of back injury if:

- There is not enough room—in particular, vertically—to carry out the activity
- The floor is uneven or is slippery, presenting tripping hazards, in relation to the worker's footwear
- The place of work or the working environment prevents the handling of loads at a safe height or with good posture by the worker
- There are variations in the level of the floor or the working surface, requiring the load to be manipulated on different levels
- The floor or footrest is unstable
- The temperature, humidity, or ventilation is unsuitable

The activity may present a risk particularly of back injury if it entails one or more of the following requirements:

- Overfrequent or over-prolonged physical effort involving, especially, the spine
- An insufficient bodily rest or recovery period
- Excessive lifting, lowering, or carrying distances
- A rate of work imposed by a process that cannot be altered by the worker

Annex II describes individual risk factors. The worker may be at risk if he or she:

- Is physically unsuited to carry out the task in question
- Is wearing unsuitable clothing, footwear, or other personal effects
- Does not have adequate or appropriate knowledge or training

The complete text of this directive may be downloaded in several languages from http://europa.eu.int/eur-lex/.

ACKNOWLEDGMENTS

This work is dedicated to the members of ISO/TC 159/SC 3/WG 4 who developed this standard, especially to my colleagues from the writing group.

NOTE

1 Taking into account the field of application and the intended user population, it will be necessary to select an alternative appropriate reference mass as shown in Table 18.2.

REFERENCES

90/269/EEC. (1990). Council Directive of 29 May 1990 on the minimum health and safety requirements for the manual handling of loads where there is a risk particularly of back injury to workers (fourth in-dividual Directive within the meaning of Article 16 (1) of Directive 89/391/EEC).

EN 614-1. Safety of machinery—Ergonomic design principles—Part 1: *Terminology and general principles.*

EN 614-2. Safety of machinery—Ergonomic design principles—Part 2: *Interactions between the design of machinery and work tasks.*

EN 1005-2. Safety of machinery—Human physical performance—Part 2: *Manual handling of machinery and component parts of machinery.*

EN 1050. *Safety of machinery—Risk assessment.*

EN-ISO 7250, *Basic list of definitions of human body measurements to technical design.*

Garg, A., Chaffin, D., & Herrin, G. D. (1978). Prediction of metabolic rates for manual materials handling jobs. *American Industrial Hygiene Association Journal, 39*(8), 661–674.

Genaidy, A. M., & Ashfour, S. S. (1987). Review and evaluation of physiological cost prediction models for manual materials handling. *Human Factors, 29*(4), 465–476.

Grieco A., Occhipinti E., Colombini D., & Molteni G. (1997). Manual handling of loads: The point of view of experts involved in the application of EC Directive 90/269. *Ergonomics, 40*(10), 1035–1056.

ISO 7730. Moderate thermal *environments—Determination of the PMV and PPD indices and specification of the conditions for thermal comfort.*

ISO 11226. *Ergonomics—Evaluation of working postures.*

ISO 14121. *Safety of machinery—Principles of risk assessment.*

ISO 1228 Ergonomics part 1: Lifting and carrying part 2: pushing and pulling part 3: Handling of law loads at high frequency.

ISO/IEC Guide 51. *Safety aspects—Guidelines for their inclusion in standards.*

Schaub, Kh., & Landau, K. (1997). Computer-aided tool for ergonomic workplace design and preventive health care. In *Human Factors and Ergonomics in Manufacturing, 7*(4), 269–304.

19 Ergonomics of Manual Handling—Part 2

Pushing and Pulling

Karlheinz G. Schaub and Peter Schaefer

CONTENTS

INTRODUCTION

RELATION TO OTHER STANDARDS, GUIDELINES, AND REGULATIONS

Probably because pushing and pulling is physically less demanding than lifting and carrying, there is a lack of risk-assessment methods on a quantitative basis by means of a formula. It is the aim of ISO 11228-2 to bridge that gap. Several approaches exist for the evaluation of pushing and pulling tasks in a worldwide overview.

In Europe, pushing and pulling, as two types of manual materials handling, are regulated by the Council Directive 90/269/EEC "on the minimum health and safety requirements for the manual handling of loads where there is a risk particularly of back injury to workers." Next to this directive, manual materials handling is addressed by the European Union (EU) machinery directive (98/37/EC)—EN 1005, a set of harmonized CEN standards that supports this directive. The next chapters offer additional information about the dual European system of health and safety at work and the manual handling directive.

THE DUAL CONCEPT OF HEALTH AND SAFETY IN EUROPE

On their way to building up a political union, the EU Member States decided to add social components to their various branches of the community that were predominantly of economic nature for many years. It was intended that these social components should also include measurements for preventive healthcare and aspects of ergonomic workplace and product design. The adoption in 1986 of the Single European Act gave new impetus to the occupational health and safety measures taken by the community. This was the first time health and safety at work had been directly included in the EEC Treaty of 1957 and was done through the new articles 100a and 118a. Article 100a requires the harmonization of national legislation. The objective is to remove all barriers to trade in the single market and allow free movement of goods and people across borders. In principle, Article 100a does not permit Member States to set higher requirements for their products than those laid down by the directives.

Of most significant importance to the level of protection in the EU Member States is that directives adopted under Article 118a lay down minimum requirements concerning health and safety at work. According to this principle, the member states may raise their level of protection, if desired, or raise it if it is lower than the minimum requirements set by the directives. Beyond this, they have the obligation to maintain and introduce more stringent protective measures than required by the directives.

Articles 100a and 118a contribute to the improvement of the Member States' working environments as well as in the equal or better protection of their workers. Directives under Article 100a are intended to ensure the placing on the market of safe products; under Article 118a, they are intended to ensure the healthy and safe use of the products at the workplace (EC 1993). Articles 100a and 118a had been reformulated as Articles 95 and 137 in the Maastricht treaty.

The Framework Directive on health and safety at work (89/391/EEC)—including the corresponding Individual Directives (e.g., Visual Display Units [VDU] work, manual materials handling, and personal protective equipment) as well as the Machinery Directive (98/37/EC; formerly 89/392/EEC)—serve the implementation of Articles 95 and 137 of the Maastricht treaty. Whereas the Framework Directive addresses employers and employees, the Machinery Directive only focuses on the designer and manufacturer of machinery.

THE MANUAL HANDLING DIRECTIVE

This Directive, which is the fourth individual Directive within the meaning of Article 16(1) of Directive 89/391/EEC, lays down minimum health and safety requirements for the manual handling of loads where there is a risk, particularly of back injury to workers. It obliges the employer to take appropriate organizational measures or use the appropriate means—mechanical equipment—to avoid the need for manual handling of loads by workers. Where this cannot be avoided, the employer shall take the appropriate organizational measures, use the appropriate means, or provide workers with such means to reduce the risk involved in the manual handling of such loads, having regard to Annex I of this directive.

Wherever the need for manual handling of loads by workers cannot be avoided, the employer shall organize workstations in a way that makes handling as safe and healthy as possible and:

- Assess—in advance, if possible—the health and safety conditions of the type of work involved and particularly examine the characteristics of loads, taking account of Annex I.
- Take care to avoid or reduce the risk—especially of back injury—for workers by taking appropriate measures, considering, in particular, the characteristics of the working environment and the requirements of the activity, taking account of Annex I.

Workers or their representatives shall be informed of all measures to be implemented, pursuant to this Directive, with regard to the protection of safety and of health.

Employers must ensure that workers or their representatives receive general indications and—where possible—precise information on the weight of a load and the center of gravity of the heaviest side when a package is eccentrically loaded.

Employers must ensure that workers receive proper training and information on how to handle loads correctly and the risks they might be open to, particularly if these tasks are not performed correctly, having regard to Annexes I and II.

Annex I describes the characteristics of the load. The manual handling of a load may present a risk particularly of back injury if it is:

- Too heavy or too large
- Unwieldy or difficult to grasp
- Unstable or has contents likely to shift
- Positioned in a manner requiring it to be held or manipulated at a distance from the trunk, or with a bending or twisting of the trunk
- Likely, because of its contours or consistency, to result in injury to workers, particularly in the event of a collision

A physical effort may present a risk particularly of back injury if it is:

- Too strenuous
- Only achieved by a twisting movement of the trunk
- Likely to result in a sudden movement of the load
- Made with the body in an unstable posture

The characteristics of the work environment may increase a risk particularly of back injury if:

- There is not enough room—in particular, vertically—to carry out the activity
- The floor is uneven—presenting tripping hazards—or is slippery in relation to the worker's footwear
- The place of work or the working environment prevents the handling of loads at a safe height or with good posture by the worker,
- There are variations in the level of the floor or the working surface, requiring the load to be manipulated on different levels,
- The floor or footrest is unstable
- The temperature, humidity, or ventilation is unsuitable

The activity may present a risk particularly of back injury if it entails one or more of the following requirements:

- Over-frequent or over-prolonged physical effort involving, in particular, the spine
- An insufficient bodily rest or recovery period
- Excessive lifting, lowering or carrying distances
- A rate of work imposed by a process that cannot be altered by the worker

Annex II describes individual risk factors. The worker may be at risk if he or she:

- Is physically unsuited to carry out the task in question
- Is wearing unsuitable clothing, footwear, or other personal effects
- Does not have adequate or appropriate knowledge or training

The complete text of this directive may be downloaded in several languages from http://europa.eu. int/eur-lex/.

GENERAL APPROACHES

This standard is under development and has reached the status of a draft. When the working group the international standard has finished the draft, it will be sent out as a committee draft Standard (ISO/CD 11228-2) for voting.

STRUCTURE OF THE STANDARD

The Standard follows the predefined typical structure of ISO Standards. After the foreword, introduction, scope, and normative references, Chapter 3 offers definitions required for that Standard. Chapter 4, as the main part of the Standard, offers recommendations for the avoidance of hazardous manual handling tasks and the execution of a risk assessment composed of two risk assessment methods. Method 1—an easily applicable checklist—is described in detail in Annex A. Method 2—a more sophisticated quantitative approach—is described in detail in this chapter. Annex E offers an example for Method 2.

Annex A describes in detail Method 1 of the risk assessment. Method 1 is an easy-to-handle checklist that allows the rating of a task for high or low risks to verbally describe the problems arising from the task and formulate possible remedial actions to lower the existing risk.

The method presented takes into consideration the following:

- The task
- The loads or objects to be moved
- The working environment
- Other factors
- Management and organizational issues

Annex C offers biomechanical considerations of pushing and pulling. Annex D deals with measures for risk reduction and covers the following:

- Avoidance of repetitive handling
- Design of the work: task, workplace, and work organization
- Design of object, tool, or material handled
- Design of the working environment
- The worker capabilities

Annex E finally offers examples for the application of the more sophisticated risk-assessment Method 2.

Basic Philosophy

A major part of the Standard aims to offer a comprehensive risk assessment for pushing and pulling tasks. Depending on the sources reviewed, the cardiovascular system might be the bottleneck for pushing and pulling tasks, as well as the muscular or the skeletal system (Schaefer, Kapitaniak, & Schaub, 2000). Especially for high loads, the initial forces required might exceed or come close to the maximal force limits, whereas the sustained forces required are on a very low level only—if the environmental (moderate ambient thermal environment, good coupling in between the feet and the floor, smooth floor without obstacles, etc.) and task conditions (no sudden accelerations, obstacles, or curves with narrow radius, etc.) are appropriate. It was believed that, contrary to lifting and carrying, pushing and pulling would not be of biomechanical relevance (Mital, Nicholson, & Ayoub, 1997). Recent investigations of Jäger (2001), however, showed that—especially for pulling tasks—high lumbar spine loads might result from pulling tasks.

When pushing and pulling objects, the posture adapted and the location of force exertion have a substantial impact on maximum voluntary contraction (MVC) (Schaub, Berg, & Wakula, 1997). The posture adopted and the location of force exertion depends on the position of the handles and the stature of the operator. So, the biomechanical load situation for a given task may vary among the operators dependent on their stature. To allow an overall individual independent risk assessment, it was decided to realize a model that considers muscular as well as skeletal bottlenecks, and that takes into consideration the effect of the stature onto maximal force limits and the postures adapted.

Data Sources

Relevant tables from DIN 33411-5—a comprehensive national German Standard (18 tables with maximal action forces for the 5th, 10th, 15th, 50th, and 95th force percentiles of a grand total of 188 different types of force exertions)—were used to describe the influence of the relative working height (gripping height as percentage of stature) onto the maximal force capabilities (see Figures 19.1 and 19.2). For a conservative assessment, the values for the 5th force percentiles were chosen.

The procedure of EN 1005-3 derives acceptable load limits out of maximal static force limits by several multipliers that consider the velocity, frequency, and action time of force exertions, the task duration, and a "risk" multiplier.

Snook's tables will be used to set up multipliers for the influence of frequency and duration for maximum limits of pushing and pulling tasks.

RISK-ASSESSMENT MODEL

Approaches

ISO 11228-2 provides two different ways of risk assessment. The first one (Method 1) is a checklist-based procedure that yields very rough and rather pessimistic estimates. Results of Method 1 may be understood in such a way that loads accommodated in the green zone are surely green. However, any accommodation in red zones may turn out to be green or yellow as well, that's why Method 1 will not be presented here. The checklists are easy to apply and may be found in the Standard. In the following, the focus is Method 2, which provides a detailed assessment and evaluation of risk.

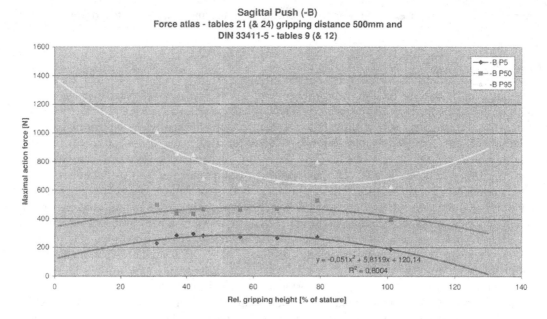

FIGURE 19.1 Maximal static push action forces in relation to gripping height and force percentiles.

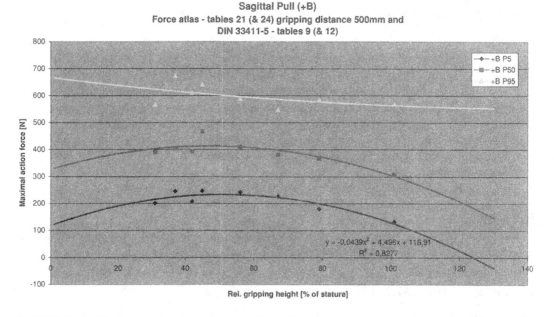

FIGURE 19.2 Maximal static pull action forces in relation to gripping height and force percentiles.

INNOVATIONS

- Method 2 introduces two characteristic elements that are absolutely new in this field. These are:
- User-oriented assessments—limits reflecting demographic profiles of any optional user population

- Skeletal load limits—in addition to muscular aspects, skeletal limits consider spinal aspects (e.g., compressive load of lumbar spine)

User-Oriented Assessment

Industrial production generally seeks to meet the demands of a wide and heterogeneous clientele as precisely as possible. Hence, as user profiles change, the design of industrial products is changing as well. To ensure good interaction between man and machine, technical aids generally should reflect major characteristics of envisaged user populations—for example, demographic profiles as described by age, gender, and stature distributions. That kind of target group orientation not only addresses classical anthropometries but also influences manipulation forces and may directly affect health protection. The general problem is how to protect different user populations by the same safety level—in other words, how to make sure that user groups all over the world are really taking comparable risks.

That's why ISO 11228-2 introduced force limits reflecting demographic profiles in particular. These characteristics are given by distributions of age, gender, and stature.

Such an evaluation procedure is absolutely new in the field of load rating. Traditional approaches either refer to pure male or pure female populations or to a fixed mix of both (e.g., NIOSH, 1981; Siemens, 1969). It is an outstanding characteristic of ISO 11228-2 that its special load limits react directly to the slightest changes in demographic profiles.

Skeletal Load Limits

This risk estimation approach adopts a multidisciplinary approach, giving suitable consideration to biomechanical, physiological, and psychophysical parameters. The biomechanical approach considers force exertions in relation to both individual strength capabilities and the risk of injury. Such injuries not only include muscular diseases but also cover skeletal diseases—that is why this risk estimation approach additionally considers lumbar spine compression in relation to lumbar spine strength for different age populations.

How the Model Works

Method 2 is a quite complex procedure and provides a breakdown of risk assessment into four major parts (see Figure 19.3):

Part 1: Muscle-Based Force Limits F_{Br}

Part 1 adopts a two-step approach:

Step 1: Basic force limits (F_B). Basic force limits (F_B) are adjustable strength limits of the intended user population, taking into account the effects of age, gender, and stature.

Step 2: Adjusted force limits (F_{Br}). Adjusted force limits (F_{Br}) reflect characteristics of the work task such as velocity, frequency, and duration. The effects of these characteristics are accounted for by a set of multipliers (m_v, m_f, and m_d):

$$F_{Br} = m_v \times m_f \times m_d \times F_B.$$

Part 2: Skeletal-Based Force Limits (F_{LS})

Part 2 provides load limits (F_{LS}) that are based on the limited compressive strength of the lumbar spine. The procedure integrates the following two steps:

Step 1: Compressive strength limit (Fc). In a first step, compressive strength limits are determined by taking into account the effects of age and gender on the foreseeable user population.

Step 2: Action force limits (F_{LS}). Step 2 introduces a relation between compressive strength limits (Fc) determined by step 1 and action forces as observed at the workplace.

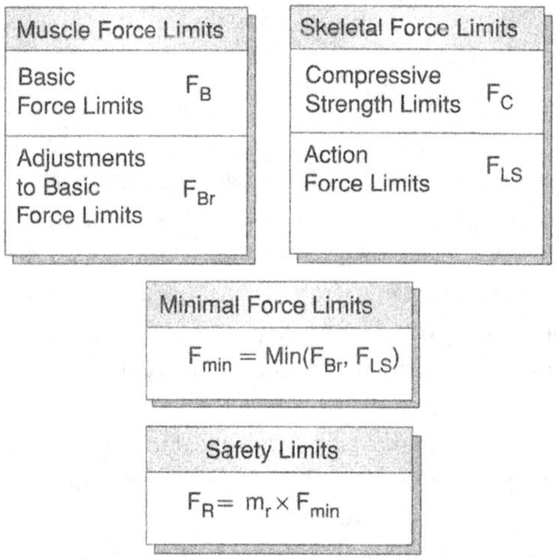

Muscle Force Limits		Skeletal Force Limits	
Basic Force Limits	F_B	Compressive Strength Limits	F_C
Adjustments to Basic Force Limits	F_{Br}	Action Force Limits	F_{LS}

Minimal Force Limits

$$F_{min} = Min(F_{Br}, F_{LS})$$

Safety Limits

$$F_R = m_r \times F_{min}$$

FIGURE 19.3 Risk-assessment procedure (Method 2) in ISO 11228-2.

This way the action force limit (F_{LS}) makes sure that compressive strength limits (F_C) of the lumbar spine generally are not exceeded at work.

Part 3: Minimal Limits (F_{min})

Part 3 selects the minimum between muscle-based force limits (F_{Br}) and skeletal-based force limits (F_{LS})

$$F_{min} = Min\left(F_{Br}, F_{LS} \right).$$

Part 4: Safety Limits (F_R)

Safety limits (F_R) are based on capacity limits (F_{min}) and suitable risk multipliers (m_r), such that

$$F_R = m_r \times F_{min}.$$

MUSCLE FORCE LIMITS

Basic Force Limits

Part 1 of the risk-assessment method specifies a calculation procedure providing basic force limits that change with changing characteristics of optional user populations. Required characteristics are given by distributions of age, gender, and stature.

The technical procedure is a step-by-step approach, as described in the following.

Objectives. The procedure calculates force limits when pushing or pulling, at selected absolute handle heights, and with specified target populations.

Input Data. To get started, the procedure first requires a set of input data (see Figure 19.4). In detail, these are the following:

- Selected absolute handle height above ground—for example, 1.2 m
- Characteristics of the envisaged user population—that is, stature distribution and distributions of age and gender
- Strength distributions from tables or experimentally found in young females (20 years ≤ age < 30 years), when pushing or pulling, and at selected relative handle heights

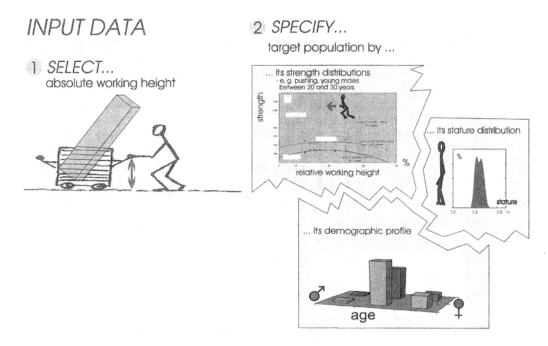

FIGURE 19.4   Input data of risk assessment model.

Predicting Statures. In the first step, the procedure predicts statures (Figure 19.5) resulting from various working situations. Generally, there is a relation between these statures on one side and relative working heights (h_{rel}) on the other when working on given absolute levels (h_{abs}):

$$S = \frac{h_{abs}}{h_{rel}}m \quad \text{with: } S: \text{stature.}$$

Practically, the procedure works as follows (see Figure 19.5):

- Select absolute working height—for example, 1.2 m above ground
- Select a set of relative working heights covering the range man is able to work. Intervals between selected heights should be reasonably spaced
- Predict statures for each relative working height selected previously—for example, when working 1.2 m above ground and relative working height is 120%, then stature is at 1.71 m

Finally, there is a set of statures allowing work at one and the same absolute working height while selecting various relative working heights.

Weighting Multipliers. The chance stature (S) observed certainly depends on the specific stature distribution to be found in an envisaged target population. On the other side, stature (S) is defined when working at a given relative height (h_{rel}) and on a given absolute level (h_{abs}). That's why the aforementioned stature probabilities describe incidence rates of those corresponding working situations in the same way. Such probabilities may define weighting multipliers to be determined for each relative height in particular (Figure 19.6).

The way these weighting multipliers will be found is explained in Figure 19.6. The major steps are:

- Get a specific stature distribution of the envisaged target population

PREDICTING STATURES

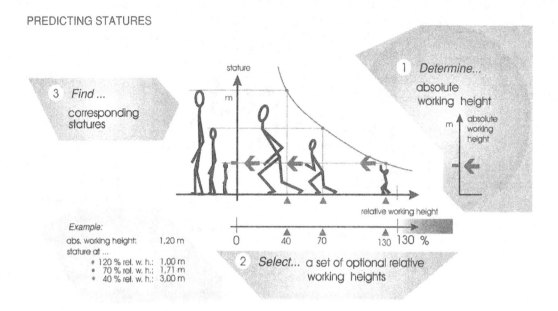

FIGURE 19.5   Statures required when working at selected relative heights.

FINDING WEIGHTING MULTIPLIERS

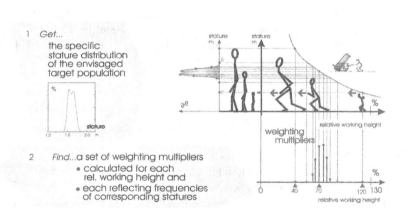

FIGURE 19.6   Multipliers weighting relative working heights.

- Select a set of relative working heights
- Predict corresponding statures
- Determine probabilities that these statures will be found in the target population
- Assign multipliers to stature probabilities

Introduction of Strength Distributions. Strength distributions may be given by percentile functions (e.g., 15th, 50th, and 85th percentiles) depending on relative working heights (see Figures 19.1 and 19.2). This is the experimental base of a set of distributions to be determined at selected relative working heights. In the end, each relative working height accommodates a specified strength distribution and an appropriate weighting multiplier (see Figure 19.7).

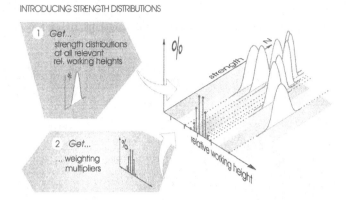

FIGURE 19.7 Distributions describing strength at selected relative working heights.

Weighting Strength Distributions. The next step adjusts reference distributions at each relative working height by the aforementioned weighting multipliers—each multiplier modifies his or her distribution at his or her relative working height (see Figure 19.8). Such modifications reflect the probabilities of these relative working heights to be found among the envisaged user population. The procedure are:

- Get strength distributions at each relative working height
- Get according weighting multipliers
- Multiply each distribution by its "own" multiplier

Demographic Fitting. Up to this point, results strictly represent reference groups only—for example, females between 20 and 30 years of age. That's why another step adjusts these reference distributions to real ones that may be observed within real user populations.

This work is done by the "Synthetic Distributions Method," a special procedure that was standardized by EN 1005-3 and was introduced into ISO 11228-2 as well. A general overview is given in Figure 19.9. The procedure simply calculates modified distributions reflecting the effects of age and gender within optional user populations. These modified distributions are no longer approximations to normal.

Combined Distribution Functions. The physical strength of any user population as a total may be described by combined distributions shaped by demographic profiles and suitable sets of relative working heights reflecting stature distributions. Those distributions may be found by a simple sum integrating weighted distributions of all relative working heights.

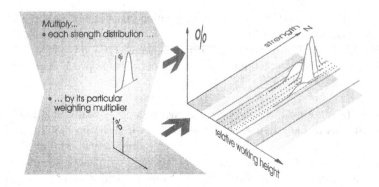

FIGURE 19.8 Adjusting distributions to demographic profiles.

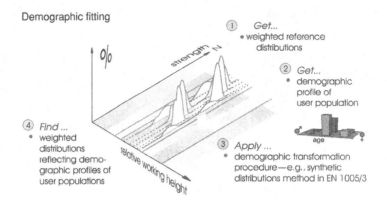

FIGURE 19.9 Adjusting distributions to demographic profiles.

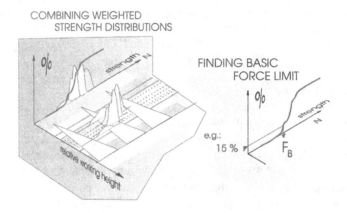

FIGURE 19.10 Finding combined strength distributions and basic force limits.

Finally, a second integration of these combined distributions with increasing strength yields a combined distribution function (see Figure 19.10).

To define load limits in a reproducible way, percentile approaches were applied. Such an approach starts by combined strength distribution functions as found within well-defined target populations at selected activities. This allows force limits to be found easily in the following way (see Figure 19.10):

- Specify particular percentages of the user population that should be able to do the activities under consideration (here: 85%)
- Determine the percentage of people not capable—here: 110—85 = 15%
- Read the corresponding force limit (F_B) directly out of appropriate distribution functions

Activity Limits

Basic force limits (F_B) are estimate short-term capacity limits. Usually, that kind of limit is far from any real activities of daily living. Hence, these short-term limits should be adjusted to real human life. In ISO 11228-2, this is done by a set of multipliers accounting for the effects of velocity, frequency, and duration (see Figure 19.11). This approach adopts the multiplier system of EN 1005/3 in European Standardization. Possibly, this particular multiplier approach in ISO 11228-2 will be replaced by another procedure based on more recent findings.

Velocity Multiplier (m_v). Human force-generating capacity is reduced with the increasing speed of concentric movements. This is realized by the velocity multiplier (m_v) as determined by Table 19.1.

Frequency multiplier (m_f). Frequently repeated actions may increasingly lead to fatigue effects resulting in reversible reductions of human strength. Principally, these effects depend on action time and cycle frequency—where each cycle duration may be broken down into its action time and its individual break. A frequency multiplier (m_f) was defined to describe fatigue effects. Details are found in Table 19.2. An overview is given in Figure 19.11.

TABLE 19.1

Velocity Multiplier (m_v) Depending on Moving Speed

m_v	1.0	0.8
Movement	No	Yes

"No" Action implies no or a very slow movement.
"Yes" Action implies an evident movement.

TABLE 19.2

Frequency Multiplier (m_f) Defined by Action Time and Cycle Frequency

<Frequency of Actions (min⁻¹)>
Action Time (min)

	<0.2	0.2–2	2–20	>20
≤0.05	1.0	0.8	0.5	0.3
>0.05	0.6	0.4	0.2	Not applicable

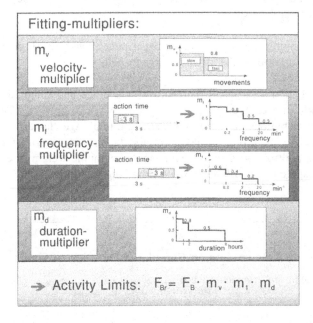

FIGURE 19.11   A set of multipliers designed to adjust basic force limits (F_B) to real activities of daily living.

Duration multiplier (m_d). Human force-generating capacity is further reduced by the overall time spent on the job. Actions that are loading muscle structures may add up, causing local fatigue. To regard those local effects, it's not only overall action time that should be considered—in particular, durations of similar actions also seem to be important. That's why "similar actions" are defined in the Standard by actions that are very similar in character (i.e. pushing and pressing).

Those duration effects are described by multiplier m_d. Here, "duration" always refers to the total duration of similar activities including interruptions.

Activity Limits (F_{Br}). Limits adjusted to real activities of daily living may be found by simple reduction of basic force limits F_B. In particular, F_B is reduced by the following set of multipliers accounting for the effects of velocity, frequency, and duration:

$$F_{Br} = F_B \times m_v \times m_f \times m_d,$$

where F_B is the basic force limit, m_v is the velocity multiplier (Table 19.1), m_f is the frequency multiplier (Table 19.2), and m_d is the duration multiplier (Table 19.3).

SKELETAL FORCE LIMITS

In addition to muscular approaches (muscle force limits F_B and F_{Br}), ISO 11228-2 adopts spinal aspects as well. However, the procedure described here is a tentative approach and probably will be replaced by a more sophisticated method. Basically, this preliminary approach has two major steps:

Step 1: Estimates compressive force limits of lumbar spine
Step 1: Finds action force limits

COMPRESSIVE FORCE LIMITS

Human spinal strength apparently depends on the effects of age and gender (see Figure 19.12; Jäger, 2001). That's why spinal limits should reflect these effects—that is, when demographic profiles change, spinal limits should change as well. To this end, this spinal approach basically adopts the same demographic philosophy that was realized in the muscle force approach (see Figure 19.9).

The major steps are as follows:

- Start out with compressive strength as an experimental base (see Figure 19.12).
- Find regressions describing the effects of age in both genders.
- Introduce age classes.
- Calculate distribution parameters (percentiles or average and standard deviation) of compressive strength in each age class.
- Generate log-distributions of compressive strength in all age classes.
- Check the demographic profile of the target population using age-intervals in Figure 19.12.
- Find weighting multipliers accounting for the demographic weight of each age class.
- Multiply each age class distribution by its weighting multiplier.

TABLE 19.3
Duration Multiplier (m_d) Depending on Cumulated Duration of Similar Actions

Duration (Working Time) in Similar Actions (hours)

Duration	≤ 1 hour	1–2 hours	2–8 hours
m_d	1.0	0.8	0.5

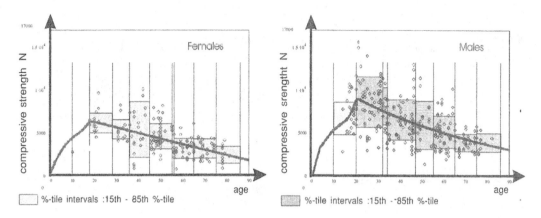

FIGURE 19.12   Compressive strength of lumbar spine (Jäger, 2001; Rosenberg, 2003).

- Sum up all weighted age class distributions to get total spinal strength distributions of both genders.
- Integrate total strength distributions with increasing strength to get the total strength distribution functions of males and females.
- Determine the 15th percentile to find the compressive strength limits of the lumbar spine.

Certainly, those lumbar spine limits change with changing user populations. In Figure 19.13, a variety of precalculated LBS (lumbar spine) limits may be found, estimating a set of preselected situations. Its characteristics are two different age groups—general EU-working population and "active" EU-seniors and a selection of specified ratios between males and females.

In fact, these limits are changing in a rather "natural" way with changing profiles of target populations (see Figure 19.13).

ACTION FORCE LIMITS

In the following step, limitations of externally applied forces are determined in such a way that compressive force limits of the lumbar spine are not exceeded. The procedure to determine those action force limits has not yet been designed in detail. That's why the following method provides a very rough approach that will be improved on in the near future.

At the moment the following procedure applies:

Compressive force limits of lumbar spine

RATIO males/females %	ALL AGES 15 - 64 y kN	"ACTIVE" SENIORS 50 - 64 y kN
0 : 100	3.0	2.4
25 : 75	3.2	2.5
59 : 41	3.6	2.9
75 : 25	3.8	3.0
100 : 0	4.2	3.4

Target population:
EU 12 (1993)—working population of the 12 EU countries in 1993

FIGURE 19.13   Precalculated LBS force limits varying with selected user populations.

- *Determine...* the average stature ($\bar{h}$) of
  the target population suggesting
  anthropometry of an average user.

e.g., $\bar{h}$ stature

- *Select...* the absolute working height h_w.

e.g., h_w

- *Predict...* the most probable working
  posture of an average user.

e.g.,

- *Find...* the shoulder joint angle (SJA).
- *Find...* the force angle (FA).

i.e.,

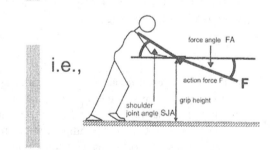

- *Check...* appropriate charts if available.
  For example, see Figure 19.14.
- *Determine...* the action force limit
  (F_{LS}) by the chart selected.

e.g.,

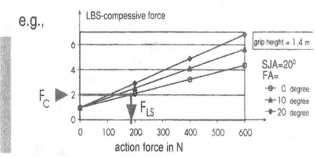

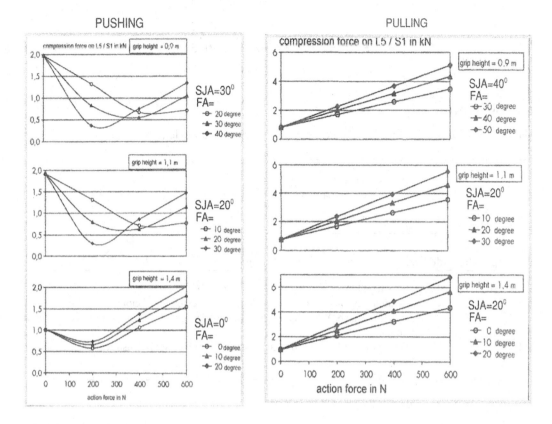

FIGURE 19.14 Compressive strength of lumbar spine depending on action forces at selected pushing or pulling activities (Jäger, 2001).

MINIMAL FORCE LIMITS

At this stage, the evaluation procedure provides two capacity limits made to assess the activities under consideration—these are muscle force limits (see Figure 19.9) and skeletal load limits (see Figure 19.12). That's why the procedure has to decide now which limit to adopt for further processing. Such a decision will be made on a safe basis. To make sure that no less than a predefined majority is protected, either way, the lower one of both limits should be selected.

That's why the procedure determines the minimum of muscular force limit F_{Br} and skeletal load limit F_{LS}:

$$F_{min} = \text{Min}(F_{Br}, F_{LS}).$$

In this way, the resultant capacity limits F_{min} make sure that a specified safety level is not undercut in the evaluation procedure.

SAFETY LIMITS

Both muscle force limits (F_{Br}) and skeletal load limits (F_{LS}) are capacity limits that allow wide majorities to do specified activities of daily living. On the other hand, that kind of force limit does not necessarily guarantee safe jobs. Certainly, health risks may be found below maximal force levels as well. Hence, above-capacity limits F_{min} should be transformed into safety limits. This is done by two multipliers m_r^{red} and m_r^{green} that define the green and red safety limits Fr:

$$F_R^{\text{red}} = m_r^{\text{red}} \cdot F_{\min} \quad \text{with } m_r^{\text{red}} = 0,7$$
$$F_R^{\text{green}} = m_r^{\text{green}} \cdot F_{\min} \quad m_r^{\text{green}} = 0,5$$

RISK ASSESSMENT

Above safety limits, F_R^{red} and F_R^{green} are drawing up a three-zone model, providing three categories to easily assess any real load (L):

Red zone:	load $L \geq F_R^{\text{red}}$
Yellow zone:	$F_R^{\text{green}} <$ load $L < F_R^{\text{red}}$
Green zone:	load $L \leq F_R^{\text{green}}$.

Interpretations depend on the zones accommodating load L, in particular:

Green zone:	No action is required.
Yellow zone:	Risk estimation shall be continued, including other relevant risk factors, followed as soon as possible by redesign. Where redesign is not possible, other measures to control the risk shall be taken.
Red zone:	Action to lower the risk (e.g., redesign, work organization, worker instruction, and training) is necessary.

EXAMPLE

To demonstrate the way ISO 11228-2 works, please see Figure 19.15. This example calculates basic force limits (F_B) when pushing. For demonstration purposes, limits were made for American and Japanese test populations widely differing in stature distributions. Further strength distributions of both nations were assumed to be identical. In each of both nationalities, limits were determined for a variety of predefined subpopulations specified in particular by a mix of age and gender.

Age Mixes

In this example, there are three age mixes arranging the general working population in three different ways:

- Junior—population concentrates in age group 1 (age < 20 years)
- All ages—equal distributions in all three age groups
- Senior—population concentrates in age group 3 (50 < age < 65 years)

Gender Mixes

Each of the aforementioned age mixes, with three different ratios between males and females, is realized:

- Mix 1: male/females = 0: 100%
- Mix 2: males/females = 50: 50%
- Mix 3: males/females = 100: 0%

A combination of these mixes in age and gender further yields a 3 × 3 array as shown in Figure 19.15.

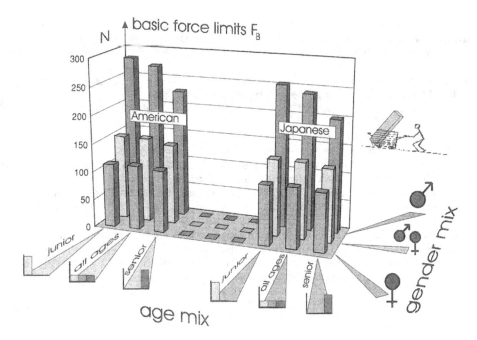

FIGURE 19.15 Adaptive load limits of ISO 11228-2 varying within and between two different user populations.

Finally, basic force limits F_B are calculated

- at the place of each element in the array
- regarding both the American and the Japanese population

Interpretations. Results in Figure 19.15 quantitatively demonstrate the wide adaptivity of the innovative ISO procedure. Its limits do not only reflect given demographic profiles—in both nations, limits are decreasing with increasing age and increasing female representation. Its limits additionally depend on "nationalities" as an effect of stature distributions in particular.

PERSPECTIVES

The risk assessment procedure in ISO 11228-2 generally demonstrates the use of what might be understood as "target group ergonomics." That kind of target group orientation in the field of ergonomics provides very flexible limits adjustable to optional demographic profiles. This makes sure that each particular user group may be protected on a specified safety level—there is little overprotection and little underprotection. Global assessments usually do not work so precisely. Above all, it must be realized that any overprotection or underprotection may produce considerable costs—apart from all kinds of individual misfits. In this more economical sense, ISO 11228-2 seems to be very cost-effective. That is why it may be an interesting perspective to introduce that kind of target group ergonomics into other fields of ergonomics as well.

REFERENCES

DIN 33411-5. Human physical strength—Maximal static action forces—Values.
90/269/EEC. (1990). Council Directive of 29 May 1990 on the minimum health and safety requirements for

the manual handling of loads where there is a risk particularly of back injury to workers (fourth individual Directive within the meaning of Article 16 (1) of Directive 89/391/EEC).

EN 1005-3. Safety of machinery—Human physical performance—Part 3: Recommended force limits for
machinery operation.

Jäger M. (2001). *Belastung und Belastbarkeit der Lendenwirbelsäule im Berufsalltag*, Fortschr.-Ber. VDI
Reihe 17 Nr. 208, VDI Verlag Düsseldorf, ISBN 3-18-320817-2.

Mital, A., Nicholson A. S., & Ayoub M. M. (1997). *A guide to manual materials handling*. 2nd edition.
London: Taylor & Francis.

NIOSH. (1981). *Work practices guide for manual lifting* (Technical Report, Publication No. 81-122). U.S.
Department of Health and Human Services. Cincinnati, OH:

Rosenberg, S. (2003). *Human lumbar spine—structural stabilities and load limits*, under publication. Inst. of
Ergonomics, TU München, Boltzmannstr. 15, 85747 Garching, Germany.

Schaefer, P., Kapitaniak, B., & Schaub, Kh. (2000). Custom made load limits shaped by age, gender and
stature distributions—A new ISO-approach. In Proceedings of the *IEA 2000/HFES 2000 Congress*,
4-349–4-351.

Schaub, Kh., Berg, K., & Wakula, J. (1997). Postural and workplace related influences on maximal force
capacities. In IEA '97 (Ed.), From experience to innovation (Vol. 4, pp. 219–221). Tampere, Finland:
The 13th Triennial Congress of the International Ergonomics Association.

Siemens. (1969). *Lastentransport von Hand, Mitteilung aus dem Labor für angewandte Arbeitswissenschaften*,
Nr. 6.

20 Recommended Force Limits for Machinery Operation
A New Approach Reflecting User Group Characteristics

Peter Schaefer and Karlheinz G. Schaub

CONTENTS

INTRODUCTION

Industrial production generally seeks to meet the demands of a wide and heterogeny clientele as precisely as possible. Hence, as user profiles change, the design of industrial products changes as well. To ensure good interaction between man and machine, technical aids generally should reflect major characteristics of envisaged user populations (e.g., demographic profiles as described by age and gender). That kind of target group orientation not only addresses classical anthropometries but also influences manipulation forces and may directly affect health protection.

The general problem is how to protect different user populations by the same safety level—in other words, how to make sure that all user groups are really taking comparable risks. The European Union (EU) is defining safe products by CEN standardization. Some of these CEN safety limits largely depend on the profile of the intended target population—so do force limits, in particular. That's why EN 1005/3 introduced force limits reflecting demographic profiles. These characteristics are given by distributions of age and gender.

Such an evaluation procedure is absolutely new in the field of load rating. Traditional approaches either refer to pure male or pure female populations or to a fixed mix of both (e.g., NIOSH, 1981; Siemens, 1969). It is an outstanding characteristic of EN 1005/3 that its special load limits react directly to the slightest changes in demographic profiles.

BASIC ELEMENTS

This particular load evaluation procedure in EN 1005/3 combines two basic elements.

PERCENTILE APPROACH

To define load limits in a reproducible way, percentile approaches were applied. Such an approach simply starts with isometric strength distributions as found within well-defined target populations and selected activities. Results may be readily displayed by strength distribution functions. Such an approach allows force limits to be found easily in the following ways:

- Specify particular percentages of the user population supposed to be capable of doing the activities under consideration (here: 85%)
- Read the corresponding force limit directly out of appropriate distribution functions

SYNTHETIC DISTRIBUTIONS

To realize target group sensibility of force limits, a way was found to synthesize strength distributions of arbitrary target populations as described by age and gender. The aforementioned percentile approach finally yields force limits adjusted to optional user populations.

ELEMENTARY STEPS

The percentile approach and the idea to synthesize strength distributions are two major elements on the way to adjustable force limits. These elements are part of a three-step evaluation procedure.

STEP 1: BASIC FORCE LIMITS

In the first step, basic force limits are calculated. To this end, a combined strength distribution function of the intended user population is synthesized. Force limits may be found by predefined percentiles of these distributions. In EN 1005/3, the 15th percentile was selected. Such a choice makes sure that 85% of the envisaged user population can do its job.

STEP 2: ACTIVITY LIMITS

The aforementioned force limits describe human strength sustained for only a few seconds, that's why such short-term capacity limits must be adjusted to real user activities. In EN 1005/3, this is done by a set of multipliers reflecting real user situations.

STEP 3: SAFETY LIMITS

Generally, capacity limits do not necessarily ensure a safe job. To realize safety limits, a special risk multiplier was introduced to transform the capability limits into safety limits (Figure 20.1).

BASIC FORCE LIMITS F_B

Basic force limits may be directly calculated when starting by human strength data. They reflect some basic short-term force-generating capacity. To this purpose, EN 1005/3 provides two alternative ways to calculate F_B.

PROCEDURE 1

Procedure 1 is a quick way to determine force limits. This procedure assumes equal representations of males and females and may be applied

- If the envisaged user profile is similar to the general European population
- If demographic details are simply not available

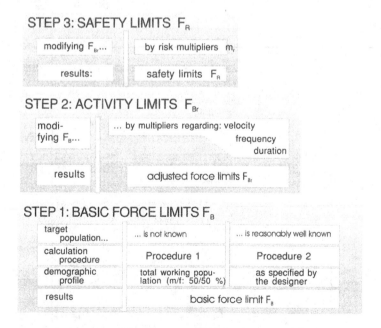

FIGURE 20.1 A three-step model to put up force limits.

PROCEDURE 2

On the other hand, procedure 2 applies if age and gender distributions of an envisaged user population are reasonably well known. In this case, it's up to procedure 2 to provide load limits that reflect demographic profiles in particular.

PROCEDURE 1

Input Parameters

In the first step, manipulations of the human operator should be analyzed to pinpoint the most hazardous activities. Such an analysis yields distribution parameters (average and standard deviation) of human strength that may be determined either by tables or measurements. These strength data ideally should represent the adult European working population. Practically, it is recommended to start with distribution parameters of female reference groups only. These female distribution parameters allow reasonably good predictions of force limits F_B including both genders. Such a procedure is integrated into EN 1005/3 (see Figure 20.2).

Approximation. If the data for female reference groups are not available, distribution parameters of young females (between 20 and 30 years) may be used approximately (see procedure 2).

Calculations

Force Distribution. Average and standard deviation defines strength distribution functions $DF(x)$. Such an approximation to normal is a good and easy way to determine force limits in most practical applications.

Logarithmic Transformation. Force limits are more reliable when shifting over to logarithmic normal distributions:

$$\bar{F}_{\ln} = \ln \bar{F} \quad \sigma_{\ln} = \ln \frac{\bar{F} + \sigma}{\bar{F}}$$

Example

$$\bar{F}_{\ln} = \ln 233 = 5.45$$

$$\sigma_{\ln} = \ln \frac{233.7 + 81}{233.7}$$

$$= 0.30$$

Activity		$\bar{F}$ [N]	σ [N]
hand work (one hand) power grip		278.0	62.2
arm work (sitting posture, one arm)			
	- upwards	58.0	18.4
	- downwards	88.6	33.2
	- outwards	65.5	26.2
	- inwards	85.6	24.6
	- pushing		
	- with trunk support	312.0	84.8
	- without trk. support	78.0	42.7
	- pulling		
	- with trunk support	246.0	45.7
	- without trk. support	67.9	33.5
whole body work (standing posture)			
	- pushing	233.7	81.0
	- pulling	164.6	44.9
pedal work (sitting posture, with trunk support)			
	- ankle action	293.4	104.7
	- leg action	542.5	156.2

FIGURE 20.2   A selection of strength distribution parameters (reference group: adult female population).

Its basic characteristics and inputs are:

- *Target population*: adult (e.g., European) working population
- *Reference group*: adult female population
- *Distribution parameters*: average force F and standard deviation σ of the reference group

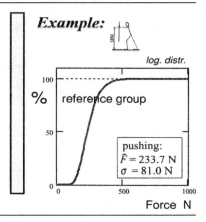

Strength Percentiles. Starting by the aforementioned distribution, parameters $\bar{F}_{\ln}$ and $\sigma_{\ln}$ logarithmic force percentiles ($\bar{F}_{\ln \%}$) may be calculated:

Example

$$\bar{F}_{\ln 15\%} = 5.45 - 0.5244 \cdot 0.30$$
$$= 5.30$$

$$\bar{F}_{\ln \%} = \bar{F}_{\ln} + z_\% \cdot \sigma_{\ln}.$$

Referring to the 15th or 1st percentile of the target group, $z\%$ amounts to

$$z_{15\%} = -0.5244$$
$$z_{1\%} = -2.0537$$

With $z_{15\%} = z_{15\%}^{\text{gen. popul.}} = z_{30\%}^{\text{females}} = -0.5244$ and $z_\% = z_{1\%}^{\text{gen. popul.}} = z_{2\%}^{\text{females}} = -2.0537$.

A simple transformation back to linear finally yields appropriate percentiles $F\%$:

$$F_\% = e^{F_{\ln \%}} \text{N}$$

Example
$$F_{15\%} = e^{5.3} = 200 \text{ N}$$
$$F_{1\%} = e^{4.84} = 127 \text{ N}$$

Results

Both percentiles $F_{15\%}$ and $F_{1\%}$ are defining basic force limits F_B:

$$F_B = \begin{cases} F_{15\%} & \text{for professional use} \\ F_1 & \text{for domestic use} \end{cases}$$

Example
$$F_B = 200 \text{ N}$$

for professional use

Activity		$\bar{F}$ [N]	σ [N]
	hand work (one hand) power grip	270.0	54.1
arm work (sitting posture, one arm)			
	- upwards	56.0	18.4
	- downwards	86.0	33.2
	- outwards	63.5	26.2
	- inwards	83.4	24.6
	- pushing		
	- with trunk support	303.0	81.0
	- without trk. support	75.5	42.7
	- pulling		
	- witht runk support	242.0	44.9
	- without trk. support	65,7	33.5
whole body work (standing posture)			
	- pushing	228.0	84.8
	- pulling	161.0	45.7
pedal work (sitting posture, with trunk support)			
	- ankle action	282.0	96.5
	- leg action	528.5	157.6

FIGURE 20.3 A selection of strength distribution parameters (reference group: females between 20 and 30 years of age.).

These force limits allow specified activities up to 85% or 99% of the adult (e.g., European) working population without exceeding their physical capacity.

PROCEDURE 2

Input Parameters

Forces. To start, procedure 2 requires distribution parameters of a defined female reference group (also see Figure 20.3).

The details are:

reference group:

- females
- $20 \leq$ age ≤ 30 years

distribution parameters:

- average force $\bar{F}$; and
- standard deviation σ of the reference group

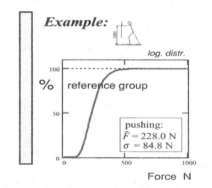

Example:

log. distr.

reference group

pushing:
$\bar{F} = 228.0$ N
$\sigma = 84.8$ N

Force N

Certainly, reference forces change with changing individual strategies and activities.

Approximation. If no data of young female reference groups are available, distribution parameters of procedure 1—adult female population—may be used approximately.

User Demography. Next, the demographic profile of the envisaged user population is specified. Results may be arranged either by age and gender:

females: $n_{f1\%}$: age < 20 years
    $n_{f2\%}$: 20 ≤ age ≤ 50 years
    $n_{f3\%}$: 50 < age ≤ 65 years
males: $n_{m1\%}$: age < 20 years
    $n_{m2\%}$: 20 ≤ age ≤ 50 years
    $n_{m3\%}$: 50 < age ≤ 65 years

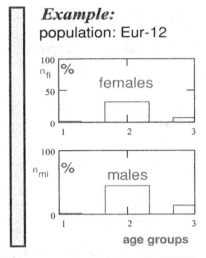

Example:
population: Eur-12

with

n_{fi}, n_{mi} percentage of subgroups as found within any user population.

$$n_{f1} = 1.6\% \quad n_{m1} = 2.0\%$$
$$n_{f2} = 31.6\% \quad n_{m2} = 43.8\%$$
$$n_{f3} = 7.6\% \quad n_{m3} = 13.4\%$$

Procedure

Procedure 2 calculates force limits adjusted to user populations as previously specified. The procedure works as follows.

Generation of Subgroup Distributions. Force averages and standard deviations of all other subgroups are simply calculated by aforementioned reference parameters ($\bar{F}$, σ) and some appropriate multipliers (a_{xx}, s_{xx}) expressing relations between age and gender (Schaefer, Rudolph, & Schwarz, 1997; Rühmann & Schmidtke, 1992):

Females
Average forces: $\bar{F}_{fi} = \bar{F} * a_{fi}$
SD:

Males
Average forces: $\bar{F}_{mi} = \bar{F} * a_{mi}$
SD: $\sigma_{mi} = \sigma * s_{mi}$

with

	Averages a_{xx}				Standard Deviations s_{xx}		
Age Groups	1	2	3	Age Groups	1	2	3
Females a_{fi}	0.96	1.00	0.93	Females s_{fi}	1.03	1.00	0.96
Males a_{mi}	1.95	2.16	1.70	Males s_{mi}	1.57	1.65	1.81

and

$i = 1 \ldots 3$:   Age groups

a_{xx}, s_{xx}:   Subgroup multipliers

 .   Average force and standard deviation of the reference group

	Example		
Age Groups	**1**	**2**	**3**
$\bar{F}_{fi}$	172.8	180.0	167.4
σ_{fi}	61.8	60.0	57.6
$\bar{F}_{mi}$	351.0	388.8	306.0
σ_{mi}	94.2	99.0	108

Logarithmic Distributions. Approaching lower force levels above approximation to normal yields increasingly poor results at lower percentiles (e.g., 1%). In this case, logarithmic distributions are more realistic. An easy transformation is calculating a new set of logarithmic distribution parameters:

$$\text{Females: } \bar{F}_{fi}^L = \ln(\bar{F}_{fi}) \quad \sigma_{fi}^L = \ln \frac{\bar{F}_{fi} + \sigma_{fi}}{\bar{F}_{fi}}$$

$$\text{Males: } \quad \bar{F}_{mi}^L = \ln(\bar{F}_{mi}) \quad \sigma_{mi}^L = \ln \frac{\bar{F}_{mi} + \sigma_{mi}}{\bar{F}_{mi}}$$

Generating Subgroup Distribution, Functions.

females:

$$DF_{fi}(x) = \frac{1}{\sigma_{fi}^L \sqrt{2\pi}} \int_{-\infty}^{\ln x} e^{-z_{fi}^2/2} dz$$

with $z_{fi} = \frac{\ln x - \bar{F}_{fi}^L}{\sigma_{fi}^L}$

x: forces

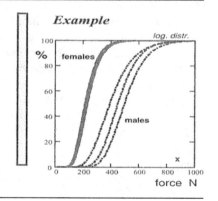

Example

Males:

$$DF_{mi}(x) = \frac{1}{\sigma_{mi}^L \sqrt{2\pi}} \int_{-\infty}^{\ln x} e^{-z_{mi}^2/2} dz$$

with

$$z_{mi} = \frac{\ln x - \bar{F}_{mi}^L}{\sigma_{mi}^L}$$

Weighting and Combining Subgroup Distributions.
$DF(x) = \sum_i (n_{fi} DF_{fi}(x) + n_{mi} DF_{mi}(x))/100$

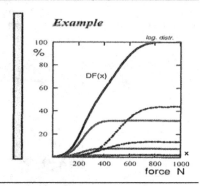

Percentiles. DF(x) is the combined distribution function of all subgroups depending on force x. So, force limits may be found by calculating the 15th or the 1st percentile of $DF(x)$:

$$DF(x) = \begin{cases} 0.15 & \text{for professional use} \\ 0.01 & \text{for domestic use} \end{cases} \Rightarrow \text{force } x$$

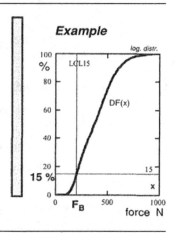

Results. Above percentile approach yields basic force limits F_B:

Example

$F_B = x$ [N]

$$F_B = 200.2 \text{ N}$$

These limits allow specified activities up to 85% or 99% of any optional user population without exceeding their physical capacity.

ACTIVITY LIMITS

Basic force limits F_B are describing short-term capacity limits. Usually, that kind of limit is far from any activities of daily living. Hence, these short-term limits should be adjusted to real human life. In EN 1005/3, this is done by a set of multipliers accounting for effects of velocity, frequency, and duration (see Figure 20.4).

Velocity Multiplier M_V

Human force-generating capacity is reduced with the increasing speed of concentric movements. This is covered by velocity multiplier *my* as determined by Table 20.1.

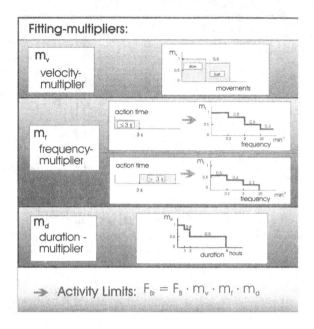

FIGURE 20.4 A set of multipliers designed to adjust basic force limits F_B to activities of daily living.

TABLE 20.1

Velocity Multiplier m_v Depending on Moving Speed

m_v	1.0	0.8
Movement	No	Yes

FREQUENCY MULTIPLIER M_F

Frequently repeated actions may increasingly lead to fatigue effects that usually make human force-generating capacity decrease. Principally, these effects depend on action time and cycle frequency where each cycle duration is made up by its action time and its individual break.

A frequency multiplier m_f was defined to describe fatigue effects. Details may be found in Table 16.2. An overview is given in Figure 20.4.

DURATION MULTIPLIER M_D

Human force-generating capacity is further reduced by the overall time spent on the job. Actions loading muscle structures may add up, causing local fatigue. To regard those local effects, it is not only overall action time that should be considered but also the duration of similar actions. That is why "similar" actions are defined in the Standard by actions that are very similar in character (i.e. pushing and pressing) (Table 20.2).

Those duration effects are described by multipliers (see Table 20.3). Here, "duration" always refers to the total length of similar activities, including interruptions.

TABLE 20.2

Frequency Multiplier m_f Defined by Action Time and Cycle Frequency

	Frequency of Actions (minute^{-l})			
Action Time (minute)	<0.2	0.2–2	2–20	>20
≤0.05	1.0	0.8	0.5	0.3
>0.05	0.6	0.4	0.2	0.3
				Not applicable

TABLE 20.3

Duration Multiplier m_d Depending on the Cumulated Duration of Similar Actions

	Duration (Working Time) in Similar Actions (hour)		
Duration	<1	1–2	2–8
m_d	1.0	0.8	0.5

ACTIVITY LIMITS F_{BR}

Limits adjusted to activities of daily living may be found by the simple reduction of basic force limits F_B. In particular, F_B is reduced by the previous set of multipliers assessing the effects of velocity, frequency, and duration:

$$F_{Br} = F_B \times m_v \times m_f \times m_d,$$

where F_B is basic force limit, m_v is velocity multiplier, m_f is frequency multiplier, and m_d is duration multiplier.

SAFETY LIMITS

Activity limits are capacity limits allowing a wide majority of specified activities of daily living. That kind of force limit does not necessarily guarantee safety on the job. Certainly, health risks may be found below maximal force levels as well. Hence, the above capacity limits should be reduced to safe limits. This is done by two different multipliers, m_r^{red} and m_r^{green}, defining green and red safety limits F_R:

$$F_R^{red} = m_r^{red} \cdot F_{Br} \quad \text{with} \quad m_r^{red} = 0.7$$
$$F_R^{green} = m_r^{green} \cdot F_{Br} \quad m_r^{green} = 0.5$$

RISK ASSESSMENT

Above safety limits F_R^{red} and F_R^{green}, define a three-zone model providing categories for assessing easily any real load L:

TABLE 20.4
Risk Multiplier m_r Defining Risk Zones

Risk Zone	m_r
Recommended	≤0.5
Not Recommended	0.5–0.7
To Be Avoided	>0.7

Red zone:	load $L \geq F_R^{red}$
Yellow zone:	$F_R^{green} < \text{load } L < F_R^{red}$
Green zone	load $L \leq F_R^{green}$

Interpretations depend on the zones accommodating load L (see also Table 20.4):

Green zone:	Recommended zone—the risk of disease or injury is negligible. No intervention is needed.
Yellow zone:	Zone is not recommended—risks of disease or injury cannot be neglected. Additional risk estimation should be done considering other risk factors. Risk analysis may yield acceptable risks even in the yellow zone. If, on the other hand, analysis is reconfirmed a risk re-design or other measures may be needed.
Red zone:	Zone to be avoided—risks of disease or injury are obvious and cannot be accepted. Further activities to lower risks are necessary.

EXAMPLE

To demonstrate the way EN 1005/3 works, an example is given in Figure 20.5. This presents selected activity limits F_{Br} when pushing, as described by the input data in Figure 20.5. These limits are allocated to a set of predefined target populations. Each of these populations is specified by a mix of age and gender.

AGE MIXES

    Mix 1: young people—all people accommodated by age group 1
    Mix 2: all ages—equal distributions in all three age groups
    Mix 3: seniors only—all people accommodated by age group 3

GENDER MIXES

Each of the previous age mixes realizes three different ratios between males and females:

    Mix 1: males and females = 0%:100%
    Mix 2: males and females = 50%:50%
    Mix 3: males and females = 100%:0%

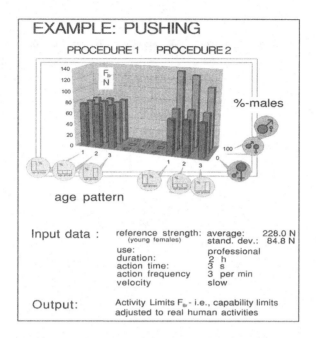

FIGURE 20.5 Activity limits calculated with selected demographic profiles.

A combination of the previous mixes in age and gender further yields a 3 × 3 array as shown in Figure 20.5. Finally, activity limits are calculated at the place of each element in the array and by procedures 1 and 2.

INTERPRETATION

Results in Figure 20.5 quantitatively demonstrate the way force limits depend on given demographic profiles. Obviously, limits decrease not only with increasing age but also when changing from male to female populations.

Such a "natural behavior" of force limits is only provided by procedure 2. It is obvious that procedure 1 is absolutely resistant to any changes in demographic profiles. Hence, these results demonstrate that procedure 2 can reflect demographic profiles quantitatively, whereas procedure 1 continues old load rating traditions.

REFERENCES

EN 1005–3. Safety of machinery—Human physical performance—Part 3: Recommended force limits for machinery operation.

NIOSH. (1981). *Work practices guide for manual lifting* (Technical Report, Publication No. 81–122). Cincinnati, OH: U.S. Department of Health and Human Services.

Rühmann H., & Schmidtke H. (1992). *Körperkräfte des Menschen, Otto Schmidt KG Köln, Dokumentation Arbeitswissenschaft*, Bd. 31. ISBN 3-504-65637-9.

Schaefer, P., Rudolph, H., & Schwarz, W. (1997). Variable force limits for optional target populations—A new approach realized in CEN—Standardization, Proceedings of the 13th IEA Conference, Tampere, 4, 533–535.

Siemens. (1969). *Lastentransport von Hand, Mitteilung aus dem Labor für angewandte Arbeitswissenschaften*, Nr. 6.

21 Guidelines for the Prevention of WMSDs
The Italian Experience

Enrico Occhipinti and Daniela Colombini

CONTENTS

INTRODUCTION

The introduction of general and more ergonomically oriented norms in Italy from 1994, derived from a series of UE Directives in OS&H at the workplace (European Directives 391/89, 269/90, 270/90), called the attention of several stakeholders (employers, trade unions, OH&S professionals, public authorities) to WMSDs prevention issues.

In particular, acknowledgment of European Directive 269/90 (1990) concerning manual handling of loads resulted in the adoption of risk assessment and management procedures that were partially renewed in 2008 (with the update of the general law concerning health and safety at work): this issue regards over four million workers usually engaged in physically heavy tasks (about 20% of the labor force).

On the other hand, the increasing reports of occupational upper limb WMSDs (by now at the first place among the most frequently reported occupational diseases) induced the national and regional authorities—in the absence of a specific regulation on the subject—to issue guidelines forassessing and managing the risk from upper limb repetitive movements potentially involving about six million workers (30% of the general workforce), mainly from the manufacturing industry.

On this basis, the major points of the following Italian guidelines are summarized and discussed.

• The guidelines for the application of European Directive 269/90 on manual handling of loads, prepared by the authors and officially adopted by the Conference of the Italian Regions (1996) and by the National Institute for Safety and Prevention at Work (ISPESL); some aspects related to the updates inserted in the new law "D.Lgs 81/08" on all aspects of Health and Safety at Work will be also presented.

• The guidelines for the prevention of UL-WMSDs connected to repetitive movements and exertions, prepared by the authors, officially adopted by the Lombardy Regional Government in 2002 with relevant updates in 2009 (Regione Lombardia, 2009) and 2014. In this view, it is worth underlining that so far, no national governmental specific guidelines have been issued; that, among Italian regions, Lombardy is the most densely populated (nine million inhabitants) and industrialized (more than four million workers); and that other regions (Veneto, Emilia-Romagna, Puglia) have substantially adopted the Lombardy guideline's approach.

CONTENTS OF THE GUIDELINES

MANUAL HANDLING OF LOADS (DIRECTIVE EC 90/269)

Exposure Assessment

Risk assessment—the individual and/or collective probability of contracting dorso-lumbar spinal disorders due to manual load handling—is one of the pillars of preventive measures required by new European regulations and intervention methodologies in the field of ergonomics.

It is worth recalling that an extremely simplified interpretation of the assessment concept has become standard practice in workplaces and in other applications—according to which, for example, load handling may be assessed solely based on the load weight (as stated in old national regulations and standards).

This was a widespread practice in Italy: in fact, a weight limit (30 kg) was initially (from 1996 to 2008) introduced in the legislation to incorporate the EC Directive in Italian law, hence, the simplification that all objects weighing less than 30 kg may be handled "safely"!

With such a scenario, when defining the appropriate tools for risk (or, better, exposure) assessment, the authors had to redefine the requirements for assessment validity and applicability, setting aside the rigors of a sophisticated scientific approach as a prerogative of research elites but also opposing the oversimplification demanded by operators in the field (often as a pretext) (Grieco et al., 1997).

In the law update in 2008, a direct reference for applying the EC Directive is made to international standards and to ISO 11228 parts 1–2 and 3, particularly (ISO, 2003, 2007a, 2007b).

Assessment of Lifting Tasks

As regards the assessment of manual lifting, the model proposed by the guideline was originally based on the Revised NIOSH Lifting Equation—RNLE (Waters et al., 1993). Obviously, this occurs via major adaptations and changes in the original model as, on the other side, suggested by ISO 11228-1 and EN 1005-2 standards (ISO, 2003; CEN, 2003).

For large-scale application of the model, it seemed (and actually was) useful to propose graphic and procedural simplifications as reported in Table 21.1, which shows a sheet for collection and processing of all the data required to calculate the lifting index. Note that in the proposed sheet, reference is made to a "reference mass" (or load constant) whose details should consider relevant aspects such as the gender and age of the working population and the correspondent level of protection as addressed in ISO 11228-1 and EN 1005-2.

It is important to underline that these documents—on account of the impossibility of achieving a univocally established reference mass or load constant (maximum recommended weight under ideal lifting conditions)—propose a range of possible constants indicating, as required, the relevant "target" population as well as the degree of its presumed protection. Considering the requirements of the Italian Law (taking into consideration aspects as gender and age) and the main reference to ISO 11228-1, the reference mass for different parts of the working population has been selected as follows: Males (18–25 years): 25 kg.; Females (18–25 years): 20 kg; Males (<18 or >45 years): 20 kg; Females (<18 or >45 years): 15 kg.

TABLE 21.1
Datasheet for the Evaluation of a Lifting Task

CALCULATION OF LIFTING INDEX

| COMPANY AREA WORK PLACE TASK | | | | | | | | | OBSERVATION DATE OBSERVER | |

LOAD CONSTANT (kg.)

	male	female	
18-45 YEARS	25	20	
<18 e >45 YEARS	20	15	

Distance of the hands from the floor at the start of lifting

(cm)	0	25	50	75	100	125	150	>175	VM	1
VERTICAL MULTIPLIER	0,77	0,85	0,93	1,00	0,93	0,85	0,78	0,00		

Vertical distance of the load between the beginning and the end of lifting

(cm)	25	30	40	50	70	100	170	>175	DM	1
DISTANCE MULTIPLIER	1,00	0,97	0,93	0,91	0,88	0,87	0,86	0,00		

X

Maximum distance between the load and the body during lifting

(cm)	25	30	40	50	55	60	>63	HM	1
HORIZONTAL MULTIPLIER	1,00	0,83	0,63	0,50	0,45	0,42	0,00		

X

Angular measure of displacement of the load from the sagittal plane

(degree)	0	30°	60°	90°	120°	135°	>135°	AM	1
ASIMMETRY MULTIPLIER	1,00	0,90	0,81	0,71	0,52	0,57	0,00		

X

Assessment of grip of the object

	GOOD	BAD	CM	1
COUPLING MULTIPLIER	1,00	0,90		

X

E

Frequency of lifts per minute and duration — FM — 1

FREQUENCY	MMH DURATION		
ACTION/MIN.	≤ 8 H (LONG)	≤ 2 H (MODERATE)	≤ 1 H (SHORT)
<0,1	1,00	1,00	1,00
<0,2 to <=0,1	0,85	0,95	1,00
0,2	0,85	0,95	1,00
0,5	0,81	0,92	0,97
1	0,75	0,88	0,94
2	0,65	0,84	0,91
3	0,55	0,79	0,88
4	0,45	0,72	0,84
5	0,35	0,60	0,80
6	0,27	0,50	0,75
7	0,22	0,42	0,70
8	0,18	0,35	0,60
9	0,00	0,30	0,52
10	0,00	0,26	0,45
11	0,00	0,00	0,41
12	0,00	0,00	0,37
13	0,00	0,00	0,00
14	0,00	0,00	0,00
15	0,00	0,00	0,00
>15	0,00	0,00	0,00
MULTIPLIERS FOR AREAS INF TO 75 CM			

	NO	YES		
G	LIFT WITH 1 UPPER LIMB	1,00	0,60	1
H	LIFT 2 OPERATOR	1,00	0,85	1

KG. LIFTED WEIGHT		RECOMMENDED WEIGHT LIMIT	

LIFTED WEIGHT		
RECOMMENDED WEIGHT		Lifting Index

Source: Adapted from Italian guidelines

It should be further stressed that adopting the RNLE model for assessments in the field of manual load lifting tasks does, however, pose some problems that can be schematically described in the following.

1. The American authors themselves (Waters et al., 1993) emphasize that the procedure is not applicable in some situations—such caution is quite understandable from a strictly scientific viewpoint—but in some cases, it may be overcome by making assumptions based on empirical data. When, for example, the load is lifted with only one arm, the EN 1005-2 proposes introducing a further multiplication factor of 0.6. If lifting is carried out by two or more operators, always in the same workplace, it was proposed to consider—as the weight actually lifted—the weight of the object divided by the number of operators, and for the recommended weight to introduce a further multiplicative factor of 0.85. Such adjustments tend to reply to widespread applicative problems that would remain unsolved. In this sense, further proposals aimed at favoring an ever-increasing practical applicability of the method are no doubt to be encouraged.

2. In many working situations, the same group of workers must carry out different lifting tasks often in the same shift. The different lifting tasks may be irregular in each period (e.g., in a warehouse with picking activities) or according to established time sequences (e.g., when an operator works every 1–2 hours on an assembly line, first loads the line, then unloads the finished products, and then packs them). In such cases, the analytical procedure for each task is not suitable to summarize the overall exposure of the individual worker (or of the group of) to lifting. Therefore, these cases require an analytical procedure for multiple tasks that is obviously more complex (Waters, Putz Anderson, & Garg, 1994). These procedures have been recently updated by the authors (Colombini et al., 2012) and included in an ISO TR (ISO, 2014) applicative of ISO 11228-1.

3. The NIOSH assessment procedure is not well suited to application in various working sectors (typically non-industrial sectors), sometimes on account of the characteristic of the lifted load, the great variability of lifting tasks, their frequent association with other manual handling tasks (trolley pulling or pushing), and the presence of other risk factors for the lumbar spine (e.g., whole-body vibrations). Agriculture, transport, and delivery of goods and healthcare for individuals who are not self-sufficient (at home or in the hospital) are typical examples. In these situations, though the NIOSH lifting index is useful—validated procedures for integrated exposure assessment were not yet available—the need for further research and proposals of specific simplified exposure assessment procedures aimed also at managing risk factors exists.

Assessment of Other Manual Handling Activities (Pulling, Pushing, Carrying)

No equally consolidated procedures based also on multidisciplinary approaches, like the NIOSH procedure for lifting, are available in the literature for the assessment of exposure to manual load handling such as pulling, pushing, or carrying. Considering this aspect, it was decided to use the data derived from the specific application of psychophysical methods summarized by Snook and Ciriello (1991). This option resulted to be substantially in agreement with the ISO 11228-2 (ISO, 2007a), which is now specifically addressed in the renewed (in 2008) Italian legislation on the matter.

This was mainly due to two reasons:

1. Such data were also expressed with reference to the percentiles of potentially satisfied (even if not necessarily protected) population. Where it was possible to select data on the "satisfaction" of 90% of the population, we were able to provide reference values that had a potential level of protection comparable to that resulting from the application of RNLE.

2. The data from psychophysical studies were expressed by Snook and Ciriello and, later, by ISO 11228-2 with reference not only to the two genders but also to structural variables (height of pushing/pulling areas, distance) as well as to organizational variables (frequency and duration of tasks). This produced a well-defined method applicable to different working situations and capable of suggesting appropriate preventive measures, when necessary.

Manual Handling Index and Its Consequences

We have seen that it is always possible to calculate—albeit with a variety of assessment procedures according to the analyzed manual handling activity—a synthetical exposure index (manual handling index = MHI), as follows:

$$MHI = \frac{\text{actually handled weight (force)}}{\text{recommended weight (force) as a function of major situation variables}}$$

Such a synthetical manual handling index, even if determined by semiquantitative assessment procedures, may become an effective tool—not so much for defining the exposure level of one worker (or group of workers) involved in manual handling—for defining the consequent preventive measures in accordance with community regulations and, more generally, with correct prevention strategies.

To reach the latter goal, it is convenient to classify MHI results at least according to a three-zone model (green, yellow, and red). This is because the level of approximation (both intrinsic and in conditions of application) of the suggested methods and procedures calls for a certain amount of caution as regards borderline results around the value of 1.

This three-zone model (or traffic light model) appeared to be useful in this sense. The MHI results could be classified as follows:

- MHI up to 0.85 = Green zone: there is no particular exposure for the working population and, therefore, no collective preventive actions are required.
- 0.86 ≤ MHI ≤ 1.00 = Yellow zone: this is the borderline zone where exposure is limited but may exist for some of the population. Prudent measures are to be taken especially in training and health surveillance of workers. Wherever possible, it is suggested to limit exposure to return to the green zone.
- MHI higher than 1.00 = Red zone: exposure exists and is significantly present. The higher the MHI value, the higher the exposure for increasing numbers of the population. MHI values may determine the priority of prevention measures that must, in any case, be taken to minimize exposure towards the yellow zone. Training and active health surveillance of workers must be undertaken in any case.

Health Surveillance Strategies

This paper is aimed at ergonomists; therefore, health surveillance problems are discussed only considering the general aspects they may be of interest to our readers.

In Europe, Council Directive 89/391/EEC requires employers to ensure that workers receive health surveillance, at regular intervals, appropriate to the health and safety risks they incur at work—including exposure to manual handling activities. Consequently, the Italian Guidelines affirm that an active health surveillance for workers exposed to manual lifting is compulsory whenever a job is found to have *a manual handling index higher than 1.*

The guideline provides reference rules synthetically reported:

- Active health surveillance should regard all thoracolumbar spinal diseases independently from work-relatedness.
- Active health surveillance of spinal WMSDs can be performed in different steps:
    a. The first step envisages, for all exposed subjects, the administration of questionnaires or anamnestic interviews according to models that are already available in the literature.
    b. The second step envisages a clinical examination of the spine only for subjects classified as positive in the previous anamnestic survey. This examination can be made by the occupational physician in the company medical department using a standardized set of specific clinical tests and maneuvers reported in the literature.

    c. The third step applies to those subjects, identified in the two previous steps, who require more specialistic (neurological, orthopaedical, etc.) or instrumental tests (image diagnostics, EMG, etc.) to complete the individual diagnostic procedure.

- The frequency of health surveillance (first + second step) checks may be established according to relative exposure indices as well as health results obtained in the latest "round" of examinations. Since health surveillance is concerned with slowly evolving chronic degenerative diseases, three-year checks are adequate in most cases.
- One of the goals of health surveillance, from a collective viewpoint, is to check whether in each working population exposed to a specific risk, the occurrence of spinal WMSDs is other than expected. To make such comparisons, adequate reference data on the whole working population are needed. The guideline report data on the prevalence of positive cases (defined according to established criteria) for cervical, thoracic, and lumbar-sacral spine in a group of workers with low, present, or past, exposure to occupational risk factors for the spine (MMH, fixed postures, WBV). Data are subdivided by gender and ten-year age classes.
- Another goal of specific health surveillance at the individual level is the earliest possible identification of subjects affected by spinal disorders for whom it would not be advisable to allow exposure levels that were defined as permissible for healthy subjects. The guideline gives detailed criteria to manage those cases.

Risk Management and Workplace (Re)design

The guideline gives full details regarding criteria and examples of task and workplace (re)design for reducing the need or the risks connected to manual handling activities. Since the issues presented are quite common in the international literature and in other specific guidelines, they are skipped in this presentation.

PREVENTION OF WMSDs CONNECTED TO UPPER LIMBS REPETITIVE EXERTIONS AND MOVEMENTS

Program for Implementation

These guidelines come, for the first time in 2003, jointly with a three-year experimental plan of the Lombardy Region involving its application in approximately 2,000 manufacturing industries of Lombardy identified based on the kind of production (mainly in mechanical, electromechanical and electronic, textile, clothing, food and meat, plastics, and rubber processing) and number of employed workers (over 50).

The plan as well as the guidelines were agreed upon between the public authority (with functions of Labor Inspectorate), employers' associations, and trade unions.

The plan was replied to in the following years, extending the field of application to other productive sectors and company dimensions and, in 2009, the guidelines were updated (Regione Lombardia, 2009) in accordance with the new Italian legislation on health and safety work and the new international standards on the matter: EN 1005-5 (CEN, 2007) and ISO 11228-3 (ISO, 2007b).

Each plan defines the goal of risk assessment and management actions application in identified companies as well as a series of actions to be carried out by the different protagonists involved. They can be summarized as follows:

- Definition of regional guidelines agreed upon between the Prevention Regional System and social partners (trade unions and employers' associations);
- Start and conclusion of an education and training program of all public operators (labor inspectors) and operators from OSH services of concerned companies;
- Assistance provided by labor inspectors in applying guidelines;
- Preliminary risk assessment and possible consequent actions in accordance with guidelines carried out by enterprises;

- Surveillance of the state of progress of the project;
- Implementation of the recording regional system of reported work-related musculoskeletal diseases (WMSDs); and
- Implementation of a regional data website on risks and injuries caused by upper limb repetitive movements and, more specifically, on preventive solutions adapted to the benefit of all potential users.

Lastly, the plan defines the process, output, and outcome indicators to check trends and results with time.

Guidelines

The guidelines, considering the general indications in European Directive 391/89, state that each employer shall also consider the risk associated with manual repetitive tasks and upper limb repetitive movements when assessing work-related risks. If such a risk is present, a specific program is to be started to reduce it.

Therefore, guidelines that refer primarily to ISO 11228-3 in the 2009 update (ISO, 2007b)—after presenting the (epidemiologic, legal, technical) state-of-the-art on this subject—provide indications on:

- Hazard identification
- Risk estimation and assessment
- Health surveillance
- Medical-legal and insurance consequences
- (re)Design of tasks, workplaces, and working facilities in view of risk reduction

The general process as indicated in the guidelines is summarized in Table 21.2.

A preliminary assessment of possible risks develop along three successive steps:

- identification of "problematic jobs"
- risk assessment
- analytical risk assessment (in selected cases).

TABLE 21.2

Signals of a Possible Exposure to Repetitive Movements and Exertions of the Upper Limbs ("Problem Job" When One or More Signals Are Present)

1. Repetitiveness: Task(s) organized in cycles lasting up to 30 seconds or requiring the same upper limb movement (or brief group of movements) every few seconds, for at least two hours in the shift.
2. Use of force: Task(s) requiring the repetitive use of force (at least once every five minutes), for at least two hours in the shift. To this, consider the following criteria: handling of objects weighing more than 2.7 kg; the handling, between the thumb and forefinger, of objects weighing over 900 g; the use of tools requiring the application of quite maximal force.
3. Bad postures: Task(s) requiring the repetitive presence of extreme postures or movements of the upper limbs, such as uplifted arms, deviated wrist, or rapid movements, or actions requiring striking movements (such as using the hand as a tool) for at least one hour continuously or two hours in the shift.
4. Repeated impacts: Task(s) requiring the use of the hand as a tool for more than ten times in an hour, for at least two hours in the shift.

Risk Assessment

As to identification of "problematic jobs" whose exposure assessment shall be carried out in the concerned working sectors, the following criteria hold valid:

- the worker/s has/have a near-daily exposure to one or more indicators of possible exposure reported in Table 21.3;
  and/or
- there are reported cases (one or more also considering the number of workers involved) of diagnosed work-related **musculoskeletal** diseases of upper limbs.

Recently, according to ISO TR 12295 (ISO, 2014), the key enters and quick assessment procedures for upper limbs repetitive movements have been adopted (see also the relevant chapter in this volume).

As to exposure estimate, all workplaces and processing already identified as "problematic" are to be analyzed first through simplified assessment tools. With this purpose, the use of appropriate investigation tools can be made, available in the literature mostly as checklists that must be filled in by specially trained staff (Figure 21.1).

The OCRA checklist is the estimation tool suggested by the guideline and it is annexed together with related instructions for use and interpretation of results (Colombini, Occhipinti, & Grieco, 2002; Occhipinti & Colombini, 2007).

As to exposure, the final score of the OCRA checklist can be interpreted according to the classification scheme (based on the so-called traffic light system) reported in Table 21.3 (Occhipinti & Colombini, 2007).

As to risk analytical assessment, it may be necessary in some specific situations. There is no precise rule fixing when a task or a workplace needs a more detailed investigation; therefore, this decision is up to the discretion and individual fortuitous requirements. Nevertheless, the decision orientation criteria are reported in the following:

- More detailed investigation can be excluded when the results and data from risk assessment are sufficiently sound, coherent with the other contextual information, and—particularly—able to address in sufficient detail the consequent actions with respect to different risk determinants.
- Risk detailed investigation should be carried out in all the cases when the risk estimate results are uncertain or do not correspond to other contextual information (e.g., WMSD occurrence), when more data are required to define the consequent preventive actions, or when it is necessary to establish a connection more precisely between risk and damage in acknowledging a UL-WWMSD as a work-related disorder.

The preferential tool for investigating the risk in detail is the OCRA Index method (Occupational Repetitive Action) (Colombini, Occhipinti, & Grieco, 2002) as also stated in ISO 11228-3.

TABLE 21.3

Classification of OCRA Checklist Results into Five Areas for Risk Exposure Level Assessment

Zone	OCRA Values	Checklist Values	Risk Classification	Suggested Actions
Green	Up to 2,2	Up to 7,5	Acceptable	None
Yellow	2,3–3,5	7,6–11	Borderline or Very Light	New Check or Improve (*)
Red-Low	3,6–4,5	11,1–14	Light	Improve+Health Surv.+Training (2*)
Red-Medium	4,6–9,0	14,1–22,5	Medium	Improve+Health Surv.+Training (3*)
Red High	More Than 9,0	More Than 22,5	High	Improve+Health Surv.+Training (4*)

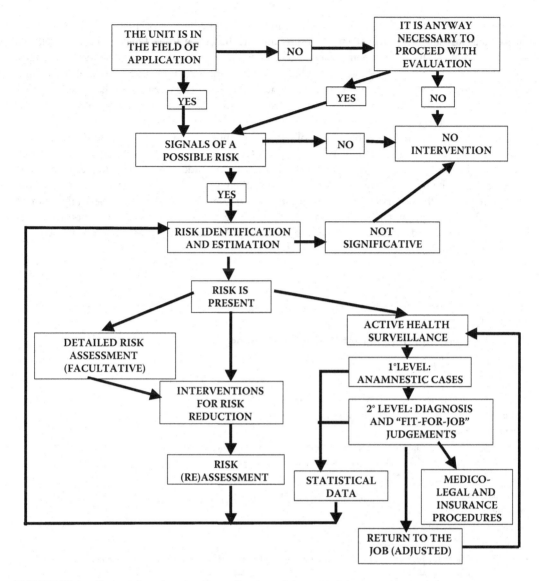

FIGURE 21.1   General flowchart for the application of the guidelines.

A special enclosure includes some considerations concerning the OCRA Index used as a probabilistic prediction tool of induced health effects (UL-WMSD) and for risk classification.

Health Surveillance Strategies

The guidelines provide detailed indications and tools for implementing and managing active health surveillance and developing all the medical-legal and insurance fulfillments resulting from the identification of fully diagnosed UL-WMSDs cases. This handbook is addressed to ergonomists and health surveillance strategies similar to those summarized for spine disorders, but details on this aspect are not reported.

Task and Workplace (Re)design

When both exposure assessment and the study of UL WMSDs have revealed a significant risk associated with repetitive and/or strenuous movements of the upper limbs, the need arises to implement specific measures aimed at re-designing tasks, procedures, workplaces, and equipments. These measures are often urgent and complex and are based on three types of coordinated and virtually simultaneous actions being carried out: structural modifications, organizational changes, and personnel training, as reported in Table 21.4. While the structural measures are almost universally accepted and widely recommended, actions involving organizational changes do not always meet with unanimous consent nor does the scientific literature provide concrete examples. The guidelines provide criteria and some concrete examples for re-designing jobs and preventing disorders caused by repetitive movements of the upper limbs. Reference is made to the three areas mentioned above, and specific indications are given for each area based on the abundant literature already available on structural modifications. A section is also devoted to the subject of possible organizational changes already investigated and applied in some field experiments and whose criteria, regarding the reduction of pace (without reducing productivity), are synthetically reported in Table 21.5. Lastly, guidelines are supplied for training programs designed to support the previous two classes of actions (i.e. structural and organizational) and devoted to workers (Tables 21.4 and 21.5).

TABLE 21.4
General Description of Different Preventive Actions

Structural modifications

- The use of ergonomic tools.
- An optimal arrangement of the workstation, furnishings, and layout.*Improve aspects related to the excessive use of force, awkward posture, and localized compressions.*

Organizational modifications

- An ergonomically designed job (pace, breaks, alternating tasks).*Improve aspects related to*:
- *movements performed frequently and repetitively for prolonged periods*;
- *absence or inadequacy of recovery periods*; and
- *job and task rotations.*

Training

- Suggestions concerning breaks.
- Appropriate information on specific risks and injuries.
- Concrete methods for performing tasks and utilizing proper techniques.*Are additional to the other interventions.*

TABLE 21.5
Brief Recommendations for Reducing the Frequency of Technical Actions (But Not Productivity)

Avoid Useless Actions: Added Arbitrarily by the Worker
⇒ Due to Manufacturing Flaws
⇒ Due to Obsolete Technologies Distribute Actions Between Both Limbs
Reduce the Repetition of Identical Actions:
⇒ By Processing Pre-Assembled Pieces
⇒ By Introducing Semi-Automatic Steps
⇒ By Replacing Manual Tasks With Hi-Tech Solutions
Reduce Auxiliary Actions:
⇒ By Creating Intersections Between the Conveyor Belt and the Work Bench

REFERENCES

CEN (2003). EN 1005-2: Safety of machinery—Human physical performance—Part 2: Manual handling of machinery and components parts of machinery.

CEN (2007). EN 1005-5: Safety of machinery—Human physical performance—Part 5: Risk assessment for repetitive handling at high frequency.

Colombini D., Occhipinti E., & Grieco A. (2002). *Risk assessment and management of repetitive movements and exertions of upper limbs: Job analysis, OCRA risk index, prevention strategies and design principles.* Ergonomics book series. Elsevier, vol. 2.

Colombini D., Occhipinti E., Alvarez-Casado E., & Waters T. (2012). *Manual lifting: A guide to the study of simple and complex lifting tasks*, Boca Raton and New York (USA): CRC Press, Taylor & Francis Group.

Conferenza dei Presidenti delle Regioni Italiane (1996). Linee guida per l'applicazione del D.LGS 626/94. *Edizioni Regione Emilia Romagna*, 359–418.

Council Directive N. 90/269 (1990). Minimum health and safety requirements for the manual handling of loads where there is a risk particularly of back injury to workers. *Official Journal of the European Communities*, N.L 156/9, 21.6.90.

Grieco A, Occhipinti E., Colombini D. & Molteni G. (1997). Manual handling of loads: The point of view of experts involved in the application of EC Directive 90/269. *Ergonomics*, *40*(10), 1035–1056.

ISO (2003). ISO 11228-1. Manual handling. Part 1: Lifting and carrying.

ISO (2007a). ISO 11228-2. Manual handling. Part 2: Pushing and pulling.

ISO (2007b). ISO 11228-3. Manual handling. Part 3: Handling of low loads at high frequency.

ISO (2014). ISO TR 12295 Ergonomics—Application document for ISO standards on manual handling (ISO 11228-1, ISO 11228-2 and ISO 11228-3) and evaluation of static working postures (ISO 11226).

Occhipinti E., & Colombini D. (2007). Updating reference values and predictive models of the OCRA method in the risk assessment of work-related musculoskeletal disorders of the upper limbs. *Ergonomics*: 50,11, 1727–1739.

Regione Lombardia (2009). *Linee guida regionali per la prevenzione delle patologie muscolo-scheletriche connesse con movimenti e sforzi ripetuti degli arti superiori* (Edizione aggiornata 2009). Decreto Dirigenziale n. 3958 del 22 aprile 2009. Available at http://www.regione.lombardia.it/

Snook S.H., & Ciriello V.M. (1991). The design of manual handling tasks: Revised tables of maximum acceptable weights and forces. *Ergonomics*, *36*(9), 1197–1213.

Waters T., Putz Anderson V., Garg A., & Fine L. (1993). Revised NIOSH equation for the design and evaluation of manual lifting tasks. *Ergonomics*, *36*(7), 749–776.

Waters T., Putz Anderson V., & Garg A. (1994). *Application manual for the revised NIOSH Lifting Equation.* Cincinnati, OH: U.S. Department of Health and Human Services.

22 Assessment of Manual Material Handling Based on Key Indicators
German Guidelines

Ulf Steinberg, Gustav Caffier, and Falk Liebers

CONTENTS

INTRODUCTION

This chapter presents a practice-based method of describing and evaluating the working conditions that prevail in the manual handling of loads. This simple method is geared to the recognition and removal of bottlenecks. Because it only covers the major activity indicators, it is called the key indicator method. This method was developed and tested from 1996 to 2001 in connection with the implementation of the European Union (EU) framework and individual directives on occupational safety and health in German national law. It consists of two independent but formally adaptable parts for lifting, holding and carrying, and for pulling and pushing. The method was drawn up in the Federal Institute for Occupational Safety and Health in close collaboration with the Committee of the Laender for Occupational Safety and Health (Länderausschuss für Arbeitsschutz und Sicherheitstechnik—LASI) with the involvement of numerous companies, scientists, accident insurance bodies, and trade unions. The method can be only used to assess risks with the aim of preventing work-related health risks from the manual handling of loads. In the six years since its first publication, this method has enjoyed a wide acceptance among possible users and a correspondingly broad application in Germany.

REQUIREMENTS FROM EUROPEAN OCCUPATIONAL SAFETY AND HEALTH LAW

German occupational safety and health law, based on the common statutory instruments of the EU, requires an assessment of the working conditions and documentation of the results of this assessment. In particular are the following Council Directives—

- **Council Directive 89/391** of June 12, 1989, on the introduction of measures to encourage improvements in the safety and health of workers at work
- **Council Directive 90/269/EEC** of May 29, 1990, on the minimum health and safety requirements for the manual handling of loads where there is a risk particularly of back injury to workers (fourth individual Directive within the meaning of Article 16 (1) of Directive 89/391/EEC)

—that have been incorporated in national law. These laws essentially contain protective goals while avoiding differentiated limits value and methodological regulations. They give the employer the responsibility for safe working conditions. For their part, the employees are obliged to conduct themselves in a safe manner.

Because in Germany about 90% of all companies are small- and medium-sized enterprises with a small workforce, a complete safety service from trained specialists is not possible. Many tasks are performed by the employers themselves or by persons within the company specially assigned for the purpose. For this reason, appropriate support for the competence of the corporate practitioners, taking due account of available resources, was a major criterion for the development of the key indicator method described in this chapter.

The application of this method is not mandatory in law. However, there is an application recommendation of the *Committee of the Laender for Occupational Safety and Health*, the body representing the authorities of the German Federal states responsible for supervising government occupational safety and health law.

ANALYSIS OF DEFICIENCIES

Risk estimates for physical load situations are, in methodological terms, very demanding and they always involve compromises. Among the three cornerstones of practicability, error of judgment, and damage model, there is an explosive area of conflict. To rationally arrive at results are particularly important in everyday judgments, for which few resources in terms of funds and time are normally estimated. As far as possible, simple algorithms are preferred, which supply a concrete result with few measurements and computation specifications. Complex relations that have a reciprocal effect and are not easy to measure are simplified. Unfortunately, the concrete figures calculated exhibit a deceptive precision that is not realistic. And unfortunately, the methodological directions do not indicate the nature and the scope of the possible error. The resulting, uncritical judgments can have far-reaching consequences.

Risk-assessment methods for physical load situations, therefore, must fulfill the following indicators in practical terms:

- Value-neutral description of the most important activity indicators
- Reliable coverage of these indicators with the lowest possible effort
- Revelation and rough quantification of relevant risks
- Indication of design bottlenecks
- Comprehensibility and traceability of the judgment by the user
- Low effort for documentation
- Calculability of assessment errors

CRITICAL CONSIDERATION OF AVAILABLE ASSESSMENT PROCEDURES

Starting with the experience gained from many years of practical ergonomic work and scientific, methodologically critical studies, many methods available worldwide were tested in a research project with the aim of making a specific application recommendation.

The following methods were evaluated, covering as they do the period between 1959–1996:

- Transport formula (Spitzer, Hettinger, & Kaminski, 1982).
- Schultetus-Burandt method (also known as Siemens-Burandt) for the determination of admissible limit values for forces and torques (Siemens AG, 1981).
- Formulae for calculating the energy conversion during physical work. (Garg, Chaffin, & Herrin, 1978)
- Work Practice Guide for manual lifting, calculation of control, and allowability limits for load weights (NIOSH, 1981).
- Manual load-displacement in standing position – determination of recommended limits values (BOSCH, 1982).
- Regulations and code of practice – Manual handling No. 8. Occupational Health and Safety Act. Melbourne (OHSA, 1988).
- Determination of maximum muscle strain when lifting and carrying (Heben und Tragen, 1987).
- Further development of the Siemens-Burandt method in the REFA chemistry experts' committee (Handhaben von Lasten, 1987).
- Simplified procedure based on the NIOSH approach of 1981 (Pangert & Hartmann, 1989).
- Der Dortmunder (computer-aided tools for the biochemical analysis of the strain on the spine when loads are manipulated) (Jäger, Luttmann, & Laurig, 1992).
- National Standard for Manual Handling (NOHSC:1001, 1990).
- Manual handling – manual handling operations regulations (HSE, 1992).
- Revised NIOSH equation for the design and evaluation of manual lifting tasks, calculation of a recommended limit load weight (NIOSH, 1981).

- ISO-CD 11228 Ergonomics, manual handling, lifting and carrying, calculation of recommended limit load weights (ISO TC 159/SC 4/WG 4: ISO CD 11228, 1994).
- ErgonLift, PC-aided calculation of biomechanical, and energy-related strain figures (Laurig & Schiffmann, 1995).
- Formula for the three-dimensional calculation of compression force (McGill, Normann, & Cholewicki, 1996).
- prEN 1005 Safety of machinery – Human physical performance, Part 2: Manual handling of machinery and component parts of machinery, calculation of recommended limit load weights when handling machines (CEN TC 122, 1993).

The titles of the methods already make clear that they were developed for different objectives. Initially, the establishment of practicability criteria had priority in connection with the humane design of work. Over the past few years, the risk estimate has become the center of attention. Against the background of the national economic relevance of persistently high sickness frequency rates with respect to the spine and the preventive notion of the EU Directives in matters of employee protection, there arose the need to conduct a prospective assessment of possible health risks.

An analysis of the studies published in the technical literature on the numerous methods revealed that these only tentatively satisfy the requirements that arise in practice. The principal problems that occurred repeatedly were that the methodological models are not comprehensible for the practical user, that methods are too often not practicable because of restrictions on their application, and that they are too laborious and that possible application errors are not defined. Apart from the high and hardly achievable effort required, this gives rise to critical application situations. The users, who are normally well practiced at their workplace, do not have a clear view of the most complicated overall system, and apply it purely schematically, if at all. This can be the cause of errors in judgment with grave consequences for personnel or the economic situation. The rejection of these methods by many of those involved is correspondingly great.

DEVELOPMENT OF THE KEY INDICATOR METHOD

MODEL APPROACH

If we examine the methods indicated in Chapter 3, checklists and analytically biomechanical or energy-related approaches dominate in the assessment of the risk from the manual handling of loads. With the biomechanical approach, the load on the lumbar spine is determined (without taking account of other joints or increasing muscular fatigue). In the energy-related approach, the aim is to minimize the strain on the heart and the circulatory system, ignoring biomechanical aspects.

With the development of the key indicator method, an attempt was made to facilitate a holistic approach and, at the same time, implement the abstract model approach for the users in concrete circumstances. The initial notions were concerned with the fact that chronic damage to the muscular and skeletal system does not progress unnoticed but emerges over an extended period in the form of a variety of complaints. Every form of constrained posture (identical muscular tension and joint positions over an extended period) leads to pain and should as far as possible be avoided by changing the posture. This discomfort becomes more evident in extreme positions of the joints. Greater physical forces, regardless of whether they involve forces of posture or action, lead to a strain on muscles, tendons, tendon roots, and ligaments. These strains are first perceived as fatigue and then if they must be applied over an extended period, as pain. From this complex unit of physical forces, postures, and positions of the joints, as well as duration and frequency, arises specific feelings of stress that result in avoidance reactions if a tolerance limit is passed. If owing to restrictions, there is no possibility of avoidance; reactions of overload can be expected.

Starting with these circumstances, which can be understood from personal experience, the key indicator method was developed based on biomechanical and energy-related knowledge and

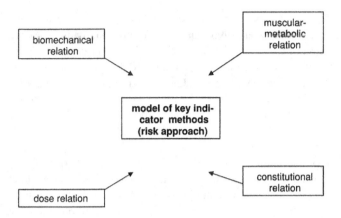

FIGURE 22.1   Model of the key indicator method.

experience. With this method, the probability of damage to the muscular and skeletal system from the manual handling of loads is evaluated. The basis for the evaluation is the acting dose. The damaging potential of load weight, posture, or working conditions thus depends on duration and frequency. The model approach of the key indicator method is based on an integrative assessment of these key indicators of the manual manipulation of loads. This approach only takes account of chronic damage and not accident-like events. The indicators of a case of damage include clinically relevant functional disturbances or pain. The direct consequences may be complaints and illnesses; the indirect consequences are work incapacity, occupational disablement, early retirement, or fluctuation.

The target variable of the method is the assessment of the risk from the manual handling of loads in the form of a risk score. This is determined by allocating a rating point to the individual key indicators according to how marked they are, and then linking them in a simple computation (see Figure 22.1).

Test of Method

To test the draft key indicator method, two different approaches were selected (revised NIOSH equation for design and evaluation of manual lifting tasks; checklist with 24 items according to the annex of the Council Directive 90/269/EEC on the minimum health and safety requirements for the manual handling of loads where there is a risk particularly of back injury to workers). They were then subjected simultaneously to a critical application test in 1995 and 1996. In 51 companies from various sectors throughout Germany, 168 workplaces were assessed in parallel using the three methods. A total of 112 individuals were involved, with different training and interests. The test encompassed the analytical-statistical part and an open discussion among all those involved on the spot.

The validity test accounted for much of the overall test. It was only possible indirectly to check whether the key indicator method can depict a largely realistic health risk. Theoretically, the frequency of diseases of the muscular and skeletal system should also increase as the rating points rise in accordance with the stress-strain formula. This linear effect was not measurable under practical conditions in Germany. Deployment of employees more in accordance with their abilities, early change of workplace, and reduction of evidently high strain rarely led to specific disorders. Instead, the "healthy worker effect" prevails. Where there are high physical strains, one encounters well-trained employees with a good constitution. The test of validity was conducted, taking data on the circumstances relating to stress-strain disorders derived from studies, expert opinions on the assessment results, and comparisons with backed-up assessments using other methods and plausibility criteria.

The result of this project was a clear decision in favor of the key indicator method. The experience using the "revised NIOSH [1993] equation for the design and evaluation of manual lifting tasks" accord with those of Dempsey (1999, 2002). By far, the largest number of activities assessed are mixed forms involving lifting, holding, carrying, and displacement of loads on the same level. Because of the limiting application criteria of this method, they cannot be assessed. The experience with the application of the checklist from the annex of the Council Directive 90/269/EEC was negative. Without any quantitative differentiation of the indicators, the evaluation was exclusively a matter of the user's own discretion. It was, therefore, just as impossible to conduct a risk evaluation as it was to ascertain the specific needs regarding action.

The development of the method was published in detail in Steinberg et al. (1998, 2000) and was discussed at numerous expert congresses. The results of the application test led to a revision of the draft key indicator method, and this was published in 1996 as a discussion proposal (LASI, 1996; Steinberg & Windberg, 1997). From 1996 to 2001, broad-based application tests were conducted. The experience with these led to a more detailed statement of the matter that was concluded for the time being with the LASI (2001) published version.

Initially, a method that takes equal account of lifting, holding and carrying, and of pulling and pushing. During the application test, it became clear, however, that this is not possible. The differences in the nature of the load and possible stresses are too great. In connection with the publication of the final version of the key indicator method in 2001, this was clearly limited to lifting, holding, and carrying. For pulling and pushing, a formally adaptable draft was published in 2002 (LASI, 2002) for public discussion and general application trials.

Error Examination

The reliability examinations conducted revealed relatively small differences in assessment for the key indicator method. If an assessment is conducted of individual persons who have precise knowledge of the working sequences being assessed, and if the applicable regulations are observed, incorrect judgments are very seldom made. When the users were compared, the lowest error proneness was found among charge-hands, foremen, and engineers.

Even so, because of the rough gradations in the load description and with correct recording, errors in the range of ±10% always had to be included in the calculation (see Figure 22.2). Assuming incorrect values like incorrect frequency, much higher error rates are possible through to totally meaningless results. The focus is, therefore. on the correct recording of indicators and a critical plausibility test. It is possible to increase the reliability of assessment by integrating it in a methods inventory (chap. 6).

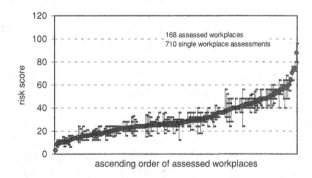

FIGURE 22.2 Mean values (filled circle) and range widths (a reference line per workplace, single values as stroke) for the point totals determined using the key indicator method, 168 workplaces, and 710 single assessments by different users (workplaces sorted in ascending order according to the level of the mean point totals).

PRESENTATION OF THE KEY INDICATOR METHOD

DESCRIPTION OF THE ACTIVITIES

The key indicator method comprises the two parts "lifting, holding, carrying loads" and "pulling, pushing loads." Both parts of the method contain descriptions of the nature and markedness of the relevant activity indicators—the so-called key indicators. This is used to define indicators that exert a major influence on the assumed effect complexes. The selection of the indicators is, therefore, geared initially to their influence on the cause-and-effect relationship. It is also important, however, that these indicators can be reliably recorded under practical conditions. The use of measured values would be desirable but it is not financially feasible because of the great expense it involves. This applies for the measurement of physical forces, postures, and working conditions. Table 22.1 gives an overview of the key indicators.

The general practical understanding of the terms has clear priority over scientific precision. This is why the term "force" is not used for pulling and pushing. The anticipated forces are roughly circumscribed instead via the indicators of weight, means of transport, positioning accuracy, and working conditions.

In the tables of the forms, the markedness of the key indicators is quantified roughly by classification into value ranges. The spectrum of value ranges largely reflects the practical conditions completely. When these value ranges are exceeded—and within the roughly incremented scales—it is possible to extrapolate as appropriate. The rough classification of the markedness of the individual indicators contained in the key indicator methods reflects two aspects. On one hand, exact and quantitative measurements of the markedness of the indicators (e.g., posture or duration) within one working day are normally only possible with great effort and time input; qualified estimates of the markedness of the indicators based on a sound knowledge of the working conditions are more effective. On the other, it is not monotonous activities that are encountered (lifting of a defined load over a certain period with a certain frequency) but mixed activities (lifting, carrying, and holding of different loads with varying frequency). Even with exact measurements via working sequence studies, relatively rough classifications are needed in such cases for an evaluation of the manual handling task.

EVALUATION

The evaluation is conducted both for lifting, holding, and carrying and for pulling and pushing in identical form as a risk rating. The probability of damage to the muscular and skeletal system is indicated—not its severity or localization. The evaluation is conducted separately from the description by multiplying the time rating by the sum of key indicator rating points. The number of points represents a measure of the prevailing risk. Between the ranges, there are fluid transitions.

The evaluation model takes account of four factors: (a) biomechanics, (b) muscular-metabolic effort, (c) acting dose, and (d) variations in constitution. Although all do not stand in isolation, they reflect the specific effects in each case.

TABLE 22.1
Overview of Key Indicators

Lifting, Holding, Carrying	Pulling, Pushing
Duration, Frequency	Duration, frequency
Load weight	Mass to be moved and transport vehicle positioning accuracy and speed of motion
Posture	Posture
Working conditions	Working conditions

The factor of *biomechanics* particularly considers the mechanical load on bones and joints from the postural and action forces applied. The forces to be transferred in the skeletal system are a measure of the internal strain and possible overstrain on individual structural elements. Possible long-term consequences are degenerative changes, and short-term overstrain may lead to fractures and cracks.

The biomechanical components are taken into consideration through the load weight, positioning accuracy, speed of motion, and posture.

The *muscular-metabolic* component relates to the activity of the muscles. Direct hazards are only possible with major cases of overload (sprains and torn muscle fiber). Otherwise, the muscles react under load situations with reversible fatigue. Increasing fatigue is, however, the cause of loss of strength and deteriorating coordination. The progress of work and, hence, the biomechanical loads, worsen. The great significance of this component is that it is perceived by the employee and that it can be measured directly (in individual cases with appropriate effort). The physiological effort under working conditions can be measured by the heartbeat and oxygen intake.

The muscular-metabolic component is taken into consideration via the duration, frequency, load weight, positioning accuracy, speed of motion, and postures.

The *dose relation* is obtained by considering the duration of the action of the biomechanical load (dose = internal force * time [Nh]) or the muscular-metabolic effort (energy = performance * time [Wh]).

The dose is taken into consideration mainly in terms of the duration.

Whereas the three components mentioned relate to activity, the *constitutional* prerequisite is considered in relation to individuals. The relationship of work strain and physical resilience must be noted. Muscular strength, endurance, physical type, and skill vary considerably. Healthy employees with sturdy bone structure and well-trained muscles are less at risk under the same load situations. The connection between individual resilience and the load situation evaluation is shown in Figure 22.3.

The evaluation is based formally on the calculated connection of the scaled activity description. Concrete figures are always calculated; they must be seen in the overall context. Two aspects are of particular importance: the fluid transitions between the ranges and the possible assessment errors (Figure 22.4).

The basic design objective should be to adhere to the range up to 25 points. This means that health damage from the manual handling of loads will largely be excluded. But it is neither physiologically necessary nor economically appropriate to assume this limit as an absolute rule. The limit is a protection for employees with low resilience. For employees able to take physically greater burdens, values around 35 are probably still acceptable (Figure 22.3). The design of work should, therefore, always take account of the efficiency present in the workforce structure. These

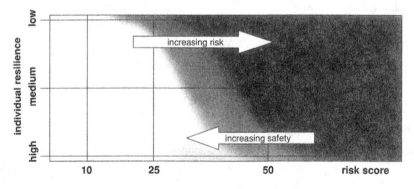

FIGURE 22.3 Consideration of the differing individual resilience in the evaluation model of the key indicator methods.

Risk range	Risk score	Description
1	< 10	Low load situation: health risk from physical overload unlikely to appear.
2	10 to < 25	Increased load situation: Physical overload is possible for less resilient persons. For this group re-design of workplace is appropriate.
3	25 to < 50	Highly increased load situation. Physical overload also possible for persons with normal resilience. Re-design of the workplace is recommended.
4	50	High load situation. Physical overload is likely to appear. Workplace re-design is essential.

FIGURE 22.4 Evaluation table per form.

ideas are based to a considerable extent on the assessment model of the *Work Practice Guide for Manual Lifting* (NIOSH, 1981).

Evaluations of the test results indicate, in this context, an occasional but serious problem— the assessment results signalize a possible risk that is not accepted as such or which cannot be eliminated with a reasonable amount of effort. Reactions are rejection of the assessment method or willful changes to the rating scales.

Nearly always, these problems arise from an excessively narrow way of looking at the matter and insufficient consideration of the assessment model. The Risk Range 2 is defined as an increased load situation, so physical overload is possible for less resilient persons. For this group, re-design of the workplace is appropriate. *Risk Range 3 is defined as* a highly increased load situation where physical overload is also possible for persons with normal resilience. Re-design of the workplace is recommended.

When terms were being drawn up, close attention was paid to these situations. The term *load situation* relates to work demands and depends on the individual. The term *strain* in this paper is the value-free reaction of an individual organism to a load as it acts. *Physical overload,* however, describes the potential negative consequences of a load situation. *Resilience* describes the ability to deal with a load situation present without any adverse stress. *Normal resilience* takes account of the very wide range of individually different physical efficiency. Muscle mass, endurance, body mass, state of health, and motivation are—alongside sex and age—some of the variables that must be considered. *Possible overload* means that damage will not always necessarily occur, but only if there is an incorrect relationship between load situation and resilience. These terms must be interpreted in the correct context.

For practical purposes—in the range of 25–50 points—this incorrect relationship must be noted. Because neither the employer nor the company doctor is in a position to make an exact estimate of resilience, an "early warning system" is of great importance. By making a sensitive record of perceived stress and health complaints on the part of employees, the load situation survey can be supplemented in terms of biomonitoring. Support by company doctors is effectively possible in the form of selected orthopedic examinations. The primary design objective, then, is not adherence to the 25-point range but an acceptable physical strain that reliably avoids damage. If the employees do not experience excessive stress, if there are no health complaints, and if there is no evidence of a higher sickness rate, there is no need for action. The prerequisite is that this must actually be verified. From 50 points, however, there is always a need for action.

Focusing on a favorable total point rating should not conceal the actual goal: the design—that is, the avoidance of bottlenecks. The focus of attention should, therefore, be on reducing high individual rating points.

It should be said here by way of warning that the simple key indicator methods primarily serve to delimit the problem. If more extensive design measures have been judged to be necessary, reasonable preparations should then always be made for investments.

FORMAL STRUCTURE

Both key indicator methods are designed in their original form as one-page worksheets. For reasons of practicality, the complete overview is important. The worksheets can be filled in directly and filed away as documentation. The reverse contains important instructions for use. A description of activities and an evaluation are drawn up in two separate stages. The description in the first stage is value-free. This means that the description remains valid without restriction later when changes may have been made to statutory regulations and assessment procedures.

The forms and the related instructions for use are attached as annexes.

LIMITS OF APPLICATION

The key indicator methods are intended only for the identification of bottlenecks and needs for action.

Evaluations of the whole work or an entire day are, thus, not possible. For this purpose, special analyses based on precise work sequence studies are needed to take account of potential synergy and compensation effects. Such studies should always remain the task of ergonomically qualified persons. The possible consequences of incorrect judgments cannot be estimated by non-specialists in ergonomics.

Use for questions with legal implications is also not permissible.

In the long term, it is also conceivable that the load situation data collected may provide the basis for more extensive data analyses. Spot-check analyses in Germany already show that the overall situation for the manual handling of loads is determined more by unfavorable postures and high time fractions than by great load weights. The efforts made over the past few decades to reduce the great load weights are already bearing fruit.

INCREASE OF ASSESSMENT RELIABILITY BY INTEGRATION IN A METHODS INVENTORY

Assessment errors can, basically, never be discounted. The assessment methods are incomplete and are subject to limitations in their application. This also applies to the key indicator methods, even though their errors are comparatively slight. The user left to his or her own resources would not, in practice, like to be confused by methodological inadequacies or to have to put in an unreasonably great effort.

One simple way out of this problem is to combine several methods that complement or check one another methodologically. For the onsite work of ergonomists and company doctors in Germany, a four-part methods inventory (Caffier & Steinberg, 2000) has been developed. It involves an average time input of less than two hours per workplace or employee to record the following:

- The objective physical strain (*key indicator method*)
- The perceived stress (*questionnaire with 47 items according to Slesina, 1987*),

- The existing health complaints (*Nordic questionnaire according to Kuorinka et al., 1987*) *and*
- Orthopedic findings (*orthopedic multistage diagnosis according to Grifka, Peters, & Bär, 2001*).

The methods can also be used singly. As with the key indicators method, the value for the corporate practitioner is not in the precise calculation and comparative evaluation but in the highlighting of relationships and the influencing factors to be considered.

DISCUSSION AND EXPERIENCE WITH APPLICATION

After seven years of experience with application, the following can be said:

- The method is currently used very widely. It has been included in numerous methodological recommendations, including German-speaking countries outside Germany itself. It is seen by many practitioners as a source of support for their professional expertise.
- The assessment results are nearly always plausible and enable one to conclude the need for action in a short time. Interpretation of the results is occasionally conducted too schematically, however, based on the classification into one of the four risk ranges and neglects the fluid transitions between the ranges.
- The proportion of relevant misapplications and misjudgments is small. The most frequent errors are assessments based on inadequate knowledge of the activity, computation errors, application to complicated work sequences, uncritical applications, and failure to give heed to the instructions. The form is frequently copied and applied without methodological explanation.
- A random compilation of the assessment results gives initial, Germany-wide overviews of the nature and scope of the physical strain involved in the manual handling of loads (Figure 22.5).
- The fears prevalent in the mid-1990s that multiple burdens would arise for companies from the obligation to assess have not become a reality. In most cases, the key indicator method is acknowledged to be an efficient tool.
- In several cases, the method has been applied far beyond its use limits for complex work designs and for appraisal purposes in legal disputes. The method is unsuitable for this purpose.
- In some cases, users have autonomously modified the method. Partly for frivolous reasons—the simple appearance is associated with the possibility of simple adaptation—and with the intention of influencing the results of the assessment. The developers of the method have not agreed and do not agree to such modifications. All experience and criticisms are collected, discussed, and decided on a consensus basis.

The prime aim in drawing up the key indicator method is to provide a suitable tool for assessing the working conditions and for justifying necessary preventive actions. A further intention of the method developers was to reduce the existing shortfall of knowledge regarding the nature and extent of work-related load situations to the muscular and skeletal system and the cause-and-effect relationship in the emergence of disorders of the muscular and skeletal system. The data routinely collected were intended to be useful when drawing up a standardized description of the load situation. At the present time, the assessment results for 747 activities involving the manual handling of loads are available in the form of a random sample (Figure 22.5). Because the distribution of the rating points for the first random sample from 1997 of 168 activities has not changed (see Figure 22.2), it can be taken as approximately representative for the load situation in Germany. The distribution is

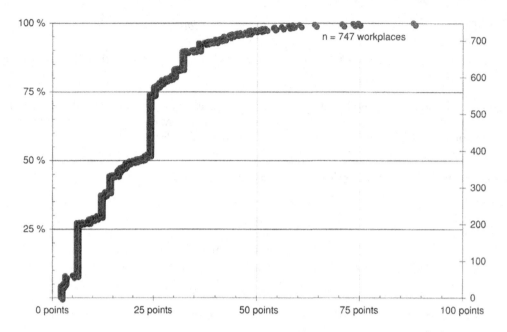

FIGURE 22.5 *Ordered distribution of risk scores in the total stock evaluated.* [Distribution of risk scores for 747 workplaces assessed using the key indicator method "Lifting, carrying, and holding loads" (in ascending order, ranges up to 25 points and up to 50 points marked)].

generally accepted as a reflection of the real situation. It documents the fact that the critical limit of 50 points is exceeded in less than 10% of the workplaces where loads are handled manually.

INTERNET PRESENTATION

Updates of the methods are accessible at the internet address http://www.baua.de. These pages also show the link with the whole methods inventory. Interactive calculation programs, partly combined with database applications, supplement the package.

REFERENCES

BOSCH. (1982). *Lastenumsetzung von Hand im Stehen—Ermittlung empfohlener Grenzwerte*. Stuttgart: Robert Bosch GmbH.

Caffier, G., & Steinberg, U. (2000). Praxisgerechtes Methodeninventar zur Umsetzung der Lastenhandhabungsverordnung. *Arbeitsschutz aktuell, 2*, 69–74.

CEN TC 122. (1993). *Ergonomics: Draft prEN 1005-2 Safety of machinery—Human physical performance. Part 2: Manual handling of machinery and component parts of machinery.*

Dempsey, P. G. (1999). Utilizing criteria for exposure and compliance assessment of multiple task manual material handling jobs. *International Journal of Industrial Ergonomics, 24*, 405–416.

Dempsey, P. G. (2002). Usability of the revised NIOSH lifting equation. *Ergonomics, 45*, 817–828.

Garg A., Chaffin, D. B., & Herrin, G. D. (1978). Prediction of metabolic rates for manual material handling jobs. *American Industrial Hygiene Association Journal. 39*, 661–674.

Grifka, J., Peters, T., & Bär, H.-F. (2001). *Mehrstufendiagnostik von Muskel-Skelett-Erkrankungen in der arbeitsmedizinischen Praxis* (Publication series of the Federal Institute for Occupational Safety and Health: Offprint, p. 62). Bremerhaven: Wirtschaftsverl. NW 2001.

Handhaben von Lasten (1987). *REFA-Fachausschusse Chemie*. 2nd edition Darmstadt: REFA-Bundesverband.

Heben und Tragen (1987). *Ermittlung maximaler Muskelbelastung nach VDI*. Daimler Benz AG.

HSE. (1992). *Manual handling—Manual handling operations regulations 1992, Guidance on regulations L23*. Sheffield: HSE Information Centre.

ISO TC 159/SC 4/WG 4: ISO CD 11228 (1994). *Ergonomics—Manual handling. Part 1: Lifting and carrying. Draft.*

Jäger, M., Luttmann, A., & Laurig, W. (1992). Ein computergestütztes Werkzeug zur biomechanischen Analyse der Belastung der Wirbelsäule bei Lastenmanipulationen: "Der Dortmunder." *Medizinische-orthopädische Technik, 112*, 305–309.

Kuorinka, I., Jonsson, B., Kilbom, A., Vinterberg, H., Biering-Sorensen, F., Andersson, G. et al. (1987). Standardisied Nordic questionnaires for the analysis of musculoskeletal symptoms. *Applied Ergonomics, 18*, 223–237.

LASI. (1996). *Handlungsanleitung zur Beurteilung der Arbeitsbedingungen beim Heben und Tragen von Lasten* (1st edition, LASI Publication 9). Länderausschuss für Arbeitsschutz und Sicherheitstechnik (Eds).

LASI. (2001). *Handlungsanleitung zur Beurteilung der Arbeitsbedingungen beim Heben und Tragen von Lasten* (4th revised edition, LASI Publication 9). Länderausschuss für Arbeitsschutz und Sicherheitstechnik (Eds.).

LASI. (2002). *Handlungsanleitung zur Beurteilung der Arbeitsbedingungen beim Ziehen und Schieben von Lasten* (LASI Publication LV29). Länderausschuss für Arbeitsschutz und Sicherheitstechnik (Eds.).

Laurig, W., & Schiffmann, M. (1995). *ErgonLIFT: Rechnerunterstützte Methodik zur Gefährdungsbewertung und Prävention beim manuellen Handhaben von Lasten*. Bielefeld: Schmidt.

McGill, S. M., Normann, R. W., & Cholewicki, J. (1996). A simple polynomial that predicts low-back compression during complex 3-D tasks. *Ergonomics, 39*, 1107–1118.

NIOSH. (1981). *Work practice guide for manual lifting* (DHHS [NIOSH] Publication, 81–122). Washington, DC: U.S. Government Printing Office.

NOHSC:1001. (1990). *National Standard for Manual Handling and National Code of Practice for Manual Handling [NOHSC: 2005 (1990)]*. In National Occupational Health and Safety Commission (Ed.). Canberra: Australian Government Publishing Service.

OHSA. (1988). *Regulations and code of practice—Manual handling Nr. 8*. Occupational Health and Safety Act 1985. Melbourne.

Pangert, R., & Hartmann, H. (1989). Ein einfaches Verfahren zur Bestimmung der Belastung der Lendenwirbelsäule am Arbeitsplatz. *Zentralblatt f. Arbeitsmedizin, Arbeitsschutz and Prophylaxe, 39*, 191–194.

Siemens AG (1981). *Ermitteln zulässiger Grenzwerte für Kräfte und Drehmomente, Arbeitsblatt*. Munich: Siemens AG.

Slesina, W. (1987). *Arbeitsbedingte Erkrankungen und Arbeitsanalyse – Arbeitsanalyse unter dem Gesichtspunkt der GesundheitsVorsorge*. Stuttgart: Enke.

Spitzer, H., Hettinger, T., & Kaminski, G. (1982). *Tafeln für den Kalorienumsatz bei körperlicher Arbeit* (6th edition). Berlin: Beuth.

Steinberg, U., Behrendt, S., Bradl, I., Caffier, G., Gebhardt, Hj, Liebers, F., et al. (2000). *Erprobung und Evaluierung des Leitfadens Sicherheit und Gesundheitsschutz bei der manuellen Handhabung von Lasten* (Publication series of the Federal Institute for Occupational Safety and Health: Research, Fb 897). Bremerhaven: Wirtschaftsverl. NW 2000.

Steinberg, U., Caffier, G., Mohr, D., Liebers, F., & Behrendt, S. (1998). *Modellhafte Erprobung des Leitfadens Sicherheit und Gesundheitsschutz bei der manuellen Handhabung von Lasten* (Publication series of the Federal Institute for Occupational Safety and Health: Research, Fb 804). Bremerhaven: Wirtschaftsverl. NW 1998.

Steinberg, U., & Windberg, H.-J. (1997). *Leitfaden Sicherheit und Gesundheitsschutz bei der manuellen Handhabung von Lasten* (Publication series of the Federal Institute for Occupational Safety and Health: Offprint, 43). Bremerhaven: Wirtschaftsverl. NW

APPENDIX 1   WORKSHEET FOR THE KEY INDICATOR METHOD FOR LIFTING HOLDING, AND CARRYING

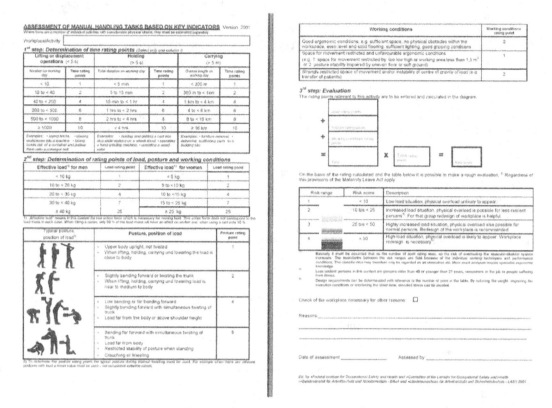

APPENDIX 2 INSTRUCTIONS FOR USE WHEN ASSESSING WORKING CONDITIONS USING THE KEY INDICATOR METHOD FOR ACTIVITIES INVOLVING LIFTING, HOLDING, AND CARRYING

Caution!

   This procedure only serves as an orienting assessment of the working conditions for lifting and carrying loads. Nevertheless, good knowledge of the manual handling task being assessed is absolutely essential when determining the time rating, load rating, posture rating, and the rating of the working conditions. If this knowledge is not present, no assessment may be made. Rough estimates or suppositions lead to incorrect results.

The assessment is basically conducted for manual handling tasks and must be related to 1 working day. If load weights or postures change within an individual activity average values must be formed. If *a number of manual handling tasks* with substantially different load manipulations arise within the overall activity, they must be *estimated* and documented *separately.*

Three steps are necessary in the assessment: First: determination of the time rating points; second, determination of the rating points for the key indicators, and third, evaluation.

In the determination of the rating points it is basically permitted to form intermediate steps (interpolation). A frequency of 40 produces the time rating point 3, for example. The only exception is the effective load of ≥40 kg for a man and ≥25 kg for a woman. These loads uncompromisingly yield a load rating of 25.

FIRST STEP: DETERMINATION OF THE TIME RATING POINTS

The time rating points are determined with reference to the table separately for three possible forms of load handling:

- For manual handling tasks characterized by the *regular repetition of short lifting, lowering, or displacement operations,* the number of operations is a determinant for the time rating points.
- For manual handling tasks characterized by the *holding* of loads, the total duration of the holding is taken: *Total duration = number of holding operations × duration of a single holding operation.*
- For manual handling tasks characterized by the *carrying* of a load, the total distance covered with the load is taken. An average speed when walking of 4 km/h, 1 m/s is assumed.

SECOND STEP: DETERMINATION OF THE RATING POINTS OF LOAD, POSTURE, AND WORKING CONDITIONS

LOAD WEIGHT

- The load rating points are determined with reference to the table separately for *men and women.*
- If, in the course of the manual handling task being assessed, different loads are handled, an *average value* may be formed where the greatest single load for men does not exceed 40 kg and for women 25 kg. For comparison purposes, peak load values can also be used. Then, however, the reduced frequency of these peaks must be taken as a basis, and in no account the total frequency.
- In the case of *lifting, holding, carrying, and setting-down activities,* the effective load must be taken. The effective load mass here is the weight force which the employee actually has to cancel out. The load is therefore not always equal to the weight of the object. When a box is tilted, only about 50% of the weight of the box acts.
- When loads are being *pushed and pulled* a separate assessment is necessary.

POSTURE

The rating points of posture are determined with reference to the pictograms in the table. The *characteristic postures during the handling of loads* must be used for the individual activity. If different postures are adopted as work progresses, an average value can be formed from the posture rating points for the manual handling task being assessed.

WORKING CONDITIONS

To determine rating points of the working conditions, the working conditions that predominate most of the time must be used. Occasional discomfort which has no safety significance will not be taken into account. Safety-relevant indicators must be documented in the text box *Check of the workplace for other reasons.*"

THIRD STEP: EVALUATION

Each task is evaluated on the basis of an *activity-related risk score* (calculation by addition of the rating points of the key indicators and multiplication with the time rating points).

- The *basis for evaluation* comprises biomechanical mechanisms of action combined with dose models. Account is taken here of the fact that the internal strain on the lower spine depends to a crucial extent on the extent to which the trunk is leaning forward and on the load weight and that it increases with increasing load duration or frequency, side bending, or twisting.
- *Summarized evaluations* are difficult with a number of manual handling tasks because they go beyond the informative scope of this orientation analysis. They normally require more extensive work analysis procedures to obtain a risk assessment.
- *Design needs that can be concluded.* From this risk estimate there is immediate evidence of design needs and approaches. Basically, the causes of high rating points must be eliminated. Specifically, these are organizational regulations in the case of high time rating points, reduction of the load weight, or the use of lifting aids in the case of high load rating points or the improvement of ergonomic conditions in the case of high posture rating points.

APPENDIX 3 WORKSHEET FOR THE KEY INDICATOR METHOD FOR PULLING, AND PUSHING

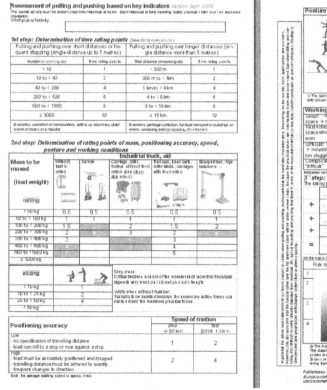

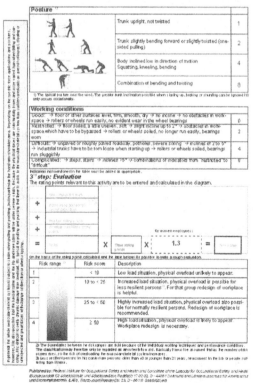

APPENDIX 4 INSTRUCTIONS FOR USE WHEN ASSESSING WORKING CONDITIONS USING THE KEY INDICATOR METHOD FOR ACTIVITIES INVOLVING PULLING, AND PUSHING

Caution!

This procedure serves for an orienting assessment of working conditions with the pulling and pushing of loads. Nevertheless, good knowledge of the manual handling task being assessed is absolutely essential when determining the time rating and the rating points for mass, positioning accuracy, speed, posture, and working conditions. If this knowledge is not present no assessment may be made. Rough estimates or suppositions lead to incorrect results.

The assessment is basically conducted for individual activities and relates to 1 working day. If load weights or postures change within an individual activity, average values must be formed. If *a number of manual handling tasks* with substantially different load manipulations arise within the overall activity, they must be *estimated* and documented *separately*.

Three steps are necessary in the assessment: First, determination of the time rating points; second, determination of the rating points for the key indicators; and third, evaluation.

In the determination of the rating points, it is basically permitted to form intermediate steps (interpolation). A frequency of 40 produces the time rating point 3, for example.

FIRST STEP: DETERMINATION OF TIME RATING POINTS

The time rating points are determined, with reference to the table, separately for pulling and pushing over short distances with frequent stopping and pulling and pushing over longer distances.

- For pulling and pushing over short distances with frequent stopping the frequency is taken as the basis.
- For pulling and pushing over longer distances the total distance is taken as the basis.

The limit value for the individual distance of 5 m should be regarded as a rough aid. In cases of doubt a decision should be taken according to which criterion arises more frequently: start-up and braking or extended pulling.

SECOND STEP: DETERMINATION OF RATING POINTS OF MASS, POSITIONING ACCURACY, POSTURE, AND WORKING CONDITIONS

MASS TO BE MOVED

The determination is conducted with reference to the table, taking account of the mass to be moved (weight of means of transport plus load) and the nature of the means of transport (industrial truck, aid). Very often drawbar-less trolleys with rollers are used. A distinction is drawn here between (steerable) steering rollers and (nonsteerable) fixed rollers.

If different loads are handled in the course of the individual activity to be assessed, an *average value* may be formed. For comparison purposes, peak load values may also be used. Then the lower frequency of these peaks must be taken as a basis and on no account the overall frequency.

Positioning Accuracy and Speed of Motion

The determination is conducted with reference to the table. The speed "fast" is equivalent to normal walking. If in special cases there are clearly faster speeds, the table can be extended as appropriate and a 4 or 8 can be given. Interpolations are permissible.

Posture

The posture rating points are determined with reference to the pictograms in the table. The *characteristic postures during the handling of loads* must be used for the individual activity. If different postures are adopted, an average value may be formed from the posture rating points for the manual handling task being assessed.

Working Conditions

To determine the rating points of the working conditions, the working conditions that predominate most of the time must be used. Occasional discomfort which has not safety significance will not be taken into account.

THIRD STEP: EVALUATION

Each task is evaluated with reference to an *activity-related risk score* (calculation by addition of rating points for the key indicators and multiplication by the time rating points). If women perform this task, the rating points are multiplied by a factor of 1.3. This takes account of the fact that women have on average about two thirds of the capacity of men.

- *The basis for the evaluation* is the probability of health damage. The nature and level of damage is not defined more closely. Account is taken of biomechanical and physiological action mechanisms combined with dose models. It is taken that the internal strain on the muscular and skeletal system depends to a crucial extent on the physical forces to be applied. These physical forces are determined by the weight of the object to be moved, the acceleration values, and the floor surface resistances. Unfavorable postures and increasing load duration or frequency increase the internal load.
  The instructions in the grey box on page 2 of the form must be adhered to.
- *Summarized evaluations* are *difficult* in the case of a number of manual handling tasks because they go beyond the informative scope of this orientation analysis. They normally require more extensive procedures of risk assessment.
- *Design needs that can be concluded.* From this risk estimate there is immediate evidence of design needs and approaches. Basically the causes of high rating points must be eliminated. Specially these are:
  - For time rating points, organizational regulations
  - For high mass rating points, reduction of load weight or use of suitable industrial trucks
  - For high rating points for speed of motion and positioning accuracy, use of wheel guides and stop buffers or reduction in workload
  - For high posture rating points, improvement in workplace design
  - The working conditions should always be "good."

23 Manual Lifting Standards

Ergonomic Assessment and Proposals for Redesign

Lilia R. Prado-León and Enrique Herrera Lugo

CONTENTS

INTRODUCTION

Low back pain (LBP) is a huge common health problem (Dionne, Dunn, & Croft, 2006; Rapoport, Jacobs, & Bell, 2004) and causes an enormous economic burden on individuals, families, communities, industry, and governments (Thelin, Holmberg, & Thelin, 2008).

In the United States, the cost of occupational injuries and illnesses is more than $170 billion annually (Occupational Safety & Health Administration, 2004). Most people who experience activity-limiting LBP go on to have recurrent episodes. Estimates of recurrence after one year ranged from 24% to 80% (Hoy, Brooks & Blythc, 2010).

The most recurrent occupational factors found were physical load, mainly in Manual Materials Handling (MMH) tasks and awkward working postures; often observed in occupations such as warehouse, blue-collar or construction work, or nursing (Zheng, Hu, & Shou, 1994; Smedley, Egger, & Cooper, 1995; Nahit et al., 2001; Plouvier, Renahy, & Chastang, 2008; Mitchell, O'Sullivan, & Burnett, 2008). Within MMH, lifting has been the task representing the greatest risk (Prado, Celis, & Avila, 2005; Xiao, Dempsey, & Lei, 2004; Mitchell, O'Sullivan, & Burnett, 2008).

Due to evidence that lifting is a risk factor for the presence of LBP, standards, and guidelines for evaluating it has been developed, along with strategies to prevent or minimize this type of risk.

CONCEPT OF MANUAL MATERIALS HANDLING

This is an occupational activity undertaken very often and in various work settings. It is characterized by the presence of one or more of the following activities that involve excess effort or overexertion and movements in extreme postures (Kumar & Mital, 1992; Pheasant, 1991; Worksafe: Travail Sécuritaire, 2011): lifting; lowering; pushing; and pulling and carrying animals, people, heavy objects, equipment, or tools without the assistance of mechanical devices.

1. NIOSH (1994) mentions that "handling means that the worker's hands move individual containers manually by lifting, lowering, filling, emptying, or carrying them" (p. 8).

From a systemic point of view, MMH includes four components (Ayoub, 1992).

- Worker
- Task
- Tools and equipment
- Setting

Table 23.1 presents the elements of the task and its indicators 1 (Ayoub & Mital, 1989).

Each MMH task requires a dynamic and static muscular effort. Damage may occur when the effort produced by these tasks exceeds the viscoelastic capacity of the ligaments, tendons, bones, and discs (Pheasant, 1991). This excess may arise from overuse, overexertion, or temporal accumulation. By overuse is understood repetitive activity (frequency) of certain MMH tasks within the workday, though this doesn't imply overload when applied to characteristics of an isolated task.

TABLE 23.1

MMH: Component Task, Its Elements, and Indicators

Elements	Indicators
Load dimensions	Measurement of mass, push/pull force requirements, mass moment of inertia, load measurement: height, width, depth
Load distribution	Localized measure of load unit (one- or two-handed)
Coupling	Measure of hand-object interface or method of gripping or fastening the load
Load stability	Localization and consistency (liquids or voluminous materials)
Layout of Workstation	Measurement of the work area's spatial properties, such as the distance of movement, obstacles, and nature of the destination
Frequency/Duration/Step	Measurement of time dimensions for the task: frequency, duration, and dynamics of the activity (short-term or long-term)
Complexity	Measurement of combined demands such as manipulation, movement requirements, activity objective, precision tolerance, and number of kinetic components

---

In other words, when a worker carries out the task of lifting 20 kg, the weight does not qualify as overload, but doing this task with a frequency of one lift per minute throughout an eight-hour workday does count as one. Overload, thus, encompasses the remaining MMH elements: load weight, lifting posture, workstation layout, complexity, load distribution, stability, etc., and finally, temporal accumulation—which refers to the total period the person has been performing MMH activities.

CONTROLS

Reducing a work activity's ergonomic risks requires taking concrete actions, called controls, which may be divided into administrative and engineering controls.

ADMINISTRATIVE CONTROLS

These refer to the organization of a task. They are considered necessary when the task evaluated presents ergonomic risk that is not severe. NIOSH (1994) mentions the following:

- Alternating heavy and light tasks
- Providing workstation variety to eliminate or reduce repetition (overuse of one particular muscle group)
- Adjusting working hours, practices, and pace
- Providing recovery time (e.g., short rest periods)
- Modifying work practices so they are carried out only in the proper zone (knees-shoulders and close to the body)
- Rotating workers

ENGINEERING CONTROLS

These are more drastic controls and, thus, are more effective. They involve engineering changes to the workstation—in other words, redesigning it. At times, this may only mean changing tools or using certain aids. Other times, it includes rearranging the location of station elements, processes, products, or materials. They are considered indispensable when the evaluated work represents high ergonomic risk.

CONCEPT OF LOW BACK PAIN

LBP refers to those pathological conditions that fall under musculoskeletal disorders (MSDs) presenting pain in the lower part of the back and are significantly related to tasks done on the job (Pheasant, 1991). The pain often radiates toward the thighs or buttocks, restricting mobility in the back, possibly causing muscular spasms due to incorrect functional use of the lumbosacral spine (Cailliet, 1990; La Dou, 1993; Crenshaw & Campbell, 1988).

LBP is frequently an incapacitating condition and many patients with acute LBP develop chronicity (Teasell & White, 1994).

From an ergonomic point of view, the principal occupational causes of MSDs are highly repetitive activities, often undertaken from inadequate postures with movement of the involved corporal segments and pressure from work equipment upon the body. Putz-Anderson (1998) also underscores the importance of non-existent or insufficient rest/recuperation.

LBP constitutes one of the most important MSDs. The initial supposition is that everyone does things that are potentially damaging to the back, but if these actions take place repetitively, a cumulative process of damage arises over weeks, months, or years. This situation causes the damage range to exceed the recuperation range, producing degenerative damage to the lumbosacral spine, which manifests in one context or another although the context may not be directly provoked by the damage but by prior antecedents (Pheasant, 1991).

STANDARDS AND GUIDELINES FOR MANUAL LIFTING

A standard is a document providing requirements, specifications, guidelines, or characteristics that may be consistently used to assure that materials, products, processes, and services are adapted to their purpose.

The International Standards Organization (ISO) develops international standards for manufacture (services as well as products), trade, and communication for all industrial branches. The ISO is a network of national standards institutes from more than a hundred countries, with a Central Ministry in Switzerland coordinating the system. Standards are developed by a technical committee made up of experts in each area (http://www.iso.org/iso/home/standards.htm).

Standards in lifting tasks are focused on providing means to achieve the objective of adapting lifting tasks to human capabilities. Only in this way may the person or team responsible for occupational health consider findings in the field of ergonomics, allowing them to limit workload to a permissible level and enhance occupational health and safety.

In this sense, it is obvious that standardization is an important and efficient way of allowing the application of ergonomic findings.

ISO 11228-1:2003 ERGONOMICS—MANUAL HANDLING

The following presents some of the most important ISO 11228-1:2003 terms and definitions pertaining to manual lifting:

Manual lifting. Moving an object from its initial position upwards without mechanical assistance
NOTE: This also includes handling people or animals.

Manual lowering. Moving an object from its initial position downwards without mechanical assistance

Ideal posture for manual handling. Standing symmetrically and upright, keeping the horizontal distance between the center of mass of the object being handled and the center of mass of the worker less than 0.25 m, and the height of the grip less than 0.25 m above knuckle height.

Unfavorable environmental conditions. Conditions that give an additional risk to the lifting or carrying task, like a hot or cold environment or slippery floor.

Ideal conditions for manual handling. Conditions that include ideal posture for manual handling, a firm grip on the object in neutral wrist posture, and favorable environmental conditions.

Repetitive handling. Handling an object more than once every five minutes.

Neutral body posture. Upright standing posture with the arms hanging freely by the side of the body.

According to ISO 11228-1:2003, risk assessment should determine the mass of the object, taking into account the lifting frequency, duration, posture, angle of asymmetry, handle design, load stability, lifting travel distance, and horizontal lifting distance.

ISO 11228-1:2003 also mentions preventive elements based on reduction of risk factors by task, object, workplace, job organization, and environmental conditions.

The first recommendation is to avoid MMH by mechanizing work tasks; the second is to redesign the job.

Recommendations for ideal MMH conditions are:

- Moderate thermal environment
- Two-handed operation
- Unrestricted standing posture

- MMH by just one person
- Gentle lifting
- Good coupling between hands and object
- Good friction between feet and floor
- Minimal incidence of MMH tasks other than lifting, as well
- The objects to be lifted not be cold, hot, or contaminated
- Vertical load-displacement should be less than or equal to 0.25 m and not below knuckle level nor above shoulder height
- Upright trunk, with no rotation.
- The load should be kept close to the body.

ISO 11228-1:2003 also mentions that a complementary—though no less important—part of prevention is providing workers with information and training.

Principal elements for evaluating a lifting task in ISO are considered in the NIOSH (National Institute of Occupational Safety and Health) equation, presenting an easier way of applying standards for those who are not specialists. The NIOSH method is also described in ISO standard 11228-1:2003—Ergonomics—Manual Manipulation—Part 1: Lifting and transport. The NIOSH Equation for evaluating lifting will be described next.

NIOSH Equation (Revised, 1991)

This tool evaluates risk for manual two-handed lifting tasks and may recommend solutions for identifying risks. It is a relatively sophisticated model with the capacity of evaluating the effect of trunk rotation, hand-coupling, significant control, and multiple tasks. It also has a wide range of work duration and lifting frequency, greater than that of the 1981 version.

Development of the equation

The equation was based on three criteria derived from scientific literature and the combined judgment of experts from the fields of work biomechanics, psychophysics, and physiology.

The criteria are:

Biomechanical. Biomechanics is an interdisciplinary field of study that integrates knowledge from biological sciences and mechanical engineering. NIOSH defines it as the study of the human body as a system operating under two sets of laws: the biological and those of Newtonian mechanics. The biomechanics approach to the problem of LBP estimates mechanical stress produced by forces acting upon the lower back by two basic measurements: (a) compressive forces generated in L5S1 of the spinal column and (b) pressure generated in the abdominal cavity.

Employing biomechanics, the NIOSH equation set 770 lbs as the maximum compression non-injury force on the disc.

Physiological. These criteria are based on the human body's metabolic and circulatory responses to various loads (Kumar & Mital, 1992). Physiology determines the rate of maximum energy cost to the entire body without fatigue to be from 2.2 to 4.7 kcal/minute.

Psychophysical. Referring to the relationship between human sensations and their physical stimuli, they focus on the subject's own perception of effort and excess effort. From the psychophysical standpoint, the maximum acceptable weight is 75% for women workers and 99% for men workers.

NIOSH methodology is based upon the equation

$$RWL = LC \times HM \times VM \times DM \times AM \times FM \times CM$$

where RWL is the recommended weight limit, LC is the constant for load weight, HM is the horizontal multiplier, VM the vertical multiplier (start of lifting), DM is the distance multiplier, AM is the asymmetry multiplier, CM is the coupling multiplier, and FM is the frequency multiplier.

Next, three (3) cases will be presented to which the NIOSH equation was applied, serving as examples for utilizing manual lifting standards for risk evaluation in the context of Mexico.

CASE STUDIES

In Mexico, many MMH—unmechanized—tasks are carried out. This country lacks precise data on the impact of LBP, but according to data provided by the Mexican Social Security Institute in 2012, dorsopathies were the second cause of disability at a national level as well as in the state of Jalisco (situated in the country's western region). There is no occupational health and safety legislation in any detailed way to address the ergonomic aspects of preventing these problems, and it is only the multinational companies that implement ergonomic programs for their workers. Applying international ergonomic standards in this context, thus, becomes particularly useful for determining the possibility of developing lumbar injuries.

Three (3) job descriptions are presented, from different businesses within the food and drink industry, all in the metropolitan area of the Mexican city of Guadalajara, in the state of Jalisco. Two of the jobs analyzed involve warehousing/merchandise handling tasks. This type of task has been documented as high risk (Gardner, Landsittel, & Nelson, 1999; St-Vincent et al., 2005). The jobs analyzed were loading dock worker (beer cartons), packer of sacks of dextrose, and loading dock worker (milk crates).

LOADING DOCK WORKER—BEER CARTONS

This worker's labor consists of placing beer cartons on a pallet, with the cartons carried to him along a conveyor belt. After the correct number of cartons has been arranged, he proceeds to secure them with PVC film, and a forklift carries the now-full pallet away.

This activity is carried out by three workers, two of whom stand along the length of the conveyor belt, taking directly from it to add to the stack; the first provides greater frequency since he takes all the cartons he can, while the second only places the cartons the first didn't manage to lift. Every half hour they rotate their posts, each worker resting after having completed the two mentioned stations and just recording the cartons stacked, while the other two continue doing the loading work (see Figure 23.1).

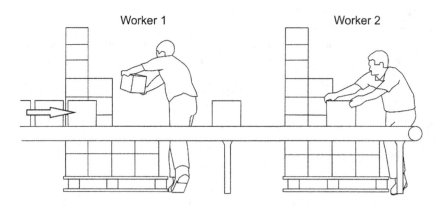

FIGURE 23.1 Loading dock workers—beer cartons.

Application of the NIOSH Equation

First contemplated for making an ergonomic evaluation of lifting this load (15 kg per carton) was a tool that would determine the compression of the worker's L5/S1 intervertebral disc; however, given the twisting present in the worker's trunk and the repetitiveness of the activity, it became necessary to use the 1991 NIOSH equation that addresses this type of task feature. Because the activity entails taking the beer cartons from a fixed (conveyor) point and placing them in an eight-level stack on a pallet (15 cm), it was decided to use the multitask variant that facilitates using the ergoweb © software.

Results

Results of the analysis showed the task to be considered is of high risk since it had a Composite Lifting Index (CLI) of 3.

In particular, results of the Single Task lifting Index (STILI) showed that placing the cartons on levels 8, 7, and 6 of the stacks, respectively, generated the greatest risk, and the first tier also represented a risk. As was expected, the levels nearest the height of the conveyor yielded lesser value and lesser risk. Still, it should be noted that all the lifting, regardless of the level at which cartons were placed, involved risks that originated from the horizontal distance between the center of the worker's ankles and the center of the hands—specifically, at the start of lifting from the conveyor belt.

Redesign Proposal

The CLI result of 3 indicated the appropriateness of implementing engineering changes to eliminate or minimize the risk, so several proposals were made aiming to reduce the impact of ergonomic risk factors on the worker's lower back. All of these incorporate the use of various technologies—some simple and others more complex, such as complete automation of the process.

Of the recommendations described in the said document, we will only mention those addressing improvements to loading dock work done by the worker:

- Using the support of a vacuum lifter elevator system so that the load weight would not be directly lifted by the worker, it is necessary to ponder this system's response speed (see Figure 23.2).
- Using a variable height turntable platform (leveler pallet rotator) so that wherever the pallet is, the stack can always be in the most advantageous position (see Figure 23.3).
- Using a deflector on the conveyor, as this directs the carton to the worker's hands and to the pallet, saving movements and effort (see Figure 23.4).

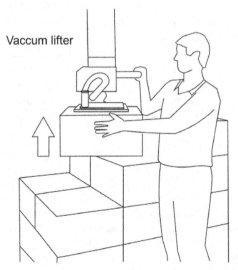

Vaccum lifter

FIGURE 23.2 Utilizing an elevator system to lift the load.

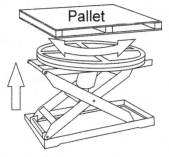

FIGURE 23.3   Using a leveler pallet rotator allows placing the stack in the most convenient position for the worker.

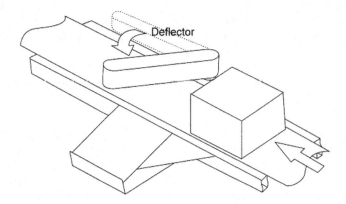

FIGURE 23.4   Using a deflector on the conveyor.

Packer—Sacks of Dextrose

This worker's responsibility is to pack powdered dextrose—a component much used in the food industry—into sacks. This job consists of placing a sack with a plastic bag inside on a scale, with one hand keeping the bag adjusted to the mouth of the feeder located at the height of the sack. This arranged, the worker proceeds to fill the sack while, with the other hand, operating the valve that allows dextrose to fall into the sack. While this is happening, he must watch the scale's digital display to ascertain when it reaches the desired weight (25 kg). To make the powder flow more freely, he has another valve that permits air to be injected, which he only activates when the material gets stuck in the ducts during the filling operation. When the weight is slightly over the limit, he stops the valve action and must then remove the excess dextrose with a spoon. Once the exact weight is obtained, he proceeds to lower the sack to the floor (see Figure 23.5) so that another worker can take the sack and sew it shut with a sewing machine. This work rotates each hour with other production line workers so that one of them prepares the dextrose mix with other additives, the just-described worker fills the sacks, the third closes them, and the fourth stacks them on the pallet.

Application of NIOSH Equation

To conduct the ergonomic evaluation of lifting sacks on this job, the first considered was the use of a tool that would determine the compression of the worker's L5/S1 intervertebral disc, since some of the biomechanical characteristics of certain movements in sack lifting turned out to be ideal for this. However, the repetitiveness with which the action was carried out (1.5 sacks per minute), as well as the worker's trunk twisting (presenting mainly when letting go of the load), supported the decision to use the 1991 NIOSH equation, which addresses these factors.

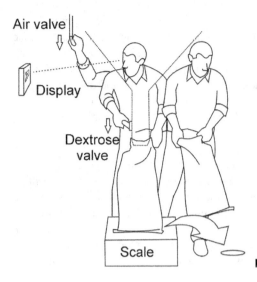

FIGURE 23.5 Packer—sacks of dextrose.

Results

Results of the 1991 NIOSH equation permitted establishing that the activity could be categorized as risky since a Lifting Index (LI) of 1.5 was obtained. Partial equation results indicated that the elements contributing most to generating risk were: first, coupling, because the worker took the sack by its upper points, hanging onto it with great effort; secondly, the lifting frequency; third, the horizontal distance between the load and the worker's body; and fourth, twisting or asymmetry at the waist while lifting.

Redesign Proposal

Obtaining an LI of 1.5 suggests the possibility of solving the problem with administrative controls alone; however, these had, in some way, already been established when implementing a job rotation among the four workers making up the production line, so it was necessary to make another suggestion. Analyzing the activity and physical characteristics of the work station brought us to the conclusion that it would be very simple to install a small roller conveyor (see Figure 23.6) or a

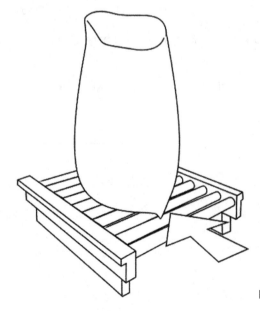

FIGURE 23.6 Roller conveyor.

smooth stainless steel platform—either of these to be at the same height—placed alongside the weighing platform to avoid completely lifting; it would be sufficient to slide the full sack to one side and, depending on the location and length of the tool, it might not need to be lifted as well by the following worker—the one sewing the shut with a sewing machine. If we were addressing the coupling multiplier as the departure point for generating a solution, we would find ourselves trying to solve the problem with a possibly complex tool or component, with the worker still manually lifting the sack. Thus, the better and more economical solution was chosen, eliminating the necessity of lifting the sack.

While the present description of the evaluation refers to lifting loads, it may be noted that at this workstation, several other worker actions were evaluated, and based on these, relevant recommendations were made. One referred to relocating the air valve that increased material flow, suggesting that it be placed at the front since its being behind the worker was what occasioned an inadequate posture for the operation, as the worker had to simultaneously hold the mouth of the sack open with the other hand.

The recommendation for the scale's digital display also consisted of relocating it to the front, since its being to the side caused the worker to turn his neck each time he wanted to see how the sack was filling.

LOADING DOCK WORKER—MILK CRATES

This worker must stack polypropylene crates containing milk cartons, with the stacks later carried by other workers to delivery trucks. Previously, in another area of the processing plant, the cartons, recently filled with milk, are placed in the crates which, once filled, are taken to the storage area with the help of a roller conveyor through an opening in the wall that connects the two areas. The first action taken in the warehouse with these crates is to stack them—five 21 kg crates high (see Figure 23.7). Once this stack is formed, the worker must transfer it to another part of the warehouse to free up space and continue stacking crates. This transfer is done with the aid of a pole with a hook on the end, which he inserts into the grid of the bottom crate, pulls the stack, walking backward to the desired point where he stops pulling and laterally pushes the stack until it is properly placed.

Application of the NIOSH Equation

In observing this job, it was directly determined that the ideal tool for ergonomically evaluating this activity was NIOSH, since the lifting task for stacking presented high lifting frequency, evident twisting at the waist, and wide horizontal distance between the hands and the center of the feet. The multitask variant of the equation was used for the various load placement levels.

Results

Of the jobs that were analyzed, this was the one showing the highest ergonomic risk to the lower back: the CLI obtained was 13.8, which indicates high risk.

FIGURE 23.7 Milk crate loader.

Upon analyzing the particular results of the equation, it was observed that the levels that mostly generated risks were the 5th, 4th, and 1st tiers, respectively. The levels providing the least risk were the 3rd and 2nd. These differences are explained by the vertical load-displacement distance being congruent with the generalized recommendation of lifting loads at waist level. However, the risk was still high at every level. Analyzing the particular multipliers determined that the riskiest factors for this task—from greatest importance—were the horizontal distance, asymmetry angle, and lifting frequency.

If we address the principle of the equation that data from lifting at the origin should be considered when there is no significant control, as was the case here, the point where trouble begins is the roller conveyor, from which the worker takes the milk crates. For the task of placing the load at the destination, the contribution to ergonomic risk is considered based upon its different vertical displacement levels.

Redesign Proposal

In this case, it is crucial to effect engineering controls, given the high lifting index obtained.

For the above-described beer carton loader, simple technical solutions are offered (see Figures 23.2–23.4); the same type of implement was also suggested for this job, but given the high risk represented, more sophisticated technologies were also recommended. At any rate, the ideal solution for this is to totally eliminate manual lifting of the load.

One of the options was to utilize an industrial robot, which—with appropriate programming—could stack cartons in the requisite manner (see Figure 23.8).

Another option is using automatic loading systems—which, for the particular nature of this case may be relatively economical (Figure 23.9).

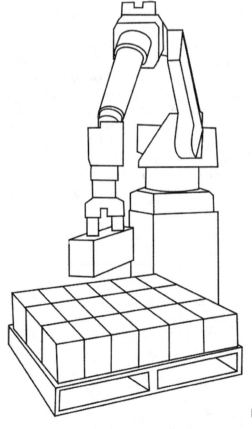

FIGURE 23.8   Robotic arm for loading.

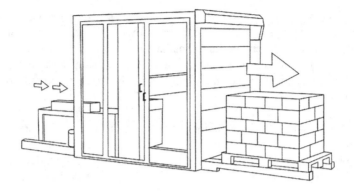

FIGURE 23.9 Automatic loading or stacking system.

CONCLUSIONS

The study cases exemplified how using international lifting standards included in the NIOSH equation makes it possible—in a relatively simple and easy way—to identify lifting task risks that may contribute to lumbar or LBP injuries. It may also be observed how the same tool offers a guideline in establishing redesign suggestions for risk prevention.

Although countries like Mexico find it difficult to completely eliminate manual lifting, these redesign suggestions have included possible effective solutions for reducing risks: handling aids such as variable height platforms, conveyor belts, or auxiliary grippers.

While evaluation of the complete job description has not been presented (for going beyond the scope of the present work), it is important to note the involvement of different tasks—not just lifting, but also other such MMH tasks such as transport and pushing-pulling—obtained that should be added to the results, along with the risks implied by the other tasks.

It is, however, necessary to note that work on the standardization of manual lifting remains to be done in the future, as current job situations are frequently not considered by the standards. An example is the case of an unusual posture we found in the food industry, in the job of product packer. Figure 23.10 shows the working posture which, according to NIOSH equation guidelines, cannot be evaluated by this tool. One may observe that the worker effects load lifting from a seated position—if the same lifts were done while standing, this would inarguably be an ideal case for evaluation with the equation.

Trying to interpolate the variables set by the NIOSH equation with this case might show each as present, but at a greater level, of risk. Horizontal and vertical localization is present at origin as

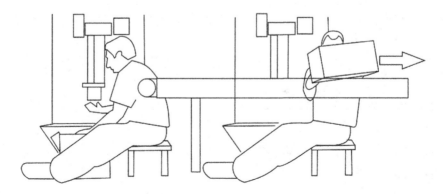

FIGURE 23.10 Packer's working posture outside NIOSH guidelines.

well as destination, as is vertical displacement and asymmetry angle albeit being seated and totally restricting movement of the legs—which in standing position would actively engage with raising by lowering the load or with turning of the body—the worker should compensate for his restrictions by displacing his upper extremities more extensively and increasing the bending and twisting of his trunk.

Other equation variables are present: time remaining seated (eight hours), repetition or frequency of the task, and (regular) coupling.

Additional workstation features could neither be evaluated with this equation. Among the notable negative characteristics are lack of seat cushioning and lumbar support. All the above clearly goes against the guidelines and standards established for injury prevention.

Thus, the analysis of this type of activity should resort to other tools, like the 3DSSPP © (see Figure 23.11), which—although analyzes static postures—makes it possible to obtain valuable information such as compressive force present in the L5/S1 intervertebral disc, forces present in different back muscles, and percentage of the population capable of doing some of the movements involved in this activity. It has the advantage of providing anthropometric data in a way that corresponds to the actual person.

To continue with atypical cases not covered in the standards, there is one-handed lifting—another situation present in actual lifting tasks that does not comply with the criteria for applying the NIOSH equation.

Lifting in teams of two or more people may also be the object of more conscientious study, which can benefit from more guidelines and standards that permit us to evaluate these types of situations.

Other elements such as vibration added to lifting—as with water delivery workers who experience whole-body vibration because their job is to not only drive the delivery truck but also distribute (raising/lowering and manually transporting the water jug)—may increase the risk of developing lumbar injury (Prado, Celis, & Avila, 2005). Thus, this is an additional factor to consider when using the NIOSH equation in estimating the risk of injury to the lower back.

As has been observed, the task of manual lifting is indubitably complex and entails a multiplicity of elements that made it more difficult to arrive at a standardization to cover all variants. However, we firmly believe that in the not-too-distant future, advances in knowledge of these factors may be included in more integral tools, allowing evaluations in the cases we have mentioned that is efficient and relatively easy.

```
Analyst: Leon;  Company: UG;  Task: Untitled; Sex: Male
Date: 11/29/14                  Analysis Summary
-----------------------------------------------------------------
  Anthropometry                   |     Percentage of Population with
  -------------                   |       Sufficient Strength Capability
Height (in)         69.7          |
Weight (lb)        165.6          |
                                  | Elbow  ####################  100
  Force on Hand    Right    Left  |
  -------------   -------  ------- | Shoulde:####################   99
Magnitude (lb)      10      10     |
Components (lb)                    | Torso  ####################    99
   X               0.0     0.0     |
   Y               0.0     0.0     | Hip    ####################    99
   Z             -10.0   -10.0     |
                                  | Knee   ####################    99
L5/S1 Disc Compression Force (lb) 1SD |
        +--------+------+          | Ankle  #                      3 SUL
  ##                      97+- 14  |
                                  |        -----------------
+--+--+--+--+|+--+--+--+|-+--+-    |         0  20 40 60 80 100
      |           |                |
    BCDL        BCUL
   Estimated Ligament Strain (%) 5.49
                      3DSSPP (3.0)
      Copyright 1995, The University of Michigan, ALL RIGHTS RESERVED
```

FIGURE 23.11 Analysis of packer posture with 3DSSPP ©.

At any rate, though current standards do not address all possible cases, examining and applying the NIOSH equation to three different jobs showed that using available lifting standards could identify risk elements and their degree of contribution to potential lower back damage, as well as help set administrative or engineering controls conducive to the prevention of musculoskeletal disorders.

Last but not any less important the is consideration of training. While acknowledging that training alone is not an ergonomic improvement, it should be utilized in conjunction with the redesign of the task or workstation (NIOSH Publication No. 2007-131, 2007; ISO 11228-1:2003, 2003). On one hand, it is necessary to focus upon training in the sense of promoting occupational education so that by applying knowledge, workers constitute their own tool for change. When someone is conscious of the damage they may do and has ready-to-hand knowledge allowing self-regulation of their own lifting tasks, their probability of modifying behavior—and of success—is better than for those given "orders" on what and what not to, without explanations as to why.

On the other, there is worker physical training. In this sense, it has been reported that gradual training in lifting can strengthen muscles, bones, ligaments, and tendons, and reduce the probability of injury (Chow, 2001). Physical training may be general or oriented to a specific task. Knapik and Sharp (1998) mention that both are effective, but general training may be useful to improve a wide range of MMH tasks while specific training offers greater gains for specific MMH tasks, such as manual lifting.

Identifying risk factors that may be modified—such as the nature of the task, its duration, and repetitiveness—by means of ergonomic standards and guidelines, could lead to establishing preventive measures that reduce the incidence of LBP, with its great individual, economic, and social repercussions. In this sense, it may be noted that epidemiologic studies have supported a conclusion that back injuries may be prevented or reduced by 33% if the workstation is redesigned (Snook, 1978; quoted by Kumar & Mital, 1992).

REFERENCES

Ayoub, M., & Mital, A. (1989). *Manual materials handling*. Bristol, PA: Taylor & Francis.

Ayoub, M. (1992). Problems and solutions in manual materials handling: The state of the art. *Ergonomics*, 35(7-8), 713–728.

Cailliet, R. (1990). *Síndromes dolorosos: Incapacidad y dolor de tejidos blandos*. México: Manual Moderno.

Chow, D. H. K. (2001). Lifting strategies. Edit. Karwowski W. *International encyclopedia of ergonomics and human factors*, Taylor & Francis: London, (Vol. 1, pp. 260–262.).

Crenshaw, A., & Campbell, A. (1988). *Cirugía Ortopédica*. 7a. edición. Argentina: Médica Panamericana.

Dionne, C. E., Dunn, K. M., & Croft, P. R. (2006). Does back pain prevalence really decrease with increasing age? A systematic review. *Age and Ageing*, 35(3), 229–234.

Gardner, L. I., Landsittel, D. P., & Nelson, N. A., (1999). Risk factors for back injury in 31,076 retail merchandise store workers. *American Journal of Epidemiology*, 150(8), 825–833.

Hoy, D., Brooks, P., & Blythc, F. (2010). The epidemiology of low back pain. *Best Practice & Research Clinical Rheumatology*, 24(6), 769–781.

ISO 11228-1:2003. (2003). *Ergonomics—Manual handling—Part 1: Lifting and carrying*.

Knapik J.J., & Sharp M. A. (1998). Task-specific and generalized physical training for improving manual-material handling capability. *International Journal of Industrial Ergonomics*, 22(3), 149–160.

Kumar, S., & Mital, A. (1992). Margin of safety for the human back: A probable consensus based on published studies. *Ergonomics*, 35(7–8) 769–781.

La Dou, J. (1993). *Medicina Laboral*. México: El Manual Moderno.

Mitchell, T., O'Sullivan, P., & Burnett, A. (2008). Low back pain characteristics from undergraduate student to working nurse in Australia: A cross-sectional survey. *International Journal of Nursing Studies*, 45, 1636–1644.

Nahit E, Macfarlane G, Pritchard C., et al. (2001). Short term influence of mechanical factors on regional musculoskeletal pain: a study of new workers from 12 occupational groups. *Occupational and Environmental Medicine*, 58(6), 374–381.

NIOSH. (1994). *Application manual for the revised lifting equation*. U.S. Department of Health and Human Services. Public. Cincinnati OH: Health Service Center for Disease Control and Prevention, National Institute for Occupational Safety and Health, Division of Medical and Behavioral Science.

NIOSH Publication No. 2007-131 (2007). *Ergonomic guidelines for manual material handling*. Cincinnati, OH: NIOSH.

Occupational Safety & Health Administration (2004). Accessed June20, 2013, from: http://www.osha.gov/publications/osha3173.pdf

Pheasant, S. (1991). *Ergonomics, work and health*. Hong Kong: McMillan Press, Scientific & Medical.

Plouvier, S., Renahy, E., & Chastang, J. (2008). Biomechanical strains and low back disorders: Quantifying the effects of the number of years of exposure on various types of pain. *Occupational and Environmental Medicine, 65*, 268–274.

Prado, L., Celis, A., Avila, R. (2005). Occupational lifting tasks as a risk factor in low back pain: A case-control study in a Mexican population. *Work, 25*, 107–111.

Putz-Anderson, V. (1998). *Cumulative trauma disorders: A manual for musculoskeletal diseases of the upper limbs*. Bristol, PA: Taylor & Francis.

Rapoport, J., Jacobs, P., & Bell, N.R. (2004). Refining the measurement of the economic burden of chronic diseases in Canada. *Chronic Diseases in Canada, 25*(1), 13–21.

M. St-Vincent, D. Denis, D. Imbeau, & M. Laberge. (2005). Work factors affecting manual materials handling in a warehouse superstore. *International Journal of Industrial Ergonomics, 35*, 33–46.

Smedley, J., Egger, P., & Cooper, C. (1995). Manual handling activities and risk of low back pain in nurses. *Occupational and Environmental Medicine, 52*, 160–163.

Teasell R., & White. (1994). Clinical approaches to low back pain. Part 1. Epidemiology, diagnosis and prevention. *Canadian Family Physician, 40*, 481–485.

Thelin, A., Holmberg, S., Thelin, N. (2008). Functioning in neck and low back pain from a 12-year perspective: A prospective population-based study. *Journal of Rehabilitation Medicine, 40*(7), 555–561.

Worksafe: Travail Sécuritaire. (2011). *Ergonomics guidelines for manual handling*. Saint John, Canada: WorkSafeNB.

Xiao, G., Dempsey, P., & Lei, L. (2004). Study on musculoskeletal disorders in a machinery manufacturing plant. *Journal of Occupational and Environmental Medicine, 46*, 341–346.

Zheng, Y., Hu, Y., & Shou, B. (1994). An epidemiologic study of workers with low back pain. *Chinese Journal of Surgery, 32*(1), 43–58.

24 ISO Technical Report 12295

Application Document for ISO Standards on Manual Handling (ISO 11228-1, ISO 11228-2, and ISO 11228-3) and Evaluation of Static Working Postures (ISO 11226)

Enrico Occhipinti and Daniela Colombini

CONTENTS

Full Name of the Standard

ISO TR 12295. Ergonomics—Application document for ISO standards on manual handling (ISO 11228-1, ISO 11228-2, and ISO 11228-3) and evaluation of static working postures (ISO 11226)

INTRODUCTION

The ISO 11228-series and ISO 11226 establish ergonomic recommendations for working postures and different manual handling tasks. These standards provide information for designers, employers, employees, and others involved in work, job, and product design such as occupational health and safety professionals.

The ISO 11228 series consists of the following parts, under the general title "Ergonomics-Manual handling":

- Part 1: Lifting and carrying (ISO, 2003)
- Part 2: Pushing and pulling (ISO, 2007a)
- Part 3: Handling of low loads at high frequency (ISO, 2007b)

ISO 11226 "Ergonomics – Evaluation of static working postures" (ISO, 2000) specifies recommended limits for static working postures without or with minimal external force exertion while considering body angles and duration.

While 11228-1, 2, and 3 and ISO 11226 are each self-contained with respect to data and methods, users may need guidance in selecting or using the standards in their specific application.

The Technical Report (TR) here presented has been published in April 2014 (ISO, 2014).

It offers a simple risk assessment methodology for small and medium enterprises and non-professional activities. For expert users, more detailed assessment methodologies are presented in the annexes.

All the contents are coherent with provisions of the main standards addressed in the title.

It is to be underlined that an ISO TR is a sort of "guideline" presenting data on the "state-of-the-art" in relation to a particular subject. The document is entirely informative in nature and does not contain matter implying that it is normative. It clearly explains its relationship to normative aspects of the subject that are dealt with in international standards related to the subject.

SCOPE OF THE TR

This application document guides the potential users of the ISO 11228-1, 2, 3 series and ISO 11226.

Specifically, it guides the potential user and provides additional information in the selection and use of the appropriate standards.

Depending on whether specific risks are present, this application document is intended to assist the user in deciding which standards should be applied. This application document has a dual scope:

1. To provide all users—particularly those who are not experts in ergonomics—with criteria and procedures:
   - to identify the situations where they can apply the standards of the ISO 11228 series and/ or ISO 11226; and
   - to provide a "quick assessment" method to easily recognize activities that are "certainly acceptable" or "certainly critical." If an activity is "not acceptable," it is necessary to complete a detailed risk assessment as set out in the standard, but it should be possible to continue with the subsequent actions. Where the quick-assessment method shows that the activity risk falls between the two exposure conditions, then it is necessary to refer to the detailed methods for risk assessment set out in the relevant standard.

This scope and approach are described in the main text of the Technical Report.

The use of the quick-assessment approach is best completed using a participatory approach involving workers in the enterprise. Such involvement is considered essential in identifying effective priorities when dealing with the different hazard and risk conditions and—where necessary—effective risk reduction measures.

2. To provide all users, especially those who are sufficiently familiar with the standards of the ISO 11228 series, with details and criteria for applying the risk assessment methods proposed in the original standards of the series. This second part of the scope is achieved through separate annexes (A, B, C) related to ISO 11228 Parts 1, 2, and 3, respectively. These provide information relevant to the practical application of methods and procedures presented in the ISO 11228-series based on application experiences of the standards. Some modifications of the methods explained in the standards are described in the present Technical Report—which is intended to be supplemental to the users—with a particular focus on applications where multiple manual tasks are performed by the same worker(s). No Annex was provided for the application of ISO 11226 on static working postures.

GENERAL APPROACH OF THE TR

All the relevant ISO standards adopt a general approach to risk assessment and management according to four basic steps:

- hazard identification
- risk estimation
- detailed risk evaluation
- risk reduction

As a consequence, the TR envisages the following levels (Figure 24.1):

FIRST LEVEL
Addressed to preliminary identification of the main hazards (or problems) associated with working condition and priority identification via specific "Key Enters."

SECOND LEVEL
Focused on risk factors for WMSDs and consisting of a "quick assessment" of identified hazards (via Key Enters).

THIRD LEVEL
As a result of the second level, recognized risk estimation tools are used (from corresponding ISO standards).

Beyond the third level, if the study (finalized to subsequent stages of preventive measure adoption) needs more details, a reference could be made to more analytical methods proposed by standards. Detailed methods, however, must be used only upon definite circumstances by skilled and trained staff.

HAZARD IDENTIFICATION BY KEY ENTERS

This step is aimed at checking the existence of a working hazard/problem (hazard identification)—in this case consisting of a potential biomechanical overload for WMSDs—and whether a further analysis is necessary.

The ISO Application Document (as well as the standards it refers to) draws some useful elements for key entries to access to subsequent assessment (second and, maybe, third level) of the different conditions of "possible" biomechanical overload.

Table 24.1 reports the "key enters" representing the first (entry) level of hazard identification for the consequent application of related ISO standards.

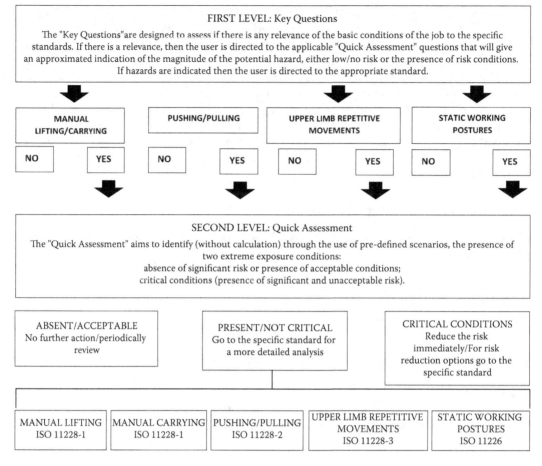

FIGURE 24.1 Flowchart of the approach to risk identification and estimation by ISO TR 12295.

QUICK ASSESSMENT

The "Quick Assessment" aims to identify, without calculation, the presence of two "extreme" exposure conditions:

- the absence of risk or acceptable risk
- the presence of a relevant risk (or the presence of extremely hazardous risk factors that are not acceptable), also labeled as critical conditions (critical code)

When either of these conditions is met, it is not necessary to make a more detailed estimation of the exposure level using the corresponding standard (the applicable standard can still provide ideas and information for the correction of the risk factors). However, when none of the two "extreme" conditions is met, it is necessary to conduct a risk assessment by methods reported in the corresponding standard.

It is emphasized that the use of the quick-assessment method is best completed using a participatory approach involving workers in the enterprise. Such involvement is considered essential to identify effective priorities when dealing with the different hazard and risk conditions and—where necessary—risk reduction measures.

Criteria for this level, especially if aimed at quickly checking acceptable conditions, are often explicitly present in the international technical standards. As an example, see steps 1 and 2 of ISO

TABLE 24.1

Key Enters to the Evaluation of Different Conditions of Biomechanical Overload as Considered in ISO 11226 and 11228 (parts 1-2-3)

THE KEY-QUESTIONS

1	*Application of ISO 11228-1*		

Is there manual lifting or carrying of an object of 3 kg or more present?   NO   YES

If NO, then this standard is not relevant, go to the next Key Questions regarding the other standards

If YES, then go to step 2 of Quick Assessment

2   *Application of ISO 11228-2*

Is there two-handed whole-body pushing and pulling of loads present?   NO   YES

if NO, then this standard is not relevant, go to the next Key Questions regarding the other standards

If YES, then go to step 2 of Quick Assessment

3   *Application of ISO 11228-3*

Are there one or more repetitive tasks of the upper limbs with a total duration of one hour or   NO   YES
more per shift?

Where the definition of "repetitive task" is: *task characterized by repeated work cycles*

or

tasks during which the same working actions are repeated for more than 50% of the time.

If NO, then this standard is not relevant, go to the next Key Question regarding the other standards

If YES, then go to step 2 of Quick Assessment

4   *Application of ISO 11226*

Are there static or awkward working postures of the HEAD/NECK, TRUNK, and/or UPPER   NO   YES
AND LOWER LIMBS maintained for more than four seconds consecutively and repeated for a
significant part of the working time?

For example:

– *HEAD/NECK (neck bent back/forward/sideways, twisted)*

– *TRUNK (trunk bent forward/sideways/, bent back with no support, twisted)*

– *UPPER LIMBS (hand(s) at or above the head, elbow(s) at or above shoulder, elbow/hand(s)*
behind the body, hand(s) turned with palms completely up or down, extreme elbow flexion-
extension, wrist bent forward/back/sideways)

– *LOWER LIMBS (squatting or kneeling) maintained for more than 4 seconds consecutively and*
repeated for a significant part of the working time

If NO, then this standard is not relevant.

If YES, then go to step 2 of Quick Assessment

11228-1 or method 1 of the EN 1005-2 standard (CEN, 2003). They both, as to manual lifting, address to quickly check the congruence of the lifted mass and related lifting frequency with reference values provided in the standards themselves. In other cases, such as in repetitive manual work, the reference to a quick acceptability assessment is maybe less explicit but can still be deducted from the standard text itself. In this sense, EN 1005-5 (CEN, 2007) provides more elements than ISO 11228-3.

On the other hand, for quick assessment of extreme/critical conditions, it is possible to apply definitions and criteria inherent in the methods recommended by standards-setting the presence of one or more extremely problematic element(s). Such are the values of weights lifted beyond the maximum recommended value, extreme lifting areas, extremely high action frequencies with upper limbs, the presence of repetitive maximal strength demands, etc.

Lifting and Carrying

A preliminary check of some adverse environmental, object, and organizational conditions is highly recommended since they could represent an additional risk in manual handling; then, one can apply questions addressed for checking an acceptable condition (Table 24.2). If all the listed conditions are present (i.e. "YES" answers), the assessed task is acceptable (Green area) and it is not necessary to continue the risk evaluation.

If any of the conditions is not met, apply ISO 11228-1, step 3. Practically, apply the Revised NIOSH Lifting Equation (Waters et al., 1993).

The "Quick Assessment" procedure could also be used in identifying critical conditions (for lifting and carrying). The term critical condition means that the manual lifting and/or carrying of objects is not recommended. If any of the conditions reported in Table 24.3 is met, a critical situation in lifting and/or carrying is present, and an ergonomic intervention is necessary to re-design the task as a high priority.

Whole Body Pushing and Pulling

A preliminary check of some adverse environmental, object, and organizational conditions is highly recommended since those conditions could represent an additional risk in whole-body pushing and pulling; subsequently one can apply questions addressed for checking an acceptable condition for pushing and pulling (Table 24.4).

The "Quick Assessment" procedure is also used in identifying critical conditions for pushing and pulling (Table 24.5).

If at least one of the conditions reported in Table 24.5 is met, a critical situation in pushing and/or pulling is present, and an urgent ergonomic intervention is necessary to redesign the task as a high priority. The critical conditions given here are indicated in ISO 11228-2.

Please note that, when considering *force magnitude*, one can consider the experience of worker (s) in terms of the perceived effort. In determining the perceived effort, the use of the CR-10 Borg scale (Borg, 1998) is suggested in estimating the force developed during pushing and/or pulling.

Repetitive Task(s) of the Upper Limbs

To establish acceptable risk, use Table 24.6. If all the listed conditions are present, then the examined task is in the Green area (ACCEPTABLE), and it is not necessary to continue the risk evaluation. If any of the conditions is not met, address to ISO 11228-3, Method 1, and, when necessary, Method 2.

A quick assessment can also be used in identifying "critical conditions." If any of the conditions are met, then a critical situation is present and an ergonomic intervention is necessary to redesign the task as a high priority (Table 24.7).

Static Working Postures

To establish acceptable risks, use Table 24.8. If any of the conditions are not met, it is necessary to apply ISO 11226.

For static working posture, a quick assessment of "critical condition" has not been provided by TR 12295, given the scarcity of applicative information on this peculiar aspect.

It is important to underline that once hazards have been identified (by key enters and quick assessment), especially in the case of "critical conditions," every effort should be aimed at reducing the risk rather than proceeding with useless—and sometimes very complex, more detailed—risk evaluations. As a consequence, simple and concrete solutions for limiting the main risk determinants, both mechanical and organizational, could be suggested according to priorities.

TABLE 24.2

Lifting and Carrying—Preliminary Factors to be Considered and Quick Assessment of Acceptable Conditions

LIFTING AND CARRYING: PRELIMINARY ASPECTS

Is the working environment unfavorable for manual lifting and carrying?

Presence of extreme (low or high) temperature		NO	YES
Presence of slippery, uneven, unstable floor		NO	YES
Presence of insufficient space for lifting and carrying		NO	YES

Are there unfavorable object characteristics for manual lifting and carrying?

The size of the object reduces the operator's view and hinder movement		NO	YES
The load center of gravity is not stable (example: liquids, items moving around inside of object)		NO	YES
The object shape/configuration presents sharp edges, surfaces, or protrusions		NO	YES
The contact surfaces are too cold or too hot		NO	YES
Does the task(s) with manual lifting or carrying last more than eight hours a day?		NO	YES

If all the questions are answered "NO," then continue the quick assessment. If at least one of the questions is answered "YES," then APPLY THE STANDARD ISO 11228-1. The consequent specific additional risks HAVE TO be carefully considered to MINIMIZE THESE RISKS.

LIFTING: QUICK ASSESSMENT—ACCEPTABLE CONDITION green code)

3 TO 5 kg	Asymmetry (e.g., body rotation, trunk twisting) is absent	NO	YES
	Load is maintained close to the body	NO	YES
	Load vertical displacement is between hips and shoulders	NO	YES
	Maximum frequency: less than five lifts per min	NO	YES
5.1 TO 10 kg	Asymmetry (e.g., body rotation, trunk twisting) is absent	NO	YES
	Load is maintained close to the body	NO	YES
	Load vertical displacement is between hips and shoulder	NO	YES
	Maximum frequency: less than one lift per min	NO	YES
MORE THAN 10 kg	Loads of more than 10 kg are not present	NO	YES

If all the questions are answered "YES," then the examined task is in the green area (ACCEPTABLE) and it is not necessary to continue the risk evaluation. If at least one of the questions is answered "NO," then evaluate the task(s) by ISO 11228-1.

CARRYING: QUICK ASSESSMENT—ACCEPTABLE CONDITION

RECOMMENDED CUMULATIVE MASS (total kg carried during the given durations for the given distance below): is the carried cumulative mass LESS than recommended values considering distances (more/less than ten m) and duration (one min; one hour; eight hours)?

Duration	Distance ≤ 10 m per action	Distance > 10 m per action		
8 hours	10,000 kg	6000 kg	NO	YES
1 hour	1500 kg	750 kg	NO	YES
1 min	30 kg	15 kg	NO	YES
	Awkward postures during the carrying are not present		NO	YES

If all the questions are answered "YES," then the examined task is in the green area (ACCEPTABLE) and it is not necessary to continue the risk evaluation. If at least one of the questions is answered "NO," then evaluate the task(s) by ISO 11228-1.

To this aim, reference could be made to the corresponding standards or to several proposals by the literature or the institutional websites of the international or national authorities dealing with occupational safety and health.

TABLE 24.3

Lifting and Carrying—Quick Assessment—Critical Condition

LIFTING AND CARRYING – QUICK ASSESSMENT: CRITICAL CONDITION.

If only one of the following conditions is present, risk has to be considered as HIGH and it is necessary to proceed with task re-design

CRITICAL CONDITION: presence of lifting/carrying task lay-out and frequency conditions exceeding the maximum suggested

VERTICAL LOCATION	The hand location at the beginning/end of the lift is higher than 175 cm or lower than 0 cm.	NO	YES
VERTICAL DISPLACEMENT	The vertical distance between the origin and the destination of the lifted object is more than 175 cm	NO	YES
HORIZONTAL DISTANCE	The horizontal distance between the body and load is greater than the full arm reach	NO	YES
ASYMMETRY	Extreme body twisting without moving the feet	NO	YES
FREQUENCY	More than 15 lifts per min of SHORT DURATION (manual handling lasting no more than 60 min. consecutively in the shift, followed by at least 60 min of break-light task)	NO	YES
	More than 12 lifts per min of MEDIUM DURATION (manual handling lasting no more than 120 min consecutively in the shift, followed by at least 30 min of break—light task)	NO	YES
	More than eight lifts per min of LONG DURATION (manual handling lasting more than 120 min consecutively in the shift)	NO	YES

CRITICAL CONDITION: presence of loads exceeding following limits

Males (18–45 years)	**25 kg**	NO	YES
Females (18–45 years)	**20 kg**	NO	YES
Males (<18 or >45 years)	**20 kg**	NO	YES
Females (<18 or >45 years)	**15 kg**	NO	YES

CRITICAL CONDITION FOR CARRYING: presence of cumulative carried mass greater than those indicated

Carrying distance 20 m or more in eight hours/Carrying distance per action 20 m or more	**6000 kg in eight hours**	NO	YES
Carrying distance less than 20 m in eight hours/Carrying distance per action less than 20 m	**10,000 kg in eight hours**	NO	YES

If at least one of the conditions has a "YES" response, then a critical condition is present. If a critical condition is present, then apply ISO 11228-1 for identifying urgent corrective actions.

TABLE 24.4

Pushing and Pulling—Preliminary Factors to Be Considered and Quick Assessment of Acceptable Conditions

PUSHING AND PULLING: PRELIMINARY ASPECTS

Working environment conditions

Are floor surfaces slippery, not stable, uneven, have an upward or downward slope or are fissured, cracked, or broken?	NO	YES
Are restricted or constrained movement paths present?	NO	YES
Is the temperature of the working area high?	NO	YES

Characteristics of the object pushed or pulled

Does the object (or trolley, transpallet, etc.) limit the vision of the operator or hinder the movement?	NO	YES
Is the object unstable?	NO	YES
Does the object (or trolley, etc.) have hazardous features, sharp surfaces, projections, etc. that can injure the operator?	NO	YES
Are the wheels or casters worn, broken, or not properly maintained?	NO	YES
Are the wheels or casters unsuitable for the work conditions?	NO	YES

If the answers for all the conditions are "NO," then continue the quick assessment. If at least one of the answers is "YES," then apply ISO 11228-2. The consequent specific additional risks HAVE TO be carefully considered to MINIMIZE THESE RISKS.

PUSHING AND PULLING: QUICK ASSESSMENT—ACCEPTABLE CONDITION

Hazard	**Force magnitude**		
	The force magnitude does not exceed approx. 30 N (or approximately 50 N for frequencies up to once per five min up to 50 m) for continuous (sustained) force exertion and approx. 100 N for peak (initial) force application. Alternatively, the perceived effort (obtained by interviewing the workers using the CR-10 Borg scale) shows the presence, during the pushing-pulling task(s), of an up to SLIGHT force exertion (perceived effort) (score 2 or less in the Borg CR-10 scale).	NO	YES
Hazard	**Task duration**		
	Does the task(s) with manual pushing and pulling last up to eight hours a day?	NO	YES
Hazard	**Grasp height**		
	The push-or-pull force is applied to the object between hip and mid-chest level.	NO	YES
Hazard	**Posture**		
	The push-or-pull action is performed with an upright trunk (not twisted or bent).	NO	YES
Hazard	**Handling Area**		
	Hands are held inside shoulder width and in front of the body.	NO	YES

If all the questions are answered "YES," then the examined task is in the green area (ACCEPTABLE) and it is not necessary to continue the risk evaluation. If at least one of the questions is answered "NO," then evaluate the task(s) by ISO 11228-2.

TABLE 24.5
Pushing and Pulling—Quick Assessment—Critical condition

PUSHING AND PULLING - QUICK ASSESSMENT: CRITICAL CONDITION.

If only one of the following conditions is present, risk has to be considered as HIGH and it is necessary to proceed with task re-design.

Hazard **FORCE MAGNITUDE**

    A. Peak initial force during push-or-pull (to overcome rest state (inertia) or to accelerate NO YES
or to decelerate an object): The force is at least 360 N (males) or 240 N (females).

    B. Continuous (sustained) push-or-pull (to keep an object in motion): The force is at least
250 N (males) or 150 N (females)Alternatively, during the pushing-pulling task(s), the
perceived effort using the CR-10 Borg scale (obtained by interviewing the workers),
shows the presence of high peaks of force (perceived effort) (a score of 8 or more on the
Borg CR-10 scale)?

Hazard **POSTURE**

The push-or-pull action is performed with the trunk significantly bent or twisted.   NO YES

Hazard **FORCE EXERTION**

The push-or-pull action is performed in a jerky manner or in an uncontrolled way.   NO YES

Hazard **HANDLING AREA**

Hands are held either outside the shoulder width or not in front of the body.   NO YES

Hazard **GRASP HEIGHT**

Hands are held higher than 150 cm or lower than 60 cm.   NO YES

Hazard **FORCE DIRECTION**

The push-or-pull action is superimposed by relevant vertical force components ("partial   NO YES
lifting").

Hazard **TASK DURATION**

Does the task(s) with manual pushing and pulling lasts more than eight hours a day?   NO YES

If one or more answers are "YES," then a critical condition is present. If a critical condition is present, then apply ISO 11228-2 for identifying corrective actions.

TABLE 24.6
Quick Assessment for Activities with Manual Repetitive Movements: Check of an Acceptable Condition (Green Area)

REPETITIVE TASK(S) OF THE UPPER LIMBS: QUICK ASSESSMENT - ACCEPTABLE CONDITION

Are either of the upper limbs working for less than 50% of the total time duration of repetitive task(s)?	NO	YES
Are both elbows held below shoulder level for 90% of the total duration of the repetitive task(s)?	NO	YES
Is there a moderate force (perceived effort = 3 or 4 in CR-10 Borg scale) exerted by the operator for no more than one hour during the duration of the repetitive task(s)?	NO	YES
Absence of force peaks (perceived effort = 5 or more in CR-10 Borg scale)?	NO	YES
Presence of breaks (including the lunch break) that last at least eight min every two hours?	NO	YES
Are the repetitive task(s) performed for less than eight hours a day?	NO	YES

If all the questions are answered "YES," then the examined task is in the green area (ACCEPTABLE) and it is not necessary to continue the risk evaluation. If at least one of the questions is answered "NO," then evaluate the task(s) by ISO 11228-3.

TABLE 24.7

Quick Assessment for Activities with Manual Repetitive Movements: Check of Presence of a Critical Condition

REPETITIVE TASK(S) OF THE UPPER LIMBS: QUICK ASSESSMENT: CRITICAL CONDITION.

Are technical actions of a single limb so fast that cannot be counted by simple direct observation?	NO	YES
One or both arms are operating with elbow at shoulder height for half or more than the total repetitive working time?	NO	YES
A "pinch" grip (or all the kinds of grasps using the fingers tips) is used for more than 80% of the repetitive working time?	NO	YES
There are peaks of force (perceived effort = 5 or more in CR-10 Borg scale) for 10% or more of the total repetitive working time?	NO	YES
There is no more than one break (lunch break included) in a shift of 6–8 hours?	NO	YES
Total repetitive working time is exceeding eight hours within a shift?	NO	YES

If at least one of the questions is answered "YES" then a critical condition is present. If a critical condition is present, then apply ISO 11228-3 for identifying urgent corrective actions.

TABLE 24.8

Static Working Postures—Quick Assessment of Acceptable Conditions

Head and trunk evaluation

Are both the trunk posture AND the neck posture symmetrical?	NO	YES
Is the trunk flexion to the front less than 20° OR, in case of backward inclination, is the trunk fully supported?	NO	YES
Is there trunk flexion between 20° and 60° to the front AND is the trunk fully supported?	NO	YES
Is neck extension absent OR in case of neck flexion to the front, is it less than 25°?	NO	YES
Is backward head inclination fully supported OR, in case of head inclination to the front, is it less than 25°?	NO	YES
If sitting, is a convex spinal curvature absent?	NO	YES

Upper limb evaluation (evaluate the more loaded limb)—Right/Left

Are awkward upper arm postures absent?	NO	YES
Are the shoulders not raised?	NO	YES
Without full-arm support, is the upper arm elevation less than 20°?	NO	YES
With full arm support, is there an upper arm elevation up to 60°?	NO	YES
Are extreme elbow flexion/extension AND extreme forearm rotation absent?	NO	YES
Is extreme wrist deviation absent?	NO	YES

Lower limb evaluation (evaluate the more loaded limb)—Right/Left

Is extreme knee flexion absent?	NO	YES
Is the knee not flexed in standing postures?	NO	YES
Is there a neutral ankle position?	NO	YES
Is kneeling or crouching absent?	NO	YES
When sitting, is the knee angle between 90° and 135°?	NO	YES

If all the questions are answered "YES," then the examined task is in the green area (ACCEPTABLE) and is not necessary to continue the risk evaluation. If at least one of the questions is answered "NO," then evaluate the task(s) by ISO 11226.

ANNEX A: APPLICATION INFORMATION REGARDING ISO 11228-1 (LIFTING AND CARRYING)

The purpose of this Annex is to provide the "expert" users of ISO 11228-1 with useful information that is needed to perform a risk assessment of manual lifting and carrying activities.

The Annex gives details on the following aspects:

- reference masses to be used when considering gender and age;
- classification of the results of risk assessment, introducing the concept of the Lifting Index (LI);
- demonstration (by an example) of a task evaluation that emphasizes the need to address work;
- organization;
- an approach (derived from the standard) for the analysis of manual lifts operated by several (two or more) workers;
- an approach for the analysis of lifts carried with one upper limb;
- carrying limits in other than "ideal conditions"; and
- the evaluation of variable lifting tasks (when different masses are lifted while holding different body postures (by considering various load placement positions) with examples for the calculation of Variable Lifting Index (VLI).

Some of the more relevant of these aspects will be briefly reported here.

REFERENCE MASS

By considering the contents of Table C.1 in ISO 11228-1 and of similar tables in other relevant standards (CEN, 2003), the following Reference Mass could be adopted in relation to the gender/age of the working population (Table 24.9).

LIFTING INDEX AND CONSEQUENT CLASSIFICATION

Once the Recommended Limit for Mass (m_R) has been computed—starting from Reference Mass (m_{ref}) reported in Table 24.9—and using procedures and equations given in step 3 of ISO 11228-1 (see also Revised NIOSH Lifting Equation by Waters et al., 1993), it is possible to compare the Actually Lifted Mass (m_A) with the resulting Recommended Limit (m_R) by computing a Lifting

---

TABLE 24.9
Suggested Reference Masses Related to Age and Gender When Applying ISO 11228-1

Working Population by Gender and Age	Reference Mass (m_{ref})
Men (18–45 years old)	25 kg
Women (18–45 years old)	20 kg
Men (<18 or >45 years old)	20 kg
Women (<18 or >45 years old)	15 kg

Note: 23 kg is the reference mass used in the U.S. National Institute of Occupational Safety and Health (NIOSH) Lift Equation, which is the source of the lifting analysis method used in ISO 11228-1. The use of 23 kg as the reference mass accommodates at least 99% of male health workers and at least 75% of healthy female workers at LI = 1.0.

TABLE 24.10

Classification of Exposure by Lifting Index Values

Lifting Index Value	Exposure Level	Interpretation	Consequences
LI ≤ 1.0	Acceptable	*Exposure is acceptable for most members of the reference working population.*	Acceptable: no consequences
1.0 < LI ≤ 2.0	Risk present	*A part of the adult industrial working population could be exposed to a moderate risk level.*	Redesign tasks and workplaces according to priorities
2.0 < LI ≤ 3.0	Risk present; high level	*An increased part of the adult industrial working population could be exposed to a significant risk level.*	Redesign tasks and workplaces as soon as possible
LI > 3.0	Risk present; very high level[a]	*Absolutely not suitable for most working population.[a]*	Redesign tasks and workplaces immediately

Notes

[a] Consider only for exceptional circumstances where technological developments or interventions are not sufficiently advanced. In these exceptional circumstances, increased attention and consideration must be given to the education and training of the individual (e.g., specialized knowledge concerning risk identification and risk reduction).

Index (LI). The Lifting Index (LI) is equal to the ratio between the actually lifted mass (m_A) and the corresponding recommended limit for mass (m_R).

When using the Lifting Index, the classification of results, coherent with the one given in ISO 11228-1 step 3, becomes:

- Lifting index ≤ 1 = acceptable condition.
- Lifting index > 1 = not recommended condition.

For a better interpretation of the resulting Lifting Index—especially when they are greater than 1 (not recommended condition) or to better address the intervention priorities with reference to relevant scientific literature on the matter—one can refer to Table 24.10.

Multitask Lifting Analysis

To this peculiar but relevant aspect from an application point of view, TR 12295 devotes several pages that are very hard to synthesize here.

Detailed procedures of analysis are given after defining the lifting task characteristics according to the criteria given below:

- MONO TASK is defined as those that involve lifting of only one kind of object (with the same load), always using the same postures (body geometry) in the same layout, origin, and destination. In this case, the "traditional" Lifting Index (LI) computational procedure could be followed as substantially reported in ISO 11228-1 and in Waters et al. (1993).
- COMPOSITE TASK is defined as tasks that involve lifting objects (generally of the same kind and mass) using different geometries (collecting and positioning from/on shelves placed at several heights and/or depth levels). Each individual geometry is a task "variant" and takes the name of "subtask." In this case, the Composite Lifting Index (CLI) computational procedure could be applied, as presented in the Applications Manual for the Revised NIOSH

Lifting Equation (Waters, Putz-Anderson, & Garg, 1994). It is to be underlined that no more than 10–12 variants or subtasks could be considered by this procedure.

- VARIABLE TASK is defined as a lifting task where both the geometry and load mass vary in different lifts performed by the worker(s) in the same period. The VLI (Variable Lifting Index) procedure is suggested in assessing these complex types of lifting tasks. The procedure is briefly illustrated in TR 12295, but reference is made to the volume "*Manual lifting: A guide to the study of simple and complex lifting tasks*" by Colombini et al. (2012).
- SEQUENTIAL TASK is defined as a job where the worker rotates between two or more mono task, composite task, and/or variable task during a work shift (each task lasting no less than 30 min consecutively). In this case, the Sequential Lifting Index (SLI) computational procedure could be followed, as addressed in Waters, Lu, and Occhipinti (2007).

ANNEX B: APPLICATION INFORMATION FOR ISO 11228-2 (PUSHING AND PULLING)

In ISO 11228-2, risks related to pushing or pulling are estimated and assessed in a multi-disciplinary approach considering physiological, psychophysical, and biomechanical capabilities. The two methods to evaluate and assess the risks from pushing-or-pulling tasks are provided:

- Method 1 represents a simple risk assessment checklist and psychophysically-based tables resulting in a two-zone risk assessment approach (risk acceptable or not acceptable).
- Method 2, which is based on a physiological and biomechanical approach, permits the determination of the level of risk via a three-zone assessment approach (risk acceptable, conditionally acceptable, or not acceptable).

Method 1 is sufficiently clear and well known; it uses the psychophysical tables for two-handed pushing and pulling.

Method 2, which should be considered a more detailed method, is rather complex to be applied considering the way it is presented in ISO 11228-2.

Annex B in TR 12295 tries to explain how simple the use of method 2 is by referring to pre-calculated values of basic force limits and skeletal compressive strength limits (by age and gender) and presenting a simplified procedure for determining the final "Safety Limit" to be compared with measured forces applied on the field.

The detailed procedure of method 2 is divided into four parts, according to:

- Part A: muscle force limits
- Part B: skeletal force limits
- Part C: permitted maximum forces
- Part D: safety limits

Part A and Part B should be determined in parallel and in Part C, the most protective limit—as derived from part A and B—should be selected.

In practice, the most protective limit is quite always derived from Part A, so apart from exceptional circumstances, one may only use a simplified procedure for Part A to determine the maximum permitted forces and safety limits.

The Annex explains how to measure force limits (Part A) in a simplified way by using data already presented in ISO 11228-2.

Part A determines force limits based on maximum isometric muscle-force measurements ("muscular static strength") and subsequent adjustments. Thereby, the maximum forces are reduced according to population characteristics (i.e. age, gender, and stature) and the requirements of the operations (i.e. frequency, duration, and distance of push/pull task). So, the muscle force limits are derived in a two-step procedure:

- Step 1: determining the "basic force limits F_B" based on muscle strength; and
- Step 2: reducing F_B due to actual population and task characteristics, resulting in "F_{Br}."

In Step 1, muscular static-strength values can easily be depicted from provided tables considering the working heights of near the floor and about 2 m, several males-to-females distribution ratios, two age profiles (all ages vs. elderly), and working experience (professional vs. domestic populations):

- Select the subgroup most similar to the intended user population.
- Select the appropriate table (pushing vs. pulling, professional vs. domestic populations).
- Read the precalculated force limit according to population subgroup profile and working height.

After this procedure in Step 1 of Part A, the basic force limit (F_B) is defined.

In Step 2, adjustments to basic force limits (F_B) are performed by considering reduction factors due to the distance and frequency of the push-or-pull tasks. For push-or-pull distances less than 5 m, the factors are based on initial forces, whereas they are based on sustained forces for longer distances. The distance-related factor depends on the gender distribution (i.e. on the male-and-female percentages) of the intended user population.

After this procedure in Step 2 of Part A, the action-force limit (F_{Br}) based on muscle strength is defined.

The procedure is depicted in Figure 24.2 by using the right part of the flowchart (standard use).

Disregarding Part B, one may use the action-force limit (F_{Br}) for the final computation of the Safety Limit (Part D).

In Part D, the safety limit is determined due to the provision of the risk multiplier m_r, which represents the limiting criteria in defining the green, yellow, and red zone of risk. That means the overall risk of injury is rated by applying a three-zone risk rating system. This risk is evaluated based on the actual resultant (measured) force F_R and the limiting force F_L (in practice F_{Br}).

Acceptable risk (green zone) when $F_R \leq 0.85 \, F_L \, (F_{Br})$

- The risk of disease or injury is negligible or is at an acceptably low level for the entire operator population.
- No action (i.e. no redesign) is required.

Conditionally acceptable risk (yellow zone) when $0.85 \, F_L \, (F_{Br}) < F_R \leq 1.0 \, F_L \, (F_{Br})$:

- There is a risk of disease or injury that cannot be neglected for the entire operator population or part of it.
- Further risk estimation by analyzing contributory risk factors and redesign shall be performed as soon as possible. If redesign is not possible, other measures to control the risk are to be taken.

Not acceptable risk (red zone) when $F_R > F_L \, (F_{Br})$:

- There is a considerable risk of disease or injury that cannot be neglected for the operator population.
- Immediate action to reduce the risk is necessary (e.g., redesign, work organization, worker instruction, and training).

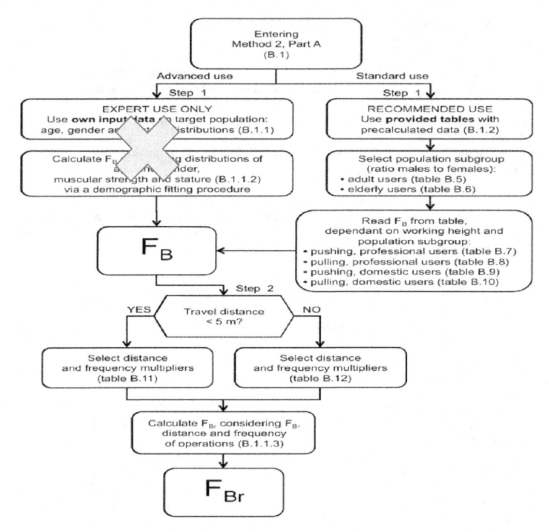

FIGURE 24.2 Derivation of action-force force limits F_{Br} based on muscular strength; follow the standard use.

ANNEX C: APPLICATION INFORMATION FOR ISO 11228-3

Annex C provides additional information relevant to the practical application of methods and procedures presented and/or recommended in ISO 11228-3. The purpose of Annex C is to provide the expert users of ISO 11228-3 with useful information when—having completed the key questions and the quick assessment—they need to apply the standard to perform a risk assessment of manual repetitive tasks.

The information is related to:

- the OCRA Index method (Preferred Method 2 in ISO 11228-3), particularly shoulder posture and some organizational relevant factors;
- presentation of the OCRA Checklist as a useful tool for simple risk assessment (Method 1 in ISO 11228-3);
- further details on "Multitask Analysis," with a particular focus on applications of the OCRA Index and checklist methods, where multiple manual repetitive tasks are performed by the same worker(s) in a shift;

- other methods suggested for a detailed risk assessment (Method 2 in ISO 11228-3); and
- brief references on other methods recently developed for the purposes of simple risk assessment (Method 1 in ISO 11228-3).

Some of the more relevant of these aspects will be briefly reported here.

ADVANCES IN THE APPLICATION OF THE OCRA INDEX

The advances regard the analysis of shoulder postures and related multipliers, the three-level classification of "repetitiveness (lack of variation or stereotype)" with corresponding multipliers, and the confirmation of the classification of daily duration of repetitive task in several scenarios (with related multipliers) from 120 min to more than 480 min with one-hour increments. Some details about these updates are reported in paragraph 13 of "Repetitive actions and movements of the upper limbs."

OCRA CHECKLIST AS A USEFUL TOOL FOR METHOD 1—SIMPLE RISK ASSESSMENT

The OCRA checklist (Colombini, Occhipinti, and Grieco 2002; Colombini, Occhipinti, & Fanti, 2005) is one of the methods/tools suggested in ISO 11228-3 Annex A for Method 1 (simple risk assessment). Since the OCRA checklist is based on the same general framework, criteria, and definition of the "Consensus Document" assumed as a reference point in the same Annex A, it was considered useful to briefly report an updated description of the tool to favor its application for ISO 11228-3 Method 1.

The OCRA checklist is useful in quickly identifying the presence of the main risk factors for the upper limbs and classifying the consequent exposure. It is, therefore, recommended for the initial screening of several workstations in an enterprise featuring repetitive tasks, while the complete OCRA Index is useful for the (re)design or in-depth analysis of workstations and repetitive tasks.

The analysis system suggested with the OCRA checklist starts by assigning scores for each of the main risk factors (frequency, force, posture) and the additional factors. For each risk factor, several scenarios are presented and for each scenario, a score is suggested (ranging from 0 to maximum as the potential risk increases). The sum of the partial scores (for each risk factor: frequency, force, posture, additional factors) obtained in this way produces a partial final score. To obtain the final exposure value, two multipliers must be applied to calibrate the partial final score, considering both the net daily duration of repetitive work and the presence of hours without adequate recovery (Figure 24.3). This procedure estimates the actual exposure in different levels (absent, borderline, light, medium, and high) (Table 24.11) (Occhipinti & Colombini, 2004, 2007).

The OCRA Checklist is synthetically presented in the Annex (also with the full paper model).

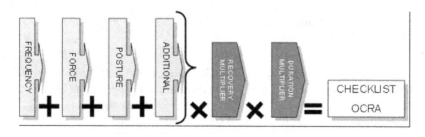

FIGURE 24.3 Computation model for the OCRA checklist final score.

TABLE 24.11

OCRA Checklist Scores and Correspondence with OCRA Index Values for Exposure Classification Purposes

OCRA Checklist Score	OCRA Index		Exposure Level
≤7.5	≤2.2	Green	No risk (acceptable)
7.5 < score ≤ 11.0	2.2 < index ≤ 3.5	Yellow	Borderline or very low risk
11 < score ≤ 14.0	3.5 < index ≤ 4.5	Red light	Light risk
14 < score ≤ 22.5	4.5 < index ≤ 9.0	Red medium	Medium risk
> 22.5	>9.0	Red high	High risk

OCRA Multitask Analysis

When computing for the OCRA Index (or the OCRA Checklist score) considering the presence of more than one repetitive task, a "traditional" procedure has been proposed both in literature and in ISO 11228-3 (main text and Annex C). This approach, which results could be defined as "time-weighted average," seems to be appropriate when considering rotations among tasks that are performed very frequently—for instance, almost once every 90 min (or shorter). In those scenarios, "high" exposures are presumed to be somehow compensated by "low" exposures that alternate very quickly between each other. As a consequence, the traditional procedure for the OCRA Index multitask analysis is confirmed when rotation among repetitive tasks is performed almost every 90 min. The index or checklist score will be defined as OCRA Multitask Average.

On the contrary, when rotation among repetitive tasks is less frequent (i.e. once every one 40 min or more), the "time-weighted average" approach could result in an underestimation of the exposure level (as it flattens the peaks of high exposures). For those scenarios, an alternative approach is based on a more realistic concept that the most stressful task is the minimum starting point. Hence, the result of this approach will be between:

- the OCRA Index (or checklist score) of the most stressful task considered for its individual longest continuous duration; or
- the OCRA Index (or the checklist score) of the same most stressful task when it is (only theoretically) considered lasting for the overall duration of all examined repetitive tasks.

A novel procedure estimates the resulting index within this range of minimum to maximum values. The consequent index will be defined as OCRA Multitask Complex.

The procedure is based on the following formula (Occhipinti, Colombini, & Occhipinti, 2008):

$$\text{OCRA Multitask Complex} = \text{OCRA}_{1(\text{Dum}1)} + (\Delta\text{OCRA}_1 \times K)$$

Where:

- *1,2,3,...,N = repetitive tasks ordered by OCRA values (1= highest; N = lowest) computed considering respective real continuous duration multipliers (Dum_i) and Rc_M(the same for all the tasks);*
- *Dum_i = duration multiplier for $task_i$real continuous duration;*
- *Dum_{tot} = duration multiplier for total duration of all repetitive tasks;*

- $\Delta\,OCRA_1 = OCRA$ of $task_1$considering Dum_{tot}—$OCRA$ of $task_1$considering Dum_1;
- $\dfrac{K = (OCRA_{1\,max} * FT_1) + (OCRA_{2\,max} * FT_2) + ... + (OCRA_N * FT_N);}{(OCRA_{1\,max})}$
- $OCRA_{i\,max} = OCRA$ of $task_i$considering Dum_{tot};
- $FT_i = Fraction$ of $Time$ (values from 0 to 1) of $task_i$with respect to the total repetitive time.

For all the operative details about OCRA Index and OCRA Checklist models and evaluations, see the site: www.epmresearch.org. A free download of related software (in excel) is available.

STUDYING (BY OCRA METHOD) MULTIPLE REPETITIVE TASKS WITH ROTATIONS ALONG WEEKS, MONTHS, OR YEAR

In industrial manufacturing sectors, tasks rotate often in a similar way every day and, consequently, the previous procedures could be easily applied. On the contrary, in some productive sectors (agriculture, construction, cleaning, retail, etc.), exposure assessment is much more complex, characterized by the presence of several tasks over periods longer than a typical working day (weekly, monthly, yearly turnover).

Studies are reported (Colombini & Occhipinti, 2008) to organize models in assessing situations where tasks rotate by weeks, months, or a year. In general, those studies are based on the use of the Checklist OCRA and on adaptations of the two multitask analysis approaches (average and complex) that have been previously presented.

The general procedure for studying such situations implies three operating stages:

- Completing a *preliminary organizational study* to establish the kind of turnover: the periodicity of the different repetitive tasks as repeated in time (daily or weekly or monthly or yearly).
- Defining the *"intrinsic" risk level in each task* using the OCRA checklist. Intrinsic level means ascribing to the repetitive task a net duration of 440 min/shift with two breaks, 8–10 min each, and a lunch break of at least 30 min.
- Applying *specific mathematical models* (adaptations of average or complex approach) considering intrinsic values as well as organizational patterns (duration, frequency, and sequences) of individual tasks under study.

The choice of the most predictive model will necessarily be based on the collection of relevant epidemiological data. The preliminary data collected seem to confirm a better validity of the OCRA Multitask Complex model.

ADVANCES ON OTHER METHODS

Short references are given to advances in "other methods" suggested by ISO 11228-3 as Strain Index and HAL/ACGIH TLV for a "detailed" risk assessment with recommendation about their use for ISO 11228-3.

Moreover, a very brief note addresses more recent methods that are not reported in ISO 11228-3 that could be used for a simple risk assessment. They are HARM (Hand Arm Risk Assessment Method), ART-Tool (Assessment of Repetitive Tasks of the upper limbs), KIM-MHO (Key Indicator Method—Manual Handling Operations), and EAWS (European Assembly Worksheet—section 4).

REFERENCES

Borg, G. (1998). *Borg's perceived exertion and pain scales*, Champaign, IL: Human Kinetic Europe.

CEN. (2003). *EN 1005-2: Safety of machinery—Human physical performance—Part 2: Manual handling of machinery and component parts of machinery.*

CEN. (2007). *EN 1005-5: Safety of machinery—Human physical performance—Part 5: Risk assessment for repetitive handling at high frequency.*

Colombini D., Occhipinti E., & Grieco A. (2002). *Risk assessment and management of repetitive movements and exertions of upper limbs.* Amsterdam: Elsevier Science.

Colombini D., Occhipinti E., & Fanti M. (2005). *Il metodo OCRA per l'analisi e la prevenzione del rischio da movimenti ripetuti.* Collana Salute e lavoro, Milano: Franco Angeli Editore.

Colombini D., & Occhipinti E. (2008). The OCRA method (OCRA index and checklist). Updates with special focus on multitask analysis. Eds W. Karkwoski and G. Salvendy. *Conference Proceedings: AHFE 2008.* Las Vegas.

Colombini D., Occhipinti E., Alvarez-Casado E., Waters T. (2012). *Manual lifting: A guide to the study of simple and complex lifting tasks*, CRC Press, Taylor & Francis Group. Boca Raton and New York (US)

ISO. (2000). *ISO 11226. Ergonomics—Evaluation of static working postures.*

ISO. (2003). *ISO 11228-1. Ergonomics—Manual handling—Lifting and carrying.*

ISO. (2007a). *ISO 11228-2. Ergonomics—Manual handling—Pushing and pulling.*

ISO. (2007b). *ISO 11228-3. Ergonomics—Manual handling—Handling of low loads at high frequency.*

ISO. (2014). *ISO TR 12295. Ergonomics—Application document for ISO standards on manual handling (ISO 11228-1, ISO 11228-2 and ISO 11228-3) and evaluation of static working postures (ISO 11226).*

Occhipinti E., & Colombini D. (2004). Metodo Ocra: Aggiornamento dei valori di riferimento e dei modelli di previsione dell'occorrenza di patologie muscolo-scheletriche correlate al lavoro degli arti superiori (UL-WMSDs) in popolazioni lavorative esposte a movimenti e sforzi ripetuti degli arti superiori. *La Medicina del Lavoro*, 95, 305–319.

Occhipinti E., & Colombini D. (2007). Updating reference values and predictive models of the OCRA method in the risk assessment of work-related musculoskeletal disorders of the upper limbs. *Ergonomics*, *50*(11), 1727–1739.

Occhipinti E., Colombini D., & Occhipinti M. (2008). Metodo Ocra: messa a punto di una nuova procedura per l'analisi di compiti multipli con rotazioni infrequenti. *La Medicina del Lavoro*, *99*(3), 234–241

Waters T. R., Putz-Anderson V., Garg A., & Fine L. J. (1993). Revised NIOSH equation for the design and evaluation of manual lifting tasks. *Ergonomics* 36(7), 749–776.

Waters T. R., Putz-Anderson V., & Garg A. (1994). *Applications manual for the Revised NIOSH Lifting Equation.* DHHS (NIOSH) Publication No. 94-110. National Institute for Occupational Safety and Health, Centers for Disease Control and Prevention. Cincinnati, OH, 45226.

Waters T. R., Lu M. L., & Occhipinti E. (2007). New procedure for assessing sequential manual lifting jobs using the revised NIOSH lifting equation. *Ergonomics*, *50*(11), 1761–1770

Section V

Standards for Human-Computer Interaction

25 Standards, Guidelines, and Style Guides for Human-Computer Interaction

Tom Stewart

CONTENTS

INTRODUCTION

Standards, guidelines, and style guides generally exist to improve the consistency of the user interface and improve the quality of interface components. They help specifiers procure systems and system components that can be used effectively, efficiently, safely, and comfortably. They also help restrict the unnecessary variety of interface hardware, software, and technology and ensure that the benefits of any variations are fully justified against the costs of incompatibility, loss of efficiency, and increased training time for users. Even standards that are still under development can have an impact on hardware and software development. The major suppliers play an active part in generating the standards and increasingly, they are incorporating the guidance on good practice into products before the standards themselves are published.

In this chapter, we describe a number of human-computer interaction (HCI) standards that have been developed internationally and explain how they can be used in conjunction with other guidelines and style guides to improve user experience.

Although designing usable systems requires far more than simply applying standards, guidelines, and style guides, they can, nonetheless, make a significant contribution by promoting consistency, good practice, common understanding, and an appropriate prioritization of user interface issues.

CONSISTENCY

Anyone who uses computers knows too well the problems of inconsistency between applications—even within the same application. Inconsistency, even at the simplest level, can cause problems. Just three examples:

- Press the <escape> key in one place and you are safely returned to your previous menu choice. In another place, you are unceremoniously "dropped" to the operating system, the friendly messages disappear, and you lose all your data.
- On the web, inconsistency is rampant. Even something as straightforward as a hypertext link may be denoted by underlines on one site, by a mouseover on a second site, and by nothing at all on a third site.
- Different and confusing keyboard layouts sit side-by-side in many offices.

Standards, guidelines, and style guides play an important part in helping address these issues by collating and communicating agreed best practices for user interfaces and for the processes by which they are designed and evaluated. They can provide a consistent reference across design teams or across time to help avoid such unpleasant experiences. Indeed, in other fields, consistency—for example, between components that should interconnect—is the prime motivation for standards. It is certainly a worthwhile target for user interface standards.

GOOD PRACTICE

In many fields, standards provide definitive statements of good practice. In user interface design, there are many conflicting viewpoints about good practice. Standards, especially International Standards, can provide independent and authoritative guidance. International Standards are developed slowly, by consensus, using extensive consultation and development processes. This has its disadvantages in such a fast-moving field as user interface design, and some have criticized any attempts at standardization as premature. However, there are areas where a great deal is known which can be made accessible to designers through appropriate standards, and there are approaches to user interface standardization, based on human characteristics, which are relatively independent of specific technologies.

The practical discipline of achieving consensus helps moderate some of the wilder claims of user interface enthusiasts and helps ensure that the resulting standards do represent good practice. The slow development process also means that standards can seldom represent the leading edge of design. Nonetheless, properly written, they should not inhibit helpful creativity.

COMMON UNDERSTANDING

Standards themselves do not guarantee good design, but they provide a means for different parties to share a common understanding when specifying interface quality in design, procurement, and use.

- *For users*, standards allow them to set appropriate procurement requirements and evaluate competing suppliers' offerings.
- *For suppliers*, standards allow them to check their products during design and manufacture and provide a basis for making claims about the quality of their products.
- *For regulators*, standards allow them to assess the quality and provide a basis for testing products.

APPROPRIATE PRIORITIZATION OF USER INTERFACE ISSUES

One of the most significant benefits of standardization is that it places user interface issues squarely on the agenda. Standards are serious business; whereas many organizations pay little regard to research findings, few organizations can afford to ignore standards. In Europe—and, increasingly, in other parts of the world—compliance with relevant standards is a mandatory requirement in major contracts.

A Note on Terminology

HCI standards, guidelines, and style guides represent three approaches in improving the usability of systems. They are not mutually exclusive categories. For example, many HCI standards simply provide agreed guidelines rather than specific requirements. Some style guides are implemented in such a way that they have become standards from which designs cannot vary.

However, generally (and in this chapter), we use the terms as follows:

- Guidelines—recommendations of good practice that rely on the credibility of their authors for their authority
- Standards—formal documents published by standards-making bodies that are developed through some form of consensus and formal voting process
- Style guides—a set of recommendations from software providers or agreed upon within development organizations to increase consistency of design and promote good practice within a design process of some kind

HCI STANDARDS

SOURCES OF HCI STANDARDS

Many standards bodies have been in existence for some time and are organized according to traditional views of technology and trade. Software is used as part of systems that involve a range of technologies. The purpose of this section is to introduce one of the key standard organizations that are working on standards relevant to HCI and describe its main activities briefly.

In most people's minds, one of the most basic and fundamental objectives of standardization is to minimize unnecessary variations. Ideally, for any product category, there is one standard that should be satisfied, and products that meet that standard give their owners or users some reassurance about quality or about what standards makers refer to as "interoperability." Thus, yachtsmen in Europe who buy a lifejacket that meets EN 396 might reasonably expect it to keep them afloat if they have the misfortune to fall overboard in the Florida Keys. Similarly, an office manager in the United States who orders A4 paper for a photocopier might reasonably expect paper that meets that standard (ISO 216:1975) to fit, even though it is not the typical size used locally.

This brings us to a rather important point. It is often difficult to achieve a single agreed-upon standard, and a common solution is to have more than one. An obvious example concerns paper size where there are the ISO A-series (A0, A1, etc.), the ISO B series (B0, B1, etc.), as well as U.S. sizes (legal, letter, etc.). Although this solves the standard makers' problems in agreeing on a single standard, it is an endless source of frustration for users of the standard—as anyone who has forgotten to check the paper source in an e-mailed document can testify.

However, there is another reason why there are more standards than one might imagine, especially when it comes to user interface design issues. The reason is that computer technology forms the basis of many different industries, and standards can have an important impact on market success.

But it is not just at the international level that there appears to be some duplication. In the United Kingdom (U.K.), the British Standards Institution mirror committee to SC4 published an early version of the first six parts of ISO 9241 as British Standard BS 7179:1990. The prime reason for this was to provide early guidance for employers of users of visual displays who wanted to use standards to help them select equipment that met the requirements in the Schedule to the Health and Safety (Display Screen Equipment) Regulations 1992. These regulations are the U.K. implementation of a European Community Directive on the minimum safety and health requirements for work with display screen equipment (90/270/EEC). Of course, as a spin-off, the British Standards Institution was able to generate revenue from selling these standards several years before the various parts of ISO 9241 became available as British Standards.

A similar process has taken place in the United States with the Human Factors and Ergonomics Society (HFES) developing HFS 100 on Visual Display Terminal Ergonomics as an ANSI-authorized Standards Developing Organization. More recently, there are two HFES standards development committees working on HFES 100 (a new version of HFS 100) and HFES 200, which addresses user interface issues. It includes sections on accessibility, voice, and telephony applications; color and presentation; and slightly rewritten parts of the software parts of ISO 9241.

International ergonomics standards in HCI are being developed by the International Organization for Standardization (ISO). The work of ISO is important for two reasons. First, the major manufacturers are international and, therefore, the best and most effective solutions need to be international. Second, the European Standardization Organization (CEN) has opted to adopt ISO standards—wherever appropriate—as part of the creation of the single market. CEN standards replace national standards in the European Union and European Free Trade Area member states.

The International Organization for Standardization (ISO) comprises national standards bodies from member states (see www.iso.ch for more information). Its work is conducted by technical and subcommittees that meet every year or so and are attended by formal delegations from participating members of that committee. In practice, the technical work takes place in working groups of experts, nominated by national standard committees but expected to act as independent experts. The standards are developed over a period of several years and, in the early stages, the published documents may change dramatically from version to version until a consensus is reached (usually within a Working Group of experts). As the standard becomes more mature (from the Committee Draft Stage onwards), formal voting takes place (usually within the parent subcommittee) and the draft documents provide a good indication of what the final standard is likely to look like. Table 25.1 shows the main stages.

TABLE 25.1

The Main Stages of International Standards Development

WI	Work Item—an approved and recognized topic for a working group to be addressing which should lead to one or more published standards.
WD	Working Draft—a partial or complete first draft of the text of the proposed standard.
CD	Committee Draft—a document circulated for comment and approval within the committee working on it and the national mirror committees. Voting and approval are required for the document to reach the next stage.
DIS	Draft International Standard—a draft standard that is circulated widely for public comment via national standards bodies. Voting and approval are required for the draft to reach the final stage.
FDIS	Final Draft International Standard—the final draft is circulated for formal voting for adoption as an International Standard.
IS	International Standard. The final published standard.

Note: Documents may be reissued as further CDs and DISs

The International Organization for Standardization (ISO)

In the late 1970s, the kind of concern about the ergonomics of visual display terminals (also called visual display units) that stimulated German standards (see Appendix 1) became more widespread, especially in Europe.

The prime concern at that time concerned the possibility that prolonged use (especially of displays with poor image quality) might cause deterioration in users' eyesight. (Note: Since then, several studies have shown that aging causes the main effect on eyesight, and because display screen work can be visually demanding, many people only discover this deterioration when they experience discomfort from intensive display screen use. This can incorrectly lead them to attribute their need for glasses to their use of display screens.)

When a new international standards work item to address this concern was proposed, the Information Technology committee decided that this was a suitable topic for the recently formed ergonomics committee ISO/TC 159. The work item was allocated to the subcommittee ISO/TC 159/SC4 Signals and Controls, and an inaugural meeting was held at BSI in Manchester in 1983. The meeting was well attended by delegates from many countries, and a few key decisions were made.

At that time, there was a proliferation of office-based systems, and SC4 decided to focus on office tasks (word processing, spreadsheet, etc.) rather than try to include computer-aided design or process control applications. It was also decided that we would need a multi-part standard to cover the wide range of ergonomics issues that it needed to address to improve the ergonomics of display screen work. A number of working groups were established to carry out the technical work of the subcommittee. Table 25.2 lists the current working groups of ISO/TC159/SC4.

Little did any of those present realize that it would be nearly seven years for the first parts of ISO 9241 to be published and that it would take until the end of the century to publish all 17 parts. Table 25.3 shows the published parts of ISO 9241 and Table 25.4 the other published standards that were part of the ISO/TC159/SC4 work program.

HCI STANDARDS UNDER DEVELOPMENT

Although the ISO 9241 standards represented a major part of the output of ISO/TC159/SC4, a number of other standards are under development at this time (December 2004); these are listed in Table 25.5.

TABLE 25.2
The Working Groups of ISO/TC159/SC4

WG1	Fundamentals of Controls and Signalling Methods
WG2	Visual Display requirements
WG3	Control, workplace, and environmental requirements
WG4	Task requirements (disbanded)
WG5	Software ergonomics and human-computer dialog
WG6	Human-centered design processes for interactive systems
WG8	Ergonomics design of control centers

How to Use ISO 9241 Standards

Although it was not made explicit at the time, SC4 had an underlying set of assumptions about HCI design activities and how the standards would support these. These activities included:

* Analyzing and defining system requirements
* Designing user–system dialogs and interface navigation
* Designing or selecting displays
* Designing or selecting keyboards and other input devices
* Designing workplaces for display screen users
* Supporting and training users
* Designing jobs and tasks

Table 25.6 shows how it was anticipated that the standards would be used to support these activities.

Revisions to 9241

Following the completion of the 17-part ISO 9241, work is underway for a major revision and restructure to incorporate other relevant standards and make the ISO 9241 series more usable. This article describes the new structure and the principles agreed for the revision process and gives the current status of the new parts (as of December 2004).

Although computer technology has changed dramatically over the period the original ISO 9241 has been under development (more than 20 years), many of the ergonomics issues remain similar. For example, when Part 14 (menu dialogs) was originally planned, menus were usually displayed on character-based screens, and choices were typically made by selecting numbered choices by keystroke. However, by the time the standard was finished, menus were a common part of graphical interfaces, and items were selected by pointing devices. Nonetheless, the guidance on menu structures—how many options should be presented, and so on—remained applicable because it related to how people make choices and interpret information that has not changed much in time.

But changes in the technology and the way we use it have made it difficult for the 9241 standards to be up-to-date. The development of flat panel displays meant that new display standards had to be developed. The rapid pace of change meant that we could not always wait for technology to stabilize before developing standards, so we developed some design process standards. And, of course, ISO itself requires standards to be reviewed after five years.

As a result, ISO/TC159/SC4 has been working to develop a new set of standards that build on the strengths of the previous work but are also easier to use by standards users.

TABLE 25.3

ISO 9241 Standards Published by ISO/TC159/SC4 Ergonomics of Human-System Interaction Including Amendments and Revisions

ISO 9241

ISO 9241-1:1997 Ergonomic Requirements for Office Work With Visual Display terminals (VDTs)—Part 1: General Introduction ISO 9241-1:1997/Amd 1:2001

ISO 9241-2:1992 Ergonomic Requirements for Office Work With Visual Display terminals (VDTs)—Part 2: Guidance on Task Requirements

ISO 9241-3:1992 Ergonomic Requirements for Office Work With Visual Display terminals (VDTs)—Part 3: Visual Display Requirements ISO 9241-3:1992/Amd 1:2000

ISO 9241-4:1998 Ergonomic Requirements for Office Work With Visual Display terminals (VDTs)—Part 4: Keyboard Requirements ISO 9241-4:1998/Cor 1:2000

ISO 9241-5:1998 Ergonomic Requirements for Office Work With Visual Display terminals (VDTs)—Part 5: Workstation Layout and Postural Requirements

ISO 9241-6:1999 Ergonomic Requirements for Office Work With Visual Display terminals (VDTs)—Part 6: Guidance on the Work Environment

ISO 9241-7:1998 Ergonomic Requirements for Office Work With Visual Display terminals (VDTs)—Part 7: Requirements for Display With Reflections

ISO 9241-8:1997 Ergonomic Requirements for Office Work With Visual Display terminals (VDTs)—Part 8: Requirements for Displayed Colors

ISO 9241-9:2000 Ergonomic Requirements for Office Work With Visual Display terminals (VDTs)—Part 9: Requirements for Non-Keyboard Input Devices

ISO 9241-10:1996 Ergonomic Requirements for Office Work With Visual Display terminals (VDTs)—Part 10: Dialog Principles

ISO 9241-11:1998 Ergonomic Requirements for Office Work With Visual Display terminals (VDTs)—Part 11: Guidance on Usability

ISO 9241-12:1998 Ergonomic Requirements for Office Work With Visual Display terminals (VDTs)—Part 12: Presentation of Information

ISO 9241-13:1998 Ergonomic Requirements for Office Work With Visual Display terminals (VDTs)–Part 13: User Guidance

ISO 9241-14:1997 Ergonomic Requirements for Office Work With Visual Display terminals (VDTs)—Part 14: Menu Dialog

ISO 9241-15:1997 Ergonomic Requirements for Office Work With Visual Display terminals (VDTs)—Part 15: Command Dialogs

ISO 9241-16:1999 Ergonomic Requirements for Office Work With Visual Display terminals (VDTs)—Part 16: Direct Manipulation Dialog

ISO 9241-17:1998 Ergonomic Requirements for Office Work With Visual Display Terminals (VDTs)—Part 17: Form Filling Dialogs

---

The agreed title for the new ISO 9241 is "The Ergonomics of Human System Interaction." This title was selected to demonstrate the broadening of the scope from office tasks and to align the standard with the overall title and scope of SC4.

We also wanted to build on the "branding" of ISO 9241 that has become recognized as a benchmark, particularly in Europe.

Table 25.7 shows the structure and the current status of the parts.

TABLE 25.4

Other Standards Published by ISO/TC159/SC4 Ergonomics of Human-System Interaction

Other Standards

ISO 11064-1:2000 Ergonomic Design of Control Centers—Part 1: Principles for the Design of Control Centers

ISO 11064-2:2000 Ergonomic Design of Control Centers—Part 2: Principles for the Arrangement of Control Suites

ISO 11064-3:1999 Ergonomic Design of Control Centers—Part 3: Control Room layout ISO 11064-3:1999/Cor 1:2002

11064-4:2004 Ergonomic Design of Control Centers—Part 4: Layout and Dimensions of Workstations

ISO 13406-1:1999 Ergonomic Requirements for Work With Visual Displays Based on Flat Panels—Part 1: Introduction

ISO 13406-2:2001 Ergonomic Requirements for Work With Visual Displays Based on Flat Panels—Part 2: Ergonomic Requirements for Flat Panel Displays

ISO 13407:1999 Human-Centered Design Processes for Interactive Systems

ISO 14915-1:2002 Software Ergonomics for Multimedia User Interfaces—Part 1: Design Principles and Framework

ISO 14915-2:2003 Software Ergonomics for Multimedia User Interfaces—Part 2: Multimedia navigation and Control

ISO 14915-3:2002 Software Ergonomics for Multimedia User Interfaces—Part 3: Media Selection and Combination

ISO/TS 16071:2003 Ergonomics of Human–System Interaction—Guidance on Accessibility for Human-Computer Interfaces

ISO/TR 16982:2002 Ergonomics of Human–System Interaction—Usability Methods Supporting Human-Centered design

ISO/PAS 18152:2003 Ergonomics of Human–System Interaction—Specification for the Process Assessment of Human–System Issues

ISO/TR 18529:2000 Ergonomics—Ergonomics of Human–System Interaction—Human-Centered Lifecycle Process Descriptions (available in English only)

TABLE 25.5

Other Main Standards Under Development by ISO/TC159/SC4 Ergonomics of Human–System Interaction

Standard	Status
11064-5—Ergonomic Design of Control Centers—Part 5: Human–System Interfaces	Delayed
11064-6—Ergonomic Design of Control Centers—Part 6: Environmental Requirements	FDIS
11064-7—Ergonomic Design of Control Centers—Part 7: Principles for the Evaluation of Control Centres	DIS
1503 (rev)—Ergonomics Requirements for Design on Spatial Orientation and Directions of Movements	WI agreed
16071 Ergonomics of Human–System Interaction—Guidance on Software Accessibility	CD due in 12/04
23973—Software Ergonomics for World Wide Web User Interfaces	DIS in preparation

STRENGTHS AND LIMITATIONS OF HCI STANDARDS

It is important to be aware of the strengths and limitations of standards. They cannot be understood (and, therefore, used effectively) in isolation from the context in which they were developed. It is important to realize that:

- Standards are developed over an extended period of time.
- It is easy to misunderstand the scope and purpose of a particular standard.

TABLE 25.6

How Parts of ISO 9241 Were Intended to Be Used in HCI Design Activities

HCI Activity	Relevant Part of ISO 9241
Analyzing and Defining System Requirements	*ISO 9241-11:1998 Guidance on usability*
Designing User–System Dialogs and Interface Navigation	*ISO 9241-10:1996 Dialog principles*
	ISO 9241-14:1997 Menu dialogs
	ISO 9241-15:1998 Command dialogs
	ISO 9241-16:1999 Direct manipulation dialogs
	ISO 9241-17:1998 Form-filling dialogs
Designing or Selecting Displays	*ISO 9241-3:1992 Display requirements*
	ISO 9241-7:1998 Requirements for displays with reflections
	ISO 9241-8:1997 Requirements for displayed colors
	ISO 9241-12:1998 Presentation of information
Designing or Selecting Keyboards and Other Input Devices	*ISO 9241-4:1998 Keyboard requirements*
	ISO 9241-9:2000 Requirements for non-keyboard input devices
Designing Workplaces for Display Screen Users	*ISO 9241-5:1998 Workstation layout and postural requirements*
	ISO 9241-6:1998 Guidance on the work environment
Supporting and Training Users	*ISO 9241-13:1998 User guidance*
Designing Jobs and Tasks	*ISO 9241-2:1992 Guidance on task requirements*

- Standard-making involves politics as well as science.
- The language of standards can be obscure.

But,

- Structure and formality can be a help as well as a hindrance.
- The benefits do not just come from the standards themselves.
- Being international makes it all worthwhile.

Standards Are Developed Over an Extended Period of Time

One of the reasons the process is slow is that there is an extensive consultation period at each stage of development, with time being allowed for national member bodies to circulate the documents to mirror committees and then collate their comments. Another reason is that Working Group members can spend a great deal of time working on drafts and reaching consensus only to find that the national mirror committees reject their work when it comes to the official vote. It is particularly frustrating for project editors to receive extensive comments (that must be answered) from countries that do not send experts to participate in the work. Of course, the fact that the work is usually voluntary means that it is difficult to get people to agree to work quickly.

However, there are some benefits that come directly from the slow pace of the process. One benefit is that when technology is moving quicker than the makers can react, it makes it clear that certain types of standards may be premature. For example, ISO 9241-14:1997 Menu Dialogues was originally proposed when character-based menu-driven systems were a popular style of dialog design. Its development was delayed considerably for all manner of reasons. But these delays meant that the final standard was relevant to pull down and pop-up menus that had not even been considered when the standard was first proposed.

TABLE 25.7

The Structure and the Current Status of the Parts (as of December 2004)

New ISO 9241 Ergonomics of Human–System Interaction

Part	Title	Status and Notes
1	Introduction	Reserved number—could be a TR to allow frequent updates
2	Job Design	Reserved number for revision and extension of old Part 2
11	Hardware and Software Usability	Reserved number for revision and extension of old Part 11 to include hardware explicitly
20	Accessibility and Human–System Interaction	Approved WI—technical work to start a meeting in Tokyo in December
21–99	Reserved Numbers	No plans to allocate at present
100	Software Ergonomics	Reserved number for series of software ergonomics standards
100	Dialog Principles	DIS (voting ends 01/05, revision of old Part 10)
200	Human System Interaction Processes	Reserved number for revision and extension of ISO 13407 and other process standards
300	Displays and display-related hardware	Reserved number for series of display ergonomics standards (will become Introduction)
301	Introduction	CD approved, DIS in preparation but will be renumbered 300
302	Terms and Definitions	CD approved, DIS in preparation
303	Ergonomic Requirements	CD approved, DIS in preparation
304	User Performance Test Methods	CD approved, DIS in preparation
305	Optical Laboratory Test Methods	CD approved, DIS in preparation
306	Field Assessment Methods	CD approved, DIS in preparation
307	Analysis and Compliance Test Methods	CD approved, DIS in preparation
400	Physical Input Devices—Ergonomics Principles	CD approved, DIS in preparation
410	Design Criteria for Products	CD expected 12/04
420	Ergonomic Selection Procedures	CD expected 12/04
500	Workplace Ergonomics	Reserved number for revision and extension of old Part 5
600	Environment Ergonomics	Reserved number for revision and extension of old Part 6
700	Special Application Domains	Reserved number for series of ergonomics standards for specific application domains, for example, process control

Another benefit is that, during the development process, those who may be affected have the opportunity to prepare for the standard. Thus, by the time the ISO 9241-3:1992 Display requirements were published, many manufacturers were able to claim that they already produced monitors that met the standard. They had not been in that position when the standard was first proposed (although some argued that they would have been improving the design of their displays anyway). Certainly, the standards provided a clear target for both demanding consumers and quality manufacturers.

It Is Easy to Misunderstand the Scope and Purpose of a Particular Standard

HCI standards have been criticized for being too generous to manufacturers in some areas and too restrictive in others. The "overgenerous" criticism misses the point that most standards are setting minimum requirements and, in ergonomics standards, makers must be very cautious about setting such levels. However, there certainly are areas where being too restrictive is a problem. Examples include:

- **ISO 9241-3:1992 Ergonomics Requirements for Work with VDTs: Display Requirements**. This standard has been successful in setting a minimum standard for display screens, which has helped purchasers and manufacturers. However, it is biased toward Cathode Ray Tube (CRT) display technology. An alternative method of compliance based on a performance test (which is technology independent) that should help redress the balance has now been published.
- **ISO 9241-9:1999 Ergonomics Requirements for Work with VDTs: Non-keyboard input devices**. This standard has suffered because technological developments were faster than either ergonomics research or standards making. Although there was an urgent need for a standard to help users be confident in the ergonomic claims made for new designs of mice and other input devices, the lack of reliable data forced the standard makers to slow down or run the risk of prohibiting newer, even better solutions.

Standards Making Involves Politics as Well as Science

Although ergonomics standards are generally concerned with mundane topics such as keyboard design or menu structures, they generate considerable emotion among standards makers. Sometimes, this is because the resulting standard could have a major impact on product sales or legal liabilities. Other times, the reason for the passion is less clear. Nonetheless, the strong feelings have resulted in painful experiences in the process of standardization. These have included:

- **Undue influence of major players**. Large multinational companies can try to exert undue influence by dominating national committees. Although draft standards are usually publicly available from national standards bodies, they are not widely publicized. This means that it is relatively easy for well-informed large companies to provide sufficient experts at the national level to ensure that they can virtually dictate the final vote and comments from a country.
- **"Horse trading" and bargaining to achieve agreement**. End-user requirements can be compromised as part of "horse-trading" between conflicting viewpoints. In the interests of reaching an agreement, delegates may resort to making political tradeoffs largely independent of the technical merits of the issue.

The Language of Standards Can Be Obscure

In ISO, the formal rules and procedures for operating seem to encourage an elitist atmosphere with standards written for standards enthusiasts. ISO has recognized this and is attempting to make the process more customer-focused, but such changes take time. These procedures and rules reinforce elitist tendencies and sometimes resulted in standards that leave much to be desired in terms of brevity, clarity, and usability. There are three contributory factors:

- **The use of stilted language and boring formats**. The unfriendliness of the language is illustrated by the fact that, although the organization is known by the acronym ISO, its full English title is the International Organization for Standardization. The language and style are governed by a set of Directives and these encourage a wordy and impersonal style.
- **Problems with translation and the use of "Near English."** There are three official languages in ISO—English, French, and Russian. In practice, much of the work is conducted in English, often by non-native speakers. The result of this is that the English used in the standards is often not quite correct—it is "near English." The words are usually correct, but the combination often makes the exact meaning unclear. These problems are exacerbated when the text is translated.
- **Confusions between requirements and recommendations**. In ISO standards, there are usually some parts that specify what has to be done to conform to the Standard. These are indicated by the use of the word "shall." However, in ergonomics standards, we often want to

make recommendations as well. These are indicated by the use of the word "should." Such subtleties are often lost on readers of standards, especially those in different countries. For example, in the Nordic countries, they follow recommendations (shoulds) as well as requirements (shalls), so the distinction is diminished. In the United States, they tend to ignore the "shoulds" and only act on the "shalls."

Structure and Formality Can Be a Help as Well as a Hindrance

One of the benefits of standards is that they represent a rather simplified and structured view of the world. There is also a degree (sometimes excessive) of discipline in what a standard can contain and how certain topics can be addressed. Manufacturers (and ergonomists) frequently make wildly different claims about what represents good ergonomics. This is a major weakness for our customers who may conclude that all claims are equally valid and there is no sound basis for any of them. Standards force a consensus and, therefore, have real authority in the minds of our customers. Achieving consensus requires compromises, but then so does life.

The formality of the standards means that they are suitable for inclusion in formal procurement processes and demonstrating best practice. In the U.K. at least, parts of ISO 9241 may be used by suppliers to convince their customers that the visual display screen equipment and its accessories meet good ergonomic practices. Of course, they can also be "abused" in this way with overeager salesmen misrepresenting the legal status of standards, but that is hardly the fault of the standards makers.

The Benefits Do Not Just Come From the Standards Themselves

There are several ways in which ergonomics standardization activities can add value to user interface design apart from the standards themselves—which are the end results of the process.

In 1997, the U.S. National Institute of Standards and Technology (NIST) initiated a project (Industry Usability Reporting [IUSR]) to increase the visibility of software usability. They were helped in this endeavor by prominent suppliers of software and representatives from large consumer organizations. One of the key goals was to develop a common usability reporting format (Common Industry Format [OF]). This is currently being processed as an ANSI standard through NCITS. OF has been developed to be consistent with ISO 9241 and ISO 13407 and is viewed by the IUSR team as "an implementation of that ISO work." This activity in itself should have a major impact on software usability (http://www.nist.gov/iusr).

In the hardware arena, many people are aware of the TCO 99 sticker that appears on computer monitors and understand that it is an indication of ergonomic and environmental quality. What they may not know is that TCO is the Swedish Confederation of White Collar Trade Unions and that ISO 9241 was used as a major inspiration for its original specification. They publish information in English, and details are available on their website at http://www.tco.se/eng/index.htm.

Being International Makes It All Worthwhile

Although there are national and regional differences in populations, the world is becoming a single market with the major suppliers taking a global perspective. Variations in national standards and requirements not only increase costs and complexity but also tend to compromise individual choice. Making standards international is one way of ensuring that they have an impact and can help improve the ergonomic quality of products for everyone. That has to be a worthwhile objective. Table 25.8 shows the member countries of ISO/TC159/SC4.

GUIDELINES

BACKGROUND

Whereas European interest in the early 1980s seemed to focus on computer hardware ergonomics, there was a growing interest in user interface software in the United States.

TABLE 25.8
Members of ISO/TC159/SC4 Ergonomics of Human-System Interaction

"P" Members	Austria	Belgium	Canada	Czech Republic
	China	Denmark	Finland	France
	Germany	Ireland	Italy	Japan
	Korea	Netherlands	Norway	Poland
	Slovakia	Spain	Sweden	Thailand
	United Kingdom	United States of America		
"O" members	Australia Tanzania	Hungary	Mexico	Romania

Computer hardware seemed like a possible target for formal standardization. The highly contextual nature of best practice made formal standards—at best—premature and—at worst—dangerous traps for perpetuating obsolete practices in this rapidly developing technology.

Rather than attempt to develop formal standards, several human-computer interaction groups and individual researchers started to assemble collections of guidelines, tips, and hints into books and compendiums. We describe three of the most influential in this section.

Smith & Mosier's Guidelines for Designing User Interface Software

In 1986, Sidney Smith and Jane Mosier published *Guidelines For Designing User Interface Software* for the U.S. Air Force. With 944 guidelines, this document remains the largest collection of publicly available user interface guidelines in existence. These guidelines draw extensively from four sources: Brown et al. (1983), Engel and Granda (1975), MIL-STD-1472C (revised; 1983), and Pew and Rollins (1975).

Their report provides user interface guidelines in six categories:

1. Data entry
2. Data display
3. Sequence control
4. User guidance
5. Data transmission
6. Data protection

One example of a guideline from this document is:

1.3 DATA ENTRY: Text

1.3/10 + Upper- and Lower-Case *Equivalent in Search*

Unless otherwise specified by a user, treat upper- and lower-case letters as equivalent in searching text

Example: "STRING," "String,", and "string" should all be recognized/accepted by the computer when searching for that word.

Comment: In searching for words, users will generally be indifferent to any distinction between upper and lower case. The computer should not compel a distinction that users do not care about and may find difficult to make. In situations when the case actually is important, allow users to specify case as a selectable option in string search.

Comment: It may also be useful for the computer to ignore such other features as bolding, underlining, parentheses, and quotes when searching text.

See also: 1.0/27 3.0/12

Although focused on character user interfaces (GUIs), many of the guidelines are still relevant to today's user interfaces—especially in websites.

Shneiderman's User Interface Guidelines

Ben Shneiderman published a landmark text in 1987, *Designing the User Interface*. This book contains many tables of guidelines and includes detailed explanations and background research to justify each guideline. He presented the "eight golden rules of dialog design," which, he explained, represents underlying principles of design that were applicable to most interactive systems. These principles were:

- **Strive for consistency**—in particular, use consistent sequences of actions and the same terminology wherever appropriate. He claimed that this was the most frequently violated yet "the easiest one to repair and avoid." Certainly, it ought to be easy to ensure consistency, but as we discuss later with respect to style guides, it can prove difficult to ensure that unhelpful inconsistencies do not creep into designs, especially where distributed teams are involved.
- **Enable frequent users to use shortcuts**—abbreviations, special keys, and macros can all be appreciated by frequent knowledgeable users. One of the traps of Windows, Icons, Mouse, and Pop-up menu (WIMP) interfaces is that they are very easy to demonstrate to senior management and superficially appear simple to use. We have seen several examples where experienced users have become extremely frustrated by mouse-intensive systems—in some cases, to the extent of suffering work-related upper-limb disorders.
- **Offer informative feedback**—visual feedback can be particularly effective. Shneiderman (1987) goes on to discuss the value of direct manipulation in this context.
- **Design dialogs to yield closure**—organizing actions into groups with a beginning, middle, and end plays to a basic psychological desire for closure and provides a sense of satisfaction when tasks are completed.
- **Offer simple error handling**—helps users not to make serious errors and ensure that error messages really help them.
- **Permit easy reversal of actions**—it may not always be possible, but there is nothing more reassuring for users than knowing that they can undo an unintended action that sometimes has dramatic consequences.
- **Support internal locus of control**—another basic psychological desire is for control and, the more systems are able to provide users with control, the more satisfIED they will be, especially for experienced users. Some modern office software packages break this rule, and many of us experience extreme frustration when a "clever" piece of software insists on reformatting what we have carefully laid out on a page.
- **Reduce short-term memory load**—although the "seven plus or minus two chunks" may be an oversimplified description of our memory limitations, many interfaces demand extraordinary feats of memory to perform simple tasks.

Nielsen's Usability Heuristics

In 1990, Molich and Nielsen (1990) carried out a factor analysis of 249 usability problems to derive a set of "heuristics" or rules of thumb that would account for all of the problems found. Nielsen (1994a) further revised these heuristics, resulting in the ten guidelines listed in the following.

1. **Visibility of system status**—The system should always keep users informed about what is going on through appropriate feedback within a reasonable time.
2. **Match between system and the real world**—The system should speak the user's language—with words, phrases, and concepts familiar to the user—rather than system-oriented terms. Follow real-world conventions, making information appear in a natural and logical order.
3. **User control and freedom**—Users often choose system functions by mistake and will need a clearly marked "emergency exit" to leave the unwanted state without having to go through an extended dialog. Support undo and redo.
4. **Consistency and standards**—Users should not have to wonder whether different words, situations, or actions mean the same thing. Follow platform conventions.
5. **Error prevention**—Even better than good error messages is a careful design that prevents a problem from occurring in the first place.
6. **Recognition rather than recall**—Make objects, actions, and options visible. The user should not have to remember information from one part of the dialog to another. Instructions for use of the system should be visible or easily retrievable whenever appropriate.
7. **Flexibility and efficiency of use**—Accelerators—unseen by the novice user—may often speed up the interaction for the expert user, such that the system can cater to both inexperienced and experienced users. Allow users to tailor frequent actions.
8. **Aesthetic and minimalist design**—Dialogs should not contain information that is irrelevant or rarely needed. Every extra unit of information in a dialog competes with the relevant units of information and diminishes their relative visibility.
9. **Help users recognize, diagnose, and recover from errors**—Error messages should be expressed in plain language (no codes), precisely indicate the problem, and constructively suggest a solution.
10. **Help and documentation**—Even though it is better if the system can be used without documentation, it may be necessary to provide help and documentation. Any such information should be easy to search, should focus on the user's task, list concrete steps to be carried out, and not be too large.

Nielsen (1994b) describes a method for structuring a guidelines-based user interface review. Nielsen's method uses Molich and Nielsen's (1990) user interface heuristics, and indeed terms this method a "heuristic evaluation." The method, however, can be applied using any set of HCI Guidelines. For further information on evaluation techniques, see Chapter 57 "Inspection-Based Evaluations" by Cockton, Lavery, and Woolrych (2003).

STYLE GUIDES

Although few organizations have the history, infrastructure, or stamina to impose and police rigid user interface standards, style guides can be developed to reduce the unnecessary variation caused by dispersed design teams and extended system development timescales.

The main stages in the process involve the following.

CHOOSING THE RIGHT GUIDELINES

We have already pointed out that there are many good sources of guidelines, including a number of proprietary style guides provided by major vendors. There is no right answer as to which guidelines to select. It depends on the specific system under development and the style of the interface.

TAILORING THE GUIDELINES INTO SPECIFIC DESIGN RULES FOR YOUR APPLICATION

For instance, a guideline that states that displays should be consistently formatted might be translated into design rules that specify where various display features should appear, such as the display title, prompts and other user guidance, error messages, command entries, and so forth. For maximum effectiveness, guideline tailoring must take place early in the design process before any actual design of user interface software. To tailor guidelines, designers must have a thorough understanding of task requirements and user characteristics. Thus, task analysis is a necessary prerequisite of guidelines tailoring.

The process of developing, reviewing, and agreeing on style guides can be a positive process in enhancing organizational communication, especially across traditional organizational barriers.

IMPLEMENTING THE STYLE GUIDE

Many style guides offer little more than general recommendations on good practice. The problem with these is that they take considerable interpretation and may, therefore, still result in different parts of a system behaving differently. The mere presence of a style guide does not ensure consistency. Designers have to choose to conform—or be disciplined to conform—to achieve the benefits.

One of the best ways of encouraging them to follow a guide is to support it with a code library and provide lots of good examples for designers to follow. Interactive demonstrations can be particularly valuable.

POLICING AND MAINTAINING THE STYLE GUIDE

In practice, it is better to motivate and encourage designers to follow good examples in style guides than to rely on post-design monitoring and policing operations. As we have pointed out earlier, international standards are slowly being developed, which address many different aspects of user interface design—both in hardware and software. These standards can be used to provide support for in-house measures. In our experience, senior managers are more likely to take style guide and user interface issues seriously if they know that there are public standards that support them. We have incorporated checklists based on parts of ISO 9241 in some style guides that we have developed. Appendix 2 shows a checklist based on the seven key principles in ISO 9241-10:1996 *Dialogue Principles*. Each principle has been rephrased as a question (in bold) with specific questions below. The "correct" answer is usually "yes," although there may be occasions where the recommendation is not applicable or not possible; for example, it may not be possible to *"undo"* a *"commit"* action. Appendix 3 shows a similar checklist for user guidance based on ISO 9241-13:1998 *User Guidance*.

The checklists can be used by designers in reviewing their own work as well as by those involved in signing-off the design.

CONCLUSION

One of the recurring themes in this chapter is what might be called Stewart's Law of Usability Standards—the easier it is to formulate the usability standard or guideline, the more difficult it is to apply in practice. By this, I mean that a simple guideline like "allow the user to control the pace and sequence of the interaction" has a great deal of backing as a general guideline. In practice, of course, there are many situations where the rule fails, where the answer is "it depends," and where the context makes it more appropriate for the system to control some part of the interaction. The designer, wishing to follow such a simple guideline, needs to interpret it in the context of the system, and this requires insight and thought. The alternative approach—where the guideline is preceded by statements defining the context, for example, "if the user is x and the task is y and the environment is z, then do ABC"—becomes so tedious and confusing that they are quickly ignored. In the ISO 9241 series described earlier, there has been an attempt to define a middle course that involves giving specific practical examples as an aid to design. In ISO 13407 *Human-Centered Design Processes for Interactive Systems,* the standard is concerned with the process itself.

ISO 13407-1999 HUMAN-CENTERED DESIGN PROCESSES FOR INTERACTIVE SYSTEMS PROVIDES

Guidance for project managers to help them follow a human-centered design process. By undertaking the activities and following the principles described in the Standard, managers can be confident that the resulting systems will be usable and will work well for their users.

The Standard describes four principles of human-centered design

- Active involvement of users (or those who speak for them)
- Appropriate allocation of function (making sure human skill is used properly)
- Iteration of design solutions (allowing time for iteration in project planning)
- Multidisciplinary design (but beware large design teams)

and four key human-centered design activities:

- Understand and specify the context of use (make it explicit—avoid assuming it is obvious)
- Specify user and organizational requirements (note there will be a variety of different viewpoints and individual perspectives)
- Produce design solutions (note plural, multiple designs encourage creativity)
- Evaluate designs against requirements (involves real user testing, not just convincing demonstrations)

To claim conformance, the Standard requires that the procedures used, the information collected, and the use made of results are specified (a checklist is provided as an annex to help). This approach to conformance has been used in a number of parts of ISO 9241 because so many ergonomics recommendations are context-specific. Thus, there is often only one "shall" in these standards that generally prescribes what kind of evidence is required to convince another party that the relevant recommendations in the Standard have been identified and followed.

We believe this is one way of ensuring that usability standards remain relevant when technology changes and offering practical help to designers and developers.

ACKNOWLEDGMENT

Some material in this chapter was also used in the *Human-Computer Interaction Handbook* edited by Julie Jacko and Andrew Sears.

REFERENCES

Brown, C. M., Brown, D. B., Burkleo, H. V., Mangelsdorf, J. E., Olsen, R. A., & Perkins, R. D. (1983, June 15). *Human factors engineering standards for information processing systems (LMSC-D877141).* Sunnyvale, CA: Lockheed Missiles and Space Company.

Cockton, G., Lavery, D. & Woolrych, A. (2003), Inspection-based evaluations. In J. A. Jacho, & A. Sears (Eds.), The human-computer interaction handbook. Mahwah, NJ: Lawrence Erlbaum Associates.

Engel, S. E., & Granda, R. E. (1975, December). *Guidelines for man/display interfaces* (Technical Report TR 00.2720). Poughkeepsie, NY: IBM.

MIL-STD-1472C, Revised. (1983, September 1). *Military standard: Human engineering design criteria for military systems, equipment and facilities*. Washington, DC: Department of Defense.

Molich, R., & Nielsen, J. (1990, March). Improving a human-computer dialogue. *Communications of the ACM, 33*(3), 338–348.

Nielsen, J. (1994a). Enhancing the explanatory power of usability heuristics. In *Proceedings of CHI'94 Conference* (pp. 152–158). Boston, MA, April 24–28, 1994.

Nielsen, J. (1994b). Heuristic evaluation. In J. Nielsen, and R. L. Mack (Eds.), Usability inspection methods. New York: Wiley.

Pew, R. W., & Rollins, A. M. (1975). *Dialog specification procedures*(Report 3129, revised). Cambridge, MA: Bolt Beranek and Newman.

Shneiderman, B. (1987). *Designing the user interface—Strategies for effective human-computer interaction.* Reading, MA: Addison-Wesley.

APPENDIX 1 HISTORY—DEUSCHES INSTITUT FÜR NORMUNG (DIN)

Thirty years ago, the German National Standards organization started to publish a series of standards which shook the computer world. These standards, DIN 66-234, were published in a number of parts and collectively addressed the ergonomics problems of Visual Display Terminals and their workplaces.

If we ask why there was widespread concern, especially from the computer manufacturers, many of whom happened to be based in the United States, then we receive two answers. One answer, which was popular at the time, was that the Standards were based on too little and too recent research. A particular issue that received such criticism was the requirement that the thickness of the keyboard should be restricted to 30 mm. A number of manufacturers reported studies disputing the importance of keyboard thickness, arguing with the proposed dimension and demonstrating that users showed preferences for quite different arrangements.

Of course, it should not be overlooked that the 30-mm keyboard thickness was a very difficult target to reach at that time. Most key mechanisms themselves required greater depth, and major manufacturers had substantial investment in tooling keyboards to quite different thicknesses.

The second answer is that the very idea of ergonomics requirements affecting sales directly was completely alien to many of the suppliers. Certainly, the large manufacturers employed ergonomists, human factors engineers, and psychologists in their research and development departments. Certainly there was a growing recognition that the human aspects of computer technology were important. But at that time, price performance was the main objective, and it came as a major culture shock for the computer industry that ergonomics standards could have such a major impact on whether a product would sell.

Note that it was not the DIN standard itself but its integration into workplace regulations (ZH 618 Safety Regulations for Display Workplaces in the Office Sector, published by the Central Association of Trade Cooperative Associations) that gave the ergonomics requirements such "teeth." Failure to comply with these regulations leaves an employer uninsured against industrial compensation claims.

DIN 66 234 also contained a number of parts that dealt exclusively with software issues. For example, Part 3 dealt with the grouping and formatting of data; Part 5, with the coding of information; and Part 8, with the principles of dialog design. Although these were more in the form of recommendations, they too were heavily criticized, particularly for their broad scope and their inhibitory effect on interface design.

Indeed, a major criticism of most early standards in this field was that they were based on product design features such as height of characters on the screen. Such standards were specific to current technology, for example, cathode ray tubes (CRT), and did not readily apply to other technologies. They may therefore inhibit innovation and force designers to stick to old solutions.

APPENDIX 2 TASK DESIGN CHECKLIST BASED ON ISO 9241-10:1996 DIALOG PRINCIPLES

	Yes	No
Is the dialog suitable for the user's task and skill level?	☐	☐
Does the sequence match the logic of the task?	☐	☐
Are there any unnecessary steps that could be avoided?	☐	☐
Is the terminology familiar to the user?	☐	☐
Does the user have the information they need for the task?	☐	☐
Is extra information available if required? (*keep dialog concise*)	☐	☐
Does the dialog help users perform recurrent tasks?	☐	☐
Does the dialog make it clear what the user should do next?	☐	☐
Does the dialog provide feedback for all user actions?	☐	☐
Are users warned about (and asked to confirm) critical actions?	☐	☐
Are all messages constructive and consistent?	☐	☐
Does the dialog provide feedback on response times?	☐	☐
Can the user control the pace and sequence of the interaction?	☐	☐
Can the user choose how to restart an interrupted dialog?	☐	☐
Does the dialog cope with different levels of experience?	☐	☐
Can users control the amount of data displayed at a time?	☐	☐
Is the dialog consistent?	☐	☐
Are the appearance and behavior of dialog objects consistent with other parts of the dialog?	☐	☐
Are similar tasks performed in the same way?	☐	☐
Is the dialog forgiving?	☐	☐
Does the dialog provide "undo" (and warn when not available)?	☐	☐
Does the dialog prevent invalid input?	☐	☐
Are error messages helpful?	☐	☐
Can the dialog be customized to suit the user?	☐	☐
Does the dialog offer users different ways of working?	☐	☐
Does the dialog provide helpful defaults?	☐	☐
Can users choose different levels of explanation?	☐	☐
Can users choose different data representations? (e.g., show files as icons or lists)	☐	☐
Does the dialog support learning?	☐	☐
Does system feedback help the user learn? (e.g., menu items that indicate shortcut key combinations)	☐	☐
Is context-sensitive help provided? (where possible)	☐	☐

APPENDIX 3 CHECKLIST BASED ON ISO 9241-13:1998 USER GUIDANCE

	Yes	No
General		
Can the user guidance be readily distinguished from other information?	☐	☐
Do system-initiated messages disappear when no longer applicable?	☐	☐
Do user-initiated messages remain until the user dismisses them?	☐	☐
Are messages specific and helpful?	☐	☐
Can the user continue while the guidance is displayed?	☐	☐
Are important messages distinctive?	☐	☐
Can users control the level of guidance they receive?	☐	☐
Wording		
Do messages describe results before actions? (e.g., to clear screen, press del)	☐	☐
Are most messages worded positively? (i.e., what to do not what to avoid)	☐	☐
Are messages worded in a consistent grammatical style?	☐	☐
Is the guidance written in a short, simple sentences?	☐	☐
Are messages written in the active voice?	☐	☐
Is the wording user oriented?	☐	☐
Prompts—To Indicate That the System Is Waiting for Input		
Do prompts indicate the type of input required?	☐	☐
Is online help available to explain prompts? (if required)	☐	☐
Are prompts displayed in consistent positions?	☐	☐
Does the cursor appear automatically at the first prompted field?	☐	☐
Feedback—To Indicate That the System Has Received Input		
Does the system always provide some form of feedback for all user actions?	☐	☐
Is normal feedback unobtrusive and nondistracting?	☐	☐
Are the type and level of feedback suitable for the skills of the users?	☐	☐
Does the system provide clear feedback on system state? (e.g., waiting for input)	☐	☐
Are selected items always highlighted?	☐	☐
If the action requested is not immediate, is there feedback that requests have been accepted (and confirmation when they are complete), for example, remote printing?	☐	☐
Does the system show progress indicators when appropriate?	☐	☐
Is system response feedback appropriate? (not too fast or too slow)	☐	☐
Status Information—To Indicate What the System Is Currently Doing		
Is appropriate status information available at all times?	☐	☐
Is status information always displayed in a consistent location?	☐	☐
Does the system always indicate when user input is not possible?	☐	☐
Are system modes clearly indicated?	☐	☐
Error Prevention and Validation		
Are the function keys consistent across the system?	☐	☐
Does the system anticipate problems and warn the user appropriately?	☐	☐
Does the system warn the user when data might be lost by a user action?	☐	☐
Is "undo" provided, where appropriate?	☐	☐
Can users modify or cancel input prior to input?	☐	☐
Can users edit wrong input (rather than have to reenter the complete field)?	☐	☐
Does the field level validation:		
Immediately indicate that there is an error?	☐	☐

(Continued)

General	Yes	No
Position the cursor at the beginning of the first incorrect field?	☐	☐
Indicate all fields in error (including cross-field errors)?	☐	☐

Error Messages

	Yes	No
Can users get more help easily if required?	☐	☐
Do error messages indicate what is wrong and what should be done?	☐	☐
Can the user tell when an error message has reoccurred?	☐	☐
Do error messages disappear when the error has been corrected?	☐	☐
Can users remove error messages prior to correction if they wish?	☐	☐
Do error messages appear in a consistent location?	☐	☐
Can error messages be moved if they obscure part of the screen?	☐	☐
Do error messages appear as soon as the wrong input has been entered?	☐	☐
Can users turn off confirmation screens?	☐	☐
Can users adjust the volume of warning tones or messages?	☐	☐

Online Help

	Yes	No
Is the online help context sensitive?	☐	☐
Is the system-initiated online help unobtrusive?	☐	☐
Can users turn off system-initiated online help?	☐	☐
Can the users initiate online help by a simple consistent action?	☐	☐
Does the system accept synonyms and close spelling matches when the user searches for the appropriate area of online help?	☐	☐
Is the online help clear, understandable, and specific to the users' tasks?	☐	☐
Does the online help provide both descriptive and procedural help?	☐	☐
Can the user easily go between the task screens and the online help?	☐	☐
Can the users configure online help to suit their preferences?	☐	☐
Can the user easily return to the task?	☐	☐
Are there suitable features to help users find appropriate help?Are there quick methods for:		
Going directly to another screen	☐	☐
Browsing	☐	☐
Exploring linkages between topics	☐	☐
Returning to the previous help page	☐	☐
Returning to a home location	☐	☐
Accessing a history of previously consulted topicsIf the online help system is large, are any of the following provided to aid search:		
String search of a list of topics?	☐	☐
Keyword search of online help text?	☐	☐
Hierarchical structure of online help text?	☐	☐
Map of online help topics?	☐	☐
If the online help system has a hierarchical structure, is there:		
An overall indication of the structure?	☐	☐
Easy access to any level in the hierarchy?	☐	☐
An obvious and consistent method of accessing more detail?	☐	☐
A quick means of accessing the main (parent) topic?	☐	☐
Are topics self-contained? that is, not dependent on reading previous sections	☐	☐
If the information is scrollable, does the topic remain clear?	☐	☐
Does the context-sensitive help provide information on:		
The current dialog step?	☐	☐

(Continued)

General	Yes	No
The current task?	☐	☐
The current applications?	☐	☐
The task information presented on the screen?	☐	☐
Does the online help explain objects, what they do, and how to use them?	☐	☐
Does the online help indicate when it is not available on an object?	☐	☐

26 International Standards in Information and Communication Technology (ICT) Accessibility and Their Application in Operating Systems

Anna Szopa

CONTENTS

INTRODUCTION

The importance of information and communication technology (ICT) accessibility is underlined in fundamental international and national acts. The first was created in 1973 in the United States: Section 508 of the Rehabilitation Act discusses the development, procurement, maintenance, or use of electronic and information technology (Section 508 of the Rehabilitation Act, U.S. General Services Administration). The next important document is Section 255 of the Communications Act created in 1996. It says that telecommunication equipment manufacturers and service providers should make their products and services accessible to people with disabilities if such access is readily achievable (Section 255 of the Communications Act, Federal Communications Commission). The third act was also created in the United States—21st Century Communications and Video Accessibility Act (CVAA). In 2010, the CVAA updated federal communications law to increase the access of persons with disabilities to modern communications (CVAA, Federal Communications Commission). In Europe, the first important activities related to accessibility was issued in 2015, the European Accessibility Act discusses the problem of removing barriers for disabled people and creating opportunities for accessible products and services (European Commission).

Nowadays, accessible technologies are built in every operating system and device, and software companies race to make them more useful and flexible. In most cases, accessibility is a company's core consideration from the earliest stages of product design through release. A big part of society faces problems in accessing ICT devices. Therefore, it is so important to design accessible user interfaces that can help all users navigate their computers (Cook, Polgar, & Encarnação, 2019). Computer accessibility features refer not only to the disability or aging but also to a personal preference, or a unique work style; therefore, information technology producers and developers have made, so far, many strides to make ICT easier to use. An important component of accessibility is auto-personalization. Computer auto-personalization means that the device can be set up exactly the way users need or prefer with the data and software they need (Szopa & Vanderheiden, 2020). The aim of the chapter is to provide a chronological presentation of international standards related to ICT accessibility and outline the latest accessibility features that are built into operating systems.

ACCESSIBILITY—INTERNATIONAL STANDARDS

Accessibility in ICT is well supported in related international standards and recommendations. The first ISO standards appeared in 2001. The ISO/IEC GUIDE 71:2001 are the guidelines for standards developers that address the needs of older persons and persons with disabilities. The document describes the requirements for usability and accessibility and analyzes the aspects that should be considered when designing products and services for the elderly and disabled people (ISO, the International Organization for Standardization, ISO/IEC GUIDE 71:2001). This standard has been revised by ISO/IEC GUIDE 71:2014 (ISO, the International Organization for Standardization, ISO/IEC GUIDE 71:2014).

Six years later, ISO created standard ISO/IEC TR 19766:2007 Information technology—Guidelines for the design of icons and symbols accessible to all users, including the elderly and persons with disabilities. The document provides recommendations relating to the design of icons to support accessibility for the elderly and people with disabilities. These recommendations assist accessible implementation of all icons for users (ISO, the International Organization for Standardization, ISO/IEC TR 19766:2007).

Standard ISO 9241-20:2008 Ergonomics of human-system interaction—Part 20: Accessibility guidelines for information/communication technology (ICT) equipment and services appeared in 2008. This document provides tables that enable standards developers to relate the relevant clauses of a standard to the factors that should be considered to ensure that all abilities are addressed. It also offers descriptions of body functions or human abilities and the practical implications of impairment and a list of sources that standards developers can use to investigate more detailed and specific guidance materials (ISO, the International Organization for Standardization, ISO 9241-20:2008).

The most important international standards are described in ISO/IEC 24786:2009 Information technology—User interfaces—Accessible user interface for accessibility settings. The standards make information technologies more accessible by ensuring that people who have problems using the computer can adjust the accessibility settings by themselves. The norm specifies requirements and recommendations in making accessibility settings accessible and provides guidance on specific accessibility settings. It presents how to activate accessibility functions access and operates accessibility setting mode. It summarizes in detail requirements for specific disabilities (low vision, blind, deaf, hard of hearing, physical disabilities, cognitive, language, and learning disabilities). The norm applies to all operating system user interfaces on computers and other devices, and it does not apply to the user interface before the operating system is loaded and active (ISO, the International Organization for Standardization, ISO/IEC 24786:2009). The norm defines the most important terms related to computer accessibility. Some of the terms can be found in the following norms: ISO/IEC 2382-1:1993, Information technology—Vocabulary—Part 1: Fundamental terms

(ISO, the International Organization for Standardization, ISO/IEC 2382-1:1993) and ISO 9241-171:2008, Ergonomics of human-system interaction—Part 171: Guidance on software accessibility (ISO, the International Organization for Standardization, ISO 9241-171:2008).

Further standard ISO/IEC 13066-1:2011 Information technology—Interoperability with assistive technology (AT)—Part 1 is standardizing requirements and recommendations for interoperability and provides a basis for designing and evaluating interoperability between IT and AT (ISO, the International Organization for Standardization, ISO/IEC 13066-1:2011).

In 2012, ISO/IEC 40500:2012 Information technology—W3C Web Content Accessibility Guidelines (WCAG) 2.0 was issued. The standard includes recommendations for making web content more accessible and usable (ISO, the International Organization for Standardization, ISO/IEC 40500:2012). The standards are available at The W3C Web Accessibility Initiative (WAI) and refer to the information in a web page or application, including natural information such as text, images and sounds, code, or markup that defines the structure, presentation, etc. (Information technology—W3C Web Content Accessibility Guidelines (WCAG) 2.0).

In the same year, ISO/IEC 29136:2012 Information technology—User interfaces—Accessibility of personal computer hardware was issued. It provides requirements and recommendations for the accessibility of personal computer hardware to be used in planning, developing, designing, and distributing these computers (ISO, the International Organization for Standardization, ISO/IEC 29136:2012).

In 2014, ISO issued ISO 14289-1:2014 Document management applications—Electronic document file format enhancement for accessibility—Part 1: Use of ISO 32000-1 (PDF/UA-1), The primary purpose of the standard is to define how to represent electronic documents in PDF format to allow the file to be accessible (ISO, the International Organization for Standardization, ISO 14289-1:2014).

And in 2015, ISO/IEC TS 20071-21:2015 Information technology—User interface component accessibility—Part 21: Guidance on audio descriptions was introduced. The norm provides recommendations in describing audiovisual content in an auditory modality (ISO, the International Organization for Standardization, ISO/IEC TS 20071-21:2015).

In 2020, ISO issued two standards ISO 24552:2020 Ergonomics—Accessible design—Accessibility of information presented on visual displays of small consumer products. The document describes how to present information on small visual displays to make the product more accessible for older people and people with low vision or color deficiency (ISO, the International Organization for Standardization, ISO 24552:2020). The second standard was ISO/TS 21054:2020 Ergonomics—Accessible design—Controls of consumer products. The document defines design principles of accessibility for controls of consumer products so that users from a population with the widest range of user needs, characteristics, and capabilities can use controls to operate and control consumer products in the same manner and ease as users without disabilities (ISO, the International Organization for Standardization, ISO/TS 21054:2020).

Besides useful recommendations and guidelines, the ISO Standards define the most important terms related to accessibility such as (ISO, the International Organization for Standardization, ISO/IEC 24786:2009):

- accessibility feature—feature (etc.) that is specifically designed to increase the usability of products for those experiencing disabilities (ISO, the International Organization for Standardization, ISO 9241-20:2008)
- accessibility setting—setting to make the user interface more accessible for people with disabilities
- accessibility setting mode—mode where the user adjusts accessibility settings
- auditory feedback—function that allows individuals to hear whether their operations (e.g., key input) have been accepted by the computer
- on-screen keyboard—software that presents a keyboard on the display screen that is operable

by a pointing device and that generates input that is identical to that which comes from a physical keyboard

- screen reader—function that reads the characters and other information on the screen aloud to the user to allow access to the information on screen without viewing the screen
- shortcut—operation which immediately invokes an action without displaying intermediate information (such as menus) or requiring pointer movement or any other user activity
- visual emphasis—function that allows users to change the visual aspects to improve visibility
- visual feedback—function that allows users to know visually whether their operations (e.g., key input) have been accepted by the computer
- voice operation—function that allows users to operate a computer with voice commands through a microphone (e.g., the voice command "Switch to Mail" activates the email application)

Outside the norms related to accessibility software and hardware, developers are supposed to follow the recommendations related to user interface as:

- ISO/IEC 18021:2002 Information technology—User interfaces for mobile tools for management of database communications in a client-server model (ISO, the International Organization for Standardization, ISO/IEC 18021:2002)
- ISO/IEC 25010:2011 Systems and software engineering—Systems and software Quality Requirements and Evaluation (SQuaRE)—System and software quality models (ISO, the International Organization for Standardization, ISO/IEC 25010:2011)
- ISO/IEC 25051:2014 Software engineering—Systems and software Quality Requirements and Evaluation (SQuaRE)—Requirements for quality of Ready to Use Software Product (RUSP) and instructions for testing (ISO, the International Organization for Standardization, ISO/IEC 25051:2014)

ACCESSIBILITY IN OPERATING SYSTEMS

Nowadays, accessibility features are often built into every operating system and device, and software producers race to implement new solutions—accessibility options to help with many different special needs. Below are the latest accessibility solutions built into Windows, Apple, and Android operating systems; presented also is the tool named Morphic auto-personalization that allows computer settings to be adjusted according to the user's needs and preferences.

WINDOWS

Windows, to assist users with vision problems, can adjust size and color, apply color filters, distinguish colors, boost contrast or get rid of color entirely, magnify the screen, and make the mouse more visible by changing the color and size of the mouse pointer. Windows navigator helps navigate the PC; offers simplified navigation and intelligent image description, making it easy to explore a page without missing a thing on the screen. The new Windows research project Microsoft Soundscape uses innovative audio-based technology to enable people with blindness or low vision to build a richer awareness of their surroundings, thus becoming more confident in navigating new environments. For users with hearing problems, a narrator lets them convert stereo sound into a single channel so everything can be heard better; displays audio alerts visually; makes notifications stick around longer; and uses closed captions to read the words that are spoken in movies, television shows, etc. Neuro-sensitive users can minimize distractions by reducing animations and turning off background images, clean up taskbar clutter, simplify the start menu and quiet notifications, and use reading view to clear distracting content from web pages. In addition, for anyone who needs writing assistance, Windows offers help in constructing sentences with text suggestions. Tell Me quickly accesses commands in several Office 365 applications without navigating the command ribbon; this feature can be used to discover the difficult-to-find capabilities. Windows gives several options to make devices easier to use; users can change the shape and color of the mouse pointer, use Mouse Keys to move the mouse pointer with the numeric keypad, and support eye control (Windows 10 Accessibility Features).

APPLE

Apple devices let users write a text without seeing the screen because VoiceOver describes exactly what is happening. In addition, users can choose from a range of color filters and fine-tune them or turn on Invert Colors. Hover Text lets you choose the fonts and colors. Improved dictation and richer text editing features help users write more efficiently, while simple vocal commands quickly open and interact with apps. Switch Control is an assistive technology that uses built-in features as well as switches, a joystick, or other adaptive devices to control the screen. The Accessibility Keyboard is fully customizable and gives users advanced typing and navigation capabilities. It also includes toolbar support as well as improved typing, autocapitalization, and word suggestions. Text-to0Speech can make learning easier by letting users hear what they are reading and writing; Speak Screen can read text from newspapers, books, web pages, or email (Apple Accessibility).

ANDROID

Android offers TalkBack. To interact with the device, it describes the user's actions and talks about alerts and notifications. If users want spoken feedback at certain times, they can turn on Select to Speak to select items on their screen to hear them read or described aloud or point the camera at something in the real world. Voice Access controls the device with spoken commands. Users can use their voice to open apps, navigate, and edit text hands-free. Switch Access interacts with your Android device by using one or more switches instead of the touchscreen. Time to take action (Accessibility timeout) chooses how long to show messages. BrailleBack connects a refreshable braille display to the user's device via Bluetooth. BrailleBack works with TalkBack for a combined speech and braille experience, allowing text to be edited and interact with the user's device. Users can use Live Transcribe to capture speech and view it as a text for users with hearing problems. Sound Amplifier, with wired headphones, amplify the sounds in the environment and Hearing Aid Support can pair hearing aids with an Android device to hear more clearly. Thanks to Real-Time Text (RTT), users can text to communicate during calls (Android Accessibility).

MORPHIC AUTO-PERSONALIZATION

The tool was designed by the team from the Trace Research and Development Center and is a part of the Global Public Inclusive Infrastructure initiative (GPII). It allows the adjustment of computer settings to suit the user's needs and preferences. It also helps discover setting options available to make the computer more accessible. Preferences are saved in the cloud and can be used on any device where Morphic is installed. Furthermore, Morphic can run on various types of operating systems, browsers, and devices that the users have. In this manner, people can have computers instantly set up for them without doing it themselves (Szopa & Vanderheiden, 2020; Szopa et al., 2019). At this moment, two versions of Morphic are available. Morphic Personal provides quick access to a fixed set of features such as text size, glare reduction, magnifier, personal reading, and volume. Morphic Plus adds the ability to personalize various accessibility features and links that are included on MorphicBar. Furthermore, Morphic Plus allows users to create a customized environment for clients or users, helping them be more productive at work or school and more connected to communities (Morphic).

SUMMARY

Accessibility is driven in part by law, regulated by standards, and supported by guidelines. In the chapter, the author presented fundamental international standards that currently shape the world of accessibility. Standards were presented chronologically and were associated with user needs, human-system interaction, user interface, accessible design, web content, electronic document file format, ICT equipment and services, and software quality.

In the second part, the latest accessibility solutions were presented. The summary shows that companies are increasingly focused on making their technology products and services more accessible and the solutions are becoming more effective. The effort brings together engineers, designers, artists, managers, and a huge group of professionals that make accessibility the future of technology. Accessibility features can be used successfully by people with a wide range of abilities and disabilities, and they help users perform better in their communities and jobs. Furthermore, the presented solutions are fundamental in ensuring a positive user experience.

By looking at the present trends in ICT development, we can say that the knowledge and awareness of accessibility will result in better design methods, tools, and products. Because we should never forget that almost every person in the world will need assistive technology eventually, accessibility should be already perceived as a crucial aspect of futuristic universe design.

REFERENCES

Android Accessibility. https://www.apple.com/accessibility/ (accessed 04/17/2020).

Apple Accessibility. https://www.apple.com/accessibility/ (accesses 04/15/2020).

Cook, A., Polgar, J., & Encarnação, P. (2019). *Assistive technologies: Principles and practice*. Elsevier Health Sciences. ISBN: 9780323523387

CVAA, Federal Communications Commission. https://www.fcc.gov/consumers/guides/21st-century-communications-and-video-accessibility-act-cvaa (accessed 06/04/2020).

European Commission. https://ec.europa.eu/social/main.jsp?catId=1202 (accessed 09/05/2020).

Information technology—W3C Web Content Accessibility Guidelines (WCAG) 2.0. https://www.w3.org/WAI/standards-guidelines/wcag/#iso (accessed 02/22/2020).

ISO, the International Organization for Standardization. ISO/IEC GUIDE 71:2001. https://www.iso.org/standard/33987.html (accessed 02/06/2020).

ISO, the International Organization for Standardization. ISO/IEC GUIDE 71:2014. https://www.iso.org/standard/57385.html (accessed 02/04/2020).

ISO, the International Organization for Standardization. ISO/IEC TR 19766:2007. https://www.iso.org/standard/42128.html (accessed 02/10/2020).

ISO, the International Organization for Standardization. ISO 9241-20:2008. https://www.iso.org/standard/33987.html (accessed 09/01/2020).

ISO, the International Organization for Standardization. ISO/IEC 24786:2009. https://www.iso.org/standard/41556.html (accessed 02/10/2020).

ISO, the International Organization for Standardization. ISO/IEC 2382-1:1993. https://www.iso.org/standard/7229.html (accessed 02/10/2020).

ISO, the International Organization for Standardization. ISO 9241-171:2008. https://www.iso.org/standard/39080.html (accessed 02/10/2020).

ISO, the International Organization for Standardization. ISO/IEC 13066-1:2011. https://www.iso.org/standard/53770.html (accessed 02/14/2020).

ISO, the International Organization for Standardization. ISO/IEC 40500:2012. https://www.iso.org/standard/58625.html (accessed 02/20/2020).

ISO, the International Organization for Standardization. ISO/IEC 29136:2012. https://www.iso.org/standard/45159.html (accessed 02/25/2020).

ISO, the International Organization for Standardization. ISO 14289-1:2014. https://www.iso.org/standard/64599.html (accesses 02/27/2020).

ISO, the International Organization for Standardization. ISO/IEC TS 20071-21:2015. https://www.iso.org/standard/63061.html (accessed 03/03/2020).

ISO, the International Organization for Standardization. ISO 24552:2020. https://www.iso.org/standard/73276.html accessed 03/03/2020

ISO, the International Organization for Standardization. ISO/TS 21054:2020. https://www.iso.org/standard/69765.html (accessed 03/03/2020).

ISO, the International Organization for Standardization. ISO/IEC 18021:2002. https://www.iso.org/standard/30806.html (accessed 03/10/2020).

ISO, the International Organization for Standardization. ISO/IEC 25010:2011. https://www.iso.org/standard/35733.html (accessed 03/10/2020).

ISO, the International Organization for Standardization. ISO/IEC 25051:2014. https://www.iso.org/standard/61579.html (accessed 03/10/2020).

Morphic. https://morphic.org/ (accessed 10/30/2020).

Section 255 of the Communications Act, Federal Communications Commission. https://www.fcc.gov/consumers/guides/telecommunications-access-people-disabilities (accessed 07/03/2020).

Section 508 of the Rehabilitation Act, U.S. General Services Administration. https://www.section508.gov/manage/laws-and-policies/state (accessed 09/02/2020).

Szopa, A., & Vanderheiden, G. (2020). The Importance of Computer Auto Personalization. Advances in Usability, User Experience, Wearable and Assistive Technology. Proceedings of the AHFE 2020 Virtual Conferences on Usability and User Experience, Human Factors and Assistive Technology, Human Factors and Wearable Technologies, and Virtual Environments and Game Design, July 16–20, 2020, USA

Szopa, A., Jordan, B., Folmar, D., Vanderheiden, G. (2019). The auto-personalization computing project in libraries. Advances in usability and user experience. Proceedings of the AHFE 2019 International Conferences on Usability & User Experience, and Human Factors and Assistive Technology, July 24–28, 2019, Washington D.C., USA

Windows 10 Accessibility Features. https://www.microsoft.com/en-us/accessibility/windows?activetab=pivot_1%3aprimaryr2 (accessed 04/02/2020).

27 Human Factors Engineering of Computer Workstations
HFES Draft Standard for Trial Use

Thomas J. Albin

CONTENTS

INTRODUCTION

BSR/HFES 100 Human Factors Engineering of Computer Workstations (HFES 100) is a specification of the recommended human factors and ergonomic principles related to the design of the computer workstation. HHFES 100 is primarily intended for fixed, office-type computer workstations for moderate to intensive computer users. These design specifications are intended to facilitate the performance and comfort of computer workers.

HFES 100 addresses the interaction between the computer user and the hardware components of the operator–machine system. It addresses input devices, output devices, and furniture, as well as the integration of these components into a human user–computer hardware system.

Specifications are given for individual components. As an example, the specifications for furniture components describe a range of adjustments necessary to accommodate users who vary in size and use multiple working postures.

HFES 100 provides guidance as to how these individual workstation components are to be integrated into a whole or system that is ergonomically sound. It is possible to ergonomically utilize well-designed computer workstation components that individually conform to the specifications of HFES 100 yet combine them inappropriately. As an extreme example, a desk only designed for standing work might mistakenly be installed in an area where all other components were intended for sitting work. The chapter on Installed Systems is intended to prevent such mismatches.

Given that the individual components and the system can accommodate a variety of users and postures, HFES 100 also provides guidance on how to accommodate an individual user. As an example, it provides guidance on how a chair that is designed to accommodate a variety of sizes of users and a variety of postures may be adjusted to fit an individual.

Finally, it provides guidance regarding desirable conditions in the environment around the computer workstation.

THE NEED FOR A DESIGN STANDARD FOR COMPUTER WORKSTATIONS

The computer is a basic and pervasive tool in all aspects of life. According to the U.S. Census Bureau, three-fourths of all households in the United States had computers in 1997. Further, according to the same source, about 71% of all children used a computer at school. Finally, U.S. census data indicate that half of all employed adults use a computer in their workplace (Newberger, 1999).

Given that half of all workers in the United States use computers, designing computer workstations to facilitate computer worker performance offers the prospect of significant benefits. Research shows a clear association between the performance of computer workers and the ergonomics of the workstation. As an example, Dainoff (1990) found that performance increased by approximately 20% when the workstation components were adjusted to accommodate the individual user.

The U.S. Census Bureau (2001) estimates that there were 105,000,000 individuals employed in all businesses in the United States in 1997 and that the total payroll for these individuals was approximately $3 trillion (U.S. Census Bureau, 2000). Because one-half of all employed individuals used computers in the workplace that year, the payroll for computer workers in 1997 was approximately $1.5 trillion. Although one would not expect such dramatic facilitation of performance in every case, as was demonstrated by Dainoff (1990), the scale is so large that even small changes are significant. The performance benefit realized from good ergonomic practices in computer workstations dictates that it is in the employers' interest to provide computer workplaces that are ergonomically well designed.

Many employers do not have ergonomics or human factors professionals available to assist them in providing ergonomically sound computer workplaces. HFES 100 is a concise resource document for employers to use in providing well-designed computer workplaces when selecting components, integrating the components into a system, and fitting a workstation to an individual user.

HFES is divided into four major chapters: "Installed Systems," "Input Devices," "Visual Displays," and "Furniture."

INSTALLED SYSTEMS

The "Installed Systems" chapter describes how to put together all the workstation components into a system that matches the capabilities of the intended user. It is intended primarily for the System Installer—that is, the individual or individuals responsible for selecting the components and installing them as an integrated system. As such, it deals primarily with issues such as hardware components, noise, thermal comfort, and lighting.

HFES 100 (Human Factors and Ergonomics Society, 2002) lists the following four key issues that the System Installer must address:

- Ensure that all components individually comply with the appropriate specifications of HFES 100
- Ensure that the selected components are compatible with one another when installed
- Ensure that the workstation properly fits the intended user
- Ensure that the users are informed about the proper use and adjustment of the workstation components

Some individuals may fall outside the design specifications for the individual workstation components given in HFES 100. As an example, furniture specifications are written to accommodate individuals ranging between the 5th percentile females and 95th percentile males. The "Installed Systems" chapter offers some guidance on how to accommodate these individuals.

INPUT DEVICES

The "Input Devices" chapter addresses issues such as physical size, operation force, and handedness in the design of the following input devices:

- Keyboards
- Mouse and puck devices
- Trackballs
- Joysticks
- Styli and light pens
- Tablets and overlays
- Touch-sensitive panels

VISUAL DISPLAYS

The Visual Displays chapter covers monochrome and color CRT and flat panel displays. It is not intended to apply to displays on items such as photocopy machines and telephones. It addresses the human factor issues related to the design of displays such as viewing characteristics, contrast, and legibility.

FURNITURE

The "Furniture" chapter describes specifications for workstation components such as chairs and desks. There are productivity and comfort benefits associated with changing posture while performing computer work, and computer workers have been observed to frequently change positions while working. Consequently, the furniture designer needs to anticipate a range of movement and change in working posture.

HFES 100 describes the various postures that a furniture designer should consider. There are four reference postures: reclined sitting, upright sitting, declined sitting, and standing. The computer worker will frequently change the position of their body parts while working in any of the Reference Postures. The expected ranges of movement of body parts are described as User Postures.

ANTHROPOMETRY

The anthropometric data utilized in the Furniture and Installed Systems chapters were developed from the database of U.S. Army personnel (Gordon et al., 1989). A description of the anthropometric methods and models used is provided as an appendix to the "Furniture" chapter.

REFERENCES

Dainoff, M. (1990). Ergonomic improvements in VDT workstations: Health and performance effects. In S. Sauter, M. Dainoff, & M. Smith (Eds.), *Promoting health and productivity in the computerized office* (pp. 49–67). Philadelphia: Taylor and Francis.

Gordon, C. C., Churchill, T., Clauser, T. E., Bradtmiller, B., McConville, J. T., Tebbets, I., et al. (1989). *1988 Anthropometric survey of U.S. Army personnel: Summary statistics interim report* (Tech. Report NATICK/TR-89-027).

Human Factors and Ergonomics Society. (2002). *Board of Standards Review/Human Factors and Ergonomics Society 100—Human factors engineering of computer workstations—Draft Standard for Trial Use*, Santa Monica, CA: Human Factors and Ergonomics Society.

Newberger, E. C. (1999). *Computer use in the United States. Population characteristics.* U.S. Department of Commerce, Economics and Statistics Administration, U.S. Census Bureau.

U.S. Census Bureau. (2000). 1997 Economic census: Comparative statistics for the United States 1987 SIC basis. U.S. Census Bureau, http://landview.census.gov.epcd/ec97sic/E97SUS.HTM

U.S. Census Bureau. (2001). Statistics about business size (including small business) from the U.S. Census Bureau. U.S. Census Bureau, http://www.census.gov/epcd/www/smallbus.html

28 Practical Universal Design Guidelines
A New Proposal

Kazuhiko Yamazaki, Toshiki Yamaoka, Akira Okada,
Sohsuke Saitoh, Masatoshi Nomura, Koji Yanagida,
and Sadao Horino

CONTENTS

INTRODUCTION

Universal design (UD) practices are gaining importance in product design. Because there have been no established universal design methods, designers often tend to use their own experiences and intuition in extracting problems and creating new universal designs. In view of these situations, the Ergo-Design Technical Group of the Japan Ergonomics Society (EDTG/JES) organized a working group for proposing integrated UD methods that could be used by designers. This working group developed new practical UD guidelines together with the members of EDTG/JES and collaborators, including ergonomics experts in universities and relevant industrial designers working in the private sectors. The working group utilized various opportunities to discuss UD needs such as the annual conferences of the Society and its Kantoh District (Tokyo area) chapter, as well as regular and special seminars organized by the Ergo-Design Technical Group. The working group, thus, examined the proposals of the members for practical UD guidelines. This chapter describes the new guidelines, including the basic design principles and the detailed methods for accomplishing new designs that meet the requirements of the proposed guidelines.

These guidelines propose ready-to-use procedures for implementing a logical and systematic design approach for universal design. They provide practical guidance for designers and others who are looking for practical design tools.

The following summarizes the basic policies for the development of practical UD guidelines and gives the contents of the guidelines. This includes concrete guidance on how to apply them. In developing the guidelines, the highest priority has been placed on methods of extracting user requirements that are considered critical in developing a universal design. The working group was organized for a specified period in the EDTG/JES. It consisted of the following members: the Chair of WG, Toshiki Yamaoka (Wakayama University); members of the WG: Akira Okada (Graduate School of Osaka City University), Sosuk Saito (Human Factor Co., Ltd.), Masatoshi Nomura (NEC Corporation), Hiroharu Yanagida (Sanyo Electric Co., Ltd.), and Kazuhiko Yamazaki (IBM Japan, Ltd.); and the Chair of EDTG/JES: Sadao Horino (Kanagawa University).

The guidelines consist of two parts—Part 1: Practical Guide and Part 2: Reference Information. These two parts cover the following topics.

Part 1: Practical Guide

1. Definition of UD and the basic approach
2. UD design processes
3. UD user segments table
4. UD matrix
5. Methods of extracting user requirements
6. Methods of the construction and design of the UD concept
7. Methods of UD evaluation

Part 2: Reference Information

1. Information for the UD user segments table and UD data
2. UD matrix samples
3. Application of the UD matrix
4. UD action checklist
5. Case studies on the use of the Practical Universal Design Guidelines
6. Related information

The Practical Guidelines are intended for direct use by designers and have been prepared in accordance with the following basic policies: (a) design can be implemented without particular preliminary knowledge of UD; (b) the Practical UD Guidelines are produced for actual UD users (business owners, and company employees) and not merely for the promotion of the UD concept alone; (c) the Practical UD Guidelines are based on a pragmatic approach; (d) the Practical UD Guidelines are based on human-centered design methods; (e) the Practical UD Guidelines are based on logical design methods; (f) following the Practical UD Guidelines should allow customization; and (g) the guidelines comply with IEC/ISO Guide 71.

OUTLINE OF THE PRACTICAL UD GUIDELINES

The following introduces the outline of the Practical UD Guidelines.

THE DEFINITION OF UNIVERSAL DESIGN AND THE BACKGROUND FOR EMERGING UD NEEDS

The term universal design is defined as follows: to design products, environments, and information with the aim of providing all users—in a fair and equitable manner—with satisfactory services that respond to their varied needs. The need for UD has emerged due to the following background situations: (a) human-centered design has become increasingly important, whereas products tend to incorporate more functions and advanced technologies; (b) companies are highly appreciated when they respond to special needs and market needs put forward by the aging society, meet various growing ISO-related standards and legal provisions of laws and regulations, and fulfill their own social functions and responsibilities; and (c) there are fundamental changes in the social environment—such as an aging population that widens the diversity in the use of products—and enhanced public expectations for efficient welfare spending.

Clearly, these situations have extensive implications for the relevant design processes of products produced for a wide range of users.

METHODS OF LID IMPLEMENTATION BY COMPANIES

To promote UD, it is necessary to disseminate the idea that "UD activities reflect the corporate philosophy of a company, represent its business operations, and signify its presence" so that the significance of UD must be recognized throughout each company. UD can bring about direct and indirect effects on a company in meeting the varied needs of the users of their products.

Direct effects include:

- Improvement of product features
- Enhancement of the corporate image
- Reduction of total costs
- Development of new businesses and products

Indirect effects include:

- Market expansion
- Dissemination of social responsibilities of corporations and other organizations.

Effective promotion of UD requires the establishment of steps (procedures) to achieve successful UD implementation and planning of aggressive activities (action) in the final stage. A company can start with whatever it can undertake without following established procedures, but it is desirable to set up suitable internal organizations (structure) and infiltrate (awareness-raising) the UD concept throughout the company members—including the executive manager and employees at all levels—to solidify unity within the company through the following: (a) Procedures: establishment of the study of successful cases, the whole design processes, design methods, and the check system for design review; (b) Action: promotion of the "top-down" design approach, active proposals for new design ideas, and dissemination of the approach to organizations and people outside the company; (c) Structure: establishment of the project, creation of a company-wide council, and selection of a department in charge of UD promotion, (d) Awareness rising: production of manuals, holding of lectures and seminars for employees, and dissemination of information concerning UD applications conducted by other companies and institutions.

UNIVERSAL DESIGN PROCESS

Universal design means the design of products and environments usable by all people to the greatest extent possible, without the need for adaptation or specialized designs. "All people" include an entire range of different ages, cultures, locations, and disabilities—but the universal design does not mean that all people can use all kinds of products or services. Some severely disabled individuals will still need specific modifications. The goal of universal design is not only to eliminate physical barriers but also to remove mental barriers considering economic factors such as availability, manufacturing costs, and price. In reality, it is not easy to adopt a universal design approach in product design because universal design means covering a wide range of users. It is very difficult for product designers to find answers to questions such as, "what should be the design requirements?" and "how should ideas and designs be evaluated?" The situation becomes overly complicated, making it hard for product designers to adopt universal design principles.

The proposed Practical Universal Design Guidelines utilize the user segments table to define user groups and a UD matrix to determine relevant design requirements, creating a new design concept, and evaluating the ideas. The user segments table consists of all user segments from the universal design viewpoint. The UD matrix consists of user groups, user tasks, and all requirements specified by user groups and user tasks.

The process of the proposed UD guidelines involves defining user groups by using the user segments table, defining user tasks by using basic user tasks, and then formulating the UD matrix for the defined user groups and user tasks. The next steps are forming a design concept by using the UD matrix with priorities for each identified requirement, creating a detailed design based on the design concept, and evaluating the ideas by using the UD matrix. The final step is the evaluation of the design proposal by real users.

DEVELOPMENT PROCESS

Before the actual design process starts, it is important to share the understanding of the importance and merits of universal design practices with all the related people involved in product development. Without such a shared understanding of universal design practices, the design process will not succeed smoothly.

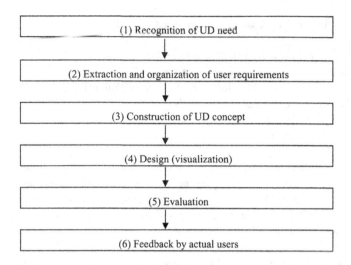

FIGURE 28.1   Universal design development process.

As shown in Figure 28.1, the typical development process is as follows:

- Recognition of the need for UD (awareness-raising and information sharing)
- Extraction and identification of user requirements
- Construction of the UD concept
- Design (visualization)
- Evaluation
- Feedback from actual users

DESIGN PROCESS

As shown in Figure 28.2, the process of the proposed Practical UD Guidelines is to define user groups by applying the user segments table, defining user tasks by examining basic user tasks, and then composing a UD matrix for the defined user groups and the defined user tasks. The next steps are forming a design concept by using a UD matrix with priorities for each of the requirements, creating a detailed design based on the design concept, and evaluating the ideas by using the UD matrix. The last step corresponds to the user evaluation of the final proposed design.

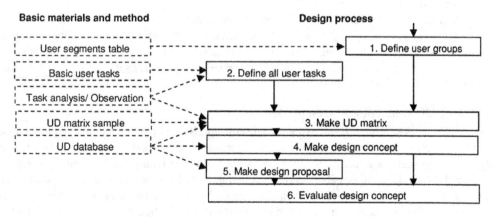

FIGURE 28.2   Design process.

Thus, the detailed design process in this sequence is as follows:

- Define target user groups by referring to the user segments table
- Define all user tasks by referring to basic user tasks
- Prepare the UD matrix by referring to the sample UD matrix
- Establish the UD concept and design method
- Make the design proposal
- Evaluate the design proposal

Define target user groups by referring to the user segments table. The user segments table is prepared to help designers find out the necessary types of users of the product considered and to put these types in some target user groups in the UD matrix. The UD user segments constitute a matrix of rows and columns. "Function 1, 2" in the columns and "User type 1, 2" in the rows on the left table correspond to "factors to be considered" and "user type" in the user segments table, respectively.

- Clarify the user attributes and user functions to be adapted to the users by referring to the user segments table.
- Define several target user groups for each product.

Define All User Tasks by Referring to Basic User Tasks

Basic user tasks are typical, and they include preparation, start, acquisition of information, cognition, judgment, understanding, operation, completion of work, and maintenance. The basic user tasks are prepared in the UD matrix.

- Define typical user scenarios for targeted products.
- Define all user tasks based on the basic user tasks.

It is possible to define some user tasks that aren't derived from basic user tasks. When adding more details, it is better to use the task analysis method or the direct observation method. It is also better to define the user tasks after selecting one of the user scenarios from among several user scenarios.

Prepare the UD Matrix by Referring to the Sample UD Matrix

UD matrix is the table where the X-axis (rows) is used for user groups and the Y-axis (columns) is for each user task; each cell represents user requirements.

- The sample UD matrix on Practical Universal Design Guidelines ensures how to create a new UD matrix.
- Determine the user scenario and the user environment.
- Select the target users—based on Step 1—in the target user section of the UD matrix template.
- Enter the user tasks from Step 2 in the user tasks section of the UD matrix template.
- Determine the requirements for each cell of the UD matrix. It is useful to use a UD database to define all the requirements.

Establishing UD Concept and Design Method

There are two ways to structure and express a UD concept. Based on extracted user requirements, a UD concept can be structured by a bottom-up system or by a top-down system (by the planning party). A UD concept is ranked into three levels ((A): shall be realized, (B): should be realized, (C): should be realized, if possible) and this ranking is utilized in the subsequent design and evaluation stages. User requirement items determined by using a UD matrix or simplified user requirement extraction method are changed into visual forms through brainstorming while taking cost and technical aspects into consideration.

Make the Design Proposal

Create ideas to meet each requirement and then consolidate all the ideas into one system design. This step should use the UD database to make the ideas.

Evaluate the Design Proposal

Evaluate each idea and the design proposals based on the design concept by designers. The design proposal should be evaluated by the team and by real users.

Compact Design Process

The compact design process is a simplified way for a designer to design a product from the universal design viewpoints. Without making a UD matrix, the designer may utilize an existing UD matrix with necessary modifications and utilize an action checklist to evaluate the design ideas.

As shown in Figure 28.3, the compact design process starts by looking at existing UD matrix samples to understand basic requirements. For this purpose, the designer needs to select an existing UD matrix from the samples of the UD matrix on the Practical Universal Design Guidelines. Alternatively, the designer may utilize the same kind of UD matrix that was created for a previous product by the designer. The rest of the process is almost the same as the standard design process. Instead of evaluation by a team, the designer will be able to use an action checklist contained in the Practical Universal Design Guidelines.

METHODS OF EXTRACTING USER REQUIREMENT

Target users are narrowed down by using a user segments table or its simplified version, and relevant user requirements are extracted based on the defined UD matrix. Other simplified methods of extracting user requirements under examination include: (a) the use of an action checklist, (b) extraction of user requirements based on user participation (workshop style), (c) making of a simplified UD matrix, (d) simplified task analysis, and (e) direct observation. However, when a UD matrix is used, there is no need to use a simplified user requirement extraction method.

UD User Segments Table

A user segments table (user segments table [Table 21.1], user segments material) is produced to extract users at UD and users requiring special considerations and organizing user types (groups) to be considered in the product design stage. There are two types of UD user segments tables: (A) user segments table and (B) user segments material (Table 28.1).

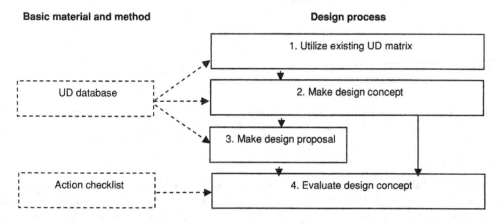

FIGURE 28.3 Compact design process.

TABLE 28.1
Structure of the User Segments Table

Factors (Seeing (Eyesight)—), Hearing—to Consider

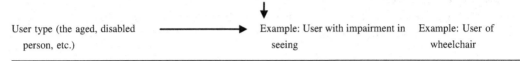

| User type (the aged, disabled person, etc.) | ⟶ | Example: User with impairment in seeing | Example: User of wheelchair |

TABLE 28.2
Structure of a UD Matrix

Three Aspects of a Product	UD Principles	Flow of the Basic User Tasks	User Tasks	User Groups (Users with Impairment in Seeing, Wheelchair Users, etc.)	
Usable	(1) Easy to acquire	(1) Preparation	Task (1)		
	information	(2) Start			
			Task (2)		
	(2) Easy to understand	(3) Acquire information	Task (3)		
	(3) Reduce load on mind and body	(4) Cognition, Judgment, Understanding	Task (4) ⟶		Requirement
	(4) Safety	(5) Operation	—		
	(5) Maintenance	(6) Completion of work	—		
		(7) Maintenance			
Useful	(1) Proper price	(specification)_____			
	(2) Ecology	(specification)_____			
	(3) Function	(specification)_____			
	(4) Performance	(specification)_____			
Desirable	(1) (Beauty)	(specification) _____			
	(2) Pleasant to use	(specification) _____			
	(3) Want to own	(specification) _____			

UD MATRIX

A UD matrix (Table 28.2) is used to effectively extract UD requirements according to different situations. User groups are indicated along the horizontal axis (rows), and individual tasks relating to the three facets of a product and operations are classified along the vertical axis (columns). Then, individual UD-related requirements for each user group are indicated in intersecting cells.

UD-related requirements vary depending on the products and target users. In addition, classification based on product aspects and tasks results in organized determination of requirements with minimal omissions. It also allows easier prioritization. As a result, the following benefits can be realized: (a) overall UD image can be understood visually, (b) prioritization of requirements can be realized for an easy establishment of the design concept, (c) solutions to problems are more easily discovered, (d) the use of required databases can be facilitated; this makes it possible to minimize undesired omissions and facilitate the checking process, (e) individual applications to different products are made possible even for designers without previous experiences, and (f) the examination of new applications and functions in a new product development is made possible.

USER SEGMENTS TABLE

APPROACH TO THE USER SEGMENTS TABLE

The user segments table is prepared to help designers identify the various types of users of a product being designed. These user types selected will be put into relevant target user groups in a UD matrix (Nomura et al., 2002). As shown in Table 28.3, this user segments table constitutes a matrix of rows and columns. The columns consist of items related to physical and mental functions (visual, acoustic, motor, cognitive, etc.), demographic (gender, economy, etc.), cultural (language, custom, etc.), and environmental factors—all of which should be considered by the designers for a given product. The rows list the basic types of users—people with disabilities, people with temporary disabilities, children, and so on. To better understand the table, some cells in the matrix list typical or general user examples and additional notes and related human characteristics data for some user examples are listed in extra tables and figures as the appendices in the Practical Universal Design Guidelines.

"Function 1, 2" in the columns and "User type 1, 2" in the rows on the left table correspond to "factors to consider" and "user type" in the user segments table, respectively. Several figures in the cells represent user examples.

IDENTIFYING USER GROUPS

As shown in Figure 28.4, designers select target users that should be considered in designing the product after referring to user examples listed in the cells or in the respective columns and rows in this table. These users are classified into user groups according to their needs that must be accommodated in the product design. At that point, user groups are applied in the UD matrix. To make it simpler to select target users, a simplified table of user segments is also proposed in the Practical Universal Design Guidelines. This simplified table consists of ready-made target user groups that are general and basic. This table will be presented in the final version of the practice guideline.

UD MATRIX

WHAT IS A UD MATRIX?

A UD matrix is a matrix to be formulated by the designer by considering all the relevant requirements for the target users concerned. It helps designers easily pinpoint the requirements for UD that may vary in different situations. The X-axis (rows) is used for user groups and the Y-axis (columns) for the three aspects of products and each user task. Thus, each cell contains individual requirements for UD.

TABLE 28.3
User Segments Table

Cognitive		Others	Demographic				Culture			Environment Non-User
J Knowledge/Decision	K Intellect	L Other abilities	M Age	N Gender	O Economy	P Qualification	Q Language	R Custom	S Nationality/Religion	T Those around user
Aged	User with a cognitive impairment	Aged								Heart pacemaker
Visitor from different culture area		Allergy		Male/Female user of product for women/me			Foreigner	Foreigner	Foreigner	
User Under Urgent Situations Infant or child		Infant or Child	Minor							Infant or Child
Novice/inexperienced user					Financially poor person	Person with no driver's license	User who cannot understand the Japanese language	Foreigner (exports)	Foreigner (exports)	Non-smoker Neighbor staff for maintenance

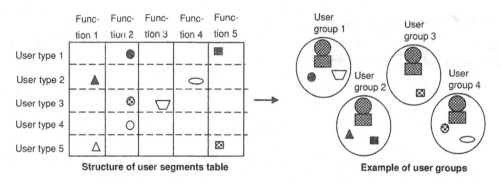

FIGURE 28.4   Selection of target user groups from the user segments table.

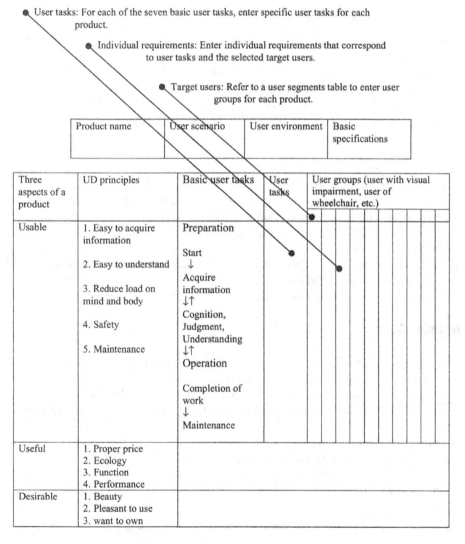

FIGURE 28.5   How to develop a UD matrix?

How to Create a UD Matrix?

As shown in Figure 28.5, in creating a UD matrix for a particular product to be designed, it is advisable to follow the procedures below in a ready-made format:

- Enter the product classification, the user environment, the user scenario as well as the basic specifications foreseen.
- For each of the seven basic user tasks, enter the user tasks for each product in the column on the right.
- Refer to the user segments table to identify user groups for each product and enter them in the columns across the top.
- Enter individual requirements in each of the matrix cells. Refer to entries on the left and above to enter minimum general requirements.
- Use a database, if necessary, when identifying individual requirements.

Three Aspects of a Product and UD Principles

The fundamental requirement in developing and supplying products is ensuring fairness. It is, therefore, necessary to consider three aspects: usable, useful, and desirable. The UD principles that correspond to these three aspects should be taken into account. The examples of related items that are of interest to the designer when examining individual requirements are shown in the following sections. The designer must follow these principles in realizing UD. These principles can be incorporated into a form of an action checklist to be used by designers.

Usability

- A product should be designed to make it easy to acquire information.
- Information about how to operate should be presented in a way that is easily identified and, where appropriate, through more than one sensory channel.
- The brightness, size, colors, and contrast of markings should be appropriate and conducive to understanding.
- Layout, clues, emphasis, and mapping of information presented should be appropriate.
- Information for users should be presented in an easily recognizable position.
- Information for users should be presented at an appropriate timing.

Easy to Understand

- Operating points, markings, and moving parts should be easily recognizable.
- Expressions and language should be easy to understand.
- Sufficient information for understanding and judgment should be provided.
- The way of using the designed product should be familiar to or felt as natural by the users.
- Appropriate feedback to the users about the results of the operation done should be provided during or immediately after the operation.

Reduce Physical and Mental Load

- Operating steps should be reduced to the minimum through appropriate omissions or automation.
- All operations should be done with one hand or finger.
- Enough space should be provided for access regardless of the user's posture or body shape.

- Controls should be within easy reach and are easily visible.
- The shape and size of the controls and mechanical stress in handling them should be designed so that the product can be operated with adequate force.
- Proper hitches and anti-slip devices should be located where the operation could be slippery.
- The product should support the part of the body or the load handled, if necessary.

Safety

- Sufficient time allowance should be provided, and due considerations should be given to the operation to prevent erroneous operation.
- Safety measures should be provided to prevent potential operational errors and avoid escalation to a system malfunction or a higher degree of danger.
- Markings and alarm sound against danger should be presented in such a way that users do not fail to understand its meaning.
- Users should not be required to take any action that might involve any serious or immediate danger.

Maintenance

- Due considerations should be given to maintenance actions that might be taken by users.
- Appropriate workspace, working posture, work time, and repairability should be secured for maintenance by specialists.

USEFULNESS

Proper Price

- The product must allow its purposeful use with its standard model alone or by adding additional simple options with no extra costs incurred.

Ecology

- The product should be well constructed in terms of durability and is recyclable.

Availability of all Necessary and Adequate Functions

- All relevant mode control mechanisms and adjustment functions should be available for different user groups.
- Multipurpose or easy to additionally operate functions should be available to meet diverse access needs.
- The number of functions should not be too many that make operations complicated.

The Product Performance Should Be Provided to the Extent Required and Should Be Sufficient Enough

- The product should allow its use without any concerns over malfunctioning.
- Sufficient levels of performance should be available to meet various access needs.
- Basic performance is not obsolete in comparison with other standard products.

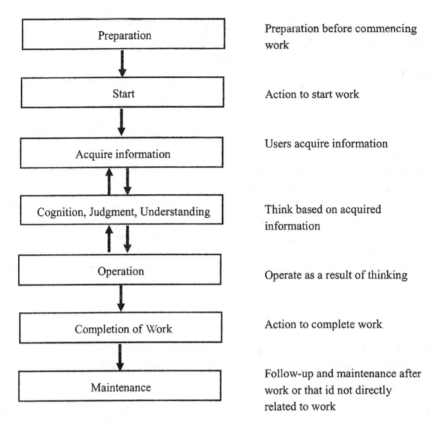

FIGURE 28.6 Basic user tasks.

ATTRACTIVENESS

Beauty

- Consideration to UD should enhance beauty.

Pleasantness in Use

- Consideration of weakened physical function should not necessarily mean any negative image.
- Consideration of UD should enhance the joy of use (Figure 28.6).

Wanting to Own

- The product should be attractive not only visually but also aurally and tactually.
- Attachment to the product should be enhanced by showing considerations for the user's characteristics.

BASIC USER TASKS

Regarding "usability" among three aspects of a product, the UD requirements for usability vary in individual user tasks. Although concrete user tasks vary in each product, the basic user tasks have something in common. Their real order may differ depending on the tasks performed, with some tasks omitted as appropriate. The theoretical order for basic user tasks is shown in Figure 28.6.

BENEFITS OF A UD MATRIX

There are significant benefits derived from applying a UD matrix. This may be confirmed in while developing one. To facilitate this stage, UD matrix samples have been arranged as examples for some products (an automatic teller machine, a cellular phone, a laptop personal computer, washing machine, etc.). Consequently, the following benefits are expected when a UD matrix is applied.

A UD matrix will:

- Provide a whole picture of UD at a glance,
- Help designers form a concept as they prioritize requirements with reference to the matrix,
- Help find clues in solving problem,
- Make it possible to use a relevant database,
- Make it easy to check requirements, since there are few omissions,
- Apply to individual examples of each product, and
- Help examine new uses and functions for new products.

DESIGN EVALUATION METHOD

The purpose of design evaluation is to verify whether a design plan is compatible with the UD concept. In a general evaluation method, omission of design examination in connection with the concept is checked, and a UD action checklist can be used to confirm the appropriateness of the overall UD process, the product, and the ease of its use. In the final stage, with the participation of target users (healthy users and physically challenged users), a design evaluation of the developed system is conducted to extract any previously undiscovered points by using a checklist and relevant tools. New items extracted through the evaluation process are examined, and the design is re-worked for improvement. To achieve a higher level of UD performance, it is necessary to establish this development cycle.

REFERENCE

Nomura, M., Yanagida, K., Yamaoka, T., Yamazaki, K., Okada, A., & Saito, S. (2002). A proposal for universal design practical guidelines (3): Using an UD matrix to pinpoint the requirements of UD, International Conference for Universal Design, 2002, 718–722.

29 Virtual Environment Usage Protocols

Kay M. Stanney, David A. Graeber, and Robert S. Kennedy

CONTENTS

INTRODUCTION

Technological advances over the past decade have laid the foundation for ubiquitous computing. One such advancement that could potentially address people's information and general needs is virtual environment (VE) technology (Stanney, 2002). Virtual environments allow users to be immersed in three-dimensional (3D) digital worlds—surrounding them with tangible objects to manipulate and venues to traverse—that they can experience from an egocentric perspective. Through the concrete and familiar, users can enact known perceptual and cognitive skills to interact with a virtual world; there is no need to learn the contrived conventions of more traditional graphical user interfaces. Virtual environments also extend the realm of computer interaction, from the purely visual to multimodal communication that more closely parallels human-human exchanges. Virtual environment users not only see visual representations but also reach out and grab objects, "feel" their size, rotate them in any given axis, hear their movement, and even smell associated aromas. Such experiences do not have to be in solitude, as VE users can take along artificial autonomous agents or collaborate with other users who also have representations within the virtual world. Taken together, this multisensory experience should afford natural and intuitive interaction.

The previous paragraph describes an ideal—unfortunately, it is not the current state of the art. In today's virtual environments, users are immersed in an experience with suboptimal visual resolution, inadequate spatialization of sound, encumbering interactive devices, and misregistration of tracking information (Durlach & Mavor, 1995; Stanney et al., 1998). These technological shortcomings engender adverse physiological effects and pose usability concerns that require additional study to identify how best to design and use VE technology. Substantial research is needed before the functionality, reliability, and appropriateness of VE technology are understood to the point of use by a wide variety of people. This research should determine how to design and utilize VE interfaces that will enable the broadest possible spectrum of citizenry to interact easily and effectively with these systems (National Research Council, 1997).

The fundamental question is: can an understanding of human physiological responses to VE technology be developed such that design guidelines and usage protocols that render VE interfaces usable by the broad spectrum of people who may wish to utilize this technology be devised? Efforts by many researchers (Cobb et al., 1999; DiZio & Lackner, 1997, 2002; Harm, 2002; Howarth & Finch, 1999; Lawson et al., 2002; Stanney, Kennedy, & Kingdon, 2002; Stanney et al., 2003; Viirre & Bush, 2002; Wann & Mon-Williams, 2002; Welch, 2002; Wilson, 1997; Wilson, Nichols, & Haldane, 1997; Wilson, Nichols, & Ramsey, 1995) have made substantial gains toward this goal. Heeding the warning of Biocca (1992) who indicated that cybersickness (McCauley & Sharkey, 1992), a form of motion sickness in VE systems, could be a snake lingering in the underbrush and threatening the widespread diffusion of VE technology, researchers have set out to overcome a number of challenging objectives. This research has focused on developing tools to measure the adverse effects of VE exposure (Kennedy & Stanney, 1996; Stanney, Kennedy, et al., 1999), examining the psychometrics of cybersickness (Graeber, 2001a, 2001b; Kennedy et al., 2001; Kennedy, Stanney, & Dunlap, 2000; Kingdon, Stanney, & Kennedy, 2001; Stanney & Kennedy, 1997a, 1997b; Stanney et al., 2003; Stanney, Lanham, et al., 1999), developing usage protocols (Stanney, Kennedy, & Kingdon, 2002), investigating system-related issues that influence cybersickness (Stanney & Hash, 1998; Stanney et al., 1998, 2003), examining the efficacy of réadaptation mechanisms in recalibrating those exposed to VE systems (Champney, Stanney, & Kennedy, under revision), and examining the influences of cybersickness on human performance (Stanney, Kingdon, et al., 2002; Stanney, Kingdon, & Kennedy, 2001), among other related pursuits.

Although tremendous progress has been and continues to be made toward these objectives, during these pursuits, a nettlesome problem has reared that could substantially limit the accessibility of VE technology. The problem is that a substantial number of people cannot withstand prolonged exposure to VE and quickly withdraw from these systems. This exodus is predictably due to the adverse effects associated with exposure, which may be as minor as a headache or as severe as vomiting or intense vertigo (Cobb et al., 1999; Howarth & Finch, 1999; Regan & Price, 1994; Stanney, Kennedy, et al., 1999; Stanney et al., 1998, 2003; Wilson, 1997; Wilson, Nichols, & Haldane, 1997). It is on rare occasions that VE exposure causes an emetic response (about 1.5%); while many of those exposed experience some level of nausea, disorientation, or oculomotor problems (Lawson et al., 2002; Stanney, Kingdon, et al., 1998, 2003; Stanney et al., 1998). Other problems, such as sleepiness and visual flashbacks, also occur (Lawson et al., 2002). Approximately 80–95% of those exposed to a VE report some level of adverse symptomatology (Stanney et al., 1998). These disturbances lead a substantial proportion to prematurely cease their interaction, impeding the general accessibility to this technology. In fact, dropout rates of as high as 50% have been found in exposures of one hour (Stanney et al., 2003). In general, dropout rates ranging from 5% to 50% have been found in a number of VE studies (Cobb et al., 1999; DiZio & Lackner, 1997; Howarth & Finch, 1999; Regan & Price, 1994; Singer, Ehrlich, & Allen, 1998; Stanney, Lanham, et al., 1999; Wilson, Nichols, & Ramsey, 1995; Wilson, Nichols, & Haldane, 1997). Based on the tendency to have dropouts, the U.S. Army Research Institute (Knerr et al., 1998) has suggested that VE exposures should be limited to 15 min—a period that may be too short for many training, educational, or analysis-based applications. An alternative is for those who utilize this technology to screen individuals who are susceptible to adverse effects, leaving it available only to those who are non-susceptible. Such an approach was considered by Chevron, who used a VE CAD tool to build 3D earth models for seismic imaging, structural mapping, and reservoir characterization (Kowalik, personal communication, August 20, 2001) and will likely be adopted by others. Yet, such accessibility limitations should not be tolerated. All children should be able to enjoy the excitement of an immersive VE learning experience (Moshell & Hughes, 2002). All employees should be able to leverage their VE design tools (Davies, 2002). All medical, military, and other personnel should be able to utilize VE training systems (Knerr et al., 2002; Satava & Jones, 2002). By developing a solid understanding of physiological responses to VE technology, limited accessibility to VE technology could potentially be circumvented. From this understanding, user-centered design principles and usage protocols for VE systems could be identified—which, if applied successfully, should render VE systems suitable for a broad spectrum of people. (Note: VE exposure should likely be avoided by certain populations, including those susceptible to photic seizures and migraines, those displaying co-morbid features of various psychotic, bipolar, paranoid, substance abuse, those who are claustrophobic or have other disorders where reality testing and identity problems are evident, and those with preexisting binocular anomalies [Rizzo, Buckwalter, & van der Zaag, 2002; Stanney, Kennedy, & Kingdon, 2002; Viirre & Bush, 2002; Wann & Mon-Williams, 2002].)

In 1997, the Life Sciences Division at NASA Headquarters, Washington, DC, funded a study with one of the primary objectives being to characterize the current state of knowledge regarding aftereffects in virtual environments (Stanney et al., 1998). Two of the most critical research issues identified by this effort were to (a) establish VE design guidelines that minimize adverse effects and (b) establish means of determining user susceptibility to cybersickness and aftereffects. This chapter seeks to address both issues so that this knowledge can be used to support more effective human-VE interaction. The realization of VE usage protocols and design guidelines are essential because recent advances in VE applications are quite impressive and have the potential to advance many fields—from education to medicine. In fact, a recent review of VE applications by Stone (2002) provides clear evidence that VE technology has advanced sufficiently to support a wide range of applications in areas such as engineering, micro- and nanotechnology, data visualization, ergonomics and human factors design, manufacturing, military, medicine, retail, and education. The development of an understanding of human physiological responses to VE technology can ensure that such applications are designed so that adverse effects on their users are minimized.

DEFINITION OF TERMS

Adaptation: Decline in the amplitude of a sensory response in the presence of a constant prolonged stimulus.

Aftereffect: Any effect of virtual environment exposure that is observed after a participant has returned to the physical world.

Cybersickness: Sensations of nausea, oculomotor disturbances, disorientation, and other adverse effects associated with virtual environment exposure.

Habituation: Gradual decline in sensory response to repeated stimulus exposure.

Virtual environment (VE): A three-dimensional dataset describing an environment based on real-world or abstract objects and data.

FULL NAME OF THE STANDARD

Virtual Environment Usage Protocols

OBJECTIVE AND SCOPE

The current chapter focuses on contributing to the advancement of VE technology by promoting a better understanding of human physiological limitations that impede the widespread use of this technology and identifying techniques that can render VE systems more usable, useful, and accessible. This is done through the investigation of factors known to affect the level of adverse effects associated with VE exposure, including system design parameters (i.e. degrees of freedom [DOF] of user control and scene complexity), usage variables (i.e. exposure duration and intersession interval), and individual characteristics (i.e. susceptibility and gender). The major objectives being addressed by this chapter and their expected significance are as follows:

- Increase exposure durations and their overall proportion completed via theorized design guidelines and usage protocols (i.e. exposure management techniques) that depress the stimulus intensity of a VE system and drive acclimation.
- Decrease dropout rates via theorized design guidelines and usage protocols that afford repeated VE exposures with systematically determined intersession intervals.
- Create equal opportunity for VE system use despite motion sickness susceptibility.

DISCUSSION OF FACTORS EFFECTING VIRTUAL ENVIRONMENT USAGE

Table 29.1 provides a synopsis of the extent and severity of the adverse effects associated with VE exposure.

Based on the studies summarized in Table 29.1, the most conservative predictions would be that at least 5% of all users—and potentially up to half of those exposed—will not be able to tolerate prolonged use of current VEs and that a substantial proportion would experience some level of adverse effects. Further, females, younger (<23 years), older (>40 years) individuals, and those highly susceptible to motion sickness may be particularly bothered by VE exposure. In addition to the problems listed in Table 29.1, the repercussions associated with such adverse effects may also include unequal opportunities for VE accessibility among the moderate to highly motion-sickness-susceptible population (Graeber, 2001a, 2001b), decreased user acceptance and use of VE systems (Biocca, 1992), decreased human performance (Kolasinski, 1995; Lawson et al., 2002; Stanney, Kingdon, et al., 2002), and acquisition of improper behaviors (e.g., reduction of rotational movements [i.e. roll, pitch, and yaw] to quell side effects; Kennedy, Hettinger, & Lilienthal, 1990).

To summarize the findings in Table 29.1, VE systems can be hampered by intense malaise, high attrition rates, limited exposure durations, the possibility of rejection of the system by the intended user population, conceivable decrements in human performance, and unequal opportunities for use. Although the exact causes of these problems remain elusive, they are thought to be a result of system design (e.g., scene content and user control strategies) and technological deficiencies (e.g., lag, distortions, and limited sensorial cues), as well as individual susceptibility (see review of factors by Kolasinski, 1995, Stanney, 1995, and Stanney et al., 1998). The most widely accepted theory is that mismatches (due to system design issues or technological deficiencies) between the sensory stimulation provided by a VE or simulator expected due to real-world experiences (and thus established neural pathways) are thought to be the primary cause of motion sickness (a.k.a Sensory Conflict Theory; Reason, 1970, 1978; Reason & Brand, 1975). From the individual susceptibility perspective, age, gender, prior experience, individual factors (e.g., unstable binocular vision, individual variations in the interpupillary distance (IPD), and susceptibility to photic seizures and migraines), drug and alcohol consumption, health status, and ability to adapt to novel sensory environments are all thought to contribute to the extent of symptoms experienced (Kolasinski, 1995; McFarland, 1953; Mirabile, 1990; Reason & Brand, 1975; Stanney, Kennedy, & Kingdon, 2002; Stanney et al., 1998). Although considerable research into the exact causes is requisite and has been ongoing for decades in its various forms (seasickness, motion sickness, simulator sickness, space sickness, and cybersickness; Chinn & Smith, 1953; Crampton, 1990; Kennedy & Fowlkes, 1992; McCauley & Sharkey, 1992; McNally & Stuart, 1942; Reason, 1970, 1978; Reason & Brand, 1975; Sjoberg, 1929; Stanney et al., 1998; Tyler & Bard, 1949; Wendt, 1968), there are currently few (if any efforts) focusing on the development of means to reduce the adverse effects by "acclimating" users to the VE stimulus. Yet, stimulus-response studies, such as the classical conditioning studies of Pavlov (1928), offer a potential paradigm through which to realize such acclimation. There is, however, a lack of understanding about the factors that drive VE stimulus intensity and acclimation (i.e. stimulus depression) versus sensitization (i.e., heightened response) to such a stimulus, such that this knowledge can be used to identify VE usage protocols that minimize adverse effects. In fact, current usage of VE technology generally treats users as if they are immune to motion sickness or possess low motion sickness susceptibility and are capable

TABLE 29.1
Brief List of Problems Associated with Exposure to VE Systems

- 80%–95% of individuals interacting with a head-mounted display (HMD) VE system report some level of side effects, with 5%–50% experiencing symptoms severe enough to end participation. Approximately 50% of those dropouts occur in the first 20 min and nearly 75% by 30 min (Cobb et al., 1999; DiZio & Lackner, 1997; Howarth & Finch, 1999; Regan & Price, 1994; Singer, Ehrlich, & Allen, 1998; Stanney, Kennedy, & Kingdon, 2002; Stanney, Lanham, et al., 1999; Stanney et al., 2003; Wilson, Nichols, & Ramsey, 1995; Wilson, Nichols, & Haldane, 1997).

- Virtual environment exposure can cause people to vomit (about 1.5%), and approximately three-fourths of those exposed tend to experience some level of nausea, disorientation, and oculomotor problems (Cobb et al., 1999; DiZio & Lackner, 1997; Howarth & Finch, 1999; Regan & Price, 1994; Singer, Ehrlich, & Allen, 1998; Lawson et al., 2002; Stanney et al., 1998, 2003; Wilson, Nichols, & Ramsey, 1995; Wilson, Nichols, & Haldane, 1997).

- With prolonged (>45-min) VE exposure, oculomotor problems may become more pronounced, whereas nausea and disorientation tend to level off (Stanney et al., 2003).

- Subjective report of sickness post VE exposure is 2.5–3 times greater than sickness reported by pilots training in motion-based and non-motion-based U.S. Navy fixed-wing and helicopter simulators (Stanney & Kennedy, 1997a; Stanney et al., 1998).

- Before the age of two, children appear to be immune to motion sickness, after which, susceptibility increases until about the age of 12, when point it declines again (Money, 1970). Those over 25 are thought to be about half as susceptible as they were at 18 years of age (Mirabile, 1990). However, older individuals (>40 years) exposed to VE systems can be expected to experience substantially more nausea, as well as greater oculomotor disturbances and disorientation as compared to younger individuals (Stanney et al., 2003). Younger individuals (<23 years) exposed to VE systems may experience more oculomotor disturbances than older populations (Stanney et al., 2003).

- Females exposed to VE systems can be expected to be more susceptible to motion sickness and to experience higher levels of oculomotor and disorientation symptoms as compared to males (Graeber, 2001a; Kennedy, Lanham, Massey, Drexler, & Lilienthal, 1995; Stanney et al., 2003). In general, females tend to adapt more slowly to nauseogenic stimulation (McFarland, 1953; Mirabile, 1990; Reason & Brand, 1975).

- Individuals susceptible to motion sickness can be expected to experience more than twice the level of adverse effects to VE exposure as compared to non-susceptible individuals (Stanney et al., 2003).

- Individuals exposed to VE systems can be expected to experience lowered arousal (e.g., drowsiness and fatigue) upon post-exposure (Lawson et al., 2002; Stanney et al., 2003).

- Flashbacks (i.e. visual illusion of movement or false sensations of movement, when not in the VE) can be expected to occur quite regularly to those exposed to a VE system (Lawson et al., 2002; Stanney et al., 2003).

- Prolonged disorientation (e.g., dizziness and vertigo) can be expected after VE exposure, with symptoms potentially lasting more than 24 hours (Stanney & Kennedy, 1998; Stanney et al., 2003).

of rapid acclimation to novel sensory environments. This is not the case, as evidenced by the intensity and extent of side effects listed in Table 29.1. Additional research is required because, although the beginnings of VE design guidelines are under development (Stanney, Kennedy, & Kingdon, 2002) and expectations of use have been identified (Stanney et al., 2003), means of prolonging exposure via acclimation or other such conditioning strategies are still required to ensure VE systems are accessible by a broad spectrum of users.

To leverage and compare conditioning strategies, a means to characterize VE stimulus intensity is required. Research focused on developing an understanding of the system design and technological and individual drivers of adverse effects associated with VE exposure can be used to make predictions on VE stimulus intensity (see Table 29.2). One can see from Table 29.2 that there are a diverse number of factors influencing this. By developing an understanding of these factors, means of acclimating users to a VE stimulus could potentially be identified.

The factors reviewed in Table 29.2 can be used to characterize the intensity of a VE stimulus. The technological factors (e.g., system consistency, lag, update rate, mismatched IPDs, and unimodal and intersensorial distortions) are system-specific and may be resolved—to some extent—as technology progresses, whereas the system design (e.g., DOF of user control, scene complexity), usage (e.g., exposure duration, intersession interval), and individual factors (e.g., susceptibility, gender) are not dependent on technological progress and will likely have an enduring influence on human–VE interaction. Among these factors, exposure durations of 15 min or more, very short (<2 days) or extended (>5 days) intersession intervals, high DOF of user control, and complex visual scenes are all known to lead to a more intense VE stimulus. In terms of user control, Held (1965) and Reason (1978) provided evidence that suggests that motion sickness can be overcome if users have control over their movements and receive an appropriate sensory response (e.g., visual, vestibular, or proprioceptive) to their actions. Essentially, in response to a neural mismatch (i.e., sensory conflict), users create a new "neural store" with which incoming sensory information is compared, eventually resulting in a new neural match (i.e. the reafference copy resulting from effector stimulation [i.e. stimuli resulting from one's own muscular activity] overwrites the efference copy [i.e. motor impulses]). However, Stanney and Hash (1998) found that high DOF of motion, although being superior to passive motion in minimizing cybersickness, may not be the best solution to the sickness problem. Under such conditions, VE users may not be able to efficiently update the neural store with the abundant amount of sensory information resulting from their unrestricted movements. Thus, by allowing users streamlined control (i.e. only those DOF necessary for supporting an activity) within a VE, the neural store may be updated quickly in response to streamlined reafference.

Individual susceptibility also plays a critical role. Stanney et al. (2003) found that susceptible individuals can be expected to experience more than twice the level of adverse effects to VE exposure as compared to non-susceptible individuals. McFarland (1953) found females and children to be five and nine times, respectively, more susceptible to motion sickness than adult males. Thus, to gauge the intensity of a VE stimulus, individual susceptibility must be considered. Kennedy and Graybiel (1965) developed a Motion History Questionnaire (MHQ) that could prove useful in assessing individual susceptibility. From its inception, the MHQ's purpose was to ascertain levels of motion sickness susceptibility to a variety of provocative motion challenges using an individual's history of motion experiences. As a result of the MHQ's utility, an array of scoring keys has been developed (Kennedy & Fowlkes, 1992) for various motion challenges (e.g., high- and low-frequency vertical motion, Coriolis stimulation, simulators, etc.), and, more recently, for circular vection (Graeber, 2001b) and virtual environments (Kennedy et al., 2001). The latter study demonstrated that the MHQ is promising as a predictive tool for VE sickness but indicated that it needs to be further refined. A recent analysis (Kingdon, 2001) indicated that the predictability of this tool could be potentially enhanced by coupling questionnaire items with a psychophysical battery test to assess the postural stability, dexterity, and ability of the human visual system to perform smooth pursuit tracking (i.e. gaze nystagmus) similar to the roadside battery test used to assess sobriety (Tharp, Burns, & Moskowitz, 1981). Such enhanced predictability of individual susceptibility could assist in fully characterizing the influences of VE stimulus intensity.

The summary in Table 29.2 suggests that the sickness-inducing characteristics of a VE stimulus could potentially be reduced by shortening the exposure duration, maintaining an intersession interval of two to five days, reducing the DOF of user movement control to only those necessary to support an activity, and simplifying visual scenes. Table 29.2 further suggests that the exact strategy that is most effective may depend on individual susceptibility. If these tactics are coupled with conditioning approaches, reductions in adverse effects and associated dropout rates can be expected.

Prior research has implemented various conditioning techniques to reduce adverse effects in, and hasten acclimation to, a variety of altered sensory environments. Conditioning regimens, including adaptation (one prolonged exposure to the full intensity of a stimulus), habituation

TABLE 29.2
Factors Influencing VE Stimulus Intensity

- Adverse effects associated with VE exposure are positively correlated with exposure duration (Kennedy, Stanney, & Dunlap, 2000). Lanham (2000) has shown that sickness increases linearly at a rate of 23% per 15 min. Dropouts occur in as little as 15 min of exposure (Cobb et al., 1999; DiZio & Lackner, 1997; Howarth & Finch, 1999; Regan & Price, 1994; Singer, Ehrlich, & Allen, 1998; Stanney, Kennedy, & Kingdon, 2002; Stanney, Lanham, et al., 1999; Stanney et al., 2003; Wilson, Nichols, & Ramsey, 1995; Wilson, Nichols, & Haldane, 1997).
- Intersession intervals of 2–5 days are effective in mitigating adverse effects, while intervals less than two or greater than five days are ineffective in reducing symptomatology (Kennedy et al., 1993; Watson, 1998).
- Repeated exposure intervals within a session less than two hours apart appear to heighten adverse effects upon reentry (Graeber, 2001a).
- As the amount of user movement control in terms of DOF and head tracking increases, so does the level of nausea experienced (Stanney & Hash, 1998; Stanney et al., 2003).
- Complete user movement control (6 DOF) can be expected to lead to 2.5 times more dropouts than streamlined control (3 DOF).
- Movements in rotational axes (e.g., roll, pitch, and yaw) may be more provocative than linear movement (Money & Myles, 1975).
- The rate of visual flow (i.e. visual scene complexity) may influence the incidence, and more so the severity of motion sickness experienced by an individual (Kennedy & Fowlkes, 1992; McCauley & Sharkey, 1992).
- Complex visual scenes may be more nauseogenic than simple scenes with complex scenes possibly resulting in 1.5 times more emetic responses; however, scene complexity does not appear to affect dropout rates (Dichgans & Brandt, 1978; Kennedy, Dunlap, et al., 1996; Stanney et al., 2003).
- Such effects may be exacerbated by a large field-of-view (FOV) (Kennedy & Fowlkes, 1992), high spatial frequency content (Dichgans & Brandt, 1978), and visual simulation of action motion (i.e. vection [Kennedy, Dunlap, et al., 1996]).
- Various technological factors thought to influence how provocative a VE includes: system consistency (Uliano, Kennedy, & Lambert, 1986); lag (So & Griffin, 1995); update rate (So & Griffin, 1995) mismatched IPDs (Mon-Williams, Rushton, & Wann, 1995); and unimodal and intersensorial distortions (both temporal and spatial [Welch, 2002]).
- Individual factors thought to contribute to an individual's motion sickness susceptibility include age, gender, prior experience, visual predispositions (e.g., unstable binocular vision, individual variations in IPD, and susceptibility to photic seizures and migraines), drug and alcohol consumption, health status, and ability to adapt to novel sensory environments (Kolasinski, 1995; McFarland, 1953; Mirabile, 1990; Reason & Brand, 1975; Stanney, Kennedy, & Kingdon, 2002; Stanney et al, 1998).
- Individuals who have experienced an emetic response associated with carnival rides can be expected to experience more than twice the level of adverse effects to VE exposure as compared to those who do not experience such emesis (Stanney et al., 2003).
- Individuals with higher preexposure drowsiness will be more likely to experience drowsiness upon post-VE exposure, and those exposed for 60 min or longer can be expected to experience more than twice the level of drowsiness as compared to those exposed for a shorter duration (Lawson et al., 2002; Stanney et al., 2003).
- As drowsiness increases, one can expect a greater severity of flashbacks (Lawson et al., 2002; Stanney et al., 2003).
- Body mass index (BMI) does not tend to be related to VE sickness symptoms; however, those with higher BMIs may be less prone to experience an emetic response (Stanney et al., 2003).

(repeated exposures to the full intensity of a stimulus), cognitive therapies (use of education and biofeedback), dual adaptation (repeated exposures until abatement of side effects and negative aftereffects), incremental adaptation (an incrementing of stimulus intensity within one exposure), and incremental habituation (an incrementing of stimulus intensity across multiple exposures) have been applied to underwater, zero gravity and acrobatic flight maneuver environments, artificial optical distortions (e.g., prism lenses and optokinetic drums), Pensacola Slow Rotation Room (SRR), and simulators with varying degrees of success (see review by Graeber, 2001a). Although such conditioning regimens have been investigated, they have primarily been curtailed to adaptation and habituation regimens despite the proven performance of other conditioning regimens to mitigate motion sickness in various sensory environments and lackluster success of adaptation regimens in reducing cybersickness in VE systems. Incremental conditioning approaches have particularly been very successful in minimizing side effects in an array of environments, including the SRR (Cramer, Graybiel, & Oosterveld, 1978; Graybiel, Deane, & Colehour, 1969), combat flight training (Bagshaw & Stott, 1985), and artificial optical distortions (Hu, Stern, & Koch, 1991) as well as being recommended as an effective method in mitigating simulator sickness (Kennedy et al., 1987) and cybersickness (Graeber, 2001a; Welch, 2002). Incremental conditioning protocols have been effective not only in reducing adverse effects but also increasing the rate (Lackner & Lobovitz, 1978) and extent (Ebenholtz & Mayer, 1968) of acclimation obtained in comparison to adaptation and habituation protocols.

The Dual Process Theory (DPT) of neural plasticity (Groves & Thompson, 1970; Prescott, 1998; Prescott & Chase, 1999) provides a theoretical means of understanding how particular conditioning regimens to various motion sickness-inducing environments may affect the magnitude of a sensory conflict (Graeber, 2001a). In the context of a VE system, the DPT would state that there are two opponent processes undertaken upon VE stimulus onset (see Figure 29.1). The two processes are depression (i.e. lessening or acclimation) and sensitization (i.e. heightening) of sickness outcome, which are carried out in parallel through different tracks. The DPT theory suggests that the depression process occurs along the stimulus-response (S-R) pathways, whereas the sensitization process is undertaken through the "state" system (i.e. the central nervous system [CNS]). During VE stimulus processing, a confluence occurs where the processes converge to yield the observed response (e.g., net outcome, a.k.a. sickness severity). The observed response is suggested to be a function of the integration of these two paths and their characteristics based on VE stimulus strength, number of VE stimulus exposures, and degree of neural plasticity in an individual.

In essence, based on the DPT theory, VE stimulus intensity should set the relationship between the opponent processes, which should then determine the proclivity of acclimation in the CNS and the degree to which each opponent process contributes to the net outcome. Concurrently, a number of exposures should co-determine the rate of acclimation in the CNS and subsequent depression of the sensitization tracks' weighting. As VE stimulus intensity increases, the sensitization opponent process should play a more dominant role in determining the net outcome—whereas the effect of the depression opponent process should wane. The aspect of DPT of concern to VE usage protocols is primarily the sensitization opponent process and—subsequently, the depression curve—how it may be shaped through behavioral modification. It is herein suggested that the sensitization opponent process' effect on net outcome determines the degree of side effects experienced (i.e. net outcome is equivalent to the degree of cybersickness). In other words, as the VE stimulus intensity increases, the influence of the sensitization opponent process also increases, thereby exerting greater control over the net outcome (i.e. heightened cybersickness). This is similar to the idea of long-term potentiation (Carlson, 2001), which is a long-term increase in the excitability of a neuron due to a particular input—especially if exposure to the input is repeated with a brief interstimulus interval.

By using an incremental approach—in which a VE stimulus is incrementally increased in intensity within one exposure (incremental adaptation) or across multiple exposures (incremental habituation)—it is suggested that the VE stimulus intensity may be low enough that the depression track would dominate the net outcome and afford more rapid acclimation to the stimulus in the CNS, thereby mitigating the outward expression of cybersickness in the net outcome. If a VE user can complete an adaptive process at each increment of stimulus intensity (i.e. depressing sensitization to preexposure levels or below), then each stepwise increase in VE stimulus intensity—assuming the increase is within the bounds of moderate to rapid CNS acclimation—should keep the magnitude and duration of sensitization to a minimum. In theory, this should afford attainment of exposure to VE stimulus intensities that would normally result in supramaximal sensitization (i.e. a VE stimulus intensity of sufficient strength to inhibit acclimation in the CNS, yielding a continued state of heightened sensitization that does not depress and predictably leads to cessation of exposure) without experiencing supramaximal sensitization. Further research is needed to verify this supposition.

According to DPT, using an incremental approach should result in both minimal sensitization that dissipates relatively quickly and near-maximal depression that will gradually dominate the net behavioral output over time (see Figure 29.1). In other words, by incrementing the intensity of the VE stimulus, depression—rather than sensitization—is facilitated, which allows one to capitalize on humans' inherent neural plasticity and maximize the rate and extent of acclimation to a VE stimulus.

Graeber (2001a) investigated the efficacy of an incremental approach to minimize motion sickness associated with exposure to a vection drum. The results suggest that by employing an incremental approach, it may be possible to substantially reduce the adverse effects and dropout rates. Graeber (2001a) found dropouts were reduced to approximately 20% among high motion sickness susceptibles, compared to a dropout rate of 50% for high susceptibles employing non-incremented protocols. Although this work demonstrated the potential efficacy of such conditioning strategies in a vection drum, further study is needed to determine their suitability in VE systems. More specifically, conditioning strategies for incrementally increasing VE stimulus intensity by manipulating both system design and usage variables need to be identified.

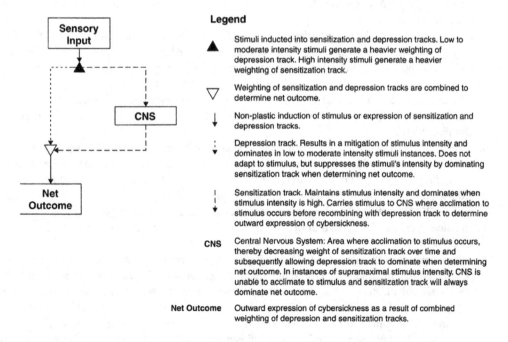

FIGURE 29.1 DPT points of induction and expression of sensitization and depression.

VIRTUAL ENVIRONMENT USAGE PROTOCOL

Based on the previous review of VE-related usage factors, it is suggested that the optimal conditions to engender acclimation to VE exposure can be identified and characterized by understanding the affects of system design (e.g., DOF of user-initiated control and visual scene complexity), usage (e.g., exposure duration and intersession interval), and individual factors (e.g., susceptibility and gender) on VE stimulus intensity. With this understanding, conditioning strategies conducive to facilitating prolonged VE exposure can be developed. Thus, a systematic VE usage protocol that minimizes risks to users can be developed. A comprehensive VE usage protocol should consider each of the following steps (Stanney, Kennedy & Kingdon, 2002).

1. Following the guidelines in Table 29.3, design the VE stimulus to minimize adverse effects.
2. Following the guidelines in Table 29.4, quantify the VE stimulus intensity of the target system and compare it to quartiles in Table 29.5. Table 29.5 provides quartiles of sickness symptoms based on 29 VE studies (Cobb et al., 1999; Kennedy, 2001; Kennedy, Jones, et al., 1996; Stanney, 2001; Stanney & Hash, 1998) that evaluated sickness via the Simulator Sickness Questionnaire (SSQ; Kennedy et al., 1993). If a given VE system is of high intensity (say the 50th or higher percentile, with a total SSQ score of 20 or higher), significant dropouts can be expected.
3. Following the guidelines in Table 29.6, identify the individual capacity of the target user population to resist the adverse effects of VE exposure.
4. Following the guidelines in Table 29.6, set exposure duration and intersession interval.
5. Provide warnings for those with severe susceptibility to motion sickness, seizures, migraines, colds, flu, or other ailments.
6. Educate users about the potential risks of VE exposure. Inform them of the insidious effects they may experience during exposure, including nausea, malaise, disorientation, headache, dizziness, vertigo, eyestrain, drowsiness, fatigue, pallor, sweating, increased salivation, and vomiting. Depending on the VE content, potential adverse psychological effects may also need to be considered.
7. Inform users about the potential adverse *aftereffects* of VE exposure. Inform users that they may experience disturbed visual functioning, visual flashbacks, as well as unstable locomotor and postural control for prolonged periods following exposure. Relating these experiences to excessive alcohol consumption may prove instructional.
8. Inform users that if they start to feel ill, they should terminate their VE interaction because extended exposure is known to exacerbate adverse effects (Kennedy, Stanney, & Dunlap, 2000).
9. Prepare users. Donning an HMD is a jarring experience (Pierce et al., 1999). Depending on the complexity of the virtual world, it can take 30–60 seconds to adjust to the new space (Brooks, 1988). Prepare users for this transition by informing them that there will be an adjustment period.
10. Adjust environmental conditions. Provide adequate airflow and comfortable thermal conditions (Konz, 1997). Sweating often precedes an emetic response, so proper airflow can enhance user comfort. In addition, extraneous noise should be eliminated, as it can exacerbate ill-effects.
11. Adjust equipment to minimize fatigue. Fatigue can exacerbate the adverse effects of VE exposure. To minimize fatigue, ensure all equipment is comfortable and properly adjusted to fit. HMDs should fit snuggly and be evenly weighted about a user's head, stay in place when unsupported, and avoid uneven loading to the neck and shoulder muscles. Many HMDs have adjustable head straps, IPDs, and viewing distance between the system's eyepieces and the user's eyes. Ensure users optimize these adjustments to obtain proper fit. Tethers should not obstruct the movements of users. DataGloves and other effectors should not induce excessive static loads via prolonged unnatural positioning of the arms or other extremities.

TABLE 29.3
Addressing System Factors That Influence the Strength of a VE Stimulus

- Ensure any system lags and latencies are stable; variable lags and latencies can be debilitating.
- Avoid high levels of user movement in terms of DOF of user control (i.e. neural store can be updated quickly in response to streamlined reafference).
- Reduce DOF to those necessary for the activity being supported (e.g., don't provide pitch unless needed)
- Minimize display and phase lags (i.e. end-to-end tracking latency between head motion and resulting update of the display).
- Optimize frame rates.
- Provide adjustable IPD.
- When large FOVs are used, determine if it drives high levels of vection (i.e. perceived self-motion); if it does, consider reducing the FOV.
- If high levels of vection are found and they lead to high levels of sickness, then reduce the spatial frequency content of visual scenes.
- Provide multimodal feedback that minimizes sensory conflicts (i.e. provide visual, auditory, haptic, and kinesthetic feedback appropriate for the situation being simulated).

TABLE 29.4
Steps to Quantifying VE Stimulus Intensity

1. Get an initial estimate. Talk with target users (not developers) of the system and determine the level of adverse effects they experience.
2. Observe. Watch users during and after exposure and note comments and behaviors.
3. Try the system yourself—particularly if you are susceptible to motion sickness—and obtain a first-hand assessment of the adverse effects.
4. Measure the dropout rate. If most people can stay in for an hour without symptoms, then the system is likely benign; if most people drop out within ten min, then the system is probably in need of a redesign.
5. Measure. Use simple rating scales to assess sickness and visual, proprioceptive, and postural measures to assess aftereffects.
6. Compare. Use Table 29.5 to determine how the system under evaluation compares to other VE systems.
7. Report. Summarize the severity of the problem, specify required interventions (e.g., warnings and instructions), and set expectations for use (e.g., target exposure duration and intersession intervals).
8. Expect dropouts. With a high-intensity VE stimulus, dropout rates can be high.

TABLE 29.5
Virtual Environment Sickness Quartiles

Quartile	SSQ Score
25th	15.5
50th	20.1
75th	27.9
95th	33.3
99th	53.1

TABLE 29.6

Factors Affecting Individual Capacity to Resist Adverse Effects of VE Exposure

- Limit initial exposures. For medium to strong VE stimuli, limit initial exposures to a short duration (e.g., ten min or less).
- Adaptation. Set inter-session exposure intervals 2–5 days apart to enhance individual adaptability.
- Adaptation and habituation. Use an incremental approach where a VE stimulus is incrementally increased in intensity within one exposure (incremental adaptation) or across multiple exposures (incremental habituation),
- Determine whether users can complete an adaptive process at each increment of stimulus intensity (i.e. depressing sensitization to preexposure levels or below); if not, lower stimulus intensity.
- Repeat exposure. Avoid repeated exposure intervals occurring less than two hours apart if adverse effects are experienced in an exposure.
- Age. Expect little motion sickness for those under age two; expect greatest susceptibility to motion sickness between the ages of two and 12; expect motion sickness to decline after 12, with those over 25 being about half as susceptible as they were at 18 years of age.
- Gender. Expect females to be more susceptible than males (perhaps as great as three times more susceptible).
- Anthropometries. Consider setting VE stimulus intensity in proportion to body weight and stature.
- Individual susceptibility. Expect individuals to differ greatly in motion sickness susceptibility and use the MHQ or another instrument to gauge the susceptibility of the target user population.
- Couple MHQ questionnaire items with the roadside battery test used to assess sobriety (Tharp, Burns, & Moskowitz, 1981) or other psychophysical test batteries (e.g., postural stability, dexterity, and gaze nystagmus) to enhance its predictability.
- Drug and alcohol consumption. Limit VE exposure to those individuals who are free from drug or alcohol consumption.
- Rest. Encourage individuals to be well-rested before commencing VE exposure. Inform individuals that if they become drowsy, they may experience a greater frequency and severity of flashbacks.
- Predisposition. Inform individuals who have experienced an emetic response associated with carnival rides that they may become ill during VE exposure.
- Ailments. Discourage those with cold, flu, or other ailments (e.g., headache, diplopia, blurred vision, sore eyes, or eyestrain) from participating in VE exposure. Encourage those susceptible to photic seizures and migraines, as well as individuals with preexisting binocular anomalies to avoid exposure.
- Fitness and overall health. Many factors (e.g., alcohol, flu, hangover, sleep loss, high stress, etc.) contribute to overall health and fitness and, thus, can be expected to exacerbate sickness in a VE; whereas each individual factor may minimally influence adverse effects if they are concatenated, then you can expect that there will be a sickness that, rather than being attributable to the VE, may be due to a combination of conditions in the individual.
- Clinical user groups. Obtain informed sensitivity to the vulnerabilities of these user groups (e.g., unique psychological, cognitive, and functional characteristics). Encourage those displaying co-morbid features of various psychotic, bipolar, paranoid, substance abuse, claustrophobic, or other disorders where reality testing and identity problems are evident to avoid exposure.

12. Avoid provocative movements. For strong VE stimuli, warn users to avoid movements requiring high rates of linear or rotational acceleration and extraordinary maneuvers (e.g., flying backward) during initial interaction (McCauley & Sharkey, 1992).
13. Monitor users. Throughout VE exposure, an attendant should be always available to monitor the user's behavior and ensure their well-being. An attendant may also have to assist users if they become stuck or lost within the virtual world, as often happens.
14. Look for red flags. Indicators of impending trouble include excessive sweating, verbal frustration, lack of movement within the environment for a significant amount of time, and less overall movement (e.g., restricting head movement). In addition, before they become ill, users will tend to become very quiet and will totally stop talking. Users who

demonstrate any of these behaviors should be observed closely, as they may experience an emetic response. Extra care should be taken with these individuals after postexposure. Note: it is beneficial to have a bag or bin located near users in the event of an abrupt emetic response.

15. Termination. Set criteria for terminating exposure. Exposure should be terminated immediately if users verbally complain of symptoms and acknowledge that they are no longer able to continue. Also, to avoid an emetic response if telltale signs are observed (i.e. sweating and increased salivation), exposure should be terminated.

16. Postexposure care. Some individuals may feel ill or unsteady upon postexposure. These individuals may need assistance when initially standing up after exposure.

17. Debriefing. After exposure, the well-being of the users should be assessed. Measurements of their hand-eye coordination and postural stability should be taken. Similar to field sobriety tests (Tharp et al., 1981), these can include measures of balance (e.g., standing on one foot, walking an imaginary line, and leaning backward with eyes closed), coordination (e.g., alternate hand clapping and finger to nose touch while the eyes are closed), and eye nystagmus (e.g., follow a light pen with the eyes without moving the head).

18. Releasing. Set criteria in releasing users. Specify the amount of time after exposure that users must remain on-premises before driving or participating in other high-risk activities. In our lab, a 2-to-1 ratio is used; post-exposure users must remain in the laboratory twice the amount of exposure time to allow recovery. Do not allow individuals who fail in debrief tests or are experiencing adverse aftereffects to conduct high-risk activities until they have recovered (e.g., have someone drive them home).

19. Follow-up. Call the users the next day or have them call to report any prolonged adverse effects.

CONCLUSIONS

The risks associated with VE exposure are real and include ill effects during exposure, as well as the potential for prolonged aftereffects. To minimize these risks, VE system developers should quantify and minimize VE stimulus intensity, identify the capacity of the target user population to resist the adverse effects of VE exposure, and then follow a systematic usage protocol. This protocol should focus on warning, educating and preparing users, setting appropriate environmental and equipment conditions, limiting initial exposure duration and user movements, monitoring users and looking for red flags, and setting criteria in terminating exposure, debriefing, and release. Adopting such a protocol may minimize the risk factors associated with VE exposure.

ACKNOWLEDGMENTS

This material is based on the work supported in part by the Office of Naval Research (ONR) under Grant N000149810642, the National Science Foundation (NSF) under Grants DMI9561266 and IRI-9624968, and the National Aeronautics and Space Administration (NASA) under Grants NAS9-19482 and NAS9-19453. Any opinions, findings and conclusions, or recommendations expressed in this material are those of the authors and do not necessarily reflect the views or the endorsement of the ONR, NSF, or NASA.

REFERENCES

Bagshaw, M., & Stott, I. R. (1985). The desensitization of chronically motion sick aircrew in the Royal Air Force. *Aviation, Space, and Environmental Medicine, 56*(12), 1144–1151.

Biocca, F. (1992). Will simulation sickness slow down the diffusion of virtual environment technology? *Presence: Teleoperators and Virtual Environments, 1*(3), 334–343.

Brooks, Jr., F. R. (1988). Grasping reality through illusion: Interactive graphics serving science. In *ACM SIGCHI Conference Proceedings* (pp. 1–11). Washington: ACM.

Carlson, N. R. (Ed.). (2001). *Physiology of Behavior* (7th edition). Boston: Allyn & Bacon.

Champney, R., Stanney, K. M., & Kennedy, R. S. (under revision). *Virtual environment réadaptation mechanisms.* University of Central Florida.

Chinn, H. I., & Smith, R. K. (1953). Motion sickness. *Pharmacological Review, 7,* 33–82.

Cobb, S. V. G., Nichols, S., Ramsey, A. D., & Wilson, J. R. (1999). Virtual reality-induced symptoms and effects (VRISE). *Presence: Teleoperators and Virtual Environments, 8*(2), 169–186.

Cramer, D. B., Graybiel, A., & Oosterveld, W. J. (1978). Successful transfer of adaptation acquired in a slow rotation room to motion environments in navy flight training. *Acta Oto-Laryngologica, 85,* 74–84.

Crampton, G. H. (Ed.). (1990). *Motion and space sickness.* Boca Raton, FL: CRC Press.

Davies, R. C. (2002). Applications of systems design using virtual environments. In K. M. Stanney (Ed.), *Handbook of virtual environments: Design, implementation, and applications* (pp. 791–806). Mahwah, NJ: Lawrence Erlbaum Associates.

Dichgans, J., & Brandt, T. (1978). Visual-vestibular interaction: Effects on self-motion perception and postural control. In R. Held, H. W. Leibowitz, & H. L. Teuber (Eds.), *Handbook of sensory physiology, Vol. VIII: Perception* (pp. 756–804). Heidelberg: Springer-Verlag.

DiZio, P., & Lackner, J. R. (1997). Circumventing side effects of immersive virtual environments. In M. Smith, G. Salvendy, & R. Koubek (Eds.), *Design of computing systems: Social and ergonomic considerations* (pp. 893–896). Amsterdam, Netherlands: Elsevier Science Publishers, San Francisco, CA (August 24–29).

DiZio, P., & Lackner, J. R. (2002). Proprioceptive adaptation and aftereffects. In K.M. Stanney, (Ed.), *Handbook of Virtual Environments: Design, Implementation, and Applications.* (pp. 791–806). Mahwah: NJ: Lawrence Erlbaum Associates.

Durlach, B. N. I., & Mavor, A. S. (1995). *Virtual reality: Scientific and technological challenges.* Washington, DC: National Academy Press.

Ebenholtz, S. M., & Mayer, D. (1968). Rate of adaptation under constant and varied optical tilt. *Perceptual and Motor Skills, 26,* 507–509.

Graeber, D. A. (2001a). *Use of incremental adaptation and habituation regimens for mitigating optokinetic side effects.* Unpublished doctoral dissertation, University of Central Florida.

Graeber, D. A. (2001b). *Application of the Kennedy and Graybiel Motion History Questionnaire to predict optokinetic induced motion sickness: Creating a scoring key for circular vection* (Tech. Report No. TR-2001-03). Orlando, FL: Naval Air Warfare Center, Training Systems Division.

Graybiel, A., Deane, E, & Colehour, J. (1969). Prevention of overt motion sickness by incremental exposure to otherwise highly stressful coriolis accelerations. *Aerospace Medicine, 40,* 142–148.

Groves, P. M., & Thompson, R. F. (1970). Habituation: A dual process theory. *Psychological Review, 77*(5), 419–450.

Harm, D. L. (2002). Motion sickness neurophysiology, physiological correlates, and treatment. In K. M. Stanney (Ed.), *Handbook of virtual environments: Design, implementation, and applications* (pp. 791–806). Mahwah, NJ: Lawrence Erlbaum Associates.

Held, R. (1965). Plasticity in sensory-motor systems. *Scientific American, 72,* 84–94.

Howarth, P. A., & Finch, M. (1999). The nauseogenicity of two methods of navigating within a virtual environment. *Applied Ergonomics, 30,* 39–45.

Hu, S., Stern, R. M., & Koch, K. L. (1991). Effects of pre-exposure to a rotating optokinetic drum on adaptation to motion sickness. *Aviation, Space, and Environmental Medicine, 62,* 53–56.

Kennedy, R. S. (2001). *Unpublished research data,* Orlando, FL: RSL Assessments, Inc.

Kennedy, R. S. et al. (1987). Guidelines for alleviation of simulator sickness symptomatology. *Aviation, Space, and Environmental Medicine, 60,* 10–16.

Kennedy, R. S., Lanham, D. S., Massey, C. J., Drexler, J. M., & Lilienthal, M. G. (1995). Gender differences in simulator sickness incidence: *Implications for military virtual reality systems. Safe Journal, 25*(1), 69–76.

Kennedy, R. S., Dunlap, W. P., Berbaum, K. S., & Hettinger, L. J. (1996). Developing automated methods to quantify the visual stimulus for cybersickness. *Proceedings of the Human Factors and Ergonomics Society 40th Annual Meeting* (pp. 1126–1130). Santa Monica, CA: Human Factors & Ergonomics Society.

Kennedy, R. S., & Fowlkes, J. E. (1992). Simulator sickness is polygenic and poly symptomatic: Implications for research. *International Journal of Aviation Psychology, 2*(1), 23–38.

Kennedy, R. S., & Graybiel, A. (1965). *The Dial test: A standardized procedure for the experimental production of canal sickness symptomatology in a rotating environment* (Rep. No. 113, NSAM 930). Pensacola, FL: Naval School of Aerospace Medicine.

Kennedy, R. S., Hettinger, L. J., & Lilienthal, M. G. (1990). Simulator sickness. In G. H. Crampton (Ed.), *Motion and space sickness* (pp. 247–262). Boca Raton, FL: CRC Press.

Kennedy, R. S., Jones, M. B., Stanney, K. M., Ritter, A. D., & Drexler, J. M. (1996, June). *Human factors safety testing for virtual environment mission-operation training* (Final Report, Contract No. NAS9-19482). Houston, TX: NASA Johnson Space Center.

Kennedy, R. S., Lane, N. E., Berbaum, K. S., & Lilienthal, M. G. (1993). Simulator sickness questionnaire: An enhanced method for quantifying simulator sickness. *International Journal of Aviation Psychology*, *3*(3), 203–220.

Kennedy, R. S., et al. (2001, September 5–7). Use of a motion history questionnaire to predict simulator sickness. In *Presented at Driving Simulation Conference 2001*, Sophia-Antipolis (Nice), France.

Kennedy, R. S., & Stanney, K. M. (1996). Postural instability induced by virtual reality exposure: Development of a certification protocol. *International Journal of Human-Computer Interaction*, *8*(1), 25–47.

Kennedy, R. S., Stanney, K. M., & Dunlap, W. P. (2000). Duration and exposure to virtual environments: Sickness curves during and across sessions. *Presence: Teleoperators and Virtual Environments*, *9*(5), 466–475.

Kingdon, K. (2001). Effects of low stereo acuity on performance, presence, and sickness within a virtual environment. Unpublished master's thesis, Orlando, FL: University of Central Florida

Kingdon, K., Stanney, K. M., & Kennedy, R. S. (2001, October 8–12). Extreme responses to virtual environment exposure. In *The 45th Annual Human Factors and Ergonomics Society Meeting* (pp. 1906–1910). Minneapolis/ St. Paul, MN.

Knerr, B. W., Breaux, R., Goldberg, S. L., & Thurman, R. A. (2002). National defense. In K. M. Stanney (Ed.), *Handbook of virtual environments: Design, implementation, and applications* (pp. 857–872). Mahwah, NJ: Lawrence Erlbaum Associates.

Knerr, B. W. et al. (1998). *Virtual environments for dismounted soldier training and performance: Results, recommendations, and issues* (ARI Tech. Rep. No. 1089). Alexandria, VA: U.S. Army Research Institute for the Behavioral and Social Sciences.

Kolasinski, E. M. (1995). *Simulator sickness in virtual environments* (ARI Tech. Rep. 1027). Alexandria, VA: U.S. Army Research Institute for the Behavioral and Social Sciences.

Konz, S. (1997). Toxicology and thermal discomfort. In G. Salvendy (Ed.), *Handbook of human factors and ergonomics* (2nd edition, pp. 891–908). New York: Wiley.

Lackner, J. R., & Lobovitz, D. (1978). Incremental exposure facilitates adaptation to sensory rearrangement. *Aviation, Space and Environmental Medicine*, *49*, 264–362.

Lanham, S. (2000). The effects of motion on performance, presence, and sickness in a virtual environment. Masters Thesis, University of Central Florida.

Lawson, B. D., Graeber, D. A., Mead, A. M., & Muth, E. R. (2002). Signs and symptoms of human syndromes associated with synthetic experiences. In K. M. Stanney (Ed.), *Handbook of virtual environments: Design, implementation, and applications* (pp. 791–806). Mahwah, NJ: Lawrence Erlbaum Associates.

McFarland, R. A. (1953). *Human factors in air transportation: Occupational health & safety*. New York: McGraw-Hill.

McCauley, M. E., & Sharkey, T. J. (1992). Cybersickness: Perception of self-motion in virtual environments. *Presence: Teleoperators and Virtual Environments*, *1*(3), 311–318.

McNally, W. J., & Stuart, E. A. (1942). Physiology of the labyrinth reviewed in relation to seasickness and other forms of motion sickness. *War Medicine*, *2*, 683–771.

Mirabile, C. S. (1990). Motion sickness susceptibility and behavior. In G. H. Crampton (Ed.), *Motion and space sickness* (pp. 391–410). Boca Raton, FL: CRC Press.

Mon-Williams, M., Rushton, S., & Wann, J. P. (1995). Binocular vision in stereoscopic virtual-reality systems. *Society for Information Display International Symposium Digest of Technical Papers*, *25*, 361–363.

Money, K. E. (1970). Motion sickness. *Psychological Review*, *50*(1), 1–39.

Money, K. E., & Myles, W. S. (1975). Motion sickness and other vestibulo-gastric illnesses. In R. F. Naunton (Ed.), *The vestibular system* (pp. 371–377). New York: Academic Press.

Moshell, J. R., & Hughes, C. E. (2002). Virtual environments as a tool for academic learning. In K. M. Stanney (Ed.), *Handbook of virtual environments: Design, implementation, and applications* (pp. 791–806). Mahwah, NJ: Lawrence Erlbaum Associates.

National Research Council. (1997). *More than screen deep: Toward every-citizen interfaces to the nation's information infrastructure*. Washington, DC: National Academy Press.

Pavlov, I. P. (1928). *Lectures on conditioned reflexes*. W. H. Gantt (Trans.), New York: International Publishers.

Pierce, J. S., Pausch, R., Sturgill, C. B., & Christiansen, K. D. (1999). Designing a successful HMD-based experience. *Presence: Teleoperators and Virtual Environments, 8*(4), 469–473.

Prescott, S. A. (1998). Interactions between depression and facilitation within neural networks: Updating the dual process theory of plasticity. *Learning & Memory, 5*(6), 446–466.

Prescott, S. A., & Chase, R. (1999). Sites of plasticity in the neural circuit mediating tentacle withdrawal in the snail Helix aspersa: Implications for behavioral change and learning kinetics. *Learning & Memory, 6*, 363–380.

Reason, J. T. (1970). Motion sickness: A special case of sensory rearrangement. *Advancement in Science, 26*, 386–393.

Reason, J. T. (1978). Motion sickness adaptation: A neural mismatch model. *Journal of the Royal Society of Medicine, 71*, 819–829.

Reason, J. T., & Brand, J. J. (1975). *Motion sickness*. New York: Academic Press.

Regan, E. C., & Price, K. R. (1994). The frequency of occurrence and severity of side-effects of immersive virtual reality. *Aviation, Space, and Environmental Medicine, 65*, 527–530.

Rizzo, A. A., Buckwalter, G., & van der Zaag, C. (2002). Virtual environment applications in clinical neuropsychology. In K. M. Stanney (Ed.), *Handbook of virtual environments: Design, implementation, and applications* (pp. 1027–1064). Mahwah, NJ: Lawrence Erlbaum Associates.

Satava, R. M., & Jones, S. B. (2002). Medical applications of virtual environments. In K. M. Stanney (Ed.), *Handbook of virtual environments: Design, implementation, and applications* (pp. 721–730). Mahwah, NJ: Lawrence Erlbaum Associates.

Singer, M. J., Ehrlich, E. A., & Allen, R. C. (1998). *Effect of a body model on performance in a virtual environment search task* (ARI Tech. Rep. 1087). Alexandria, VA: U.S. Army Research Institute for the Behavioral and Social Sciences.

Sjoberg, A. A. (1929). Experimental studies of the eliciting mechanism of sea sickness. *Acta Oto-Laryngologica, 13*, 343–347.

So, R. H., & Griffin, M. J. (1995). Effects of lags on human operator transfer functions with head-coupled systems. *Aviation, Space, and Environmental Medicine, 66*, 550–556.

Stanney, K. M. (1995). Realizing the full potential of virtual reality: Human factors issues that could stand in the way. *Virtual Reality Annual International Symposium '95* (pp. 28–34). Los Alamitos, CA: IEEE Computer Society Press.

Stanney, K. M. (2001). *Unpublished research data*. Orlando, FL: University of Central Florida.

Stanney, K. M. (Ed.). (2002). *Handbook of virtual environments: Design, implementation, and applications*. Mahwah, NJ: Lawrence Erlbaum Associates.

Stanney, K. M., & Hash, P. (1998). Locus of user-initiated control in virtual environments: Influences on cybersickness. *Presence: Teleoperators and Virtual Environments, 7*(5), A41–A59.

Stanney, K. M., & Kennedy, R. S. (1997a, September 22–26). Cybersickness is not simulator sickness. In *Proceedings of the 41st Annual Human Factors and Ergonomics Society Meeting* (pp. 1138–1142). Albuquerque, NM.

Stanney, K. M., & Kennedy, R. S. (1997b). The psychometrics of cybersickness. *Communications of the ACM, 40*(8), 67–68.

Stanney, K. M., & Kennedy, R. S. (1998, October 5–9). Aftereffects from virtual environment exposure: How long do they last? In *Proceedings of the 42nd Annual Human Factors and Ergonomics Society Meeting* (pp. 1476–1480). Chicago, IL.

Stanney, K. M., Kennedy, R. S., Drexler, J. M., & Harm, D. L. (1999). Motion sickness and proprioceptive aftereffects following virtual environment exposure. *Applied Ergonomics, 30*, 27–38.

Stanney, K. M., Kennedy, R. S., & Kingdon, K. (2002). Virtual environments usage protocols. In K. M. Stanney (Ed.), *Handbook of virtual environments: Design, implementation, and applications* (pp. 721–730). Mahwah, NJ: Lawrence Erlbaum Associates.

Stanney, K. M., Kingdon, K., Graeber, D., & Kennedy, R. S. (2002). Human performance in immersive virtual environments: Effects of duration, user control, and scene complexity. *Human Performance, 15*(4), 339–366.

Stanney, K. M., Kingdon, K., & Kennedy, R. S. (2001). Human performance in virtual environments: Examining user control techniques. In M. J. Smith, G. Salvendy, D. Harris, & R. J. Koubek (Eds.), *Usability evaluation and interface design: Cognitive engineering, intelligent agents and virtual reality* (Vol. 1 of the *Proceedings of HCI International 2001*, pp. 1051–1055). Mahwah, NJ: Lawrence Erlbaum Associates.

Stanney, K. M., Kingdon, K., Nahmens, I, & Kennedy, R. S. (2003). What to expect from immersive virtual environment exposure: Influences of gender, body mass index, and past experience. *Human Factors*, *45*(3), 504–522.

Stanney, K. M., Lanham, S., Kennedy, R. S., & Breaux, R. B. (1999, September 27–October 1). Virtual environment exposure drop-out thresholds. In *The 43rd Annual Human Factors and Ergonomics Society Meeting* (pp. 1223–1227). Houston, TX.

Stanney, K. M. et al. (1998). Aftereffects and sense of presence in virtual environments: Formulation of a research and development agenda. Report sponsored by the Life Sciences Division at NASA Headquarters. *International Journal of Human-Computer Interaction*, *10*(2), 135–187.

Stone, R. J. (2002). Applications of virtual environments: An overview. In K. M. Stanney (Ed.), *Handbook of virtual environments: Design, implementation, and applications* (pp. 827–856). Mahwah, NJ: Lawrence Erlbaum Associates.

Tharp, V., Burns, M., & Moskowitz, H. (1981). *Development and field test of psychophysical tests for DWI arrest* (DOT Final Rep. ODT HS 805 864). Washington, DC.

Tyler, D. B., & Bard, P. (1949). Motion sickness. *Physiological Review*, *29*, 311–369.

Uliano, K. C., Kennedy, R. S., & Lambert, E. Y. (1986). Asynchronous visual delays and the development of simulator sickness. *Proceedings of the Human Factors Society 30th Annual Meeting* (pp. 422–426). Dayton, OH: Human Factors Society.

Viirre, E., & Bush, D. (2002). Direct effects of virtual environments on users. In K. M. Stanney (Ed.), *Handbook of virtual environments: Design, implementation, and applications* (pp. 581–588). Mahwah, NJ: Lawrence Erlbaum Associates.

Wann, J. P., & Mon-Williams, M. (2002). Measurement of visual aftereffects following virtual environment exposure. In K. M. Stanney (Ed.), *Handbook of virtual environments: Design, implementation, and applications* (pp. 731–749). Mahwah, NJ: Lawrence Erlbaum Associates.

Welch, R. B. (2002). Adapting to virtual environments. In K. M. Stanney (Ed.), *Handbook of virtual environments: Design, implementation, and applications*. Mahwah, NJ: Lawrence Erlbaum Associates.

Wendt, G. R. (1968). *Experiences with research on motion sickness* (NASA Special Publication No. SP-187). In *Fourth Symposium on the Role of Vestibular Organs in Space Exploration*. Pensacola, FL.

Wilson, J. R. (1997). Virtual environments and ergonomics: Needs and opportunities. *Ergonomics*, *40*(10), 1057–1077.

Wilson, J. R., Nichols, S., & Haldane, C. (1997, August 24–29). Presence and side effects: Complementary or contradictory? In M. Smith, G. Salvendy, & R. Koubek (Eds.), *Design of computing systems: Social and ergonomic considerations* (pp. 889–892). Amsterdam, Netherlands: Elsevier Science Publishers, San Francisco, CA.

Wilson, J. R., Nichols, S. C., & Ramsey, A. D. (1995). Virtual reality health and safety: Facts, speculation and myths. *VR News*, *4*, 20–24.

Watson, G. S. (1998). The effectiveness of a simulator screening session to facilitate simulator sickness adaptation for high-intensity driving scenarios. *Proceedings of the 1998 IMAGE Conference*, The IMAGE society, Chandler, AZ.

30 Location and Arrangement of Displays and Control Actuators

Robert W. Proctor and Kim-Phuong L. Vu

CONTENTS

INTRODUCTION

The performance of an operator is highly dependent on the location and arrangement of displays and controls. A well-designed interface considers the positioning and grouping of displays and controls and the relationships between them. The display arrangement should allow the operator to detect and identify critically displayed information with minimal effort. This information needs to be easily associated with the appropriate control that affects the system function captured by the display. The control should be located where it is readily accessible and can be operated comfortably. To accomplish these goals, the display and control arrangements need to be designed according to the principles of perception, response selection, motor control, anthropometrics, and biomechanics.

Displays, controls, and their relations have been studied since the earliest days of Human Factors and Ergonomics as a distinct field. Consequently, much is known about factors that influence performance with different display-control configurations, and numerous guidelines that reflect this knowledge have been developed. In addition, there has been considerable research on stimulus-response compatibility and related effects from which recommendations can be made—such as response-effect compatibility—in recent years, but this has not yet been incorporated into specific standards and guidelines.

The objective of this chapter is to provide a thorough description of existing standards and guidelines regarding the location and arrangement of displays and controls. In addition, we will discuss recommendations that can be made based on current knowledge. In all cases, we will discuss the findings that underlie the standards, guidelines, and recommendations.

DEFINITION OF RELEVANT TERMS

a. Accessibility index—A quantitative measure of control layout that considers the ranked frequency of use of the controls and their distance from the operator.

b. Acuity—The ability to perceive and distinguish details in a visual image. Acuity is highest in central vision and decreases dramatically in peripheral vision.

c. Fitts' law—A law used to estimate movement time (MT) as a function of the target width (W) and distance (D). $MT = a + b \log_2 (2D/W)$, where $\log_2 (2D/W)$ is the index of difficulty, a is a constant for the time to initiate movement, and b is a constant that accommodates the rate of movement.

d. Functional groupings—Using grouping principles to organize displays or controls so that those relating to common functions are grouped together.

e. Gesture control—Hand movements used to signal controlling actions, especially for touch screen devices.

f. Grouping principles—Organizational principles that determine which parts of a display or control arrangement are perceived as belonging together and are separate objects.

g. Head-up displays—Head-up displays superimpose a virtual image of the information display on the central area where the outside world is viewed. The purpose is to allow the operator to check displays without looking away from the outer environment.

h. Line of sight—A hypothetical line from the eyes to the point on which they are fixated. The farther away from the line of sight a display is located—that is, the greater the visual angle between the line of sight and the display—the more difficult it is to identify details of the display.

i. Link analysis—An analysis that can be applied to display panel design to determine the locations in which components of a display panel should be placed. In a link analysis, displays with a high frequency of use or importance are placed in central locations, and displays that are scanned in sequence are placed in adjacent locations.

j. Mental model—A dynamic representation of an event or scenario that reflects the person's understanding of the situation and can promote accurate situation awareness. Mental models direct the comprehension of new information, reasoning, solution of problems, and decisions made under uncertainty.

k. Population stereotypes—Expected relationships within a given population between movements of controls and their effects on a display or system.

l. Reach envelope—The distance from the operator within which controls should be for a certain percentage of the population (typically 95%) to be able to reach them easily. For desktop computer entry, the controls should be within reach of the extended forearm and hand.

m. Response-effect compatibility—Differences in the speed and accuracy with which responses can be selected as a function of the effects that they produce on a display or system.

n. Saccadic eye movement—A rapid, ballistic eye movement made to a pre-determined location either voluntarily or reflexively to the onset of a stimulus or a change in some property.

o. Spatial compatibility—Performance is better when the spatial relations in a display are mapped in a corresponding manner to the spatial relations of a response panel or control.

p. Stimulus-response compatibility—Differences in the speed and accuracy with which responses to stimuli can be selected for different stimulus-response arrangements and mappings. This is sometimes called display-control compatibility.

q. Touch screen displays—Displays for which the controls are represented by activation through touching of specific areas of the display itself.

r. Useful field of view (UFOV) —UFOV is a measure of visual attention that specifies the range of the visual field where people can select an object from multiple stimuli. UFOV allows assessment of visual processing speed, divided attention, and selective attention. It decreases with age and is a useful predictor of driving accidents.

s. Visual angle—The angle at the eye subtended by the image of an object. The visual angle is computed as $\tan^{-1}$(size/distance).

t. Visual field—The area in which visual stimuli can be detected. Acuity varies across the visual field, being highest in foveal vision.

FULL NAME OF THE STANDARDS (GUIDELINES)

a. ANSI/HFES 100-2007, Human Factors Engineering of Computer Workstations.

b. DoD Military Standards—1472D Human Engineering Design Criteria for Military Systems Equipment and Facilities, Section 5.

c. Ergonomics of Design and Use of Visual Display Terminals (VDTs) in Offices: Specification for Keyboards (BS 7179: Part 4: 1990).

d. Ergonomic Design of Control Centres (ISO 11064).

e. Ergonomic Requirements for the Design of Signals and Control Actuators—Part 1: Human Interactions with Displays and Control Actuators (ISO/CD 9355-1).

f. Ergonomic Principles for the Design of Signals and Control Actuators—Part 2: Displays (ISO/CD 9355-2).

g. Ergonomic Requirements for Office Work with Visual Display Terminals (VDTs) (ISO 9241).

h. Ergonomic Requirements for Work with Visual Displays Based on Flat Panels (ISO 13406-1,2).

i. Guidelines for Using Anthropometric Data in Product Design (HFES 300).

j. Human Factors Engineering of Software User Interfaces (HFES 200).

k. Human Integration Design Handbook (HIDH) (NASA/SP-2010-3407/REV1).

l. Medical Electrical equipment—Medical image display systems—Part 1: Evaluation methods (IEC 62563-1:2010).

m. Nuclear power plants—Control rooms—Application of visual display units (VDUs) (BS IEC 61772:2009).

n. Safety of Machinery; Ergonomics Requirements for the Design of Displays and Control Actuators; Part 1: General Principles for Human Interactions with Displays and Control Actuators (CEN EN 894-1; PREN 894-1; BSI BS EN 894-1).

o. Safety of Machinery—Ergonomics Requirements for the Design of Displays and Control Actuators—Part 2: Displays (CEN EN 894-2; PREN 894-2; BSI BS EN 894-2).

p. Safety of Machinery—Ergonomics Requirements for the Design of Displays and Control Actuators—Part 3: Control Actuators (CEN EN 894-2; PREN 894-3; BSI BS EN 894-3).

q. Visual display requirements (BS EN 29241-3:1993; ISO 9241).

JUSTIFICATION OF STANDARDS

LOCATION AND ARRANGEMENT OF DISPLAYS

When designing display panels, the placement of the components is crucial because of the relation of acuity to the location in the visual field. A visual display must be located within the visual field if it is to have any chance of being detected and responded to. For young adults, the field of view is approximately 180°, but it decreases to about 140° in older adults. Within the visual field, acuity varies drastically. It is highest in central, foveal vision (approximately 1° of visual angle) and decreases sharply as the stimulus location moves further into the visual periphery. Surrounding the fovea is a somewhat larger region of 5°, called the parafovea, where acuity is still high but not as high as in foveal images.

The central 30° of the field is the focal vision where images can be brought through saccadic eye movements. The remainder of the visual field is called ambient vision. It plays a role in guiding movement and maintaining spatial orientation, but head movements are required for images in this region to be brought into the foveal vision. Studies have shown that the useful field of view is reduced considerably when multiple stimuli are present and need to be processed.

When the head and eyes are oriented directly ahead, the line of sight is horizontal. However, this posture is uncomfortable because it requires the neck and eye muscles to be tense. It is more comfortable to relax the muscles for both the head and the eyes. The head is in the most comfortable position when it is inclined slightly forward, making the normal line of sight relative to the head 10–15° below the horizontal. When the eyes are also relaxed, the normal line of sight is 25–30° below the horizontal.

Because time and effort are required to move the head and eyes, it is important to minimize the amount of movement that is necessary to process the information from a display arrangement. A link analysis can be used to help determine where the individual components of a display panel should be placed. As applied to display design, components are arranged according to their importance, frequency, and sequence of use, with adjacent components between the strongest sequential scanning links.

The distance of displays from the operators is also a factor to consider. Both the accommodation of the lens and the vergence angle of the eyes vary as a function of distance from the operator for distances up to approximately 200 cm. The need for re-accommodation and change in vergence angle when switching among display components or between displays and the external environment should be minimized. One benefit of a head-up display—where the displays are located on the windshield—is that it reduces the differences in accommodation and vergence when switching between the outside world and the display panel, relative to a head-down display. However, this benefit has its associated costs such as creating clutter, being difficult to see against the environmental background, and reducing the visibility of objects in the outer environment.

Glare is a factor with many visual displays, including CRTs and LCDs. A distinction is often made between discomfort glare—where the glare mainly reduces visual comfort—and disability glare—where the glare from higher intensity light sources reduces visibility. Glare on display screens can be reduced by positioning the display away from light sources, using shades and screen filters, and so on.

Although most displays are two-dimensional, 3D displays that use the depth cue of binocular disparity are relatively common for entertainment. In the workplace, though, most 3D displays are presented on 2D screens, where monocular depth cues form a 3D image on the screen. Although 3D displays can provide a more natural depiction of elements in natural environments, they compress information along certain axes on a 2D screen, making it difficult for observers to accurately extract distance information along the compressed dimensions. What people typically refer to as 3D displays use polarized glasses or some other means to produce the cue of binocular disparity in the images at the eyes. Special consideration needs to be given to the implementation of such displays.

LOCATION AND ARRANGEMENT OF CONTROLS

Because controls often need to be seen to select the correct one, the visual concerns that apply to displays also apply to control panels. However, anthropometric and biomechanical factors about the placement of controls are more important. If a control is placed where it cannot be reached and operated easily, this may present problems for the operator. When designing for reach, the 5th percentile reach values are often used because they accommodate all but the very smallest users. A distinction is also often made between the reach envelopes of males and females since males are of larger physical stature on average. The immediate reach envelope refers to the region where the controls can be reached without bending, and the maximum reach envelope refers to the region in which controls can be reached by bending. The exact values for the reach envelope vary as a function of the task, the height at which the control is placed, and whether the operator is standing or sitting. For example, at table level, Sengupta and Das (2000) estimated the maximum straight reach when seated to be 58.3 and 64.5 cm for females and males, respectively. These values varied as a function of height from the tabletop and degrees of angle horizontally from straight ahead. The reach envelope for standing operators was approximately 4 cm longer for males and 6 cm for females.

In addition to ensuring that an operator can reach a control, it is important that the operator not confuse the controls. If an operator must select one among several controls, they should be separated and distinct from one another to minimize the possibility of activating the incorrect one. As with displays, failure to place controls that are used in sequence together may result in excessive movements of the operator and cause fatigue.

Touchscreens have become popular over the last decade because they provide a natural interaction method where operators can use their hands or fingers directly to manipulate objects and controls on the screen. The display is highly reconfigurable and flexible. Moreover, because a separate input device is not needed, touchscreens can save space in a work environment. The main disadvantages of touchscreens are (a) the hands can occlude the display, (b) touchscreens are not ideal for tasks requiring the operator to point to an exact location on the screen, and (c) the input of text is slowed down by virtual keyboards compared to physical keyboards.

Gesture input refers to single movements or combinations of movements that command a response. Gestures are typically used to input commands for touchscreen displays and interactive systems (such as the Wii system from Nintendo). The spatial properties of movements are often mapped to the resulting consequences—such as when a leftward hand swipe on a touchscreen makes the right part of a display visible on the screen or when gestures are used to control spatial actions in a virtual reality display. A benefit of gestural responses is that they can make interaction with the device or virtual world more natural.

DISPLAY-CONTROL RELATIONS

Most standards focus on the placement and arrangement of displays and controls, but less on their relation. However, the process of determining what action to take in response to the displayed information is a major component of response time. It has been known, since at least the time of Fitts and Seeger's (1953) classic study on stimulus-response compatibility, that the time for response selection is minimal when spatial compatibility is high. Performance is best when displays and their corresponding controls are configured in similar arrangements and each display is mapped to the spatially corresponding control. If display-control arrangements are not highly compatible, the operator may be delayed in responding, or may even take the wrong action.

Response-effect compatibility refers to the relation between a response and its effect on the system or environment it is operating in. The prior example of a gesture being mapped to a spatially corresponding shift in the visual display is an example of a compatible response-effect relation. Response-effect compatibility is important because the anticipation of the consequences of a response plays a significant role in the selection and control of action.

DISCUSSION OF SPECIFIC STANDARDS AND GUIDELINES

ARRANGEMENT AND LOCATION OF DISPLAYS

Display Position

Displays should be positioned according to the following guidelines:

- *Identification:*
  The display needs to be positioned where it can be detected and identified readily.
- *Visibility:*
  The operator should be able to see all the essential displays from his normal workplace position.
    - Primary instruments should be positioned within 30° of the line of sight to avoid excessive movements of the eyes, head, or body.
    - Visual warning signals should also be positioned within 30° of the line of sight to allow the operator to see the signal readily even if it is not often used.
    - Auditory warning signals do not have to be positioned within the normal line of vision since they can be detected independent of where the operator is looking. Auditory warning signals can be used to direct where an operator should look.
    - Secondary instruments should be positioned within 60° of the line of sight. This will allow the operator to read the displays by moving the eyes, without having to change the position of the head.
    - Infrequently used instruments do not need to lie within the normal line of vision.

- *Derivatives of Change:*
  When different derivatives of change need to be displayed, displays should be positioned from left to right in order of increasing derivatives.
- *Viewing Distance:*
  The viewing distance should be at least 40 cm for many office tasks, with 50 cm being preferable. Closer distances may cause eye strain.
- *Maximum Line of Sight Angle:*
  The line of sight should be no more than 60° above or below the horizontal. For a display to be legible, the size and distance need to be considered as well as the location in the visual field relative to the line of sight.
- *Glare and Reflection:*
  The display should be positioned in a location that minimizes glare and reflection from surrounding objects.

Display Layout

- *Related Information:*
  Displays of highly related information sources should be located close to each other. There are many ways in which different sources of information can be highly related. For example, displays can provide correlated information or information pertinent to the same system function.
- *Importance:*
  Displays that are the most important should be placed in foveal vision. Important displays should compose a "central zone" on the display panel. This zone should ideally be within 30° of visual angle.
- *Frequency:*
  Frequently used displays should be in the primary visual field and adjacent to each other.

- *Sequence:*
Displays that are typically viewed sequentially should be placed close together, in the order in which they are scanned. People have a bias to scan along the horizontal dimension more than the vertical dimension.
- *Consistency:*
When using different display panels or screens for similar or related tasks, the same elements should be positioned in the same locations. This will improve performance by allowing the operator to direct attention to the desired locations effortlessly, minimizing visual search time and memory load.
- *Perceptual Grouping:*
The Gestalt perceptual organizational principles should be used to group related displays together and separate them from other groupings. Commonly used principles are:
  - Similarity: Display elements intended to be perceived as grouped together should be similar in appearance.
  - Spatial proximity: Elements intended to be perceived as a group should be placed close together.
  - Common fate: Elements intended to be perceived as a group should move in the same direction and speed.
  - Connectedness: Elements intended to be perceived as a group should be connected by lines. This is sometimes called the use of sensor lines.
  - Common region: Elements intended to be perceived as a group should be within a common boundary. This is often used for over-the-counter medications, where a single dose is surrounded by a perforation boundary.

- *Display Integration:*
For settings in which information about complex systems at different levels of abstraction needs to be displayed, integrating across different views is a challenge. For tasks such as fault diagnosis that require problem-solving, presenting the information integrated in a single display yields better performance than presenting the information for the separate views in distinct windows.
- *Clutter Avoidance:*
There should be adequate spacing between all display components on a display panel. Avoid excessive placements of display elements within a small area. Clutter is of particular concern for head-up displays, where each new display element or symbol that is added increases the difficulty of perceiving the outer environment.

ARRANGEMENT AND LOCATION OF CONTROLS

Position of Controls

- *Identification:*
The control needs to be positioned where it can be detected and identified readily.
- *Visibility:*
The operator should see all the essential controls from the normal workplace position.
- *Accessibility:*
Controls should be in a position that allows easy and comfortable access by the operator. For example, the slope of a keyboard should be positioned between 0° and 25°, and the operator should be able to make additional adjustments. The keyboard should also be placed in a stable position that is independent of the display.
- *Expectancy:*
Controls should be positioned where people would expect them to be located. Population stereotypes can help determine the expectancy for the targeted users.

Control Layout

- *Importance:*
  Controls that are most important (i.e. primary controls) should be located close to the operator.
- *Frequency:*
  The most frequently used controls should be located close to the operator, inside the immediate reach envelope where they can be reached without bending. Infrequently used controls can be placed at further distances.
- *Sequence of Use:*
  Controls operated in sequence should be placed close together and preserve the sequential relation.
- *Accessibility:*
  All controls should be arranged within the reach envelope to allow easy access. Avoid placing controls where operators must make awkward movements.
- *Tactile landmarks:*
  Controls should be located near tactile landmarks. This is especially needed for operators with visual disabilities.
- *Grouping:*
  Related controls should be grouped together, and unrelated controls should be separated for the operator to easily identify groups of controls. Examples of grouping include:
  - Group controls with similar functions together on the control panel
  - Make related controls equal in size

- *Movement Space*:
  The layout of the controls should not cause the operator to experience any discomfort when operating a control or moving between controls.
- *Movement Time:*
  Movement time between controls should be minimized. Movement time is an increasing function of movement distance and a decreasing function of target width. It can be estimated using Fitts' Law.
- *Corresponding Control Labels:*
  The labels on control panels should be larger than usually recommended so that both younger and older adults can see them. If the guidelines for control labels were based on young adults, the characters will need to be a minimum of 20% larger recommended to accommodate older adults. Labels should also have a good contrast, with a 10:1 contrast ratio of labels with their background recommended.
- *Symmetry:*
  Design the panel to be as symmetrical as possible.
- *Expectancy:*
  The control should be in a position that the operator expects to avoid excessive search for the control. Population stereotypes can help determine the expectancy for the targeted users.
- *Consistency:*
  When more than one control panel is used or when operators must switch between different control panels, the specific controls should be located consistently across panels.
- *Clutter Avoidance:*
  Adequate distances should be maintained between controls.

DISPLAY-CONTROL RELATIONS

Although many guidelines and standards are established for the location and arrangements of displays and controls independently, the display-control layout should also optimize the interrelationship between display-control pairs.

- *Stimulus-Response Compatibility:*
Stimulus-response compatibility refers to the natural tendency to respond faster and more accurately with some mappings of stimuli to responses than to others. When designing display-control layouts, it is important to map displays with their associated controls in a compatible manner. Maintaining compatibility is particularly important when workload or stress is high because performance with incompatible mappings will deteriorate more than with compatible mappings. However, maintaining compatibility is not as easy as it sounds because it varies as a function of tasks and their goals. Below are specific compatibility relations:
  - Relative Location: it is important to maintain the relative location of displays and their associated controls. Compatibility effects occur with respect to relative locations, not just absolute locations. For example, when four burners are arranged in a rectangle on a stovetop and are mapped to four linearly arranged controls, ambiguity arises as to which control operates a specific burner. However, if the burners are staggered so that they are in four distinct locations on the horizontal dimension, then it is easy to determine based on relative location which burner corresponds to each control.
  - Frames of Reference: compatibility effects occur with respect to many frames of reference. These include body midline, direction of attention, and location relative to environmental cues and objects. When multiple displays or controls are organized into groups, locations can be coded at both the global and local levels.
  - Parallel versus Orthogonal: parallel mappings (e.g., displays in left-right locations mapped to controls in left-right locations) yield better performance than orthogonal mappings (e.g., displays in top-bottom locations mapped to controls in left-right locations). Designers should try to align displays and controls along the same axis.
  - Orthogonal Mappings: when parallel mappings are not possible and displays are mapped orthogonally to their associated controls, the mapping of "top" with "right" and "bottom" with "left" often yields better performance than the mapping of "top" with "left" and "bottom" with "right." However, the preferred mapping is affected by the location of the controls relative to the display and to other controls.
  - Pure versus Mixed Mappings: performance is better when the same mapping is used for all display-control pairs than with different mappings. The most ideal layout is one where all display-control mappings are compatible. When mappings are mixed, performance often suffers more for the compatible relation than the incompatible relation.
  - Rules: when displays cannot be mapped compatibly to their spatially corresponding controls, one that produces a systematic relation (e.g., opposite) is better than a random mapping. For example, performance is much better with a mirror-opposite relation between displays and controls than with a random one. This is because the individual associations between displays and controls do not have to be remembered.
  - Prevalence Effects: when the display-control configuration can be coded along two dimensions at once, a prevalence effect may occur where coding, with respect to one dimension, dominates the other. The dominant dimension is likely to be the one made salient by both the display and control environments.
  - Simon Effects: spatial correspondence effects occur when the location is irrelevant to the task. For example, when stimuli and response vary in left-right locations and responses made to the stimulus attribute is nonspatial (e.g., identity: S or H), responding is faster when the location of the stimulus and response corresponds spatially than when it does not.
  - Intentions and Task Goals: compatibility effects are not an automatic consequence of physical relations. Rather, they depend on the operators' intentions and the task's goals.

- Dimensional Overlap: compatibility effects occur not only for spatial stimuli and responses but for any situation in which the stimulus dimension overlaps with the response dimension. This overlap may involve conceptual similarity (when the display and control sets can be categorized along the same dimension), perceptual similarity (when the display and control sets are physically similar), and structural similarity (e.g., when the display and control sets maintain an ordered relationship).

- *Proximity:*
  Displays should be close to their associated controls.
- *Movement:*
  The direction of movement for a control should be compatible with the direction of movement for both the feedback indicator and the system movement. The following are movement principles based on population stereotypes:
  - Clockwise-to-Right-or-Up: clockwise control movement is expected to move a pointer toward the right for horizontal displays and upward for vertical displays.
  - Warrick's Principle: the pointer is expected to move in the same direction as the side of the control that is nearest the display.
  - Clockwise-to-Increase: clockwise control is expected to cause an increase in the reading of the display.
  - Clockwise-Away: clockwise control is expected to cause the pointer to move away from the control.

- *Practice:*
  Performance with an incompatible display-control configuration improves with practice but is typically worse than it would be with the same amount of practice if the configuration were compatible.
- *Learning and Transfer:*
  If an operator has previous experience with a related display-control configuration, the previously learned relationship may transfer to the current situation. This transfer may facilitate performance—if the display-control relations conform to the operator's expectations—or otherwise, interfere.
- *Display-Control Configurations:*
  When the display and its associated control are on the same panel, the control should be placed under the display if possible. If this arrangement is not possible, the control should be placed to the right of the display; because most people are right-handed, the display will not be obstructed when the operator reaches for the control. When multiple displays need to be monitored while adjusting a single control, the control should be placed under the display in the middle in a manner that will allow the control to be operated without obscuring the displays.
- *Redundancy:*
  The display-control configuration should contain redundant coding of information for important relations.
- *Controllability:*
  The display-control configuration should help guide the operator through the tasks and allow the operator direct control over the system.
- *Concurrent use:*
  Displays that are monitored simultaneously while manipulating a control should be placed where the operator can easily see them.
- *Mental model:*
  A designer should be able to predict which display-control mappings would be most compatible based on knowledge of the operators' mental representation of the task and system.

EXAMPLES OF APPLICATION

- *Multifunction Displays:*
  In a multifunction display, operators proceed through a hierarchy of information presented on a computer screen by pressing or clicking buttons. With such displays, how to map the hierarchy to the buttons for successive screens is a significant human factor problem. Several general design principles mentioned above—such as frequency and sequence of use—are applicable. Moreover, search time can be reduced dramatically by minimizing the average distance between buttons for successive screens and maximizing repeated selections of the same buttons. Optimization methods can be applied to select the best mapping relative to a defined cost function. For more details, see Francis (2000).
- *Aircraft Display-Control Layout:*
  A British Midland Airways Boeing 737-400 aircraft crashed on January 8, 1989 when the functioning right engine was shut down instead of the failing left engine. The investigators noted that one factor contributing to this error was the layout of the engine displays relative to the controls. The engine displays were grouped into left and right rectangular blocks arranged on the cockpit instrument panel. The left grouping contained the primary instruments for both engines; the right grouping, the secondary instruments for both engines. Within each grouping, the dials for the left engine were in the left column and those for the right engine in the right column. The throttle controls for the left and right engines were aligned with the primary and secondary groupings, respectively. Thus, left-right display-control compatibility was maintained relative to the columns within each group, but not to the left and right groups themselves. With this configuration, spatial compatibility was maintained within the local instrument group, but not for the global grouping. Because the global blocks were aligned with the controls, it was probably more important to maintain global compatibility than local compatibility in this case. For more details, see Learmount and Norris (1990).
- *Stovetop Designs:*
  A common arrangement of the four burners on a stovetop is rectangular. There are numerous ways in which the controls can be arranged and mapped to the burners. Not too surprisingly, when the control arrangement is isomorphic to the burner arrangement, a compatible mapping of the burners to controls produces fast and accurate performance because the relative location of the display configuration matches that of the control configuration. However, performance with this arrangement suffers dramatically with an incompatible burner-control mapping. The cost associated with this incompatibility can be reduced by connecting each burner to its control with sensor lines. More commonly, the controls are arranged linearly along the horizontal dimension, and the control configuration is not isomorphic with the display arrangement. In this case, there is no natural mapping of burners to controls. This ambiguity can be reduced, and the performance improved, by staggering the burners so that they are in four distinct locations on the horizontal dimension. This produces burner-control correspondence along the horizontal dimension based on relative location. Alternatively, performance can be improved by using sensor lines to connect the burners in the rectangular arrangement to their controls in the horizontal arrangement. This example illustrates that there are many alternative ways to address potential problems of incompatibility. For a reference, see Chapanis and Yoblick (2001).

REFERENCES FROM WHICH THE STANDARDS AND GUIDELINES REVIEWED ABOVE ARE BASED

Chapanis, A., & Yoblick, D. A. (2001). Another test of sensor lines on control panels. *Ergonomics, 44,* 1302–1311.

Fitts, P. M., & Seeger, C. M. (1953). S-R compatibility: Spatial characteristics of stimulus and response codes. *Journal of Experimental Psychology, 46*, 199–210.

Francis, G. (2000). Designing multifunction displays: An optimization approach. *International Journal of Cognitive Ergonomics, 4*, 107–124.

Learmount, D., & Norris, G. (1990). Lessons to be learned. *Flight International*, 31 October–6 November, 24–26.

Sengupta, A. K., & Das, B. (2000). Maximum reach envelope for the seated and standing male and female for industrial workstation design. *Ergonomics, 43*, 1390–1404.

LIST OF OTHER RELEVANT STANDARDS/GUIDELINES WITH FULL REFERENCE

a. Man-Machine Interface (MMI)—Actuating Principles (IEC 447, CENELEC EN 60447)

b. Visual Display Requirements (BS EN 29241-3)

c. Deutsches Institut für Normung: 666234: VDT Workstations, Part 3: Grouping and formatting of data.

d. Location and Direction of Motion of Operator's Controls for Agricultural Tractors and Self-Propelled Agricultural Machines (AS 1246)

e. Earth-Moving Machinery—Instrumentation and Operator's Controls—Part 5: Zones of Comfort and Reach for Controls (ISO 6682, AS 2956.5)

f. Flight Deck Panels, Controls, and Displays, Part 8: Flight Deck Head-up Displays (ANSI/SAE ARP 4102/8)

g. Earth-Moving Machinery—Instrumentation and Operator's Controls—Part 0: General Introduction and Listing (AS 2956.0)

h. Operator Controls and Displays on Motorcycles, Recommended Practice (SAE J107)

i. Instruments and Controls in Motor Truck Cabs, Location and Operation of (SAE-J680)

j. Human Engineering Considerations in the Application of Color to Electronic Aircraft Displays (ANSI/SAE ARP 4032)

k. Construction and Industrial Equipment, Instrument Face Design, and Location For (SAE J209)

l. Ergonomics Aspects of Indicating Devices; Types, Observation Tasks, Suitability (DIN 33413 PT 1)

m. Photometric Guidelines for Instrument Panel Displays that Accommodate Older Drivers, Information Report (SAE J2217)

n. Earth-Moving Machinery—Instrumentation and Operator's Controls—Part 2: Operating Instrumentation (ISO 6011, AS 2956.2

o. Remote Control and Display Unit Design (ARINC 561-11 SEC 10.0)

p. Control Panel, Aircraft, General Requirements for (MIL-C-81774A-1)

q. Display, Head-Up, General Specification for (MIL-D-81641 NOTICE 2)

r. Control Panel, Aircraft, General Requirements for (MIL-C-81774A-1)

s. Human Engineering (MIL-STD-1472F)

t. Man-Systems Integration Standards (NASA-STD-3000 REV B VOL II)

u. Aircrew Station Control Panels (NATO STANAG 3869 ED 1 AMD 4)

v. Design Objectives for CRT Displays for Part 23 Aircraft (SAE ARP 4067)

31 Maximizing the Value of Enterprise Human-Computer Interaction Standards
Strategies and Applications

Wei Xu

CONTENTS

INTRODUCTION

Human factors/ergonomics (HFE) standards are not only a useful reference for experienced HFE practitioners but are also guidance for organizations that are inexperienced in HFE design practice. HFE standards can give credibility to the value of introducing user-centered methods (Bevan, 2001). As computing technologies advance, knowledge of HFE has spread to the computing-related work, and the field of human-computer interaction (HCI) has grown rapidly. Accordingly, HCI standards have evolved for guiding practice. There is a great deal of literature concerned with the development and practice of international standards (e.g., International Organization for Standardization/ISO) and national (e.g., The American National Standards Institute/ANSI) HFE or HCI standards, but little on the practice of HCI standards at an enterprise level. This chapter will focus on the practice of HCI standards there. It intends to assess the challenges of enterprise HCI standards from strategy, development, and governance perspectives. Specifically, we discuss the practices at Intel Corporation, a high-tech enterprise environment.

Hierarchy of HCI Standards

Like HFE standards, the HCI standards system is a pyramidal multi-level structure model. From top to bottom, this multi-level model includes international standards, regional/national standards, and enterprise standards.

The highest level in the HCI standards hierarchy is the relevant international standards issued by the ISO Human-System Interaction Committee (ISO Technical Committee/TC 159). The TC 159 has four sub-technical committees that published 143 standards in the areas of HCI, user experience (UX) and usability, etc. Most of the published standards are related to HCI/UX. For example, in the well-known ISO 9241 series of standards, each includes a sub-series of standards. Within the ISO 9241 series, ISO 9241-210 specifies human-centered design (HCD) methods and principles; the standard ISO 9241-220 is a document for HCD processes (to be published); and ISO 9241-230 are the standards for evaluation methods in HCD (currently being drafted).

The second level in the hierarchy is the regional/national HFE standards, most of which are related to HCI. For example, in the U.S., national standards include nearly 40 government standards—published by the National Aeronautics and Space Administration (NASA), the Federal Aviation Administration (FAA), etc.—and non-government standards published by the Human Factors and Ergonomics Society (HFES), ANSI, etc. Among these, HFES has published two HCI-related standards, including Human Factors Engineering for Computer Workstations (ANSI/HFES 100, 2007) and Human Factors Engineering of Software User Interfaces (HFES 200, 2008).

The third level is the HCI standards that are published by a variety of enterprises and organizations, which are called enterprise HCI standards herein. These enterprise standards specify the HCI standards at a detailed level across the industry and organization domains. The standards may be applicable to the external or internal audience of an enterprise. For example, Microsoft has issued a detailed user interface (UI) design standard for the Microsoft Windows family of software products. Apple has issued UI design standards for applications across the iOS technology platform. These standards are for an external audience, thus ensuring that developers across industries follow the UI design standards when developing specific products. In addition, many corporations, such as Intel, have published a series of HCI standards for internal use. On one hand, these internal HCI standards ensure internal organization and project teams adopt uniform design standards across internal product lines, while on the other, they also enable and facilitate the HCI design practices within a corporation.

Generally speaking, the contents of the HCI standards across the three layers reflect the relationship between inheritance and consistency. The highest level of international standards defines consensus in all aspects—such as the guiding principles and design principles—rather than the specific design requirements, reflecting the flexibility of specific design within a certain scope. Meanwhile, the lowest level of enterprise standards is more detailed for specific design within specific domains/platforms.

VALUE AND CHALLENGES OF INTERNATIONAL/NATIONAL HCI STANDARDS

As a foundation, the higher-level (international/national) HCI standards do provide benefits. As an example, the approach to software development in the ISO 9241 series is based on detailed guidance and principles for design, rather than precise interface specifications, thus permitting design flexibility while avoiding constraints on design (Bevan, 2001). As a result, these ISO standards often describe principles—not specific solutions—for implementation.

Also, these higher-level HCI standards provide potentially complementary approaches to the assurance of usability and UX in solutions during the HCI practice when used in combination with lower-level HCI standards, such as individual enterprise HCI standards (Bevan, 2001). From a software user interface (UI) design perspective, the use of ISO 9241 and ANSI/HFES 100 standards in combination with a style guide based on individual corporation's branding requirements is useful for detailed and consistent branding design on the UI. As a lesson learned, the attempt by the IEEE to develop a standard for the drivability of UI (IEEE, 1993) eventually failed to achieve consensus (Bevan, 2001). This approach does not, in itself, ensure usability.

However, these higher-level HCI standards face challenges in practice. Enterprise HCI standards have been more influential than ISO, and the ISO standards have not been widely adopted (Bevan et al., 2016). The challenges are outlined below.

Availability of Best-Known Methods

To demonstrate the value of the higher-level HCI standards, one must successfully implement them in product development within an organization's environment. However, in the HFE and HCI communities, there is little information about how to effectively apply these HCI standards at an organization level (e.g., an enterprise environment) and how to develop own enterprise HCI standards and effectively manage the governance of compliance in practice.

Low Perceived Value of HCI Standards

Influenced by the organizational culture on HFE/HCI, challenges occur because HFE/HCI professionals' expertise and contributions to design can be undervalued (Green, 2002). As a result, HCI standards are difficult to establish and may not be adopted in practice.

Challenge of Conformance

Very few of the higher-level HCI standards specify the UI design precisely, instead defining general principles from which appropriate UI design and procedures can be derived. This gives the standards authority for good professional practice but makes it more difficult to demonstrate conformance (Bevan, 2001).

Intended Audience

As guided by a human-centered design philosophy, there is no perfect set of guidelines in HCI standards: different audiences have different needs (Bevan, 2005). It is difficult for the higher-level

HCI standards to be comprehensive enough to serve all purposes of all types of audience across domains. After identifying the intended audience, HCI professionals in enterprises must specify the scope and the depth of guidelines in the HCI standard to include in a specific domain/platform.

Timing Upgrade

Today, the pace of technological change continues to accelerate, and while all ISO standards are reviewed at least once every five years (Bevan, 2001), higher-level HCI standards can quickly become out of date. This makes it more difficult to assess the conformance of a product to the standards as documented (Reed et al., 1999). Standards are supposed to contain a single requirement specifying what kind of information shall be provided to demonstrate that the relevant recommendations in the standard have been identified and followed. This approach cannot be guaranteed at these higher-level standards, as so many recommendations are context-specific.

Domain-Specific Issues

It is difficult to find standards relevant to given projects from the myriad domain-specific standards published by many organizations and agencies (Swaminathan & Rantanen, 2014). There are also very few comprehensive standards available for adoption across the industry, and this may not even be feasible, as there are a variety of domains across the industry.

THE NEEDS FOR ENTERPRISE HCI STANDARDS

As discussed above, there are many challenges in implementing international and national HCI standards to meet specific enterprise needs within a corporate environment. Thus, there is a need to develop more specific HCI standards at the enterprise level that are guided by these higher-level HCI standards. Specifically, in our practice at Intel, we recognize that at least two types of HCI standards are needed when delivering both external and internal digital software solutions: design standards and methodological standards.

HCI Design Standards

With the development of new technologies, there have been efforts to develop HCI design standards in the areas of voice, touch, and gestures. However, most of the current standards focus on the design of visual human-machine interfaces; thus, the discussion in this section also focuses on this aspect. Based on our practices over years as a typical use case in the visual UI platform, HCI design standards for enterprises should include the following categories:

- *UI Design Standards:* the standards should adopt the design guidance and principles inherited from related ISO standards and ANSI/HFES-100 standards. In addition, the standards should specifically reflect the enterprise brand requirements for UI design, specific and usable UI design patterns consistently to be used across the enterprise, and the HCI design guidelines across platforms.
- *Visual Design Resource Library:* this library should provide a visual design resource based on the enterprise's branding requirements so projects can quickly reuse the assets for faster and consistent design that will naturally comply with the enterprise UI design standards.
- *Conceptual UI Design Pattern Library*: this library should provide various conceptual UI design patterns (i.e. wireframe/low-fidelity prototypes) to guide UI prototyping work across project teams. The purpose is to encourage projects to carry out UI prototyping and reuse appropriate patterns for usable UI design.
- *Code-Based UI Component Library*: driven by both the visual design resource library and the conceptual UI design pattern library, this library should include code-based UI components. A process should be set up to ensure that UI components can be accumulated to the library based on project code work over time. The developers of future projects can directly reuse code-ready UI components, saving development costs and ensuring usable design.

HCI Methodology Standards

To enable the development of organizational maturity in HCI, methodological standards are also needed, which also helps the adoption of the design standards.

- *User-Centered Design (UCD) Process Standards*: the higher-level HCI standards, such as ISO 9241-210/220, can be leveraged to communicate the needs for UCD activities across programs/projects. The enterprise's UCD process standards should define required activities/ methods at key stages of the organization's internal product development process and checkpoints for activities from a governance perspective.
- *UX Quality Standards*: the UX quality standards should specify standardized indicators to be used in the organization, the benchmark value of the indicators, the method of verification, the tracking method, and reporting procedures.
- *Organizational UX Maturity Model*: the model should define the levels of maturity, the dimensions (factors) for the evaluation of maturity, and specific checklists.
- *Holistic UX Design Standards*: the UI design standards only cover the UI portion. For best HCI practice, a holistic UX approach must be taken so that additional factors that impact the UX can be addressed, including application performance, workflow, UI writing, UI internationalization, user help/training, etc.

REASONS FOR ENTERPRISE HCI STANDARDS

Promoting Consistent Design and Saving Costs

Consistent design means there is no need to set separate design standards for each product. One standard can cover a range of products. From a UX ecological perspective, consistent design means delivering a unified UX across platforms and products. By standardizing design, a central HCI team in an organization (if it exists) can work with development teams to develop UI assets that can be reused across projects (e.g., code-based UI components), helping developers to quickly generate UI, saving time and costs in development, repair, and future upgrades.

Improving Product UX

The detailed UX design specifications defined by enterprise standards are based on previous UCD activities across projects, so these have a guarantee of usability/UX for a family of products. Furthermore, design based on principles of consistency and usability can help the users (for the same product family) match the mental models they have built using previous products when learning to use new products, thereby reducing learning time, improving the UX, and reducing user support costs.

Promoting Product Brands

Any product has a certain brand, and its design is crucial for UX and marketing. The design of product brands is often closely related to the UI, so the standardization of UI design is conducive to standardizing the unified product brand design. For example, a page template that reflects the product brand is available across all pages of an enterprise's website.

Evangelizing UX Practice Within the Enterprise

The process by which developers refer to UI design standards is an opportunity by itself for them to learn UX. UX practice has been integrated into the standardized UCD development process, which is a very good process for the development team to learn UCD and coordinate with UX personnel. UI design standards and UCD process standards also help the team promote UX, while standardized metrics and benchmarks help the organization's management track UX quality across products.

Promoting More Efficient UCD Activities

UX artifacts—such as a series of personas and user journey maps that are built using standardized user research methods (including user interviews, questionnaires, etc.)—can be used to build a cross-project reusable UX artifact repository for reuse across projects in the future. If necessary, new project teams can update these artifacts based on updated user requirements. This will save time for future UCD activities.

Helping Outsourcing Contract and Business Acquisition

Driven by cloud-computing technology and cost control, many enterprises enlist third-party development contractors to complete some product development. Enterprise HCI design standards facilitate the communication of HCI design requirements and acceptance criteria with contractors. In addition, when an organization acquires other companies, a clear established UI design standard is conducive in providing guidance on the HCI design for the products of the acquired company.

Facilitating the Growth of the UX Organization

In a large and medium-sized corporation, the experience of conducting HCI work across internal organizations may be uneven. These organizations need help from the enterprise's central HCI group in methodology and design resources. The development of HCI standards, including design and methodology, will help the dispersed UX personnel adapt to growth. In addition, the process of developing these standards is a learning and growth process for the central group, which is also conducive to enhance the influence and leadership of the central HCI group in relation to the diverse business product lines.

DEVELOPING STRATEGY

UX Strategy

Today, enterprise resource planning (ERP) systems are commonly used in corporations. ERP is a complex computer-based digital environment of solutions that enterprise employees rely on to do their daily jobs. These solutions include customer relations management (CRM), supply chain management (SCM), human capability management (HCM), finance systems, business intelligence (BI), product support systems, and utility and productivity solutions, among others. The huge set of ERP solutions are mixed with home-grown and vendor solutions. Over the years, these were deployed to internal employees with a variety of business and legal requirements, as well as technological requirements. New technology, such as cloud-based SaaS solutions, also adds more vendor solutions. As a result, there can be inefficiency or lack of integration across data, business processes, and UI.

From a UX perspective, enterprise users daily apply a set of solutions to accomplish their tasks. They interact with these solutions through a variety of user touchpoints, including user awareness/marketing, devices/installation, access/security, solution UI, business process, user support, and user context using data, etc. Therefore, UX is not simply influenced by a single solution or an individual touchpoint, but instead by how all these solutions and touchpoints are combined to provide a *unified experience* for end-users. If any of these touchpoints were to break down, it would create a negative experience. Intel is no exception, and a lack of good UX in many individual apps and across apps due to siloed solutions requires users to use multiple siloed apps to accomplish their job (e.g., SaaS and homegrown solutions). Thus, the hybrid environment of today's ERP systems poses several serious HCI and UX challenges to the delivery of a unified experience.

Based on extensive research, a unified experience strategy has been proposed to address the UX issues along with corporation-wide digital transformation efforts (Xu, 2014; Xu et al., 2019). To support the unified experience strategy, enterprise HCI standards, processes, and an organization maturity model were developed within the IT organization. Below, we focus our discussion on enterprise HCI standards.

OWING ORGANIZATION OF THE ENTERPRISE HCI STANDARDS

To support the implementation of the unified experience strategy, the Intel IT Cross-Domain HCI/ UX Technical Working Group (TWG) was set up as a horizontal capability and leadership team to drive the HCI standards and governance. The HCI/UX TWG is embedded within the overall IT enterprise architecture and technical governance body.

The TWG's charter is to: (1) drive HCI design strategy; (2) develop and maintain HCI standards; and (3) lead and manage the governance of HCI standards. The HCI/UX TWG is chaired by a principal human factors researcher/architect, and meetings are held bi-weekly. The membership includes representatives of HFEs who are working in different business units and representatives from software architects, developers, interaction designers, business architects, and so on. One of the major tasks is to develop/maintain enterprise HCI standards, review/audit the HCI design of cross-platform projects, and approve proposed HCI standards and design assets. Also, the HCI/UX TWG chair serves as representative of the TWG for multiple technical committees, which helps with the alignment and evangelization of HCI/UX across organizations.

EXPERIENCE OF USING ENTERPRISE HCI STANDARDS

A few years earlier, we developed and published a few HCI-related standards, from both design and methodology perspectives. To further understand the gaps, we conducted a UX study with 20 software developers. The data collected were mapped out into the user journey map with major pain points and needs identified (see Figure 31.1). As shown in Figure 31.1, there are many pain points along the developer's end-to-end experience of using the HCI standards, including "Be aware" (e.g., "I know that there are HCI standards available for us"), "Access" (e.g., "I know where I can find the standards"), "Understand" (e.g., "I understand the content of the standards," "I understand which parts are mandatory vs. recommended"), "Apply" (e.g., "I know how I can apply the standards to my development work"), and "Be governed" (e.g., "I know what governance process we need to follow for compliance"). The major gaps identified and the actions to be taken across the end-to-end experience journey are summarized as follows.

- *Low Awareness of HCI Standards*: many people did not know HCI standards or knew only some of them. For those who did, they only knew the basic UI design standards documents (i.e. the look and feel design standards for applications).
- *Not Sure Where to Find HCI Standards*: many people do not know where to locate these HCI standards or had difficulty finding them.

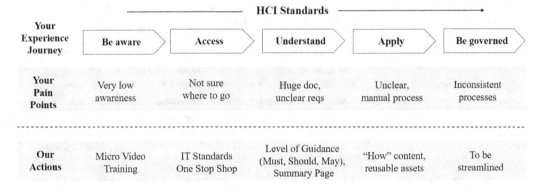

FIGURE 31.1 The gaps identified and actions to be taken.

- *Unclear About the Content of HCI standards*: most people did not fully understand the content of HCI standards documents. In particular, they were not sure whether the statements defined in the standards were requirements (mandatory) or recommendations (optional) from the compliance perspective.
- *Not Sure How to Apply the HCI standards:* the lack of reusable assets made it hard for developers to apply the requirements defined in the standards to their development work.
- *Inconsistent Governance Process*: varied versions of review processes were found across different meetings/forums, and the execution of the review/audit to a large extent relied on the expertise of project technical leaders.

DEVELOPING AN ACTION PLAN

An action plan was developed to specifically address the gaps identified (see Figure 31.1). The action plan was developed through a process. First, all the possible actions to address a specific gap were defined and then prioritized in terms of resources needed and impact. Second, the prioritized actions were presented to the stakeholders. Finally, the action plan was approved by the management, and resources were assigned for implementation.

SOLUTIONS AND EXECUTION

TRANSFORMING HCI DESIGN ASSETS

"HCI design assets" broadly refers to HCI related standards, guidelines, tools, and reusable UI resources—that is, anything that positively support HCI work. Based on the strategy and action plan defined, we started a transformation of the existing HCI design assets. Prior to the transformation, there were only a few paper-based HCI standards documents, such as the Master Look and Feel Design Standards for digital solutions, along with some ISO standards (e.g., SIO 9241-100). Table 31.1 summarizes the "before" and "after" states of various items that were transformed, as well as the benefits, obtained.

TABLE 31.1
Transforming UX Assets

Before	After	Benefits
Static documents (Word, PDF, etc.) distributed across places	An online one-stop shop consolidating all documents in digital formats	Easy to access
Descriptive guidelines, only covering "do" and "don't" statements	Reusable UI assets, allowing developers/ designers to reuse the assets for compliance (e.g., visual, design pattern, UI components)	Reusable, agile
A UI-centric approach standards and assets	A UX-centric approach by publishing non-UI standards (e.g., app performance, app writing, design guide for persona & user journey map)	Holistic UX approach to minimize UX risk
Unclear content of the standards (mandatory vs. recommendation)	Each requirement/statement assigned with RFE labels (must, should, may)	Clarified level of guidance for compliance of standards
A "one-way" push model, only allowing the audience to access UX assets	A "two-way" model allowing developers/ designers to contribute, rate, and comment on UX assets	More interactive and participatory for developers/designers

From Distributed Documents to an Online One-Stop Shop

In the past, our enterprise HCI standards were distributed to the whole organization in a Word/PDF document, but this approach didn't deliver a good UX as it was hard for users to find and access them. To address this point as documented in the user journey map (see Figure 31.1), we have built an online one-stop shop that consolidates all the HCI standards documents in the form of digital content. An enterprise collaboration (social media) platform is also used to facilitate the participatory approach (see Table 31.1).

From a "Farm of Links" to Searchable, Filterable, and Sortable Content

Although the one-stop shop consolidated all the existing HCI standards documents in a central location, it was primarily a farm of links connecting individual standards documents. It was not easy for users to find the content within or across documents. To address the problem, the contents of all the HCI standards documents were converted to a web-based format with meta tags defined across documents to help the implementation of search/filtering/sorting capabilities (see Figure 31.2). This helped users easily find the content more efficiently and effectively.

From Descriptive Guidelines to Reusable UI Assets

The traditional way of delivering UI-related HCI standards is based on descriptive guidelines such as "do" or "don't." For instance, "don't use x color over x color for low contrast;" "use x UI design pattern in scenario x." This approach is difficult for developers/designers to follow and it is not efficient, leading to low adoption and difficult compliance from a governance perspective. To overcome the issues, we created three reusable UI assets libraries:

- *Visual Design Library*: includes HTML-based UI style sheets based on the Intel Corporate brand requirements and the enterprise UI design standards (e.g., Bootstraps framework-based); an images/icons library with over 500 icons with different formats per corporate brand requirements.
- *Conceptual UI Design Pattern Library*: includes over 200 UI design concepts (wireframe) in responsive design across desktop, mobile, and tablet platforms. The catalog includes page layout, navigation, various UI elements, and comprehensive page design. All the design patterns were selected based on usability validation through projects or the industry's best-known methods.
- *Code-Based UI Component Library*: includes many code-based UI components built by UI technology (e.g., Angular JavaScript, React). These UI components are based on some of the

Title	Guidance	Status	Domain	Area	Solution Type	App Layer
Decision Framework For UI Selection	Should Use	Ratified	Application	Development	COTS, SaaS, Cust...	UI
HP ALM	Should Not Use	Ratified	Application	Tools	COTS	UI/UX
IT Application Internationalization Guidelines	Should Use	Ratified	Application	Development	COTS	SaaS	Cu...	UI/UX
IT Application Performance Standards	Must Use	Ratified	Application	Development		UI/UX
IT Axure MLAF Based UI Widget Library	Should Use	Ratified	Application	Development	COTS	SaaS	Cu...	UI/UX
IT Master Look And Feel Standards	Must Use	Ratified	Application	Development	COTS	SaaS	Cu...	UI/UX
IT Touch UI Design Guidelines	Should Use	Ratified	Application	Development	COTS	SaaS	Cu...	UI/UX
IT UI Design Guidelines For Info Visualization, BI Reports, & Dash...	Should Use	Ratified	Application	Development		UI/UX
IT UI Responsive Web Design Guidelines	Should Use	Ratified	Application	Development	COTS	SaaS	Cu...	UI/UX
IT UX Governance Standards	Must Use	Ratified	Application	Development		UI/UX
Intel Icon Library	Must Use	Ratified	Application	Development		UI/UX
Intel Writing Style Guide And Editorial Standards For User Interface	Should Use	Ratified	Application	Development	COTS	SaaS	Cu...	UI/UX
MEAP Any	Must Not Use	Ratified	Application	Platforms		
Mobile UI Design Standards And Guidelines	Must Use	Ratified	Application	Development		UI/UX
Plugins Native Application	Must Not Use	Ratified	Application	Development		
Plugins Web Browser	Must Not Use	Ratified	Application	Development		

FIGURE 31.2 A screenshot illustrating the filterable and sortable UI.

conceptual UI design patterns with visual design (style sheet, icons, font, etc.) applied. Thus, developers can easily reuse the UI components during development, which will guarantee usable UI components with compliance to Intel brand requirements by default. Over time, we expect increasingly reusable UI components to be added by various project teams (Figure 31.3).

From UI-Centric to Holistic UX Approach

Driven by our unified UX strategy, we have taken a holistic UX approach by developing more UX-centric HCI standards beyond the originally narrow scope of UI design. Over time, more UX-related HCI standards have been (or will be) developed and published. These include:

- *Application Performance Standards:* these specify the pass/fail criteria of the performance (page load/response) test for IT applications.
- *Software Accessibility Standards:* these define the requirements for accessibility of software, driven by relevant ISO standards.
- *Application Writing Guide:* this provides guidelines for writing on the UI of applications to make content easy to understand, in a consistent style necessary to achieve the corporation brand.
- *UX Guidelines in Business Process Reengineering (planned):* these guidelines will define the UX guiding principles and methodology that can be applied to the process of redesigning business processes from a UX perspective.
- *UX Analytics Guide (Support tools):* this guide provides the instructions of using application usage/UX analytics capabilities that help projects to collect users' usage and behavioral data (e.g., click streams) after its release. The collected data will help projects identify UX gaps for improvements.
- *UX indicators (planned):* we plan to standardize key UX metrics (e.g., productivity, user satisfaction scores, user support volume/cost) across organizations. The next step is to integrate these UX metrics into the overall operation dashboard of the IT organization, providing higher visibility to management.

From Generic UX to Specific UX

The development of HCI standards should meet the needs of the organization's business and technology strategy. Besides the "generic UX" related HCI standards (e.g., UI design, methods, metrics, etc.), we also develop specific HCI standards as follows:

- *Technology-Driven Approach*: as the pace of technological change continues to accelerate, people are looking for HCI design guidance, and we have published timely and relevant HCI

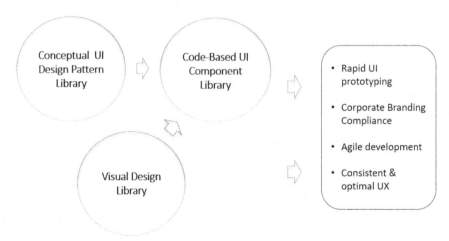

FIGURE 31.3 Usable and reusable UI assets (the three libraries).

standards, such as UI Design Standards for Mobile Solutions and UX Design Guide for Web Responsive Design.

- *Business Strategy-Driven Approach:* cloud computing and cost control enables a business to seek SaaS/vendor solutions. To minimize the UX risk, we published "UX Design Guidelines for SaaS/Vendor Solutions," where key activities, such as, "projects must incorporate UX into the vendor selection scorecard" and "projects must conduct a UX assessment by following a standardized UX assessment template" are defined. Some UX/HCI design requirements can also be directly cited in the purchase contracts to help minimize UX risk.

From a One-Way Push Model to a Two-Way Participatory Model

In the past, HCI standards were published in a push model; that is, the HCI organization published standards as Word/PDF documents, with the assumption that the content of the standard would be understandable and easily adopted by the developers/designers. Over time, we realized that there were problems with this one-way push model, and gradually deployed the following two-way user participatory model. The new approach has helped the adoption of the HCI standards and allowed the owners of the standard to receive feedback for improvement.

- *A User Feedback Mechanism*: a mechanism was set up by leveraging the internal social media platform, where developers/designers can make comments on a standard item or a design pattern and provide rating scores.
- *A User Contribution Mechanism:* for the UI Conceptual Design Pattern Library and the Code-Based UI Components Library, developers and designers can contribute their own assets once validated through project usability activities (e.g., usability testing), and the HCI TWG will send a reward for the contributions. This approach has helped the growth of the libraries and reduced the need for development resources.

ENHANCING GOVERNANCE OF THE HCI STANDARDS

To a large extent, the value of HCI standards will only be realized if they are applied to design and development, which largely relies on whether a rigorous governance process is set up. There is an industry-wide consensus that a governance model enables an organization to achieve a desired objective (Kiskel, 2011).

Level of Compliance

As shown in Figure 31.1, one of the typical challenges in governing HFE standards is either the developers misunderstand the content of the standards (requirements vs. recommendations) or the author of the HFE standards typically fails to clarify the level of guidance (Reeds et al., 1999). As a result, the developers/designers don't know whether they must or should comply with the standards, making it more difficult. We have adopted the RFC 2119 system that standardizes the keywords to indicate the requirement levels of standards, to address the challenge (Bradner, 1997), specifically on the following tag system illustrated in Figure 31.2:

- *Must Use*: means that the standard/guidance is required to be used
- *Must Not Use:* means that the standard/guidance should not be used
- *Should Use:* means that the standard/guidance is recommended, but not enforced
- *Should Not Use:* means that the standard/guidance is not recommended to be used
- *May Use:* means that the standard/guidance is optional

Governance Tracking Process

One of the successful strategies is to follow the requirements defined in the standards throughout the design and development process, don't wait until the end to think about standards (Reeds et al.,

1999). We have integrated the governance of these HCI standards into the existing development process with a tracking process defined. The tracking process of these HCI standards requires projects to be checked or audited at a few key checkpoints along the development process to ensure their conformance. However, in reality, it is not easy to implement all at once. Over time, progress has been made within the organization—from a manual process, followed by a semi-automated tracking process for certain domains, and then transition to a fully automated governance tracking process (which is still not fully realized at this point).

- *Manual Review/Audit Process*: the HCI TWG creates an HCI standards compliance checklist (for the "must use" requirements defined, refer to Figure 31.2) used in Solutions Architecture Review Meetings (weekly). The architecture review meetings are held based on a mature software governance process, which helps facilitate the governance of HCI standards.
- *Semi-Automated Tracking Process*: for a few specific domains/platforms (e.g., mobile solutions), a semi-automated workflow-based tracking process is used. This process sets up three domain-specific reviewers for decision-making prior to release, including the domains of security, architecture, and UX/HCI. All mobile projects must activate their accounts of this tracking process at the beginning of development. All three review experts across the three domains must finish their respective reviews, document, and sign off their approval before the project can move to the next steps until release to production. The entire flow is facilitated through email notifications among the three parties.
- *Strategic Review*: besides the tactical reviews discussed above, the HCI TWG chair attends IT Technical Review Committee (TRC) meetings as a standing member. This is to ensure that major technical decisions across the whole organization will include HCI/UX considerations.

Handling Exceptions and the Waiver Process

There are always exceptions and waiver requests from project teams. We have set up a sub-process to handle these. Per the process, a project team can request a waiver for compliance with a mandatory requirement specified in these HCI standards by filling out a request form with their supporting reasons and an expiration date for the waiver. The leader of the technical governance committee reviews and dispositions the request (approval and the expiration date if a waiver is granted, or disapproval).

Integrating UX Architecture Artifacts into the BDAT Framework

In the enterprise architecture (software) world, The Open Group Architecture Framework (TOGAF) model has been followed for practices with a relatively good governance process implemented (TOGAF, 2019). The TOGAF model defines the architecture and governance across the business, data, applications, and technology (BDAT) domains. Leveraging off the TOGAF model has helped the HCI community integrate HCI architecture artifacts into the BDAT domains.

- *Defining UX Architecture Artifacts:* key UX architecture artifacts are defined, including the following three:
  - "Experience Gaps and Needs" defines the user's pain points when using the "as is" solutions and the needs for the "to be" solutions. The content can be delivered in a format of personas and/or user journey maps. The UX architect provides input as requirements into Use Case Diagram and Process Transition Roadmap architects of the business architecture domain.
  - "UX prototype" is used as an approach to facilitate the project's iterative activities through UX/UI prototyping, usability testing, and design improvement to minimize the UX risk in solutions.

- "UX Roadmap" defines the user's needs over a certain period in terms of capabilities or usages. We plan to integrate this UX artifact for greater influence on strategy.

- *Integrating UX Artifacts into BDAT Architecture Governance:* technical architecture governance is mature. For major programs (e.g., digital transformation programs), the UX architecture artifacts are integrated into the Business Architecture (i.e. Experience Gaps and Needs, UX Roadmap) and Application Architecture domains (i.e. UX Prototype).

LEVERAGING OF AN ORGANIZATION UX MATURITY MODEL

To support the unified UX strategy, we built an organization UX maturity model with two purposes: (1) a roadmap-based guidance for an organization (program-level) aspiring for a UX-driven organization and (2) an assessment tool of UX capabilities for an organization (program-level) to identify the existing gaps and needs for further steps. Table 31.2 shows a high-level view of the UX maturity model; detailed requirements have been documented as checklists within each cell of the model. Specifically, the UX governance of the HCI standards—along with the organization's UX strategy, user-centered process, and UX metrics—are included in the model. This is intended to reinforce the implementation of the enterprise HCI standards. There is still more work needed to fully realize the benefits of the UX maturity model.

CONCLUSIONS

In summary, the development of enterprise HCI standards that complement the international/national HCI standards and will maximize the value of the HCI standards across all levels is greatly needed. Specifically, our practices in developing and governing enterprise HCI standards demonstrate that strategic approaches are needed to effectively maximize the influences of enterprise HCI standards through the transformation of HCI design assets, enhanced governance, and the maturation of the UX.

As we move forward, we face the following challenges. First, we are entering the era of artificial intelligence (AI) technology while almost all the HCI standards across international, national, and enterprise levels have been developed specifically for non-AI-based products. There is a need to develop HCI standards in design and methodology, especially at the enterprise level (Xu, 2019).

Secondly, an agile development process has been widely adopted in practice, but the existing HCI standards were developed primarily based on the traditional product development lifecycle

TABLE 31.2

The Summary of the Organization UX Maturity Model

	L1: Exploration	L2: In Transition	L3: Sustainable	L4: Proficient	L5: Ideal
Organization UX Strategy	None or limited	Developing	Gaining Commitment	Influencing	Transforming
UX Governance	Optional & voluntary	Growing	Practicing	Proactive	Collaborative
User-Centered Design Process	Limited & Inconsistent	Learning	Adopting	Improving	Integrated
UX Metrics	None or inconsistent	Project-specific	Org. centric & standardized	Continuous improvement	Automated & holistic

(e.g., waterfall). Although people claim that the flexible and iterative nature of ISO 9241-210 makes it a good basis for both UX design and an agile development process (Maguire, 2017; ISO, 2017), we do observe the difficulty of integrating HCI process/methods into an agile process, and better-known methods are needed.

Finally, to further integrate HCI/UX architecture artifacts into the enterprise architecture framework (i.e. the BDAT framework) to maximize the HCI work benefits, we feel more work is needed—such as building a solid relationship between UX artifacts and existing key enterprise architecture artifacts like business use case diagram, application functional requirement, and solution architecture—so that the gaps and needs identified in the UX artifacts (e.g., personas) can be traceable and trackable to ensure that they are addressed in the solution.

REFERENCES

ANSI/HFES 100. (2007). *Human factors engineering for computer workstations.* https://www.hfes.org/publications/other-publications/ansihfes-100–2007-human-factors-engineering-of-computer-workstations

Bevan, N. (2001). International standards for HCI and usability. *International Journal of Human Computer Studies,* 55(4), 533–552

Bevan, N., Carter, J., Earthy, J., Geis, T., Harker, S. (2016). New ISO standards for usability, usability reports and usability measures. In M. Kurosu (Ed.): *Proceedings of International Conference of Human Computer Interaction 2016, Part I, LNCS 9731,* Toronto: Springer, pp. 268–278, 2016. DOI: 10.1007/978-3-319-39510-4_25.

Bradner, S. (1997). *Key words for use in RFCs to indicate requirement levels.* https://www.ietf.org/rfc/rfc2119.txt

HFES 200. (2008). *Human factors engineering of software user interfaces.* https://www.hfes.org/resources/technical-standards/hfesansi-standards

Bevan, N. (2005). Guidelines and standards for web usability. In *Proceedings of HCI International 2005,* Lawrence Erlbaum, Mahwah, NJ.

Reed, P., Holdaway, K., Isensee, S., Buied, E., Fox, J., Williams, J., Lund, A. (1999). User interface guidelines and standards: progress, issues, and prospects. *Interacting with Computers,* 12: 119–142.

Green, P. (2002). Why safety and Human Factors/Ergonomics standards are so difficult to establish. In D. de Waard, K.A. Brookhuis, J. Moraal, and A. Toffetti (2002), *Human factors in transportation, communication, health, and the workplace.* Maastricht, the Netherlands: Shaker Publishing.

IEEE (1993). *Recommended practice for graphical user interface drivability.* P1201.2 Balloting Draft 2.

ISO (2017). *ISO 9241-210: Ergonomics of human-system interaction – Part 210: Human-centred design for interactive systems.* Geneva: International Organisation for Standardisation.

Kiskel, J. (October 2011). *Governance model: Defined.* Cognizant 20-20 Insights.

Maguire, M. (2017). Using human factors standards to support user experience and agile design. *UAHCI'13: Proceedings of the 7th International Conference on Universal Access in Human-Computer Interaction: Design Methods, Tools, and Interaction Techniques for eInclusion – Volume Part I,* Las Vegas, NV, July 21–26, 2013. pp. 185–194. DOI: 10.1007/978-3-642–39188-0_20

Swaminathan, A., & Rantanen, E. (2014). Usability of human factors standards. *Proceedings of the Human Factors and Ergonomics Society Annual Meeting,* 58, 591–594. DOI: 10.1177/1541931214581125

TOGAF (2019). *The TOGAF Standard, Version 9.2 overview.* https://www.opengroup.org/togaf

Xu, W. (2014). Enhanced ergonomics approaches for product design: A user experience ecosystem perspective and case studies. *Ergonomics,* 57(1): 34–51.

Xu, W. (2019). Toward human-centered AI: A perspective from human-computer interaction. *ACM Interactions,* 26(4): 42–46. doi.org/10.1145/3328485.

Xu, W., Furie, D., Mahabhaleshwar, M., Suresh, B., & Chouhan, H. (2019). Applications of an interaction, process, integration, and intelligence (IPII) design approach for ergonomics solutions. *Ergonomics,* 62(7): 954–980. doi.org/10.1080/00140139.2019.1588996.

Section VI

Management of Occupational Safety and Health

32 The Benefits of Occupational Health and Safety Standards

Denis A. Coelho, João C. O. Matias, and João N. O. Filipe

CONTENTS

INTRODUCTION

In this chapter, occupational health and safety is approached from the viewpoints of both Human Factors and Ergonomics and standardization, including management systems standards. The field of Occupational Health and Safety (OHS) is summarily characterized, and a historical perspective of its evolution is given, in relationship to Human Factors and Ergonomics (HFE). While several components of what makes the field of OHS up today pre-existed as specialized fields before, the foundation of the discipline of HFE (following World War II)—and some of the genesis of HFE, OHS, and HFE—have developed as two distinct, although somewhat overlapping, areas of activity. Links and commonalities can be found today between the two. Environmental conditions (such as noise, climate, or lighting) are an example of an area that is dealt with within the discipline of HFE and is also a central concern of OHS. While inadequate postures and movements at work and psychosocial factors impinge on OHS, HFE is equipped with knowledge and methods to perform the design of work and to deal with these work design factors. Regarding legislation, regulations, and standardization, OHS has gained a head start compared to HFE (the Occupational Safety and Health Act of 1970 in the U.S.A., the ILO R164 recommendation of 1981 at an international level, and the directive 89/391 on safety and health at work of 1989 in the EU), which supports the inclusion of this discussion on OHS standards within the realm of HFE standards, given the intersecting interests of the two disciplines. Regarding economic considerations, some HFE interventions have been the object of cost-benefit analyses reported in the literature. In what concerns OHS, the literature has focused on the cost of occupational accidents, injuries, illnesses, and fatalities. Estimates of costs and benefits of complying with (or adopting) OHS standards are also available in the form of literature dealing with the mandatory U.S. OSHA standards.

Central to the implementation of OHS is the assessment of hazards and risks and their resolution or reduction, besides the consideration of cost-benefit analyses, which may provide an important input in decision-making regarding alternative risk control strategies and measures. Despite the different legal statuses of specific OHS standards (whether legal or voluntary), statistical overviews of reported occupational fatalities, injuries, and illnesses in the U.S.A. and EU support the consideration of OHS as a paramount issue in organizations. This chapter focuses on OHS legislation, regulations, and voluntary standards in the EU and U.S.A., as well as management systems. Great emphasis is given to OHSAS 18001 as a specification for an OHS management system. It is seen as an effective tool for guiding the design of a continuously improved OHS management system in an organization. Lessons learned from quality management systems reinforce the interest in a management system in the OHS area of an organization. Emphasis is also given to the advantages of integrating management systems within an organization (quality, environmental, and OHS).

OCCUPATIONAL HEALTH AND SAFETY

Occupational Health and Safety (OHS) is a field of activity in organizations where several professional specialties collaborate. Safety engineering, occupational medicine, and ergonomics are principal specialties involved in the field of OHS. Assuring health and safety in the workplace is typically a joint effort of a number of professional activities. The purpose of OHS may be seen as assuring the good health and safety of people at work, thus preventing fatalities, injuries, and illnesses brought about by exposure to work-related health and safety hazards. Assuring OHS is accomplished by creating a controlled work environment. In the words of the EU Commission's (2002) strategy on health and safety at work between 2002 to 2006, this entails developing an approach that is both global and preventive, geared towards promoting well-being at work and going beyond the mere prevention of specific risks.

EVOLUTIONAL PERSPECTIVE OF OCCUPATIONAL HEALTH AND SAFETY

Work hazards to safety and health have been acknowledged throughout history. The loss of hearing by workers processing stone, bronze, and iron was already recorded in the Roman Empire 2000 years ago (Tytyk, 2004), but it was only during modern times—after the processes of industrialization in Europe and the United States—that occupational health and safety concerns began to be dealt with systematically. While ergonomists and human factor specialists only came onto the industrial scene after World War II, industrialization was to bring the advent of a set of professions whose activities would foster improvements to the health and safety conditions of workers. The principles of work organization, introduced by Frederic W. Taylor (1856–1915) in American Industry, promoted the division of industrial labor into monotonous tasks with repetitive movements, alienating the operators from their work (Saha, 1998). At that time, factory owners often viewed workers as unreliable, inefficient, and requiring supervision and force, while children, along with women and men, labored long hours in unsafe, unhygienic conditions (Tytyk, 2004). A number of new industrial professions developed as a response to the need for productivity improvements and profit gains. The professions of work hygienists, illumination technicians, and occupational physicians were some of the novel specialties that appeared during industrialization. The activity of these professionals promoted improvements to the otherwise appalling work safety and health conditions of factory workers.

At an international level, the consciousness of the importance of preventing occupational injuries and illnesses led to action starting in the early 20th century. Since 1919, the International Labour Conference has been issuing, for ratification by member countries, a set of conventions and recommendations concerning occupational safety and health and the working environment. Starting with a fragmented, patchy coverage of specific hazards in selected sectors of activity (e.g., white lead in paint—1921, marking of weight for packages transported by vessels—1929, or safety provisions in building—1937), the scope of the instruments issued gradually enlarged. The occupational safety and health recommendation of 1981 (ILO R164: 1981) encourages a systematic approach to prevention, extended to all branches of economic activity and all categories of workers. Currently, the International Labour Organization (ILO, the first specialized agency of the United Nations, established in 1946) collaborates with the International Ergonomics Association (IEA) in developing the forthcoming ILO instrument on ergonomics and the prevention of musculoskeletal disorders. The two institutions are also working together in developing the IEA/ILO publications, "Checkpoints on ergonomics" on a cross-section of industry application areas (Caple, 2004). This ongoing collaboration provides evidence of widespread recognition of the fact that effective application of ergonomics in work design while promoting a balance between worker characteristics and task demands provides the worker safety and physical and mental well-being.

Relationship Between Human Factors and Ergonomics and Occupational Health and Safety

As a scientific discipline and a professional specialty, HFE is somewhat younger than others involved in OHS. This notwithstanding, HFE professionals are specially requested to deal with specific OHS problems of the working environment. In an article about changes in OHS education, Dijk (1995), coming from an occupational medicine background, mentions safety experts, occupational physicians, occupational nurses, occupational hygienists, work and organization specialists, and occupational physiotherapists as the professionals collaborating, at that time in the Netherlands, in the activities of providing OHS services to organizations. Ergonomists and work psychologists were expected to shortly join the aforementioned professionals in applying their selves to OHS problems, particularly those dealing with work-related disorders like musculoskeletal disorders and mental breakdown, burn-out, and the delayed consequences of psychic trauma as an effect of aggressive or depressing events at work. Assertively, in a review of occupational risk management, Zimolong (1997) acknowledged that a workplace design that does not take ergonomic principles into account is likely to lead to an increase in errors and accidents.

In a report about OHS in Spain, Sesé et al. (2002) consider a number of OHS aspects of the workplace that are used to characterize the working environment: physical exposure (noise, vibration and thermal stress), postures and movements (lifting and handling heavy objects, repetitive movements, and pain or tiring postures), exposure to chemical substances, carcinogens, neurotoxic chemicals, and infectious biological agents. They also considered psychosocial factors (e.g., work rate, violence at work, or monotonous work), as well as accidents and occupational diseases (produced by chemical agents, skin diseases produced by substances and chemical agents, produced by inhalation of substances, infectious and parasitic occupational diseases, produced by physical agents and systemic diseases). A number of these categories are HFE themes and objects of study, with an array of ergonomic methods of intervention and design applicable in these areas.

These include, from the OHS categories listed, the physical environment, postures and movements, and psychosocial factors. Interestingly, standards concerned with the physical environment have generally not been produced under the heading of ergonomics. This has often led to a technical or engineering approach, where standards have concentrated on the physical aspects of the environment with little detail concerning human responses or ergonomics methods (Parsons, 1995). In this regard, it is important to emphasize that HFE is different from most bodies of knowledge that have supported the discipline. While the primary purpose of anthropology, cognitive science, psychology, and sociology is to understand and model human behavior, the main purpose of ergonomics is design (Helander, 1997). Still, research in ergonomics can be both applied and basic, since it is also often necessary to engage in basic research to seek support for design activities—in doing so, ergonomics has created a sizable body of knowledge, one that is constantly being enlarged (Coelho, 2012). The involvement of HFE in assuring OHS at the workplace, whether at the design stage or in post-design interventions, is hence, appropriate and desirable given the relevance of the knowledge and methods of HFE to many important dimensions of OHS.

Within organizations, opportunities for collaboration between OHS and HFE may be many. Given the closeness of the two disciplines and their overlapping activities, the links between them can serve as the basis upon which to build synergistic operational programs within organizations. In a study discussing ergonomic design principles and programs in terms of a loss management viewpoint, Amell, Kumar, and Rosser (2001) show how an Occupational Injury and Illness surveillance program may be used in determining whether an ergonomic intervention is required and evaluating the intervention. Notwithstanding a preventive approach to ergonomics and to OHS, the approach Amell, Kumar, and Rosser (2001) present is admittedly suited to workplaces and work activity where ergonomic principles were not employed in the design stage.

Economics of Interventions

The economic worth of HFE and OHS interventions has been the object of analysis. Beevis and Slade (1970) identified a dozen cases where ergonomic benefits had been expressed in financial terms and resulted in cost savings. The authors stated that the examples available at the time were scarce and that few of those published included economic data. In their opinion, this reflected the difficulty of converting the usual ergonomic criteria of human performance into costs—for example, when assigning a monetary value to a reduction in the incidence of mistakes or accidents. Publishing on the same theme 33 years later, Beevis (2003) verifies that assigning costs and benefits to ergonomic applications remains very arduous. Nevertheless, ergonomists should be able to discuss its potential benefits with clients or project managers. To that end, Beevis (2003) points out that, when ergonomics is considered as part of a risk-reduction strategy, it may be possible to estimate the reduction in the risks of costs or to estimate the reduction of the risk of loss of anticipated benefits. In the process-safety field, cost-benefit analysis of risk-reduction systems is done by examining the reduction in expected annual loss from the implementation of those systems and comparing this with the annual cost of implementing such systems to determine whether the cost is justified (Antes, Miri, & Flamberg, 2001). This approach to cost-benefit analysis is quantitative and, as such, does not consider intangible factors like the public and personnel perception of alternative risk-reduction systems (Antes, Miri, & Flamberg, 2001). Moreover, the result of any cost-benefit analysis is only as accurate as the assumptions made to determine the costs of implementation and the anticipated benefits. In establishing the economic case for HFE or OHS interventions, estimating the economic value of the anticipated benefits is bound to be more troublesome than estimating the cost of implementing the applications or recommendations.

In assessing the cost/benefits of human factors, Rouse and Boff (1997) consider the variation in aspects—such as human performance, accidents, or absenteeism—as determinants of the likelihood of benefits resulting from improvements in health and safety at the workplace. Translating those determinants into economic terms, in the context of a specific organization, could be a daunting task, especially if overall processes of the organization have not been characterized and work tasks are not standardized. A literature search on the theme of costs and benefits in respect to OHS measures returned a few studies that focused, however, on the cost of occupational accidents, injuries, illnesses, and fatalities (Dembe, 2001; Leigh, Cone, & Harrison 2001; Weil, 2001; Rikhardsson & Impgaard, 2004). In these studies, the analysis of those costs (to society at large, to companies, or to individuals) is used to establish a case for the implementation of preventive measures to safeguard health and safety at the workplace. The existence of occupational fatalities, injuries, and illnesses also raises moral questions, especially when the former is viewed as an element of cost-benefit analyses. Another study (Seong & Mendeloff, 2004) deals with the methods used to estimate the benefits of the U.S. OSHA (mandatory) safety standards. The controversial OSHA Ergonomic Program Standard of the United States (it was several years in the making and generated controversy at almost every step along the way, being overturned by the U.S. Congress in 2001) is the theme of an article (Biddle & Roberts, 2004) that deals with the cost-effectiveness of the program. In these two last studies, the starting point is the comparison between the costs incurred by employers at large in implementing the standards (at a nation-wide level in the U.S.) and the cost savings and prevention of injuries and fatalities brought about by the implementation. The values presented in both studies are based on OSHA estimates and the methods used to arrive at those estimates are also analyzed. In preparation for the promulgation of a new mandatory occupational safety and health standard, OSHA seeks information on the costs and benefits of various potential preventive strategies intended to face the particular hazards involved in the standard (US-DOL, Fact Sheet No. OSHA 92-14). While establishing the cost-effectiveness of ergonomic and OHS mandatory standards is not a straightforward task, in the U.S., endeavors are carried out to that end by taking nationwide statistical values into account to produce cost and benefit estimates for OSHA standards. These estimates corroborate the cost-effectiveness at a macroeconomic level of the regulations intended for compulsory implementation. Nevertheless, in the U.S., OSHA's estimates of the benefits of new mandatory standards are the object of debate because alternative methods and sources of compiled data can be used to establish those projections.

Large-scale losses in organizations, such as those arising from major fires or explosions, or otherwise involving loss of life, make the accidents immediately behind those losses very visible. The losses accruing from these accidents may, accordingly, be very thoroughly costed. Less well understood, however, is the nature and extent of loss from accidents of a more routine nature. Most accidents in industry typically injure but do not kill people, while damaging and interrupting production processes (EU Commission, 2004). The costs of this type of accident can often be extended to sick pay, increased insurance premiums, or maintenance budgets. Few organizations have the mechanisms to identify these extended costs separately and fewer still actually identify and examine the costs of accidents systematically (Rikhardsson & Impgaard, 2004). Many employers mistakenly believe that they are covered by insurance for most of the costs arising from accidents, but often, uninsured costs far exceed insured costs (EU Commission, 2004). Additionally, there are also intangible costs due, for example, to loss of business image, customer satisfaction, employee morale, or goodwill (Antes, Miri, & Flamberg, 2001). These intangibles are hard, if not impossible, to quantify in financial terms. For these reasons, accident costs are usually incomplete. Thus, it may be difficult to demonstrate the financial benefits of accident reduction measures, in some cases.

RISKS TO OCCUPATIONAL HEALTH AND SAFETY

A central goal in the implementation of OHS is the resolution or reduction of risks, which should follow a thorough identification of workplace hazards and the assessment of their degree of risk. Besides being hazards at the workplace, some events and situations may also represent threats to property integrity or have the potential to cause environmental damage. Hence, safety engineering is another discipline with intersecting activities in OHS, besides HFE. The hazards with which OHS is primarily concerned are those that can directly affect the well-being (health- and safety-wise) of people at work. A hazardous event or situation can have negative consequences to people at work, including illnesses (whether acute or chronic) and accidents causing physical injury (non-fatal or fatal), and may also have psychological impacts on people (e.g., anxiety, stress, or depression). An array of hazards can be present in the workplace. A provisional categorization of hazards to health and safety at work is shown in Table 32.1.

Risks may accrue differently from exposure to single hazards than with the interaction of hazards (e.g., interaction among dangerous substances and agents, or between these and physical exposures, as well as interactions between inadequately set factors in the working environment, whether of a physical or psychosocial nature). Moreover, many of the hazards are not directly perceivable and their negative consequences have to be inferred from knowledge, personal experience, or reliance on documentation (Zimolong, 1997). Preventive action is, hence, necessary, and organizations should handle both obvious and non-obvious hazards systematically and continuously. In planning for ongoing activities of hazard identification, risk assessment, and risk resolution or reduction (control), organizations should establish and maintain procedures pertaining to activities and facilities, which include, according to the requirements set out in the OHSAS 18001 specification (BSI-OHSAS 18001:1999BSI-OHSAS 18001 1999; 2007):

- routine and non-routine activities,
- activities of all personnel that has access to the workplace (including contractors and visitors), and
- facilities at the workplace, whether provided by the organization or others.

Procedures provide a description that can be used as an external memory in guiding activities according to the principles safeguarding OHS and, as such, are instruments for the management of risk. In general, measures taken within organizations for the management of risk should follow the principle of elimination of hazards where practicable, followed in turn by risk reduction (either by reducing the likelihood of occurrence or the potential severity of injury or damage), with the adoption of personal protective equipment (PPE) only as a last resort (BSI-OHSAS 18002:2002). The magnitude of risks can be assessed using the definition presented by Rodrigues and Guedes (2003). It estimates risk by considering it as the combination of the probability of a hazardous event or situation occurring and its severity in terms of seriousness of consequences. Defined in this way, risk is directly proportional to both probability and severity. The greater the probability of occurrence of a hazardous event or situation, or the greater the

TABLE 32.1

Provisional Categorization of Workplace Hazards to Health and Safety

Type of Hazard	Hazards Falling in This Category
Dangerous substances and agents	• exposure to chemicals, carcinogens, toxins, and infectious and parasitic biological agents
Physical exposures	• noise, vibration, radiation and inadequate illumination, and thermal environment • hazards associated with other (physical) ergonomic factors, such as working postures, workload, or materials handling
Integrity threats	• fire, explosion, and reactivity hazards • threats to physical integrity in the form of hazards that can result in cuts, falls, projections, crushing, asphyxiating, or electrocution
Psychosocial factors	• harassment and violence at work • inadequate job design in terms of work rate, job demands and content, repetitive work, mental workload, or interacting technology

severity of the consequences, the greater the risk. Likewise, the smaller the probability or the severity, the smaller the risk. Once an organization is able to appreciate all significant OHS hazards and to estimate their probability and severity, risks can then be classified, identifying those that must be eliminated as well as those that should be controlled by specific measures within the organization. According to Antes, Miri, and Flamberg (2001), organizations have many different methods for analyzing hazards and risks, with the most common being what-if analysis, checklist reviews, hazard and operability (HAZOP) procedures, and failure modes and effects analysis (FMEA). Once risks are identified and categorized, the organization must decide which preventive measures or mitigation strategies to implement. Cost-benefit analysis of alternative risk control measures can help organizations select the most effective means of risk resolution or reduction.

Risk control measures in OHS entail changes in the workplace and the work environment. These changes typically involve setting up signs, disseminating guidance materials, providing incentives, setting performance goals, establishing training programs, intervening in work design, and performing ergonomic (re)design of workplaces and work environments. The control of new and emerging OHS risks of a psychosocial nature, linked to social change, may necessitate intervention in other spheres, namely, human relations—both within the organizations and in a broader context. Emerging disorders such as stress, depression, anxiety, violence at work, harassment, and intimidation are becoming increasingly prominent in Europe in education, health, and social services (EU Commission, 2002). These disorders cannot be linked solely to exposure to a single particular hazard but are likely to result from the interaction of a wide set of factors, including the degree of acceptance of human diversity within organizations, hierarchical relations, work design, working time arrangements, and computer-related fatigue.

THE NEED FOR OCCUPATIONAL HEALTH AND SAFETY

OHS has progressed significantly in recent decades. The U.S. Occupational Safety and Health Administration stated in 2003 that it had, since its creation in 1970, driven the work-related fatality rate down by 62% and promoted a reduction in overall injury and illness rates of 42% (OSHA, 2003). Ten years later, from a reported 2013 U.S. fatalities count of 4,405 starting from an estimated 140 hundred in 1970, the annual fatality rate is cut by over two-thirds (this signifies a reduction from 38 to 12 fatal injuries per day, according to OSHA, 2014). An estimated annual cost of 170 billion U.S. dollars for occupational injuries and illnesses (OSHA, 2003) was irrefutable evidence that occupational accidents and work-related health problems are expensive in modern society; the estimate of this annual burden was updated to 250 billion U.S. dollars in 2011 (Leigh, 2011). According to Dupré (2001), Eurostat estimated that every year, occupational accidents in the EU resulted in 150 million lost working days and that a further 350 million days were lost because of work-related health problems (these figures are from 1998 and 1999 and a 15-country EU; EU27 data from Eurostat for 2007 indicated an estimated minimum of 367 million lost working days due to sickness). In economic terms, this meant a projected 2.6%–3.8% of the collective EU gross national production (GNP) was lost every year (EU OSHA, 2001). Statistical data on occupational accidents and illnesses in the U.S.A. (for the year 2002) and the EU (for the year 2000), shown in Table 32.2, provides insight into the dimension of the problem.

More than 5,000 people lost their lives at work every year in the European Union, and in the United States, fatalities directly related to work had a similar magnitude. In addition, OSHA (2003) estimated that perhaps as many as 50,000 workers died in the U.S. each year from illnesses where workplace exposures are a contributing factor. In the EU, statistics indicated that every year, more than 75,000 people were so severely disabled that they could no longer work (Saari, 2001). Injury insurance systems are designed to protect those injured and their dependants, but the costs of work-related accidents and illnesses extend not only to those directly affected (financially and emotionally) but also to employers and organizations (lost productivity and negative impact on the

TABLE 32.2

Overview of Occupational Accident/Injury and Health Problems/Illness Statistics for the EU and the USA

EU (15 Countries, 2000)[1]	USA (2002)[2]
Total number of employees covered in data	
142.2 million	137.7 million
Total accident-related deaths at work/Fatal work injury count	
5237	5525
Occupational accidents leading to more than three days absence from work: 4.8 million	Work-related injuries and illnesses that resulted in days away from work: 1.44 million
Total number of accidents, including those that did not involve absence from work: 7.6 million	Total non-fatal work-related injuries: 4.4 million
Total work-related health problems: 7.7 million	Total non-fatal work-related illnesses: 0.29 million

Notes

1 Source: Eurostat, European Social Statistics (EU-15 countries: Austria, Belgium, Denmark, Finland, France, Germany, Greece, Ireland, Italy, Luxembourg, Netherlands, Portugal, Spain, Sweden, and United Kingdom).
2 Source: US Department of Labor, Bureau of Labor Statistics (the figures shown for work-related injuries and illnesses only concern private industry).

public's perception) and to countries, given their significant impact on the economy. Ultimately, these costs fall on the citizens at large, as taxpayers and consumers.

Countries, organizations, and employees have a myriad of tools at hand for promoting health and safety at the workplace. Some of these are laws and regulations, which are mandatory and enforced by government agencies. Methods and standards that can guide organizations in complying with legislation, or in attaining performance levels in OHS above minimum legal requirements—including OHS management systems geared to continuous improvement—are also available. The sections of this chapter that follows focus on existing standards (both mandatory and voluntary) in the area of health and safety in the workplace in the United States and the European Union. The OHSAS 18001 specification, supporting the implementation of a system for the management of OHS within organizations, is also discussed.

OCCUPATIONAL HEALTH AND SAFETY LEGISLATION, REGULATIONS AND VOLUNTARY STANDARDS

Normative documents in OHS may be divided into three main types. Those that are legislative or regulatory are mandatory; standards that are intended for voluntary implementation by organizations; and internal procedures developed within organizations. In this section, legislation and voluntary standards will be dealt with in more detail than internal procedures. Still, the latter warrants some attention as possible precursors to standards, especially those developed within organizations of significant size and encourage prevention as a suitable approach to OHS. The benchmarking of "best in class" organizations may be at the origin of the diffusion of some internal procedures to other organizations, depending on the results evidenced by their original application and their usefulness. Internal OHS procedures may assume great importance in those cases where they become adopted as voluntary standards, although in a somewhat modified version. This version is usually prepared by a specialized technical committee within a standards publishing organization (for instance, European standards in ergonomics are produced by working groups of CEN TC 122 "Ergonomics"). The creation of voluntary standards may also take place in reaction to specific legislation, providing guidance in interpreting and implementing it. Mandatory standards (legislative and regulatory) in the EU and in the U.S., standards for voluntary implementation, and OHSAS 18001 constitute the focus of this section.

LEGISLATION AND REGULATIONS

The principle underlying the development of legislation and regulations in OHS is that all workers are entitled to work in healthy, safe, and hygienic conditions. Mandatory standards and legislation are developed to guarantee these rights, which assist each worker. The range of the success and benefits of implementation of mandatory standards and legislation may depend, however, on the thoroughness of inspection activities. Inspection can work as a motivation for the implementation of the minimum requirements set forth in legislation and regulations, but to that end, it ought to be effective not only at the initial licensing stage of businesses but also in continuity. Inspection can be decisive in reaching the objectives underlying the development of laws and regulations in OHS. If employers expect that they will not be inspected if they do not have a rooted culture of health and safety at work in their organization, they may be tempted to neglect OHS regulatory or legislative requirements. Health and safety at the workplace are also deeply rooted in an organization's culture. In this regard, it is recognized that inspectors have a crucial role to play as agents of change to promote better compliance—first through education, persuasion, and encouragement, then through increase in enforcement activities, where necessary (EU Commission, 2004).

Employers must apply all the minimum mandatory requirements to satisfy their obligations of assuring basic health and safety conditions to their employees. Employers have the duty to evaluate and control risks, not only in terms of safety but also regarding the workers' health. Therefore, they should adopt measures, give instructions, inform the workers or their representatives, and organize health and safety activities at work. Employees have the right to communicate to labor inspectors the situations or conditions where minimum OHS requirements have not been met. Besides their own rights, legal and regulatory provisions on OHS enunciate that employees also have duties, which include executing the instructions and recommendations of their employer, in addition to demonstrating a collaborative attitude in assuring OHS conditions.

Legislation and regulatory mandatory standards in the occupational health and safety domain deal with a wide range of areas. The aspects covered include the organization of prevention services, the organization of work, the protection of more vulnerable worker groups, the prevention of exposure to specific hazards, the prevention of industrial accidents, fire protection, and industry-specific preventive measures. In what follows, OHS legislation and regulation in the EU and the U.S. are presented separately. For the EU, the focus will be drawn on the EU directives on OHS, which are disseminated and adopted in the internal law of the EU member states. For the U.S.A., the role of OSHA in developing mandatory standards promulgated by the US Congress will be presented.

Legislation on Safety and Health at Work in the EU

Prevention is the guiding principle for occupational health and safety legislation in the European Union. According to a communication from the EU Commission (2004), the goal of instilling a culture of prevention rests on the double foundation that the minimum requirements provide a level playing field for businesses operating within the large European domestic market and provide a high degree of protection to workers, avoiding pain and suffering and minimizing the income foregone for enterprises by preventing occupational accidents and diseases. Hence, to avoid accidents and occupational diseases, EU-wide minimum requirements for health and safety protection at the workplace have been adopted. In the member states of the European Union, OHS legislation comes from two different sources. One is the legislation developed by each member state in its internal law. The other is the obligatory adoption (transposition) of EU directives to the internal law of every member state. Because the legislative systems covering safety and health at the workplace differed widely among the individual member states and needed improvement, the Council of the European Union adopted, by means of directives, minimum requirements for encouraging improvements to guarantee the protection of the health and safety of workers (e.g., directives 89/391/EEC, 94/33/EC). The adoption of these directives was not intended to permit any reduction in the levels of protection already achieved in individual member states prior to the transposition of the directives. They aimed, rather, at harmonizing national provisions on the subject, which often included technical specifications and self-regulatory standards, resulting in different levels of safety and health protection, and permitting competition at the expense of safety and health (directive 89/391/EEC). Hence, EU countries are committed to encouraging improvements in and harmonizing OHS conditions across member states.

Directive 89/391/EEC on the introduction of measures to encourage improvements in OHS is the European OHS framework directive. As such, it served as a basis for other (individual) directives—which, together with the framework directive, form a coherent whole. Directive 89/391/EEC lays down the principles for the introduction of measures to encourage improvements in the safety and health of workers and provides a framework for specific workplace environments, developed in individual directives. The OHS framework directive and the individual directives that spring from it are listed in Table 32.3. Besides the framework and its individual directives, there are still other complementary directives in the OHS domain, of which examples are given in Table 32.4.

The transposition of framework directive 89/391/EEC and of the individual directives on OHS to the internal law of EU member states has led to the harmonization of the legislative framework in OHS across the EU. The shift of paradigm represented by the EU health and safety legislation (which envisaged moving beyond a technology-driven approach to accident, injury, and illness prevention to a policy much more focused on the personal behavior and on organizational structures) had a major impact on the national OHS systems (EU Commission, 2004). The impact was greater in those countries that had either less developed legislation in the field or legislation based on corrective principles rather than on a preventive approach to assuring OHS (EU Commission, 2004). For the Nordic EU member states (Denmark, Finland, and Sweden), transposition did not require major adjustments since they already have rules in place that were in line with the OHS EU directives. In Austria, Belgium, France, Germany, the Netherlands, and the United Kingdom, the framework directive served to complete or refine existing national legislation. In the EU-15 countries of the South (Greece, Italy, Portugal, and Spain) as well as in Ireland and Luxembourg, the framework directive had considerable legal consequences because prior to its adoption, these countries possessed antiquated or inadequate legislation on the subject. Enlargement in the year 2004 brought in new member states to an EU—comprised by 25 countries then—and was further enlarged to EU-27. In many of the new member states, the prevention culture has yet to be rooted (EU Commission, 2004). Transposition of the EU directives on OHS to their internal law will constitute a decisive step toward instilling a generalized preventive approach to OHS. Member states that already benefited from the EU legislation to modernize their occupational health and safety rules emphasize the following innovative aspects of the framework directive (EU Commission, 2004):

- the broad scope of application, including the public sector,
- the principle of objective responsibility of the employer,
- the requirement that a risk assessment shall be drawn up and documented,
- the obligation to establish a prevention plan based on the results of the risk assessment,
- the recourse to prevention and protective services, and
- workers' rights to information, consultation, participation, and training.

Some member states also reported on difficulties in enforcing the legislation adopted as a result of transposing the EU directives. Although framework directive 89/391 recognizes the need to avoid imposing administrative, financial, and legal constraints that would hold back the creation and

TABLE 32.3
EU Framework Directive on Occupational Health and Safety and Its Individual Directives

Content	Directive
Framework: introduction of measures to encourage improvements in the safety and health of workers at work	89/391/EEC
Workplaces: minimum health and safety requirements for the workplace	89/654/EEC
Use of work equipment: minimum safety and health requirements for the use of work equipment by workers at work	2009/104/EC ([1])
Use of personal protective equipment: minimum health and safety requirements for the use by workers of personal protective equipment at the workplace	89/656/EEC ([2])
Manual handling: minimum health and safety requirements for the manual handling of loads where there is a risk particularly of back injury to workers	90/269/EEC
Work with screen display equipment: minimum safety and health requirements for work with screen display equipment	90/270/EEC
Carcinogens: protection of workers from the risks related to exposure to carcinogens at work	90/394/EEC ([3])
Biological agents: protection of workers from risks related to exposure to biological agents	2000/54/EC
Safety signs: minimum requirements for the provision of safety and/or health signs at work	92/58/EEC
Pregnant workers: introduction of measures to encourage improvements in the safety and health at work of pregnant workers and workers who have recently given birth or are breastfeeding	92/85/EEC
Mineral-extracting industries (drilling): minimum requirements for improving the safety and health protection of workers in the mineral-extracting industries through drilling	92/91/EEC
Mineral-extracting industries: minimum requirements for improving safety and health protection of workers in surface and underground mineral-extracting industries	92/104/EEC
Fishing vessels: minimum safety and health requirements for work onboard fishing vessels	93/103/EC
Chemical agents: protection of the health and safety of workers from the risks related to chemical agents at work	98/24/EC
Explosive atmospheres: minimum requirements for improving the safety and health protection of workers potentially at risk from explosive atmospheres	99/92/EC
Physical agents—vibration: minimum health and safety requirements regarding the exposure of workers to the risks arising from physical agents (vibration)	2002/44/EC
Physical agents—noise: minimum health and safety requirements regarding the exposure of workers to the risks arising from physical agents (noise)	2003/10/EC
Temporary or mobile construction sites: implementation of minimum health and safety requirements at temporary mobile or construction sites	92/57/EEC
Transport activities: minimum safety and health requirements for transport activities and workplaces on means of transport	–([4])

Notes
1 Replaces 89/655/EEC (modified by directives 95/63/EC and 2001/45/EC).
2 Amended by directives 93/68/EEC, 93/95/EEC, and 96/58/EC.
3 Modified by directive 97/42/EC and amended by directive 99/38/EC.
4 This is not a directive but an amended proposal for a council decision.

development of SMEs, problems in implementing this legislation in SMEs in practice have been reported (EU Commission, 2004). These are ascribed to specific administrative obligations, formalities, and financial burdens, as well as the time required to develop appropriate measures, which gave rise to a certain negative reaction from SMEs, especially in Belgium, Denmark, Germany, the Netherlands, Sweden, and the United Kingdom. The overriding objective of framework directive

TABLE 32.4

Examples of Other EU Directives in the Domain of OHS (the Directives Shown Here Are Not Individual Directives of Framework Directive 89/391/EEC)

Content	Directive
Temporary workers: supplements the measures to encourage improvements in the safety and health at work of workers with a fixed-duration employment relationship or a temporary employment relationship	91/383/EEC
Medical treatment onboard vessels: minimum safety and health requirements for improved medical treatment onboard vessels	92/29/EEC
Young People: protection of young people at work	94/33/EC
Electrical equipment for use in potentially explosive atmosphere in mines susceptible to firedamp	98/65/EC

89/391 and that of its individual directives is achieving a high level of protection of the safety and health of workers. This goal can only be met if all actors involved (employers, workers, workers' representatives, and national enforcement authorities) cooperate to pursue the efforts that are necessary to attain a comprehensive, effective, and correct application of the content of the EU directives on OHS. EU directives also acknowledge that technical standards constitute an indispensable reference in adopting procedures and measures to meet legal requirements in OHS.

Mandatory Standards on Safety and Health at Work in the USA

In the U.S.A., the Occupational Safety and Health Act of 1970 authorizes the Secretary of Labor, through the Occupational Safety and Health Administration (OSHA), to set mandatory occupational safety and health standards applicable to businesses affecting interstate commerce (US-DOL, Fact Sheet No. OSHA 92-14). In its website (http://www.osha.org), OSHA states that its mission is to assure the safety and health of American workers by setting and enforcing standards; providing training, outreach, and education; establishing partnerships; and encouraging continual improvement in workplace safety and health. OSHA regulations are mandatory standards designed to reduce on-the-job injuries and limit the risk of developing occupational diseases. According to Fact Sheet No. OSHA 92-14 of the U.S. Department of Labor, the impetus to develop a new standard can come from several sources, including public petitions, the U.S. Congress, information from governmental departments, referral from the Environmental Protection Agency, OSHA's own initiative or requests from OSHA advisory committees. In preparing a new standard, OSHA seeks information to determine the extent of a particular hazard or group of hazards, investigates currently used and potential protective measures, and estimates the costs and benefits of various protective strategies. OSHA uses several sources of information, including surveys, meetings with employers and employer groups, and focus group discussions with workers from many plants and industries across the U.S.A. OSHA also consults with the interest groups potentially affected by an envisaged new standard, including industry and labor, which meet to hammer out agreements serving as the basis for a proposed rule.

Individual states in the U.S.A. are encouraged to establish and maintain their own job safety and health programs only if they are "at least as effective" as federal standards; these are subject to federal approval. States are also allowed to develop standards covering areas or issues not regulated by OSHA's (federal) standards, but only if they are "required by compelling local conditions" and will not "unduly burden interstate commerce" (US-DOL, Fact Sheet No. OSHA 92-14). Interestingly, during the first two years following its creation, the OSHA was authorized to promulgate national consensus standards and other federal standards as OSHA standards. National consensus standards came from voluntary standards developed by organizations such as the

American National Standards Institute and the National Fire Protection Association (US-DOL, Fact Sheet No. OSHA 92-14).

Most OSHA regulations (standards-29 CFR) cover hazards that may be present in a wide set of industries. These standards are compiled as the OSHA General Industry Standards (standards-29 CFR—part 1910—Occupational Safety and Health Standards). The non-reserved subparts of these standards are listed in Table 32.5. OSHA also promulgated other standards in OHS, some of which apply to a single industry, with examples shown in Table 32.6. Besides Part 1910 and the industry-specific parts shown in Table 32.6, OSHA has produced a number of other regulations that are relevant to its activity. Examples are shown in Table 32.7.

OSHA is not limited to creating and enforcing standards. Because the agency was allegedly burdening employers with rules, inspections, and penalties, it decided to adopt an alternative approach that includes providing on-site consultative assistance to employers who want help in establishing OHS conditions, offering a choice for employers. According to OSHA (2003), voluntary, cooperative relationships among employers, employees, unions, and the OSHA can be a useful alternative to traditional OSHA enforcement and an effective way to reduce worker deaths, injuries, and illnesses. The OSHA also provides guidelines and other voluntary standards to assist in implementing regulations and improving OHS conditions beyond minimum regulatory requirements.

VOLUNTARY STANDARDS IN OCCUPATIONAL HEALTH AND SAFETY

Voluntary standards play a crucial role in the adoption and implementation of procedures and measures required by applicable legislation and regulations in the OHS domain. Budgetary limitations and consequent resource and staff shortages are at the root of the difficulty the national organisms responsible to

---

TABLE 32.5

Subparts of OSHA's General Industry Standards, Excluding Reserved Subparts (Standards-29 CFR—part 1910—Occupational Safety and Health Standards)

Subpart	Heading
A	General
B	Adoption and Extension of Established Federal Standards
D	Walking-Working Surfaces
E	Means of Egress
F	Powered Platforms, Manlifts, and Vehicle-Mounted Work Platforms
G	Occupational Health and Environmental Control
H	Hazardous Materials
I	Personal Protective Equipment
J	General Environmental Controls
K	Medical and First Aid
L	Fire Protection
M	Compressed Gas and Compressed Air Equipment
N	Materials Handling and Storage
O	Machinery and Machine Guarding
P	Hand Portable Powered Tools and Other Hand-Held Equipment
Q	Welding, Cutting, and Brazing
R	Special Industries
S	Electrical
T	Commercial Diving Operations
Z	Toxic and Hazardous Substances

TABLE 32.6
Some Industry Specific OSHA Standards

Title	Standards—29 CFR
Construction: Safety and Health Regulations for Construction	Part 1926
Agriculture: Occupational Safety and Health Standards for Agriculture	Part 1928
Maritime sectors:	
Occupational Safety and Health Standards for Shipyard Employment	Part 1915
Marine Terminals	Part 1917
Safety and Health Regulations for Longshoring	Part 1918

TABLE 32.7
Other OSHA Regulations (Standards—29 CFR)

Title	Standards—29 CFR
State Plans for the Development and Enforcement of State Standards	Part 1902
Inspections, Citations, and Proposed Penalties	Part 1903
Recording and Reporting Occupational Injuries and Illness	Part 1904
Rules of Procedure for Promulgating, Modifying, or Revoking OSHA Standards	Part 1911
Advisory Committees on Standards	Part 1912
Safety Standards Applicable to Workshops and Rehabilitation Facilities	Part 1924
Safety and Health Standards for Federal Service Contracts	Part 1925
Discrimination against Employees under the OSHA Act of 1970	Part 1977
Identification, Classification, and Regulation of Carcinogens	Part 1990

produce laws, and rulings must accompany scientific and technological development. While the adoption of voluntary standards in OHS does not automatically confer immunity from legal obligations, some voluntary standards serve as guidelines for interpretation and consolidation of specific OHS regulations or legislation. This leads to a complementarity between legislation/regulations and voluntary standards in several activities. Voluntary standards are sound technical solutions to safety and health problems without creating additional cost and operation burdens to governments (McCabe, 2004). Legislators and regulators impel voluntary safety and health standards because they may also be introduced more easily and quickly than laws and regulations. Despite the voluntary adhesion to this kind of standard, these are procedures, rules, and measures recognized and accredited by specialized organisms. Voluntary standards are accredited by national and international organisms, such as ANSI (American National Standards Institute), AFNOR (Association Française de Normalisation), BSI (British Standards Institute), DIN (Deutsches Institut für Normung), ISO (International Standards Organization), or CEN (European Committee for Standardization). These organizations—and others of the same kind acting in other countries and regions of the world—develop standards internally and adopt standards with a national, foreign, or international origin. There are many voluntary standards applicable to OHS. Their coverage spans from personal protective equipment to fire and explosion, safety of machinery, evaluation methodologies (vibration, acoustics, noise), signals and safety signs, limit-values of exposure, and management of OHS within organizations. The development and accreditation of voluntary standards in OHS in the US and the EU are discussed in the following separate sections. Examples of voluntary standards in OHS are also given.

Voluntary Standards in Occupational Health and Safety in the US

In the U.S.A., no single government agency has control over standards. Some bodies of the U.S. government require companies to use certain voluntary standards, but there is no federal government agency that controls them. The standards are created by many different organizations and are not ratified or governed by any agency. It is the American National Standards Institute (ANSI), a private, non-profit organization, which administers and coordinates the U.S. voluntary standardization and conformity assessment system. ANSI is not a government agency or a regulatory body. According to the information presented on its website (http://www.ansi.org), ANSI itself does not develop American National Standards, but it provides all interested U.S. parties with a neutral venue to come together and work toward common agreements. ANSI approves the process by which standards become accepted. Accreditation by ANSI signifies that the procedures used by a standards developing body satisfy the essential requirements for American National Standards. ANSI also states on its website that it has currently around 270 accredited standard developer groups representing approximately 200 distinct organizations in the private and public U.S. sectors who cooperatively work to develop voluntary national consensus standards and American National Standards. In the domain of safety, health, and protection of the workers, several developers are active. Some examples of ANSI-accredited standards within the OHS domain are listed in Table 32.8, which also shows the origin organization of each of the standards listed.

Voluntary Standards in Occupational Health and Safety in the EU

There are three European standard organizations that are recognized as competent to adopt harmonized standards at the level of the EU. The European Committee for Standardization (CEN), the European Committee for Electrotechnical Standardization (Cenelec), and the European Telecommunications Standards Institute (ETSI) are the three bodies that meet the requirements expressed in EU directives. Some European Standards were developed in support of the policy, adopted in 1985, which stipulated that legislative harmonization in the form of EU directives should be limited to the essential requirements and that writing of the detailed technical specifications necessary for the implementation of directives was entrusted to the European voluntary standards organizations. On behalf of governments, of the European Commission, or of the EFTA Secretariat, the European standards organizations can be requested to develop standards. In the more recent EU directives, such as directive 98/37/EC (safety of machinery), voluntary standards accredited by CEN and Cenelec are explicitly recommended as a valid tool in implementing the requirements set forth in the directives. These voluntary standards are listed for each of the

---

TABLE 32.8

Examples of ANSI Accredited Standards in the Domain of OHS, with an Indication of Their Original Developer

Standard	Developer
ANSI Z87.1-2003—Occupational and Educational Personal Eye and Face Protection Devices	ASSE—American Society of Safety Engineers
ANSI N13.15-1985—Dosimetry Systems, Performance of Personnel Thermoluminescence	HPS—Health Physics Society
NFPA 70E-2004—Standard for Electrical Safety Requirements for Employee Workplaces	NFPA—National Fire Protection Association
UL 2351-2004—Standard for Safety for Spray Nozzles for Fire Protection Service	UL—Underwriters Laboratories

applicable EU Directives in the European Commission's website that is dedicated to enterprises (http://europa.eu.int/comm/enterprise/). Compliance with harmonized standards does not, however, translate into immunity from legal obligations. Caution should be exercised since CEN and Cenelec harmonized standards may not cover all health and safety requirements of a given directive. Thus, lawful obligations should always be assessed independent from the analysis of the requirements of the voluntary standards.

According to the information posted on its website (http://www.cenorm.be), CEN aims to draw up voluntary European standards and promote corresponding conformity of products and services in areas other than electrical and telecommunications, which are covered by the activity of Cenelec and ETSI. CEN's mission is to promote voluntary technical harmonization of standards in Europe in conjunction with worldwide bodies and its European partners, having as purpose to diminish trade barriers, promote safety, allow interoperability of products, systems, and services, and promote common technical understanding. In this way, EU member states and EFTA countries subsequently adopt the standards adopted by CEN. Table 32.9 presents some European standards on OHS and examples of corresponding national standards resulting from the adoption of the former.

The exchange of standards at an international level, with the consequent adoption of voluntary standards developed by organisms other than the European Standards Organizations, is evident in some of the European standards. As an example, standard EN ISO 12100–2:2003 (safety of machinery—basic concepts, general principles for design—part 2: technical principles) was adopted as a European standard from an original ISO standard. The same standard is adopted by European countries and transposed to their national standardization systems (e.g., BS EN ISO 12100–2:2003, in the U.K.). The national standards institutions of European countries also produce their own, which do not come from the harmonization process led by CEN, Cenelec, or ETSI, nor do they proceed from ISO (e.g., German standard DIN 58214:1997—Eye-protectors—Helmets—Terms, forms, and safety requirements). National standards institutions also adopt standards not previously drawn up by any of the three European standards organizations, directly from international standards.

MANAGEMENT OF OCCUPATIONAL HEALTH AND SAFETY

Traditionally, occupational health and safety management entailed a little more than reacting to accidents and work injuries as part of a reactive culture in workplace health and safety.

TABLE 32.9

Examples of CEN Accredited European Standards in the Domain of OHS and Corresponding National Standards Resulting from Their Transposition in France, Germany, and the UK

European Standard	France: AFNOR	Germany: DIN	UK: BSI
EN 54–12:2002—Fire detection and fire alarm systems—Part 12: Smoke detectors—Line detectors using an optical light beam	NF EN 54-12	DIN EN 54-12	BS EN 54–12:2002
EN 169: 2002—Personal eye-protection—Filters for welding and related techniques—Transmittance requirements and recommended use	NF EN 169	DIN EN 169	BS EN 169: 2002
EN 294:1992—Safety of machinery—Safety distance to prevent danger zones being reached by the upper limbs	NF EN 294	DIN EN 294	BS EN 294:1992
EN 13861:2002—Safety of machinery—Guidance for the application of ergonomics standards in the design of machinery	NF EN 13861	DIN EN 13861	BS EN 13861:2002

However, over time, it was possible to verify improvements in OHS conditions, especially because of legislation and regulations that have been developed in this domain. In the EU, legislation has clearly had a positive influence on OHS conditions. It has contributed to instilling a culture of prevention, despite some flaws—which, according to the EU Commission (2004), still holds back the achievement of the full potential of this legislation. Throughout the EU, attitudinal changes concerning the behavior and the awareness of people must take place at a large scale to fully implement in practice the concept of prevention in the management of OHS within organizations. In the U.S., OSHA has been developing and enforcing mandatory OHS standards for more than 40 years, promoting during this time a very significant reduction in work-related fatalities, injuries, and illnesses. Despite these achievements, the turn of the millennium U.S. statistics spelled out a stark reality, as did the statistics from the EU (Table 32.2). An impetus for further improvements could come from further disseminating a culture of systematic approach to OHS, where all the people in the organization are included, in a process of continual improvement. These are undoubtedly necessary, especially in moving beyond the traditional reactive approach to a preventive systematic approach in the management of OHS in organizations. Differences between a systematic approach to the management of OHS in organizations and the traditional reactive approach are shown in Table 32.10, after Bottomley (1999). An OHS management system that puts a systematic approach to prevention into practice can be part of the management systems of any organization. Such a management system can provide a set of procedures and tools that make the management of risks to health and safety at work more efficient and enable organizations to improve performance by controlling their risks to OHS.

In implementing an OHS program based on a policy of prevention and continuous improvement, a fundamental issue to be dealt with is, according to Zimolong (1997), how to provide the long-term commitment of employees and management. Participative management and a suitable leadership style may be useful to ensure the cooperation of all members, promoting commitment and involvement in OHS activities over time. There are several factors that may motivate organizations to implement an OHS management system in such a way that leads to acquiring a culture of prevention and continuous improvement. OHS regulations and legislation in the U.S. and in the EU encouraged prevention and continuous improvement through periodic revision of hazards and resolution or control of risks to workplace safety and health. Organizations may also have learned lessons from implementing quality management and environmental management systems, through ISO 9001 and ISO 14001 certification, respectively. Benefits of ISO 9001 certification have been

TABLE 32.10

Differences Between a Reactive and a Systematic Approach to the Management of OHS in Organizations, Adapted from Bottomley (1999)

A Reactive Approach to the Management of OHS in Organizations Entails That...	A Systematic Approach to the Management of OHS in Organizations Entails That...
Hazards are dealt with reactively	Hazards are identified preventively
Risk controls are dependent on individuals	Risk controls are described in procedures
Risk controls are not linked to each other	Risk controls are linked by a common method
OHS activity happens but is not planned	OHS activity is planned
Controls are monitored and reviewed regularly	Controls are reviewed after an incident
Responsibilities are not defined	Responsibilities are defined for everyone
Focus is only given to risks on the organization's own "backyard"	Public and supplier risks are managed in a planned way
There is no company policy on OHS to communicate	Company policy on OHS is communicated

reported in literature (Bryde & Slocock, 1998; Lipovatz, Steno & Vaka, 1999; Neergaard, 1999). Important benefits motivated by certification to ISO 9001 include disseminating a systematic approach that promotes consistency in operations and controlled documentation, encourages teamwork, personnel training, and improved motivation and organization. Some problems linked to ISO 9001 certification were also reported, including its high cost, the extra documentation that is necessary, and the resistance to change of some people in the organization (Quazi & Padibjo, 1998).

A review of empirical research, which was conducted in the U.K. by Wright (1998), identified what motivates managers to proactively manage OHS (using a planned and systematic approach). The review suggested that there are two decisive factors that motivate both SMEs and large organizations to initiate OHS improvements. These factors are the fear of losing credibility and a belief that it is necessary and morally correct to comply with OHS regulations. Other factors identified were the aim to improve staff morale and productivity, the extension of modern concepts such as Total Quality Management to all areas of responsibility, and the integration of OHS into quality and environmental management systems. The need to avoid or reduce the immediate costs associated with ill health and injuries could also be a very strong motivating factor. In the EU, many enterprises excuse the absence of an OHS management system because of its implementation costs (EU Commission, 2004). But this should not be the case because EU legislation (EU framework directive 89/391/EEC) does not call for sophisticated management systems. It simply encourages applying basic management principles in the field of OHS. The size of the organization and its economic sector of activity are not critical in determining the potential use of an OHS management system, although according to Bottomley (1999), these may determine the manner and style of implementation.

The implementation of management systems based on standards and specifications in organizations at a worldwide level had an incremental development over the years. Quality management systems based on the ISO 9001 standard were very much disseminated. A similar path was followed in certification to ISO 14001 (environmental management system) by organizations. The OHS management system based on the Occupational Health and Safety Assessment Series 18001 was also implemented in organizations at a worldwide level. In many of these organizations, quality and environmental management systems were already in place or were implemented simultaneously with the implementation of the OHS management system. An OHS management system based on OHSAS 18001 provides a set of procedures and tools that enable an efficient management of OHS, promoting continual improvements in organizational performance through the ongoing processes of hazard identification, risk assessment, and control of risks to OHS.

OCCUPATIONAL HEALTH AND SAFETY MANAGEMENT SYSTEMS: SPECIFICATION AND GUIDELINES FOR IMPLEMENTATION

The demand for a recognizable OHS management system standard, against which management systems could be assessed and certified, was at the origin of the development of the OHSAS (Occupational Health and Safety Assessment Series) 18001 specifications. OHSAS 18001 was developed collaboratively by a group representing standards institutions and other organizations with concerns on OHS, with an international and national base (Germany, Ireland, Japan, Mexico, Norway, Singapore, South Africa, Taiwan, and the U.K.). It was issued in several countries and languages, in some cases as a standard and in others as a specification.

The OHSAS 18001 specification (BSI-OHSAS 18001:2007) and its accompanying guidelines for implementation (BSI-OHSAS 18002:2000, 2008) entail a preventive and proactive approach to the management of OHS. Emphasis is given to ensuring the identification, evaluation, and control of risks to OHS in a continuously improved manner. In what follows, the applicability and envisaged benefits of the OSHAS 18001 specification for OHS management systems are illustrated. An overview of the requirements set out in the specification is given and the compatibility of an OHS management system with other management systems in organizations is also discussed.

Applicability and Envisaged Benefits

The OHSAS 18001 specification is deemed applicable to organizations that may wish to establish an OHS management system to eliminate or minimize risk to employees and other interested parties who may be exposed to OHS risks associated with its activities (BSI-OHSAS 18001:2007). This OHSAS specification is applicable to the implementation, maintenance, and continuous improvement of an OHS management system. An organization may also implement it to assure itself of its conformance with its stated OHS policy or demonstrate such conformance to others (BSI-OHSAS 18001:2007). Organizations seeking certification/registration of their OHS management systems by an external organization, or willing to make a self-determination and declaration of conformance with this OHSAS specification, may also choose to implement it. Although the requirements in OHSAS 18001 are intended to be incorporated into any OHS management system, the extent of their application will depend on the OHS policy of the organization, the nature of its activities, and the risks and complexity of its operations (BSI-OHSAS 18001:1999, 2007).

The specification does not state specific OHS performance criteria nor does it give detailed specifications for the design of a management system, but the advantages of certification in OHSAS 18001 demonstrate the business and social benefits of health and safety improvements. The British Standards Institute produced some case studies that are posted on its website (http://emea.bsi-global.com/OHS/CaseStudies/). The U.K. Health and Safety Executive has also published on its website (http://www.hse.gov.uk) a series of case studies setting out the business case for good health and safety management. The case studies prepared by these two institutions from the U.K. cover a variety of industry sectors, including pharmaceutical, construction, manufacturing, petrochemical, aerospace, defense, utilities, food, and service companies, as well as organizations in the public sector. The benefits that are reported in the case studies spring from improvements in operational efficiency, productivity, and public image, to a reduction in lost working days, accidents, and medical claims; from recognition by insurers and heightened worker satisfaction to involvement in assuring OHS.

Overview of Requirements

There are 18 requirements set out in the OHSAS 18001specification. Table 32.11 presents a summary of all the requirements in the OHSAS 18001 specification. The guidelines for the implementation of OHSAS 18001 (BSI-OHSAS 18002:2000, 2007) explain the requirements, detailing their intent and the typical inputs, process, and outputs that can be used in implementing each of the requirements. Two of the 18 requirements portrayed in OHSAS 18001 are discussed in this section. There is a requirement for defining an occupational health and safety policy statement. This requirement establishes the overall sense of direction and sets the principles of action for an organization. This policy must be authorized by the top management and should clearly state overall health and safety objectives and a commitment to improving health and safety performance. It should also be consistent with other management policies of the organization (e.g., pertaining to quality or environmental aspects). The OHS policy statement of the organization should be reviewed periodically to ensure that it remains relevant and appropriate to the organization. The policy should be communicated to all employees and to other groups or individuals affected by the OHS performance of the organization. In many countries, OHS legislation and regulations demand consultation and participation of employees in their organization's OHS management systems.

Another requirement set out in the OHSAS 18001 specification concerns establishing and maintaining procedures for the ongoing identification of hazards, the assessment of risks, and the implementation of necessary control measures. The OHSAS 18001 specification and the guidelines

TABLE 32.11

Overview of Requirements Set Out in the OHSAS 18001 Specification, According to OHSAS 18001:2007

OHSAS 18001 Requirement on	Some Elements of the Requirement
(4.1) General requirements	The organization shall establish and maintain an OHS management system (…)
(4.2) OHS policy	The top management shall define and authorize the organization's OHS policy and ensure that within the defined scope of its OHS management system (…)
(4.3) Planning	
(4.3.1) Planning for hazard identification, risk assessment, and risk control	The organization shall establish, implement, and maintain a procedure(s) for the ongoing hazard identification of risk assessment and determination of necessary controls. (…)
(4.3.2) Legal and other requirements	The organization shall establish, implement, and maintain a procedure(s) for identifying and accessing the legal and other OHS requirements that are applicable to it. (…)
(4.3.3) Objectives and program (s)	The organization shall establish, implement, and maintain documented OHS objectives, at relevant functions and levels within the organization. (…)
(4.4) Implementation and operation	
(4.4.1) Resources, roles, responsibility, and accountability	The top management shall take ultimate responsibility for the OHS and the OHS management system (…)
(4.4.2) Competence, training, and awareness	The organization shall ensure that any person(s) under its control performing tasks that can impact on OHS is (are) competent on the basis of appropriate education, training, or experience, and shall retain associated records. (…)
(4.4.3) Communication, participation, and consultation	With regard to its OHS hazards and the OHS management system, the organization shall establish, implement, and maintain a procedure(s) for. (…)
(4.4.4) Documentation	The OHS management system documentation shall include (…) a. the OHS policy and objectives; b. description of the scope of the OHS management system; c. description of the main elements of the OHS management system and their interaction, and reference to related documents; d. documents, including records, required by this OHSAS Standard.
(4.4.5) Control of documents	Documents required by the OHS management system and by this OHSAS Standard shall be controlled (…)
(4.4.6) Operational control	The organization shall determine those operations and activities that are associated with the identified hazard(s) where the implementation of controls is necessary to manage the OHS risk(s) (…)
(4.4.7) Emergency preparedness and response	The organization shall establish, implement, and maintain a procedure(s): a. to identify the potential for emergency situations; and b. to respond to such emergency situations. (…)

(Continued)

TABLE 32.11 (Continued)

OHSAS 18001 Requirement on	Some Elements of the Requirement
(4.5) Checking	
(4.5.1) Performance measurement and monitoring	The organization shall establish and maintain procedures to monitor, and measure OHS performance on a regular basis (...)
(4.5.2) Evaluation of compliance	Consistent with its commitment to compliance [see 4.2c)], the organization shall establish, implement, and maintain a procedure(s) for periodically evaluating compliance with applicable legal requirements (see 4.3.2). The organization shall evaluate compliance with other requirements to which it subscribes (see 4.3.2). The organization may wish to combine this evaluation with the evaluation of legal compliance referred to in 4.5.2.1 or to establish a separate procedure(s) (...)
(4.5.3) Incident investigation, nonconformity, corrective action, and preventive action	The organization shall establish, implement, and maintain a procedure(s) to record, investigate, and analyze incidents. (...) The organization shall establish, implement, and maintain a procedure(s) for dealing with actual and potential nonconformity (ies) and for taking corrective action and preventive action (...)
(4.5.4) Control of records	The organization shall establish and maintain records as necessary to demonstrate conformity to the requirements of its OHS management system and of this OHSAS Standard, and the results achieved (...)
(4.5.5) Internal audit	The organization shall ensure that internal audits of the OHS management system are conducted as planned (...)
(4.6) Management review	The top management shall review the organization's OHS management system, at planned intervals, to ensure its continuing suitability, adequacy, and effectiveness. (...)

for its implementation do not make recommendations on how these activities should be conducted but establish principles by which the organization can determine whether its processes are suitable and sufficient. The hazard identification, risk assessment, and risk control processes form the core of the whole OHS management system, enabling the organization to continuously identify, evaluate, and control its OHS risks. These processes should be carried out with respect to normal operations within the organization, and in what concerns abnormal operations. The latter include periodic or occasional operations such as plant cleaning and maintenance, as well as plant start-ups and shut-downs. Potential emergency conditions should also be covered in hazard identification, risk assessment, and risk control processes. Depending on the competencies that exist within the organization, it may be necessary for the organization to seek external advice or services in support of conducting these processes. Following the performance of the processes of hazard identification, risk assessment, and risk control, any corrective or preventive actions identified as necessary should be monitored to assure their timely completion. The results of the processes of hazard identification, risk assessment, and risk control should also be considered as input for the establishment of revised or new OHS objectives in the organization.

Compatibility with Other Management Systems

The OHSAS 18001 specification was developed to be compatible with ISO 9001 and ISO 14001 standards. Reviews are to be done whenever new revisions to either are made to ensure continuing compatibility and facilitate the integration of quality, environmental, and OHS management systems in organizations. In the year 2000, a revision to ISO 9001 was done (the earlier version was dated from 1994), which promoted a greater approximation to the Total Quality Management philosophy. Emphasized, among others, was continuous improvement, which is required in a formal manner by the three sets of standards. In this revision1, emphasis was also given to the management of resources, through the comprehensive treatment of elements such as information, communication, infrastructure, and work environment. In what concerns the latter, a change was made in the 2000 version of the standard for a quality management system toward emphasizing human resources within the organization, with the introduction of a new concept of working environment. The importance of human resources and their working environment for the quality of products is explicitly emphasized. There are several requirements that are common to the three management systems, such as prevention, conformance, continuous improvement, process management, or leadership. For these reasons, organizations running certified ISO compliant quality and environmental management systems—or, in the process of implementing these—can more easily implement a system of management of occupational health and safety based on the OHSAS 18001 specification.

Some organizations may benefit from having integrated management systems; others may prefer to adopt different systems based on similar management principles. Given the absence of a single management system standard or specification that brings together the currently separate standards and specifications for quality management, environmental management, and occupational health and safety management, an alternative way to reap the benefits of a simultaneous approach to the management of the three systems could be pursued. It consists of integrating these systems into practice inside the organization (Matias & Coelho, 2002, 2011; Coelho & Matias, 2010). The benefits of such integration should be greater than the sum of the partial benefits of the independently managed systems to be of worth for organizations. Disadvantages should also be relatively smaller. There may also be barriers to such integration. Given that an organization has functioned with the systems as separate entities, fear of change may exist and manifest itself as opposed to the integration, which will affect the company organization. This situation is clearly more relevant for organizations that have previously attained success with separate management systems. Favorable arguments for the integration of the different management systems spring from economies of scale in integrating information channels and documentation. Information bottlenecks can pre-exist due to communication problems occurring in disperse management systems. These disperse systems in the organization may, however, share common goals, such as continuous improvements or prevention of nonconformance and accidents. With a unified management system, the interests of all the organization's stakeholders (employees, customers, shareholders, suppliers, and society) would be more conveniently satisfied (Figure 32.1).

CONCLUSION

Assuring OHS can be accomplished by creating a controlled work environment, where risks to health and safety have been resolved or are under control. The involvement of HFE in assuring OHS at the workplace—whether at the design or in post-design interventions—is appropriate and desirable, given the relevance of the knowledge and methods of HFE to many important dimensions of OHS. Within organizations, opportunities for collaboration between OHS and HFE may be many. Effective application of ergonomics in work design, while promoting a balance between worker characteristics and task demands, promotes worker safety and physical and mental well-being. In establishing the economic case for HFE or OHS interventions, estimating the economic value of the anticipated benefits is bound to be more troublesome than estimating the cost of implementation. Similarly, while establishing the cost-effectiveness of ergonomic and OHS mandatory standards is not a straightforward task, endeavors are carried out to that end in the U.S. by taking nationwide statistical values into account to produce cost and benefit estimates for OSHA standards. At the level of organizations, the costs of accidents are usually incompletely assessed,

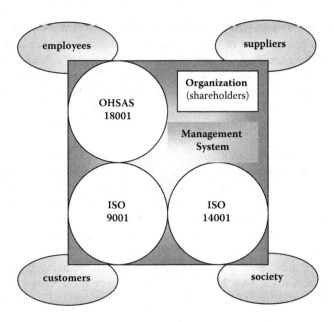

FIGURE 32.1 The organization's stakeholders and the integration of certified management systems within the organization.

when comparing them with the costs of accident prevention to determine the benefits of accident reduction. The costs of work-related accidents and illnesses extend not only to those directly affected but also to organizations and countries, given their significant impact on the economy. Ultimately, these costs fall on the citizens at large, as taxpayers and consumers. Preventive action in OHS is undoubtedly necessary, and organizations should handle both obvious and non-obvious hazards systematically and in a continuous manner.

OHS regulations and legislation in the U.S.A. and EU encourage prevention and continuous improvement through periodic revision of hazards and resolution or control of risks to workplace safety and health. Through the adoption of EU directives on OHS to their internal law, EU countries are committed to encouraging improvements in OHS conditions and harmonizing OHS conditions across the EU. In the EU, legislation has clearly had a positive influence on OHS conditions. It has contributed to instilling a culture of prevention, although further disseminating a culture of a preventive and systematic approach to OHS is still needed, moving beyond the traditional reactive approach. Inspectors have a crucial role to play as agents in promoting compliance to OHS legislation and regulations, not only through enforcement activities but also through education, persuasion, and encouragement. People also have a very important role in promoting their safety and health at their workplace. If people can agree about something, then they are more likely to do it than if they are simply told to. A proactive attitude toward OHS should not only be seen as a concern of management or as a law or regulation that must be obeyed, but it is recommended for everyone in an organization.

Voluntary standards assume an important role in the development and improvement of work conditions, because they complement mandatory standards and legislation, supporting their implementation. EU directives acknowledge that technical standards constitute an indispensable reference in adopting procedures and measures to meet legal requirements in OHS. In the U.S., OSHA provides guidelines and other voluntary standards to assist in implementing regulations and improving OHS conditions beyond minimum regulatory requirements set forth in OSHA mandatory standards.

An OHS management system based on OHSAS 18001 provides a set of procedures and tools that enable an efficient management of OHS, promoting continual improvements in organizational performance through the ongoing processes of hazard identification, risk assessment, and control of risks to OHS. The advantages of certification in OHSAS 18001 demonstrate the business and social benefits of health and safety improvements. These include improvements in operational efficiency, productivity and public image, reduction in lost working days, accidents and medical claims, recognition by insurers, and heightened worker satisfaction and involvement in assuring OHS.

The acceptance and dissemination of ISO 9001 and ISO 14001 certified management systems—and the benefits they promoted in organizations—opened the path for the implementation and certification of other management systems in organizations. The OHSAS 18001 specification was developed to be compatible with ISO 9001 and ISO 14001 standards, with several requirements that are common to the three management systems, including prevention, continuous improvement, and process management. Organizations running ISO-compliant quality and environmental management systems—or are in the process of implementing them—can, thus, more easily implement a system of management of occupational health and safety based on OHSAS 18001. This set of three management systems can be integrated within the organization, with benefits springing from integrating information channels and documentation.

REFERENCES

Amell, T. K., Kumar, S., & Rosser, B. W. J. (2001). Ergonomics, loss management, and occupational injury and illness surveillance. Part 1: elements of loss management and surveillance. A review. *International Journal of Industrial Ergonomics*, *28*, 69–84.

Amended proposal for a Council Decision concerning the minimum safety and health requirements for transport activities and workplaces on means of transport—Individual Directive within the meaning of Article 16 of Directive 89/391/EEC, COM/93/421FINAL—SYN 420, Official Journal C 294, 30/10/1993, p. 4.

ANSI N13.15-1985 – *Dosimetry systems, performance of personnel thermoluminescence*, HPS—Health Physics Society, McLean, VI.

ANSI Z87.1-2003 – *Occupational and educational personal eye and face protection devices*, ASSE - American Society of Safety Engineers, Des Plaines, IL.

Antes, M. K., Miri, M. F., & Flamberg, S. A. (2001). Selection and design of cost-effective risk reduction systems. *Process Safety Progress*, *20*(3), 197–203.

Beevis, D., & Slade, I. M. (1970). Ergonomics – Costs and benefits. *Applied Ergonomics*, *1*(2), 79–84 (republished in Applied Ergonomics, 34 (2003), 413–418).

Beevis, D. (2003). Ergonomics – Costs and benefits revisited. *Applied Ergonomics*, *34*, 491–496.

Biddle, J., & Roberts, K. (2004). More evidence on the need for an ergonomic standard. *American Journal of Industrial Medicine*, *45*, 329–337.

Bottomley, B. (1999). *Occupational health and safety management systems: Strategic issues report*. National Occupational Health and Safety Commission, Canberra, Australia. Available online at http://www.nohsc.gov.au

Bryde, D. J., & Slocock B. (1998). Quality management systems certification: A survey. *International Journal of Quality and Reliability Management*, *15*(5), 467–480.

BS EN ISO 12100-2:2003, *Safety of machinery - Basic concepts, general principles for design - Part 2: Technical principles*, British Standards Institute, London, UK.

BSI-OHSAS 18001:1999, *Occupational health and safety assessment series: Occupational health and safety management systems – Specification, Incorporating Amendment No. 1*, British Standards Institution (13 December 2002), London, UK.

BSI-OHSAS 18001:2007, *Occupational health and safety assessment series: Occupational health and safety management systems – Requirements*, British Standards Institution (first published July 2007), London, UK.

BSI-OHSAS 18002:2000, *Occupational health and safety management systems – Guidelines for the implementation of OHSAS 18001*, British Standards Institution (13 December 2002), London, UK.

BSI-OHSAS 18002.2007, *Occupational health and safety management systems – Guidelines for the Implementation of OHSAS 18001*, British Standards Institution (revision first published in 2007), London, UK.

Caple, D. C. (2004). IEA international development. In: *Proceedings of the Ergonomics Congress of Portuguese Speaking Countries*. Funchal, Portugal, July 26–28, 1994. CD-ROM, APERGO – Portuguese Ergonomics Association, Lisbon, Portugal.

Coelho, D. A. (2012). Inaugural editorial: A new human factors and ergonomics journal for the international community is launched. *International Journal of Human Factors and Ergonomics*, *1*(1), 1–2.

Coelho, D. A., & Matias, J. C. (2010). Innovation in the organisation of management systems in Portuguese SMEs. *International Journal of Entrepreneurship and Innovation Management*, *11*(3), 324–329.

Council Directive 94/33/EC of 22 June 1994 concerning the protection of young people at work, Official Journal L 216, 20/08/1994, pp. 12–20.

Council Directive 96/58/EC amending for the third time Council Directive 89/686/EEC, Official Journal L 236, of 18/9/1996, p. 44.

Commission of the European Union Directive 98/65/EC of 3 September 1998 adapting to technical progress Council Directive 82/130/EEC on the approximation of the laws of the Member States concerning electrical equipment for use in potentially explosive atmospheres in mines susceptible to firedamp, Official Journal L 257, 19/09/1998 pp. 29–34.

Council of the European Communities Directive 89/391/EEC of 12 June 1989 concerning the introduction of measures to encourage improvements in the safety and health of workers at work, Official Journal L 183, 29/06/1989, pp. 1–8.

Council of the European Communities Directive 89/654/EEC of 30 November 1989 concerning the minimum safety and health requirements for the workplace, Official Journal L 393 of 30/12/1989, pp. 1–12.

Council of the European Communities Directive 89/655/EEC of 30 November 1989 concerning the minimum safety and health requirements for the use of work equipment by workers at work, Official Journal L 393, 30/12/1989, pp. 13–17.

Council of the European Communities Directive 89/656/EEC of 30 November 1989 concerning the minimum health and safety requirements for the use by workers of personal protective equipment at the workplace, Official Journal L 393, 30/12/1989, pp. 18–28.

Council of the European Communities Directive 90/269/EEC of 29 May 1990 concerning the minimum health and safety requirements for the manual handling of loads where there is a risk particularly of back injury to workers, Official Journal L 156, 21/06/1990, pp. 9–13.

Council of the European Communities Directive 90/270/EEC of 29 May 1990 concerning the minimum safety and health requirements for work with display screen equipment, Official Journal L 156 of 21/06/1990, pp. 14–18.

Council of the European Communities Directive 90/394/EEC of 28 June 1990 concerning the protection of workers from the risks related to exposure to carcinogens at work, Official Journal L 196 of 26/07/1990, pp.1–7.

Council of the European Communities Directive 91/383/EEC of 25 June 1991 supplementing the measures to encourage improvements in the safety and health at work of workers with a fixed-duration employment relationship or a temporary employment relationship, Official Journal L 206, 29/07/1991, pp. 19–21.

Council of the European Communities Directive 92/29/EEC of 31 March 1992 on the minimum safety and health requirements for improved medical treatment on board vessels, Official Journal L 113, 30/04/1992, pp. 19–36.

Council of the European Communities Directive 92/57/EEC of 24 June 1992 concerning the implementation of minimum safety and health requirements at temporary or mobile construction sites, Official Journal L 245, 26/08/1992, pp. 6–22.

Council of the European Communities Directive 92/58/EEC of 24 June 1992 Concerning the minimum requirements for the provision of safety and/or health signs at work, Official Journal L 245 of 26/08/1992, pp. 23–42.

Council of the European Communities Directive 92/85/EEC of 19 October 1992 on the introduction of measures to encourage improvements in the safety and health at work of pregnant workers and workers who have recently given birth or are breastfeeding, Official Journal L 348, 28/11/1992, pp. 1–8.

Council of the European Communities Directive 92/91/EEC of 3 November 1992 concerning the minimum requirements for improving the safety and health protection of workers in the mineral-extracting industries through drilling, Official Journal L 348 of 28/11/1992, pp. 9–24.

Council of the European Communities Directive 92/104/EEC of 3 December 1992 concerning the minimum requirements for improving the safety and health protection of workers in surface and underground mineral-extracting industries, Official Journal L 404 of 31/12/1992, pp. 10–25.

Council of the European Union Directive 93/68/EEC amending the Council Directive 89/686/EEC, Official Journal L 220, of 30/8/1993, p. 1.

Council of the European Union Directive 93/95/EEC amending for the second time Council Directive 89/686/EEC, Official Journal L 276, of 9/11/1993, p. 11.

Council of the European Union Directive 93/103/EC of 23 November 1993 concerning the minimum safety and health requirements for work on board fishing vessels, Official Journal L 307 of 13/12/1993, pp. 1–17.

Council of the European Union Directive 95/63/EC of 5 December 1995 amending Directive 89/655/EEC, Official Journal L 335 of 30/12/1995, pp. 28–36.

Council of the European Union Directive 97/42/EC of 27 June 1997, Official Journal L 179 of 08/07/1997, pp. 4–6, (amended in Official Journal C 123 of 22/04/1998, p. 21).

Council of the European Union Directive 98/24/EC of 7 April 1998 on the protection of the health and safety of workers from the risks related to chemical agents at work, Official Journal L 131, 05/05/1998, pp. 11–23.

Council of the European Union Directive 99/38/EC of 29 April 1999 amending Directive 90/394/EEC, Official Journal L 138, 01/06/1999 pp. 66–69.

Dembe, A. E. (2001). The social consequences of occupational injuries and illnesses. *American Journal of Industrial Medicine*, 40, 403–417.

Dijk, F. J. H. van (1995). From input to outcome: changes in OHS education. *Safety Science*, 20, 165–171.

DIN 58214: 1997 – *Eye–protectors – Helmets – Terms, forms and safety requirements*, Deutsches Institut für Normung (DIN), Berlin, 1997.

Directive 98/37/EC of The European Parliament and of The Council of 22 June 1998 on the approximation of the laws of the Member States relating to machinery, Official Journal of the European Communities L 207, 23/07/1998, pp. 1–46.

Directive 2003/10/EC of the European Parliament and of the Council of 6 February 2003 on the minimum health and safety requirements regarding the exposure of workers to the risks arising from physical agents (noise), Official Journal L 042, 15/02/2003, pp. 38–44.

Dupré, D. (2001). Statistics spell it out. *Magazine of the European Agency for Safety and Health at Work*, 4, 5–7. http://osha.eu.int

EN 54-12:2002 – *Fire detection and fire alarm systems – Part 12: Smoke detectors – Line detectors using an optical light beam*, European Committee for Standardization (CEN), Brussels, Belgium.

EN 169:2002 – *Personal eye–protection – Filters for welding and related techniques – Transmittance requirements and recommended use*, European Committee for Standardization (CEN), Brussels, Belgium.

EN 294:1992 – *Safety of machinery – Safety distance to prevent danger zones being reached by the upper limbs*, European Committee for Standardization (CEN), Brussels, Belgium.

EN 13861:2002 – *Safety of machinery – Guidance for the application of ergonomics standards in the design of machinery*, European Committee for Standardization (CEN), Brussels, Belgium.

EN ISO 12100-2:2003 – *Safety of machinery – Basic concepts, general principles for design – Part 2: Technical principles*, European Committee for Standardization (CEN), Brussels, Belgium.

EU Commission. (2002). Commission of the European Communities COM(2002) 118 final. Adapting to change in work and society: a new Community strategy on health and safety at work 2002–2006 (Communication).

EU Commission. (2004). Commission of the European Communities COM(2004) 62 final. Communication on the practical implementation of the provisions of the Health and Safety at Work Directives 89/391 (Framework), 89/654 (Workplaces), 89/655 (Work Equipment), 89/656 (Personal Protective Equipment), 90/269 (Manual Handling of Loads) and 90/270 (Display Screen Equipment).

EU OSHA. (2001). Economic Impact of Occupational Safety and Health in the Member States of the European Union. European Agency for Safety and Health at Work.

European Parliament and Council of the European Union Directive 99/92/EC of 16 December 1999 on minimum requirements for improving the safety and health protection of workers potentially at risk from explosive atmospheres, Official Journal L 23, 28/01/2000, pp. 57–64.

European Parliament and Council of the European Union Directive 2000/54/EC of 18 September 2000 concerning the protection of workers from risks related to exposure to biological agents at work, Official Journal L 262, 17/10/2000 pp. 21–45.

European Parliament and Council Directive 2001/45/EC of the 27 June 2001 amending Council Directive 89/655/EC, Official journal L 195, 10/07/2001 p. 46–49.

European Parliament and Council Directive 2002/44/EC of the of 25 June 2002 concerning the minimum health and safety requirements regarding the exposure of workers to the risks arising from physical agents (vibration), Joint Statement by the European Parliament and the Council, Official Journal L 177, 06/07/2002 pp. 13–20.

Helander, M. G. (1997). The human factors profession. In G. Salvendy (Ed.). *Handbook of human factors and ergonomics*, 2nd Edition (pp. 3–16). New York: John Wiley & Sons, Inc.

ILO R164: 1981 – *Recommendation R 164 concerning occupational safety and health and the working environment*. International Labour Organization. Available online at http://www.ilo.org

ISO 9001: 2000 – *Quality management systems: Requirements*, International Organization for Standardization (ISO), Geneva, Switzerland.

ISO 14001: 1996 – *Environmental management systems – Specification with guidance for use*, International Organization for Standardization (ISO), Geneva, Switzerland.

Leigh, J. P., Cone, J. E., & Harrison, R. (2001). Cost of occupational injuries and illnesses in California. *Preventive Medicine, 32*, 393–406.

Leigh, J. P. (2011). Economic burden of occupational injury and illness in the United States. *Milbank Quarterly, 89*(4), 728–772.

Lipovatz, D., Steno F., & Vaka, A. (1999). Implementation of ISO 9000 quality systems in Greek enterprises. *International Journal of Quality and Reliability Management, 16*(6), 534–551.

Matias, J. C. O., & Coelho, D. A. (2002). The integration of the standards systems of quality management, environmental management and occupational health and safety management. *International Journal of Production Research, 40*(15), 3857–3866.

Matias, J. C. O., & Coelho, D. A. (2011). Integrated total quality management: Beyond zero defects theory and towards innovation. *Total Quality Management & Business Excellence, 22*(8), 891–910.

McCabe, J. (2004). *Voluntary safety standards: In the public interest?* CFA Consumer Assembly 2004. Washington DC, March 11. http://www.ansi.org.

Neergaard, P. (1999). Quality management: A survey on accomplished results. *International Journal of Quality and Reliability Management, 16*(3), 277–289.

NFPA 70E-2004 - *Standard for Electrical Safety Requirements for Employee Workplaces*, NFPA—National Fire Protection Association, Quincy, MA, USA.

Occupational Safety and Health Act of 1970. Public Law 91-596. 84 STAT. 1590. 91st Congress, S.2193. December 29, 1970, as amended through January 1, 2004. http://www.osha.gov

OSHA (2003). *All about OSHA occupational health and safety administration*. U.S. Department of Labor. Information booklet available online at http://www.osha.gov

OSHA (2014). *Worker fatalities Reported to federal and state OSHA*. U.S. Department of Labor. https://www.osha.gov/dep/fatcat/dep_fatcat.html

Parsons, K. C. (1995) Ergonomics of the physical environment – international ergonomics standards concerning speech communication, danger signals, lighting, vibration and surface temperatures. *Applied Ergonomics, 26*(4), 281–292.

Quazi, H. A., & Padibjo, S. R. (1998). A journey toward total quality management through ISO 9000 certification. *International Journal of Quality and Reliability Management, 15*(5), 489–508.

Rikhardsson, P. M., & Impgaard, M. (2004). Corporate cost of occupational accidents: An activity-based analysis. *Accident Analysis and Prevention, 36*, 173–182.

Rodrigues, C., & Guedes, J. F. (2003). *Linhas de orientação para a interpretação da norma OHSAS 18001/NP 4397 [Guidelines for the interpretation of the OHSAS 18001/NP 4397 standard]*. Porto: APCER – Associação Portuguesa de Certificação. http://www.apcer.pt

Rouse, W. B., & Boff, K. R. (1997). Assessing cost/benefits of human factors. In G. Salvendy (Ed.). *Handbook of human factors and ergonomics*, 2nd Edition (pp. 1617–1633). New York: John Wiley & Sons, Inc.

Saari, J. (2001). Accident prevention today. *Magazine of the European Agency for Safety and Health at Work, 4*, 3–5. http://osha.eu.int

Saha, A. (1998). Technological innovation and western values. *Technology in Society, 20*, 499–520.

Seong, S. K., & Mendeloff, J. (2004). Assessing the accuracy of OSHA's projections of the benefits of new safety standards. *American Journal of Industrial Medicine, 45*, 313–328.

Sesé, A., Palmer, A. L., Cajal, B., Montaño, J. J., Jiménez, R., & Llorens, N. (2002). Occupational safety and health in Spain. *Journal of Safety Research, 33*, 511–525.

Standards-29 CFR – Part 1902 – *State Plans for the Development and Enforcement of State Standards*, Occupational Safety and Health Administration (OSHA), U.S. Department of Labor.

Standards-29 CFR – Part 1903 – *Inspections, Citations, and Proposed Penalties*, Occupational Safety and Health Administration (OSHA), U.S. Department of Labor.

Standards-29 CFR – Part 1904 – *Recording and Reporting Occupational Injuries and Illness*, Occupational Safety and Health Administration (OSHA), U.S. Department of Labor.

Standards-29 CFR – Part 1910 – *Occupational Safety and Health Standards*, Occupational Safety and Health Administration (OSHA), U.S. Department of Labor.

Standards-29 CFR – Part 1911 – *Rules of Procedure for Promulgating, Modifying or Revoking OSHA Standards*, Occupational Safety and Health Administration (OSHA), U.S. Department of Labor.

Standards-29 CFR – Part 1912 – *Advisory Committees on Standards*, Occupational Safety and Health Administration (OSHA), U.S. Department of Labor.

Standards-29 CFR – Part 1915 – *Occupational Safety and Health Standards for Shipyard Employment*, Occupational Safety and Health Administration (OSHA), U.S. Department of Labor.

Standards-29 CFR – Part 1917 – *Marine Terminals*, Occupational Safety and Health Administration (OSHA), U.S. Department of Labor.

Standards-29 CFR – Part 1918 – *Safety and Health Regulations for Longshoring*, Occupational Safety and Health Administration (OSHA), U.S. Department of Labor.

Standards-29 CFR – Part 1924 – *Safety Standards Applicable to Workshops and Rehabilitation Facilities*, Occupational Safety and Health Administration (OSHA), U.S. Department of Labor.

Standards-29 CFR – Part 1925 – *Safety and Health Standards for Federal Service Contracts*, Occupational Safety and Health Administration (OSHA), U.S. Department of Labor.

Standards-29 CFR – Part 1926 – *Safety and Health Regulations for Construction*, Occupational Safety and Health Administration (OSHA), U.S. Department of Labor.

Standards-29 CFR – Part 1928 – *Occupational Safety and Health Standards for Agriculture*, Occupational Safety and Health Administration (OSHA), U.S. Department of Labor.

Standards-29 CFR – Part 1977 – *Discrimination against Employees under OSHA Act of 1970*, Occupational Safety and Health Administration (OSHA), U.S. Department of Labor.

Standards-29 CFR – Part 1990 – *Identification, Classification, and Regulation of Carcinogens*, Occupational Safety and Health Administration (OSHA), U.S. Department of Labor.

Tytyk, E. (2004). Evolutional background for humanizing technology. *Human Factors and Ergonomics in Manufacturing, 14*(3), 307–319.

UL 2351-2004 – *Standard for Safety for Spray Nozzles for Fire-Protection Service*, UL - Underwriters Laboratories, Research Triangle Park, NC, USA.

US-DOL, U.S. Department of Labor. Setting occupational safety and health standards. Program Highlights, Fact Sheet No. OSHA 92-14. http://www.osha.gov

Weil, D. (2001). Valuing the economic consequences of work injury and illness: A comparison of methods and findings. *American Journal of Industrial Medicine, 40*, 418–437.

Wright, M. S. (1998). Factors motivating proactive health and safety. Prepared by ENTEC UK Ltd for the Health and Safety Executive (Contract Research Report 179/1998), UK. http://www.hse.gov.uk.

Zimolong, B. (1997). Occupational risk management. In G. Salvendy (Ed.). *Handbook of Human Factors and Ergonomics*, 2nd edition (pp. 989–1020). New York: John Wiley & Sons, Inc.

33 Occupational Health and Safety Management Systems
Trends, Analysis, and Effectiveness

P. K. Nag and Anjali Nag*

CONTENTS

INTRODUCTION

The concept of occupational health and safety management system (OHSMS) has become increasingly popular among the stakeholders to assure compliance with the requirements of OHS controls. The emphatic endeavor is visible in applying management framework to implement cost-effective OHS and assisting governments, employers, and employees to better target their priorities in the prevention of workplace injuries and diseases (Robson et al., 2007; Fernández-Muñiz, Montes-Peón, & Vázquez-Ordás, 2009; Stolk et al., 2012). Revolving around the families of International Standards Organization (ISO) management standards—such as ISO 9000 quality management, ISO 14000 environmental management, ISO 26000 social responsibility, ISO 22000 food safety management—the systems approach has drawn the attention of the standards organizations, accreditation bodies, and the national regulatory agencies in formalizing, implementing, and evaluating the OHS management framework. Several national standards, proprietary, and sectoral schemes on OHSMS have been deployed, aiming at building a documented approach to OHS improvement in the enterprises and bringing visibility to its business objectives (Nag & Nag,

2003; Saksvik & Quinlan, 2003; Cambon, 2007). This contribution elucidates the core characteristics of OHSMS, the outlines and updates of national and sectoral OHSMS models, key features of system documentation requirements for certification and/or contractual obligations, trends of country-specific acceptance of OHSMS models, and analyses of the effectiveness and challenges of the management systems in diverse sectors of employment.

WHAT IS AN OHS MANAGEMENT SYSTEM?

For a long time, there has been debate and endeavor in defining and characterizing the concept of OHSMS (Frick & Wren, 2000; Stolk et al., 2012). It is aptly understood that a technical or scientific definition—to be useful—must be directed towards the intended population so that the broad meaning of the definition and its characterization is interpretable. Often, the concept of OHSMS has been vaguely or broadly defined, allowing shades of difference to get a unified understanding (Robson et al., 2005). This contribution compiles the documented views of the nature, meaning, and significance of OHSMS. As commonly understood, an OHSMS is a planned, documented, and verifiable process of managing hazards and risks of health and safety at workplaces, increasing productivity by reducing the direct and indirect costs associated with accidents and improving the quality of products and services. What makes it a management system is the deliberate linking and logical sequencing of processes and elements of the system that govern the way for a repeatable and identifiable way of managing OHS. In other words, an OHSMS provides a direction to OHS activities in accordance with the organizational policies, regulatory requirements, industry practices, and standards—including negotiated labor agreements. Ideally, the system of management of OHS is distinguished from traditional OHS programs by being more proactive, process-based, and better internally integrated with the core functions of the organization. Therefore, the purposes of establishing OHSMS are to:

Align OHS objectives with business objectives of an organization; integrate OHS programmes/systems into the business systems; establish a logical framework upon which to establish an OHS programme; devise a set of policies, targets, plans and procedures; provide an auditable reference for performance benchmarking, and establish a continual improvement framework towards organizational effectiveness.

The International Occupational Hygiene Association (IOHA, 1998) analyzed 13 publicly available management system documents on OSHMS, environmental management systems, and quality assurance and management systems and models, and identified the common characteristics of a universal OHSMS assessment instrument (Redinger & Levine, 1998). This comprises the primary elements as (a) communication, (b) evaluation, (c) continuous improvement, (d) integration, and (e) management review. The structure includes:

Initiation (OHS inputs)—(management commitment and resources, regulatory compliance and system conformance, accountability, responsibility and authority, and employee participation);

Formulation (OHS process)—(occupational health and safety policy, goals and objectives, performance measures, system planning and development, baseline evaluation and hazard/risk assessment, and OHSMS manual and procedures);

Implementation/Operations (OHS process)—(training System, technical expertise, personnel qualifications, hazard control, process design, emergency preparedness and response, hazardous agent management system, preventive and corrective action, and procurement and contracting);

Evaluation (Feedback)—(communication system, document and record management, evaluation, auditing and self-inspection, incident investigation and root cause analysis, and health/medical program and surveillance);

Improvement/Integration (Open system elements)—(continual improvement, integration and management review).

A close resemblance of the structure of management elements was also found in the information paper of Bottomley (1999) to the Australian National Occupational Health and Safety Commission below:

Organization, responsibility, accountability— (senior manager involvement, line manager/supervisor, specialist personnel, management accountability, performance measurement, company OHS policy);

Consultative arrangements— (health and safety representatives, issue resolution—employee, employer and OHS representatives, joint OHS committees and employee participation);

Specific programme elements— (health and safety rules and procedures, training program, workplace inspections, incident reporting and investigation, hazard prevention, data collection and analysis/record keeping, OHS promotion and information provision, purchasing, design, emergency procedures, medical and first aid, monitoring and evaluation, work organization issues).

Apart from the elements included, their linkages are critical to form a management system. Following extensive review and analysis of literature, Robson et al. (2005) abbreviated an OSHMS as "... *an integrated set of organizational elements involved in the continuous cycle of planning, implementation, evaluation, and continual improvement, directed toward the abatement of occupational hazards in the workplace. Such elements include ... the organizations' OSH-relevant policies, goals and objectives, decision-making structures and practices, technical resources, accountability structures and practices, communication practices, hazard identification practices, training practices, hazard controls, quality assurance practices, evaluation practices, and organizational learning practices.*"

OHSMS—NATIONAL AND SECTORAL MODELS

Enterprises of all kinds are under constant pressure to comply with OHS regulations and perform in line with their policy. Inevitably, various OHSMS models and schemes have been in use now, with relative applicability to diverse sectors. A brief outline of different national and international OHSMS standards, sectoral models, and country-specific standards are included in Tables 33.1–33.3. The management elements of many models and schemes (e.g., BS 8800; OHSAS 18001; ILO OSH 2001) have been explicitly ascribed to Edwards Deming's PDCA (Plan-Do-Check-Act) management cycle, as depicted in Figure 33.1. In the "Plan" phase, an enterprise establishes the objectives and processes necessary to deliver results in accordance with its OHS policy; in the "Do" phase, the enterprise shows the implementation of the stated processes. The "Check" phase refers to the monitoring, measuring, and reporting of results, and verifying whether the planned objectives have been met. For evaluation, in the "Act" phase, suitable corrective measures are implemented to further improve results.

BRITISH HSE GUIDANCE AND BS 8800

One of the earliest OHS system guidance—HS (G) 65 (1993) brought out by the British Health and Safety Executive (HSE)—described the issues that organizations need to address to effectively manage health and safety. Together with the approved codes of practice on applicable regulations (e.g., the British Health and Safety Act), the HS (G) 65 approach provided employers with a basis in managing health and safety (Figure 33.2).

TABLE 33.1

Brief Outlines of the OHS Management Systems

BS 8800—ISO (1996)	OHSAS 18001 (2007)	ILO-OSH (2001)	SA 8000 (2008)	ANSI Z10 (2012)
HS(G)65 approach 4. OHSMS initial status review 4.1 **OHS policy** 4.2 **Organizing** (responsibilities, organizational arrangements, documentation) 4.3 **Planning and implementing** (risk assessment, legal and other requirements, management arrangements) 4.4 Measuring performance 4.5 Audit 4.6 Periodic status review **ISO 14001 approach** 4. OHSMS initial status review 4.1 **OHS policy** 4.3 **Implementation and Operation** (structure and responsibility, training, awareness and competence, communications, OHSMS documentation, and operational control, emergency preparedness and response) 4.4 **Checking and corrective action** (monitoring and measurement, corrective action, records, audit) 4.5 **Management review**	1–3. Scope, references, terms & definitions 4 OH&S management system requirements 4.1 General requirements 4.2 **OH& S policy** 4.3 **Planning** (Hazard identification, risk assessment and determining controls, legal and other requirements, objectives and program (s) 4.4 **Implementation and operation** (resource, roles, responsibility and authority, competence, training and awareness, communication, participation and consultation, documentation, control of documents, operational control, emergency preparedness, and response) 4.5 **Checking** (performance measurement and monitoring, evaluation of compliance, incident investigation, nonconformity, corrective and preventive action, control of records, internal audit) 4.6 **Management review** ANNEX A: Correspondence between OHSAS 18001:2007, ISO 14001:2004 and ISO 9001:2000	2. National frameworks for OSH-MS (national policy and tailored guidelines) 3. **OSHMS policy** (OSH policy, employee participation) **Organizing** (responsibility, accountability, competence and training, documentation, communication) **Planning and implementation**(initial review, system planning, development and implementation, OSH objectives, hazard prevention, control measures, management of change, emergency prevention, preparedness and response, procurement, contracting) **Evaluation** (performance monitoring and measurement, investigation of work-related injuries, OSH performance, audit, management review) **Action for improvement** (preventive and corrective action, continual improvement)	**Social accountability requirements** 1. Child labor 2. Forced and compulsory labor 3. Health and safety 4. Freedom of association & right to collective bargaining 5. Discrimination 6. Disciplinary practices 7. Working hours 8. Remuneration 9. **Management systems** 9.1 Policy 9.2 Management representative 9.3 SA8000 Worker representative 9.4 Management review 9.5–9.6 Planning and implementation 9.7–9.10 Control of suppliers/subcontractors and subsuppliers 9.11–9.12 Addressing concerns and taking corrective action 9.13–9.14 Outside communication and stakeholder engagement 9.15 Access for verification 9.16 Records **Note:** SA 8000 management system is included herewith, considering its proximity and relevance to OHSMS.	1. Scope, purpose & application 2. Definitions 3. **Management leadership & employee participation** 3.1. Management Leadership (OHSMS, policy, responsibility, and authority) 3.2. Employee participation 4. **Planning** 4.1. Initial and ongoing reviews 4.2. Assessment and prioritization 4.3. Objective 4.4. Implementation plans and allocation of resources 5. **Implementation and operation** 5.1. OHSMS operational elements (risk assessment, hierarchy of controls, design review, management of change, procurement, contractors, emergency preparedness) 5.2. Education, training, and awareness 5.3. Communication 5.4. Document and record control 6. **Evaluation and corrective action** 6.1. Monitoring and Measurement 6.2. Incident Investigation 6.3. Audits 6.4. Corrective and Preventive Actions 6.5. Feedback to the Planning Process 7. **Management review** (review process and review outcomes) **Appendices (A–O):** A—Policy statements B—Roles and responsibilities C—Employee participation D—Planning-identification, assessment, and prioritization

ANNEX B: Correspondence
between OHSAS 18001, OHSAS
18002 and ILO-OSH:2001
Guidelines on occupational safety
and health management systems.

E—Objectives/implementation plans
F—Risk assessment
G—Hierarchy of control
H—Management of change
I—Procurement
J—Contractor safety and health
K—Incident investigation
L—Audit
M—Management review process

TABLE 33.2
Brief Outlines of the Sectoral OHSMS Schemes

MASE/DT78 (2004)	GEHSE (2004)	Responsible Care Management System (UK) (1998)	E&P Forum (oil & gas companies) (1998)
1. **Policy of SHE** (responsibility for the direction, Policy of SHE, Indicators and expected dissemination, Planning of SHE, Organization of resources, Animation, Communication, analysis of causes of accidents and incident)	2. **Policy and personnel** (HSE policy, training specifications, Empowerment of stakeholders, etc.)	2–3. **Leadership commitment, Policy**	1–2. **Leadership commitment, Policy, and strategic objectives**
2. **Human resources** (training and professional qualification, New entrants, appointment of personnel, competency, skills empowerment, system appropriation)	3. **Planning and operation** (Planning details, Plan of prevention, identification, and analysis of risk, Treatment of Modifications, Work Permit, Emergency preparedness,	4. **Identifying requirements** (regulatory and other requirements, risk assessment, defining program)	3. **Organization, resources, and documentation** (structure and responsibilities, resources, competence, contractors, communication, documentation control)
3. **Operation** (preparation, organization intervention, material and facilities, plan of prevention, matters of subcontractors, management documentation)	4. **Subcontractors/outsourcing**	5. **Planning** (HS&E objectives, setting targets for improvement, planning for control, performance criteria, emergency preparedness)	4. **Evaluation and risk management** (identification, evaluation, recording of hazards and effects, objectives and performance criteria, risk reduction measures)
4. **Controls** (audits, inspection, verification, etc.)	5. **Measures of performance** (safety, medical surveillance, revisit of experience)	6. **Organization** (structure and responsibility, MR, resources, competence, documentation, communications)	5. **Planning** (asset integrity, procedures, work instruction, management of change, contingency, emergency planning)
5. **Continual improvement, management review**	6. **Audits and controls** (Inspection and audits, Corrective Actions and anomalies, Management review)	7. **Implementation and control** (people, purchasing, contractors, manufacturing, storage, transportation, distribution, management of change)	6. **Implementation and monitoring** (activities, monitoring, records, non-compliance, corrective action, incident reporting, follow-up)
		8. **Monitoring** (measurement, checking, and inspection, internal audit, improvements, records)	7. **Auditing and reviewing**
		9. **Management review**	

The British Standards Institution released the BS 8800 (1996) guidance on OHSMS to improve OHS performance in organizations and help the management of OHS integrate with other aspects of business to build a responsible image within the marketplace. Structurally, the BS 8800 standard is based on two approaches (i.e. the HSE guidance—HS (G) 65 and ISO 14001 standards [Table 33.1]). The BS 8800 contains annexes on the practical guidance, including planning and implementing, risk assessment, and measuring performance. The standard explicitly states that it should not be quoted as if it were a specification and that it should not be used for certification purposes.

OHSAS 18001:1999/2007

A specification standard—the Occupational Health and Safety Assessment Series (OHSAS 18001:1999)—was developed by a consortium of national standards and certification bodies. The national standards and proprietary certification schemes, which were used to create the specification of the standard, are given in BS-OHSAS 18001 (1997). Although OHSAS 18001 is not an ISO standard, this specification has wide global acceptance from diverse businesses, and a large number of the certification bodies certify according to this standard. OHSAS 18001 has been revised in July 2007, making it significantly closer to the international standard ILO-OSH 2001.

The schematic of the system elements of OHSAS 18001 requirements is illustrated in Figure 33.3. An accompanying publication—OHSAS 18002—is a guidance document that directly

TABLE 33.3

Brief Outlines of the Country Adopted OHS Management Systems

US OSHA program	AS/NZS 4801 (Australia/New Zealand) (1997)	SafetyMAP (1999)	Spain (UNE 81900)	Norway, Royal Decree (1992)	The Netherlands (NPR 5001)	Ireland (KS NR 11–97) (1997)	Japan (JISHA 1997)
VPP (1988) a. Management commitment and planning b. Hazard assessment c. Hazard correction and control d. Safety and health training e. Employee participation **Safety and Health Program (1989)** a. Scope and application b. Basic obligation c. Management leadership and employee participation d. Hazard assessment and control e. Information and training f. Program evaluation g. Dates h. Definitions	4.1. Commitment and policy (leadership and commitment; initial OHS review, OHS policy) 4.2. Planning (hazards identification, assessment and control of risk, legal and other requirements, objectives and targets, performance indicators, management plans) 4.3. Implementation (ensuring capability, support action, risk assessment and control, contingency preparedness and response) 4.4. Measurement and evaluation (inspection, testing and monitoring; audits, corrective, and preventive action) 4.5. Review and continual improvement.	1. Building and sustaining commitment 2. Documenting strategy 3. Design and contract review 4. Document control 5. Purchasing 6. Working safety by system 7. Monitoring standards 8. Reporting and correcting deficiencies 9. Managing movement and materials 10. Collecting and using data 11. Auditing of management systems 12. Developing skills and competencies	4.1. OHS policy 4.2. OHS management system 4.3. Responsibilities (management and staff responsibility and resources, management review) 4.4. Risk evaluation (record of legal, regulatory, and other requirements, hazard identification, Risk evaluation and control, maintenance of risk control measures) 4.5. Prevention planning (OHS management program) 4.6. OHS handbook and documents 4.7. Performance control (active control, verification, ensuing control, non-conformity, and corrective actions) 4.8. Safety and health records 4.9. Evaluation of OHSMS (audits, review of OHSMS)	4.1. External requirements (stakeholder, legal requirements) 4.2. Management responsibility (leadership commitment, policy, and objectives, system documentation) 4.3. Management system structure (organization, documentation) 4.4. Communication (internal and external) 4.5. Planning (human resource management, establishing targets, management program, management of change, loss prevention, contingency, and emergency preparedness) 4.6. Operational and functional management (marketing, development, procurement, processing) 4.7. Measuring and evaluation (process performance, evaluation, assessment, audit, review, records) 4.8. Improvement (non-conformance handling, preventive and corrective action, improvement)	4.3 Maintenance and development of OHSMS 5 First and periodic review of OHS status (initial and periodic management review) 5.3 Risk assessment (initial and periodic risk assessment) 6 Decision-making and planning (OHS policy, plan of action) 7 Organization and implementation 7.1 Organization (responsibilities, authority, resources, OHS targets, consultation with staff, representatives, qualification, training and information, expert assistance, information management, internal/external information and communication) 7.2 Implementation of plan of action 8 Performance measurement (review OHSMS, evaluation of plan of action)	4. OHSMS principles 5. Initial status review 6.1 OHS policy 6.2 Planning (legal and other requirements, objectives and targets, planning for OHS) 6.3 Implementation and operation (structure and responsibility, consultation, training, awareness and competence, communication, documentation, operational control, management and control of contractors, emergency preparedness and response) 6.4 Checking and corrective action (monitoring and measurement, accidents and incidents of non-compliance with OHSMS—corrective and preventive action, audit, records) 6.5 Management review	1. Corporate policy and CEO's policy 2. Establishment's policy, director's Policy 3. OHS management programs 4. OHS management manuals 5. OHS management organization system 6. OHS committees 7. Guidance to subcontractors 8. OHS education 9. OHS daily activities 10. Machinery and equipment safety 11. Chemical safety 12. Working environment 13. Work management 14. Medical surveillance 15. Promotion of physical and mental health 16. Comfortable workplaces 17. Measures for workers in need of special care 18. Traffic safety of commuters 19. Analysis of accident causes 20. Emergencies

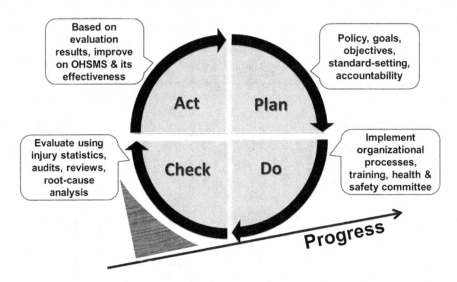

FIGURE 33.1   Plan-Do-Check-Act management cycle.

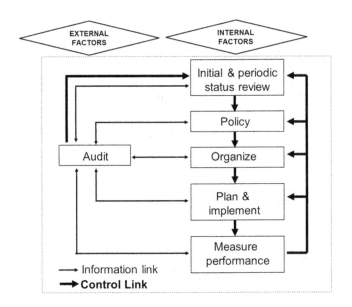

FIGURE 33.2   Element of OHSMS in HS (G) 65 approach.

corresponds to the specifications. The specification standard sets out the minimum requirements (Table 33.1) for the best OHSAS practice and applies to organizations of all types and sizes, accommodating various geographical, cultural, and social conditions. The British Standards Institution states that the creation of OHSAS was necessary to reduce marketplace confusion and meet the demands to create a global framework to apply in any organization that wishes to:

Establish an OHSMS to eliminate or minimize risk to employees and other interested parties who may be exposed to OHS risks associated with its activities; implement, maintain, and continually improve an OHSMS; assure itself of its conformance with its stated OHS policy; demonstrate such conformance

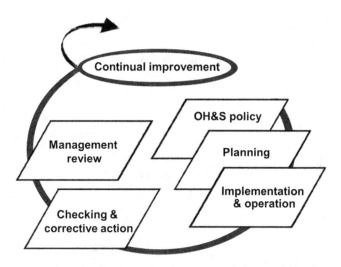

FIGURE 33.3 Element of OHSMS in the OHSAS 18001 approach.

to others; seek certification of its OHSMS by an external organization; or make a self-determination and declaration of conformance with the OHSAS specification.

ILO-OSH MS:2001

Recognizing the positive impact of introducing OHSMS at the enterprise level, the International Labor Office brought out a consensus model (ILO-OSH 2001) according to the principles defined by the ILO's tripartite constituents (governments, employers, and workers). The readers may note that the ILO-OSH model is referred to as the occupational safety and health management system (OSH-MS). The ILO-OSH model reflects the ILO values that advance the objectives of labor conventions, ratified by member nations. Some related conventions are 155 occupational safety and health (1981), 161 occupational health services (1985), 162 asbestos (1986), 167 safety and health in construction (1988), 170 chemicals (1990), 174 prevention of major industrial accidents (1993), 176 safety and health in mines (1995), and others.

Structurally, the ILO-OSH model is distinctively different from the OHSAS 18001 by the former's unique focus in developing national OHS policy and guidelines, including tailored enterprise-level management framework (Figure 33.4). For enterprise-level application, the ILO-OSH model is constituted by 16 elements which are grouped under five sections—policy, organizing, planning and

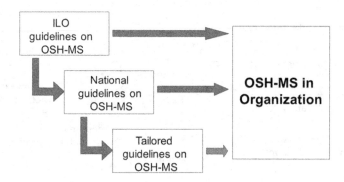

FIGURE 33.4 ILO-OSH national framework for OHSMS implementation.

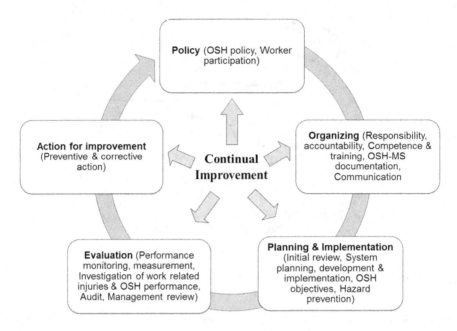

FIGURE 33.5   Element of ILO-OSH guidelines on the OHS management system.

implementation, evaluation, and action for improvement (Figure 33.5). This model has emphasis on dimensions such as employee participation, training at all levels, injury investigation, and health programs (Table 33.1).

SA 8000 (SOCIAL ACCOUNTABILITY STANDARD)

The SA 8000 (2008), a social accountability management standard (Figure 33.6), seeks to guarantee the basic rights of workers and workplace conditions. The requirements of workers—child labor, forced labor, health and safety, right to collective bargaining, discrimination, disciplinary practices, working hours, and remuneration—have been defined with respect to national laws, international human rights norms, and the principles of the tripartite ILO conventions. The structured management system enables the enterprises of diverse sectors and sizes to develop and enforce policies and

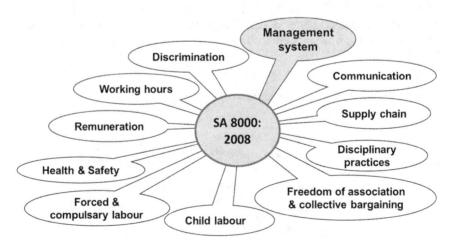

FIGURE 33.6   Elements of SA 8000—social accountability standard.

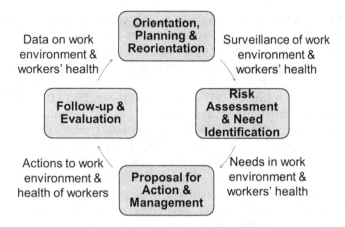

FIGURE 33.7 Logical cycle of activities for basic occupational health services.

procedures in order to manage issues that it can control and influence. It is an auditable framework for independent verification of the ethical production of goods and services. For integration of management systems in an enterprise, the SA 8000 deserves merit to be categorized as an OHSMS. Besides a recognizable well-structured system framework, the SA 8000 requires compliance of the provisions of several related ILO conventions and recommendations on OHS. Likewise, the basic occupational health services (BOHS) systems framework was launched jointly by the WHO/ILO/ICOH (Rantanen, 2005a, 2005b), which is recognized as a step towards implementing ILO conventions on occupational health delivery. The logical flow of the BOHS scheme is shown in Figure 33.7.

ANZI Z10 (2005/2012)

The ANZI Z10 standard (2005, revised 2012) provides a new blueprint of OHSMS for widespread benefits in health and safety, productivity, financial, performance, quality, and other business objectives in the U.S. This is basically a performance standard, general process, and system guidelines that apply to organizations of all sizes and types. On the primary template of Deming's PDCA cycle of continual improvement, the Z10 standard includes seven sections, as described in Table 33.1. The framework includes the conventional system elements of BS 8800. 15 appendices elucidate the dimensions of the system elements, such as policy statements, roles and responsibilities, assessment, audit information, and others. Under the implementation and operation of OHSMS, the hierarchy of controls sets enterprises to take classic risk reduction steps through elimination, substitution of less hazardous materials/process/operation/equipment, engineering controls, warnings, administrative controls, and the use of personal protective equipment. The management review covers the management review process, review outcomes, and follow-up—which should be conducted at least once a year to ensure that proper actions are taken following evaluation, and improvements are continuously done.

UPDATES ON ISO 45001—A NEW INTERNATIONAL STANDARD FOR OHSMS

Since the first time OHSAS 18001:1999 (revision 2007) was introduced as a model of OHSMS, the revised standard has been widely practiced in as many as 127 countries. Often, concerns on the international credibility of tailored OHSMS schemes and models have been voiced since these models were not created through the formal standards development process. The voluntary forms of development of the specification documents are wide open, leading to considerable confusion in establishing the credibility of the OHSMS certification across the sectors of business.

The vivid need for synchronizing OHSMS to an internationally recognized standard led to the initiative of a new standard (ISO 45001) to improve the OHS for local, national, regional, and international levels in both developed and developing countries. Active involvement of ILO in the development of the new standard will dispel earlier debates and arguments of relevance, context of application, and conflict of interest (if any).

The first committee draft—ISO PC 283 (October 2013)—on ISO/CD 45001 is a single document with specifications and an interpretation guidance annex. The draft clauses follow the same high-level framework (Annex SL) as ISO 9001 and ISO 14001, such as (1) scope, (2) normative references, (3) terms and definitions, (4) context of the organization, (5) leadership, (6) planning, (7) support, (8) operation, (9) performance evaluation, and (10) improvement. The intent behind Annex SL is to maintain alignment with other ISO management systems, to help organizations establish integrated management systems covering multiple disciplines. Those familiar with the specifications of different OHSMS models will recognize that the core of the ISO/CD 45001 primarily draws from OHSAS 18001:2007, ANSI Z10:2012 standard, and other specifications. However, the practitioners may recall that the OHSAS 18000 series contains two documents (18001 as requirements and 18002 as guidance).

The proposed ISO/CD 45001 further emphasizes leadership and management roles and introduces a new risk-based approach. It insists that no OHSMS can be effective unless OHS aspects receive visibility and are embodied in the overall management system of the organization. However, the leadership role as regard to OHS performance and outcomes requires further elucidation. Also, in Clause 6 (planning), there are inconsistencies in the use of certain terms such as "hazard" and "incident." Clause 7 highlights the requirements on resources, competence, and awareness, including documented information, communication, participation, consultation, and representation. Clause 8 is centered on the hierarchy of control, management of change, outsourcing, procurement, contractors, emergency preparedness, and response. Clause 9 refers to monitoring, measurement, analysis, and the evaluation of compliance, with concomitant challenges on the aspects of internal audit and management review. Clause 10 is mostly concerned with incident investigation, nonconformity, corrective action, and continual improvement. Annex A of the standard aims to give guidance on the interpretation of ISO 45001. Regarding Clause 3, questions have been raised related to definitions such as "worker/employee" and "workplace," and about changing the term "contractor" to "external provider."

The final standard is expected to be released in 2016 and will replace OHSAS 18001:2007. The challenge of the new standard is to define the functions of an organization to look beyond its immediate OHS issues (e.g., internal employees, contractors, and suppliers) and how their work might affect the neighbors in the surrounding area. There is a complete neglect of mention to the environmental and biological monitoring and diagnostic criteria of occupational diseases and disorders. In other words, the new standard should allow organizations to determine the breadth of application pertaining to them, without deliberately excluding certain areas of operation.

SECTORAL OHSMS MODELS

Several sector-specific auditable and non-auditable OHSMS frameworks have been evolved for application in chemical, petrochemical, construction, oil, and gas installations. While the models have a wide difference in the structure and content and in the order of presentation of system elements, as shown in selected widely practiced schema (Table 33.2), they have a common resemblance with the PDCA framework in managing the workplace environment, products as well as health and safety.

Responsible Care: the U.S. Chemical Manufacturers Association designed Responsible Care (1992) as the health and safety code of management practice. With the broad outline on program management, identification and evaluation, prevention and control, communication, and training,

Responsible Care is a multidisciplinary means to enable member enterprises to promote OHS among employees, contractors, and the public, and to protect the environment.

The British Responsible Care management system (1998) represents the International Council of Chemical Association's Responsible Care program. In contrast to the U.S. Responsible Care code, the British document appears more comprehensive (Table 33.2). It includes a guidance and self-assessment document and describes the key requirements of OHS and the environmental management system that can be incorporated into the management systems of an enterprise. The Responsible Care does not explicitly address medical program and surveillance; however, the guidance includes the requirements of British HSE—HS (G) 65 (1993), Eco-Management and Audit Scheme, EMAS (1993), ISO 14001 (1996/2004), and BS 8800 (1996).

E&P Forum (oil and gas companies) and other business Charter: the International Exploration and Production Forum (E&P Forum) guidelines (1998) of the international association of oil and gas companies try to develop and apply OHSMS and environmental management systems in their exploration and production operations worldwide. A comprehensive management system addresses core structural issues for programs and procedure development, including a controlled documentation system. The structure and content of the E&P guidelines (Table 33.2) are sufficiently generic to make them adaptable to different types of enterprises, systems, and cultures.

MASE/DT 78 and GEHSE: the MASE (2004)—Manuel d'Amélioration Sécurité Enterprise—originated in the 1990s, led by Exxon, to better structure the security approach and industrial hygiene in petrochemical sites. Currently, MASE is used by other industrial sectors such as steel and nuclear. MASE defines the minimum requirements (Table 33.2) for the establishment of a prevention system revolving around (a) the commitment of the management of the company, (b) competence and professional qualifications of staff, (c) the preparation and organization of work, (d) controls, and (e) continuous improvement. The companies meeting the audit criteria required by MASE are certified for one or three years, in the territory of France.

The security agreement between the Union of Chemical Industries (France) and employee organizations requires that the intervener company/contractors involved in construction, maintenance, and logistics facilities must set up a system of management in OHS (DT 78, June 2004), covering (a) policy and general organization, (b) competence and training, (c) preparation and organization of work, (d) control of outsourcing, (e) evaluation and monitoring, and (f) tracking. The external auditor assesses the compliance documentation and conditions and grants authorization for 1–3 years, according to the level and scale of implementation.

Since September 2008, the MASE and DT78 standards are one in their goal: to put in place a common approach for safety improvement among companies, business users, and social partners, especially in the chemical sector—in France and at the European level—and avoid regulatory constraints. The authorized management system will benefit a larger number of employees—beyond those working on Seveso and high threshold sites; facilitate integration of the management systems; and enhance organizational performance (security, competitiveness, etc.). Corollary to the MASE certification, the GEHSE (2000/2004) is the commitment guide on hygiene, safety, and environment for outside contractors involved in hydrocarbon deposits or small oil facilities, including service stations. The specification of GEHSE (Table 33.2) was designed based on the industry agreement on safety within the oil institutions. It is generally present in parts of the French territory not subject to MASE.

KEY FEATURES IN OHSMS DOCUMENTATION

An organization has been advised to implement OHSMS for the systematic identification, evaluation, and prevention or control of workplace hazards. The written guidance or documentation depends on the size and complexity of the operations of an enterprise. Generally, the OHSMS documentation might be followed consistent with the ISO 9001/ISO 14001 systems, with three levels of documentation in a pyramidal structure; that is,

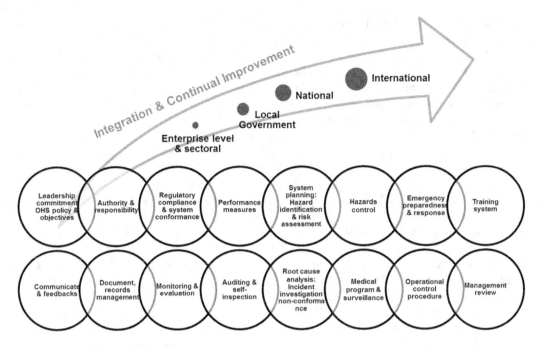

FIGURE 33.8 Dimensions for OHSMS documentation.

a. *The top-level is the OHSMS manual, referring to the scope and policy of the management system. It depicts the orderly arrangement of interdependent activities that define the processes in an organization.*

b. *The second level is made up of the standard operating procedures (SOPs)—task procedures, health and safety procedures, risk assessment, and specific instructions to the processes and products.*

c. *The third level is referred to as records management that essentially means the documented evidence of OHS performance.*

The OHSMS variables and dimensions that are central to a systems approach are thematically shown in Figure 33.8, and a brief description of the variables and documentation needs, for example, are given in Table 33.4. The structural content of the OHSMS documentation may be organized depending on the model or scheme advocated (Nag, 2002).

GLOBAL ACCEPTANCE AND COUNTRY SPECIFIC ADOPTION OF OHSMS MODELS

There has been unanimity that safety, health, and environmental issues are integral to other priorities of enterprise development. The revelation of alarming statistics of work-related accidents or diseases (~2.3 million people died in 2013, as per ILO) reaffirms the necessity of a strong OHSMS across organizations to mitigate health risks and accidents, avoid legal entanglements, reduce insurance costs, and allure a culture of positivity at workplaces. Despite the apparent distinction between the traditional OHS aspects and the concepts of OSHMS, one recognizes that there exists a lot of regulations, as well as auditable and non-auditable OHSMS guidelines that have been differently advocated in the enterprises. Limitations exist that there is no consensus on a globally harmonized strategy on OHS management. Concerns are vivid on the international credibility of the practiced OHSMS schemes and models. Anticipation is alive that the release of the new ISO 45001 by 2016 might receive wider visibility and acceptance among stakeholders for synergized application across organizations in developed and developing countries.

TABLE 33.4
The OHSMS Dimensions for Documentation

Dimensions	Description	Documentation (Indicative)
Leadership commitment, OHS policy, and objectives	Commitments to OHS development are operationally defined. The formal OHS policy reflects the leadership commitment in setting out measurable goals and objectives towards meeting the OHS performance expectations of the organization.	Identification and approaches to improving the core elements of OHSMS on a time horizon (e.g., improving existing OHS features; reduction in risk levels; frequency of undesired incident(s); recognizing disorders and diseases).
Accountability, responsibility, and authority	The organization defines the roles of personnel who are involved in OHSMS functions. Executive ownership and accountability are critical (e.g., budget, health and safety committee, medical services, emergency preparedness, training).	A management representative is designated, making him accountable for the establishment of prevention and health promotion programs through anticipation, recognition, evaluation, and control of OHS hazards. Roles and responsibilities shall be documented and communicated to employees and other relevant parties.
Regulatory compliance and system conformance	Government and non-government regulations/ standards to which an organization subscribes impose requirements on OHSMS. Broadly, both regulatory compliance and conformance to other system requirements lay the groundwork for a systematic management approach.	Emphasis is on the framework of action on OHS in recognizing the regulatory requirements, disclosure of information, back-up enforcement, and prosecution. The documented procedure elucidates how well the organization complies with legal OHS requirements. • recording the results of legal OHS compliance evaluations; • identifying activities that might be affected by applicable legal and other requirements; and • monitoring the control interventions, consequent to changes in OHS legislation.
Performance measures	For valid and reliable OHS performance measurements, the indicators, variables, measurement units, and their logical relationships are required to be established. A distinction has been made between the leading and trailing indicators.	A documented procedure is required to monitor and measure OHS performance. It outlines the parameters that determine, for example, whether the OHS policy and objectives are achieved, risk controls have been implemented, lessons are learned from OHSMS failures (accidents and illnesses), the effectiveness of training, communication, and consultation programs, and relevant information for management review.
System planning for hazard identification and risk assessment	The organization requires a documented OHS management program (strategies and plan of actions) encompassing all significant OHS hazards in its domain. Essentially, the hazard is defined as a source or situation with the potential to cause harm, such as injury or ill health, damage to property, and workplace environment.	• identification of hazards and determination of the level of risks associated with the hazards; • description of the measures to monitor and control the risks, particularly those that are not tolerable; • OHS objectives and actions to reduce identified risks, and follow-up monitoring activities; and • identification of core competency and training requirements to implement the control measures.
Hazard control	The methods of hazard control systems are generally recognized as administrative controls, engineering controls, or by using personal protective equipment (PPE). The hazard control covers the management of hazardous agents (e.g., radioactive materials, noise, heat, cold, lasers, hazardous wastes).	ILO-OSH (2001) indicates the implementation of preventive and protective measures to control hazards and risks, in an order of priority. Other OHSMS models are not explicit to hazard control, such as the approaches to managing hazardous agents. OHSAS 18001:2007 emphasizes hazard identification and risk assessment. The hazard control should reflect the principles of eliminating hazards and reducing risks (reducing the likelihood of occurrence or potential severity of injury or damage).

(Continued)

TABLE 33.4 (Continued)

Dimensions	Description	Documentation (Indicative)
Emergency preparedness and response	This refers to the manner in which the organization prepares for and responds to emergencies and accidents. It develops plans and procedures to cope with them, test its planned responses, and seeks to improve the effectiveness of emergency responses.	• identification of potential accidents and emergencies; • personnel (e.g., fire-fighters, first-aid staff, toxic spillage specialists); • documented emergency plans, including evacuation procedures; • hazardous materials and emergency action required; • interface and communication with external emergency responders; • availability of vital information (e.g., plant layout drawings, hazardous material data); • emergency equipment list (alarm systems, means of escape, fire-fighting equipment, first aid equipment, test records); and • records and review of the practice drills, and recommended actions arising from the reviews.
Operational control procedure	The organization establishes operational control procedures to identify risks (including those introduced by contractors or visitors) and documents the instances where any failure might lead to accidents or deviation from the OHS policy and objectives. The procedures are to be reviewed periodically.	• *Purchase or transfer of goods and services and use of external resources* (e.g., purchase or transfer of hazardous materials; instructions for handling of machinery and materials; OHS competence of contractors; OHS provisions for new plant or equipment). • *Hazardous tasks* (e.g., identification of hazardous tasks; pre-determination of working methods; pre-qualification of personnel; permit-to-work systems to control exposures to hazardous tasks). • *Hazardous materials* (e.g., inventory identification, storage provisions, and control of access; MSDS). • *Maintenance of safe plant and equipment* (e.g., maintenance of plant and equipment; segregation and control of access; testing of OHS related equipment; local exhaust ventilation systems, medical facilities, PPE, guarding; shutdown, fire suppression; radiological sources, safeguards, and monitoring devices).
Training system	The training is an integral component of OHS management, having qualified personnel to provide OHS services to the organization.	• an effective procedure ensures competency requirements for individual roles; • analysis of training needs and programs for employees; and • training and evaluation records. ILO-OSH (2001) is emphatic that training should be provided to all participants at no cost and should preferably take place during working hours.
Communication system/feedback channels	An effective communication system with defined feedback channels facilitates participation in good OHS practices. A viable communication system should identify how, and to whom, information on the functioning of the OHSMS will be transmitted.	A defined procedure is required for worker participation and consultation with contractors and other parties, for appropriate communication and feedback. • formal management for employee consultations through OHS committees; • employee involvement (and employee OHS representatives) in hazard identification, site inspection, risk assessment, incident investigation, and control; • OHS briefings for employees and other interested parties; and • notice boards containing OHS performance data, newsletter, etc.

Dimensions	Description	Documentation (Indicative)
Document and records management	A well-functioning document and record management system is a key indicator of whether the OHSMS is currently in conformance; that is, documentary evidence that the OHSMS operates effectively, and the processes have been carried out under safe conditions.	All documents and data critical to the operation of the OHSMS should be identified and controlled. The document control procedure should define the controls for the identification, approval, issue, and removal of OHS documentation, together with the control of OHS data.
Monitoring and evaluation	The monitoring and evaluation of an OHSMS encompasses performance monitoring, auditing, and self-inspection, incident investigation, root cause analysis, the availability of medical surveillance and health programs, management review, etc.	• procedure(s) for monitoring and measuring, and evidence of the results of implementing OHSMS; • measuring equipment lists, measurement procedures; critical equipment lists, including equipment inspection checklists; • calibration records, maintenance activities, and results; and • OHSMS checklists, medical surveillance programs.
Auditing and self-inspection[a]	The planned auditing (internal and external) allows reviewing the organization's OHSMS effectiveness and conformity to the specification it subscribes to. ILO-OSH (2001) recommends consultation on the selection of auditors. OHSAS 18001:2007 requires that the personnel undertaking the audits should be in a position to do so impartially and objectively.	• OHSMS audit plan and procedures; • OHSMS audit reports, including non-conformance reports, corrective action requests; • closed-out non-conformance reports; and • evidence of the reporting of the audit results to top management.
Incident investigation, non-conformance—root cause analysis	Incident investigation refers to the activities in determining the origin and cause(s) of accidents and incidents. The organization shall have a documented procedure for reporting and evaluating/investigating accidents, and non-conformances. Root cause analysis allows in moving up the causal chain in detecting, analyzing, and eliminating potential causes of nonconformities.	incidents investigation reports, including hazard identification, risk assessment, and risk control reports; • non-conformance register and reports; • evidence of evaluation of corrective and preventive actions taken; and • management review inputs.
Medical surveillance and health programs	This refers to the activities associated with providing occupational health services and developing the operation of a health surveillance and health promotion program. The OHSMS models evolved on the ISO 14001 templates do not explicitly mention the elements of the medical program; however, these are aspects that provide feedback to the hazard control system.	For medical surveillance, the professional may consider the ILO list of occupational diseases, with reference to the exposure characteristics of workers. The WHO (2007) proposes a national policy framework to deal with all aspects of workers' health, including primary prevention of occupational hazards, protection and promotion of health at work, employment conditions, and a better response from health systems to the workers' health.
Management review	Periodic management review involves evaluating the ability of OHSMS to meet the needs of the organization, its stakeholders, employees, and regulating agencies.	• minutes of the review meeting; • revisions to the OHS policy and objectives in light of audit results, changing circumstances, and commitment to continual improvement; • corrective/improvement actions, with assigned responsibilities and targets for completion; review of corrective action; and • emphasis in the planning of future internal OHSMS audits.
Continual Improvement	The OHSMS builds on the concept for continual improvement in the system to eliminate or minimize the potential risks of workplace injury and illness.	The enterprise demonstrates, through appropriate documentation, the processes of OHSMS to achieving continual improvement.
Integration	Integration can be viewed from both macro -as well as micro-level perspectives. The macro perspective emphasizes a systems approach to	An enterprise certified to process-based models of ISO 9001:2000 or ISO 14001 has already placed the generic framework to integrate health and

(Continued)

TABLE 33.4 (Continued)

Dimensions	Description	Documentation (Indicative)
	viewing OHSMS as a cost-effective operational framework, within the scope of the business. The micro-level perspective considers integration based on the model elements. Besides, the relative effectiveness of OHSMS models depends on their compatibility with other management systems, in practice.	safety, and other management systems. An OHSAS 18001:2007 compliant enterprise can be reassured that their system will also be compatible with the ILO-OSH guidelines and the forthcoming ISO 45001 standard on health and safety. Primarily, the ILO-OSH (2001) provides guidance to enterprises for the integration of OHS elements into their overall policy and management arrangements. OHSAS 18001 specifies requirements to enable enterprises to control risks and to improve OHS performance.

Note

a The OHSMS models are a mix of auditable and non-auditable standards. One may opt to have its OHSMS audited against a certification standard. The process of certification offers independent verification and auditing that an organization has become more self-regulatory in promoting OHS. During the three years validity period of the OHSAS 18001 certification, annual surveillance audits ensure that the certified organization continues to comply with the OHSAS 18001 requirements. A full reassessment for the renewal of the certificate shall be conducted every three years. Many OHSMS standards (e.g., ILO-OSH 2001) are not intended for certification purposes.

The broad outlines of the existing OHSMS models and schemes can be schematically grouped into the following categories:

- Mandatory OHSMS with regulatory and advisory measures;
- Voluntary OHSMS standards with and without national level certification support;
- Good OHS practice guides (not intended for certification);
- Self-regulatory OHSMS schemes and promoted through statutory bodies; and
- Sectoral models/schemes for outside companies/contractors and business users.

Mandatory OHSMS With Regulatory Measures

Many countries favored the mandatory OSHMS that arose from government legislation, advisory, enforcement, and inspection on a set of OHS management principles, to be implemented by the employers. The relative utility of this approach is to make the regulations uniformly applicable to all types of enterprises, including small workplaces. For example, the European Union Directive 89/391/EEC (Walters, Wilthagen, & Jensen, 2002) defines employer responsibilities in OHS management, including the evaluation of risks to the health and safety of workers, the implementation of preventive measures, and the integration of measures into the activities of the enterprises at all hierarchical levels.

Generally, the Scandinavian countries make it obligatory for all enterprises and public institutions to integrate preventive actions of OHS into the management system of the enterprise. Internal Control legislation is perhaps the most comprehensive among the existing mandatory OHSMS(s), which was regulated in the Norwegian offshore industries in the early 1980s and was further extended to all onshore industries in 1992. The Labor Inspectorate may adopt suitable inspection methods and strategies from the traditional on-site inspections to a more system-oriented approach. The Norwegian regulation recognizes that a higher quality in working life requires the involvement and active participation of the top management.

Norges Standardisengsforbund 96/402803 (1996) facilitates OHS operation within the framework of statutory requirements and evaluation of an organization's management against the established and recognized international standards, as appropriate. It covers multiple issues in a single document, such as process quality and its influence on the quality of products and services, protection of the environment against pollution and waste, OHS, and safety of products and services.

The Brazilian Ministry of Labor guidelines (NR-9, 1997) promulgates the obligatoriness for implementation of an environmental risks prevention program (PPRA) by employers that have workers under their responsibilities; however, the regulation may not be recognized as a true OHSMS due to the lack of core system elements such as management commitment, OHS goals and objectives, performance measures, programs and procedures, documentation, records management, and auditing. The Brazilian standards (NBR 10004) cover issues on safety and occupational medicine. The Workwell audit in Ontario (Canada) is a special type of mandatory OHSMS intervention that is administered through a legislated agency, the Workplace Safety and Insurance Board (Bennett, 2002). This is required for workplaces with poor performance in terms of worker compensation claims or legislative compliance.

In Australia, the OHS Act 2000 (supplemented by OHS Regulation 2001, New South Wales) requires employers to identify, assess, and eliminate or control risks, and consult with the employees on OHS matters to provide a safe workplace. The OHS Act is implemented by WorkCover, New South Wales, as referred to in the Workplace Injury Management and Workers Compensation Act 1998. Through WorkCover, industry codes of practice are made to provide practical guidance and advice on how to achieve the standards required by the OHS Act.

Indonesia and Singapore are unique in making OHSMS mandatory in specified undertakings. The Indonesia Ministry Regulation (1996) implements OHSMS in enterprises with more than 100 workers with hazardous potential in the workplace risk. Regulation PER.18/MEN/XI/2008 on the Implementation of an Audit for OHSMS (2008) includes a direction to regulate the procedures for registration as auditors and to conduct an audit of occupational health and safety at least once every three years in a facility. The Ministry issues certificates to industries, recommended by an independent audit body. In Singapore, the mandatory implementation of OHSMS applies to shipyards, specified construction worksites, and three classes of factories in the manufacturing sector.

In Hong Kong, under the Safety Management Regulation (1999), the "general duty provisions" are imposed, which places the maintenance of OHS as a continuous legal responsibility on all those who have control over the workplace environment. The proprietors or contractors of industrial undertakings are obliged to develop, implement, and maintain a safety management system, including 14 process elements (e.g., management commitment, training of workers, and evaluation of hazards, safety procedures, and programs). The OHS Council as a statutory body is responsible for safety audit schemes, including auditor accreditation.

Voluntary OHSMS Standards With and Without Certification Support

Until December 2013, over one million management systems conforming to ISO 9001:2008/ISO 14001:2004 standards have been implemented and certified worldwide. The top five countries, in terms of the total number of certifications of the management standards, are China, Italy, Germany, Japan, and the U.K. China alone represents nearly 43% of the total number of certifications issued up to date.

The voluntary OHSMS standards or guidelines, with formalized specifications (e.g., BS 8800, OSHAS 18001), were first promoted by commercial organizations and industry associations, targeting large enterprises to implement good corporate citizenship and employee welfare. The OHSAS 18001 specification was designed to align and integrate with the ISO 9000/ISO 14000 standards. This auditable standard has widely been adopted for certification purposes. The ISO 45001 standard is in the making and, upon release, may replace the OHSAS standard.

National voluntary OHSMS(s) exist in many countries, with or without certification purposes (Table 33.3). No official estimates are available as regards the certifications of different voluntary OHSMS models; however, the certifying bodies that represent OHSAS 18001:1999/2007 have about greater than three-fourth of the marketplace certification in ISO 9001/ISO 14001. This suggests the scope and enormity of the application of the voluntary OHS management systems. Further, the applications of OHSMS are also aligned with the national regulatory requirements.

The Australian government has been active in advancing national OHS strategies by creating a mixture of regulatory mandates and voluntary adoption of OHSMS by the enterprises (Saksvik & Quinlan, 2003; Gallagher, Underhill, & Rimmer, 2001). Audit systems have been introduced, incorporating the essential elements of OHSMS into their criteria (e.g., WorkSafe Plan, Western Australia; TriSafe Management Systems Audit, Queensland; Safety Achiever Business System, South Australia; SafetyMAP, Victoria). The SafetyMAP is used in Health and Safety Organization (HSO, Victoria) certification program at three levels—initiation, transition, and advanced. The joint accreditation system of Australia and New Zealand (JAS-ANZ) controls the certification of AS/NZS 4801 and 4804 standards on OHSMS.

The U.S. Occupational Safety and Health Administration (OSHA) encourages states and territories to develop their own safety and health plans under the Occupational Safety and Health Act (1970). The Voluntary Protection Program (VPP) was promulgated to recognize and approve worksites with exemplary OHS management programs (U.S. Federal Register, 1988). The VPP levels (Star, Merit, and Demonstration) are designed to recognize the level of achievements by the enterprises in incorporating OHS programs into their total management system. There is an emphasis that an average VPP worksite has a lost workday incidence rate of 52% below average for its industry. The Safety and Health Program Management guidelines (US Federal Register, 1989) identify four general elements—management commitment; employee involvement; worksite analysis; and hazard prevention and control, safety, and health training. The guidelines aimed at maintaining an organizational program that provides systematic policies, procedures, and practices to protect employees, allowing them to recognize and control workplace hazards, specific job hazards, and potential hazards. Corresponding to the ISO 9001:1994 version of the quality assurance system, the American Industrial Hygiene Association (AIHA, 1996) designed the ANSI accredited guidance, comprised of 20 primary sections in designing, implementing, and evaluating OHS management systems. The new ANZI Z10 (2012) is a voluntary consensus standard on OHSMS (Table 33.1) and this performance standard applies to organizations of all sizes and types.

The Netherland's Normalisatie-Instituut guide NPR 5001 (1997) functions to (a) guarantee OHS management through a planned approach to risks prevention, (b) encourage people to control risks, build relationships with stakeholders, and meet the statutory OHS obligations, and (c) promote organizations for continual OHS improvement and implement plans to integrate into the business management. Every employer is obliged to have a contract with a certified OHS Advisory Service (Arbodienst) in executing a risk assessment and implementation plan and carrying out periodic evaluation of the OHS situation. The Dutch Labour Inspectorate enforces actions directed on the failures at the workplace level and in the OHSMS. Like the Dutch standard, the Irish code of practice on OHSMS (National Standards Authority of Ireland, NSAI, 1997) is not intended for certification purposes. This code of practice may be implemented in support of compliance with OHS legislation, which is regulated by the Irish Health and Safety Authority under the Safety, Health, and Welfare Act (1989).

Asociacion Espanola de Normalizacion y Certificacion, UNE 81900 (1996) documented general rules for implementation of an OHSMS. Three documents included are the auditable specification standard, audit process, and a vocabulary document. In the overall analysis, the Spanish certification standard is sufficiently generic to be adaptable to any organization regardless of its size or the nature of its activities. The Polish national program, PL 9407 (1996), included a collection of best practices and models to deal with OHS management in small and medium-sized enterprises.

The Polish standardization authority PN-N-18001/18002 standards define requirements of OHSMS and guidelines on risk assessment.

Utilizing internationally recognized OHSMS frameworks, the British Safety Council (2013) developed a unique voluntary Five Star Audit model to objectively evaluate the OHS system(s) and arrangements against current best practice techniques. The specification covers 66 system elements under five sections of the audit (policy and organization—750 points; strategy and planning—1,125 points; implementation and operation—1,500 points; performance measurement—1,125 points; evaluation and review—500 points), totaling a maximum numerical value of 5,000 points. Two management indicators (leadership and continual improvement) are evaluated as the elements and as scoring areas within other elements. The cumulative scoring of an enterprise-level management system is expressed in a numerical safety grading system on 100 percentage points.

Good OHS Practice Guides

Certain OHS practice or service guides are essentially management systems that assist organizations and industrial enterprises to establish a responsible image within the marketplace. For example, the BS 8800 guidance on OHSMS stands apart to show how OHS performance can be improved, and the system of management integrated with other aspects of business. The specifications of the British Standard (Table 33.1) are not used for certification purposes. The influence of the system elements and the framework of BS 8800 is visible in OHSAS 18001 and ILO-OSH (2001). ISO 14001 has been adopted as a generic template to develop these OHSMS models; however, the key dimensions—such as employee participation, medical surveillance, and health programs—are embedded in ILO-OSH (2001) model, as described in Table 33.3. The ILO-OSH model is not intended for certification purposes.

Most OHSMS models are primarily directed to large enterprises. The situations in SME and informal sectors are excruciating (Nag & Nag, 2003; Hasle & Limborg, 2006; Hasle, Kines, & Andersen, 2009)—particularly in populous nations like China and India—and their low commitment towards OSH management is generally obvious (Micheli & Cagno, 2010; British HSE, 2007) due to compelling circumstances (e.g., large employment share, lack of economic and technological resources). The prevailing national OHS laws and management models do not exclude these sectors; however, there is an absence of links between the constituent players (e.g., regulatory agencies, employers/contractors, workers) in structuring contextual programs and systems.

The system designed as the BOHS has received wide publicity for its applicability in linking and sequencing of processes in managing workplace hazards in small and micro enterprises (Rantanen, 2005b). Also, the SA 8000 (2008) and auditable social accountability management standard may be potentially useful through the national OHS strategies, since the system explicitly addresses the basic rights of workers and workplace conditions and requires the compliance with provisions of international human rights norms and related ILO conventions and recommendations.

Self-Regulatory OHSMS Schemes and Promoted Through Statutory Bodies

Self-regulatory OHSMS schemes, with the hybridization of voluntary motives and legislative requirements, have also been adopted in many countries and promoted through statutory bodies. Based on the good practice of the ILO-OSH (2001) model—the State Administration of Work Safety—the People's Republic of China (2001) issued OHSMS guidelines, the guidance committee of the State Commission of Economy and Trade Trial Standards developed national guidelines, and the accreditation organizations and auditors are certified by the OHSMS guidance committee. Both the Occupational Disease Control Act and the Work Safety Act came into force in 2002.

In India, the Factories Act (1948, amendment in 1988) and other legislations brings provisions concerning the enabling steps to improve OHS management in the registered enterprises. The rule provisions cover hazardous industries and processes, OHS surveys, notifiable diseases, accidents, employee participation in safety management, etc. The directorate general of factory advice service and labor institutes (DGFASLI) advises the central and state governments on the administration of the Factories Act and coordinates the inspections. The efforts go towards the inclusion of OHSMS elements in defining OHS goals, objectives, and performance measures. The public safety standard IS 18001:2007 on OHSMS is made available by the Bureau of Indian Standards; simultaneously, there is a dramatic rise in the OHSAS 18001 certification through different accredited bodies.

The Malaysian OHS Act (1994) extends the legal basis of protecting workers to all sectors of economic activity. Under the Act, the National Council for Occupational Safety and Health is a tripartite forum that develops and reviews national OHS strategies and practical programs. The certifications of OHSAS 18001 as private programs are also available. In Japan, under the Industrial Safety and Health Act (1972), the Ministry of Health, Labor, and Welfare are responsible for establishing the industrial accident prevention programs. The Ministerial ordinance (1997) published by the Japan Industrial Safety and Health Association (JISHA) provides guidance that helps enterprises establish OHSMS (Kogi & Kawakami, 2002). These apply to enterprises of all scales and types of industry, and there is no official certification of OHSMS; it only certifies those who are trained in OHSMS. The Korean Occupational Safety and Health Agency (KOSHA), under the Industrial Safety and Health Act (1990, amendment 2000), promotes OHSMS and certification in line with the OHSAS 18001, and assists enterprises. The government ensures OHS regulations and promotes self-management to reflect effective OHS practices (International Labor Office, 2003).

SECTORAL MODELS AND SCHEMES

A quick scan of the Fortune 1000 listed corporations around the world reveals a vast variety of unique OHS programs that have evolved and implemented and successfully integrated with other business functions in the enterprises. This contribution included an analysis of a selected auditable OHSMS schema (e.g., Responsible Care, E&P Forum, MASE/DT 78, GEHSE systems) that are widely practiced in chemicals, petrochemicals, oil, and gas installations. The E&P Forum (1998) is comprehensive in addressing the traditional OHS issues and core management system approaches, such as leadership commitment, resource allocation, management review, continual improvement, and integration with other organizational systems. With relative flexibility to comply with regulatory obligations in France, thousands of French enterprises have voluntarily decided to adopt sector-specific OHSMS schemes, such as MASE/DT 78 to better deal with occupational risks linked to their activities (INRS, 2004; Drais, Favaro, & Aubertin, 2002; Cambon, 2007).

The OHSMS model(s) is the enterprise-level framework(s), and the implementation of a system is a means of meeting applicable regulatory and legislative obligations. It may be recognized by those concerned that they are not substitutes to the statutory OHS laws in the context of its application. It is not within the scope of this handbook to bring an account of the individual/proprietary OHS management systems; however, the effectiveness of OHSMS(s) with respect to different sectors of employment is briefly elucidated in the next section.

EFFECTIVENESS, PROSPECTS, AND CHALLENGES OF OHSMS PROGRAMMES

The systems approach to OHS management is increasingly being used worldwide by organizations of all types and sizes. On one hand, the standards bodies, commercial organizations, and industry associations adopted national as well as sectoral OHSMS frameworks, and on the other, UN agencies like ILO and WHO proposed global strategies on OHS development (International Labor Office, 2003; World Health Organization, 2007). Many national governments formulated national OHS strategies and programs, with the option of a mix of mandated legislative requirements and enterprise's voluntary initiatives.

For an OHSMS to remain useful and authoritative, it is important that the system is effective and reflects demonstrated needs. The analysis of impacts of OHSMS(s) on OHS management and interventions demand comparison of objective and subjective outcome measures on a reasonable time horizon (Verbeek, Pulliainen, & Kankaanpää, 2009; Tompa et al., 2010). Robson et al. (2007) compared the voluntary and mandatory OHSMS(s) and distinguished its effects on (a) implementation (i.e. change in workplace OSHMS), (b) intermediate outcomes (e.g., awareness and better safety climate, increased hazards reporting by employees and their increased involvement on OHS issues), (c) final OHS outcomes (e.g., decline in work-related injury/illness, reduction in loss of working hours), and (d) socioeconomic outcomes (e.g., decrease in costs related to disability, such as workers' compensation costs or short and long-term disability costs; firm insurance premiums or workplace productivity). Hovdena et al. (2008) investigated the conditions of employee influence on OHS matters in the Norwegian offshore oil and gas sector. The results indicate that the safety representatives perceived the climate of participation and collaboration as less conducive to overall OHS regulations in the enterprise; on the other hand, the employers had more reliance on the formal OHSMS. Therefore, any lack of consistency in perception might manifest in the measures of intervention.

Fernández-Muñiz, Montes-Peón, and Vázquez-Ordás (2009) undertook a literature review to formulate hypotheses—to identify good practices in safety management—and further tested these in 455 Spanish enterprises. The enterprises with long experiences in OSHMS(s) revealed more positive feedbacks in terms of OHS and economic outcomes, irrespective of voluntary and/or mandatory OSHMS interventions. Further study in 131 OHSAS 18001-certified enterprises (Fernández-Muñiz, Montes-Peón, & Vázquez-Ordás, 2012) noted that the top management's commitment and communication influences the safety behavior, safety performance, employee satisfaction, and the organization's competitiveness. By investigating 1 OHSAS 18001-certified PCB manufacturing industries in Taiwan, Chena et al. (2009) also emphasized that the top management's commitment and support was critical for success in OHSAS implementation, whereas the reason for its failure was poor collaboration among personnel in the industry. The level of commitment of the top management, the frequency of unsafe conduct by employees, subcontractor rule violations, the completion rate of corrective and preventive measures, and the level of fire-fighting system were significant performance indicators in OHSMS performance evaluation. A questionnaire survey (Vinodkumara & Bhasib, 2011) at eight chemical industries in Kerala (India) revealed that the employees had different perceptions about the safety variables in industries with OHSAS 18001, ISO 9001, and no certification. Step-wise regression analysis yielded a set of safety management practices that well predicted safety behavior in OHSAS 18001 certified units, whereas the safety rules and procedures emerged as common predictors of safety behavior in all three categories of industries.

There is an increasing trend among the companies and the certifying bodies in integrating and implementing more than one management systems that are mutually compatible and better aligned. Different independent management systems are integrated with linkages so that similar processes are seamlessly managed and executed without duplication. Since no integrated management systems (IMS) standard exists to date, these are primarily structured endeavors in integrating audit programs of multiple management systems (e.g., combining ISO 9001, ISO 14001, OHSAS 18001 [Nag, 2002]). The connections between different categories of management systems are not that real, apart from some structural resemblance to the logical flow of management elements. Therefore, integration of the systems has perceptible limitations from an operational point of view.

Salomone (2008) investigated sample Italian companies as regards to the potential of IMS, from the analysis of common aspects, such as real motivations (company image, costs saving, etc.), obstacles (unclear regulations, lack of financial support, etc.), driving forces, and external pressures that companies meet when implementing the management systems. Pheng and Kwang (2005) conducted a survey in 30 construction companies in Singapore, those who were certified to IMS, and observed that the larger companies reaped more benefits and incurred lesser costs. The majority of the companies, who strived for IMS certification voluntarily, expressed that the benefits of

IMS outweighed its costs. Also, a small number of companies strived for IMS because of regulatory requirements.

Research studies are scanty on the measures of OHSMS performance and the effectiveness of the OHSMS models and schemes in securing healthy and safe workplaces. On statistical consideration, the available studies do not permit reliable generalization as to the effectiveness of OHSMS models, and any likelihood of success of the systems in different industrial contexts. However, the evidence favors that an OSHMS in an enterprise brings better awareness of OHS goals and risk identification and yields increased corrective measures and positive socioeconomic and health outcomes. Reduced incidence of accidents and work-related diseases decreased absenteeism and higher productivity are the indicators of the system compatibility between worker protection and corporate competitiveness.

NOTE

* Former Scientist, National Institute of Occupational Health (Indian Council of Medical Research), Ahmedabad, India

REFERENCES

American Industrial Hygiene Association (AIHA). (1996). *Occupational health and safety management system: An AIHA guidance document*, No. AIHA OHSMS 96/3/26, 31 pp.

ANSI. (2012). *American national standard for occupational health and safety management systems (ANSI Z10-2012)*.

Australia, New South Wales, The Occupational Health and Safety (OHS) Act 2000 (supplementary OHS Regulation 2001).

Australia, Victoria, Health and Safety Organization (HSO) (1999). *Safety management achievement programme (SafetyMAP)*.

Australia, Workplace injury management and workers compensation Act 86 (1998).

Bennett D., Health and safety management systems: liability or asset? *Journal of Public Health Policy*, 23: 153–171, 2002.

Bottomley B., *Occupational health & safety management systems: Information paper*, National Occupational Health & Safety Commission, Commonwealth of Australia, (1999).

Brazil, Ministry of Labor guidelines. (1997). NR-9 (PPRA)—Environmental Risk Prevention Programme.

British Health and Safety Executive. (1993), *Successful health and safety management: HS (G) 65, HMSO*. ISBN 07176 0425 X.

British HSE, *Health and safety in the small to medium-sized enterprise — Psychosocial opportunities for intervention*, Health and Safety Executive, Sudbury, UK, 2007.

British Safety Council. (2013). *Five star occupational health and safety audit, Specification document*, pp.12.

British Standards Institution. (1996). *BS 8800, Guide to occupational health and safety management systems*.

British Standards Institution. (2007, orig. ver., 1999). *BS-OHSAS 18001:2007, Occupational health and safety management systems—Requirements*.

Cambon J. (2007). *Vers une nouvelle methodologie de mesure de la performance des systemes de management de la sante-securite au travail. Sociology*. Ecole Nationale Superieure des Mines de Paris.

Chena C. Y., Wub G. S., Chuangc K. J., Mac C. M., A comparative analysis of the factors affecting the implementation of occupational health and safety management systems in the printed circuit board industry in Taiwan, *Journal of Loss Prevention in the Process Industries.*, 22: 210–215, 2009.

China, The State Administration of Work Safety. (December 2001). *OHS management system guidelines*.

Drais E., Favaro M. and Aubertin G., *Les systèmes de management santé-sécurité en entreprise: caractéristiques et conditions de mise en oeuvre*, Institut National de la Recherche et de la Sécurité, Paris, 2002

DT 78 (June 2004). *Manuel UIC d'habilitation des entreprises extérieures de l'industrie chimique* (Union des Industries Chimiques), France.

European Union. (1993). *The Council of the European Communities Regulation 1836/93, No. L 168—Community Eco-Management and Audit Scheme (EMAS)*.

Fernández-Muñiz B., Montes-Peón J. M., Vázquez Ordás C. J., Relation between occupational safety management and firm performance, *Safety Science*, *47*: 980–991, 2009.

Fernández-Muñiz B., Montes-Peón J. M., Vázquez-Ordás C. J., Safety climate in OHSAS 18001-certified organisations: Antecedents and consequences of safety behaviour, *Accident Analysis & Prevention*, *45*, 745–758, 2012.

Frick K. and Wren J., Reviewing occupational safety and health management: Multiple roots, diverse perspectives and ambiguous outcomes, Frick K, Jensen PL, Quinlan M and Wilthagen T (eds), *Systematic occupational safety and health management: Perspectives on an international development*, Emerald Group Publishing Ltd., Bingley, 2000.

Gallagher, C., Underhill, E., & Rimmer, M. (2001). *Occupational health and safety management systems: A review of their effectiveness in securing healthy and safe workplaces*. National Occupational Health and Safety Commission (Commonwealth of Australia). ISBN 0642 709815.

GEHSE, *Guide d'Engagement Hygiène, sécurité, environnement pour les entreprises extérieures intervenant dans les dépôts d'hydrocarbures ou les petits établissements pétroliers ou les stations-service*, Association GEHSE, France 2000 (rev., February 2004).

Hasle P. and Limborg H. J., A review of the literature on preventive occupational health and safety activities in small enterprises, *Industrial Health*, *44*: 6–12, 2006.

Hasle P., Kines P. and Andersen L. P., Small enterprise owners' accident causation attribution and prevention, *Safety Science*, *47*, 9–19, 2009.

Hong Kong, Factories and Industrial Undertakings (Safety Management) Regulation, November 1999.

Hovdena, J., Lieb, T., Karlsenc, J. E., & Alterend, B. The safety representative under pressure: A study of occupational health and safety management in the Norwegian oil and gas industry, *Safety Science*, *46*: 493–509, 2008.

1988. Ministry of Labour, The Factories Act. 1948 (amendment 1988).

Indonesia, The Ministry Regulation (PER.05/MEN/1996), Implementation of Occupational Health and Safety Management System (1996); Regulation (PER.18/MEN/XI/2008), Implementation of an Audit for OHSMS, November 2008.

INRS, *De l'évaluation des risques au management de la santé et de la sécurité au travail*, Institut National de Recherche et de Santé, Paris, 2004.

1998. International Exploration and Production Forum (E&P Forum). (1998). Report No. 6.36/210—*Guidelines for the development and application of health, safety and environmental management systems*.

International Labor Office. (2001). *Guidelines on occupational safety and health management systems (ILO-OSH)*. Geneva. ISBN: 92-2-111634-4).

International Labor Office. (2003). *Promotional framework for occupational safety and health, Report IV (1)*. Geneva.

International Occupational Hygiene Association (IOHA). (1998). *Occupational Health and Safety Management Systems, Review and Analysis of International, National, and Regional Systems and Proposals for a New International Document* (prepared for the International Labour Office, Geneva; Dalrymple, H., Redinger, C., Dyjack, D., Levine, S., Mansdorf, Z.).

International Standards Organization. (2004). *ISO 14001:2004 (draft rev. 2015) Environmental management systems—Requirements with guidance for use*. Geneva.

International Standards Organization. (2008). *ISO 9001:2000 (rev. 2008) Quality management systems—Requirements*. Geneva.

International Standards Organization. (2013). *The ISO survey of management system standard certifications – 2013*. Geneva.

Ireland, National Standards Authority of Ireland. (1997). *NSAI SR 320, Recommendation for an occupational health and safety (OH and S) management system*.

Ireland, Safety, Health and Welfare at Work Act (1989/2005).

Japan, Industrial Safety and Health Regulation (Regulation No. 52), Tokyo: Ministry of Health, Labour and Welfare; 2005 (Ordinance on OHSMS guidance, Japan Industrial Safety and Health Association (JISHA), March, 1997).

Kogi K. and Kawakami T. Trends in occupational safety and health management systems in Asia and the Pacific, *Asian-Pacific Newsletter—Occupational Health and Safety*, *9*: 42–47, 2002.

Korea, Republic of, Ministry of Labor. (1990, amended 2000). Industrial Safety and Health Act, *Chapter II – Safety and health management systems*.

Malaysia, Occupational Safety and Health Act, 1994.

MASE, *Manuel d'Amélioration Sécurité Entreprise*, Association MASE (association d'entreprises utilisatrices et enterprises extérieures), France (June 2004).

Micheli G. J. L. and Cagno E., Dealing with SMEs as a whole in OSH issues: Warnings from empirical evidence, *Safety Science*, 48, 729–733, 2010.

Nag P. K. and Nag A., A national priority on occupational health and safety management system, *ICMR Bulletin*, 33: pp. 15, 2003.

Nag P. K., The management systems—Quality, environment, health and safety: ISO 9000:2000, ISO 14001, OHSAS 18001, pp. 683 (Mumbai: Quest Publications), 2002.

Nederlands Normalisatie-Instituut. (1997). *NPR 5001, Dutch Technical Report—Guide to an occupational health and safety management system.*

Norges Standardisengsforbund 96/402803. (August 1996). *Management principles for enhancing quality of products and services, occupational health & safety, and the environment.*

Norway, Royal Decree. (1992). *Regulations relating to systematic health, environmental and safety activities in enterprises* (Internal Control Regulations).

Pheng, L. S., & Kwang, G. K.. ISO 9001, ISO 14001 and OHSAS 18001 management systems: Integration, costs and benefits for construction companies, *Architectural Science Review*, 48, 145–151: 2005.

Poland, Labour Inspectorate. (November 1996). *Worker Protection Programme PL 9407, Safety and health management in SME's: Best EU practices regarding safety and health management in small and medium enterprises (SME's).*

Rantanen J. *Basic occupational health services: A WHO/ILO/ICOH/FIOH guideline.* 2nd edition Helsinki, Finland: Finnish Institute of Occupational Health, p 19, 2005a.

Rantanen J. Basic occupational health services—Their structure, content and objectives. *SJWEH Supplements* no 1: 5–15, 2005b.

Redinger C. F. and Levine S. P.. Development and evaluation of the Michigan occupational health and safety management system assessment instrument: A universal OHSMS performance measurement tool, *American Industrial Hygiene Association Journal*, 59: 572–581, 1998.

Robson L., Clarke J., Cullen K., Bielecky A., Severin C., Bigelow P., Irvin E., Culyer A., and Mahood Q., *The effectiveness of occupational safety and health management systems: A systematic review*, Institute for Work & Health, Toronto, 2005.

Robson L. S., Clarke J. A., Cullen K., Bielecky A., Severin C., Bigelow P. L., Irvin E., Culyer A. and Mahood Q., The effectiveness of occupational safety and health management system interventions: A systematic review, Safety Science, 45: 329–353, 2007.

SA 8000, *International Standard, Social Accountability International (SAI), USA* (2008).

Saksvik PØ and Quinlan M., Regulating systematic occupational health and safety management: Comparing the Norwegian and Australian experience, *Industrial Relations*, 58: 33–59, 2003.

Salomone R, Integrated management systems: Experiences in Italian organizations, *Journal of Cleaner Production*, 16: 1786–1806, 2008.

Spain, Asociacion Espanola de Normalizacion y Certificacion (December 1996). *UNE 81900, Prevention of occupational risks: General rules for implementation of an occupational safety and health management system.*

1997. Standards Australia/Standards New Zealand. (1997). *AS/NZS 4804/4801, Occupational health and safety management systems – Specification with guidance for use.*

Stolk C. V., Staetsky L., Hassan E., and Kim C. W., *Management of occupational safety and health: An analysis of the findings of the European survey of enterprises on new and emerging risks*, pp. 58, (Luxembourg: Office of the European Union), 2012.

Tompa E., Verbeek J., van Tulder M. and de Boer A., Developing guidelines for good practice in the economic evaluation of occupational safety and health interventions, *Scandinavian Journal of Work, Environment & Health*, 36: 313–318, 2010.

U.K. Chemical Industries Association. (1998). *Responsible care management system* (3rd edition).

USA, Chemical Manufacturers Association. (1992). *Employee Health and Safety Code—Responsible Care: A resource guide for the employee health and safety code of management practice.*

USA, Occupational Safety and Health Administration (OSHA). (1988). Federal Register, 12/4/1988; Voluntary Protection Programs.

USA, Occupational Safety and Health Administration (OSHA). (January 26, 1989). *Safety and Health Program Management Guidelines, Standard*—Federal Register, 29 (CFR), 1926 Subpart C, 54(18): 3094–3916.

Verbeek J., Pulliainen M. and Kankaanpää E., A systematic review of occupational safety and health business cases, *Scandinavian Journal of Work, Environment & Health*, 35: 403–412, 2009.

Vinodkumara M. N. and Bhasib M. A study on the impact of management system certification on safety management, *Safety Science*, 49: 498–507, 2011.

Walters D, Wilthagen T and Jensen P.L., Introduction 'Regulating health and safety management in the European Union', Walters D. (ed.), *Regulating health and safety management in the European Union—A study of the dynamics of change*, Presses Interuniversitaires Européennes, Brussels, 2002.

World Health Organization. (2007). Workers' health: Global plan of action. In *60th world health assembly*. WHA60.26.

34 European Union's Legal Standard on Risk Assessment

Kaj Frick

CONTENTS

RISK ASSESSMENT TO BE UNDERSTOOD WITHIN ITS WIDER SETTING

The European Union's (EU's) standard on Risk Assessment is formally a simple regulation (in the following, "Risk Assessment" refers to this standard as opposed to all other forms of "risk assessment"). It consists of two paragraphs in the EU Council Directive 89/391/EEC of June 12, 1989 "on the introduction of measures to encourage improvements in the safety and health of workers at work" (usually known as the Framework Directive). Article 6.3(a) states that "*the employer shall... evaluate the risks to the safety and health of workers*" and article 9.1(a) adds, "*the employer shall be in possession of an assessment of the risks of safety and health at work.*"

However, this is not a real standard. Neither is it an internationally- nor nationally-accepted definition of a procedure. It is also not a mandatory, legal regulation. All EU directives are addressed to the member states, not their citizens. Thus, the Framework Directive had to be transposed into national laws before its requirements came into legal force (at different times in the various member states). The transposition processes depend very much on domestic politics and their results may—and do—vary in important aspects. This is also the case for the Risk Assessment (see, further, Walters, 2002a, on the background, comparison, and analysis of the Directive, its transposal, and partly its implementation in seven member states).

That the Framework Directive is not a proper standard does not diminish its importance. It is indeed the framework for regulating workplace occupational health and safety (OHS) within the EU. Its regulation of Risk Assessment is, therefore, fundamental for the management of ergonomic and other risks at work. However, the brief and principal requirement in the Directive makes it necessary to place Risk Assessment in its wider setting to understand the risks and how they are to be assessed. This is more important as the requirements of the Framework Directive, as a whole, are very demanding.

To clarify the EU requirement on Risk Assessment, this chapter:

- Discusses the concept of risk assessment and its development
- Describes how the new process-oriented OHSM strategy, notably Directive 89/391/EEC, has evolved from this development and from other changes in OHS policies
- Describes the type and content of the Framework Directive's Risk Assessment, especially the risks and how they are to be assessed
- Details the slow implementation in—and outlines some principles of—the interpretation and application of the Risk Assessment "standard" even in small firms

SCIENTIFIC AND POLITICAL ROLES OF RISK ASSESSMENT

WHAT IS RISK ASSESSMENT?

Although the formulations vary much between many writers (see following), there is a large degree of principal consensus that *risk assessment is a combination of analysis and evaluation of risks of the use of a certain process, substance, or technology.* It consists of the following steps:

Risk Analysis

1. Define the object that is to be assessed—for example, a substance, a work process, or a production system—and the circumstances and conditions of its use. (Some writers call this "formulate the alternatives," or "system description.")
2. Identify the possible health and safety risks of the object. (According to the EU's official "Guidance on Risk Assessment at Work," this consists of: "Collect information," "Identify hazards," and "Identify those at risk;" European Commission, 1996, p. 13.)
3. Estimate the level of this risk by systematic qualitative or quantitative means, such as exposure measurements or statistics on the probability of accidents (EU guidance: "Identify patterns of exposure of those at risk").
4. Summarize 2 and 3 into how many and how serious diseases or injuries are likely to be caused by the risk object (EU guidance: "Probability of harm/severity of harm in actual circumstances").

Risk Evaluation

5. The expected risk of harm is compared to the relevant criteria of acceptable risk. At the workplace level, this can be regulations or internal health and safety goals. Societal risk evaluations often try to compare the risk of harm to the social gains by using a substance or technology.

The identification of the risks and their evaluation ends the risk assessment. For many writers, this is the first half of "health and safety (or risk) management," the second half is the "risk control" or "risk management." Although it is not expressed in formal risk management terms, this is also the case with the EU Framework Directive's standard on Risk Assessment (see section under Risk Assessment and the New OHS Management Strategy). Such a control starts with an evaluation of alternative methods to prevent, eliminate, or reduce the risks to the acceptable level, often by means of a cost-benefit analysis. After the optimal measures have been chosen, these must be implemented. Then their effects have to be monitored and, if necessary, new measures must be taken. Finally, the whole process of risk management should be audited and improved. Risk assessment is a technique to assess and manage the risks—a continuous or repeated process (when any conditions of the assessment are changed).

RISK ASSESSMENT AS SCIENCE

The aforementioned systematic procedure is a modern construction. However, risks are an eternal aspect of human life. We, therefore, always have to assess them—usually unconsciously—to minimize harm to our health, families, crops, businesses, or whatever is important to us. Our split-second grasp of the traffic—number, distance, speed of cars, slipperiness, visibility, width of the street, obstacles, and so forth—as we cross a street is a daily example. Since ancient times, thinkers have also sought to draw general conclusions from such assessments. The Roman writer Vitruvius advised against the use of lead pipe for drinking water, as he noted that exposure to lead fumes was linked to blood diseases (British Medical Association, 1992, p. 1). Ramazzini is perhaps the most famous example. In his work, "on the diseases of workers" (1700), he carefully observed and assessed the health risks of some 50 vocations.

But explicit and systematic methods of risk assessment are much more recent. What started as risk analysis around 1970 (Otway, 1985; Renn, 1985, p. 113) has since evolved into a broad array of instruments. This development is illustrated by Ogden's (2003) example of how the analysis and regulation of asbestos risks has evolved considerably since 1968. We now have methods to assess, analyze, evaluate, estimate, and control or manage the problems of risks, safety, hazards, or exposures in modern production. The professionalization of this methodology was marked by the formation of a Society for Risk Analysis in the United States in 1980 and in Europe in 1986 (Dwyer, 1991, p. 255).

There are now more than 100 titles on risk assessment in the library of the Swedish National Institute for Working Life alone (http://arbline.arbetslivsinstitutet.se) and close to another 100 if we add the similar concepts of "hazard" and of "evaluation" and "analysis." The most important differences are in the types of risks these works discuss and what level of analysis they emphasize—that is, the balance between risk assessments as advice to societal decisions versus to workplace improvements. On one hand, chemical risks dominate; on the other, accident risks are also rampant. Many are medical-toxicological works on how to assess the risks of diseases and environmental damages of hazardous substances (e.g., Sadhra & Rampal, 1999; Vollmer et al., 1996). Exposures to other work environment risks—such as noise, radiation, and heat—are assessed with the same dose-response principles (e.g., Holmér, 2000).

Technically oriented risk assessments of sequences of events, which may turn into accidents, also abound. They include special assessments of machine safety (IEC, 1995; ISO, 1999) and risks of disasters in chemical factories, nuclear power plants, offshore oil platforms, and other high-risk facilities (Ansell & Wharton, 1992; Cox & Cox, 1996; Stewart & Melchers, 1998; HSE, 1998; Harms-Ringdahl, 2001). These are sometimes developed into mathematical models of probabilistic risk assessment (Kumamato & Henley, 1996; Mosleh & Bari, 1998). Of the assessments of special risks, the ergonomic ones may combine the mechanical-sequential logic of analyzing work systems with the dose-response logic of exposure to different types, durations, and severities of musculoskeletal strain (Colombini, Occhipinti, & Grieco, 2002; HSE, 1994). A similar combination of perspectives, often with the help of checklists, is used in risk assessments of special tasks or industries (e.g., Krüger et al., 1998; Stem, 1980). There are also writers with a broad perspective on business risks in general, who combine, for example, chemical disasters with fraud and fire (e.g., Reason, 1997).

If we instead cut them along the level of analysis, HSE (1999), Roberts-Phelps (1999), and Jeynes (2002) give very hands-on advice on how to assess health and safety risks in workplaces.

Many checklists are even more concrete in structuring what risks to look for and how to assess the results (see www.pk-rh.com/en/index.html and www.prevent.se/english/). Others have a general perspective on the varying risks they discuss but also aim to help workplaces to assess (and control) these (e.g., Cox & Cox, 1996; IEC, 1995; Harms-Ringdahl, 2001; Holmér, 2000). Several writers—like Otway and Peltu (1985), Ansell and Wharton (1992), Hansson (1993), Kumamato and Henley (1996), Stewart and Melchers (1998), Vollmer et al. (1996), and Sadhra and Rampal (1999)—mainly focus on the societal principles and politics of risk analysis and assessment.

With such a variety of problems and social levels, it is not surprising that we lack an international standard to define risk assessment. The influential guideline, BS 8800 (1996, p. 4; on OHSM systems), only contains a brief definition of this as "the overall process of estimating the magnitude of risk and deciding whether or not the risk is tolerable or acceptable." According to the EU guidance (European Commission, 1996, p. 11), risk assessment is "the process of evaluating the risk to health and safety of workers while at work arising from the circumstances of the occurrence of a hazard at the workplace." Although this guidance describes seven steps of the assessment procedure—and another seven for the management of identified risks—these are still too general, as exemplified earlier. (See also the similar definition in the ILO Guidelines on OHS management systems from 2001, which are discussed later.)

There are definitions of risk assessment for special purposes (e.g., in ISO, 1999, on machine safety). There is also a general one, that is, "the systematic use of available information to identify hazards and estimate the risk to individuals or populations, property or the environment" (IEC, 1995). Other works give more exhaustive, but also quite varied, descriptions of what risk assessment is, as in Wharton (1992, p. 7), British Medical Association (1992, p. 19), Cox and Cox (1996, p. 32), Rampal and Sadhra (1999, p. 19), and Harms-Ringdahl (2001, p. 43). See also the comparisons by Rakel (1996) and by Boix and Vogel (1999, p. 20). Nevertheless, there is a large principal consensus in these works behind the description of risk assessment given earlier.

RISK ASSESSMENT AS A RESPONSE TO PUBLIC CONCERNS

Risk analysis and assessment, as systematic instruments for scientists and other experts, are largely responses to the public's growing concerns of the risks to health and safety and to the environment of advanced industrial production. Carson's (1962) alarm on the death of nature was a major turning point in the debate. The slow thalidomide disaster of the early 1960s gave another jolt to public trust in how those in power and their experts handled risks. Especially from the 1970s onwards, it has been revealed that the production or use of diverse substances—such as vinyl chloride and asbestos—may be (and often is) very dangerous (Markowitz & Rosner, 2002). In 1972, the United Nations Conference on the Human Environment in Stockholm, therefore, "placed risk management issues on the same high level on the international agenda as peace, trade, finance, and economic development" (Majone, 1985, p. 41).

Likewise, human development and the use of advanced technologies has, from the beginning, been accompanied by disasters—mines, buildings, bridges, and dams collapsed, ships sank, (later) steam boilers exploded, and trains and airplanes crashed. In industrial production, munitions and explosives factories have regularly exploded (which, e.g., gave rise to the DuPont safety philosophy; Mottel, Long, & Morrison, 1995). Later, the growth of the chemical industry resulted in, for example, the IG Farben explosion in Germany in 1921—which killed 550 people—and a series of chemical fires and explosions in Texas in 1947, killing 561 people (Perrow, 1984, pp. 120, 105). Safety in the nuclear power industry has long been contested, and even more so after the Three Mile Island became a near disaster in 1979 and the Chernobyl incident in 1986. With the gradual development of international media, these and other disasters increasingly became public knowledge. Seveso in 1976, Bhopal in 1984 (which killed some 10,000 people; Eckerman, 2004), Challenger in 1986, and Piper Alpha in 1988 are some of the most well known. This resulted in a debate on high-risk technologies (Perrow) or even on the modern industrialized world as a risk society (Beck, 1992).

Otway (1985, p. 3) summarizes the development and how the experts reacted:

"As contemporary societies began to produce a dazzling array of consumer goods, an awareness was forming that they were also producing a variety of toxic new chemicals, polluting the environment, and destroying vital natural resources. The public impacts of industrial accidents were also without precedence in size and kind."

"Many scientists and engineers were puzzled by lay group challenges to informed expert opinion. They believed that regulatory decisions would be less controversial if they could be given a firmer 'factual' basis. Their attempts to be more technically rigorous led to the emergence of the new 'science' of risk analysis, the use of available data, supplemented by calculation, extrapolation, theory, and expert judgment, to define the risks to people due to their exposure to hazardous materials or operations."

RISK ASSESSMENT AS CONTESTED POLITICS

Otway (1985) covers the two main aspects of risk assessment: on one hand, it has a scientific, expert nature of a systematic method to identify and evaluate risks, and, on the other, such assessments have a political setting. As they are instruments to guide decisions on risks, they are inherently political. The debate on and practice of risk assessment, thus, also concerns issues of power, competence, trust, and legitimacy among the stakeholders of risk. The latter are primarily:

- The risk producers—for example, owners-managers of nuclear or chemical plants
- The risk exposed—for example, workers or consumers
- The politicians who may regulate between risk-producers and the risk exposed but who may be risk-producers themselves
- The scientists-experts who advise on how to handle or minimize the risks.

Perrow (1984, pp. 306–328) and Dwyer (1991, pp. 237–269) further analyze the dual functions—the risk—of risk analysis and assessment. On one hand, these are tools to advise the risk-producers on how to control the risks. On the other, they are part of a strategy to limit the reactions and regulations of such technologies. Advanced calculations proving the risks to be acceptable, and methods of correct risk communication, are means to reassure the public and their political representatives. However, risk analysis and evaluation are difficult, and the experts' methods and outcomes are contested. Attempts to construct "criteria (for risk acceptance) which are usable, transparent, and agreed by all parties concerned" (Cox & Cox, 1996, p. 34) are often rejected (e.g., by Perrow, pp. 309–310). Hansson (1993) criticizes all the assumptions of risk analysis, including that they define their decision problems too narrowly, that they cannot obtain reasonably accurate probability estimates of the resulting risks, and that their evaluations of the alternatives (i.e. risk acceptance criteria) are biased (cf. also, the discussions in Otway & Peltu, 1985).

The need to analyze and evaluate risks is not contested nor is the need for good methods to do so. Risks can be diminished but rarely abolished. They often must be balanced against each other or against various gains. Systematic methods to guide risk decisions are usually better than intuitive impressions, especially in complex technical systems. The issue is, these, not the use as such but the "risk" of poor quality and of practical misuse of risk assessments. A comprehensive discussion of this is outside our scope (see Hansson, 1993). However, some clarification is necessary, as the EU's standard on Risk Assessment is also a political decision with varying and debated applications (Bercusson, 1996, chap. 23; Boix & Vogel, 1999; Karageorgiou et al., 2000).

With linking issues, risk assessments are criticized because of their:

1. Reliability: are their results accurate that the production and use of this substance or the technology entails the predicted level of risk?
2. Validity: whose knowledge is relevant in assessing the risks?
3. Sincerity: are the assessments used in good faith to minimize risks, or are they means to promote the interests of risk producers (or their opponents)?
4. Responsibility: are risks assessments (and control) to be voluntary—for risk producers, advised by experts—or are they to be regulated?

1. The reliability claims of "systemic safety [including risk assessment] to be able to plan production to reduce risks of accidents to an absolute minimum" (Dwyer, 1991, p. 244) have been questioned on, at least, three grounds:
   • Theoretically, the evidence on which to assess the risks is mainly retroactive data—for example, systematic statistics on past incidences. Yet, the assessment aims to predict the likelihood of future risks. It, therefore, requires a considerable element of fantasy, which makes it more or less inexact. The definition of the risk object is also, by necessity, a subjective selection of factors and variables, whereas all others are disregarded in the assessment (cf. Hansson, 1993).
   • Empirically, experts often disagree in their assessments—that is, in what type of risk a certain object entails—even when these are based on the same evidence. One contentious issue is the assessment's level of uncertainty and how to handle this (Hansson, 1993). The input data are also often unreliable (see Collinson, 1999, on systematic nonreporting of oil rig incidents).
   • Risk predictions do not work if their conditions are not met. Technical changes may be made, outer circumstances may vary, and maintenance may deteriorate—all without much notice of their risk effects. How possible changes should be included in the definition of the risk object is an open issue. But disaster inquiries (e.g., Cullen, 1990; Hopkins, 2000) demonstrate that, even in technically advanced organizations, safety claims and safety practices are often not the same.
2. This uncertainty also questions the validity of the risk expertise. Like other professionals, risk experts try to capture their market by emphasizing the unique importance of their competence (Dwyer, 1991, pp. 255–260). However, with demonstrated shortcomings in their assessments, risk-exposed and other laypersons have—often successfully—asserted that assessments should include their experience and views of the production and its risks. Assessments are also criticized for a poor validity in their risk evaluations, notably a too narrow view of the costs and benefits involved (e.g., Dorman, 1996; Dwyer, p. 238; Hansson, 1993; Perrow, 1984, pp. 308–310; Ruttenberg, 1981a).

3. The sincerity of an assessment is, in turn, related to the validity of practical competence. Not only is the involvement of the risk exposure in the assessment contested, but a tradition of keeping them uninformed of the risks also exists. The French state was not alone in limiting public knowledge of the operation of high-risk technologies (Dwyer, 1991, p. 295). However, growing media and public interest have made such policies by governments and risk producers less tenable (Dwyer, p. 260). Secrecy can turn into a public scandal, with grave political consequences. For example, fear of another scandal was a major reason behind the recent shift in French policies on OHS risks, due to the "asbestos crisis" (Rivest, 2002, p. 100). Professional methods for "risk communication" have instead been developed to convince the public that the risk assessments of the experts, usually on behalf of the risk producers, are correct. But even special communication methods often must work against distrust when it is revealed that the public has often been actively deceived of the nature and severity of the risks to which it has been exposed (Brodeaur, 1974; Markowitz & Rosner, 2002; cf. also, the tobacco companies' denial of the addiction risks of nicotine).

4. Finally, the responsibility for the risk assessment is also contested. Many risk experts have claimed that proper application of their advanced methods was the best way to control risks and have, therefore, opposed public regulation (Dwyer, 1991, p. 245; Perrow, 1984, p. 307). "Professionalism with strong ethics is better than laws" (Concha, 1983; in Dwyer, p. 292). However, they—and the risk producers who opposed regulation—have largely been unsuccessful. The development of "scientific" risk assessment has generally not been a strategy of "self-regulation" where the risk producers could avoid having external regulations imposed on their operations. Instead, regulations have, in some cases, been demonstrated to not only reduce risk levels but also promote technological and organizational innovations within risk-producing industries (Olsen, 1992; Ruttenberg, 1981b; Ashford & Caldart, 1997, chap. 10). See the previous section on the discussion of regulation versus voluntary standards under Risk Assessment as a Response to Public Concerns.

Likewise, the development of modern risk assessment is strongly linked to the growth of regulation (Otway & Peltu, 1985; Rampal & Sadhra, 1999, p. 4). The assessment of a widening array of risks have been mandated by increased regulations in both in the United States and the European Union (EU) as a standard tool to minimize societal harm from various risks (Ashford & Caldart, 1997, chaps. 2 and 4; Otway, 1985; and Walters, 2002a, chap. 2). Within the EU, "the current occupational health and safety legislation... depends on a risk assessment approach to managing and controlling hazards" (Jeynes, 2002, p. 13).

RISK ASSESSMENT AND THE NEW OHS MANAGEMENT STRATEGY

RISK ASSESSMENTS OF PRODUCTS AND OF PRODUCTION IN EU

The evolvement of Risk Assessment as the mandated method to control risks at work is part of the general development of OHS regulations within the EU. The common market was, from the beginning, also a social project. Occupational health and safety were included in the Founding Rome Treaty of 1957 (article 118). Like other EU policies, those on OHS started slowly in the 1960s, with the Directive 67/540/EEC on dangerous substances as an early example. An important step was the so-called Seveso Directive (82/501/EEC) on major-accident hazards involving dangerous substances (modernized as 96/82/EC). It requires manufacturers to identify the (major accident) hazards of their operations, though this was more linked to the environment than to the OHS debate.

As part of the EU activism after the Single European Act of 1986, new forms were developed to regulate the work environment (a term that intentionally marks a broad definition of the OHS concept; Walters, 2002a, p. 43). This resulted in the parallel and overarching Framework Directive (89/391/EEC) and Machinery Directive (89/392/EEC). Especially during the late 1980s and early 1990s, these were followed by several directives on special risks, which were issued in accordance with the former. The one on Material Handling (90/269/EEC) is one of several examples but other directives on conditions of work were also adopted—for example, on Working Time (93/104/EEC; see further Vogel, 1994; and Walters, 2002a, on the background and principles of the Framework Directive).

The OHS Directives have the dual purpose of avoiding risks and promoting the integration of the single market. The Directives are standards in the market sense. Harmonized EU requirements on the prevention of risks of the use of goods are to abolish the varying national regulations, which act as trade barriers. Local risk assessments of production are also to promote the single market by equalizing the terms of competition among the member states. At the same time, the upward harmonization of national regulations is also a social policy to improve occupational health and safety, other working conditions, and the environment within the EU.

The concept of Risk Assessment became prominent with the Framework Directive. However, in various formulations, such assessments have been fundamental also in earlier OHS directives (e.g., in 83/477/EEC on protection against asbestos risks). The EU legislation on risk assessment can be divided into either general assessments of specific risks or local assessments of general risks:

- The use of specific products—mainly machinery and chemical substances—is to be assessed as general risk sources. To secure that all products placed on the market are safe—and, therefore, shall be allowed free trade—the producers are to assess (control, inform about, etc.) the risks of all foreseeable use of these risk sources. With some changes in the terminology, such product risk assessments have been required since the Directive on dangerous substances in 1967. A large number of amendments and new directives—notably the Machinery Directive in 1989—have gradually made them stricter, both in coverage of risks products and in the required assessments and controls. In the case of substances, the common EU requirements are about classification, labeling, packaging, marketing, and other border-crossing trade aspects. However, worker exposures to such substances are determined by the local conditions of production. The corresponding regulations of occupational exposure limits (OEL) are predominantly set at national levels, but with the growing EU, attempts to coordinate the definition of acceptable risks—that is, the levels of OELs (see 98/24/EC and 2000/39/EC). Machinery, on the other hand, is more to be used as delivered. Thus, sufficient protection against all possible risks—both risk information and the material safeguards—are to be part of the harmonized EU regulations. The specification of these protections is to be determined through European standards, which has resulted in a large standardization process (Walters, 2002a, p. 44).

- To secure safe and sound conditions in individual production units, those responsible are to assess (and control or at least reduce) their general risks. The Seveso Directive launched the principle that manufacturers with potentials for major accidents should assess these risks. The Framework Directive expanded this into a general principle that employers shall assess and manage all OHS risks of their operations (including the use of chemicals, as guided by national OELs). It has since been included in a number of daughter and other directives on working conditions. These types of Risk Assessments may interact with the risk assessments of machinery. For example, changes in machinery at a workplace may require a new risk assessment according to the Machinery Directive, while the surrounding safety and health is to be assessed according to the Framework Directive (see further Walters, 2002a, pp. 41–44, on this interaction).

WORKPLACE RISK ASSESSMENTS ARE TOOLS WITHIN OHS MANAGEMENT

The mandatory workplace Risk Assessments place the responsibility to detect and abate the problems on the employers as risk producers. They aim to make employers assume the responsibility to prevent work-related ill health—which they have since the beginning of OHS regulation. According to quality control principles, the employers—and their managers—can much more effectively prevent the health risks than can the authorities as representatives of the workers. It is, therefore, not enough for risk-producing employers to passively await the authorities' assessments and ensuing prescriptive orders, which define the risks and how to abate them. Such a strategy of external control of OHS risks never worked well. With modern high-risk technologies, it has become even more untenable (as discussed earlier).

As mentioned, the evolvement of Risk Assessment was largely a result of growing regulation. The regulation—especially of workplace risks—is itself part of the broader development of OHS management, as a process-oriented prevention strategy. This emphasizes that OHS risks are much more effectively and efficiently controlled at their sources—in the planning and management of the operations causing the risks. To manage the prevention, employers, therefore, must set up appropriate quality control procedures—not only Risk Assessments but also training and proper distribution of responsibilities and authorities.

Since the late 1980s, there has been a development and dissemination of such methods to manage OHS. This has taken the form of active marketing of voluntary OHSM systems (especially in Anglo-Saxon countries), of national and international standards and guidelines for such systems, and of hybrid implementation of OHSM mixing voluntary and mandatory means (see also later the ILO Guidelines on OHS management systems). Above all, OHSM regulations (rarely constructed as complex systems, as they are also to be applicable also to small firms) have been adopted by the EU and many other OECD countries—with the United States still a notable exception—and by increasingly developing nations (see Frick et al., 2000, on the background, purposes, and problems of regulated OHSM).

CONTESTED IMPLEMENTATION OF OHSM

The spread of OHSM marks a shift from specified, prescriptive risk-regulations to generic, performance-oriented ones (Gunningham & Johnstone, 2000). This is not only a functional issue. As mentioned, earlier, politics and policies of regulation also affect the balance of power over the risk assessments and the resulting controls. Most controversial is probably in how far the increased risk assessment by the employers will be accompanied by a corresponding decrease in the authorities' regulation and enforcement. The producers of equipment already largely certify themselves—according to the Machinery Directive—to comply with the risk regulations on the product market. It is feared that a too lenient interpretation and implementation of regulated OHSM—including the Framework Directive's required Risk Assessment—may result in the same development in the labor market. Employers may certify themselves to be following the OHS regulations, with little or no external control of labor inspections (Nichols & Tuckers, 2000; cf. also, the debate on voluntary OHSM programs in the United States in Needleman, 2000).

In other words, is the requirement that employers assess and control "their" risks a functional decentralization to increase the effectiveness of OHSM (as was discussed earlier)? Or is the state shedding its responsibility to protect its citizens—the workers—under the (OHS) law? In the latter case, it is argued that letting the risk producers assess themselves is placing an excessive trust in the capability and interests of employers who operate under the economic pressure of market competition (Nichols & Tucker, 2000; Dorman, 2000; cf. the critique of the machinery standardization process in TUTB, 1996; and in Koukoulaki & Boy, 2003).

THE FRAMEWORK DIRECTIVE IS THE FOREMOST OHSM REGULATION

The balance between success (improved prevention) and sham (deregulation, with little control of prevention) in the practice of the OHSM strategy remains an open question. The answer largely depends on the varying conditions of its implementation, as discussed in Frick et al. (2000). However, Risk Assessment is unavoidably part of this open and contested implementation. This is especially so, as the Framework Directive is the foremost example of these political OHSM regulations. OHS policies all over the EU are to be governed according to the principles of this directive.

Within the Directive, the mandatory Risk Assessment is to be the starting point for its intended comprehensive OHSM. Employers are to chart and assess all risks of their production to take on the responsibility of controlling them. A thorough Risk Assessment is to both enable and motivate them (and their managers) to improve the work environment. The regulations, therefore, more of mandate or guide employers on how to assess and control the risks and focus less on what these risks are. Information on possible risks, and on suitable prevention measures, is instead to be gathered as part of the OHS management process.

TYPE AND CONTENT OF THE FRAMEWORK DIRECTIVE'S RISK ASSESSMENT

NOT A NORMAL ERGONOMIC STANDARD

The EU Framework Directive is indeed a framework, with "general principles concerning the prevention of occupational risks... as well as general guidelines for the implementation of the said principles" (Article 1.2). It is an ambitious modernization of OHS policies in the EU but also a political compromise. Some of its requirements are intentionally ambiguous, and they all have to be adapted to the varying national structures and traditions (see, further, Walters, 2002a).

The following discussion aims to clarify the Risk Assessment intended by the Framework Directive. It is illustrated with some national varieties, but these form no exhaustive analysis of the differences between the EU states (more are presented in the national chapters and in the comparison by Walters, 2002a, pp. 284–288; and in the analysis of the national transposition by Vogel, 1994, 1998). To know what employers must comply with, legal scholars and OHS practitioners have to study the national regulations and the various guidance on these.

The paragraphs on Risk Assessment earlier are also of principal nature, not guidance on how to do it. This requirement must instead be understood within its setting of the Directive. Formally, this is addressed to the member states. These must not only bring into force laws and regulations to comply with its requirements (Article 18.1) but also ensure adequate control and supervision of its workplace implementation (Article 4). The obligation of workers is mainly to "take care as far as possible of his own safety and health and that of other persons affected by his acts or omissions" (Article 13). Above all, the Directive contains principles on how employers in all economic sectors shall secure a good work environment. Some of their major duties are to:

- Ensure the safety and health of workers in every respect related to the work (Article 5.1).
- Provide the organization and means necessary for this (Article 6.1).
- Evaluate and prevent all OHS risks, including organizational ones (Article 6.2–3).
- Implement this through a prevention hierarchy, starting by avoiding the risks and ending with personal protectives and safety training (i.e. safe behavior) as last resort (Article 6.2).
- Either they possess sufficient competence to comply with the Directive's requirement or enlist such competence (Article 7).
- Organize effective worker involvement in the OHSM through information, training, consultation, and participation (Articles 10–12). However, this and other OHS measures may in no circumstances involve the workers at any cost. Participation must be extensive but free from costs to the workers (Article 6.5).

The Risk Assessment is an integrated aspect of this prescribed OHSM. As such, it differs in many respects from the majority of ergonomic standards in this book. Most of these also focus on their material content, for example, as quantitative anthropometrical or biomechanical measures (e.g., many ISO and CEN standards; cf. also Fallentin et al., 2001). Risk Assessment (and OHSM in general), on the other hand, is not quantitative at all. It deals with the process of OHS improvement, not the content of the risks to be prevented. Although most ergonomic standards are voluntary (at least formally), the Framework Directive and its Risk Assessment is the foremost mandatory OHS regulation within the EU (Walters, 2002a).

The difference is not absolute. The wide variety in standards is discussed in the earlier chapters (and in the comparative evaluation by Fallentin et al., 2001). Some ergonomic standards are mandatory by law, not by agreements between experts or the concerned parties (see earlier section on regulatory versus voluntary standards and their different logics). Ergonomic standards may also include process elements, such as programs of assessment and intervention, especially to counter complex problems like repetitive work (see earlier discussion on material versus process—or program—standards). The EU Directive on Manual Handling (90/289/EEC) is an example of a regulated process standard (in fact a mini version of the Framework Directive). Yet, it is illuminating to contrast its type and content to material and voluntary ergonomic standards. The following table is simplified—and, thus. slightly exaggerated—but should still illuminate their principal differences (Table 34.1).

Only Risks Within the Employment Relation

The differences in type between mandatory and voluntary ergonomic standards and the process regulations have already been discussed, and the means and level of compliance with the Risk Assessment is dealt with later in this chapter. However, before we go into the content of the Risk Assessment, it is important to note both the width and the limits of its application. On one hand, it applies to all employers, not only to some employers with special risks or to those who choose to use it. The European Court of Justice made this clear in its ruling against Germany (ECJ C-5/00). By exempting

TABLE 34.1
Differences Between Mandatory Risk Assessment and Voluntary Ergonomic Standards

Aspects of Standard	Standard	
	Mandatory Risk Assessment	Voluntary Ergonomic
Type		
Aim	Prevention process	Ergonomic result
Logic behind	Political-legal process	Experts' analysis
Applicability	All employers	Special risks/voluntary
Compliance through	Inspection-enforcement	Market/certificates
Content		
Risks		
Scope of Risks	All = organization & technology	Technology (usually)
Level of Risks	Ensure safety & health	Acceptable risk
Individualization	Individual adaptation	Human averages
Assessment		
Who Assesses	Social partners	Experts
Knowledge of Risks	Mandatory competence	Ignorance accepted?
Goal of Assessment	Prevention hierarchy	Safe person before safe place?

employers with ten or fewer workers from the duty to keep documents containing the results of a Risk Assessment, the German legislation had not properly transposed the Framework Directive.

On the other, the focus on the employment relationship is also a growing limitation. As with nearly all labor law, the Framework Directive is based on 19th-century industrial relations. To assess and prevent risks at work is a duty for the employers, but only to their workers. Yet, in modem production, both sides of the employment relationship wither. Companies produce more in networks by, for example, outsourcing or splitting into separate legal entities (Castells, 1996, chap. 3; Larsson, 2000). On the labor side, casual workers replace long-term employees (Quinlan & Mayhew, 2000). The production—and its work environment—is, thus, increasingly controlled by contract relations under trade law instead of employment relations under labor law, and contract partners are rarely obliged to assess the OHS risks of the producing workers.

There are some regulatory and voluntary possibilities to assess risks to workers outside of the employment relationship. Producers-employers in the UK must prevent injuries also to third parties such as visitors (HSE, 1997, p. 45; see also Johnstone, 1999, for an interpretation of the British and Australian duty of care to cover contract relations). "Voluntarily," large customers also increasingly use contracts as a market mechanism to spread—require—systematic OHSM from their contractors and suppliers (Walters, 2001, pp. 345–355). As with any voluntary standard, risk assessment can be applied more freely on the market than by regulation, but then there is no assurance that it conforms to the requirements of the Framework Directive.

WHAT RISKS TO ASSESS?

Scope of the Risks

These requirements are strict. The risks include everything that may harm the workers. The Framework Directive integrates the Scandinavian broad definition of the work environment (Walters, 2002a, p. 43), which explicitly includes organizational risks. Article 6:2(g) states that, "the employer shall... develop a coherent overall prevention policy which covers technology, organization of work, working conditions, social relationships and the influence of factors related to the work environment," and Article 6.3(a) mandates employers to "evaluate the risks to the safety and health of workers, inter alia in the choice of work equipment, the chemical substances or preparations used, and the fitting-out of workplaces."

The European Court of Justice has upheld this wide risk definition. It ruled that working time influenced the health of workers. It was, therefore, legitimate for the EU to regulate this under Article 118a of the EEC Treaty (Directive 93/104/EEC; ECJ case C-84/94). The court confirmed that every risk that may affect the health and safety of workers is part of the work environment (and, thus, shall be assessed). In this, it followed the advice of the advocate general who—among other things—stated that the base—the preventive aim of the Framework Directive—is *far from a view in which the protection of workers is limited to physical or chemical factors.* He also referred to the EU's previous acceptance of WHO's wide health concept as a base for its OHS directives (e.g., in Directive 92/85/EEC on the rights of pregnant women). The court confirmed that all risks must be assessed when it ruled against Italy (in ECJ C-49/00) that the list of examples in Article 6.3 (a)—that have been translated into Italian Law—was not enough, as inter alia had been excluded. Finally, it noted, *that the occupational risks which are to be evaluated by employers are not fixed once and for all, but are continually changing in relation, particularly, to the progressive development of working conditions and scientific research concerning such risks.*

The inclusion of all risks has important consequences. The Risk Assessment must also identify and evaluate organizational risks of (for example) social relations at work, management style, gender segregation, organization and distribution of work tasks, pay systems, working time schedules, worker autonomy in performing tasks, violence and threats, and mental and physical workload. All of these—and more—organizational factors have been documented to affect the health of workers (see, e.g., Johnson & Johansson, 1991; and Marklund, 2001). Likewise, the more traditional technical-chemical risks that are objects of national technical regulations or lists of OELs are (only) examples of such risks. The Risk Assessment may well have to consider other "traditional" risks (as is made clear in, e.g., the German law, Schaapman, 2002, p. 138).

The wide work environment concept is included in the regulations on Risk Assessment in most EU countries or in their general OHS laws, like in the Netherlands (Arbeidsomstandighedenwet, 1999) and France (Rivest, 2002, p. 98). In the Danish Workplace Assessment (Jensen, 2002) and the Swedish Systematic Work Environment Management (Frick, 2002, p. 231), the inclusion of organizational prevention was even a major aim. The scope of risks to assess in the U.K. is possibly more restrictive (see, e.g., Walters, 2002a, pp. 251–258). The formal transposition (Management of Health and Safety at Work Regulations; HSE, 1992) only orders employers to carry out a suitable and sufficient assessment. The OHS authority is ambiguous in its information on how to interpret this. On one hand, the general OHSM information includes organizational factors and lists several of these (in HSG-65; HSE, 1997, p. 13). On the other, the guide on Risk Assessment (HSE, 1999, p. 6) informs employers that they need to show that they "dealt with all the obvious significant hazards," which may exclude a duty to assess less obvious but possibly serious risks (see following on the requirement to have sufficient competence to identify all risks).

Risks to Individual Workers

The risks to individual workers—not only workers, in general—must be assessed. Article 6.2 (d) states that the prevention shall be based on a number of general principles, including "adapting the work to the individual;" Article 9.1(a), that the Risk Assessment shall include risks "facing groups of workers exposed to particular risks"; and "particularly sensitive risk groups must be protected against dangers which especially affect them" (Article 15). The risks to pregnant women, immigrant workers with poor language, older or inexperienced workers, and any other sensitive-exposed group must be specially assessed. But the work environment must also be compared to workers' individual sensitivities, competencies, capabilities, and so forth, and the preventive measures necessary for each worker must be taken. (In this adaptation to individual workers, the Framework Directive links into the later EU policies of "employability" [i.e. to promote a high level of employment as a major means for economic development within the union; Walters, 2002a, pp. 54–56].)

Levels of Risk Beyond Specified Regulations

Risk to individuals and levels of risk are related. To ensure the safety and health of sensitive groups and individuals discussed earlier, the employer may well have to assess and control risks beyond adherence to (i.e. are lower than) OELs and other specified national regulations. The limits to how low or improbable the employer has to investigate are (partly) defined by national regulations and by the individual cases. However, thorough Risk Assessments must bear in mind that OELs and other specified measures-limits are temporary regulations on minimum requirements and do not guarantee against harm (as is demonstrated in many studies, e.g., in Larsson, 1991, p. 208). National regulations—and the Framework Directive—usually aim at the lowest possible level. Employer organizations also advised their members the same—for example, when the Swedish Engineering Confederation and the Metal Workers Union jointly set 25% of the OEL as a maximum limit to aim for—in engineering plants (Verkstadsföreningen, 1978).

"To ensure the safety and health of workers" (Article 5:1) is also an absolute requirement. Economic relativizations of this duty, such as "so far as is reasonably practicable" (HSE, 1997, p. 46), are common, for example, in British OHS regulations. The U.K. transposition has been criticized for falling short of the Framework Directive, including the Risk Assessment (e.g., by Bercusson, 1996; Walters, 2002a, pp. 251–255) but the European Court of Justice has not yet tried the issue. However, the limitations in the British employers' duties pertain mainly to the actions they must take to prevent the risks (HSE, 1997). This should not be confused with assessing the risks—which must all be carefully identified and evaluated, whether they are reasonably practical to prevent or not.

WHO ASSESSES AND HOW

Not Only a Risk-Management Problem

To this broad risk definition should be added a similarly wide assessment process. Employers unequivocally have to assess all risks. However, the Risk Assessment and its result is not a pure management issue that is up to them alone. Many works on risk assessment fail to properly understand—or at least communicate—the crucial difference between economic business risks and social OHS risks. They describe OHS as (only) a matter of successful management without including the political aspect of state regulation and worker rights.

In doing so, their advice on Risk Assessment may fall short of the Framework Directive, which is clear in this respect. As the major stakeholder in healthy workplaces, workers or their representatives are to be consulted "on all questions relating to health and safety at work" (Article 11:1)—also on the Risk Assessment. Their participation is to be supported by, for example, "all the necessary information" that shall be provided by the employers (Article 10:1) and by various rights for—for example—training and paid time off, as specified in national regulations, and may not involve them in any costs (Article 6.5).

However, participation is not only a worker's right to look after their OHS interest against the business interests of the employer but is also fundamental to make the employers' OHSM effective and efficient. A thorough assessment must include the worker's experience of the risks, which cannot be fully identified by experts alone (cf. section under Risk Assessment as Contested Politics: and Gustavsen, 1980). Their practical production competence is also often needed to develop workable preventive measures. The importance of local cooperation between the social partners made Denmark mandate "Workplace Assessment" instead of assessments of individual risks in their transpose of the Framework Directive (Jensen, 2002). It also motivated Swedish Systematic Work Environment Management (Frick, 2002, p. 221). Worker participation in Risk Assessment (and in the rest of the OHSM) also largely follows national industrial relation traditions in other EU countries (with changes, though, in several cases due to the Framework Directive). For example, Italy and France extended such participation rights. The U.K. did so only after losing similar cases in the European Court of Justice (ECJ C-382/92 and C-383/92). Yet, in the many non-unionized British workplaces, employers still have much discretion in how to consult their workers (Walters, 2002a, pp. 256–258; see further Boix & Vogel, 1999; and Walters & Frick, 2000, on the motives for and difficulties of participation in Risk Assessment and in OHSM in general).

Competence Is Mandatory

In carrying out the Risk Assessment, the social partners are to be advised by OHS experts. The preamble of the Framework Directive states that "*employers shall be obliged to keep themselves informed of the latest advances in technology and scientific findings... concerning the inherent dangers in their undertakings.*" Article 7 also requires employers to either assign internal personnel with enough OHS competence to carry out all the obligations required by the Framework Directive or enlist external expertise. This article has been especially open to varying national interpretations in its transposition (see Hämäläinen et al., 2001). The so-called preventive services remain a contested issue in several EU states. Sweden transposed Article 7 as late as 2001; Denmark still debates how to comply, whereas France is reluctant to give up its system of company doctors (see the national chapters and the comparison in Walters, 2002a). The U.K. regulation largely leaves it to employers to decide what competence they may need (Walters, 2002a, pp. 255–256, and 290–293). Recently (May 2003; case C-441/01), the EU court found the Dutch transposal of Article 7 incompliant with the Directive, as it permitted employers to delegate the Risk Assessment to the external consultants of OHS services (i.e. not only to be advised by their competence). However, in whatever way it is organized, the Framework Directive requires employers to have enough OHS competence to identify and evaluate all risks (including organizational ones) that are known by experts to potentially harm workers. The competence needed, thus, depends on each workplace's possible risks.

Upstream Assessment for Primary Prevention

The assessment is to guide the prevention, as mandated by Article 6.2. According to this, risks are primarily to be completely avoided and then combated at the source. Only as a last resort shall workers be protected by personal protectives and instructions in safe behavior (among other things, as repeated research has demonstrated "safe behavior" to be much less effective than "safe place"; see, e.g., Gallagher, 1997; Gallagher, Ritter, & Underhill, 2001). The prevention hierarchy may be hard to comply with. Caution—to follow rules of safe behavior in risky situations—remains unavoidable in many production situations. However, whenever behavior control is an important part of the OHSM, the burden of proof—of compliance with the Framework Directive—rests with the employers. They should be able to demonstrate that everything possible has been done to avoid the risks—that workers are not placed in situations where caution is necessary (on worker behavior versus upstream prevention in OHSM, see Frick & Wren, 2000; Nichols & Tucker, 2000; and Wokutch & VanSandt, 2000).

To support the prescribed prevention, the Risk Assessment must, therefore, focus upstream to find the organizational and technological roots of the risks. For example, in ergonomics, macroergonomic risk assessments (and changes) of the work organization may often be more important than microergonomic improvements of workstations and tools to avoid strenuous tasks (Neumann, 2001). When "safe behavior" is unavoidable, the Risk Assessment must search for any detrimental influence on this behavior—say, by payment systems that encourage cutting corners or by working times, workloads, or noise that may hamper a safe behavior. How the Risk Assessment can identify these and other upstream risks is largely determined by its scope and competence (as discussed earlier).

THE INTERPRETATION AND APPLICATION OF RISK ASSESSMENT AS A STANDARD

Risks Are Poorly Assessed in EU Workplaces

The EU member states have interpreted and transposed the Framework Directive in line with their general political differences (Walters, 2002a). The U.K. gives employers much freedom to decide which risks to assess, how workers shall participate in this, and what competence they need in the assessment. At the other end, the Nordic model of Denmark, Finland, and Sweden encompasses the broad risk definition of the work environment based on local cooperation among the social partners and emphasizes the need for advice by a holistic OHS competence (despite ongoing debates on how to organize it). The other EU states are often somewhere in between these aspects of transposing the Directive (with numerous national varieties due to each country's tradition and structure in OHS policies).

These differences may be larger in theory (law in the book) than practice (law in action). We know little about how risks are assessed at EU workplaces. Despite the intended fundamental role of the Framework Directive, very few serious evaluations have been done of its implementation, including of the Risk Assessment. The figures we have are mainly from surveys to employers—self-reported compliance. Some scattered results—pertaining to the "old" 15 member states between 1996 and 2000—on this are:

- Sweden: 55% of the employers assessed the risks (AV, 2001, Appendix, p. 68).
- Greece: very few employers had assessed the risks (Karageorgiou et al., 2000, p. 267).
- Denmark: some 30% have started and another 15% finished (Karageorgiou et al., p. 270).
- The Netherlands: around 25% to 30% assessed the risks (Karageorgiou et al., p. 273).

- Germany: about half of the employers assessed the risks (Schaapman, 2002, p. 141).
- U.K.: around one-third of the employers assessed the risks (Karageorgiou et al., p. 268).
- Norway (which also must comply with 89/391/EEC): some 45% of the employers implemented the OHSM regulation of "internal control," which includes Risk Assessment (Gaupset, 2000, p. 340).
- Spain: 46% to 86% of the employers (depending on their size) assessed the risks (Walters, 2002b, p. 89).

There are both quantitative and qualitative problems with these data. Self-reported compliance is always likely to be overstated compared to the labor inspection's evaluations—possibly even more compared to expert studies (as was the case in Sweden; Frick, 2002, p. 228). The widespread lack of competent OHS advice and other survey answers (e.g., "knows the regulation") indicates that employers often do not fully understand what is required of their Risk Assessments. They are, therefore, even more likely to overstate their compliance. As an example, two-thirds of the Swedish local communities rated themselves to comply with the OHSM ordinance (i.e. with the Framework Directive; AV, 2001, p. 14). Most probably, they had organized and documented some Risk Assessment. Yet, at the same time, sickness absenteeism was exploding among their personnel, and case studies demonstrated structural obstacles to their OHSM (Larsson, 2000, pp. 209–212).

Whether one conducted a Risk Assessment or not depends on how it is defined. The Framework Directive is very ambitious. All risks must be assessed in a thorough process. Case studies indicate that few employers fully satisfy these demands. Often, organizational aspects are excluded, worker participation is poor, or the assessments do not go upstream to the roots of the problems (see Boix & Vogel, 1999, pp. 7–27; Frick & Wren, 2000, pp. 34–42; Walters, 2002a, pp. 285–288). The implementation of the Risk Assessment intended by the Framework Directive is, therefore, probably less—maybe much less—than what the employers have reported.

Risk Assessments are mandatory also in small firms. As mentioned, the European Court of Justice confirmed that all employers, irrespective of size, must have written assessments. Details on how to conduct and document them may be simplified, but all should anyhow be adapted to the unique local production and work situation—that is, may and should vary according to many variables other than size. Yet by far, the strongest and most pervasive difference in the surveys above is by size. Among firms with fewer than 50 workers, some 15%–50% report that they have assessed their OHS risks, and even less of the really small ones have done so. Judging from studies of the levels and problems of implementing OHSM in small firms (e.g., Eakin, Lamm, & Limborg, 2000; Frick & Walters, 1998; Walters, 2001, 2002b), their reported compliance is extra likely to be exaggerated.

WHY IS THE IMPLEMENTATION SO DIFFICULT?

Why is it so difficult to implement the Risk Assessment? One reason is that integrated OHSM requires changes in how employers manage their own businesses. On one hand, even small firms should be able to handle their OHS risks more systematically than they usually do. They know their products. To survive, they also have to assess and handle their other business problems in a systematic manner. On the other, the mandated OHSM was rare to start with. Some large factories have traditions of Safety Management, but these rarely take in the organizational causes (how, e.g., management structures affect accident rates or mental strain) or are supported by a genuine participation. The organizational development needed for a thorough Risk Assessment had to be achieved in and by management structures with limited capacities. All organizations are full of problems, conflicts, failures, and mistakes. Extra demands—in this case, to systematically improve the work environment—are not easy to fulfill

The interest in the organizational development needed to integrate OHSM properly into the rest of management may also be lacking. Employer-managers may be reluctant to develop their organizations to systematically identify all risks—possibly even new ones that may well be costly to correct. How much the economic and the occupational health and safety interests are opposing or overlapping is much debated. For example, Dorman (2000) claims that practice demonstrates the conflict to be strong, whereas the British HSE instead promotes voluntary compliance with the regulated OHSM as a "business case" of "safety pays" (HSE, 1997). Others see it as depending on the circumstances of production, often with opportunities for OHS competence to increase the joint interests (Frick, 1997; European Agency, 1999). Nevertheless, many managers regard OHS measures more as costs than investments. With usually tight budgets, they should be less interested to systematically assess the work environment, as they may incur new costs to abate the OHS risks. Ergonomists and others who promote Risk Assessments can rarely avoid the issue of how managers perceive the economic interests in improving the work environment.

The implementation of thorough Risk Assessments (and the rest of the OHSM) at the workplaces is obstructed by limited management capacities and severe economic conflicts. However, it also is also hampered by changes in the structure of production (as discussed under Only Risks Within the Employment Relation) in which responsible employers are harder to track and large ones are fragmenting into (at least formally) independent small firms. On the other hand, the implementation is promoted by what may be called the OHS infrastructure. This consists of both the resources and the policies of OHS authorities and other OHS-actors, such as the social partners, OHS experts (like health services), and insurance companies. Compared to the broad and deep-going changes needed to implement the Framework Directive and its Risk Assessment, their resources are quite limited (see, further, Walters, 2002a, on how national OHS infrastructures have promoted this implementation).

THE PURPOSE IS MORE IMPORTANT THAN THE PROCEDURE IN RISK ASSESSMENT

These are some answers to why the implementation has been so poor. However, a more important question is: what to do? How shall ergonomists and other OHS actors apply the EU standard of Risk Assessment? This section does not aim to be a practical guide on how to assess risks at work. As mentioned (under Risk Assessment as Science), there is a large literature of both practical guides and analytical works oriented to a wide variety of workplace and policy issues. It is also important that applications of the Risk Assessments comply with the varying national regulations (on which there also is ample information and guidance issued by the OHS authorities and others). Here, we shall instead discuss some of the principal issues in the application.

First, ergonomists, other OHS practitioners, and employers cannot completely rely on the existing guidance material (though much of it is useful). To adhere to its advice may not be enough to secure compliance with the regulations. This material is often sold on a market where employer-managers may prefer messages that do not interfere too much with how they run their operations. The writers, who commonly produce it, are more skilled in complex risk analysis techniques than in, for example, health effects of organizational factors or on the industrial relations of risk assessments—both of which are covered by the regulations. However, for example, Boix and Vogel's guide is useful not only for unionists (for which it is primarily intended) but also for those who organize Risk Assessment as a starting point of a broad workplace development.

Second, how to assess the OHS risks depends much on why it is done. Is it primarily in compliance with the OHS authorities' external demands, or are there also strong internal motives to improve the work environment, as part of the business development? The local answer to this defines the basis for the interaction among the concerned actors: management, workers, ergonomists, and other OHS experts. It will greatly influence what Risk Assessments one tries to implement—especially if they look into the production and its problems. OHS experts often have a know-how that can help managers combine economic and OHS interests (Oxenburgh, Marlow, & Oxenburgh, 2003).

This does not mean that conflicts of interests disappear with such competence, rather that the implementation of the Risk Assessment must focus as much on its wider setting—why it is done and how its results are going to be used—as on the formal procedures of identifying and evaluating the risks. If the motive is predominantly one of legal compliance to external demands, then ergonomists should note the broad nature of these demands and try to secure that all of them are included in the Risk Assessment. If the Risk Assessment is to support the internal development of work and production, the same broad issues have to be included, but it is easier to discuss and organize the changes in management required for a good implementation of all the OHSM. However, in both cases, ergonomists and other OHS practitioners have to secure enough competence in the broad organizational and industrial relations issues of the Risk Assessment. This is usually not included in their training.

Usability of OHSM Standards in Small Firms

If and how small firms can use Risk Assessments and other OHSM standards also depends on how they interpret and apply them. As for other employers, the choice is mainly between a legalistic, minimalist, and a more integrated developmental interpretation of the standards. On one hand, small firms also have to obey the law, and all EU states require fairly extensive Risk Assessments of employers. However, if these requirements are perceived and applied as purely external obligations from the authorities, they will also be regarded as "paperwork," foreign to their informal management. With no internal motivation, the bureaucratic Assessments will have to be implemented in small firms mainly through enforcement. As these are some 90% of all workplaces, the authorities are hardly able to implement the Risk Assessment (and other OHSM) in small firms through inspections, orders, and sanctions alone.

The alternative is not purely voluntary compliance. At least some level of integration of Risk Assessment into how they run their businesses is even more important in small firms. The more Risk Assessment is seen as a tool that can (also) support—for example, a more efficient production or improved workplace relations—the more the necessary adaptation of their management will be achievable—that is, acceptable to the owner-managers. However, small firms usually need support to be able to use OHSM standards as tools for their internal development from whatever OHS actors there are. These external actors need to help translate the formal regulations into workable methods in the individual firm. This—again—requires competence not only in OHS but also in organizational issues (see Frick & Walters, 1998; Walters, 2001, 2002b).

Ergonomists who see Risk Assessment as mainly a tool for workplace development may be important actors. They, other OHS actors, and the few small firm managers who directly try to implement the Risk Assessment can find abundant guidance material—including the checklists, and so forth—supported by the European Agency for Safety and Health at Work. However, for small firms, it is essential that these and other tools are indeed used as development support, not as "tick-in-the-box" formal checklists. The purpose of assessing risks and what comes next must be discussed for the tools to become helpful for a small firm.

REFERENCES

Ansell, J., & Wharton, F. (Eds.). (1992). *Risk: analysis, assessment and management.* Chichester, John Wiley.

Arbeidsomstandighedenwet. (1999). *Wet van 18 maart 1999, houdende bepalingen ter verbetering van de arbeidsomstandigheden.* The Hague: Staatsblad van het Koninkrijk der Nederlanden.

Ashford, N., & Caldart, C. (1997). *Technology, law and the working environment.* Washington, DC: Island Press.

AV. (2001). *Ett aktivt arbetsmiljöarbetel Rapport 2001:12.* Solna: Arbetsmiljöverket.

Beck, U. (1992). *Risk society: Towards a new modernity.* London: Sage.

Bercusson, B. (1996). *European labour law.* London: Butterwick.

Boix, P., & Vogel, L. (1999). *Risk assessment at the workplace — A guide for union action*. Brussels: TUTB.

British Medical Association. (1992). *Living with risk—The British Medical Association guide*. Chichester: Wiley.

Brodeaur, P. (1974). *Expendable Americans*. New York: Viking Press.

BS 8800. (1996). *Guide to occupational health and safety management systems*. London: British Standard Institution.

Carson, R. (1962). *Silent spring*. Boston: Houghton Mifflin.

Castells, M. (1996). *The rise of network society*. Oxford: Blackwell.

Collinson, D. (1999). Surviving the rigs: Safety and surveillance at North Sea oil installations. Organization Studies, 20(4), 579–600.

Colombini, D., Occhipinti, E., & Grieco A. (2002). *Risk assessment and management of repetitive movements and extertions of upper limbs*. Oxford: Elsevier.

Concha, S. (1983). Editorial. *Hazard Prevention*, 19(2):2.

Cox, S., & Cox, T. (1996). *Safety, systems and people*. Oxford: Butterworth-Heinemann.

Cullen, L. (1990). *The public inquiry into the Piper Alpha disaster*. London: Departement of Energy.

Dorman, P. (1996). *Markets and mortality—Economics, dangerous work and the value of the human life*. Cambridge: Cambridge University Press.

Dorman, P. (2000). If safety pays, why don't employers invest in it? In Frick et al. (Eds.), *Systematic occupational health and safety management—Perspectives on an international development*. Oxford: Elsevier.

Dwyer, T. (1991). *Life and death at work—Industrial accidents as a case of socially produced error*. New York: Plenum Press.

Eakin, I., Lamm, R, & Limborg, H. J. (2000). International perspective on the promotion of health and safety in small workplaces. In Frick et al. (Eds.), *Systematic occupational health and safety management—Perspectives on an international development*. Oxford: Elsevier.

Eckerman, I. (2004). *The Bhopal saga—Causes and consequences of the world's largest industrial disaster*. Hyderabad, India: Universities Press Private Ltd.

European Agency. (1999). *Health and safety at work—A question of costs and benefits?* Magazine No. 1, European Agency for Safety and Health at Work. Bilbao. http://agency.OHSa.eu.int/publications/magazine/en/magl.html

European Commission. (1996). *Guidance on risk assessment at work*. Luxemburg: Office for Official Publication of the European Communities.

Fallentin, N., Viikari-Juntura, E., Worsted, M., Kilbom, Å. (2001). Evaluation of physical workload standards and guidelines from a Nordic perspective. *Scandinavian Journal of Work, Environment & Health*, 27 (suppl. 2), 1–52.

Frick, K. (1997). Can managers see any profit in health and safety?—Contradictory views and their penetration into working life. *New Solutions*, 7(4), 32–40.

Frick, K. (2002). Sweden: Occupational health and safety management strategies from 1970–2001. In D. Walters (Ed.), *Regulating health and safety management in the European Union*. Brussels: P.I.E. Peter Lang.

Frick, K. P., Jensen, L., Quinlan, M., & Wilthagen, T. (Eds.). (2000). *Systematic occupational health and safety management—Perspectives on an international development*. Oxford: Elsevier.

Frick, K., & Walters, D. (1998). Worker representation on health and safety in small enterprises: Lessons from a Swedish approach. *International Labour Review*, 137(3), 367–389.

Frick, K., & Wren, J. (2000). Reviewing occupational health and safety management—Multiple roots, diverse perspectives and ambiguous outcomes. In Frick et al. (Eds.), *Systematic occupational health and safety management—Perspectives on an international development*. Oxford: Elsevier.

Gallagher, C. (1997). *Planned approaches to health and safety management*. Sydney: National Occupational Health and Safety Commission.

Gallagher, C., Ritter, M., & Underhill, E. (2001). *Occupational health and safety management systems: A review of their effectiveness in securing healthy and safe workplaces*. Sydney: National Occupational Health and Safety Commission.

Gaupset, S. (2000). The Norwegian Internal Control Reform. In Frick et al. (Eds.), *Systematic occupational health and safety management—Perspectives on an international development*. Oxford: Elsevier.

Gunningham, N., & Johnstone, R. (2000). The legal construction of OHS management systems. In Frick et al. (Eds.), *Systematic occupational health and safety management—Perspectives on an international development*. Oxford: Elsevier.

Gustavsen, B. (1980). Improving the work environment: A choice of strategy. *International Labour Review*, 119(3), 271–286.

Hämäläinen, R. M., Husman, K., Räsänen, K., Westerholm, P., & Rantanen, J. (2001). *Survey of the quality and effectiveness of occupational health services in the European Union and Norway and Switzerland.* Helsinki: Finnish Institute of Occupational Health.

Hansson, S. O. (1993). The false promises of risk analysis. *Ratio*, 6, 16–26.

Harms-Ringdahl, L. (2001). *Safety analysis—Principles and practice in occupational safety.* London: Taylor & Francis.

Holmér, I. (2000). Assessment of cold exposure. *International Journal of Circumpolar Health*, 60, 413–421.

Hopkins, A. (2000). *Lessons from Longford—The ESSO gas plant explosion.* Sydney: CCH Australia.

HSE (1992). *Management of health and safety at work.* Sudbury: Health & Safety Executive.

HSE (1994). *Upperlimb disorders: Assessing the risks.* Sudbury: Health & Safety Executive.

HSE (1997). Successful health and safety management—HSG65. Sudbury: Health & Safety Executive.

HSE (1998). *Assessment principles for offshore safety cases.* Sudbury: Health & Safety Executive.

HSE (1999). *Five steps to risk assessment.* hse.gov.uk/pubns/indg163.pdf.

IEC (1995). *Dependability management: Risk analysis of technological systems (IEC 300-3-9).* Geneva: International Electrotechnical Commission.

ISO (1999). *ISO 14121: Safety in machinery—Principles of risk assessment.* Geneva: International Standard Organisation.

Jensen, P. L. (2002). Assessing assessment—The Danish experience of worker participation in risk-assessment. *Economic & Industrial Democracy*, 23(2), 201–227.

Jeynes, J. (2002). *Risk management: 10 principles.* Oxford: Butterworth-Heinemann.

Johnson, J., & Johansson, G. (Eds.). (1991). *The psychosocial work environment: Work organization democratization and health.* Amityville: Baywood.

Johnstone, R. (1999). Paradigm crossed? The statutory occupational health and safety obligations of the business undertaking. *Australian Journal of Labour Law*, 12, 73–112.

Karageorgiou, A., Jensen, P. L., Walters, D., & Wilthagen, T. (2000). Risk assessment in four member states of the European Union. In Frick et al. (Eds.), *Systematic occupational health and safety management—Perspectives on an international development.* Oxford: Elsevier.

Koukoulaki, T., & Boy, S. (2003). Globalizing technical standards. *Impact and challenges for occupational health and safety.* Brussels: TUTB-Saltsa.

Krüger, D., Louhevaara, V., Nielsen, J., & Schneider, T. (1998). *Risk assessment and preventive strategies in cleaning work.* Bremerhaven: Wirtschaftsverlag.

Kumamato, H., & Henley, E. (1996). *Probabilistic risk assessment and management for engineers and scientists.* New York: IEEE Press.

Larsson, T. (1991). *Arbetsmiljöns styrning—"Kristinehamnsmodellen."* Uppsala: SAMU.

Larsson, T. (2000). The diffusion of employer responsibility. In Frick et al. (Eds.), *Systematic occupational health and safety management—Perspectives on an international development.* Oxford: Elsevier.

Majone, G. (1985). The international dimension. In H. Otway & M. Peltu (Eds.), *Regulating industrial risks—Science, hazards and public protection.* London: Butterworths.

Marklund, S. (Ed.). (2001). *Worklife and health in Sweden 2000.* Stockholm: National Institute for Working Life.

Markowitz, G., & Rosner, D. (2002). *Deceit and denial: The deadly politics of industrial pollution.* Berkeley: University of California Press.

Mosleh, A., & Bari, R. (Eds.). (1998). Probabilistic safety assessment and management, PSAM 4. *International Association for Probabilistic Safety Analysis and Management.* London: Springer.

Mottel, W., Long, J., & Morrison, D. (1995). *Industrial safety is good business—The DuPont story.* New York: Van Nostrand.

Needleman, C. (2000). OHSA at the crossroad: Conflicting frameworks for regulating OHS in the United States. In Frick et al. (Eds.). *Systematic occupational health and safety management—Perspectives on an international development.* Oxford: Elsevier.

Neumann, P. (2001). *On risk factors for musculoskeletal disorder and their sources in production systems design.*, Lund: Department of Design Sciences, Lund University.

Nichols, T., & Tucker, E. (2000). OHS management systems in the UK and Ontario, Canada: A political economy perspective. In Frick et al. (Eds.). *Systematic occupational health and safety management—Perspectives on an international development.* Oxford: Elsevier.

Ogden, T. (2003). The 1968 BOHS Chrysotile Asbestos Standard. *Annals of Occupational Hygiene*, 47(1), 3–6.

Olsen, P. B. (1992). Six cultures of regulation—Labour inspectorates in six European countries. Copenhagen: Handelshøjskolen.

Otway, H. (1985). Regulation and risk analysis. In H. Otway, & M. Peltu (Eds.). *Regulating industrial risks—Science, hazards and public protection*. London: Butterworths.

Otway, H., & Peltu, M. (Eds.). (1985). *Regulating industrial risks—Science, hazards and public protection*. London: Butterworths.

Oxenburgh, M., Marlow, P., & Oxenburgh, A. (2003). *Increasing productivity and profit through health and safety*. London: Taylor & Francis.

Perrow, C. (1984). *Normal accidents—Living with high-risk technologies*. New York: Basic Books.

Quinlan, M., & Mayhew, C. (2000). Precarious employment, work re-organisation and the fracturing of OHS management. In Frick et al. (Eds.), *Systematic occupational health and safety management—Perspectives on an international development*. Oxford: Elsevier.

Rakel, H. (1996). *Workplace risk assessment: A comparative analysis of regulatory practices in five EU member states*. Norwich: Environmental Risk Assessment Unit, Norwich University.

Ramazzini, B. (1700). *De Morbis Artifcum*. Modena.

Rampal, K., & Sadhra, S. (1999). Basic concepts and developments in health risk assessment and management. In S. Sadhra, & K. Rampal (Eds.), *Occupational health—Risks assessment and management*. Oxford: Blackwell.

Reason, J. (1997). Managing the risks of organizational accidents. Aldershot: Ashgate.

Renn, O. (1985). Risk analysis: Scope and limitations. In H. Otway, & M. Peltu (Eds.), *Regulating industrial risks—Science, hazards and public protection*. London: Butterworths.

Rivest, C. (2002). France: From a minimalist transposition to a full scale reform of the OHS system. In D. Walters (Ed.), *Regulating health and safety management in the European Union*. Brussels: P.I.E. Peter Lang.

Roberts-Phelps, G. (1999). *Risk assessment—A Gower health and safety workbook*. Aldershot: Gower.

Ruttenberg, R. (1981a). Why social regulatory policy requires new definitions and techniques for assessing costs and benefits: The case of occupational safety and health. *Labor Studies Journal*, 114–131.

Ruttenberg, R. (May-June 1981b). *Regulation is the mother of invention* (pp. 42–47). Working Paper.

Sadhra, S., & Rampal, K. (Eds.). (1999). *Occupational health—Risks assessment and management*. Oxford: Blackwell.

Schaapman, M. (2002). Germany: Occupational health and safety discourse and the implementation of the framework directive. In D. Walters (Ed.), *Regulating health and safety management in the European Union*. Brussels: P.I.E. Peter Lang.

Stem, R. (1980). *Introduction to risk assessment. Risk assessment in the welding industry: Part 1*. Copenhagen: The Danish Welding Institute.

Stewart, M., & Melchers, R. (Eds.). (1998). *Integrated risk assessment—Applications and regulations*. Rotterdam: Balkema.

TUTB. (1996). Trade union participation in European standardisation work: TUTB network sounds the alarm. *TUTB Newsletter*, 13–14.

Verkstadsföreningen. (1978). *Se om miljön*. Stockholm: Sveriges Verkstadsförening & Svenska Metallindustriar-betareförbundet.

Vogel, L. (1994). Prevention at the workplace. *An initial review of how the 1989 Community Framework Directive is being implemented*. Brussels: TUTB.

Vogel, L. (1998). *Prevention at the workplace. The impact of the Community Directives on preventive systems in Sweden, Finland, Norway, Austria and Switzerland*. Brussels: TUTB.

Vollmer, G, Giannoni, L., Sokull-Klütgen, B. & Kracher, W. (Eds.). (1996). *Risk assessment—Theory and practice*. Luxemburg: Environment Institute, Office for Official Publication of the European Communities.

Walters, D., & Frick, K. (2000). Works, participation and the management of occupational health and safety: Rethinking or conflicting strategies. In Frick et al. (Eds.), *Systematic occupational health and safety management—Perspectives on an international development*. Oxford: Elsevier.

Walters, D. (2001). *Health and safety in small enterprises: European strategies for managing improvement*. Brussels: P.I.E. Peter Lang.

Walters, D. (Ed.). (2002a). *Regulating health and safety management in the European Union.* Brussels: P.I.E. Peter Lang.

Walters, D. (2002b). *Working safely in small enterprises in Europe: Towards a sustainable system for worker participation and representation.* Brussels: TUTB.

Wharton, F. (1992). Risk management: Basic concepts and general principles. In J. Ansell & F. Wharton (Eds.), *Risk: Analysis, assessment and management.* Chichester: Wiley.

Wokutch, R., & VanSandt, C. (2000). OHS management in the United States and Japan: The DuPont and Toyota Models. In Frick et al. (Eds.), *Systematic occupational health and safety management—Perspectives on an international development.* Oxford: Elsevier.

35 ILO Guidelines on Occupational Safety and Health Management Systems

Daniel Podgórski

CONTENTS

BEGINNINGS AND DEVELOPMENTS IN STANDARDIZATION OF OSH MANAGEMENT SYSTEMS

FIRST NATIONAL STANDARDS

It is commonly recognized that the first national standard regarding the occupational safety and health management systems (OSH-MS) was the voluntary British standard BS 8800 (British Standard Institution [BSI], 1996). This standard was developed following the popularity of a systematic and standardized approach to the management systems growing worldwide. Initially, this approach applied to the quality management systems (ISO 9000) and, subsequently, the environmental management systems (ISO 14000). BS 8800 standard contains guidelines concerning the design and implementation of OSH-MS in the way, allowing its integration with the general enterprise management system. The OSH-MS proposed in this standard is based on the continual improvement cycle PDCA (Plan-Do-Check-Act), also known as Deming's cycle. This model is compliant with the management system model applied in the ISO 14001 standard (International Organization for Standardization [ISO], 1996) as well as with the model adopted in 2000 within the ISO 9001 standard regarding quality management (ISO, 2000).

Similar standards intended for voluntary application have also been worked out and established in other countries leading to the promotion of the systematic approach to OSH management. One should mention here the Dutch standard NPR 5001 (Nederlands Normalisatie-Institut [NNI], 1996), the Australian guide SAA HB53 (Standards Australia, 1994) intended for application in the construction industry, and the joint standard AS/NZS 4804 (Standards Australia/Standards New Zealand [SA&SNZ], 1997/2001) developed by the joint Australia and New Zealand technical committee. These documents constitute the guidelines and may not serve as the basis for management system certification, as it is in the case of the ISO 9001 and ISO 14001 standards. The AS/NZS 4804 standard, however, created grounds for elaboration and introduction of the AS/NZS 4801 standard (SA&SNZ, 2001) containing specifications that may be used for the conformity assessment processes regarding such systems, performed by a third party.

Standards and draft standards regarding systematic OSH management have also been prepared and published in Spain, where they appeared in six documents (Abad, Mondelo, & Llimona, 2002). The two most important standards within this series include UNE 81900 (Spanish Association of Standardization and Certification [AENOR], 1996a) and UNE 81901 (AENOR, 1996b). Activities aimed at OSH-MS standardization were also started in Poland in 1998 by establishing—within the framework of the Polish Standardization Committee (PKN)—the Technical Committee No. 276 for OSH Management. The works of this Committee have resulted in establishing the PN-N-18000 series of standards containing, at present, three documents: PN-N-18001 (PKN, 1999, 2003), PN-N-18002 (PKN, 2000), and PN-N-18004 (PKN, 2001). The first of them contains specifications for the OSH-MS and may be used as the basis for certification, whereas the others are the practical guidelines supporting occupational risk assessment procedures and OSH-MS implementation in organizations.

In several other countries, a systematic approach to OSH management is popularized based on national law rather than voluntary standards. Such a situation exists, among others, in the United States, Scandinavian countries, and Japan. In the United States, the system of so-called Voluntary Protection Programs (VPP) has been in force since 1982 under the supervision of the Occupational Safety and Health Administration (OSHA, 1982). Under this system, companies implement OSH management programs based on the rules, periodically updated and published by OSHA (OSHA, 2000a). Companies that participate in the VPP—there are some 900 of them at present—are relieved from routine and programmed OSHA inspections, replaced by periodic audits. Experience generated by the first years of VPP functioning has been used for the elaboration of guidelines for OSH-MS implementation in organizations, currently applied for wide popularization and promotion of OSH systematic management in the United States. In 1996, these activities were supported by the American Industrial Hygiene Association (AIHA, 1996), which developed and published guidelines for the OSH-MS based on the quality management system concept contained in the ISO 9000 standards.

In the Scandinavian countries, a systematic approach to OSH management was initiated in the 1990s by establishing mandatory legal provisions regarding so-called "internal control of work environment." Implementation guidelines for such systems were established and published in Norway in 1991 (Kommunaldepartementet, 1991) and in Sweden in 1992 (Swedish National Board of Occupational Safety and Health [SNBOSH], 1992). After a few years of experience, both countries verified regulations governing the internal control of the work environment and introduced their new versions in 1997 (Kommunaldepartementet, 1996; SNBOSH, 1997), and Sweden, again in 2001 (Swedish Work Environment Authority, 2001). These guidelines ensure practical implementation, in organizations, of provisions of the EU Framework Directive (89/391) regarding the introduction of measures for improvement of employees' health and safety at work (European Union, 1989).

In Japan, OSH-MS guidelines were put in force in the Ordinance of the Minister of Labor in April 30, 1999 (Japanese Ministry of Labour, 1999). Both the structure and the content of provisions in this document are in line with the AS/NZS 4804 and BS 8800 standards and other regulating documents in this field. Regardless of their legal provisions status, Japanese guidelines are intended for voluntary application by employers who aim at improved effectiveness of measures related to better work conditions in their enterprises.

Still another situation regarding regulation of the approach to OSH-MS exists in Germany, where the guidelines for implementation of OSH-MS have been established at the regional level. Such solutions are exemplified by the ASCA program, initiated by the Hessian Ministry of Social Matters, Employment, and Women (Hessisches Ministerium für Frauen, Arbeit und Sozialordnung, 1996). Guidelines for the design, implementation, and integration of OSH-MS in organizations were established by the Bavarian Ministry of Labor, Social Matters, Family, Women, and Health (Bayerisches Staatsministerium für Arbeit und Sozialordnung, Familie, Frauen, und Gesundheit, 1997).

DEVELOPMENT OF OSH MS STANDARDIZATION AT INTERNATIONAL LEVEL

In view of the interest in the development and establishment of OSH-MS international standards growing in various countries worldwide, the International Standardization Organization (ISO) analyzed in 1996 the need to initiate OSH-MS standardization process on an international level (Zwetsloot, 2000). Nevertheless, based on discussions held during an international workshop on the need to work out international OSH-MS standards—organized by ISO in Geneva in September 1996, with the participation of the International Labor Organization (ILO) as well as after an analysis of results of the voting held among national standardization bodies—the ISO Technical Management Board decided in February 1997 to discontinue further works for ILO on the standardization of OSH-MS specifications. The main reason for this decision consisted of considerable differences in methods and culture of OSH between highly developed and developing countries, expressed in, first of all, different solutions of legal systems covering the OSH issues.

Another attempt by ISO to undertake works related to OSH-MS international standards was made in 1999 by the proposal of BSI. However, after the official voting carried out among the standardization bodies—ISO members—this proposal was rejected in April 2000. This situation was caused mainly by the lack of support from developed countries for the concept of certification of OSH-MS conformity with requirements of relevant international standards (Lambert, 2000) by ILO's works, undertaken in the meantime, regarding its guidelines in this field (ILO, 2001a), and by the related conviction that ISO would not be an appropriate organization to lay down appropriate requirements for relations between employers and employees, which are the basis of efficient OSH management.

ISO's resignation to undertake activities aimed at the development of standards for OSH management generated other international initiatives in this area. The lack of standards that might create a basis for OSH-MS certification encouraged an action by some private consulting companies and certification organizations, operating internationally and specialized so far in certification of quality and environmental management systems. Such organizations—searching for new areas of business development, working out their own documents containing OSH-MS specifications, and publishing them—started to offer OSH-MS certification. Seeing a need to assure a uniform character of the approach to OSH-MS on an international level, a dozen or so various certification and standardization institutions—both private and government, representing various countries and international certification system—created a consortium whose objective was to generate a series of documents containing specifications and guidelines for OSH-MS. These activities, carried out under the leadership of BSI, resulted in the preparation of the OHSAS 18001 (BSI, 1999) and OHSAS 18002 (BSI, 2000) documents. OSH-MS models, adopted within these documents as well as other provisions, are compatible with ISO 9001 and ISO 14001 standards, assuring a possibility for OSH-MS integration with quality management systems and environmental management. Nevertheless, these documents have not been generated within the formal standardization process; therefore, they are not recognized as international standards. Although published by BSI, neither are the British standards. While preparing OHSAS 18001, it was assumed that this document would be adopted by some of the institutions participating in the consortium as the basis for their OSH MS certification activities and would replace other documents, hitherto applied by these institutions.

ILO AND ITS ROLE IN OSH MANAGEMENT SYSTEMS INTERNATIONAL STANDARDIZATION PROCESS

The International Labor Organization (ILO) was set up concurrently with the League of Nations at the Versailles Congress in 1919. Since 1946, ILO has become a specialized agency of the United Nations Organization (UNO). Its objective is to disseminate social justice rules to contribute to universal, sustainable peace. ILO acts on the principle of tripartite representation—which is unique among the UNO affiliated agencies, as the ILO Administrative Council consists of representatives of employers and employee organizations as equal partners of the government parties. These three parties are active participants of meetings organized by the ILO. The technical secretariat for ILO is provided by the International Labor Office in Geneva.

ILO has established and published for application by the Member States a number of commonly respected international conventions and recommendations related to labor and regarding—the freedom of associations, employment, social policies, conditions at work, social security, industrial relations, and labor administration, among others. ILO also carries out advisory activity and provides Member States with technical support through a network of offices and multidisciplinary teams in over 40 countries. Such support may consist of advice and training in the area of labor law, employment, development of entrepreneurship, project management, social security, occupational safety, and health and employee education (ILO, 2001b).

In view of the aforementioned competencies, empowerment, and scope of activities of ILO, it should have been expected that this organization would be more appropriate to elaborate and popularize the OSH management systems standards on an international scale. This job was undertaken in 1998 in the Department of Working Conditions of the International Labor Office (currently "SafeWork"). A ready draft of ILO guidelines and its subsequent improvement was then subject to verification by international experts (ILO, 2001a). Verified draft guidelines were also opinionated by ILO Member States and, in April 2001, served as a basis for discussions of the International Experts Forum, representing employers, employees, and governmental parties.

The final ILO guidelines text was approved on April 27, 2001 and submitted to the ILO Administrative Board that—on June 22, 2001—accepted the document for publication as ILO-OSH 2001 (ILO, 2001b). The ILO-OSH 2001 guidelines were initially published by ILO in English, French, and Spanish, but in many ILO Member States, guidelines were translated into other languages and published as local versions. An example is provided by the publication of the guidelines in the Polish language made by the Central Institute for Labor Protection (CIOP, 2001).

As it is stated in the introduction to the ILO guidelines, they are addressed to all persons responsible for OSH management. Application of these guidelines is not legally mandatory for ILO Member States or for enterprises located on their territory. Guidelines are not intended to replace provisions of the law or voluntary standards existing in these countries. On the other hand, provisions regarding OSH management contained in this document consider the main ILO conventions related to OSH—particularly Convention No. 155 regarding occupational safety and health (ILO, 1981) and Convention No. 161 regarding healthcare service for workers (ILO, 1985).

ILO GUIDELINES' ROLE IN PROMOTING OSH-MS AT THE NATIONAL LEVEL

APPLICATION LEVELS OF ILO GUIDELINES ON OSH MS

The ILO-OSH 2001 guidelines are intended for application at two levels: the national and organizational (enterprise). This constitutes the principal difference and advantage of this document compared with other standards regarding OSH-MS, which relate exclusively to the level of organizations. This specific feature of the guidelines is reflected in the document structure, consisting of three parts. The first part contains general objectives of the guidelines; the second one, provisions to be applied at the national level; and the third, guidelines regarding OSH-MS addressing the level of the organization.

NATIONAL POLICY ON OSH-MS

Provisions of the ILO guidelines intended to be applied at the national level related to the creation and functioning of national structures responsible for the promotion of the systematic approach to OSH management. It is recommended that, as far as possible, such activities should be supported by respective provisions of the national law. These provisions particularly include the following:

a. Nomination of a *competent institution* to formulate and implement national policy concerning establishment and promotion of OSH-MS
b. Formulation of coherent *national policy* on OSH-MS
c. Development of *national and tailored* guidelines concerning voluntary implementation and maintenance of OSH-MS in the organizations

The competent institution mentioned under point A should be nominated in the Member States in agreement with organizations representing the interests of employers and employees, as well as other organizations and institutions competent in the field of OSH. For many countries, government agencies for social affairs and labor law or R&D institutions involved in the OSH area are the potential candidates to be nominated as such bodies.

ILO guidelines specify the scope of recommended national policy on OSH MS. The most important aspects to be covered by such policy include:

- Promotion of OSH-MS as an integral part of the overall management of the organization
- Promotion of voluntary measures directed to systematic activities for OSH improvement
- Avoiding unnecessary bureaucracy, administration, and costs related to OSH-MS
- Support for OSH-MS activities provided by labor inspection and OSH services

NATIONAL AND TAILORED GUIDELINES ON OSH-MS

As mentioned under point C, ILO guidelines recommend development and establishment, in Member States, the national guidelines adjusted to the law and practice existing in a specific country, as well as respective tailored guidelines. National guidelines should be intended for voluntary application by organizations and should be based on the OSH-MS model provided in the ILO-OSH 2001 document. On the other hand, tailored guidelines should contain the generic elements of the national guidelines, consider the overall ILO guideline objectives, and should primarily reflect specific conditions and needs of the organization or a group of organizations, in particular:

- Their size (small, medium, and large) and organizational structure
- Types of existing hazards and the level of respective occupational risks

The aforementioned provisions of ILO guidelines relating to the national level provide possibilities for various solutions in the area of transposition, promotion, and practical application of the OSH-MS concept in various countries and organizations. These provisions indicate, in principle, three main ways of using the guidelines at the organization level, shown in Figure 35.1 (ILO, 2001b).

Taking into consideration the ILO-OSH 2001 recommendations regarding establishing the national guidelines, various actions in this area have been undertaken in various countries. In Germany, for example, the transposition of this document was made by publishing of the national guidelines for OSH management systems (Bundesanstalt für Arbeitsschutz und Arbeitsmedizin, 2002).

In Poland, on the other hand, it was assumed that the ILO guidelines would be transposed at the national level through verification and amendment of national standards of the PN-N-18001 series. This kind of approach has already resulted in the adoption in 2003 of the new versions of the

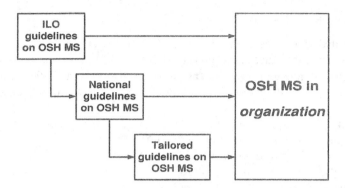

FIGURE 35.1 Principles of application of ILO guidelines at the organization level (ILO, 2001b).

PN-N-18001 standard (PKN, 2003) harmonized with the ILO guidelines. Another method of ILO-OSH 2001 transposition at the national level was applied in the Slovak Republic, where the National Labour Inspectorate developed and published voluntary guidelines on OSH MS (Narodný Inšpectorat Práce, 2002). The model of OSH-MS adopted in Slovak guidelines is based on the ILO model, whereas the practical recommendations for OSH are in line with the requirements of OHSAS 18001, BS 8800, and the Swedish law on internal control of the work environment.

It should be expected, however, that in many countries, the ILO guidelines will not be initially transposed at the national level by establishing new regulations or standards; they will rather be applied directly in their original version or the language version of specific countries.

A subsequent stage in the transposition of the ILO guidelines will consist in the preparation of respective tailored guidelines in specific countries. In some of them, such documents already exist or are being prepared, particularly in relation to special industries such as the chemical and machine industry. Nevertheless, there are still significant problems with developing a model of systematic approach to OSH management, which would be approved and practically applied by small- and medium-size organizations (SMEs). Some solutions in this area already exist, such as a guide for OSH-MS implementation in SMEs published in Germany by Länderausschuss für Arbeitsschutz und Sicherheitstechnik (LASI, 2001) and the toolset for occupational risk management in SMEs, worked out by the Institution of Occupational Safety and Health (2002) based on the concept of the Technical Research Center of Finland (VTT).

ILO GUIDELINES MODEL AND PROVISIONS REGARDING OSH-MS IN ORGANIZATIONS

The third part of the ILO guidelines, addressed to organizations, starts with provisions regarding obligations and responsibilities of employers with respect to assuring occupational safety and health, including compliance with the requirements resulting from respective legal regulations in a specific country. Additionally, the employer should provide commitment and leadership in the OSH activities and introduce organizational solutions aimed at the implementation of OSH-MS in the organization, in line with the specific model. These provisions indicate the OSH-MS model based on the PDCA continual improvement cycle and consists of five main elements. The graphic presentation of this model is shown in Figure 35.2 (ILO, 2001b).

According to other provisions of the ILO guidelines, an employer should aim at OSH-MS integration with the overall management system in the organization, including its subsystems focused on quality management or environmental management for instance. An adopted OSH-MS model facilitates meeting such a requirement as it is philosophically consistent with management system models defined in the ISO 9001 and ISO 14001 standards. It is also consistent with OSH-MS models adopted in other normative documents, such as PN-N-18001 or OHSAS 18001.

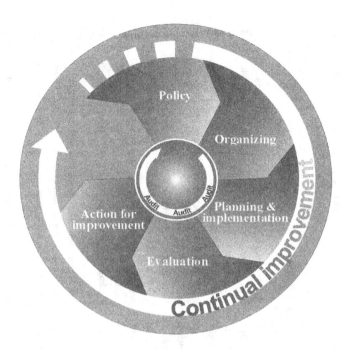

FIGURE 35.2 OSH management system model adopted in the ILO guidelines.

Provisions of the ILO guidelines relating directly to the organization level are presented in this document in the order compliant with the sequence of the main OSH-MS elements shown in Figure 35.2. Their structure and summaries are presented in Table 35.1 (based on ILO, 2001b).

NEEDS FOR FURTHER ACTIONS IN INTERNATIONAL STANDARDIZATION OF OHS-MS

Based on the analysis of ILO-OSH 2001 guidelines content discussed earlier in this chapter, one may conclude that this document constitutes a unique approach to standardization of requirements for OSH management systems on an international scale. Through a built-in mechanism assuring transposition of its provisions to the national guidelines level—and, subsequently, to the tailored guidelines level, in many countries—processes have been initiated to create new normative documents adjusted to the needs of such countries, as well as to various sectors and sizes of organizations.

Nevertheless, due to the extent of potential applications, it seems particularly urgent to develop international guidelines regarding systematic OSH management in SMEs. As shown under the ILO Guidelines' Role in Promoting OSH-MS at the National Level, such documents may be worked out and established in the developed countries that possess respective competent institutions and sustained experience in OSH-MS implementation. On the other hand, in the case of other countries, the creation of such a document may appear difficult and may require support in the form of a model that would be commonly approved at the international level. Such a model does not exist yet, and to work it out would require further research studies regarding its concept, particularly from the point of view of adaptability to the needs of organizations of various sizes (from very small to medium), various sectors typical for such organizations, as well as the needs of various organizations, institutions, and services supporting employers, and acting for the benefit of OSH (e.g., employers organizations, trade unions, insurance institutions, labor inspection, and other government agencies and consulting organizations). To ensure the useful character and wide application of this model, it should be practically verified by pilot implementations in a large group of

TABLE 35.1

Review of ILO-OSH 2001 Provisions Relating to the Organization's Level

Element	Number	Title	Summary
Policy	3.1	Occupational Safety and Health Policy	Requires the establishment of the OSH policy by the employer in consultation with workers and their representatives. Defines the required content of such a policy and indicate the need to integrate OSH-MS with other systems in the organization.
	3.2	Worker's Participation	Underlines that workers participation is the essential factor ensuring OSH-MS effectiveness. Requires consulting the workers regarding OSH activities and the introduction of solutions that encourage participation of workers in OSH-MS, including a safety and health committee.
Organizing	3.3	Responsibilities and Accountability	Indicates the necessity to define responsibilities, duties, and empowerment related to the implementation and functioning of OSH-MS. Requires designation of a member of the top management responsible for the OSH-MS implementation and performance as well as promotion of workers participation in this system.
	3.4	Competence and Training	Indicates the requirements related to competence in the field of OSH and introduction of arrangements assuring that all persons will have the competence meeting such requirements. Defines requirements regarding OSH training programs and methods of their implementation.
	3.5	OSH Management System Documentation	Defines requirements regarding the establishment and maintenance of OSH-MS documents and records, indicating their content as well as their identification, reviews, updating, publication, accessibility, and storage.
	3.6	Communication	Requires introduction of organizational arrangements and procedures assuring appropriate receiving and responding to internal and external OSH information as well as their flow among various levels of the organization. Also requires reception and responding to concerns, ideas, and other OSH-related inputs coming from workers.
Planning and Implementation	3.7	Initial Review	Recommends completion of the initial review of OSH management as the basis to create OSH-MS in the organization. Recommends that the review should be carried out in consultation with workers; defines its scope, requires documenting, and indicates the results to determine the level of reference and decision making relating to the OHS-MS implementation.
	3.8	System Planning, Development, and Implementation	Requires organizational arrangements that support planning the activities related to the implementation and maintenance of all OSH-MS elements. Recommends defining the priorities for activities and quantifiable measures of targets regarding OSH, as well as preparation of plans of tasks to achieve each objective, including the criteria to verify task implementation and needed resources.

(Continued)

TABLE 35.1 (Continued)

Element	Number	Title	Summary
	3.9	OSH Objectives	Requires establishing measurable objectives in the area of OSH and defines recommendations for their formulation and implementation.
	3.10	Hazard Prevention	This chapter contains important provisions regarding hazard identification and occupational risk management as the key element of the OSH systematic management concept.
	3.10.1	Prevention and Control Measures	Requires arrangements and procedures of ongoing identification of hazards and occupational risk assessment and introduction of adequate preventive and protective measures. Defines required hierarchy of such measures that requires them to comply with national law and good practice, to take into consideration the current state of knowledge, including information and reports of labor inspection and other services, and to be subject to periodic reviews and updates.
	3.10.2	Management of Change	Requires an assessment of the impact of all internal and external changes on OSH, undertaking adequate preventive measures before the introduction of such changes providing information and training for workers subject to those changes. Also requires carrying out identification of hazards and assessment of occupational risks before each modification or introduction of new work methods, materials, processes, and machinery.
	3.10.3	Emergency Prevention, Preparedness, and Response	Defines requirements for organizational arrangements in the area of emergency prevention, preparedness, and response including internal communication and coordination, providing first aid, firefighting actions, evacuation, training, and exercises. Requires coordination of such arrangements with external emergency services and communication with such services, authorities, and neighbor organizations.
	3.10.4	Procurement	Requires establishing the procedures to ensure conformity of goods and services purchased by the organization with applicable OSH requirements.
	3.10.5	Contracting	Requires introduction of organizational arrangements assuring contractors' compliance with the same OSH requirements and management rules as those applied by the organization. These arrangements should include the selection of contractors, communication, and cooperation methods; rules for registration of accidents; occupational diseases and incidents; and OSH training and monitoring contractors for OSH.
Evaluation	3.11	Monitoring and Measurement	Requires establishing monitoring methods for selected OSH aspects including quantitative and qualitative indicators. Indicates the role of proactive and reactive monitoring in OSH MS improvement, particularly in the area of arrangements regarding hazard identification and occupational risk assessment; also defines the main OSH aspects that should be proactively and reactively monitored.

(Continued)

TABLE 35.1 (Continued)

Element	Number	Title	Summary
	3.12	Investigation of Work-Related Injuries, Ill-Health, diseases, and Incidents, and Their Impact on Safety and Health Performance	Defines requirements regarding investigations of the causes of work-related injuries, ill-health, diseases, and incidents to identify any failures in OSH-MS. Requires the investigations to be carried out by competent persons, with the participation of workers, documenting their outcome and taking into consideration the reports of external inspection bodies, communicating results to the safety and health committee, as well as undertaking adequate corrective actions.
	3.13	Audit	Requires carrying out periodic internal audits of OSH-MS according to established policy and program. Indicates that all system elements should be subject to audits and requires the implementation of audits by competent persons and consulting the auditing process and results with workers.
	3.14	Management Review	Requires periodic OSH management reviews, carried out by the top management of the organization. Defines objectives for such reviews and the required range of reviewed factors. Also requires documenting the reviews and submission of their outcomes to the safety and health committee, workers and their representatives, and persons responsible for the relevant element of OSH-MS.
Action for Improvement	3.15	Preventive and Corrective Action	Requires the introduction of organizational arrangements regarding preventive and corrective actions, which should include identification and analysis of the root causes of any nonconformities and initiating, planning, implementation, checking the effectiveness, and documenting preventive and corrective actions.
	3.16	Continual Improvement	Requires the introduction of organizational arrangements ensuring the continual improvement of all OSH-MS elements and defines the range of factors that should be taken into account in such arrangements. Also requires comparing OSH activities and their effects with activities and effects of other organizations.

organizations representing various sectors and countries, including different levels of legislation, applied management practices, and safety cultures.

Another area where there are needs and potential possibilities for international standardization is the management of programs for ergonomic improvement at workplaces (for simplification called the "ergonomics management" or the "ergonomic program management"). Standardization in this area has been already undertaken first in the United States, in the first place to prevent work-related musculoskeletal disorders that cause the biggest social and economic losses (McSweeney et al., 2002). One should mention here the establishment by OSHA (1991) of guidelines for ergonomic program management in meat-processing plants. Further steps in this area included—among others—publishing the guidelines regarding ergonomic programs by the National Institute for Occupational Safety and Health (1997), the establishment of the obligatory standard by OSHA (2000b) in this field (subsequently rescinded in 2001), and the establishment of the standard regarding ergonomics in nursing homes (OSHA, 2003).

The aforementioned documents that define ergonomic improvement management principles prove the need for further conceptual and standardization works in this area both on the national and international level. The OSH-MS model adopted in the ILO-OSH 2001 guidelines and the tripartite representation principle adopted in the process of their creation and establishment may constitute an appropriate model for activities in this new area of international standardization.

REFERENCES

Abad, J., Mondelo, P. R., & Llimona, J. (2002). Towards an international standard on occupational health and safety management. *International Journal of Occupational Safety and Ergonomics (JOSE), 3*, 309–319.

AENOR. (1996a). *Guidelines for the implementation of an occupational safety and health management system* (Standard No. UNE 81900:1996 EX). Madrid, Spain: Spanish Association of Standardisation and Certification.

AENOR. (1996b). *Guidelines for the assessment of occupational safety and health management systems. Audit process* (Standard No. UNE 81901:1996 EX). Madrid, Spain: Spanish Association of Standardisation and Certification.

AIHA. (1996). *Occupational health and safety management system: An AIHA guidance document.* Fairfax, VA: American Industrial Hygiene Association.

Bayerisches Staatsministerium für Arbeit und Sozialordnung, Familie, Frauen, und Gesundheit. (1997). *Modell zur Entwicklung, Gestaltung, Einführung/Integration eines Managementsystems für Arbeitsschutz und Anlagensiche-heit (Occupational Health and Risk Management System).* München, Germany: Bayerisches Staatsministerium für Arbeit und Sozialordnung, Familie, Frauen, und Gesundheit.

BSI. (1996). *Guide to occupational health and safety management systems* (Standard No. BS 8800:1996). London, UK: British Standards Institution.

BSI. (1999). *Occupational health and safety management systems—Specification.* Occupational Health and Safety Assessment Series (Document No. OHSAS 18001:1999). London, UK: British Standards Institution.

BSI. (2000). *Occupational health and safety management systems—Guidelines for the implementation of OHSAS 18001* (Document No. OHSAS 18002:2000). London, UK: British Standards Institution.

Bundesanstalt für Arbeitsschutz und Arbeitsmedizin. (2002). *Leitfaden für Arbeitsschutzmanagementsysteme.* Dortmund, Germany: Bundesanstalt für Arbeitsschutz und Arbeitsmedizin (Federal Institute for Occupational Safety and Health). Retrieved March 31, 2003, from http://www.baua.de/prax/ams/leitfaden_ams.pdf

CIOP (2001). *Wytyczne do systemów zarządzania bezpieczeństwem i higieną pracy. ILO-OSH 2001 (Polish version of the ILO Guidelines on occupational safety and health management systems. ILO-OSH 2001).* Warsaw, Poland: Central Institute for Labour Protection.

European Union. (1989). *Council Directive of 12 June 1989 on the introduction of measures to encourage improvements in the safety and health of workers at work (89/391/EEC). Official Journal of the European Communities, No. L 183,* 29 June 1989, pp. 1–8.

Hessisches Ministerium für Frauen, Arbeit und Sozialordnung (1996). *ASCA: New directions in state occupational safety.* Wiesbaden, Germany: Hessisches Ministerium für Frauen, Arbeit und Sozialordnung (HMFAS).

ILO. (1981). *Occupational safety and health convention (No. 155)*. Geneva, Switzerland: International Labour Organization.

ILO. (1985). *Occupational health services convention (No. 161)*. Geneva, Switzerland: International Labour Organization.

ILO. (2001a). *Guidelines on occupational safety and health management systems (ILO-OSH, 2001)*. Geneva, Switzerland: International Labour Office. Retrieved March 31, 2003, from http://www.ilo.org/public/ english/ protection/safework/managmnt/index.htm

ILO. (2001b). *Guidelines on occupational safety and health management systems (ILO-OSH 2001)*. Geneva, Switzerland: International Labour Office.

Institution of Occupational Safety and Health. (2002). *SME risk management toolkit*. The Grange, UK: Institution of Occupational Safety and Health (IOSH). Retrieved March 31, 2003, from http://www.pk-rh.com/en/index.html

ISO. (1996). *Environmental management systems—Specification with guidance for use* (Standard No. ISO 14001:1996). Geneva, Switzerland: International Organization for Standardization.

ISO. (2000). *Quality management systems—Requirements* (Standard No. ISO 9001:2000). Geneva, Switzerland: International Organization for Standardization.

Japanese Ministry of Labour. (1999). *Guideline for occupational safety and health management systems, Ministry of Labour Notification No. 53, April 30, 1999*, Tokyo, Japan: Ministry of Labour.

Kommunaldepartementet. (1991). *Internkontroll. Forskrift med veiledning* (Internal control. Ordinance with guidelines). Oslo, Norway: Kommunaldepartementet (the former Ministry of Local Government and Labour).

Kommunaldepartementet (1996). *Forskrift om systematisk helse-, miljø- og sikkerhetsarbeid i virksomheter m. fl. (Intemkontrollforskriften)* (Regulation on systematic health, environment and safety activities in enerprises, The Internal control regulation). Oslo, Norway: Kommunaldepartementet (the former Ministry of Local Government and Labour).

Lambert, J. (2000). The German position with respect to standardisation of OH&S management systems. In D. Podgórski & W. Karwowski (Eds.), *Ergonomics and safety for global business quality and productivity. Proceedings of the Second International Conference ERGON-AXIA 2000, Warsaw, Poland, May 19–21, 2000* (pp. 315–318). *Warsaw, Poland*: The Central Institute for Labour Protection.

Länderausschuss für Arbeitsschutz und Sicherheitstechnik (LASI). (2001). *Arbeitsschutzmanagementsysteme. Hand-lungshilfe zur freiwilligen Einführung und Anwendung von Arbeitsschutzmanagementsystemen (AMS) für kleine und mittlere Unternehmen (KMU), LV 22*. Saarbrücken, Germany: Länderausschuss für Arbeitsschutz und Sicher-heitstechnik (LASI). Retrieved March 31, 2003, from http://lasi.osha.de/ publications/lv/lv22.pdf

McSweeney, K. P., Craig, B. N., Congleton, J. J., & Miller, D. (2002). Ergonomie program effectiveness: Ergonomic and medical intervention. *International Journal of Occupational Safety and Ergonomics (JOSE)*, 4, 433–449.

Narodný Inšpectorat Práce. (2002). *System riadenia bezpečnosti a ochrany zdravia pri práci. Návod za za-vedenie systému (Occupational safety and health management system. Guideline for system implementation)*. Bratislava, Slovak Republic: Narodny Inspectorat Práce.

National Institute for Occupational Safety and Health. (1997). *Elements of ergonomics programs, a primer based on workplace evaluations of musculoskeletal disorders*. DHHS (NIOSH) Publication No. 97-117. Cincinnati, OH: National Institute for Occupational Safety and Health.

NNI. (1996). *Nederlandese Praktijkrichtlinjn NPR 5001, Model voor een Arbomanagementsysteem (International version, 1997, Tech. Rep. NPR 5001: Guide to an Occupational Health and Safety Management System)*. Delft, The Netherlands: Nederlands Normalisatie-Institut.

OSHA. (1982). *Voluntary protection programs*. Federal Register, 47, 29025. Washington, DC: Occupational Safety and Health Administration.

OSHA. (1989). *Safety and health program management guidelines: Issuance of voluntary guidelines*. Federal Register, 54, 3904–3916. Washington, DC: Occupational Safety and Health Administration.

OSHA. (1991). *Ergonomics program management guidelines for meatpacking plants*. Washington, DC: Occupational Safety and Health Administration, Retrieved March 31, 2003, from http://www.ergoweb.com/resources/reference/guidelines/meatpacking.cfm

OSHA (2000a). *Revision to the voluntary protection programs to provide safe and healthful working conditions*. Federal Register, 65, 45649–45663. Washington, DC: Occupational Safety and Health Administration.

OSHA. (2000b). *Final ergonomics program standard*. November 14 (repealed March 8, 2001). Washington, DC: Occupational Safety and Health Administration. Retrieved March 31, 2003, from http://www.ergoweb.com/resources/reference/standards/standard.cfm

OSHA (2003). *Guidelines for nursing homes. Ergonomics for the prevention of musculoskeletal disorders.* Washington, DC: Occupational Safety and Health Administration. Retrieved March 31, 2003, from http://www.osha.gov/ergonomics/guidelines/nursinghome/final_nh_guidelines.pdf

PKN. (1999). *Systemy zarzgdzania bezpieczeństwem i higieng pracy—Wymagania* (Standard No. PN-N-18001:1999 Occupational safety and health management systems—Requirements). Warsaw, Poland: Polski Komitet Normal-izacyjny (Polish Standardization Committee).

PKN. (2000). *Systemy zarzgądzania bezpieczeństwem i higieng pracy—Ogólne wytyczne do oceny ijzyka zawodowego* (Standard No. PN-N-18002:2000 Occupational safety and health management systems—General guidelines for assessment of occupational risk). Warsaw, Poland: Polski Komitet Normalizacyjny (Polish Standardization Committee).

PKN. (2001). *Systemy zarządzania bezpieczeństwem i higieną pracy—Wytyczne* (Standard No. PN-N-18004:2001 Occupational safety and health management systems—Guidelines). Warsaw, Poland: Polski Komitet Normaliza-cyjny (Polish Standardization Committee).

PKN. (2003). *Systemy zarządzania bezpieczeństwem i higieng pracy—Wymagania* (Standard No. PN-N-18001:2003 Occupational safety and health management systems—Requirements). Warsaw, Poland: Polski Komitet Normal-izacyjny (Polish Standardization Committee).

Standards Australia. (1994). *A management system for occupational health, safety and rehabilitation in the construction industry (handbook)* (Standard No. SAA HB53-1994). Homebush, NSW, Australia: Standards Australia.

SA&SNZ. (2001). *Occupational health and safety management systems—Specification with guidance for use* (Standard No. AS/NZS 4801:2001). Homebush, NSW, Australia: Standards Australia and Wellington, New Zealand: Standards New Zealand.

SA&SNZ. (1997/2001). *Occupational health and safety management systems—General guidelines on principles, systems and supporting techniques* (Standards No. AS/NZS 4804:1997 and AS/NSZ 4804:2001). Homebush, NSW, Australia: Standards Australia and Wellington, New Zealand: Standards New Zealand.

SNBOSH. (1992). *Ordinance (AFS 1992:6) Internal Control of the Working Environment.* Statute Book of the Swedish National Board of Occupational Safety and Health, 1992, Solna, Sweden: Swedish National Board of Occupational Safety and Health, Publishing Services.

SNBOSH. (1997). *Ordinance (AFS 1996:6) Internal Control of the Working Environment.* Statute Book of the Swedish National Board of Occupational Safety and Health, 1997, Solna, Sweden: Swedish National Board of Occupational Safety and Health, Publishing Services.

Swedish Work Environment Authority. (2001). *Systematic work environment management (AFS 2001:1).* Swedish Work Environment Authority, Solna, Sweden. Retrieved March 31, 2003, from http://www.av.se/english/legislation/afs/eng0l0l.pdf

Zwetsloot, I. J. M. (2000). Developments and debates on OHSM system standardization and certification. In K. Frick, P. L. Jensen, M. Quinlan, & T. Wilthagen (Eds.), *Systematic occupational health and safety management—Perspectives on an international development* (pp. 391–412). Oxford, UK: Elsevier Science Ltd.

36 The Selected Elements of Shaping the Occupational Health and Safety Culture in a Company Based on a Polish PN-N-18001 Standard

Radosław Wolniak and Anna Gembalska-Kwiecień

CONTENTS

INTRODUCTION

Security culture is a state that must be developed during mutual relations and actions taken in the organization's environment. Whether it is created quickly and easily or reluctantly and problematically, it mostly depends on the employees because their psychological, emotional, social, and economic conditions depend on whether they adapt without any problems to organization requirements.

The task of the management—that wishes not only to increase the efficiency of work but also to maintain safety standards—is to educate employees with a high level of safety culture and enforce its manifestations among the crew. It is not a simple task and it requires a lot of time and effort from both the management and executive staff. It is not easy to create a security culture, but it is much harder to change the consciousness and subconsciousness of the crew.

Therefore, the guidelines included in the Polish PN-N-18001:2004 standard play a useful role and help shape a culture of occupational health and safety within the company.

THE ESSENCE OF WORK SAFETY CULTURE

The concept of safety culture in an enterprise has appeared in the literature in the 1980s; however, its beginnings date back to the 1920s. It was then noticed that the workplace is a social organization that develops its own norms, values, and procedures. In the early 1980s, companies began to be treated as organizations with a culture-specific to themselves. The term was used for the first time in a report drawn up by a commission specifically established after the Chernobyl disaster (1986). This document refers to the concept of organizational culture. As the primary cause of the Chernobyl disaster and the following disasters (explosion on the oil platform of Piper Alpha in the North Sea, the Clapham Junction railway disaster in London, etc.) was low safety culture in those plants. Since that time, this term has been increasingly appearing in studies on occupational health and safety in enterprises (Studenski, 2000; Stemn et al., 2019; Ünal et al., 2018).

Similarly, as in the case of organizational culture, you can meet many definitions of security culture in the literature on the subject (Lejko, 2010; Lis, 2013; Sujova & Cierna, 2013; Skurjat, 2014; Wyrwicka, 2014; Găureanu et al., 2019). There can be some distinguished definitions related to safety culture in security sciences and approaches closer to management sciences that understand safety culture as related to occupational health and safety (Gembalska-Kwiecień, 2012, 2017a, 2017b, 2018; Pagieła, 2014; Pacana & Stadnicka, 2017; Samarth Ramprasad & Kumar, 2018; Aburumman, Newman, & Fildes, 2019; Stoffregen, Giordano, & Lee, 2019).

The term "safety culture" was first used in a document drawn up as a result of the Chernobyl disaster—it identifies a set of human traits and attitudes that relate to the safety culture (Kowalski, 2018).

The culture of safety in the first of these trends can be defined as the total material and non-material property of a person, which serves his universally understood defense. It serves to maintain (cultivate), recover (when lost), and increase the level of security of a given entity. It consists of three dimensions: mental-spiritual, organizational-legal, and material (Piwowarski, 2012).

The culture of safety in the second of these trends can be defined as a set of features and attitudes of organizations and people that make safety matters a top priority. According to a definition, the safety culture refers to both the attitudes of employees as well as the structure of a given organization. It deals with issues related to meeting all safety requirements.

Another definition is the one presented by the British Health and Safety Commission (HCS). According to it, safety culture is a product of individual and group values, attitudes, perceptions, competencies, and behavior patterns that define the commitment and style of managing the organization's occupational health and safety (International Atomic Energy Agency, 1991; Pagieła, 2016; Kim et al., 2018; Vierendeels et al., 2018; Găureanu et al., 2019).

The safety culture can be recognized in two layers: managerial and behavioral, which are summarized in Table 36.1.

TABLE 36.1
Layers of Safety Culture

Managerial Layer	Behavioral Layer
Patterns	Behaviors
Rules and norms introduced and fixed	The way people behave
Assumptions	Convictions
Expected behaviors	The set of values and principles that are considered right
Operating activities	Attitudes
Daily procedures and rules of conduct	Values presented by teams

Source: Kowalski (2018).

The following components of the safety culture are distinguished in the literature (Studenski, 2000):

- attitude toward occupational health and safety regulations,
- trust in a preventive role of occupational health and safety regulations,
- attitude of superiors towards the subordinates who are taking risks,
- attitude of employees to the colleagues who are taking risks,
- attitude of superiors to employees who need to meet occupational health and safety standards,
- attitude towards occupational health and safety training,
- prestige of the services and the rank of the occupational health and safety employees, and
- involvement of employers in the organization of safe working conditions.

SHAPING A CULTURE OF WORK SAFETY

The safety culture is created by members of a given group who repeat their behavior according to the pattern appropriate for the safety culture norms. All definitions indicate that shaping the safety culture is a continuous and multidimensional process that all social groups must undergo to achieve a satisfactory level. One-off actions can be effective, but only for a short time.

The high value attributed to health and life is the main feature of the desired safety culture. Forming a safety culture requires creating new attitudes and values with the participation of all group members to persuade employees to eliminate excessive risk in the workplace. It is also an important part of behavior aimed at health and life protection (Booth & Lee, 1993; Bird & Germain, 1996; Ghahramani & Salminen, 2019; Nite, 2019).

Those activities are not easy because it is difficult to influence people who have already developed personality traits and behavior patterns. The safety culture in the enterprise is created by the employers who formulate the occupational health and safety policy; define threats and the standards of conduct in the situation of known threats; and promote vigilance against the possibility of new threats. Employees must know that the employer is genuinely and visibly involved in matters of occupational safety and the protection of the workers' health and life. The safety culture in an enterprise is shaped by the involvement of all employees, through group work and a sense of belonging to the company as well as proper education (Gong, 2019; Reiman et al., 2019).

Shaping the desired safety culture must be a continuous process because the actual cultural change is difficult and takes time. The one-off action taken to improve the safety culture brings only apparent results, as it only reaches the most superficial layer of culture—artifacts, as Schein called them (Makin, Cooper, & Cox, 2000; Erdal, Isik, & Fırat, 2018). For the safety culture to be of a high level, it is also necessary to reach deeper layers of culture (i.e. recognized values and basic assumptions).

When shaping the safety culture, special attention should be paid to three elements (CIOP, 2007; Podgórski & Pawłowska, 2004):

- Physical work environment (tools, machines, work organization)
- Employee behavior (compliance with occupational health and safety regulations, information, and cooperation; demonstration of care for occupational safety beyond the obligations)
- Internal characteristics of employees (knowledge, skills, motivation)

Those elements are depicted in the model of shaping the safety culture shown in Figure 36.1.

As it appears from the presented model, the effective formation of a high safety culture depends not only on the physical environment of the work but also on the behavior of employees and on their individual characteristics. Therefore, to achieve the degree of the desired safety culture, it is necessary to ensure safety that working conditions (efficient machines, devices, safe organization

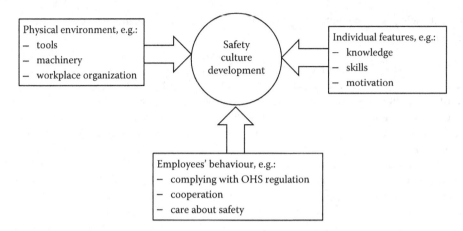

FIGURE 36.1   The model of shaping the safety culture.
Source: Podgórski and Pawłowska (2004).

of workplaces, etc.) shape safe employee behavior (compliance with occupational health and safety regulations, cooperation between employees and between employees and supervisor, etc.) and concentrate on the characteristics of individual employees (conducting training, appropriate motivation for safe conduct, etc.).

The conditions for effective shaping of the safety culture are also (Milczarek, 2001, 2002; Podgórski & Pawłowska, 2004):

* An open and honest communication related to informing employees about risks.
* The management's commitment and expression of personal interest and care for their employees' safety, compliance with occupational health and safety regulations, and treatment of security matters as equal with production problems.
* Employee participation (i.e. employee participation in the development of internal standards and documents in the field of occupational health and safety) in the activities and decisions taken within the company.
* Occupational health and safety education, or occupational health and safety training adapted to the specifics of the work and to the needs of employees.
* Analysis of accidents, reporting, and analysis of accidental and potentially accidental events that took place in the workplace; identification of causes of problems; and undertaking preventive actions.
* Motivating and strengthening safe behaviors, expressing approval and recognition to employees who act safely, and engaging in activities aimed at improving safety at the workplace.
* Cooperation between employees and an atmosphere of understanding and trust between the management and employees, and between employees from various departments and organizational levels.

Also, shaping the safety culture is favored by:

* Strengthening and shaping the employees' sense of belonging and pride in working in the company.
* Enabling professional development and realizing the professional ambitions of an employee within the company.
* Stress management by means of assessment and monitoring of stress levels among employees, job satisfaction, and pro-health prophylaxis.
* Promoting safe behaviors outside of work (on the road, at home, during rest).

All these activities help shape a high-level safety culture. However, they cannot be taken once because their effect will be short-lived. The culture of safety requires constant improvement.

Making changes in the work organization or in an applied technology, without changing the cultural conditions of the proceedings aimed at increasing the level of security, usually do not lead to the desired effects because the beneficial effect of changes in technology is overcome by taking risky behaviors. The level of safety culture has a direct impact on the employees' attitudes in terms of occupational safety. Shaping the desired safety culture, through changes in the system of values and attitudes of employees, improves the occupational health and safety of the company and increases the motivation and commitment of employees and their satisfaction with the work performed. Thanks to this, employees will comply with occupational health and safety regulations and the number of accidents will decrease. Creating the desired safety culture is the basis for ensuring worker safety and reducing the number of accidents. Particular attention should be paid to the working environment, behavior, and personality traits of the employees.

The development of a safety culture requires continuous research, which will allow assessing its level. Thanks to this research, it is possible to determine which elements of this culture needs improvement and which actions should be taken to achieve its high level.

OCCUPATIONAL HEALTH AND SAFETY MANAGEMENT SYSTEM AS A WAY TO IMPROVE SAFETY

A tool that can help improve safety and working conditions is the occupational health and safety management system, which, by systematic actions, minimizes the risks resulting from hazards in the work processes. From the end of the 1980s, guidelines regarding the systemic approach to employee safety were published in various countries. In 1999, BS OHSAS 18001:1999 Occupational Health and Safety Management Systems – Specification (updated in 2007) and PN-N 18001:2007 Occupational Health and Safety Management Systems – Requirements (updated in 2004) were issued. Those requirements were the base for enterprises in building their own occupational health and safety management systems. On March 12, 2018, the International Organization for Standardization (ISO)—after many years of efforts—published the new ISO 45001:2018 standard – Occupational Health and Safety Management Systems – requirements with guidance of use—which is the first international standard in this field. ISO 45001:2018 has been developed considering the guidelines of the International Labor Organization and national requirements, with broad international arrangements. The standard is built in a high-level structure, so it has an analog layout to the previously published norms regarding quality management (ISO 9001:2015) system and environmental management system (ISO 14001:2015). This makes it easier to integrate the occupational health and safety management system with the integrated management system ISO 45001:2018.

These documents were not created as a result of a formal standardization process implemented by ISO, but they are international documents with guidelines for the implementation and improvement of occupational health and safety management systems (Heras-Saizarbitoria et al., 2019; Pignaton et al., 2019). In the introduction to the second document, there is a note stating that the practical recommendations included in the guidelines are addressed to all people responsible for occupational health and safety management systems. They do not have legal force nor are they intended to replace national law, regulations, or standards (Palačić, 2018). The employer is responsible for the organization of activities in the field of occupational health and safety. The introduction of occupational health and safety management is an effective way to fulfill this obligation (ILO-OSH:2001; Lafuente & Abad, 2018; Kwegyir-Afful et al., 2018).

In many countries, the above-mentioned documents have been translated or national standards have been developed on their basis.

In Poland, the Central Institute for Labor Protection translated the ILO guidelines into Polish language and published them in the publication Guidelines for Occupational Health and Safety

Management Systems (CIOP, 2004) The Polish Committee for Standardization (PKN) had already initiated standardization works in 1998 to formulate requirements and guidelines in the area of implementation and improvement of occupational health and safety management systems. In February 1998, the Standardization Problem Committee No. 276 was established within the PKN for the management of occupational safety and health issues.

As a result of the work carried out, a set of PN-N-18000:2004 standards were established, including four documents:

1. PN-N-18001:2004 Occupational Health and Safety Management Systems. Requirements.
2. PN-N-18002:2004 Occupational Health and Safety Management Systems. General guidelines for risk assessment.
3. PN-N-18004:2004 Occupational Health and Safety Management Systems. Guidelines.
4. PN-N-18011:2006 Occupational Health and Safety Management Systems. Audit guidelines.

Formalized occupational safety and health management systems constitute a set of principles and interrelated elements for the general business management system that ensure the achievement of the organization's objectives to improve the safety conditions of both the employees and the environment.

The PN-N-18001:2004 standard contains requirements for a workplace safety management system that can be subjected to external verification by independent accredited institutions to obtain confirmation of compliance with the requirements of the standard and a certificate of compliance with requirements. Other standards are discussing tools for improving the occupational safety management system. They do not constitute the basis for the certification procedure. The PN-N-18004:2004 standard is a specification of the PN-EN-18001:2004 standard; it is a comprehensive, supplemented with numerous practical guidelines of the standard describing how to implement the system. Enterprises have a task to achieve compliance with the requirements—recommendations of the PN-N-18004:2004 standard to improve the existing system. The PN-N-18011:2006 standard is intended for use by auditors during audits of the occupational health and safety management system. It presents guidelines for the audit process and defines the auditors' competencies and the scope of knowledge necessary to ensure the proper conduct of audits.

The occupational health and safety management system, whose requirements have been specified in the Polish PN-N-18001:2004 standard, ensures minimizing the risk of hazard at the workplace by eliminating threats at the source.

The requirements of the above standards are a set of elements of the management system—the implementation of which should facilitate the proper conduct of each type of activity in the aspect of occupational health and safety. These are not only the requirements of the standard but above all, they are guidelines and recommendations that are useful for practical activities in the field of occupational health and safety, compliance with legal provisions in this area, and active creation of correct conditions for employees in the work environment (Kowalkow, 2004).

SELECTED GUIDELINES (ELEMENTS) FOR SHAPING THE SAFETY CULTURE INCLUDED IN THE PN-N-18001 STANDARD

Occupational health and safety culture in an enterprise can be shaped using the occupational health and safety management system presented in the PN-N-18001:2004 standard. When implementing the occupational health and safety management system, enterprises should pay special attention to elements that shape the safety culture in the long term. Due to the time perspective, standardized systems affect the improvement of the working environment and equipment at workplaces in the short term and ensure the formation of a high level of safety culture in the company in the long term.

The main objective of the PN-N-18001:2004 standard is to support activities to improve occupational health and safety by presenting requirements for an effective OHS management system. The introduction of the standard indicates that the success of implementing the OSH management system depends on the involvement of all services at all levels of the organization—particularly, the top management—and ensuring a wide participation of employees at the stage of planning, implementing, and maintaining all elements of this system.

The PN-N-18001:2004 standard can be used by enterprises whose aim is:

- implementation, maintenance, and improvement of the occupational health and safety management system;
- proceeding in accordance with the occupational health and safety policy established in-house;
- declaration of acting in accordance with the requirements of this standard; and
- attempting to obtain the confirmation of the external compliance of the occupational health and safety management system with the requirements of this standard in the organization (PN-N-18001:2004).

The standard applies to those occupational health and safety factors that the organization can oversee and may be affected by. To implement and maintain an effective occupational health and safety management system, it is recommended that the organization adopt a continuous improvement model.

The design, implementation, and functioning of the occupational health and safety management system compliant with the requirements of the PN-N-18001:2004 standard includes, in particular:

- identifying hazards and assessing the occupational risk occurring at the workplace;
- identifying and updating of legal requirements regarding occupational health and safety, development of occupational health, and safety at work policy;
- defining general and specific objectives in the field of occupational health and safety at work consistent with the occupational health and safety policy;
- developing of plans to ensure the achievement of the adopted objectives and compliance with legal requirements of occupational health and safety;
- appointing a person responsible for the implementation and maintenance of the occupational health and safety management system;
- developing and implementing training programs aimed at raising the level of employee awareness in the field of occupational health and safety;
- designing and implementing an internal communication system between various levels and units of the organization;
- designing and implementing a communication system with interested parties;
- developing the documentation required by the standard (plans, procedures, policy);
- developing a readiness and a response procedure for accidents at work; monitoring of occupational health and safety at work;
- implementing a system of internal audits to determine whether the occupational health and safety management system is consistent with the planned activities and is properly implemented and maintained; and
- conducting periodical reviews of the system by the top management aimed at establishing the system's compliance with the requirements of the standards and assessing the effectiveness of its functioning.

The PN-N-18001 standard presents individual elements of the occupational health and safety management system that are the basis for the system assessment by independent external institutions. The organizational unit that strives to achieve confirmation of compliance of the system with the requirements of the standard should meet all the obligations included in its provisions.

The next part of the chapter discusses selected requirements and guidelines related to shaping the occupational health and safety culture within the enterprise.

INVOLVEMENT OF TOP MANAGEMENT AND EMPLOYEES AS A PART OF THE WORK SAFETY CULTURE

Top management, by definition, is "a person or group of people who manage and supervise the organization at the highest level" (PN-EN ISO 9000:2015).

According to the PN-N 18001:2004 standard, an effectively implemented and properly functioning system exists only when "the highest management of an organization demonstrates strong and visible leadership and commitment to activities for occupational health and safety at work" (PN-N-18001:2004).

The basic manifestations of the management's involvement are:

- providing the necessary resources to design, implement and operate the occupational health and safety management system;
- determining and updating the policy and objectives of occupational health and safety in the organization; and
- carrying out inspections of the occupational health and safety management system.

The responsibility to shape the right organizational safety culture is based on the representatives of the top management.

In the area of safety culture activities, the most important issues accepted by the highest management are:

- standards, norms, and rules of conduct;
- behavioral patterns (behavioral artifacts);
- the way they communicate with employees;
- enforcement among employees (language artifacts);
- provision of physical security conditions (physical artifacts).

These activities will significantly shape the culture of the organization's occupational health and safety. This is because the culture of work safety of individual employees—representatives of top management—shapes the organizational culture of occupational health and safety within the organization (Ejdys, 2010).

Another very important element regarding the culture of occupational health and safety management is defining the OHS policy that is the organization's declaration regarding its intentions and principles referring to the general effects of occupational health and safety at work, defining the OHS operational framework.

The responsibility of establishing this policy is assigned to the highest management of the organization, who also bears responsibility for its implementation, communicating it to employees and ensuring that the employees understand the policy.

This declaration contains a commitment to:

- prevent accidents and occupational diseases at work;
- the organization's efforts to constantly improve the occupational health and safety at work;
- meet legal regulations and other organizational requirements;
- continuous improvement of occupational health and safety at work;
- provision of adequate resources to implement this policy; and
- additional qualifications and consideration of the role of employees and their involvement in activities for occupational health and safety at work.

Occupational health and safety policy as a document is a very important element of the safety culture. Assigning the appropriate rank to this document in the structure of strategic documents of the enterprise will ensure that the records contained therein will shape the future behavior of employees. Both the process of exposing occupational health and safety policy as a strategic document and that of enforcing the provisions contained in documents seem to be crucial in shaping the attitudes and behavior of employees.

However, it should be emphasized that the involvement of the management should not only be to formulate an occupational health and safety policy supported by appropriate measures but to also express personal interest and care for their employees' safety, comply with occupational health and safety regulations, and treat security matters as equal to the tasks performed by individual organizational units. The key to the success of an implemented and effectively functioning occupational health and safety management system is the strong leadership and commitment of the top management in activities for occupational health and safety at work. It is very important that employees feel that the threat in their workplace is not only their problem. The sense of belonging to the team, as well as the feeling that problems related to the safety of their health and life are also important for their superiors and employers, reinforce compliance with the set rules and have a real impact on activities in the area of raising the safety culture within the company.

According to PN-N-18001:2004, employee participation is one of the most important elements of the occupational health and safety management system. The participation of employees in the design, implementation, and functioning of the occupational health and safety management system is a very important element in shaping an appropriate occupational safety culture within the organization. According to the PN-N-18001:2004 requirements, "as part of the health and occupational health and safety policy, the top management should ensure consultation with employees and their representatives and their information in all aspects of occupational health and safety issues." In addition, it should introduce organizational solutions so that employees and their representatives have the time and resources to actively participate in the planning, implementation, maintenance, checking, corrective and preventive actions, and other activities for continuous improvement under the occupational health and safety management system (PN-N-18001:2004). Therefore, there is a large variety of participatory solutions that can be applied in enterprises implementing this system. It should be emphasized that, from a psychological point of view, the more an individual gets involved in making decisions, the more likely it is that he or she will obey these decisions.

The full cooperation of employees in the scope of the above-mentioned tasks is a guarantee for the effectiveness of the implemented system, and an important element in the proper shaping of the occupational safety culture.

TRAINING, AWARENESS, COMPETENCE OF THREATS, MOTIVATION

The PN-N-18001:2004 standard requires the creation of a documented procedure concerning organized trainings in the field of occupational health and safety. The training program should consider the training needs of the employees adapted to the nature of individual employee groups (Tappura et al., 2019). Education in the field of occupational health and safety covers all types of training and practical skills exercises and is an important element in shaping the culture of occupational safety within the organization. The management should ensure that all employees—at all positions—are competent in the implementation of the tasks assigned to them and have been properly trained in this area.

The organized trainings aim at making the employees aware of (PN-N-18001:2004):

- the types of threats occurring within the organization and at individual work positions and related occupational risk;

- the benefits for employees and organizations from the elimination of risks and limiting occupational risk;
- their tasks and responsibilities in achieving compliance of the action with the occupational health and safety policy and the procedures and requirements of the occupational health and safety management system, including the requirements regarding readiness and response to accidents at work and serious accidents; and
- the potential consequences of non-compliance with established procedures.

As we can see from the above guidelines, training programs should be tailored to the needs of individual groups of employees. The company should develop a system for reporting training needs for employees. Employees should be encouraged to participate in training that increases their competencies and qualifications. Support from the management for grassroots initiatives that raise employees' awareness will be susceptible to future employee behavior.

The proper occupational safety culture of the organization will lead to overcoming undesirable behavior of employees who are afraid, for various reasons (i.e. receiving a penalty), to inform about the lack of knowledge or qualifications enabling them to safely perform their duties. It should be emphasized that various types of educational programs and campaigns in the field of healthcare not covered by the periodic training system are a good practice.

Every employer is obliged to systematically analyze the cause of accidents at work and to apply appropriate preventive measures based on their results. The analysis of accidents should indicate the current situation in the area of accidents in the plant; the type of work, technologies, and positions with a special accident risk; and the scope of activities aimed at removing these hazards. Various methods of obtaining information and analyzing the causes of accidents at work are used to investigate accidents. These methods often use predefined occupational injury models or constitute a specific procedure to identify the causes of those accidents.

Motivating for safety at work is a very important guideline of the PN-N-18001:2004 norm affecting the culture of workplace safety. Motivation can be defined as a specific state of the human psyche. The employee's perception of the need to adapt to the work methods and patterns of behavior prescribed by law is called motivation for safe work.

This need is the result of a belief compatible with existing methods and will reduce or eliminate the risk of loss of life or health. An impact on employees, which intends to induce in them the need to adapt to safety patterns, is called motivating to safe conduct. This type of motivation consists of several factors (Gembalska-Kwiecień, 2017a):

- informing about accidents, their causes, and their losses,
- giving a good example,
- putting goals for the subordinate to obtain,
- promoting safe-conduct,
- assessing supervisors and employees, and
- rewarding and punishing.

The knowledge of accidents allows us to state that employees learn little from mistakes. In the event where someone had lost their life or health because of their mistakes, others would repeat the same action that previously had led to the accident. This situation indicates a lack of occupational safety culture in a given organization. In this situation, it is important to examine each accident and identify the causes that led to its occurrence. It is very important to properly examine and explain the accident and its causes. What has been done against the rules and why it happened should be investigated. Knowledge gained in post-accident investigations should be brought to the attention of the employees. Descriptions of accidents that have been presented to the employees during periodic training for self-diagnosis of causes teach them to predict the accidental effects of risk-taking. At a later stage, one should consider how to improve the organization's safety culture.

Another factor is to set a good example, especially by the supervision and management of the company. Each manager can influence the state of security by organizing technical prophylaxis, communicating and affecting the consciousness of the supervisors and employees, and giving a good example. Communicating awareness of the value of security and giving a good example are manifested by:

- demonstrating high value to security issues by management
- putting safety goals ahead of production goals
- expressing the conviction that accident-free work is a real goal possible to achieve
- striving for compliance of decisions, assessments, and conduct with proclaimed safety rules and preferences
- incorporating safety issues into production management

Every employee, regardless of their function, is obliged to follow safety regulations. This applies to both regular employees and members of the management. When lower-ranking employees see that the safety culture in a given organization shows that supervisors are complying with the regulations, then it reassures them that it is necessary to keep work safety rules (Gembalska-Kwiecień, 2017a).

The next motivating factor is giving subordinates the goals to achieve. Each supervisor should inform his team, as well as individual subordinates, that the primary objective is to perform tasks in accordance with the applicable security regulations. To achieve this goal, it is necessary to first examine the conditions and threats before the start of the task and apply the necessary safeguards. Then, they should follow it carefully and in accordance with the applicable standards of conduct in subsequent operations and activities.

Promoting safe conduct is another element of occupational safety culture regarding motivating for safe work. The factors that affect the behavior of a person are the attitudes in which he is located, his competencies and experience, and attitudes that determine how to proceed in a given situation.

A necessary element in motivating people to maintain safety is the assessment of supervision and employees. In every work establishment, department, and team—as well as in relation to management and all employees, at least once a year—an assessment should be made in terms of meeting safety standards. The following are usually considered as the assessment criteria: accident rate, occupational disease incidence, inspection results, and achieved goals, as well as special achievements in the field of improving safety. The assessments should be recorded and then considered during the revisions and promotions. All employees of the plant should be guided by the belief that the occupational safety culture in the organization shows that the condition for promotion, especially for managerial positions, is to obtain high results in the safety assessment.

The last factor of the occupational safety culture influencing the motivation for safe work is rewarding and punishing employees. Recognizing, rewarding, and praising employees for safe work by supervisors is the best incentive to maintain a high level of security in the company.

The improvement of the motivational system in the enterprise—such as creating and raising the motivation level for safe work among employees—requires managers to analyze the existing methods and instruments of motivation comprehensively and thoroughly in the enterprise, including their possible adjustment and adaptation to the constantly changing technical, organizational, and social conditions. This process is necessary to raise the culture of occupational safety (Gembalska-Kwiecień, 2017a).

COMMUNICATION

The effectiveness and efficiency of the occupational health and safety management system is guaranteed by the creation of an appropriate safety culture of the organization's work, as well as the development of an internal and external communication system. According to PN-N-18001:2004, the company should "establish and maintain procedures regarding:

- internal communication between various levels and organizational units as well as employees and their representatives;
- receiving and providing information on occupational health and safety, documenting them and their reaction in the process of communication with interested external parties;
- providing relevant information on the risks associated with the organization's activities—and the resulting occupational health and safety requirements and practices—to all subcontractors, customers, and other people who may be exposed to them; and
- accepting and analyzing comments, ideas, and information related to occupational health and safety originating from employees and their representatives and providing them with appropriate answers."

The communication system plays a very important role in the process of shaping the safety culture within the organization. The internal communication system should include the promotion of correct, safe behavior of employees.

Every employee in the company should be able to inform about occurring threats and other events affecting occupational safety, potentially accidental occurrences, as well as reporting their proposals to improve occupational health and safety.

An example of the solutions that can be used are:

- interpersonal contacts (personal, telephone)
- correspondence (mail, CD, fax)
- management meetings with selected employees
- general meetings with all employees
- posters and information boards located in the main buildings where all the most important information regarding the functioning of the company and the occupational health and safety management system are posted and regularly updated
- leaflets presenting the issues discussed in a concise and clear way
- boxes intended for reporting own improvement solutions and observed irregularities, (management should respond to this type of information without leaving them unresolved)
- trainings informing employees about hazards and preventive measures in place at the workplace

Open and honest communication is based on communication with others—teaching, listening, speaking, and reaching a compromise. Communication applies to all employees at all levels of the organizational structure. In the area of occupational health and safety, it should include reliable and systematic information about existing threats, and protection measures as well as desirable behaviors that will minimize the risk involved. Open and honest communication easily convinces employees to change. At the same time, the quality of communication in the organization has a direct impact on an employee's motivation, sense of belonging, job satisfaction, commitment and contribution of energy, efficiency, and effectiveness.

Employee participation is also directly related to good communication. In the absence of information, it is difficult to expect the employee to show interest and involvement in activities aimed at increasing safety—even the occupational safety work in the enterprise.

CONTINUOUS IMPROVEMENT

According to the PN-N-18001:2004 standard, the organization should "introduce and maintain organizational solutions for the continuous improvement of individual elements of the occupational health and safety management system and the system as a whole" (PN-N-18001:2004).

The general need to improve occupational health and safety management systems results from the following premises:

- continuous improvement is a formal requirement of the PN-N-18001:2004 standard that is the basis for their certification;
- standardized systems are open systems requiring constant adaptation to changes taking place inside the organization and in the environment; and
- system assumptions make it impossible to achieve the ideal state.

The general requirements formulated in the PN-N-18001:2004 standard give enterprises freedom in adapting them to the existing reality and organizational conditions within the framework of improvement processes.

The improvement process, as a formal requirement of the PN-N-18001:2004 standard, is important to improve the organization's safety culture. It has been narrowed down to two areas. The first area of improvement is the implementation of occupational health and safety objectives defined within the system. The second concerns situations related to non-compliance with legal and other requirements, occurrence of non-compliance, and appearance of problem situations or emergency situations. In the above approach, the attention is focused on the internal premises of improvement processes.

The improvement processes can be considered in two perspectives:

- narrow—in which the aim of improving processes is to eliminate weaknesses and disadvantages of applied solutions and to strengthen and stress internal elements determining success; and
- broad—in which the goal of improving processes is to use existing opportunities in the environment, determined by development trends and management concepts.

In the narrow sense, improvement—as a fundamental principle of PN-N-18001:2004, standardized management systems involving the gradual improvement of processes and products—can be treated as a process of solving problems. The problem can be considered in the following categories (Hamrol, 2008):

- discrepancies between what is required (assumed) and what has been achieved
- searching for opportunities to increase the efficiency and effectiveness of the processes carried out
- emergency situations, unexpected deviating from the assumed assumptions

To solve the problem, you need to know about the process or object that the problem concerns, the tools and methods of conduct, and the skills to properly use them. Particularly valuable in the process of solving problems is the ability to perceive problems correctly. The organization's occupational health and safety culture should be conducive to reporting problems that employees notice.

In a broad sense, the processes of improving standardized OHS management systems should consist of focusing on the issues of designing for occupational health and safety at work and methods of measuring the safety culture.

CONCLUSION

One of the more popular definitions of occupational safety culture developed by the British Health and Safety Executive (HSE) indicates that occupational health and safety is the result of individual and group values, attitudes, perceptions, competencies, behavioral patterns and style, and the quality of security management within the organization. Organizations with a positive culture of safety are characterized by communication based on mutual trust, the common perception of the importance of security, and trust in the effectiveness of preventive actions (Horbury & Bottomley, 1997).

The presented study deals with the selected guidelines from the PN-N-18001:2004 standard, indicating the role of individual system elements in shaping the appropriate organizational occupational health and safety culture of the organization.

The elements in shaping the occupational safety culture concern, in addition to typical organizational activities, the psychological sphere of the company's employees. The implementation and shaping of rules regulating safety problems in the organization itself will be a very ineffective activity if the support and cooperation of persons directly affected by these regulations are not obtained.

The accumulation of efforts to ensure high safety culture should be the responsibility of the top management especially because they are responsible for security throughout the enterprise.

REFERENCES

Aburumman M., Newman S., Fildes B. (2019). Evaluating the effectiveness of workplace interventions in improving safety culture: A systematic review. *Safety Science, 118*:376–392.

Bird F.E., Germain G.L. (1996). *Damage control. A new horizon in accident prevention and cost improvement*. New York: American Management Association.

Booth R.T., Lee T.R. (1993). The role of human factors and safety culture in safety management. *Proceedings of the Institution of Mechanical Engineers, Part B: Journal of Engineering Manufacture, 209*(5): 393–400.

CIOP (2004). *Central Institute for Labor Protection*, https://www.ciop.pl/.

CIOP (2006). *Stan bezpieczeństwa i higieny pracy w 2005 roku.*: http://www.ciop.pl/14541.html (Accessed March 21, 2007).

Ejdys J. (2010). *Kształtowanie kultury bezpieczeństwa i higieny pracy w organizacji*. Białystok: Oficyna Wydawnicza Politechniki Białostockiej.

Erdal M., Isik, N.S., Fırat S. (2018). Evaluation of occupational safety culture in construction sector in the context of sustainability. *Lecture Notes in Civil Engineering, 1*:245–254.

Găureanu A., Draghici A., Dufour C., Weinschrott H. (2019). The organizational safety culture assessment. *Advances in Intelligent Systems and Computing, 876*:728–734.

Găureanu A., Draghici A., Weinschrott H. (2019). Increasing the quality of occupational safety and health implementations through awareness training for those involved in implementing the safety observation report. *Quality – Access to Success, 20*:141–146.

Gembalska-Kwiecień A. (2012). Kształtowanie kultury bezpieczeństwa w przedsiębiorstwie. *Zeszyty Naukowe Politechniki Śląskiej, Organizacja i Zarządzanie, 63*:189–198.

Gembalska-Kwiecień A. (2017a). *Czynnik ludzki w zarządzaniu bezpieczeństwem pracy w przedsiębiorstwie. Wybrane zagadnienia*. Gliwice: Wydawnictwo Politechniki Śląskiej.

Gembalska-Kwiecień A. (2017b). Fundamentals of an effective corporate safety culture. *Ekonomia i Prawo, 4*:401–411.

Gembalska-Kwiecień A. (2018). Zarys problematyki kultury bezpieczeństwa pracy na przykładzie wybranego przedsiębiorstwa. *Promotor BHP, 5*:22–27.

Ghahramani A., Salminen S. (2019). Evaluating effectiveness of OHSAS 18001 on safety performance in manufacturing companies in Iran. *Safety Science, 112*:206–212.

Gong Y. (2019). Safety culture among Chinese undergraduates: A survey at a University. *Safety Science, 111*:17–21.

Hamrol A. (2008). *Zarządzanie jakością z przykładami*. Warszawa: Wydawnictwo Naukowe PWN.

Heras-Saizarbitoria I., Boiral O., Arana G., Allur E. (2019). OHSAS 18001 certification and work accidents: Shedding light on the connection. *Journal of Safety Research, 68*:33–40.

Horbury, C. R., & Bottomley D. M. (1997). *Research into health and safety in the paper industry*. Health & Safety Laboratory, IR/RAS/98/2.

ILO-OSH:2001 *Guidelines on occupational safety and health management system*. International Labour Organization.

International Atomic Energy Agency (1991). *Kultura bezpieczeństwa*. Safety Series No. 75-INSAG-4, Vienna: International Atomic Energy Agency.

ISO 45001:2018 *Occupational health and safety management systems – Requirements with guidance of use*. International Organization for Standardization.

Kim Y.G., Kim A.R., Kim J.H., Seong P.H. (2018). Approach for safety culture evaluation under accident situation at NPPs; an exploratory study using case studies. *Annals of Nuclear Energy*, *121*:305–315.

Kowalkow A. (2004). Norma PN-N 18001:2004 naturalną i skuteczną metodą doskonalenia systemów zarządzania bezpieczeństwem i higieną zgodnych z wytycznymi Międzynarodowej Organizacji Pracy, [In:] Gierzyńska-Dolna M., Kondyba-Szymański B. Eds. *Doświadczenia i efekty funkcjonowania systemów zarządzania jakością w przedsiębiorstwach*. Częstochowa: Wydawnictwo Politechniki Częstochowskiej.

Kowalski P. (2018). *Kultura bezpieczeństwa*. http://www.kulturabezpieczenstwa.pl/bezpieczenstwo/835-kultura-bezpieczenstwa (Accessed July 17, 2018).

Kwegyir-Afful E., Kwegyir-Afful E., Addo-Tenkorang R., Kantola J. (2018). Effects of occupational health and safety assessment series (OHSAS) standard: A study on core competencies building and organizational learning. *Advances in Intelligent Systems and Computing*, *594*:395–405.

Lafuente E., Abad J. (2018). Analysis of the relationship between the adoption of the OHSAS 18001 and business performance in different organizational contexts. *Safety Science*, *103*:12–22.

Lejko Z. (2010). Kultura bezpieczeństwa w przedsiębiorstwie. *Atest*, 8:36–48.

Lis K. (2013). Kultura i klimat bezpieczeństwa pracy. *Studia Oeconomica Posnaniensia*, 7:7–16.

Makin P., Cooper C.L., Cox Ch.J. (2000). *Organizacja a kontrakt psychologiczny*. Warszawa: Wydawnictwo Naukowe PWN.

Milczarek M. (2000). Kultura bezpieczeństwa w przedsiębiorstwie – Nowe spojrzenie na zagadnienia bezpieczeństwa pracy. *Bezpieczeństwo Pracy – Nauka i Praktyka*, 10:17–20.

Milczarek M. (2001). Ocena poziomu kultury bezpieczeństwa w przedsiębiorstwie. *Bezpieczeństwo Pracy – Nauka i Praktyka*, 5:17–19.

Milczarek M. (2002). *Kultura bezpieczeństwa pracy*. Warszawa: CIOP.

Nite D.K. (2019). Negotiating the mines: The culture of safety in the Indian coalmines, 1895–1970. *Studies in History*, 1:88–118.

Pacana A., Stadnicka D. (2017). *Nowoczesne systemy zarządzania jakością zgodne z ISO 9001:2015*. Rzeszów: Wydawnictwo Politechniki Rzeszowskiej.

Pagieła J. (2014). Bezpieczeństwo behawioralne jako element kształtujący kulturę bezpieczeństwa. *Zeszyty Naukowe Politechniki Śląskiej, Organizacja i Zarządzanie*, 73:471–480.

Pagieła J. (2016). Diagnoza kultury bezpieczeństwa w wybranych przedsiębiorstwach krajów UE w świetle badań pilotażowych. *Zeszyty Naukowe Politechniki Śląskiej, Organizacja i Zarządzanie*, 93:381–392.

Palačić D. (2018). Effects of OHSAS 18001 norm implementation on the improvement in safety and health at work performances in Croatia. *Sigurnost*, 3:209–223.

Pignaton A.P., Oliveira M., Carneiro C., Duques A. (2019). OHSAS 18001 certification as leadership commitment factor for improvement of the safety management performance in a mining company. *Advances in Intelligent Systems and Computing*, 791:361–369.

Piwowarski J. (2012). Kultura bezpieczeństwa i jej trzy wymiary, [In:] Hrynicki W., Piwowarski J. Eds. *Kultura bezpieczeństwa. Nauka, praktyka, refleksje*. Kraków: Wyższa Szkoła Bezpieczeństwa Publicznego i Indywidualnego "Apeiron" w Krakowie.

PN-EN ISO 9000:2015 *Systemy zarządzania jakością. Terminologia*. Polski Komitet Normalizacyjny.

PN-N-18001:2004 *Systemy zarządzania bezpieczeństwem i higieną pracy. Wymagania*. Polski Komitet Normalizacyjny.

Podgórski D., Pawłowska Z. (2004). *Podstawy systemowego zarządzania bezpieczeństwem i higieną pracy*. Warszawa: Wydawnictwo CIOP-PIB.

Reiman A., Pedersen L.M., Väyrynen S., Sormunen E., Airaksinen O., Haapasalo H.A., Räsänen T. (2019). Safety training parks – Cooperative contribution to safety and health trainings. *International Journal of Construction Education and Research*, 1:19–41.

Samarth Ramprasad K., Kumar P. (2018). 3 factor – Hot (human, organizational and technical) model for construction safety culture. *International Journal of Civil Engineering and Technology*, 8:530–541.

Skurjat K. (2014). Kultura bezpieczeństwa pracy w organizacji. *Logistyka*, 5:1368–1374.

Stemn E., Bofinger C., Cliff D., Hassall M.E. (2019). Examining the relationship between safety culture maturity and safety performance of the mining industry. *Safety Science*, *113*:345–355.

Stoffregen S.A., Giordano F.B., Lee J. (2019). Psycho-socio-cultural factors and global occupational safety: Integrating micro- and macro-systems. *Social Science & Medicine*, 226:153–163.

Studenski R. (2000). Kultura bezpieczeństwa pracy w przedsiębiorstwie. *Bezpieczeństwo Pracy*, *9*:1–4.

Studenski R. (2003). Techniczne, organizacyjne i psychologiczne uwarunkowania przyczynowości wypadkowej, Wydawnictwo Wyższej Szkoły Zarządzania Ochroną Pracy w Katowicach, Katowice.

Sujova E., Cierna H. (2013). Corporate culture as a tool to improve safety culture. *Management Systems in Production Engineering*, *3*:49–52.

Tappura S., Teperi A.M., Kurki A.L., Kivistö-Rahnasto J. (2019). The management of occupational health and safety in vocational education and training. *Advances in Intelligent Systems and Computing*, *785*:452–461.

Ünal Ö., Akbolat M., Amarat M., Tilkilioğlu S. (2018). The role of the human factor in occupational safety and health performance. *International Journal of Occupational Safety and Ergonomics*, *3*:1–6. doi: 10.1080/10803548.2018.1554932.

Vierendeels G., Reniers G., van Nunen K., Ponnet K. (2018). An integrative conceptual framework for safety culture: The Egg Aggregated Model (TEAM) of safety culture. *Safety Science*, *103*:323–339.

Wyrwicka M.K. (2014). Kultura przedsiębiorstwa a odczucie bezpieczeństwa. *Zeszyty Naukowe Politechniki Poznańskiej, Organizacja i Zarządzanie*, *63*:195–208.

37 Standards in Architectural and Building Acoustics

Joanna Jablonska and Elzbieta Trocka-Leszczynska

CONTENTS

INTRODUCTION

Human beings are continuously exposed to diverse sound field scenarios—starting from living conditions daily to the humming of devices and ducts, talking, traffic outside, and the appliances; passing through offices with open plan, hospitals, industrial machinery halls, where all noises are magnified; to the clubs, pubs, restaurant, shopping malls, concert halls, theatres, or parks. Some of the acoustic phenomena in the aforementioned places can be found desirable and favorable, while others are interpreted as noise, occasionally disturbing or even threatening physical and mental health. According to the National Institute for Occupational Safety and Health (Centers for Diseases Control and Prevention, 2018) in the United States of America (U.S.A.) in 2018, about 22 million workers were exposed to hostile noise levels at their places of employment, while the Institution of Occupational Safety and Health (IOSH) (2018) showed that around 250 million of workers were affected by this negative phenomenon globally. Since 1999, about 30% of Europeans have already been exposed to loud traffic noise, 10% to rail noise, and probably about 10% to air traffic noise (Traffic noise: Exposure and annoyance, 2001).

Prolonged exposure to excessive noise intensity or sound pressure—specific frequency or pitch of sound, depending on individual features and other factors—creates a valid threat to a person's well-being. It affects not only temporary or permanent hearing loss, tinnitus (meaning subjective perception of ringing in ears), or a temporary raising in the hearing threshold (for unborn children) but also cardiovascular disease, high blood pressure, nervous system disorders, nausea, and—under certain circumstances—internal organ vibration or even death (Everest & Pohlmann,

2014; IOSH, 2018; Nathanson & Berg, 2019; World Health Organization Europe [WHO Europe], 2009).

It must be stressed, that noise is an effect of our civilization development similar to the air, water, and soil pollution and excessive overuse of electric light or natural resources. With the increasing contamination, unwanted sound emission also rises. Moreover, documents and regulations aiming at noise limitations are still quite a new domain; the first of them were initially formulated around 50 in the 21st century (Rasmussen, 2018). Thus, it is so important to deal with these threats by creating accurate and faceable standards as well as following them in the professional work of architects, interior designers, urban and spatial planners, etc.

AIM AND PLAN

The connection of health and human well-being toward acoustic parameters of daily sound environment is undisputed; therefore, the main purpose of this chapter is to promote and highlight the importance of building and architectural acoustic standards. Moreover, the knowledge presented will be organized in a clear manner to explain the purpose and aim of documents, standards, and norms. Hence, the main plan of the chapter is as follows:

- short presentation and clarification of definitions—necessary, for different countries where diverse notions are used
- elaboration on a range of standards and fields of knowledge that they cover
- standards for build environment and within architectural and room acoustics
- purposes for the need of the formulation of the aforementioned document
- evaluation on the applicability of documents and indication of fields still requiring additional elaboration
- construction of functional areas that require individual acoustic solutions
- short conclusion for the listed issues

It must be stressed that acoustics is a very complex domain, strongly dependant on the individual features of people, organic phenomenon of human hearing, and sound perception by both the ears and the brain. Therefore, to explain certain definitions and notions, reasonable simplifications had to be introduced. It is necessary not to prolong these considerations to the technical arcana of acoustics (there are suitable and elaborate publications on this subject).

Moreover, this study is focused mainly on the practices of architecture, building, and urban planning aiming at reviling the most important aspect of this activity—targeting at creating a safe and comfortable environment for a human being. Also, this article is not focused on building protection from vibration for it is another, very broad field of research and expertise.

DEFINITIONS

To understand the nature of acoustic disturbances in the daily environment, it is crucial to start from a basic notion—**sound**—which arises as a result of periodic vibrations of air molecules, caused by vibrating objects, called sound sources (i.e. instrument strings, falling items, vocal cords). As an effect, the air pressure changes—the acoustic wave occurs (with a certain frequency), which later is interpreted by a hearing mechanism and the brain as singing, music, speaking, noise, etc. A **noise** is understood as an "unwanted sound" that occurs when the aforementioned changes of air pressure are rapid, and the movement of air particles is not periodic. However, the notion of noise is also subjective and can be interpreted differently by particular listeners (Driscoll, OSHA technical manual; Everest & Pohlmann, 2014; International Partnership ArAc-Multibook, 2015; Massalski & Massalska, 2013).

There are several basic issues connected directly toward a description of sound, and important for this elaboration is **sound intensity level**—which is logarithmic and retains a reference value. It is dependent on the relation between the certain differences in sound pressure, called **sound pressure level,** and is **described in decibels dB**. The range of sound intensity level is measured between a threshold of hearing of 0 dB and 130 dB—the so-called threshold of pain. This parameter has logarithmic and reference character; thus, they are not heard by humans in a linear manner but based on the **frequency** of acoustic wave—**expressed in hertz (Hz)**. Therefore, in acoustics, a correction according to "A" hearing curve is used to gain positive values that are as objective as possible that compare and manage this complex phenomenon. Moreover, the sound level can be measured in different ways due to varied outcomes required—for example, measurements used for industry in the USA are performed according to the sound measurement **weighting networks "A"** and then are **expressed in dBA**, need for evaluation of human risk or with special attention toward low frequencies (Driscoll, OSHA technical manual; Wroclaw Spatial Information System, 2017).

All measurements should be done according to established standards and, in this area, a lot of documents can be found. In Europe, starting from ISO 266:1975 Acoustics—Preferred frequencies for measurements; ISO 1683:1983Acoustics—Preferred reference quantities for acoustic levels; going through ISO 2204:1979Acoustics—Guide to International Standards on the measurement of airborne acoustical noise and evaluation of its effects on human beings; ISO/FDIS 20270; Acoustics—Characterization of sources of structure-borne sound and vibration—Indirect measurement of blocked forces; and toward the important specifics in ISO/CD 22955 Acoustics—Acoustic quality of open office spaces (ISO, the International Organization for Standardization, 2019). In the U.S., referring to Standard Number: 1910.95 (Part Number: 1910, Part Number Title: Occupational Safety and Health Standards)—highly detailed and broadly elaborated document—aiming at both provision of occupational safety and covering environment control indicators (Occupational Safety and Health Standards No. 1910.95 – Occupational noise exposure). Also, in particular counties like Poland: N-Z-01338: 2010—Acoustics—Measurement and assessment of infrasound noise at workplaces (self trans. from Polish), PN-ISO 1996-3: 1999 Acoustics—Description and measurement of environmental noise—Guidelines on permissible noise levels (self trans. from Polish), PN-EN ISO 9614-2: 2000 Acoustics—Determination of sound power levels of noise sources based on sound intensity measurements— "Sweep" method (introduces: EN ISO 9614-2: 1996 [IDT], ISO 9614-2: 1996 [IDT], [self-translation from Polish]), and others (Kłosak, 2015; Polski Komitet Normalizacyjny [PKN, Polish Normalization Committee]).

Another variable crucial in architectural and building acoustics is **reverberation time RT**, which is the time, in seconds, needed for the impulse of sound to drop by 60 dB compared to its maximum level reduced by 5 dB. In practice and in great simplification, interiors with long reverberation time, like churches, will be spaces where speech is illegible, yet organ music would sound properly; rooms with short reverberation time will be suitable for hearing conversation or leading language class. A too-short parameter value will be characteristic for places with "dry" sound, which will be unfavorable for staying (International Partnership ArAc-Multibook, 2015). Optimized RT is highly important for many spaces like interiors in day-to-day use; rooms where good speech understanding is crucial, like in classrooms and school halls, call centers, patient rooms in hospitals, etc.

DISCUSSION

RANGE OF DOCUMENTS

Documents that will be recalled in this chapter are created and implemented on different levels, starting from world-based, going through the continent level, then national regulations, and finishing at regional or city instructions. Only such structure, which allows to examine and formulate recommendations at both country and local levels, allow occurrence of real noise prevention. Also, a national or worldwide overview of the situation is important for it is enabling a broader summary of noise pollution reduction and a tighter control over this process (Centers for Diseases Control and Prevention, 2018; IOSH, 2018; Polski Komitet Normalizacyjny [PKN, Polish Normalization Committee]; Prawo budowlane Dz.U, 1994; Regulation of the Infrastructure Minister 2002; Traffic noise: Exposure and annoyance, 2001; WHO Europe, 2009).

As was aforementioned, all prepared documents need to cover a specific field of building, architectural, and environmental acoustics. The documents can be divided into several groups:

- defining noise threats and indicating prevention
- acoustic safety in open-air spaces (not elaborated in this article)
- acoustic safety in buildings
- buildings with additional acoustic safety requirements:
  - residential
  - higher standard residential
  - industrial
  - public
- specific public buildings rooms with additional acoustic safety requirements:
  - conference rooms
  - offices
- acoustic safety of workplaces
- workplaces with additional acoustic safety requirements:
  - open offices
  - industrial halls (Figure 37.1)

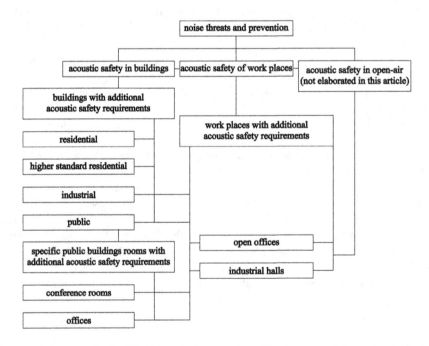

FIGURE 37.1   A Specific Field of Building, Architectural, and Environmental Acoustics [self-elaboration].

A glimpse of these documents and their purpose will be shown in the following paragraphs.

DESIGN ACCORDING TO FUNCTION

Crucial towards document selection before the design is the evaluation of building architectural-acoustical performance and the elaboration of occupation safety plans to define what sound threats are—problems that can be experienced—and the purpose of the object or room functions to be adapted. When discussing norms and standards, it must be stressed that each room and function in the building—as in each structure type—requires different sound field parameters (i.e. RT and isolation materials solutions) like exterior isolation and interior absorbing finishing—especially for acoustically active equipment and furniture. To give an example: the schoolroom, where young children need to understand each word of a teacher, needs totally other conditions than a church where pipe-organ music is played; a theatre, where each coloration of voice must be heard is different from a hospital room where patients are recovering from an operation; also, in places of administrative work or open space offices, it is important to provide communication in group work, as well as maintain the privacy of data by limiting the audibility of phone conversations. Therefore, each acoustic-ergonomic standard must be targeted at selected room or specific area functions. On some occasions, the basic start-up conditions may change throughout time (i.e. a road will be built nearby, or the air-traffic channel will be changed) so it is useful to foresee both external conditions as well as future functioning changes.

NOISE SOURCES

In contemporary open spaces of cities and villages as well as their closed areas (buildings), there are a number of noise sources—outdoor and indoor—that should be identified before any investment. As with outdoor noises sources, these can be listed as:

- traffic:
  - road communication
  - trams
  - trains
  - airplanes
  - boats

- buildings:
  - industrial
  - public
  - residential

- metro stations and tubes
- technical stations
- others

As for indoor noise sources, these can be:

- devices:
  - air conditioning
  - mechanical ventilation
  - electric appliances
  - gas appliances
  - boilers
  - machines

- ducts
- tubing
- others

To identify such threats, city noise maps will be very helpful. According to the European Parliament and the Council 2002/49/EC (European Parliament and the Council, 2002), such documents must be prepared and administrated daily. Official papers can contain different data—usually, average noise emitted for particular clusters of sources—during all 24 hour-days or nights of the year, exposing zones of overruns. Data allow designers to understand the acoustic situation of a particular spot of interest. For example, in Wroclaw (Poland, Europe), the measurements are given by the following: long-term average A-weighted sound pressure level in dB and parameter characteristic for other documents that can be found across the world. With such information, there can be indicated zones of protection, like terrains of residential, educational, recreational, health services function (Wroclaw Spatial Information System, 2017).

Hence, there are other initiatives in this manners, and maps of threats can be created in different ways, just to give an example: in New York, a part of an official policy implemented into documents is the so-called "311," a Customer Service Center where residents can complain about excessive sounds in their neighborhood. In such a way, information on urban noise is not just about average, weighted measurements, but the data is much more connected to the everyday city users and, therefore, more accurate and informative for designers and administration. After the success of this program in 2016, also public surveys were also created and launched among the city's residents (DiNapoli, Office of the New York State Comptroller, 2018).

The world or continent-level documents cannot be sufficient or oversaw without national or regional documents connected to city governed administrations. It should be mentioned that all reliable resources on acoustic city maps should be created always according to the measurements obtained in the standardized processes, with the devices allowed by low or appropriate norms, with certificates and in processes in line with appropriate documents (i.e. European Norms: ISO/NP 11819-1 Acoustics—Measurement of the influence of road surfaces on traffic noise—Part 1: Statistical Pass-By method; ISO 11819-2:2017, Acoustics—Measurement of the influence of road surfaces on traffic noise—Part 2: The close-proximity method (ISO, the International Organization for Standardization, 2019); or exemplary on national level, Polish Norms: PN-N-01341: 2000 Environmental noise—Methods for measuring and assessing industrial noise (self-translation from English), PN-EN ISO 11819-1: 2004—Acoustics—Measurement of the effect of road surfaces on road noise—Part 1: Statistical method of measurement during the ride (introduces: EN ISO 11819-1: 2001 [IDT], ISO 11819-1: 1997 [IDT]), (self-translation from English), and others (Polski Komitet Normalizacyjny [PKN, Polish Normalization Committee])).

BUILDING ACOUSTICS

Acoustic waves access the buildings either by air, as airborne sounds, or by ground and building elements—when the waves are transmitted inside of solids—as material-borne or structure-borne sound. Thus, the architect's aim is to, first, isolate the building from external noise and create a quiet living or working environment in the particular interiors. This part of the design is contained in the range of building acoustics. The second designer's responsibility is to create optimal conditions for sound propagation inside the rooms of different volumes, with beneficial sound field parameters for their particular functions, like the aforementioned RT. This domain is covered by architectural acoustics. It must be highlighted that special focus was always put on housing—especially multistory, for it is the basic human environment (Rasmussen, 2018)—and public buildings where there are organized open spaces arranged with workplaces.

General building codes and practice require some elements of acoustic isolation, which is combined with building thermal isolation and respective structural requirements (i.e. massive walls). Particular norms and standards are meant to defining other elements of acoustic building protection (not only isolation) and will be found in the aforementioned ranges. It must be stressed, however, that only the use of a combination of standards and building code requirements allows proper complex building or architectural acoustic design. Here is a recall of documents designated to general rules of isolation: in Europe ISO 717-1:2013—confirmed in 2018—titled: Acoustics—Rating of sound insulation in buildings and of building elements—Part 1: airborne sound insulation, relates to walls, floors, doors, and windows and takes into account both external sounds, like traffic noise, as well as sound sources in the premises, describing the methods for translating the acoustic measurements and their rating into recommendations for building codes (ISO, the International Organization for Standardization, 2019). A good example of such a norm related to the national building code would be Polish Norm PN-B-02151-3:2015-10—Building Acoustics—Protection from noise in buildings: Part 3: Requirements for sound insulation of building partitions and building elements (self-translation from Polish). It is focused on the isolation of external and internal walls and slabs from both airborne and material sounds that can filter inside of residential and public buildings (obligatory from 2018). So, from complex measurements, evaluation, and rating, a standard understandable by non-acoustic professionals that can be successfully applied in building and architectural design takes place. The next norm in this respect is the PN-B-02151-2 Building acoustics—Protection against noise in buildings—Part 2: Requirements for the permissible sound level in rooms (self-translation from Polish) that describes permissible sound levels A (according to the measurements already recalled) in rooms of residential, temporary residential housing, as well as public buildings. So, only the combined use of PN-B-02151-3:2015-10 and PN-B-02151-2 gives data for building design. At this point, it is worth mentioning specific norm types that are issued only for a selected range of building industry, like Building acoustics—Protection against noise in buildings—Part 5: Requirements for residential

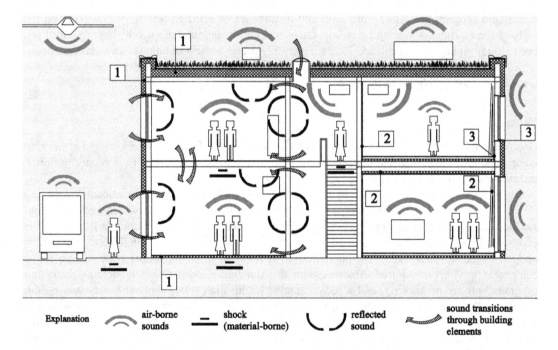

| Explanation | ⁀ air-borne sounds | — shock (material-borne) | ⌣ reflected sound | ⟿ sound transitions through building elements |

FIGURE 37.2 The levels of acoustic building phenomenon and protection: 1—isolation elements from building codes and general design rules, 2—additional elements for isolation, systems (i.e., floating floor) and sound absorbance based on norms, 3—additional isolation details from additional study and scientific elaborations [self-elaboration].

buildings with a higher acoustic standard and instructions for their rating (self-translation from Polish), (Dulak, 2017; Kłosak, 2015; Polski Komitet Normalizacyjny [PKN, Polish Normalization Committee]; PN-B-02151-2, 2018), (Figure 37.2).

Also, recall that the RT was a subject of norms in Poland and PN-B-02151-4:2015-06 Building acoustics—Protection against noise in buildings—Part 4: Requirements for reverberation conditions and indoor speech intelligibility and test guidelines (self-translation from English), which aims at limitations of reverberation noise in residential and public buildings, excluding assembly, concert, performance halls, and similar rooms. In this respect, there were two notions that introduced minimal permitted acoustic absorbance or maximal permitted RT, and both are focused on the necessary use of sound-absorbing materials, together with proper distribution around the room. Moreover, depending on the situation, the measurements on-site are connected toward providing user safety, consistent of calculations on isolation between rooms, slabs, of external facades. Material-borne or airborne sounds are also regulated by norms (i.e. in Poland, it is PN-EN ISO 10052:2007 and as it is already visible as an ISO norm (Polski Komitet Normalizacyjny [PKN, Polish Normalization Committee]; Świerczewska, 2018) [Figure 37.2]).

RESIDENTIAL BUILDINGS

A special interest should be focused on specific guidelines that are being targeted at a certain problem—like the reduction of noise during nighttime by the World Health Organization Europe (WHO Europe) designated to be included in the policies of the Member States. As it is highlighted in the document, guidelines are not standards; however, while discussing the noise reduction process, these documents are still important. Moreover, through scientific expertise, they can become standards in the near future (WHO Europe, 2009).

It is needless to say that proper sleep is crucial for good daytime functioning and health protection, so the first element of these guidelines was establishing a length of night. Moreover, the correlation between noise threshold level and disturbance of selected activity (i.e. waking up too early in the morning or during nighttime, environmental insomnia, complaints, hypertension) has been listed, giving clear guidance for necessary nighttime noise limitations. This has been contained in the following parts of the document and supported by scientific evidence because a direct connection was not observed in some cases.

WORK SAFETY

One of the main areas that must be covered by norms is the safety of workers, aimed at limiting noise exposure, indicating hearing protection methods, and guiding through the proper building, architectural, and administration solutions.

Again, the documents are targeted at specific fields—like the U.S. Occupational Safety and Health Standards no. 1910.95—Occupational noise exposure, of OSHA, which shows the hazardous noise level in sound measurement in dBA, depending on frequency and daily permissible exposure. If this allowable exposure is exceeded, the administration or engineering control is obliged to reduce such a negative phenomenon. If it is not possible, proper personal protective equipment provided is required. Among others, this standard also involves "the hearing protection program," proper monitoring, and a recommended health plan (Occupational Safety and Health Standards No. 1910.95 – Occupational noise exposure). It must be stressed that this standard is very precise, giving suitable data to follow (i.e. in industries). A similar document was also formulated in Poland: PN-ISO 1999:2000 Acoustics—Determining occupational exposure to noise and estimating noise-induced hearing damage, Introduces: ISO 1999: 1990 [IDT] (self-translation from Polish), (Polski Komitet Normalizacyjny [PKN, Polish Normalization Committee]).

Similarly, the target document in Europe would be European Directive No. 2003/10/EC—noise (European Parliament and the Council, 2003), which is focused on maximal admissible sound pressure and the daily and weekly noise exposure in a work environment. According to the document, there are required measurements—decreasing risk to possible minimums if needed—and personal protection of hearing. What is more, in all aforementioned documents there is a detailed description of the types of measurements—ways of inducing protection, and so on—pointing, on many occasions, to additional documents, norms, and standards that are practically issued toward selected topics. For example, popular in modern and contemporary buildings, open spaces, shopping malls, offices, and in university and school buildings: European ISO 3382-3:2012 currently under review and is planned to be replaced by the new ISO 3382-3 currently in development, which focuses on the increase of workplace quality with the use of specialized equipment and furnishing. Such a solution enables the adjustment of selected workplaces to different functions toward the provision of data privacy and improvement of intellectual tasks where it is crucial to limit sound emission and emission, creating an acoustic environment empowering concentration (Muratorplus, 2003; Polski Komitet Normalizacyjny [PKN, Polish Normalization Committee]).

Moreover, in spaces where communication is crucial, understanding speech is the most important factor. Both high—which occur while speaking—and low frequencies—even those at and below the threshold of hearing—must be removed, aiming to improve the hearing quality. Also, as it was already recalled, RT ought to be shortened with the use of wideband absorbance of frequencies between 125 to 4000 Hz. The aforementioned issues are especially important in spaces where telephone calls are conducted (i.e. call centers where they use highly sound-absorbent flooring, wall, and ceiling materials, and furniture against noise). Considering the diversified demands toward acoustic parameters of contemporary offices, it is required to select optimal finishing and equipment for the specific functions described in the design. Useful in this process might be additional standards, indicating noise that can be emitted by particular office devices and equipment (i.e. PN-ISO 9296:1999—Acoustics—Declared noise values for computer and office equipment (introduces: ISO 9296: 1988 [IDT]), (self-translation from Polish), (International Partnership ArAc-Multibook, 2015; Muratorplus, 2003; Polski Komitet Normalizacyjny [PKN, Polish Normalization Committee]; SonoPerf Office, 2014)).

CONCLUSIONS

EVALUATION OF STANDARDS SUFFICIENTLY

The standards are important and **crucial** for the protection of the health of all city and village users and special groups of architecture consumers—patients at healthcare facilities and the users of recreation open-air terrains or factory employees. Documents should be understood, propagated, developed, and used by administration professionals and common residents so they could demand proper and healthy everyday conditions. One of the documents of high importance in this respect is acoustic maps of the cities (i.e. indicating areas of protection from noise)—a tool very useful in planning new investments and reviling terrains where noise is exceeded and requires intervention. Such maps can serve as a reference item—for architectural, urban, or building design—in drawing out optimized conclusions and the way they should be hierarchized and structured in space, before placing new functions and urban areas.

Building safety norms for residential, public use and occupational safety must be followed in everyday architectural building design, usage, maintenance, and administration for they are a set of clear conditions allowing professionals, managers, and owners to create a good, favorable, and healthy environment. Such regulation needs to be given in areas of external and internal isolation, with reverberation noise limitation and RT indications. Interestingly enough (i.e. in Poland), not all the norms have an obligatory character (only selected obligations were introduced in 2018), so the use of them is dependent on a designer or company that can state that their estate was designed according to PN-B-02151-5:2017-10, so it is distinguished by a higher acoustic standard (Prawo budowlane Dz.U, 1994; Regulation of the Infrastructure Minister, 2002). What is more, "weaknesses" in norms and standards themselves are pointed out—like in the publication by Sabato, Sabato, and Reda (2014) where the comparison between EU and U.S. standards for occupational noise protection was made—and authors pointed out the lack of focus on direct ear stress by particular acoustical exposures. At the same time, Rasmussen (2018) also gives attention to yet another problem that is crucial while discussing standards—a vast number of buildings erected long before these documents occurred and none of them follows the recommended guidelines.

In the end, it is crucial that designers follow technical product cards that have certificates of ensuing specific (also normalized) technical laboratory research usually carried out by government institutes (depending on legislation) proving that these products provide declared technical parameters (i.e. for sound isolation). Post-construction usage will be for required comfort and safety, and the parameters declared in the specifics of a product will be actual implementation factors.

The next problem following norms and standards in practice is poor manufacturing, incorrect construction techniques, and lack of further maintain process. During design documentation, all slabs, walls, and foundations may follow proper isolation standards but if the building is raised with unpredicted technological gaps or is not air-tight, the desirable requirement will not be met. Similarly, if the finishing materials of the interiors are not installed and follow different parameters for acoustical absorption, the limitation of possible reverberation noise will miss the planned level. Thus, the construction process—and not only the design—is so important in creating quiet environmental conditions (Figure 37.2).

The differences in noise protection between properly sound isolated and air-tight structure—where correct external and between-room compartment solutions have been created—was shown in the drawing (Figure 37.2). For comparison, there is also a simulation of building without following the aforementioned recommendations. Moreover, graphic elaboration where rooms included with additional sound-absorbing materials, in comparison to interiors without these solutions. The improper examples were marked on the left of the drawing, while proper ones are placed on the right. Also, amounts of acoustic energy were symbolically shown, which clearly reveals how proper, standardized solutions in respect of building and architectural acoustics create a safer and healthier, beneficial environment (Figure 37.2).

ART OF SOUND AND ARCHITECTURE DESIGN

The number of documents listed in this article gives a notification that presented standards cannot be used for each designed case (i.e. in already recalled polish norm PN-B-02151-2 (PN-B-02151-2, 2018) there is clear information that presented norms are not set for rooms that require advanced acoustical parameters, like recording studios and laboratories for acoustical research). Also, it should be added that standards and norms will not replace the experience and general knowledge of the designers in terms of the need for acoustician consultations during the project. A great example of such cooperation would be a concert hall design where every note and its intensity, color, and brilliance matters. Another case would be a theater where the isolation from external noise is crucial, but a proper geometry of the room is also needed for specific acoustic wave propagation from an actor to a listener. In hospital design, proper sound-function-structuring (like not situating patient rooms near loud technical spaces) as well as an expert on how to mute areas for unwell people, without cutting them from beneficial sounds of nature, is needed.

In the design of a specialized type, reverberation time is important, but a number of other sound field parameters are also substantial—some subjective examples include the initial time delay gap or interaural cross-correlation while objective examples are intimacy, coloration, richness, etc.). These are the variables that are decided and designed individually for each room and no norm or standard can define them.

Summing up, similar to architecture expression in building or painting quality in art, sound field can be treated as an artistic design, especially when large-scale concert halls, theaters, or performance venues are created.

SUMMARY

In everyday environments, a lot of hazards, dangers, and noise pollution are observed despite a number of acoustic standards, norms, and regulations. Though world and national organizations and governments have been trying to counteract adverse phenomena for many years, there is still a huge demand for the formulation, revision, and escalation of knowledge on noise-reduction standards in the built environment. Moreover, there is still a need to reinforce control and pay attention to the inclusion of such norms in widely understood design.

Promoting knowledge on standards and following it through professional and administrative work will help lower the noise and reverberation noise in the built environment, both external and in interiors. It will positively affect health and well-being, bringing us closer to the realization of the WHO, EU, OSHA, IOSH standards, norms, and guidelines that increase safety and prevent acoustic pollution of human surroundings.

It is also important that the actions taken in the aforementioned countries—particularly by organizations—aiming for the increase of acoustical quality are becoming an inspiration and basis for countries where elaborated regulations were neither obligatory nor stated a law.

REFERENCES

Centers for Diseases Control and Prevention. 2018. In *Noise and hearing loss prevention*. National Institute for Occupational Safety and Health. https://www.cdc.gov/niosh/topics/noise/default.html (accessed September 12, 2019).

DiNapoli, T. P., Office of the New York State Comptroller. 2018. *Noise in New York City neighborhoods. Assessing risk in urban noise management*. https://osc.state.ny.us/reports/ (accessed September 17, 2019).

Driscoll, D. P. *OSHA technical manual*. United States Department of Labour. https://www.osha.gov/dts/osta/otm/new_noise/ (accessed September 11, 2019).

Dulak, L. *Izolacyjność akustyczna i związane z nią wymagania dotyczące budynków* (Sound insulation and related requirements for buildings – Self-Translation from Polish). (May 29, 2017). http://www.inzynierbudownictwa.pl/technika,materialy_i_technologie,artykul,izolacyjnosc_akustyczna_i_zwia-zane_z_nia_wymagania_dotyczace_budynkow,10050 (accessed October 25, 2019).

European Parliament and the Council. Directive 2002/49/EC. (June 25, 2002). *Declaration relating to the assessment and management of environmental noise – Declaration by the Commission in the Conciliation Committee on the Directive relating to the assessment and management of environmental noise, Official Journal L 189 18/07/2002 P. 0012-0026* https://eur-lex.europa.eu/legal-content/EN/TXT/HTML/?uri=CELEX:32002L0049&from=EN (accessed September 4, 2019).

European Parliament and the Council. Directive 2003/10/EC. (February 06, 2003). *Noise, Seventeenth individual Directive within the meaning of Article 16(1) of Directive 89/391/EEC*. European Agency for Safety and Health at Work. https://osha.europa.eu/en/legislation/directives/82 (accessed August–October, 2019).

Everest, A. F., Pohlmann, C. K. 2014. *Master handbook of acoustics*. McGraw-Hill Education TAB.

Institution of Occupational Safety and Health (IOSH). 2018. *Noise* https://www.iosh.com/resources-and-research/our-resources/occupational-health-toolkit/noise/ (accessed September 12, 2019).

International Partnership ArAc-Multibook. 2015. *ArAc-Multibook – Multimedial handbook for architectural acoustics*. https://arac-multibook.com/ (accessed September 17, 2019).

ISO, the International Organization for Standardization. https://www.iso.org/home.html (accessed September 17, 2019).

2003. *Klasy akustyczne mebli do biur open plan – akustyka pomieszczeń biurowych* (Acoustic classes of furniture for open plan offices – Acoustics of office rooms – Self-Translation from Polish). (March 11, 2003). Muratorplus. https://www.muratorplus.pl/technika/izolacje/klasy-akustyczne-mebli-do-biur-open-plan-akustyka-pomieszczen-biurowych-aa-xoJe-cxuZ-j4DR.html (accessed October 11, 2019).

Kłosak, A. K. *New regulations on acoustic properties of buildings*. June 15, 2015. Izolacje no. 5/2015. http://www.izolacje.com.pl/artykul/id1838,nowe-regulacje-w-zakresie-akustyki (accessed October 25, 2019).

Massalski, J., Massalska, M. 2013. *Fizyka dla inżynierów* (Physics for engineers, self-translation from Polish). WNT.

Nathanson, J. A., Berg, R. E. (2019) Noise pollution. In: *Encyclopaedia Britannica*. https://www.britannica.com/science/noise-pollution, (accessed January 19, 2019).

*Occupational Safety and Health Standards No. 1910.95 – Occupational noise exposu*re. OSHA. https://www.osha.gov/laws-regs/regulations/standardnumber/1910/1910.95 (accessed August–October, 2019).

Polski Komitet Normalizacyjny (PKN, Polish Normalization Committee), *System Cyfrowej Sprzedaży Produktów i Usług* (The system of digital sale of products and services). http://sklep.pkn.pl/ (accessed August–October, 2019).

PN-B-02151-2. (2018). *Polish Norm, Building Acoustics. Protection from noise in building. Part 2: requirements on permissible sound level in room* (self-translation from Polish).

Prawo budowlane Dz.U. (1994). nr 89 poz. 414 (Building Code Diary Act 1994 no 89 item 414 – Self-Translation from Polish).

Rasmussen, B. 2018. Building acoustic regulations in Europe – Brief history and acoustical situation. In *Proceedings of the Baltic-Nordic Acoustics Meeting 2018*, Reykjavik, Island, https://vbn.aau.dk/ws/portalfiles/portal/274779598/Rasmussen_KeynotePaper_BNAM2018_BuildingAcousticRegulationsEurope_HistorySituation.pdf (accessed August–October, 2019).

Rozporządzenie Ministra Infrastruktury z dnia 12 kwietnia 2002 r. w sprawie warunków technicznych, jakim powinny odpowiadać budynki i ich usytuowanie. Dz.U. 2002 nr 75 poz. 690 z późniejszymi zmianami (Regulation of the Infrastructure Minister form April 2002, 12 on the technical conditions to be met by buildings and their location. Diary Act 2002 no. 75 item 690, with later amendments – Self-Translation from Polish).

Sabato, A., Sabato, A., Reda, A. 2014. Protection of workers from risks caused by loud fields. Comparison between the European and the United States Standards. In *Inter-noise 2014 – 43rd International Congress on Noise Control Engineering*. Melbourne, Australia. https://www.acoustics.asn.au/conference_proceedings/INTERNOISE2014/papers/p652.pdf (accessed August–October 2019).

SonoPerf Office. 2014. *Kluczowe rozwiązania dla zapewnienia świetnej akustyki.* (Key solutions to ensure great acoustics – Self-translation from Polish). ANDRITZ Fiedler GmbH. Regensburg, Deutschland. ANDRITZ Fiedler Hradec Králové, Česká republika. http://andritz-fiedler.cz/getFile.php?q=AF%20SonoPerf_office_print_PL_FINAL.pdf (accessed October 25, 2019).

Świerczewska, M. February 01, 2018. *Akustyka ma głos* (Acoustics has vioice – Self-Translation from Polish). Otodom. https://www.otodom.pl/wiadomosci/prawo/akustyka-ma-glos-id7555.html (accessed October 25, 2019).

Traffic noise: Exposure and annoyance. 2001. European Environment Agency. https://www.eea.europa.eu/data-and-maps/indicators/traffic-noise-exposure-and-annoyance/noise-term-2001 (accessed September04, 2019).

World Health Organization Europe (WHO Europe). 2009. *Night noise guidelines for Europe*. Ch. Hurtley (Ed.), Denmark http://www.euro.who.int/__data/assets/pdf_file/0017/43316/E92845.pdf (accessed October 25, 2019).

Wroclaw Spatial Information System, *Acoustic map (with description)*. 2017. http://geoportal.wroclaw.pl/mapy/akustyczna/ (accessed August–October 2019).

38 Development of Medical Devices

Between Regulatory and Standards Requirements Versus Adopted Models

Omar Kheir, Alexis Jacoby, and Stijn Verwulgen

CONTENTS

INTRODUCTION

The advancement in information and communication technology (ICT) has made its way into medical devices while we witness their increased demand and supply, as in wearable devices and home-use medical devices. This visible increase and regulated nature of medical devices trigger the need to spread some awareness on the regulations and methods involved in their development lifecycle. This chapter serves as an introduction to the domain of medical device development, which is heavily controlled and requires proof of compliance prior to launching by the respective authorities. For this purpose, a sample of relevant regulations that are applied in the United States and the European Union will be discussed.

ISO standards related to medical devices inspire regulations set by countries—such as the U.S.—and regions—such as the EU. These standards are subject to ongoing constant updates to enhance the safety and usability controls associated with medical device development. Therefore, aside from their development efforts and concerns, developers and manufacturers have a constant growing regulatory and standards burden as well—this is where development models can be applied to support them.

Several models and standards have addressed the various phases and requirements in the product development process. Requirements deal with the general aspects of medical device development (Kinsel, 2012), specifically in their related quality, usability, and risk management (Janß, Plogmann, & Radermacher, 2016). These are governed by regulatory bodies such as health organizations and ISO standards. Medical devices should adhere to strict regulations set by the different countries where manufacturers are willing to launch their products. Most important are the European Commission (EC) Medical Device Directives (e.g., 93/42/EEC) and the U.S. Food and Drug Administration (FDA) regulations (Money et al., 2011). This has led to an abundant set of generated rules and requirements that can be difficult for interested groups and start-ups who may have very innovative ideas that may not see the light because of these burdens. Moreover, regulations and procedures required by the agencies describe *what* the documentation should contain but not *how* the design or development process should be undertaken (Ogrodnik, 2013)

Additionally, if we compare the medical devices production industry to any other, complexities and challenges are much higher as any simple usability mishap can project a significant impact and lead to serious injuries. By considering regulatory requirements applicable for the envisioned end product at the early design stages, the right choices can be made at the start to enhance the economic viability and reduce resources and efforts (Pitkänen, 2017). For example, 75% of U.S.-based medical device start-ups fail due to several reasons—on top of which is not complying with regulations (Reilly, 2015). This figure resembles the challenges and the many killing factors that start-ups in the medical device development industry encounter.

Therefore, this chapter will briefly bring some of the applicable major standards and regulations in the U.S. and the EU while emphasizing the applicable usability requirements, related development models, and the encountered challenges among them.

REGULATORY REQUIREMENTS IN THE DEVELOPMENT OF MEDICAL DEVICES

Medical devices encompass a wide range of products used in a variety of settings for the diagnosis, prevention, monitoring, or treatment of illness or disability. In line with the World Health Organization (WHO), a medical device can take the form of any article, instrument, apparatus, or machine that is used in the prevention, diagnosis, or treatment of illness or disease, or for detecting, measuring, restoring, correcting, or modifying the structure or function of the body for some health purpose (World Health Organization, 2019).

The purpose of standards in any domain is to assure consumers that their products are safe, reliable and of good quality (International Standards Organization [ISO], 2019). Standards resemble specifications, rules, guidelines, etc. that would ensure the product's optimal quality and customer satisfaction accordingly, without limiting product innovation.

Therefore, respective standards regularly undergo rigorous review and management by competent health authorities assigned by governments across the globe to regulate the development of medical devices. This chapter will focus on the regulations and standards executed in the EU and the U.S.—considering that they have been in the practice with minor reviews and restructures, unlike the standards and regulations in Asia that have been recently under extensive reformation (Wu et al., 2016).

On one hand, the national competent authorities in the EU regulate the applicable regulations and standards to outline the requirements in the form of directives for medical devices to operate. These directives categorize medical devices into four classes (I, IIa, IIb, and III) based on the risk levels associated with their intended use. To launch any medical device, the developing entity shall undergo a conformity assessment to demonstrate that requirements pertaining to the developed device's class are met to ensure that it is safe and performs as intended. This assessment is performed by accredited Notified Bodies that the EU member state designates to conduct conformity assessments (European Medicines Agency, 2019). Once a medical product passes this conformity assessment, it can be labeled with the CE mark and is ready to be released.

Also, the European Commission in the EU delegated the Medicines and Healthcare products Regulatory Agency (MHRA) in the U.K. to regulate medical devices. There are three separate major EU directives on medical devices: (1) most devices are regulated under directive 93/42/EC;

(2) implantable devices that are regulated under directive 90/385/EC; and (3) in vitro diagnostic devices (i.e. used on substances produced by the body) that are regulated under 98/79/EC and is mostly applicable to class III devices (Campbell, 2013).

Each EU country is managed by a government body called a Competent Authority that handles all matters related to medical device approval—for example, the French Agency for the Safety of Health Products. Low-risk devices can be declared right away to the Competent Authority that can decide whether to conduct inspections to confirm manufacturing standards and review the technical file for the device. As for the approval for more complex devices, Class II and above, it is directly handled by the Notified Bodies composed of independent companies that are specialized in product assessment, including medical devices for CE marks as designated by the Competent Authorities. In line with the EU directives on medical devices, the CE marking of a device in any member state is, by default, mutually recognized across all other EU member states. For example, if a manufacturer gets approval for the CE marking for a new implantable lens from a Notified Body in any European country, this medical device can be legally marketed in all EU member states accordingly.

First, a manufacturer of a device selects a properly designated Notified Body in a country of the manufacturer's choice. For approval by a Notified Body, devices are subject to performance and reliability testing linked to the risks of their intended use. For most devices, the standard is met if the device successfully performs as intended—in a way that benefits outweigh the expected risks. The specific requirements for pre-marketing clinical studies are relatively vague and not specific, and details of trials are typically not made available to the public. Therefore, even though clinical data are required for high-risk devices, guidelines for the nature and the coverage of these studies are not binding within either the manufacturers or Notified Bodies.

Following the product launch, developers shall submit any arising product flaw to the competent authorities. All information related to medical devices developed within EU boundaries—including clinical studies and approvals as well as the mentioned arising flaws—are being logged by developers themselves in the European Databank on Medical Devices (EUDAMED). On the other hand, the FDA in the U.S. is granted the single-handed power to provide reasonable assurance in the safety and effectiveness of medical devices in a centralized mechanism. This can lead to delays in issuing approvals, as witnessed in some cases where medical devices receive the marketing permit in the U.S. late after the EU. For example, a distal protection system for coronary-artery interventions received a CE mark after a single-group study involving 22 subjects showed that the device worked as intended. However, in the U.S., FDA approval for the same device came several years later based on a randomized study involving 800 subjects where a clinical endpoint of major adverse cardiac events was used (Baim et al., 2002).

Differences in the approval requirements between the U.S. and the EU are minimal when it comes to medical devices that do not require a Pre-Market Approval (PMA). Devices that are exempted from the premarket submission process—mainly class I devices in the U.S. and the EU—do not require clinical trials, which can be very controversial. As for the higher classed devices—class II and III devices—the situation is different. When it comes to transparency, the FDA has several mechanisms in making its decision-making process accessible given that open presentations to advisory committees describe particularly novel, complex, or high-risk devices, and committee panelists can publish their views (Laskey, Yancy, & Maisel, 2007). In the European Union, Notified Bodies have no obligation to publish their decision-making process or to publish any evidence on which their assessments were based (Thompson et al., 2011).

The delegated structure in the EU has proved to be faster so devices are delivered to the market earlier than the U.S. in some cases (Baim et al., 2002), as illustrated earlier. However, given the described lack of transparency and considering that data submitted to Notified Bodies pertaining to high-risk devices are classified as "commercially confidential" and not available to the public (Wild, Erdos & Zechmeister, 2014), there may be inconsistencies in the approval and reporting processes among Notified Bodies as this delegated structure allows some subjectivity. Even though the competent authority is auditing Notified Bodies, there might be differences in approaches to applying the European directives (Chen et al., 2011). This observation is understandable as the Notified Bodies can assess in favor of the manufacturers who are paying for their services (Cohen, 2012). For example, Notified Bodies can be reluctant to disapprove of medical devices for fear of losing an ongoing relationship with a manufacturer who hires them.

USABILITY REQUIREMENTS IN THE DEVELOPMENT OF MEDICAL DEVICES

Usability is also referred to as Human Factors Engineering (HFE), Ergonomics, Human Engineering (HE), Usability Engineering (UE), or Human-Computer Interaction (HCI). HFE involves the use of behavioral science and engineering methodologies in support of design and evaluation (Hegde, 2013). In the process of Medical Devices Development (MDD), much attention goes to usability requirements, heuristic evaluation, and ergonomics standards. For example, human error in operating a device can be a major cause of patient death or injury (Cafazzo & St-Cyr, 2012). Hence, manufacturers have the capacity as well as the responsibility to enhance patient safety by putting greater emphasis on the human factors engineering process in the design of medical devices (Lin, Vicente, & Doyle, 2001). During the third quarter of 2012 in the U.S., 407 medical devices were recalled—which is equivalent to more than 26.5 million units being withdrawn from the market (Borsci, 2014). While in the U.K., between 2006 and 2010, there was a 1220% increase in the number of safety problems reported, the number of recalls in Germany increased from 721 in 2010 to 1075 in 2013.

Medical devices have become more diverse in their capabilities and are used with increased frequency in busy environments with even new distractions, as with the case in devices operated by patients at home. Changing contexts induce requirements for specialized training to reduce potential operational risks. In addition to that, considering the transfer of patient care to private homes or public environments, usability for less skilled or even unskilled users—including patients and caretakers—must be addressed (Buttron, 2017).

The major benefits of HFE in medical devices is that they increase safety, reduce potential errors, decrease training and increase the ease of use, improve task performance and optimize device use, enhance user satisfaction, improve patient outcomes, lower product liability risks, facilitate the regulatory approval process, and increase the chances of commercial success (Hegde, 2013).

The general European directive on medical devices (Medical Device Directive 93/42/EEC) mentions that the devices must be designed and manufactured in such a way that, when used under the conditions and for the purposes intended, the device will not compromise the safety of the patients, or the safety and health of users. These safety risks can be in the form of **P**erception, **C**ognitive, or **A**ction risk as per the error scheme defined in the IEC 62366 (IEC 62366-1:2015) and as defined by the usability specialists referred to as "**PCA analysis.**"

Usability Standards

The FDA has issued its final version of the HFE document in 2016 following a draft version that was released back in 2011. The released final version intends to assist the industry in the human factors and usability engineering processes to maximize the likelihood that new medical devices will be safe and effective for the intended users and operating environments. The recommendations in this guidance document are intended to support manufacturers in improving the design of devices to minimize potential use errors and resulting harm. The FDA believes that these recommendations will enable manufacturers to assess and reduce risks associated with medical device use (FDA, 2016).

ISO 14971 (Medical Devices – Application of risk management to medical devices) comes in harmony with FDA guidelines (FDA, 2016), as manufacturers are expected to run risk analysis as part of their design control process, where safety measures and controls are embedded across the development process, which shall start with the early design phase and should continue throughout the detailed design and development lifecycle (Money et al., 2011).

Additionally, the main European harmonized standard (IEC 62366-1:2015) on the application of usability engineering to medical devices specifies a process for a manufacturer to analyze, specify, develop, and evaluate the usability of a medical device. Since it relates to safety, it references the ISO risk management standard as the base for the identification of usability matters via a risk management process. Several clauses of both standards come in common, such as clause 4.2 related to the identification of characteristics related to safety and clauses 4.3 - 4.4 related to the identification of hazards and sequence of events leading to hazardous events. If the results of the risk analysis indicate that risk evaluation—which is based on the risk probability of occurrence and potential harm—could lead to serious injury to the patient or the device user, then the manufacturer should have suitable human factors or usability engineering processes control, which is in line with the FDA HFE guidance document too. This is applicable to newly developed devices or existing devices that are already available in the market and subject to design correction due to arising deficiency. Any hazard that may result from the product misuse due to error performed by the user or caused by the operating environment shall be logged in the risk register and mitigated with suitable controls and will be subject to human validation test to alleviate the risks level to tolerable measures. Since this validation process shall be ongoing, the generated risk register shall be accordingly dynamic to accompany the entire production stages. Since it is practically impossible to eliminate all risks, some residual risk—which represents a risk that has been already subject to mitigation or treatment control—will keep existing no matter how extensive the mitigation plans are. Hence, these should be thoroughly analyzed to assure that further reduction of the errors' likelihood is not feasible (i.e. the benefits of the device outweigh the residual risks). The usability engineering file that compiles all usability risks could be an independent file, a part of the risk management file, or part of the product design file. It is mandatory, considering that it facilitates auditing of the design and the entire development process, providing pointers and references to all required documents in the comprehensive design file.

A famous technique that is adopted to tackle such risks is the failure mode effects analysis approach (FMEA). It can be established by an analysis team—including individuals with experience in using the device, such as a patient or a clinical expert—with a design engineer and a human factors specialist. The role of this team is to ensure that the multiple viewpoints on potential use errors and the harm that could result are being comprehensively logged and tackled.

The FMEA team can perform brainstorming sessions to identify possible use scenarios that could result in a failure mode and consider the potential harm along with the proper mitigation plan pertaining to each. Many other techniques are also available, such as the "formative evaluations" or "contextual inquiry," for the designated team to select from to serve their exercise better. Nevertheless, any approach adopted by the manufacturer to identify usability risks will not be complete unless actual user testing takes place to validate the work done and the risk mitigation efforts (McCurdie et al., 2012).

INVOLVING USERS IN THE DESIGN

The usability validation of any medical device requires the inclusion of end-users throughout the device development process during the concept and pre-concept stages to ensure that the device being produced is the right device with the right components that will meet an un-met or poorly met need. Users are not sufficiently involved in the design process (Money et al., 2011). Only very few manufacturers in medical device production use formal user-centered design—one out 11 manufacturers (Money et al., 2011). The study performed by Money et al. (2011) also mentions that UE/HFE practice is constrained by the reliance on limited numbers of senior healthcare staff, instead of users' inclusion, to confirm usability controls in place. This is possibly due to the developers' fear that including users in the design phase would cause changes to the relatively stabilizing requirements and could clash with the quality and risk identification efforts (i.e. usability solution may lead to new risks and violate or require new quality requirements). Even among manufacturers who tend to include users in their design, users are generally not brought into the developmental process until after the design brief for a new product has been produced (Martin et al., 2008). This could be since medical devices are often technology-driven rather than resulting from an identified unmet need (McCarthy et al., 2015).

Nevertheless, it is critically important to consult a wide range of specialists with various levels of experience during the design and development process to address clinical needs, human error, and patient safety. Identifying which stakeholders and parties to contact—including doctors, nurses, technicians, maintenance staff, patients, and carers—and how often they should be contacted is a thorny issue. Though, it can be as simple as performing communication within the same organization, between the organization and its suppliers, or out to regulators, users, consultants, or research providers (Vincent, Li, & Blandford, 2014).

Various techniques can be adopted for usability testing such as heuristics, cognitive walkthrough, focus groups, or the Delphi technique (Martin et al., 2008). Each can be beneficial in a way depending on the available users and the nature of the device on the table. A combination of techniques can be adopted subsequently or simultaneously to capture user requirements during medical device development. The developer will have to reach a decision regarding which technique to adopt depending on several factors such as the stage of the development of the device, the type of users involved, the expertise of the research, the type of information required, and—most importantly—the time and money available (Martin et al., 2008).

It is the class of the device under development that corresponds to how risky it is and its environment of use—such as hospitals or ambulance or patient's home. In addition to the primary operator, such as nurses or doctors or the patients themselves, the party paying for the device—such as the insurance company or the patient themselves—also determines the level of required usability validation. Hence, it is the manufacturer's responsibility to find out the best usability testing method that could be customized to meet the intended outcome.

HFE FOR STARTUPS IN MEDICAL DEVICES

Like many manufactured products, a huge proportion of medical devices are developed by small companies and start-ups who are unlikely to have the resources to employ ergonomic consultancies or staff with specific ergonomic expertise. This should not be a barrier in adopting ergonomics principles. The role of the ergonomics community should be to provide resources and knowledge that allow product developers, engineers, academics, and clinicians to make decisions with the best use of their resources for studying user requirements during the development of their product. Even if these resources are extremely limited, it is preferable that a small, well-planned, and well-conducted study is performed rather than no or minimal attempt to capture user requirements—which may be the realistic alternative. Research and development units within manufacturing bodies are more likely to adopt ergonomic principles if they believe that doing so will benefit them in terms of improved sales, a better-quality product, or a more efficient development pathway. "While pressure can be put on the industry to encourage it to focus on user-centered design practice, it is unlikely to respond to abstract directives or inducements. What is needed, therefore, is a body of exemplar case studies and demonstration projects that show how such an approach can lead to better and more competitive products."

RISK MANAGEMENT IN THE DEVELOPMENT OF MEDICAL DEVICES

Medical device failure may lead to catastrophic events to both patients and other involved users, including injuries or death. It has been estimated that medical error is among the top leading causes of death in the U.S., ranking third after heart disease and cancer, and accounting for over 250,000 deaths annually (Makary & Daniel, 2016 Makary 2016). Product recalls continuously take place due to arising risks and injuries. A recent example is the recall of 4.7 million units of the Rock'n Play Sleepers by Fisher-Price[R][TM] in April 2019 after reports of more than 30 infant fatalities occurred (rolling over while unrestrained) since the product was introduced in 2009 (Syracuse, 2019). Additionally, according to the Institute of Medicine report "To Err is Human," between 4,400 to 9,800 people die annually throughout the world in hospitals from preventable medical errors (Burton, McCaffery, & Richardson, 2006). The report also says that more people die every year because of medical errors than motor vehicle accidents, breast cancer, or even AIDS.

It is, therefore, understandable why regulators from across the world are imposing extensive measures on medical device manufacturers through regulations and standards—to demonstrate reasonable devotion toward the areas of hazard analysis and risk management (RM). Given the liability and the sensitivity of this domain, it is the responsibility and obligation of every medical device manufacturer to have a strict risk management procedure to tackle various risks associated with their products. In this respect, researchers and RM specialists established various standards, RM frameworks, and guidelines to ease this task on developers and manufacturers.

Additionally, the FDA in the U.S. performed a related analysis based on the data available in their repositories and interviews conducted with the thought leaders. It reported that medical device companies lack expertise in developing risk assessment and mitigation plans during the product development phases. Even more, risk assessment tools like design and process FMEAs are often not developed or appropriately applied (Kirkire, Rane, & Jadhav, 2015). This significantly impairs their ability to monitor and control quality throughout the manufacturing and the post-production phases.

APPLIED RISK MANAGEMENT STANDARDS AND METHODS

Within the medical device industry, one predominant standard—available in two versions—has emerged in performing risk-based activities—ANSI/AAMI/ISO 14971 (ISO 14971). The first is the international version ISO 14971:2007 and the second is the European normative version: EN ISO 14971:2012, which is intended to identify any gaps between the international standard and the requirements of the applicable European directives.

ISO 14971 is a mandatory requirement for medical device manufacturers (Chan, Ip, & Zhang, 2012) where they should establish a system for risk management and a system for reporting incidents and field safety corrective actions. The FDA in the U.S. has published a guideline in 2006 for applying risk management in medical devices—"Guidance for Industry: Q9 Quality Risk Management." Similar for the EU Regulation 2017/745 on "medical devices, amending Directive 2001/83/EC, Regulation (EC) No. 178/2002, and Regulation (EC) No. 1223/2009 and repealing Council Directives 90/385/EEC and 93/42/EEC" has been published in 2017 inspired by ISO 14971 with respect to risk management in medical devices.

In line with the above listed FDA guideline and EU regulations, the risk management process shall start by defining the systematic processes designed to coordinate, facilitate, and improve science-based decision-making with respect to risks. The core of implementing risk management is unfolded in the risk assessment process in which all risks shall be identified through the identification of hazards and the analysis and evaluation of risks associated with exposure (FDA, 2006). It starts with defining the problem description or risk question. To perform this, three questions can aid the process: what might go wrong? What is the likelihood (probability) that it will go wrong? What are the consequences (severity)? This will constitute the risk management's primary and most critical phase called "Risk Identification" and will have to be followed by two primary phases, "Risk Analysis" and "Risk Evaluation." This exercise will generate a register of risks that are evaluated and prioritized, primarily based on the likelihood of occurrence and impact using a risk score (Likelihood vs. Impact). The manufacturer needs to control those risks by setting the treatment plan relevant to each risk in the register to shrink the risk value to an acceptable level.

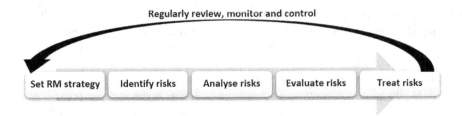

FIGURE 38.1   RM framework essential phases (ISO 14971:2007).

This register will often have to be reviewed during the entire development phases and through an ongoing monitoring strategy for the risks and the benefit-risk ratio so that risk levels are maintained and arising risks are logged. In line with the above, Figure 38.1 below sketches the essential phases for a fruitful risk management framework.

ISO 14971:2007 and the EN 2012 version define five different risk management supporting tools, each with a different approach for risk identification in their annexes to aid users throughout the risk management process. In addition to the FMEA tool, which was described earlier and is the commonly used risk management tool, there is the Preliminary Hazard Analysis (PHA), Fault-tree Analysis (FTA), Hazard and Operability Study (HAZOP), and Hazard Analysis and Critical Control Point (HACCP). Involved stakeholders (i.e. manufacturers, Notified Bodies, auditors, etc.) are expected to be trained to adopt these tools and apply them to the particular product on hand. The newest version of ISO 14971 that shall highlight more interesting and further extensive risk management requirements and tools is currently under development and is supposed to be released in late 2019 (ISO, 2019). It is worth highlighting here that there is a positive effect of risk management on project success (Olechowski et al., 2016; de Bakker, Boonstra, & Wortmann, 2010) but this insight is not yet operational due to a lack of synergy between those making purchasing decisions and the actual users of the devices.

DEVELOPMENT MODELS FOR MEDICAL DEVICES

Given that the medical devices development process is strictly regulated, and devices must receive the respective required certifications, they must comply with a set of applicable standards beforehand to prove compliance. Also, as mentioned earlier, standards pertaining to safety and risk management, whether national or international, are equally strict to ensure safety for the patients and users. Accordingly, developers of medical devices should put considerable effort to comply with all these directives, standards, and guidelines so their products could reach the market. Even more, some developers attempt to supersede the requirements of those standards to reach wider markets and further gain the trust of the end-users. That's why requirement verification and validation, in addition to various types of thorough testing, play a critical role during every phase of the development process.

Accordingly, developers are expected to allocate a sufficient amount of time and money to ensure that safety measures are applied to their developed product. The development model, equipped with verification and validation stop points, can be applied to provide developers with a process to adopt and help them achieve their target. Nevertheless, selecting the right model—which can be off-the-shelf or customized by the developer—can be challenging. The adopted model and related tools can follow the agile trend, yet it must be selected very carefully to avoid jeopardizing safety and usability.

In the next part of the chapter, we will identify the major steps that should be considered in the development of medical devices. A brief of major adopted models will be also discussed to expose the available medical devices' development techniques and methods.

FIGURE 38.2   Development primary phases (Ogot & Okudan-Kremer, 2004).

MEDICAL DEVICES DEVELOPMENT CORE PHASES

Ogot and Okudan-Kremer (2004) conducted a comparative analysis based on relevant literature and defined five phases for the normative design process, which highlight the basic phases without embedding other crucial requirements such as standards and regulations inclusion. These phases contain needs assessment; conceptualization; preliminary design and evaluation; detailed design and testing; and production. These phases are sketched in Figure 38.2:

The primary phase of any development cycle must start with the needs assessment—also referred to as requirements engineering—which is the process that focuses on discovering, analyzing, documenting, and managing requirements (Pandey & Pandey, 2012). Developers are expected to dedicate time and effort for this phase for the requirements to be accurate, traceable, and complete, as the cost of amending a requirement later in the development stage is much higher than the cost of properly identifying, analyzing, and reviewing the needed requirements at the early development stage.

Through the involvement of various professionals and end-users, requirements can be collected and divided before being grouped into smaller subsystems or modules, and subsystems from earlier projects can be re-used (Becker, 2009). Amendment to the initial requirements at a late development stage, for any given reason, can incur extensive costs and lead to development deadlock and wastage of valuable resources. Therefore, various stakeholders should take part in the requirements phase to ensure that they are comprehensively collected. Additionally, if the requirements were misinterpreted or the specifications were incorrect, the developed product can be irrelevant to what is exactly desired. About three-quarters of the development costs are caused by the 43% of changes related to user requirements according to Becker (2009). Consequently, it is of extreme importance that the requirements identification phase is given the utmost attention and needed resources, including requirements thorough verification testing, to reach a successful project.

To achieve this, requirements clauses shall be wisely identified using the available techniques, such as Entity Relationship Diagrams or Data Flow Diagrams, and accompanied with verification as well as validation controls to each clause to ensure that the product under development matches the objectives of the requirements. This will allow testing for those requirements, by specialized testing professionals along with end-users, including the maintenance, installation, training, and safety teams—each from a different perspective—and eventually lead to smoother production phases. This tested and prioritized requirements register will pave the way toward establishing the system architecture specification document that can be subdivided into component or sub-system specification documents, after which the conceptualization and further implementation phases can commence—which should be accompanied with test cases using the various available testing and verification tools or models.

MAJOR MEDICAL DEVICES DEVELOPMENT MODELS

The success rate of products depends on processes used by the organizations (Pandey & Pandey, 2012). In the field of medical device development, empirical research has proven the importance of following a complete process showing a significant correlation between the number of stages followed in the development process and the success of the new devices (Rochford & Rudelius, 1997). Also, according to Bresciani and Eppler (2007), graphic representations are recommended to enhance communication within knowledge-intensive teamwork settings. Having a model adopted throughout the development lifecycle provides a reference tool or a guideline for the developing teams, which can include novice or experienced designers, who are eager to conduct a complete end-to-end efficient process in compliance with the various regulations and standard requirements. Additionally, the adopted model can serve as training material so that all involved teams can be informed and eligible to take part throughout the entire development lifecycle. Unfortunately, the currently published models in the subject domain are not comprehensive as they either address the development of medical devices with a specific focus on regulations or any other focus area, such as risk management or intellectual property rights (IP), leading to fragmented approaches—as will be highlighted later in this section with examples—without being able to combine all focus areas (Medina, Kremer, & Wysk, 2013). Even more—although it is evident how impactful regulations are on the development of medical devices as discussed earlier—related models in places have addressed the inclusion of specific regulatory requirements without pointing out the mechanism to do so during the stage with which they should be considered. They simply correspond to a specific process point of view, mostly including graphical presentations exemplified in a series of steps not comprehensive enough for developers who should cater to the various requirements, including usability and risk management. Some of the essential and interrelated success factors that developers in medical devices should cater to to have a comprehensive production model can be sketched below Figure 38.3.

Several off-the-shelf development models exist, each with emphasis on one or more success factors. Pietzsch et al. (2009) define the stage-gate development process by professionals that are involved in the development, commercialization, and regulation processes of medical devices. This process model divides the development phases into stages that are separated with validation gates before the developer can proceed with further development stages. Also, there is another model by Alexander and Clarkson (2002) that thoroughly focuses on the medical devices validation aspect and provides a respective design validation framework. Panescu (2009) r-views the design process and focuses on issues related to obtaining IP rights for newly developed devices. Finally, Das and Almonor (2000) establish an attribute-driven concurrent engineering approach for the development of medical devices that includes functional, regulatory, and customer attributes.

FIGURE 38.3 Interrelated development success factors.

CONCLUSION

Manufacturers of medical devices are held to a higher standard than manufacturers of many other products due to the potential severity of the consequences of introducing inferior or unsafe products to the marketplace (McAllister & Jeswiet, 2003 McAllister 2003). Consequently, the process of medical device development shall be performed in adherence to the strict regulatory requirements enforced by relevant authorities, which unfold additional constraints for the development, manufacture, market, and continuous improvement of medical devices (Medina, Kremer, & Wysk, 2013).

Although there are currently several models to support developers in the various development phases, these are not comprehensive for developers—who may have diverse backgrounds and experience levels—to rely on, as they lack fundamental aspects on the "what," "how," and "when" to embed the set regulations and standards in the development lifecycle. Similarly, in the research conducted by Lamph (2012), it was determined that there is no medical device regulatory template that fits all products, which can put developers in a dilemma while struggling to execute the various requirements. This burden, triggered by the sensitivity of the innovative production domain and the respective lack of an appropriate development model, can lead to production delay, wastage of resources, or even failure while considering the stringent regulatory processes and potential litigations. Therefore, there is an evident thirst to have further advanced development models that would precisely embed the related regulatory and standards requirements at each development phase to aid and guide developers, whether they are startups or experienced in the domain, throughout their complex and elusive compliance endeavor.

REFERENCES

Alexander, K. and Clarkson, P. J. 2002. A validation model for the medical devices industry. *Journal of Engineering Design* 13(3):197–204.

Baim, D. S., Wahr D., George B., et al. 2002. Randomized trial of a distal embolic protection device during percutaneous intervention of saphenous vein aorto-coronary bypass grafts. *Circulation* 105:1285–1290.

Becker, U. (2009). Model based development of medical devices. In Proceedings of the Computer Safety, Reliability, and Security, 4–17.

Borsci, S., Macredie, R. D., Martin, J. L., and Young, T. 2014. How many testers are needed to assure the usability of medical devices? *Expert Review of Medical Devices* 11(5): 513–525.

Bresciani, S. and Eppler, M. J., 2007. Usability of diagrams for group knowledge work: Toward an analytic description. Proceedings of the *7th International Conference on Knowledge Management and Knowledge Technologies (I-IKNOW)* 1–8.

Burton, J., McCaffery, F., and Richardson, I. 2006. A risk management capability model for use in medical device companies. Proceedings of the *2006 International Workshop on Software Quality – WoSQ '06*.

Buttron, S. 2017. The importance of human factors and usability engineering in medical devices. European Market Regulatory, NAMSA. https://www.namsa.com/european-market/human-factors-usability-engineering-in-medical-devices/ (Accessed March 22, 2019).

Cafazzo, J., St-Cyr, O. 2012. From discovery to design: The evolution of human factors in healthcare. *Healthcare Quarterly (Toronto, Ont.)* 15:24–29.

Campbell, B. 2013. Regulation and safe adoption of new medical devices and procedures. *British Medical Bulletin* 107:5–18.

Chai, J. Y. 2000. Medical device regulation in the United States and the European Union: A comparative study. *Food and Drug Law Journal* 55:57–80.

Chan, S. L., Ip, W. H. and Zhang, W. J. 2012. Integrating failure analysis and risk analysis with quality assurance in the design phase of medical device development. *International Journal of Production Research* 50(8):2190–2203.

Chen, C. E., Dhruva, S. S., Bero, L. A. and Redberg, R. F. 2011. Inclusion of training patients in US Food and Drug Administration premarket approval cardiovascular device studies. *Archives of Internal Medicine* 171:534–539.

Cohen, D. 2012. How a fake hip showed up failings in the European device regulation. *BMJ* 345, e7126.

Das, S. K. and Almonor, J. B. 2000. A concurrent engineering approach for the development of medical devices. *International Journal of Computer Integrated Manufacturing* 13(2):139–147.

de Bakker, K., Boonstra, A. and Wortmann, H. 2010. Does risk management contribute to IT project success? A meta-analysis of empirical evidence. *International Journal of Project Management* 28(5):493–503.

European Medicines Agency. 2019. Guideline on the quality requirements for drug device combinations. https://www.ema.europa.eu/en/documents (Accessed November 2, 2019).

FDA. 2016. Applying human factors and usability engineering to medical devices. https://www.fda.gov/media/80481 (Accessed October 20, 2019).

FDA. 2006. Guidance for Industry: Q9 Quality Risk Management. https://www.fda.gov/media/71543/ (Accessed October 20, 2019).

Hegde, V. 2013. Role of human factors/usability engineering in medical device design. In *Proceedings of 2013 Annual Reliability and Maintainability Symposium (RAMS)* (pp. 1–5). Orlando, FL.

International Standards Organization. 2019. Benefits of standards. https://www.iso.org/benefits-of-standards.html (Accessed November 2, 2019).

Kinsel, D. 2012. Design control requirements for medical device development. *World Journal for Pediatric and Congenital Heart Surgery* 3(1):77–81.

Kirkire, M. S., Rane, S. B., and Jadhav, J. R. 2015. Risk management in medical product development process using traditional FMEA and fuzzy linguistic approach: A case study. *Journal of Industrial Engineering International* 11(4):595–611.

Janß, A., Plogmann, S., and Radermacher, K. 2016. Human-centered risk management for medical devices – New methods and tools. *Biomedical Engineering/Biomedizinische Technik* 61(2), 165–181.

Lamph, S. 2012. Regulation of medical devices outside the European Union. *Journal of the Royal Society of Medicine* 105:12–21.

Laskey, W. K., Yancy, C. W., and Maisel, W. H. 2007. Thrombosis in coronary drug-eluting stents: Report from the meeting of the Circulatory System Medical Devices Advisory Panel of the Food and Drug Administration Center for Devices and Radiologic Health. *Circulation* 115:2352–2357.

Lin, L., Vicente, K. J., and Doyle, D. J. 2001. Patient safety, potential adverse drug events, and medical device design: A human factors engineering approach. *Journal of Biomedical Informatics* 34(4):274–284.

Makary, M. and Daniel, M. 2016. Medical error – The third leading cause of death in the US. *BMJ*, 353: i2139.

McCurdie, T., Taneva, S., Casselman, M., Yeung, M., McDaniel, C., Ho, W., and Cafazzo, J. 2012. mHealth consumer apps: The case for user-centered design. *Biomedical Instrumentation and Technolog* 46(2):49–56.

Martin, J. L., Norris, B. J., Murphy, E., and Crowe, J. A. 2008. Medical device development: The challenge for ergonomics. *Applied Ergonomics* 39(3):271–283.

McAllister, P., and Jeswiet, J. 2003. Medical device regulation for manufacturers. *Proceedings of the Institution of Mechanical Engineers, Part H: Journal of Engineering in Medicine* 217(6), 459–467.

McCarthy, A. D., Sproson, L., Wells, O., and Tindale, W. 2015. Unmet needs: Relevance to medical technology innovation? *Journal of Medical Engineering and Technology* 39(7):382–387.

Medina, L. A., Kremer, G. E. O., and Wysk, R. A. 2013. Supporting medical device development: A standard product design process model. *Journal of Engineering Design* 24(2):83–119.

Money A., Barnett J., Kujis J., Craven M., Martin J., Young T. 2011. The role of the user within the medical device design and development process. Medical device manufacturers' perspective, *BMC Medical Informatics and Decision Making* 11: 15, https://doi.org/10.1186/1472-6947-11-15

Ogot, M. and Okudan-Kremer, G. 2004. *Engineering design: A practical guide.* Victoria, Canada: Trafford Publishing.

Ogrodnik, P.J. 2013. *Medical device design: Innovation from concept to market.* Oxford: Elsevier

Olechowski, A., Oehmen, J., Seering, W. and Ben-Daya, M. 2016. The professionalization of risk management: What role can the ISO 31000 risk management principles play? *International Journal of Project Management* 34(8):1568–1578.

Pandey, D. and Pandey, V., 2012. Importance of requirement management: A requirement engineering concern. *International Journal of Research and Development – A Management Review (IJRDMR)* 1 2319–5479.

Panescu, D. 2009. Medical device development. *31st Annual International Conference of the 31st IEEE Engineering in Medicine and Biology Society (EMBS)* 5591–5594.

Pietzsch, J. B., et al. 2009. State-gate process for the development of medical devices. *Journal of Medical Devices* 3:1–15.

Pitkänen, H. 2017. *The challenges of a MedTech startup in the land of scattered regulatory information.* MedTech Views. http://www.medtechviews.eu/article/challenges-medtech-startup-land-scattered-regulatory-information (Accessed March 22, 2019).

Reilly, K. 2015. 7 reasons why new medical devices don't make it to market (or fail when they do). MedTech Engine. https://medtechengine.com/article/why-new-medical-devises-dont-make-it-to-market-or-fail/ (Accessed March 22, 2019).

Rochford, L. and Rudelius, W. 1997. New product development process: Stages and successes in the medical products industry. *Industrial Marketing Management* 26:67–84.

Syracuse. 2019. 4.7 million Fisher Price Rock'n Play Sleepers recalled after infant deaths. https://www.syracuse.com/product-recalls/2019/04/47-million-fisher-price-rock-n-play-sleepers-recalled-after-infant-deaths.html (Accessed November 2, 2019).

Thompson, M., Heneghan, C., Billingsley, M. and, Cohen, D. 2011. Medical device recalls and transparency in the UK. *BMJ*, 342: d2973.

Vincent, C. J., Li, Y., and Blandford, A. 2014. Integration of human factors and ergonomics during medical device design and development: It's all about communication. *Applied Ergonomics* 45(3): 413–419.

Wild, C., Erdos, J. and Zechmeister, I. 2014. Contrasting clinical evidence for market authorisation of cardio-vascular devices in Europe and the USA: A systematic analysis of 10 devices based on Austrian pre-reimbursement assessments. *BMC Cardiovascular Disorders* 14:154.

World Health Organization. 2019. Medical devices. https://www.who.int/medicaldevices/definitions/ (Accessed November 2, 2019).

Wu, Y. H., Li, F. A., Fan, Y. T. and Tu, P.W. 2016. A study of medical device regulation management model in Asia. *Expert Review of Medical Devices* 13(6):533–543

Section VII

Safety and Legal Protection Standards

39 Printed Warning Signs, Tags, and Labels

Their Choice, Design, and Expectation of Success

Charles A. Cacha

CONTENTS

INTRODUCTION

An authority or an owner who has charge and control of a premise and its physical contents has a moral and legal responsibility toward individuals who occupy the premises. Similarly, an authority who designed, manufactured, and sold an object has moral and possibly legal responsibility toward the ultimate owner and user of the object. These obligations are usually to protect individuals from physical injury or disease. In some instances, this may apply to protecting personal property as well. Some examples of the responsibility of the authority are (a) the obligation of a landlord to tenants of a dwelling, (b) the obligation of a landlord to a visiting member of the public, (c) the obligation of an employer to an employee, and (d) the obligation of a designer and manufacturer of an object to the final owner and user of the object. These obligations may be discharged by removing any hazards that could cause injury or illness. These may also be permitted to remain, provided the authority or his or her subordinates meticulously and strictly supervise the premises or the object.

In consideration of the use of supervision, it is apparent and commonly understood that immediate constant supervision by the authority is (a) extremely impractical and (b) extremely expensive. It is also commonly agreed that the authority cannot "be every place at once." Thus, a state-of-the-art procedure known as warning is unavoidably employed. The warning is very often one of a semantic or ideographic nature and is used to appraise individuals of a particular hazard related to a premise, the contents of the premise, or an object owned by an individual. Usually, the warning is of a printed nature and is composed of paints, inks, and so forth applied to paper, cardboard, plastic, wood, or metal.

OBJECTIVE AND SCOPE

In consideration of the universal and unavoidable use of printed warnings, this chapter deals with the effective use and design of printed warnings so that minimal injuries, illnesses, and property damage occur during exposure to a particular hazard. Printed warnings are generally categorized as (a) signs that appear on walls, equipment, or other relatively large supporting structures; (b) tags that are signs of small dimensions and placed usually temporarily on an object; and (c) labels that are small permanent signs placed on an object. All these categories are dealt with while adhering to the following discussion format.

1. When is a printed warning appropriate or not appropriate?
2. What is the most effective physical and semantic design of a printed warning? (Figures 39.1 and 39.2).

It is important to note that in the choice and design of printed warnings both of the previous questions should be employed in a 1,2 sequence. This sequential application will assure effective meaningful design. It should finally be noted that non-printed warnings such as those of a visual (e.g., lights and blinkers) or an auditory nature (e.g., buzzers and bells) are not included in this chapter but are covered elsewhere in this book.

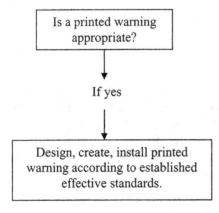

FIGURE 39.1 Sequence in decision making and design of printed warnings.

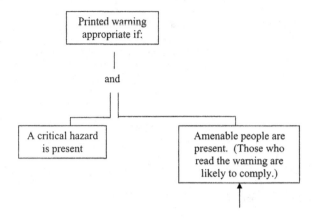

FIGURE 39.2   Decision tree for determining the appropriateness of a printed warning.

APPROPRIATENESS OF A PRINTED WARNING

Logic indicates that a warning of any kind may be successfully employed only when a proximal critical hazard exists or when individuals exposed to the critical hazard are likely to comply with the warning. A warning displayed in the absence of a critical hazard is illogical and a waste of resources. Similarly, a printed warning in the proximity of a critical hazard has only minimal value if most of those individuals associated with the hazard are not inclined to comply with the warning. Thus, the two primary requirements for the use of a printed warning are (a) the proximate presence of a critical hazard and (b) the presence of individuals who are amenable and likely to comply with the warning.

THE CRITICALITY OF A HAZARD

As was discussed, a printed warning should be used only in the presence of a hazard, particularly in the presence of a critical hazard. A minor hazard that has little consequence need not and should not bear a printed warning because of distractions and confusions that may occur from its presence among many other printed warnings in the environment. The criticality of a hazard is defined by four important major variables: (a) a potential for producing many injuries or illnesses or property damage events (frequency), (b) a potential for producing serious injuries or illnesses or property damage events (severity), (c) necessity for or inexpendability of the hazard that makes its presence unavoidable, and (d) an inability to mitigate, contain, or enclose the hazard to cancel its harmful effects (Figures 39.3 and 39.4).

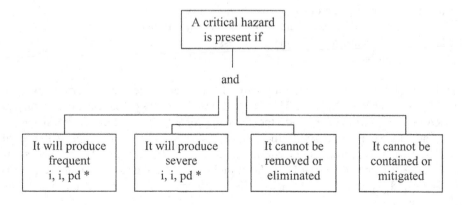

FIGURE 39.3   Criteria of a critical hazard. *Note:* *injury, illness, property damage.

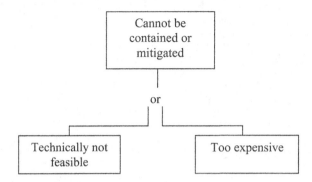

FIGURE 39.4   Reason for non-containment of a hazard.

FREQUENCY AND SEVERITY

The concepts of frequency and severity are most often applied in Safety and Risk Management to overall injuries and illnesses experienced by a large organization or subdivisions of a large organization. Procedures for calculating the frequency and severity rates of an organization appear in another section of this book. The concepts of frequency and severity may also apply, however, to an individual hazard to determine if is of a critical nature. This analysis may be an empirical ex post facto method, wherein observations reveal the number of past injuries and illnesses that have occurred (frequency) or the financial cost or length of disability time (severity) related to the hazard. Hazard analysis may also be of an a priori nature where competent, experienced person(s) estimate the frequency and severity related to a particular hazard. This analysis may then determine the hazard criticality which, in turn, partially helps decide the appropriateness or inappropriateness of a printed warning. Military Standard 882B provides a method of hazard analysis that may be employed to partially determine the appropriateness of a printed warning. Frequency (synonymous with probability) based on prior incidents or intuitive judgments is categorized as:

A. Frequent: likely to occur frequently
B. Probable: will occur several times in the life of a facility or system
C. Occasional: likely to occur sometime in the life of a facility or system
D. Remote: unlikely but possible to occur sometime in the life of a facility or system
E. Improbable: so unlikely it can be assumed that occurrences may not be experienced

Additionally, severity based on prior incidents or intuitive judgments is characterized as:

I. Catastrophic: may cause death or loss of a facility.
II. Critical: may cause severe injury or illness or major property damage.
III. Marginal: may cause minor injury or illness or minor property damage.
IV. Negligible: probably no effect on people or property.

Table 39.1 indicates a matrix method for interfacing frequencies and severities to determine the criticality of a particular hazard. This method, known as Risk Assessment and Acceptance Coding (RAAC), will not only provide the criticality of a hazard but also suggest appropriate managerial actions when dealing with the hazard. As a general rule, the greater the frequency (probability) of occurrences due to a hazard and the greater the severity of occurrences due to a hazard, the more critical is the hazard and the more important is the need for management to take remedial actions. This action would include the use of printed warnings if no engineering solutions are practicable.

TABLE 39.1

Probability and Severity Matrix

Probability Categories	Severity Categories			
	I Catastrophic	II Critical	III Marginal	IV Negligible
A Frequent	IA[a]	IIA	IDA	IVA
B Probable	IB	IIB	MB	IVB
C Occasional	IC	IIC	IIIC	IVC
D Remote	ID	IID	HID	IVD
E Improbable	IE	HE	HIE	IVE

Notes

a Interpretations: IA, IB, IC, IIA, IIB, IDA, Unacceptable Situation (needs immediate emergency action); ID, IIC, IID, MB, IIIC, Undesirable Situation (quick management decision needed); IE, HE, HID, HIE, IVA, IVB, Acceptable Situation (management review needed); IVC, IVD, IVE, Acceptable Situation (no management review needed).

ENGINEERING CONTROLS

Determining the criticality of a hazard also involves an understanding of basic engineering principles that can remove, mitigate, or contain the hazard. This understanding, plus knowledge of the unacceptable or undesirable circumstances related to frequencies and severities, will lead to an ultimate decision to create and install a printed warning or not. An aid to this process is a hierarchical decision tree initially promulgated by William Haddon and further described by Roger L. Brauer (1990) as the Energy Theory. This theory is based on the concept that unwanted events can occur only in the presence of a transfer of energy. This understanding provides a sequential procedure dealing with a hazardous source of energy. The theory and its sequences have been reported in varying formats. A condensed form created by the author of this chapter follows:

When encountering a potentially injurious source of energy:

1. Eliminate the energy source by removing it from the premises or from any object that people or property are exposed to.
2. If the energy source cannot be eliminated as above, then retain the source by keeping it in remote locations so that the effect of its expenditure is diminished to an extent that people and property are not injured or damaged.
3. If the energy source cannot be remotely located as above, then place adequate barriers such as guards, shields, or walls around it so that an energy expenditure will not cause injury to people or damage to property. Example: machine guards, protective partitions, and so forth.
4. If barriers cannot be placed around the energy source, then place barriers on or around people of property. Example: personal protective equipment such as goggles, steel tip shoes, and so forth.
5. If barriers cannot be placed around people or property then (a) provide a warning that will prevent people or property from contacting the energy expenditure when heeded and (b) train, educate, and motivate people exposed to the energy source to comply with the warning.

AN EXAMPLE OF ENERGY THEORY DECISION MAKING

A company that manufactures and sells plastic cooking utensils considers manufacturing of a new line of metal cooking utensils. This would require purchasing metal stamping machines that are potentially capable of injuring machine operators.

Ask the following sequential questions:

1. Can the metal stamping operation be excluded from the premises? No. Not if it is essential for business survival that the company open this new line of business.
2. If this metal stamping operation must be present on premises, can the operation be placed in a remote part of the plant away from employees? No. Machine operators must work in the proximity of the machines to successfully manufacture the product.
3. If the stamping machines cannot be isolated and kept away from people, can barrier guards be placed at the point of operation where the stamping occurs? No. Guarding the point of operation will not effectively allow the workpiece to be loaded into the machine by the operator.
4. If the stamping operation cannot be guarded, can personal protective equipment for the operator's hands be provided? No. Protective gloves would not withstand the forces of the ram in the stamping operation.
5. If personal protective equipment would be ineffective, then can a warning sign be placed in the proximity of the stamping operation, which will appraise operators of the hazard and advise them to avoid it? Yes. In addition to placing a warning, training, education, and motivation for those exposed to the hazard should be provided.

As may be concluded from the previous example, a warning used in safety engineering is a "last resort" and is used only when hierarchical engineering alternatives are not feasible and cannot be technically or economically applied.

ECONOMIC FEASIBILITY

In addition to engineering controls and their technical feasibilities, an additional input into decision making involves economics. Frequently, higher levels of engineering solutions are technically feasible and will give excellent protection to people and property and will consequently eliminate the need for a warning related to a hazard. These engineering solutions, however, must not be economically prohibitive. There are several techniques in Risk Management that will determine the economic viability or inviability of a safety engineering solution for controlling a hazard. The following procedure is reported by Harold E. Roland and Brian Moriarty (1983) in their book *System Safety Engineering and Management*, and by Charles A. Cacha (1997) in his book *Research Design and Statistics for the Safety and Health Professional*.

Case Study: Present Worth of Money Technique. A safety officer and the treasurer of a corporation are working together as a team to determine the financial feasibility of installing an expensive presence sensing device on all their metal stamping machines. After researching some sources of information within and outside of the corporation, they determine the following essential facts that are needed to form an opinion.

1. Cost of installation by a competent contractor: $250,000
2. Life expectancy of the stamping machines and of the presence sensing device: 30 years
3. Estimated annual savings in workmen's compensation losses and workmen's compensations premiums: $25,000
4. Rate of interest return the corporation expects for investing their excess monetary assets: 8%

The previously cited information is inserted into the following calculation:

Proposed expenditure (PE) = 250, 000

life expectancy (L) = 30

Annual savings (S) = 25, 000

Value of money (V) = .08

Present worth of money (PW_M)

Present worth of savings (PW_S)

Formulas:

$$PW_M = \frac{(1+V)^L - 1}{V(1+V)^L}$$
$$PW_S = PW_M(S)$$

Computation:

$$PW_M = \frac{(1+.08)^{30} - 1}{.08(1+0.8)^{30}}$$

$$PW_S = \frac{9.06}{0.81} = 11.18$$

$$PW_S = 11.18(25, 000) = \$279, 500$$

$$PE = \$250, 000$$

$$PW_S(\text{of } \$279, 000) > PE (\text{of } \$275, 000)$$

Because the Present Worth of Savings is larger than the Proposed Expenditure, it becomes financially advantageous and feasible for the corporation to install the presence sensing device. If the Present Worth of Savings were substantially less than the Proposed Expenditure, the corporation management might have reservations about investing in the presence sensing device. Nevertheless, despite the financial disadvantage, the presence sensing device may still be installed for the humanitarian purpose of guaranteeing safety to employees. It is advisable that a procedure such as this be applied in situations where there may be large safety expenditures. There is one apparent limitation in applying this procedure and that is the question of subjectivity. Methods of predicting Life Expectancy and estimating Annual Savings, which are only partially tempered with limited amounts of research, involve a good deal of judgment. In summary, this method—if it determines economic viability for a safety engineering device—will support the utilization of the device and will obviate the need for a printed warning. Similarly, this method—if it determines a lack of economic viability for a safety engineering device—will indicate that a printed warning coupled with training and education and motivation of people is in order.

COMPLIANCE WITH PRINTED WARNINGS

As discussed earlier, printed warnings are a last resort when engineering alternatives cannot be applied to protect people or properties from unwanted events that cause injury or damage. Printed warnings, however, are frequently and universally employed and displayed upon premises, equipment, and various objects. Since those in authority are aware of the unavoidability of using these warnings, they should also be aware of the possibility that those people exposed to these warnings may ignore these even though they are adequately presented and designed. It is important that the authorities understand that there are social-psychological variables among people that will

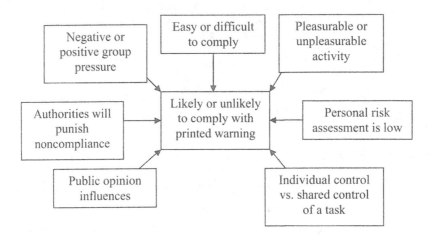

FIGURE 39.5 Factors influencing compliances or non-compliance with printed warnings.

prompt compliance or non-compliance with a printed warning. This understanding, if applied through competent training, education, and motivation, will increase the probabilities that people will comply with a warning and will therefore avoid an unwanted event. Some of the most important sociopsychological variables are listed in the following sections. Many of these are described in Volumes I and II of the Human Factors Perspectives on Warnings by Laughery, Wogalter, and Young (1991Cacha 1997, Laughery 1991; Figure 39.5).

Ease or Difficulty in Compliance. A printed warning may involve an instruction requiring an action, inaction, or an evasion by the recipient who reads it. In either case, a warning requiring minimal physical or mental effort is most likely to be complied with. A warning requiring great physical effort, complex mental processes, and inordinate inputs of time will cause non-compliance (Figure 39.5).

Group Membership. If the individual reading the warning is part of a group that is also reading it, the individual will be influenced by the general response of the group. A person, except for the most individualistic, will heed or ignore a warning according to the influence of group pressures ("When in Rome, do as the Romans do."). Similarly, an individual—though alone while reading a warning—may still feel obligated to a political, fraternal, or religious group even though that group is not present while the warning is read.

Sanctions From Authority. An understanding that punitive actions may be applied by government or organizational authorities such as employers may increase the likelihood that a printed warning will be complied with. There may be a range of responses in this area depending on how assiduously the authority enforces the warning and how submissive or independent the attitudes of the recipient of the warning are.

Pleasurable Versus Unpleasurable Activity. A printed warning may particularly succeed if it warns against an unpleasant action. Similarly, the probability of non-compliance with the warning is increased when a warning is implemented against a pleasurable activity, particularly if the activity is of a hedonistic nature.

Personal Risk Assessment. The design and placement of a written warning by those in authority is predicated upon the authority forming a risk evaluation of the hazard in question. When the warning is finally presented to an individual recipient of the warning, however, the recipient may not agree with the risk evaluation of the authority and may consider the hazard more or less hazardous than the risk evaluation originally promulgated by the authority. If the recipient believes the hazard to be less risky than the warning implies, the recipient may ignore the warning. If the recipient agrees with the risk level implied by the warning or believes the risk level is even higher, the recipient will heed the warning. Judgments by the recipient of risk levels may be based on past personal experiences and evaluation of immediate circumstance, or current events on a public or local level that report unwanted events (particularly of a catastrophic nature) related to the hazard at hand.

Personal Control. A recipient, when observing a written warning, may act individually upon a task or perform a task in conjunction with another individual. When acting alone and with full control of a task, the recipient may ignore a warning because he or she has understanding and confidence in his or her capabilities. When relying fully or partially on the action of other people, the recipient will be more likely to comply with a warning because he or she is not certain of the capabilities of other people. For the final summary for the warning design strategy, see Figure 39.6.

STANDARDS FOR PRINTED WARNING DESIGNS

After the authority responsible for a premise, equipment, or a product finally and competently decides to use a printed warning, the authority must design the warning in the most effective manner possible. This effective design will most basically deal with the physical characteristics of the warning as well as its semantic content. There are a large number of published standards providing characteristics of an effective warning. The major publications such as the American National Standards Institute (ANSI), International Organization for Standardization (ISO), Underwriters Lab (UL), and Military Standards of Department of Defense (Mil. Std.) are summarized in the following section:

A LISTING OF PROMINENT WARNINGS STANDARDS

ANSI 2535.1-2002: Standard for Safety Color Code

This publication emphasizes the importance of the uniform use of colors on signs, labels, tags, equipment, walkways, and so forth within or among all organizations. This use will enhance warnings about hazardous conditions. It is also emphasized that any warning, regardless of color, is no substitute for the reduction or elimination of a hazard. It is also emphasized that excessive use of many colors at a particular location may lead to confusion and minimize the effect of a warning. The standard indicates that there should be adequate illumination on the sign that will allow

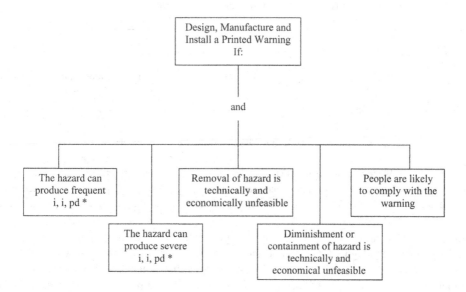

FIGURE 39.6 Requisites for a printed warning. *Note:* *injury, illness, property damage.

differentiation of colors. The specifications of various colors are in terms of the Munsell Notation System as described by the American Society of Testing Materials (ASTM) Standard ASTM D1535. It should be noted that earlier ANSI standards, such as this specification, do not appear in the current 2002 standard. Colors have been traditionally used as specified by 2535.1-1998 in the following manner.

Safety Red:	Flammable liquids
	Emergency stop bars on machines
	Emergency stop buttons on machines
	Fire protection equipment and apparatus
	Background for DANGER SIGNS
Safety Orange:	Intermediate level of hazard
	Used on hazardous machine parts
	Used on moveable guards or transmission guards
	Background color for WARNING SIGNS
Safety Yellow:	Physical hazards causing slips, trips, falls, or being caught in between
	Corrosive containers and storage cabinets for flammables
	Background color for CAUTION signs
Safety Green:	Used for emergency egress
	Used for first aid equipment
	Used for safety equipment
Safety Blue:	Used to identify safety information
	Used for mandatory signs related to personal protective equipment

ANSI Z 535.2-1998: Standard for Environmental and Facility Safety Signs

This publication is described as a visual alerting system to aid potential hazards known to exist in the environment. Major examples of this system are safety signs used in fixed locations in the environment, such as industrial facilities, commercial establishments, places of employment, and real estate properties.

The major categories of environmental and facility signs are:

Signal Word	Colors	Use
DANGER	White letters on Safety Red background	Imminent hazardous situation may result in serious injury or death
WARNING	Black letters on Safety Orange background	Potentially hazardous situation may result in serious injury or death
CAUTION	Black letters on Safety Yellow background	Potential hazardous situation may result in minor or moderate injury
NOTICE	White italic letters on Safety Blue background	Statement of policy related to safety
GENERAL Safety	White letters on Safety Green background	General safe practices and procedures

Sign design and layout illustrations in the publication are of two major categories: (a) "portrait"—in which the sign's longest dimension is vertical—or "landscape"—in which the longest dimension is horizontal; and (b) the number of panels may be three (signal word, message, and symbol or pictorial) or two (signal word and message, but not a symbol and pictorial). Generally, DANGER, WARNING, CAUTION, and NOTICE signs may make use of these formats (Figure 39.7).

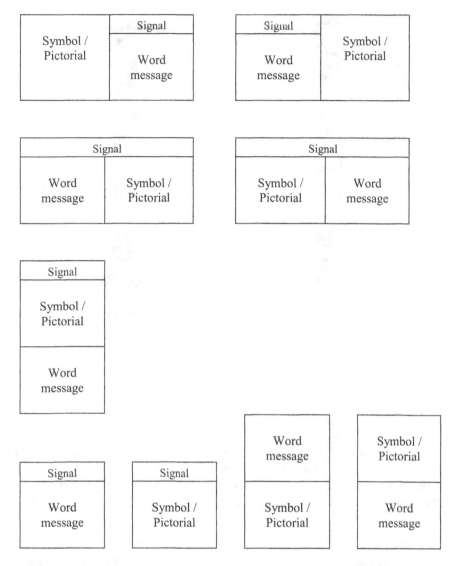

FIGURE 39.7 Sign layouts.

Letter style on signs is required to be sans serif and in upper case for signal words (Figure 39.8). Message panel lettering is a combination of upper- and lower-case sans serif. Letter styles are of a large variety: Arial, Arial Bold, Folio Medium, Franklin Gothic, Helvetica, Helvetica Bold, Meta Bold, News Gothic Bold, Poster Gothic, and Universe.

Legibility is considered to be influenced by the variables of letter height, width, spacing, and stroke width. The size of lettering depends on the message length and distance from which the sign can be easily read. Letters must be adequately spaced and not crowded.

A required minimal letter height for signal words is one unit of height for every 150 units of safe viewing distance and one in 300 for letters in the message panel.

Sign finish requires durable material. Sign placement should be a location that gives sufficient time to take evasive action from the hazard. The sign must be legible, non-distractive, and not hazardous in and of itself, and it should not be vulnerable to various motions of objects in the environment that could destroy the sign. Signs should be lit by adequate reliable illumination.

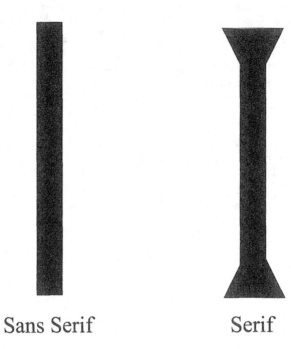

FIGURE 39.8 Letter style.

Semantically, the message in the sign should contain, in concise words, (a) the signal word, (b) the description of the hazard, (c) the consequences related to the hazard, and (d) the related evasive action. The sequence of b, c, and d may vary according to judgment.

Letter characteristics are described in Appendix B and briefly summarized here.

The stroke width/ratio of a black letter on a white background should be from 1:6 to 1:8. The ratio of a white letter on a black background should be 1:8 to 1:10. The amount of spacing (lead) between lines should be 120% of the height of letters. A table is provided for recommended message letter heights and minimum safe viewing distances (Figure 39.9). This table is based on the following formulas:

Favorable Reading Conditions	Unfavorable Reading Conditions
MSVD[a] (inches)/300	MSVD[a](feet) × 0.084

[a]Minimal Safe Visual Distance.

ANSI Z 535.3-2002: STANDARD FOR CRITERIA FOR SAFETY SYMBOLS

This standard recognizes that the recipients of a safety warning may be multicultural and multilingual. This multilingual circumstance will not allow the total success of a written warning because only a portion of the recipient population is fluent in the warning language. Symbols, however, may be comprehensible across all languages and may also provide a supplemental emotional impact on the written warning that will make it more effective.

The safety symbol is defined as an image—with or without a surrounding geometrical shape—that conveys a message without the use of words. The surrounding geometrical shape is defined as the surround shape and is usually a square or a circle.

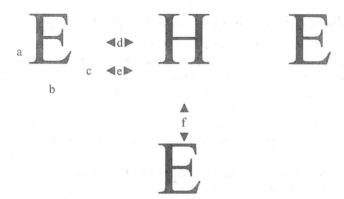

FIGURE 39.9 Lettering, a, letter height; b, letter width; c, stroke (width of letter's lines); d, separation between words; e, separation between letters; f, separation between lines (lead).

The major categories of safety symbols are:

Hazard Alert. If a surrounding shape is used, the symbol should be drawn within a yellow equilateral triangle.

Mandatory action. If a surround shape is used, the image is white within a solid safety blue or safety black circular surround.

Information. If a surround shape is used, it should be placed within a rectangle.

Prohibition. A surround shape is mandatory. The shape is a circle with a 45-degree slash from the upper left to the lower right. The image is black with safety black or safety red circle and slash (Figure 39.10).

Additional requirements are realistic proportions of figures and avoidance of narrow forms, symmetry of figures vertically and horizontally, the direction of focus of the image congruous with the physical layout of the environment, solidly drawn forms rather than an outline form, simplicity of detail with minimal embellishments, adequate symbol size to achieve legibility, and placement of the warning within the field of view and near the hazard.

The standard annexes provide procedures for drawing standardized stick figures and testing group responses to proposed new safety symbol designs.

FIGURE 39.10 Prohibition: Smoking.

ANSI Z 535.4-2002: Standard for Product Safety Signs and Labels

This standard indicates that a product safety sign or label is a sign, label, cord tag, or decal affixed to a product that provides information about that product. A product safety sign or label can be of a permanent nature—not easily removed and warns of a perpetual inherent hazard—or of a temporary nature and warns of a temporary hazard related to the product. The temporary sign is removed when the hazard no longer exists.

The principles of Z 535.2-1998 generally coincide with this standard. There is additional information related to requirements for letter height of printed material used on a small or larger sign in this standard.

Recommended Letter Height for Favorable Reading Conditions
2 ft. or less: Viewing Distance (in.)/150
>2–20 ft.: Viewing Distance (ft.-(2) × 0.03 + 0.16)
over 20 ft.: Viewing Distance (ft.)/28.6

Recommended Letter Heights for Unfavorable Reading Conditions
Viewing Distance (ft.) × 0.084

ANSI Z 535.5-2002: Standard for Safety Tags and Barricade Tapes

This standard is intended to provide warnings about temporary hazards. These warnings are not intended to remain in place as permanent but must be removed when the hazard is neutralized. These warnings, which are usually in the form of attachable tags and barrier tapes, graphically follow the principles found in Z 535.2 and Z 535.4, with the color implementation according to Z 535.1. Typically, a tag or tape contains a signal word and a message panel that contains a hazard description and/or an evasive action. Tags must be hooked as close to the hazard as possible and must not detract from other tags. Tapes should be placed in a way that gives adequate evasion time to the viewer. A table for letter heights is provided. The requirement in this table approximates the letter height requirements of Z 535.2.

ANSI Z 129.1-2000: Hazardous Industrial Chemical Precautionary Labeling

This standard provides guidelines for the preparation of precautionary information about hazardous chemicals used in the industry. This information is in the form of labels placed upon containers.

To provide warnings related to immediate and delayed physical and health hazards, a number of label characteristics are expected to be considered:

Identification of the Chemical Product or Critical Components

Signal word: DANGER, WARNING, CAUTION for immediate hazard
NOTICE and ATTENTION for delayed hazard statement of hazard(s)
Precautionary measures
Instructions in case of contact or exposure
Antidotes
Notes to physician
Instruction in case of fire
Instruction in case of spill or leak
Instructions for container handling and storage
References: also, additional label, bulletins, and material safety data sheets
Name, address, and telephone number of the manufacturers

UNDERWRITERS LAB (UL) 969-1995: MARKING AND LABELING SYSTEMS

This standard does not deal with the graphic aspects of a warning; it deals with the physical manufacturer and testing of adhesive attached labels.

ISO (INTERNATIONAL ORGANIZATION FOR STANDARDS) 3864-1984

The stated purpose of this standard is to prescribe safety colors and safety signs to prevent accidents and health hazards.

Safety colors required are red for stop, prohibition, and firefighting equipment; blue for mandatory action; yellow for caution or risk of danger; and green for safe conditions.

Contrasts in color are: safety red/white, safety blue/white, safety yellow/black, and safety green/white.

The geometric shapes of signs are circle for prohibition or mandatory action; equilateral triangle for warnings; and square or rectangle for information and instructions.

Prohibition signs, in addition to having a circular configuration, must have a 45° slash from 11 o'clock to 5 o'clock. Warning signs that do not have a suitable image that can be created should instead have an exclamation point.

The effective viewing distance (the greatest distance that a safety sign can be understood) is defined as

$$A > \frac{L^2}{2000},$$

where A is the minimum area of the sign in square meters and L is the distance in meters.

U.S. FEDERAL LEGISLATION: CONSUMER PRODUCT SAFETY COMMISSION (CPSC)

Various U.S. laws related to labels and warnings exist. 16 CFR Section 1500.121—which was created by the CPSC—appears to provide the most comprehensive design standards for labeling. This section repeats requirements from the Federal Hazardous Substances Act which are:

Signal Word: DANGER, CAUTION, WARNING
Statement of Hazard
Name of Substance
Name and address of manufacturer and supplier
Precautions to follow
Instructions
Special handling and storage
First aid instructions

The section calls for conspicuity, legibility, and contrast in type. Cautionary text must be aligned parallel to the base of the container. The label area must be adequate. A table is included, which provides the type size required in relation to the area of the label. Letter height should be no more than three times the letter width.

REFERENCES

Brauer, R. L. (1990). *Safety and health for engineers*. New York: Van Nostrand Reinhold.
Cacha, C. A. (1997). *Research design and statistics for the safety and health professional*. New York: Van Nostrand Reinhold.

Laughery Sr., K. R., Wogalter, M. S., & Young, S. L. (1991). *Human factors perspectives on warnings.* Santa Monica, CA: Human Factors and Ergonomics Society.

Roland, H. E., & Moriarty, B. (1983). *System safety engineering and management.* New York: Wiley.

APPENDIX I DEFINITIONS

Authority. Those in charge of a process, property, or organization who would devise a warning if so required.

Recipient. An individual who reads and is expected to respond to a warning.

Hazard. A physical circumstance that is capable of causing injury, illness, or property damage. Risk. A degree of probability that a hazard will cause an injury, illness, or property damage. Frequency. The number of unwanted events caused by a hazard over a specified period of time. Severity. The degree of injury, illness, or property damage resulting from an unwanted event. Warning. A stimulus conveying a communication to a recipient that the recipient is exposed to a proximate or long-range hazard.

APPENDIX II STANDARDS

Z 535.1-2002	Z 129.1-2000
Z 535.2-1998	UL 969-1995
Z 535.3-2002	ISO 864-1994
Z 535.4-2002	16 CFR SECTION 1500.121
Z 535.2-2002	

40 Record Keeping and Statistics in Safety and Risk Management

Charles A. Cacha

CONTENTS

INTRODUCTION

Medium- to large-size organizations usually establish a safety or risk management program that is designed to control the incidence of injuries and illnesses to employees. The Safety or Risk Managers who operate this program implement procedures such as safety training, safety motivation, safety inspections, and hazard abatement as integral parts of the program. By necessity, the Safety or Risk Managers are concerned about whether their program is effectively protecting the organization's employees. The criteria for testing the validity and success of a Safety and Risk Management Program have traditionally been the number (frequency) of injuries and illnesses, as well as the severity of injury and illnesses occurring to employees over a given period. This chapter describes the traditional statistical method used to validate an existing Safety and Risk Management Program and prognosticate its future success.

SCOPE AND OBJECTIVES

First, several forms of record keeping and statistics that have been traditionally used by safety and risk management professionals are described. Second, the currently prescribed method of record-keeping and statistics, as required by the U.S. Department of Labor, is presented. The traditional and current methods are similar, but a distinction is made between them. Finally, some statistical techniques in making comparisons between incident frequencies over time are provided.

Injuries and illnesses that occur within an organization are manifested by incidents known as unwanted events. These unwanted events evidenced by injuries and illnesses are usually open and obvious, detectable by the management, and are mutually exclusive of each other and may, thus, be enumerated as they occur over time. The enumeration is subdivided by units of time—generally one year, although larger or smaller units such as months may also be used. It is generally agreed that the safer an organization is, the smaller the number of unwanted events that will occur within a given unit of time. This small number also reflects the successful or unsuccessful efforts of the Risk Management Professionals. The number of injuries and illnesses occurring over a unit of time is referred to as frequency. In addition to the concept of frequency, another concept is applied—severity. Severity indicates the overall extent of seriousness of injuries across all unwanted events occurring within a unit of time. The seriousness is usually measured temporarily by the length of incapacitance of all employees who are injured or sickened. The formulas for frequency and severity follow.

FREQUENCY

Logically, to calculate injury and illness frequency, three variables should be considered: (a) the number of injuries and illnesses that occurred, (b) the amount of temporal exposure of all workers within the organization to their various job duties, and (c) an interval of time during which the organization had been operating.

The number of injuries and illnesses is generally determined by counting the number of reports generated by supervisors describing injury and illness occurrences that require medical aid beyond the ordinary perfunctory first aid treatment. This number may apply to the entire organization or to subdivisions of the organization.

The amount of exposure to workers is determined by the total number of hours expended by all workers while working for the organization and performing their required functions over an interval of time. The interval of time is a temporal segment, usually a year, which is uniformly observed in the present and into the future and may be also observed from existing records of the past.

The formula for frequency is

$$\text{Frequency} = \frac{\text{Number of Injuries and Illnesses}}{\text{Total Hours Expended by all Employees}}$$

For example, an organization experienced a total of 21 injuries and illnesses over the course of one year. Its employees worked a total of 90,000 hours during that year. Its frequency rate is

$$\frac{21}{90,000} = 0.00023.$$

SEVERITY

Although the frequency statistic may provide considerable evidence related to safety or risk management efforts, additional data collection may provide even stronger indications. Statistics may be calculated, which indicates how critical unwanted events have been. This criticality is measurable by the number of days all workers have lost. This measure is based on the consideration that the more time lost, the more severe the injury (e.g., a worker with a broken leg will most likely lose more time from the job than will a worker with a broken finger):

$$\text{Severity} = \frac{\text{Number of Days Lost by all Workers}}{\text{Total Hours Expended by all Workers}}$$

For example, an organization experienced a total of 68 days lost from the job among all injured employees during the year. Its employees worked a total of 90,000 hours. The severity rate is

$$\text{Severity Rate} = \frac{68}{90,000} = 0.00076.$$

Multiplier

The aforementioned two techniques have, in the past, been exposed to a multiplier to create a non-decimal result and relate them to realistic circumstances. This multiplier was originally 1,000,000, which is derived from 500 workers working 50 weeks a year at 40 hours per week. This multiplier was recommended by the American National Standards Institute (ANSI) Z16.1. The multiplier is currently 200,000, which is derived from 100 workers working 50 weeks a year at 40 hours per week.

The following formulas have been or are currently being used:

Prior Method

$$\text{Frequency:} \frac{\text{Number of Injuries and Illnesses}}{\text{Total Hours Worked by All Workers}} \times 1,000,000.$$

$$\text{Severity:} \frac{\text{Number of Days Lost All Workers}}{\text{Total Hours Worked by All Workers}} \times 1,000,000.$$

Present Method

$$\text{Frequency:} \frac{\text{Number of Injuries and Illnesses}}{\text{Total Hours Worked by All Workers}} \times 2,000,000.$$

$$\text{Severity:} \frac{\text{Number of Days Lost by All Workers}}{\text{Total Hours Worked by All Workers}} \times 2,000,000.$$

Prior Method

Examples from prior formulas are:

$$\text{Frequency:} \frac{21 \text{ Injuries and Illnesses}}{90,000 \text{ hours}} \times 1,000,000 = 233.$$

$$\text{Severity: } \frac{68 \text{ days lost}}{90,000 \text{ hours}} \times 1,000,000 = 756.$$

Current Method

$$\text{Frequency: } \frac{21 \text{ Injuries and Illnesses}}{90,000 \text{ hours}} \times 2,000,000 = 47.$$

$$\text{Severity: } \frac{68 \text{ days lost}}{90,000 \text{ hours}} \times 2,000,000 = 151.$$

FREQUENCY RELATED TO SEVERITY

These two concepts have dissimilar characteristics, but they are related. Some professionals may argue that frequency has greater importance because all unwanted events have the potential to be a severe event. Other professionals may argue that severity has greater significance because severity reflects a large amount of human suffering and large financial losses to the organization. Attempts have been made to integrate frequency and severity rates into one:

$$\text{Severity/Frequency Ratio} = \frac{\text{Severity Rate}}{\text{Frequency Rate}}$$

An example from prior examples is:

$$\text{Severity/Frequency Ratio: } \frac{151}{47} = 3.2.$$

Whereas frequency and severity rates are used for an entire organization or major subdivisions of an organization, the severity/frequency ratios are often calculated for smaller divisions of the organization. The severity/frequency ratio determined for a small division, if high, is a signal to the safety or risk management department that serious occurrences with severe consequences are happening and prompt safety engineering practices should be applied for the purpose of hazard abatement.

RECORD-KEEPING STANDARDS

In the past, the appropriate standards were promulgated by the American National Standards Institute (ANSI). These standards follow:

ANSI Z16.1-1967: A METHOD OF RECORDING AND MEASUREMENT
WORK INJURY EXPERIENCE

This standard deals with traditional methods of calculating frequency and severity rates. Rates are based on 1,000,000 worker hours. It also provides schedules of lost days for deaths and permanent injuries that are calculated for severity rates and may be used by private organizations; however, this has been superseded by OSHA prescribed procedures.

ANSI Z16.2-1962: METHOD OF RECORDING BASIC FACTS RELATING TO THE NATURE AND OCCURRENCE OF WORK INJURIES

This standard deals with procedures for analyzing and categorizing incidents and accidents so that managerial or engineering controls may abate hazards. Procedures are still applicable in current safety and risk management practices. Major categories in analyzing incidents and accidents are:

- Nature of Injury: cuts, lacerations, etc.
- Part of Body Affected: finger, head, etc.
- Source of Injury: machine, tool, vehicle, etc.
- Accident Type: struck by, struck against, falls, etc.

ANSI Z16.3-1973: METHOD OF RECORDING AND MEASURING THE OFF-THE-JOB DISABILITY ACCIDENTAL INJURY EXPERIENCE OF EMPLOYEES

This method analyzes injuries of employees off the job across three categories: transportation, home, and public places. The analyzes are presented as a percentage of total off-the-job injuries:

$$\frac{\text{Number of Injuries in the Category} \times 100}{\text{Total Number of Off}-\text{the}-\text{Job Injuries}} = \%$$

An overall frequency rate for off-the-job injuries:

$$\text{Frequency: } \frac{\text{Number of Injuries} \times 200,000}{312 \times \text{Average Number of Employees} \times \text{Number of Months}}$$

The number 312 is an estimate of the number of hours, monthly, spent off the job but not sleeping.

ANSI Z16.4-1977: UNIFORM RECORD-KEEPING FOR OCCUPATIONAL INJURIES AND ILLNESSES

This standard tracks the current OSHA record-keeping requirements that will be described next.

CURRENT REQUIREMENTS

Presently, the U.S. Occupational Safety and Health Administration (OSHA) requires employers with more than ten employees to use, among other forms, OSHA Form 300 in recording injuries. OSHA 300 is a log that requires the identity of the employee, job title, date of incident, a brief description of injury or illness, part of body affected, and object causing injury. Also required are the indication of whether the worker died, the days away from work, and job transfer due to injury or illness. A line entry for each injury or illness is made. An entry (also known as a recording) is made only if an injury or illness resulted in death, loss of consciousness, days away from work, restricted work activity, or job transfer or medical treatment beyond first aid. The required method in calculating the frequency rate (referred to by OSHA as the incidence rate) is to add up all line entries on the log, multiply by 200,000, and then divide by the hours worked by all employees. For example:

$$\text{Incidence Rate} = \frac{\text{Total Injury and Illness Line Entries} \times 200,000}{\text{Hours Worked By All Employees}}$$

$$\text{Example:} \quad \frac{\text{Total Entries: } 28 \times 200{,}000}{\text{Hours Worked By All Employees: } 100{,}000} = 56.$$

Severity Rate (referred to by OSHA as DART) is based on injury or illnesses that caused two or more days away from work, or which ended in two or more days of restricted job activity or job transfer:

$$\text{Example:} \quad \frac{\text{Days Lost Cases (8)} + \text{Restricted Work Cases (2)} \times 200{,}000}{100{,}000} = 20.$$

The details for this OSHA record-keeping procedure are explained in the 29 CFR Part 1964. Considering issues of legality, this procedure should be regarded as the future state in record keeping.

There are a number of statistical techniques that may help infer any substantial changes in frequencies or severities currently or over time. One of these is demonstrated here.

CONTROL CHARTS

These instruments are used to track and observe changes in frequencies over time. Observing any particularly large beneficial or detrimental change in frequency will aid the safety or risk manager to determine the success or lack of success of his or her safety or risk management program. To decide whether a change is large to a point of significance will depend on the application of a statistical technique. The following technique was acquired from the Techniques of Safety Management by Dan Petersen (1978). A sequence of computational steps follows. It is recommended that at least 20 time intervals be available for study and calculation.

Step 1. Add up all frequencies (at least 20 time intervals) ΣX.
Step 2. Add up all hours worked (at least 20 time intervals) ΣH.
Step 3. Drive a percentage from above: $\frac{\Sigma X}{\Sigma H} = P$.

$$\text{Upper Control Limit (UCL)} = P + 2.576\sqrt{\frac{P(1-P)}{\Sigma H}} \text{ at 99\% Confidence}$$

$$\text{Lower Control Limit (LCL)} = P - 2.576\sqrt{\frac{P(1-P)}{\Sigma H}} \text{ at 99\% Confidence}$$

$$\text{Upper Control Limit (UCL)} = P + 1.96\sqrt{\frac{P(1-P)}{\Sigma H}} \text{ at 95\% Confidence}$$

$$\text{Lower Control Limit (LCL)} = P - 1.96\sqrt{\frac{P(1-P)}{\Sigma H}} \text{ at 95\% Confidence}$$

Judgment will determine the choice of the 99% or 95% confidence level. This technique is usually performed monthly. Any month's entry on the chart that exceeds the UCL should cause concern to the safety or risk manager who should thereupon apply remedial actions to the physical or managerial environment. Likewise, any month's entry that appears below the LCL should indicate to the safety or risk manager that injury and illness control efforts have been particularly effective and that such efforts should be examined and repeated.

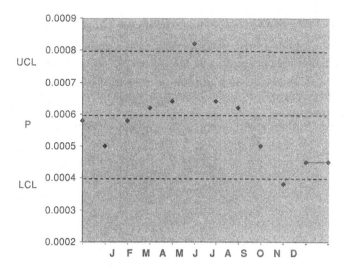

FIGURE 40.1   Year 2001 annual control chart accident frequencies.

For example: over the last 20 months, an organization has been record keeping according to OSHA requirements. A total of 60 injuries and illnesses occurred during that time. Also, in the last 20 months, the employees worked a total of 100,000 hours over 20 months.

Step 1. Add up all injuries and illnesses over 20 months: 60.
Step 2. Add up all hours worked over 20 months: 100,000.
Step 3. Derive a percentage 60/100,000:0.0006.

$$UCL = 0.0006 + 2.576\sqrt{\frac{0.0006(1 - 0.0006)}{100,000}} = 0.0008.$$

$$LCL = 0.0006 - 2.567\sqrt{\frac{0.0006(1 - 0.0006)}{100,000}} = 0.0004.$$

Enter on the control chart UCL as 0.0008 × 200,000 and 0.0004 × 200.000 as LCL The average is computed by LCL − UCL0004 × 200,000 as LCL. The average is computed by LCL-UCL÷2 × 200,000 ÷ 2 × 200.000 (Figure 40.1).

See the following figure as an example of future entries. One month (June) demonstrates a particularly undesirable circumstance and another, (October) a particularly desirable one.

REFERENCE

Petersen, D. (1978). *Techniques of safety management.* New York: McGraw Hill.

APPENDIX I: DEFINITIONS

Frequency. The number of unwanted events caused by a hazard over a specified period
  Severity. The degree of injury or illnesses resulting from an unwanted event
  Record Keeping. A methodical recording of unwanted events and their characteristics as they occur
  Control Chart. A series of graphic plots representing unwanted events over time

APPENDIX II: STANDARDS

ANSI Z16.1-1967: A method of recording and measuring work injury experience

ANSI Z16.2-1962: Method of recording basic facts related to the nature and occurrence of work injuries

ANSI Z16.3: Method of recording and measuring the off the job disabling accidental injury experience of employees

ANSI Z16.4-1997: Uniform record keeping for occupational injuries and illnesses

41 User Needs in Standards—Older People and People With Disabilities

Anne Ferguson

CONTENTS

INTRODUCTION

Older people make up an increasing percentage of the world population. A consequence of becoming older is a decrease in various functional abilities such as hearing, vision, and mobility. Younger people may also have functional impairments either from birth or as a result of accidents or illnesses. In a society that recognizes the importance of equal opportunities, it follows that products, services, workplaces, and environments should be accessible to a wide variety of users and must be designed to consider a range of user needs, including those of older people and people with disabilities.

In the standards world in the past, the needs of people with disabilities have predominantly been considered in the development of specific standards in the area of assistive technology, such as in the international standards for wheelchairs (ISO, various). The awareness of the need to build accessibility for a wider population has also been covered by various national and international standards, such as *Building Construction—Needs of Disabled People in Buildings—Design Guidelines* (International Organization for Standardization ISO, 1994). However, the needs of people with functional impairments, including older people, have not been adequately addressed in the writing or revision of standards for everyday products and services.

National legislation, such as the American with Disabilities Act ADA. (1991) in the United States (U.S.) or Disability Discrimination Act DDA. (1995) in the United Kingdom (U.K.), has been instrumental in increasing awareness of needs. Some national standards organizations, such as the Japanese Standards Association (Japanese Standards Association JSA, 2000a, 2000b, 2000c), have made progress in this area with guidelines on, for example, the usability of consumer products.

The international ISO Guide 71 (2001)—*Guidelines for Standards Developers to Address the Needs of Older Persons and Persons with Disabilities*—is intended to help standard writers ensure that all cater to the widest possible group of users, particularly older people and those with disabilities. It is, thus, an essential ergonomics standard and useful tool for designers and evaluators of products and services. It is one of several guides produced that are also of relevance to designers, ergonomists, and others. These include ISO/IEC Guide 51:1999 *Safety Aspects—Guidelines for Their Inclusion in Standards*, which covers the concept of safety and tolerable risk in fairly general terms; ISO/IEC Guide 50:2002 *Safety Aspects—Guidelines for Child Safety*, which describes a hazard-based approach to ensuring standards take account of child safety; and ISO/IEC Guide 37:1995 *Instructions for Use of Products of Consumer Interest*, which sets out the basic criteria for consumer instructions and includes, in an informative annex, a method for assessing instructions together with "non-comprehensive" compliance and evaluative checklists.

This chapter briefly describes the process which resulted in Guide 71. A description of the contents follows and then an illustration of how Guide 71 may be used to promote and ensure accessible design in products and services is given.

DEFINITION OF RELEVANT TERMS

Accessible design—guidelines for the design of consumer products to increase their accessibility for persons with disabilities or who are aging (Trace R & D Center of University of Wisconsin-Madison, 1992)

Alternative format—different presentation or representation intended to make products and services accessible through a different modality or sensory ability (ISO Guide 71)

Assistive technology—equipment, product system, hardware, software, or service that is used to increase, maintain, or improve the functional capabilities of individuals with disabilities (ISO Guide 71)

Universal design—the design of products and environments usable by all people, to the greatest extent possible, without the need for specialized design (North Carolina State University Center for Universal Design)

FULL NAME OF THE STANDARDS (GUIDELINES)

Guidelines for standards developers to address the needs of older persons and persons with disabilities. ISO/IEC Guide 71:2001

BACKGROUND

In May 1998, the Japanese Standards Association proposed to the ISO Committee on Consumer Policy [COPOLCO] that guidelines should be drawn up to address the needs of older people and people with disabilities. A policy statement was prepared by an international group in a series of meetings in Tokyo, Geneva, Washington DC, Toronto, and Paris and accepted by the Council of The International Organization for Standardization [ISO] and the International Electrotechnical Commission [IEC] in 2000. This statement (ISO/IEC, 2000) draws attention to the commercial benefits as well as to the gains for the Society in improving access to products, services, and environments. It sets out the three basic principles required to achieve the inclusion of the needs of older persons and people with disabilities in standards production and revision work. These are:

- Universal or Accessible Design
- Consumer representation of older persons and people with disabilities
- Relevant information exchange

The policy statement describes how these principles may be achieved through raising awareness, particularly by taking account of the subsequently prepared Guide 71, ensuring links between research programs and standardization, and increasing the availability of the standards in accessible formats.

ISO/IEC Guide 71

Guide 71 was drafted by a working group comprised principally of experts in the fields of disability and human factors, with subsequent input from standards organizations to ensure a document usable by standard writers. It was published in 2001 and has been widely accepted. For example, in Japan, technical criteria to be incorporated in IT equipment, software, and digital home appliances have been prepared, which are consistent with Guide 71. Also, two European standardization bodies, CEN and CENELEC, adopted and re-published Guide 71 as CEN/CENELEC Guide 6 as the prime means of enacting a mandate by the European Commission (Mandate M/283 Safety and usability of products by people with special needs), which required the production of a guidance document in the fields of safety and usability of products by people with special needs, such as the elderly and disabled people.

The third European standards body obliged to enact EC Mandate M/283 is the European Telecommunication Standards Institute [ETSI], which will use Guide 71 as a reference document. The IEC technical committee that deals with domestic electrical appliances (TC 61: Household and Similar Electrical Appliances—Safety) has now included a reference to Guide 71 in its Strategic Policy Statement; it is envisaged that other Technical Committees will also take similar steps.

ISO Guides are normally available, in printed form, in English and French. However, in keeping with the "accessible design" principles of the Guide, an English Braille version can be obtained—at the same price as the printed document—from the ISO Central Secretariat in Geneva. This is the first time an ISO publication has been made available in Braille.

Content

The introduction to Guide 71 recognizes that standards bodies have, for many years, addressed the needs of persons with disabilities in the development of specific standards around assistive technology and accessible building design, but draws attention to the fact that the needs of older people and people with disabilities are not being adequately met when standards for everyday products or services are written or revised. Thus, the Guide lays down a process that can be followed to help do this.

The very first definition in the Guide is for ergonomics or human factors, which uses the widely quoted words of Christensen, Topmiller, and Gill (1988): "that branch of science and technology that includes what is known and theorized about human behavioral and biological characteristics that can be validly applied to the specification, design, evaluation, operation, and maintenance of products and systems, to enhance safety, and effective and satisfying use by individuals, groups, and organizations."

Later, the Guide makes clear that ergonomic factors should be addressed when developing solutions to possible conflicts between safety and usability, using the example of child-resistant closures on medicines, which may make it more difficult to open for older persons who no longer see as well or have reduced strength or dexterity.

The Guide has four key sections. The first of these, Clause 6 *Developing Standards—Issues to Consider During the Standards Development Process*, suggests a process standard writers may use to address the needs of older persons and persons with disabilities when drafting a new standard or at each revision of an existing one. This requires definition of the standards project—in terms of factors such as the potential end-users—ensuring that the committee is well equipped with relevant background information and experts, in relation to awareness of aging and disability issues, injury data, and so forth, and making sure that the process of preparing the standard, its review, and eventual publication are all in a form accessible to the group under consideration—that is, older people and people with disabilities.

Clause 7, *Tables of Factors to Consider to Ensure Standards Provide for Accessible Design*, provides seven tables to help identify factors that will affect the use of a product, service, or environment and consider their significance for persons with different abilities. Each table identifies a clause or section typically found in international standards: these are Information,

Factors to consider in standards clauses on information (labelling, instructions and warnings)	Human abilities				
	9.2 Sensory				
	Seeing 9.2.1	Hearing 9.2.2	Touch 9.2.3	Taste/smell 9.2.4	Balance 9.2.5
8.2 Alternative format	▓	▓	▓		
8.3 Location/ layout	▓				▓
8.4 Lighting/glare	▓				
8.5 Colour /contrast	▓				
8.6 Size/ style of font	▓				
8.7 Clear language	▓				
8.8 Symbols/drawings	▓				
8.9 Loudness/pitch		▓			
8.10 Slow pace		▓			
8.11 Distinctive form	▓		▓		
8.12 Ease of handling	▓				▓
8.13 Expiration date marking	▓			▓	
8.14 Contents labelling	▓			▓	
8.15 Surface temperature			▓		
8.16 Accessible routes	▓				▓

FIGURE 41.1 Extracted from Table 1 of Guide 71 on "Factors to consider in clauses in information."

Packaging, Materials, Installation, User Interface, Maintenance, Storage and Disposal, and Built Environments (buildings). Figure 41.1 provides an extract from Table 1 of the Guide, showing how the factors are displayed against sensory ability. In the Guide, all seven tables have columns for sensory ability as well as physical and cognitive abilities, together with a column for "allergy." Although not typically recognized as a "disability," allergies can impose limitations on an individual's activities and, in some cases, be life-threatening.

Within each table, the user may look up the factors—such as "lighting/glare" or "surface finish"—that are particularly significant for users with particular disabilities, and then use the remaining key sections, Clauses 8 and 9, to find out more on why particular factors are important or—in general terms—what action can be taken. For example, if "ease of handling" was identified as a factor to consider—as it is in column 1, under "seeing" (see Figure 41.1)—it could be looked up under Clause 8 *Factors to Consider* to find guidance, spelling out the need to select effective color combinations for the presentation of information or consider the size, shape, and mass of a product in relation to the ease of handling.

Clause 9, *Detail About Human Abilities and the Consequences of Impairment*, provides a brief definition and description of different abilities, together with information on the effects of aging and the practical implications of impairment. Examples of hazards from which older persons and people with disabilities are more at risk because of their functional limitations are given.

For example, under "seeing," the section on effects of aging lists changes such as "loss of visual acuity," "reduced field of vision," and "sensitivity to light." Under risk of hazards, "sharp points" and "hot surfaces" appear.

The final section of Guide 71 is a Bibliography, which offers a list of sources that standard writer can use to investigate more detailed and specific guidance materials with respect to accessible design. Within Europe, a more extensive Bibliography on this topic was prepared, with funding from Mandate 283 (1999) referred to earlier. This covers Healthcare; Personal Care and Protection; Personal Mobility; Housekeeping/Household Equipment; Furnishings; Machinery and Tools, Handling Products and Goods; Communication, Information, and Signalling (ICT); Buildings and Interiors; Outdoor Environment; Traffic and Transportation; Recreation; and general information on accessibility and ergonomics.

Application of Guide 71

The Guide is intended to be used when a standard is first prepared or revised. Some work has already been undertaken by consumer representatives in the U.K. using Guide 71 to assess the effectiveness of existing standards in ensuring access for people with disabilities (BSI-CPC, 2003). The review was initially carried out in two parts of the electrical safety standard covering particular requirements for toasters and dishwashers—BS EN 60335 (1990, 1996) and their IEC equivalents—and has since been extended to other areas. The use of the tables in Section 7 of Guide 71, together with the products themselves, assisted in identifying the significant factors to consider in relation to different disabilities and possible problems in the current versions of the standards that could be considered in future revisions. For example, a possible change to the dishwasher standard resulting from the exercise would be to make explicit that dishwashers should be fitted with an "off" switch to control normal operation so that users—including those who are partially sighted—could confirm that the machine has stopped. The requested changes have been presented to the U.K. Technical Committee and it is hoped that, in due course, the European and international standards will be modified.

Technical committees and working groups are already beginning to consider Guide 71 in the production of standards and guidance. For example, reference to the Guide by the relevant international working group (ISO/TC 176 SC3 WG10, 2002a) helped ensure that the need for alternative formats for consumer information was made clear in the drafting of guidance on complaints handling. Also, the European working group (CEN TC 224 WG 6: Man-Machine Interface, 2003) compiling guidance on design for accessible card-activated devices, such as public banking and ticket machines, is incorporating definitions and concepts from Guide 71.

The probable consequence of Guide 71 and a range of national legislation in the area of disability is that more standards that consider the needs of particular groups—such as older people and people with disabilities—will follow in the future.

REFERENCES

American with Disabilities Act (ADA). (1991). Washington DC, USA.

British Standards Institution (BSI). (1990). *Specification for safety of household and similar electrical appliances. General requirements.* UK: BS 3456-201:1990, EN 60335-1:1988.

British Standards Institution (BSI) (1996a). *Specification for safety of household and similar electrical appliances. Particular requirements. Dishwashers.* UK: BS EN 60335-2-5:1996.

British Standards Institution (BSI). (1996b). *Specification for safety of household and similar electrical appliances. Particular requirements. Toasters, grills, roasters and similar appliances.* UK: BS EN 60335-2-9:1996.

British Standards Institution Consumer Policy Committee (BSI CPC). (2003). *Personal communication.*

Christensen, J. M., Topmiller, D. A., & Gill, R. T. (1988). Human factors definitions revisited. *Human Factors Society Bulletin, 31,* 7–8.

Disability Discrimination Act (DDA). (1995). UK.

European Committee for Standardization (CEN). (2003). CEN TC 224 WG 6: Man-Machine Interface. *Personal communication.*

European Committee for Standardization (CEN) and European Committee for Electrotechnical Standardization (CEN-ELEC). (2002). *Guidelines for standards developers to address the needs of older persons and persons with disabilities.* CEN/CENELEC Guide 6.

International Organization for Standardization (ISO). (various). *Wheelchairs.* Switzerland: ISO 7176 series.

International Organization for Standardization (ISO). (1994). *Building construction—Needs of disabled people in buildings—Design guidelines.* Switzerland: ISO/TR 9527:1994.

International Organization for Standardization (ISO) and International Electrotechnical Commission (IEC). (1995). *Instructions for use of products of consumer interest.* Switzerland: ISO/IEC Guide 37:1995.

International Organization for Standardization (ISO). (2002a). *ISO/TC 176 SC3 WG10 Complaints handling. Personal communication.*

International Organization for Standardization (ISO). (2002b). *Technical aids for disabled persons—Classification.* Switzerland: ISO 9999:2002.

International Organization for Standardization (ISO) and International Electrotechnical Commission (IEC). (1999). *Safety aspects—Guidelines for their inclusion in standards.* Switzerland: ISO/IEC Guide 51:1999.

International Organization for Standardization (ISO). (2000). *Addressing the needs of older persons and people with disabilities in standardization work.* Switzerland: ISO/IEC Policy Statement. 2000.

International Organization for Standardization (ISO) and International Electrotechnical Commission (IEC). (2001). *Guidelines for standards developers to address the needs of older persons and persons with disabilities.* Switzerland: ISO/IEC Guide 71:2001.

International Organization for Standardization (ISO) and International Electrotechnical Commission (IEC). (2002). *Safety aspects—Guidelines for child safety.* Switzerland: ISO/IEC Guide 50:2002.

Japanese Standards Association (JSA). (2000a). *Guidelines for all people including elderly and people with disabilities—Usability of consumer products.* Japan: JIS S 0012:2000.

Japanese Standards Association (JSA). (2000b). *Guidelines for all people including elderly and people with disabilities—Marking tactile dots on consumer products.* Japan: JIS S 0011:2000.

Japanese Standards Association (JSA). (2000c). *Guidelines for all people including elderly and people with disabilities—Packaging and receptacles.* Japan: JIS S 0021:2000.

Section VIII

Military Human Factors Standards

42 The Role of Human Systems Integration Standards in the Modern Department of Defense Acquisition Process

Joe W. McDaniel[1] and Gerald Chaikin[2]

CONTENTS

INTRODUCTION

Standardization reform was the cornerstone of acquisition reform. Many believe that all military specifications and standards are gone or that they cannot be used. However, the human engineering standardization documents that were streamlined and consolidated during the process of standards reform were revalidated as important to military acquisition.

Because standards and guidelines should be used with the understanding of when and how they were developed, this chapter discusses the history of the human engineering standardization documents—primarily specifications, standards, handbooks, and data item descriptions—and how they evolved into today's forms and how the currently approved ensemble can best be used.

STANDARDS REFORM

By the early 1990s, the high-tech industry and Department of Defense (DoD) viewed military specifications and standards as pariahs. They were seen as adding cost to government purchases and burdening industry to the point of compromising competitiveness nationally and internationally. Examples of low-cost, highly capable consumer products include rapidly evolving electronic and computer products. Although the capabilities of such products are growing at explosive rates, the human operator and maintainer are not evolving. The military standards for human factors engineering (HFE) and human systems integration (HSI) remain valid and effective.

THE CASE AGAINST MILITARY STANDARDS

Spearheaded by the electronics industry, a consortium of defense industry groups succeeded in persuading the DoD to avoid using military specifications and standards or to require special permission from high-level officials when using. How and why did this happen? State-of-the-art digital electronics technology has been undergoing a rapid and dramatic evolution. The last quarter of the 20th century saw a progression from single transistors to chips containing ten transistors, and then 1,000, and then one million, and so forth—with no end in sight. This evolution in technology has been so rapid that the military specifications and standards through which the government bought these products had become obsolete and imposed a severe burden on the electronics industry. In many cases, electronic companies required separate assembly lines for their commercial and military products. Moreover, the worldwide market for consumer electronic products now demanded more capability at a lower cost.

Back in the heyday of the space program, the government led R&D in microelectronics and was, by far, the largest customer. Today, the demands of the world marketplace are so vast that the government is no longer a driving force in the technology. The portion of U.S.-made semiconductors bought by the government declined from about 75% in 1965 to 1% in 1995. To make matters worse, the military standards for soldered circuit boards required the use of chlorinated fluorocarbons (CFCs)—known ozone-depleting chemicals—to clean the flux off the boards. Clearly, the electronic specifications and standards were obsolete, burdensome, and required the use of banned substances.

THE REPORT OF THE PROCESS ACTION TEAM

Responding to complaints from the electronics industry, DoD formed a Process Action Team (PAT) to study the problem and make recommendations. The PAT (1994) reported that the Pentagon "*does not have the ability to subsidize increasingly inefficient defense operations that do not have a self-sustaining market base.*" The PAT called for revisions to "*permit reliance on commercial products, practices, and processes.*" Describing the problems with military standards, the PAT said, "the difference between DoD and other major buyers, however, is that the military specifications and standards do not always stop at specifying what is required. Frequently, they also describe how to make a product, indeed, the one acceptable way to make it!" It recommended favoring the use of non-government over military specifications and standards.

The PAT recommended that military needs be stated as performance specifications (tell what you want, not how to do it). Stating this would solve the technical obsolescence problem and save $500 million in over two years. The PAT favored the use of state-of-the-art products and processes, including the best commercial practices and technologies, to achieve lower acquisition costs. It suggested the following test for a military standardization document: does it impede or facilitate modem manufacturing processes? Does it allow us to do things cheaper?

The DoD has always used non-government standards (NGSs). The 1994 DoD Index of Specifications and Standards (DODISS) included over 5,600 NGSs among 49,000 documents. Military standards also incorporated by reference about 5,000 NGSs.

POLICY CHANGES: WHAT THE SECRETARY OF DEFENSE DID

The Post-Cold War era has reduced the inflation-adjusted military acquisition budget from 40% to 66% (percentage varies according to source), with further reductions possible. DoD officials said that they cannot carry on business as usual, but that radical changes were needed to match the economic and technical realities of modem times. The DoD had been paying higher prices when lower-cost commercial alternatives existed and felt that they must reduce the cost of buying while preserving the defense-unique core capabilities. DoD-unique product and process specifications and standards were often identified as barriers to the industry doing business with the DoD. Generally accepting the recommendations of the PAT, (then) Secretary of Defense Dr. William J. Perry issued a policy memorandum on June 29, 1994 that stated:

> Performance specifications shall be used when purchasing new systems, major modifications, upgrades to current systems, and non-developmental and commercial items, for programs in any acquisition category. If it is not practicable to use a performance specification, a non-government standard shall be used. Since there will be cases when military specifications are needed to define an exact design solution because there is no acceptable non-governmental standard or because the use of a performance specification or non-governmental standard is not cost effective, the use of military specifications and standards is authorized as a last resort, with an appropriate waiver.

The June 29, 1994 Perry Memorandum introduced the following policy changes:

- Priority for use:
  1. Performance specifications
  2. Non-government standards
  3. Military specifications and standards as a last resort, with an appropriate waiver

- Waivers must be approved by the Milestone Decision Authority (usually the service secretary).
- Encourage contractors to propose non-government standards and industry-wide practices.

- Standards cited by other standards are automatically downgraded to non-binding guidance documents.
- Deactivate the management and manufacturer specifications and standards.
- Identify and remove obsolete specifications.
- Replace military standards with NGSs.
- Reduce direct government oversight.
- Identify and reduce or eliminate toxic pollutants.

To promote interoperability among the services, the DoD is advocating using open systems specifications and standards. Some have questioned whether this is contrary to the policy that restricts the use of detailed military specifications and standards. However, Specification and Standards Reform favors performance specifications and NGSs in solicitations and requires a waiver in using detailed specifications and some kinds of military standards (e.g., interface standards can be used without a waiver, but a design criteria standard requires a waiver before it can be used in a solicitation). This preference for performance specifications and NGSs presumably encourages contractors to propose specifications, standards, and products that are used in the private sector. This preference paves the way for contractor proposals using industry solutions consistent with the open systems approach that is widely accepted—standard products available from multiple suppliers at competitive prices. Thus, the "Open Systems" approach and Standards Reform are very consistent as to purpose and effect.

The military HFE community tried unsuccessfully for years to persuade the Defense Standards Improvements Council (DSIC) to re-designate key HFE standards from Design Criteria Standards to Interface Standards because the former cannot be cited as contractually binding without a high-level waiver. In general, only the following standardization documents may be used without a waiver:

- Performance specifications (identified by the "MIL-PRF-" designation)
- Guide specifications (e.g., JSSG 2010 Aircrew Systems)
- Commercial item descriptions (CIDs)
- Interface standards (e.g., MIL-STD-1787B *Department of Defense Interface Standard: Aircraft Display Symbology*)
- Standard practices (e.g., MIL-STD-882, System Safety Program Requirements)
- Military handbooks[3]
- Non-government standards

As a result of the MILSPEC Reform, the number of military specifications and standards was reduced by 38%—from 45,531 in 1994 to 28,326 in 1999. As a part of this streamlining initiative, the number of HFE standardization documents was reduced from 21 to 11. Only six of these are standards[4] (Chaikin, 1998). The major impact was that most of the retained HFE standardization documents lost their influence by re-designating them as non-binding guidance documents (handbooks) or as design criteria standards that require a waiver.

HISTORY OF THE MILITARY HUMAN FACTORS ENGINEERING STANDARDS

HFE design criteria standards began as responses to accidents resulting from human error. Currently, human error is still the leading cause of all accidents. The so-called "lessons learned" are merely frequently occurring errors that we hope not to repeat. Jehan (1994) discussed the rationale for having military standards and how they exist because of bona fide requirements:

> Not created for economic reasons, they exist because, as history has shown, they were required to reduce combat risk. Simply put, they represent dollars paid now to save lives later.

World War II provided the disastrous accidents that motivated what we know today as human factors engineering. The rush to build war matériel resulted in many "horror stories." Military pilots were required to fly different types of aircraft and, in those days, there was no standard control arrangement in cockpits. For example, one twin-engine bomber had the engine controls stacked vertically, with primer on top, mixture between, and throttle on the bottom. A twin-engine cargo plane also had them arranged vertically, but the order was different: throttle, primer, and mixture. Another twin-engine cargo plane had a horizontal arrangement with yet another order: mixture, throttle, and primer running from left to right. Planes crashed because pilots erroneously reverted to previous behavior patterns and operated the wrong control. The human engineering solutions included standardizing a single arrangement for engine controls and using distinct shapes for the control handles (shape coding). These standards all but eliminated this type of accident.

An important additional reason for the development of HFE standards emerged in 1958 and is best expressed when one contractor human factors manager pressured the Army to provide such design criteria, using the following reasoning: *"Testing our deliverables and identifying human factors problems without giving us your requirements and guidelines beforehand doesn't help anyone and is wasteful. If you would provide design criteria standards before our system is designed, we will know what you will test (among other things) and can design to comply with your needs."* The Army started its HFE design criteria standards efforts as a direct result of this expressed industry need.

Starting in World War II, the military began to realize the importance of acquiring systems that could be operated and maintained effectively, efficiently, cheaply, and safely. Each of the services[5] generated a host of specs and standards covering specific systems, subsystems, and classes of systems (e.g., ballistic missiles, and ground vehicles). As these proliferating documents became more numerous and costly to maintain, each of the services began to consolidate into more general-purpose specs and standards.

But it was the missile and space programs of the late 1950s and early 1960s that provided the impetus to elevate organizational human factors standards to military standards. Perhaps the most visible U.S. ballistic missile and space programs (before NASA was formed in October 1958) was run by the Air Force Ballistic Missile Division in Inglewood, CA, and the Army Ballistic Missile Agency (ABMA) of the Army Ordnance Missile Command (AOMC) at Redstone Arsenal, AL. Later AOMC became the Army Missile Command (MICOM).[6] In 1967, MICOM was selected as DoD's Lead Standardization Activity (LSA) for the Human Factors (HFAC) standardization to consolidate the principal service-peculiar human engineering specifications and standards into one tri-service specification and one tri-service standard.

Working together over the last 40 years, human factors engineers from the services, industry, and technical societies jointly developed a small set of consensus-type military standards that embody accumulated HFE knowledge. The two key documents were a military specification providing human engineering program requirements and guidelines (MIL-H-46855) and a military standard providing human engineering design criteria (MIL-STD-1472).

The first true human factors military standard was AFBM[7] 57-8A *Human Engineering Design Standards for Missile System Equipment* (November 1, 1958) that superseded a policy exhibit 57-8 dated August 1, 1957. This standard had the following major sections:

General requirements
Visual displays
Controls
Physical characteristics (components)
Ambient environment
Workplace Characteristics (anthropometry)
Hazards and Safety

The material in AFBM 57-8A was drawn from a number of technical reports, many of which eventually became chapters in the *Joint Services Human Engineering Guide to Equipment Design*[8] With some minor changes, AFBM Exhibit 57-8A was reformatted as an MIL-STD and released as MIL-STD-803 (USAF, November 5, 1959) *Human Engineering Criteria for Aircraft, Missile, and Space Systems, Ground Support Equipment*. MIL-STD-803 then evolved into a three-volume set: MIL-STD-803A-1 ((January 27, 1964) *Human Engineering Design Criteria for Aerospace System Ground Equipment*), MIL-STD-803A-2 (December 1, 1964) *Human Engineering Design Criteria for Aerospace System Facilities and Facility Equipment*, and MIL-STD-803A-3 (May 1967) *Human Engineering Design Criteria for Aerospace Vehicles and Vehicle Equipment*.

In March 1960, the Army approved ABM A XPD-844, *PERSHING Weapon System Human Factors Engineering Criteria*. In October 1961, this was updated and expanded to include all missile systems as ABMA-STD-434, *Weapon System Human Factors Engineering Criteria*. Typical source documents for ABMA-STD-434 were the same as those used for MIL-STD-803. The Army's MIL-STD-1248, *Missile Systems Human Factors Engineering Criteria* (January 20, 1964) was essentially an MIL-STD-formatted version of ABMA-STD-434A.

The MIL-STD-803A series, together with MIL-STD-1248, were the seminal documents for the original tri-service MIL-STD-1472 (February 9, 1968) *Human Engineering Design Criteria for Military Systems, Equipment, and Facilities*. MIL-STD-1472 has survived the test of time and remains the key U.S. standard for military human engineering, and—other than software ergonomics design—may still be the most cited human engineering standard in technical literature worldwide.

A more detailed history of MIL-STD-1472 from 1957 through 1978 (pre-standard predecessor documents and sources, development of first standards, conversions into military standards, consolidations of service military standards, technical changes in each version and revision, and approaches of the three services) can be found in Chaikin (1978).

THE GENERAL FAMILY OF DOD HUMAN FACTORS DOCUMENTS

An important outcome of the standardization reform initiative of the late 1990s was the cancellation of most of the single-service standards and the consolidation of their materials in a few DoD standards and handbooks. Because of the criticality of aircraft design, there continue to be two primary categories of human factors documents: *general* (MIL-STD-1472 and related handbooks) and *aircraft* (JSSG 2010 and related handbooks). The general family of DoD human factors documents includes:

> MIL-STD-1472F (August 23, 1999). *Department of Defense Design Criteria Standard Human Engineering*. Design data and information were removed from MIL-STD-1472E and inserted in MIL-HDBK-759. Some material from the canceled MIL-STD-1801 User/Computer Interface (USAF) was added to 1472.
> MIL-STD-1474D (Notice, August 1, 1997). *Department of Defense Design Criteria Standard: Noise Limits*. Implementing the policies of standardization reform, this document was updated as a tri-service design criteria standard. MIL-STD-1474 was first issued March 1, 1973 as an Army standard on noise limits, based on HEL STD S-1-63C. Since then, it's been extensively revised and expanded. As a result of recent consolidations, MIL-STD-1474D now serves as the *DoD Design Criteria Standard on Noise Limits* that is used by all the services.
> MIL-HDBK-46855A (May 17, 1999). *Human Engineering Program, Process, and Procedures*. This handbook was extensively updated to include MIL-F-46855 and DoD-HDBK-763 *Human Engineering Procedures Guide*. The superseded DoD-HDBK-763 (canceled on July 31, 1998) covered human engineering methods *and* tools. MIL-HDBK-46855A adopted or revised only those traditional methods in DoD-HDBK-763 that have remained stable over time; the section does not describe currently available automated human engineering tools, which have rapidly evolving names and features, but refers the reader to the DSSM (Directory of Design Support Methods) on the MATRIS Web site (http://dtica.dtic.mil/ddsm/).

MIL-HDBK-1908B (August 1999). *Department of Defense Handbook: Definitions of Human Factors Terms*. This handbook (previously a standard but converted to a handbook in accordance with standardization reform) is the single source of definitions for all documents in the HFAC standardization area. This avoids conflicting definitions of the same terms in human factors documents as each is developed or revised.

MIL-HDBK-759C, Notice 2 (March 31, 1998). *Department of Defense Handbook: Human Engineering Design Guidelines*. This is a companion to MIL-STD-1472 and provides design data and extended guidelines. It includes data removed from MIL-STD-1472E.

DoD-HDBK-743A (February 1991). *Anthropometry of US Military Personnel*. This contains statistics from about 40 military surveys,[9] including the 1988 Army ANthropometric SURvey (ANSUR) of 1774 men and 2208 women with more than 132 measures.

Design Criteria Standard Human Engineering—MIL-STD-1472F

Specific HFE design criteria are found in MIL-STD-1472F (August 23, 1999) *Department of Defense Design Criteria Standard Human Engineering*. MIL-STD-1472 has incorporated, by reference, ANSI/HFS 100 on Visual Display Terminal (VDT) Workstations and defers to JSSG-2010 on issues relating to aircraft crew stations, including aircraft passenger accommodation. Because JSSG-2010 does not address aircraft maintainability, MIL-STD-1472 is the appropriate guidance on design for maintainer issues for all systems, including aircraft.

The purpose of MIL-STD-1472 is to achieve mission success, system effectiveness, simplicity, efficiency, reliability, and safety of system operation, training, and maintenance. It contains a mix of requirements and guidelines to facilitate achieving required human performance and ensuring that the design is compatible with the human characteristics of operators and maintainers. The standard is divided into the following major sections:

Control/display integration	Design for the maintainer
Visual displays	Design for remote handling
Audio displays	Small systems (portable)
Controls	Operational and maintenance
Labeling	ground/shipboard vehicles
Physical accommodation	Hazards and safety
Workplace design	User–computer interface
Environment	VDT Workstations (ANSI/HFS 100)

MIL-STD-1472 provides time-tested design limits and guidance for systems, equipment, and facilities that warfighters, other operators, and maintainers can use effectively. A frequently cited MANPRINT[10] success story is how the T-800 engine in the Army's Comanche helicopter can be maintained with a nine-piece tool kit. This is far from a new idea, rather applying guidance that has been in the standard since the original MIL-STD-1472 (February 9, 1968),[11] 5.9.10.1 General. The number and diversity of fasteners used shall be minimized and commensurated with stress, bonding, pressurization, shielding, thermal, and safety requirements.

The services revised MIL-STD-1472 many years ago to incorporate provisions to ensure that women would be able to operate and maintain military systems, equipment, and facilities. Such design limits (e.g., dimensions and forces) were adjusted for compatibility with size, strength, and other characteristics of the female military population. Accordingly, MIL-STD-1472 is the primary technical tool used by DoD to ensure that women in the services are not inappropriately excluded from opportunities merely because of design-induced incompatibilities with the systems to which they might otherwise be assigned. In other words, MIL-STD-1472 can be a highly significant and effective Equal Employment Opportunity (EEO) tool.

Because of the increasing complexity of new military systems, deficiencies in the human-system interface are frequently cited in accidents. To increase the emphasis on HFE, DoD acquisition policy, circa 1993, specifically named MIL-STD-46855 and MIL-STD-1472 as "Key Standards."

HUMAN ENGINEERING PROGRAM, PROCESS, AND PROCEDURES—MIL-HDBK-46855A

The other key document was MIL-H-46855 (February 16, 1968) *Human Engineering Requirements for Military Systems, Equipment, and Facilities*—a military specification. This defined the requirements for applying human engineering to the development and acquisition of military systems. It covered the tasks to be performed by contractors in conducting a human engineering effort, including:

Defining and allocating system functions Equipment selection Analysis of tasks
Preliminary system and subsystem design
Studies, experiments, and laboratory tests (mockups, simulation, etc.)
Equipment detail design drawings
Work environment, crew stations, and facilities design
Human engineering in performance and design specifications
Equipment procedure development
Human engineering in test and evaluation
Failure analysis

MIL-H-46855 was originally a consolidation of one Army, two Navy, and one Air Force specifications—conducted simultaneously with the MIL-STD-1472 consolidation effort that was completed in February 1968. On May 26, 1994, pursuant to a re-definition of the term "standard," MIL-H-46855B was revised and converted to a military standard, MIL-STD-46855. On January 31, 1996, as part of standardization reform, MIL-STD-46855 was redesignated as MIL-HDBK-46855, *Human Engineering Guidelines for Military Systems, Equipment, and Facilities*. Because MIL-HDBK-46855 and its companion guidelines, DoD-HDBK-763,[12] were now both handbooks, it was decided to consolidate them into a new handbook. MIL-HDBK-46855A *Human Engineering Program, Process, and Procedures*—the surviving document—guides DoD and contractor program managers and practitioners regarding analysis, design, and test aspects of the human engineering program.

Key selections from MIL-HDBK-46855A

4. PROGRAM TASKS
    4.2. Detailed guidelines
        4.2.1. Analysis [Definition and allocation of system functions]
        4.2.2. HE in design and development
        4.2.3. HE in test and evaluation

5. THE SIGNIFICANCE OF HE FOR PROGRAM ACQUISITION
    5.1. HE support in system acquisition
        5.1.1. Total system approach
        5.1.2. HE and Human Systems Integration (HSI)
        5.1.5. Manpower, personnel, and training interactions and implications
        5.1.6. Scope of HE concerns
            5.1.6.1. Operators and maintainers
            5.1.6.2. Nonhardware issues
    5.2. HE activity areas
    5.3. The value of HE

6. HE PROCEDURES FOR DoD ORGANIZATIONS

6.2. Application of HE during system acquisition
6.3. Program planning, budgeting, and scheduling
    6.3.2. Work breakdown structure (WBS)

6.4. Coordination [Participation in IPTs]
6.5. Preparation of the Request for Proposal (RFP)
6.6. Proposal evaluation
6.7. Contractor monitoring

7. HE PROCEDURES FOR CONTRACTORS

7.1. HE design standards and guidelines
7.2. Program organization and management
7.3. Application of HE during system acquisition
7.4. General contractor considerations

8. HE METHODS AND TOOLS

8.1. Methods and tools section overview
8.2. HE during analysis efforts
8.3. HE analysis methods
8.4. HE during design and development
8.5. HE design and development methods
8.6. HE during test and evaluation
8.7. HE T&E methods

Over the years, human factor researchers and practitioners have developed many powerful methods to aid in HE work. Section 8 (previous) provides information regarding a number of the methods that can be applied by HE practitioners during system acquisition. The focus of Section 8 is on HE methods that are stable over time. Automated tools, however, are not included because they typically have rapidly evolving names and features. Therefore, descriptions of currently available HE tools can, instead, be found at the website for the Manpower and Training Information System (MATRIS) Office (http://dticam.dtic.mil/). The MATRIS homepage lists MIL-HDBK-46855A tools that implement many of the methods that continue to appear in the handbook. The basic description and point of contact information for the tools are available from this website. Additional data collected regarding the tools are available for DoD HFE TAG[13] members and approved DoD contractors from the Standardization office at the U.S. Army Aviation and Missile Command, Redstone Arsenal, AL.

All pre-handbook versions of MIL-HDBK-46855 also contained Data Item Descriptions (DIDs) that specified the deliverable documentation of the HFE work done on a program. In the past, calling out MIL-STD-46855 in a contract and citing one or more tailored DIDs constituted the formal HFE reporting requirements for an acquisition program. The HFAC DIDs are:

Human Engineering Program Plan
Human Engineering Progress Report
Human Engineering Dynamic Simulation Plan
Human Engineering Test Plan
Human Engineering Test Report
Human Engineering System Analysis Report
Human Engineering Design Approach Document—Operator
Human Engineering Design Approach Document—Maintainer
Critical Task Analysis Report

Downgrading MIL-STD-46855 to a handbook meant that the DIDs, which were correlated with the provisions of MIL-STD-46855, became stand-alone documents that are now rarely, if ever, contractually cited.

HUMAN SYSTEMS INTEGRATION IN AIRCRAFT

When standardization reform placed all specifications and standards in jeopardy, the military aviation community, led by the Joint Aeronautical Commanders Group,[14] reorganized and completely replaced the entire system of specifications and standards with a new system of Joint Service Specification Guides. These JSSGs cover all aspects of military aviation systems, not just human systems. However, designating "Crew Systems" as one of the ten top-level domains gives HSI unprecedented visibility in the aviation development. The following list shows the architecture of the JSSG system.

Joint Service Specification Guides Master Index	Approved
JSSG-2000B Air System	September 21, 2004
JSSG-2001B Air Vehicle	April 30, 2004
JSSG-2002 Training	Incomplete
JSSG-2003 Support Systems	Incomplete
JSSG-2004 Weapons	Incomplete
JSSG-2005 Avionics	October 30, 1998
JSSG-2006 Structures	October 30, 1998
JSSG-2007A Engines	January 29, 2004
JSSG-2008 Air Vehicle Control & Management	October 1, 2003
JSSG-2009 Air Vehicle Subsystems	October 30, 1998
JSSG-2010 Crew Systems	October 30, 1998

The concept of the "specification guides" evolved out of the Air Force's MIL-PRIME initiative. These documents have two major parts: one is a draft specification (e.g., JSSG 2010) with key numbers and requirements replaced by blanks. The second part is a set of 14 handbooks (e.g., JSSG 2010-1 through JSSG 2010-14) that discuss the issues for filling in the blanks. The actual filling in of the blanks can be a joint decision of the military and contractors. Once filled in, these guide specs become a binding part of the contract. The JSSG, then, avoided the problems of getting a waiver by not being a standard, yet becomes contractually binding in the final form. Because some of the data in the JSSG series are restricted, it was decided to limit the distribution of all JSSGs to DoD and DoD contractors.[15]

JOINT SERVICE SPECIFICATION GUIDES CREW SYSTEMS, JSSG-2010

JSSG-2010 summarizes a unified process for applying the required disciplines to the development, integration, test, deployment, and support of military aircraft crew systems. This document supports a human-centered crew station approach to the acquisition process, where the platform is designed around the human and human-generated requirements for performance as the driving force. JSSG-2010 has 14 accompanying handbooks as follows:

- JSSG-2010-1 Systems engineering guidance for the ***design of crew stations*** in fixed and rotary-wing aircraft.
- JSSG-2010-2 Guidance for the ***development requirements and verifications*** for crew systems.
- JSSG-2010-3 Guidance for the criteria to ***optimize cockpit/crew station! cabin designs*** without hindering the development of new, improved systems, including fixed and rotary wing.

- JSSG-2010-4 Guidance for the design and verification of *aircrew alerting systems*.
- JSSG-2010-5 Guidance for the development requirements and verifications for *interior and exterior airborne lighting* equipment, including specific requirements for interior lighting compatible with type I or II and class A or B night-vision-imaging systems).
- JSSG-2010-6 Guidance for the design and test information for *sustenance and waste management systems* for the support of aircrew and passengers.
- JSSG-2010-7 Guidance for the development requirements and verifications for occupant crash *protection* and for crash protective aspects of seating, restraint, and crew station, and passenger/troop station design.
- JSSG-2010-8 Rationale, guidance, lessons learned, and instructions for the *Energetic Systems* [explosive actuators] section.
- JSSG-2010-9 Guidance for the development requirements and verifications for *aircrew personal protective equipment*.
- JSSG-2010-10 Guidance for the development requirements and verifications for an *aircraft oxygen system* and its components.
- JSSG-2010-11 Guidance for the development requirements and verifications for *aircraft emergency escape systems*.
- JSSG-2010-12 Guidance for the development requirements and verifications for *deployable aerodynamic decelerator* (DAD) system or subsystem. [Parachutes are DADs]
- JSSG-2010-13 Guidance for the development requirements and verifications for an *airborne survival and flotation system* and its components. This includes many provisions for emergency egress, life support, descent, and land and water survival for extended time periods until recovery.
- JSSG-2010-14 Guidance for the performance, development, compatibility, manufacturability, and supportability requirements and verification procedures for an *aircraft windshield/canopy system* and its components.

The JSSG-2010 is incomplete in two areas: (a) aircraft maintainability that is covered in the general-purpose MIL-STD-1472 and (b) aircraft symbology that is covered by MIL-STD-1787C (January 5, 2001) *Department of Defense Interface Standard: Aircraft Display Symbology*. MIL-STD-1787B was originally planned to be one of the handbooks included in JSSG 2010 but was approved as an interface standard (that may be cited without a waiver) before the JSSG series was finished. Because an interface standard has more authority, it was decided to leave it as a stand-alone document.

RELATION TO OSHA STANDARDS

The HFE practitioner should be familiar with both OSHA (Occupational Safety and Health Administration) and military HFE-related standards on SOH (Safety and Occupational Health). The policy is to apply the "more stringent" of those available, as far as practicable.

APPLICATION OF OSHA VERSUS DoD SAFETY AND HEALTH STANDARDS

Because the OSHA statutes were not written to apply to military systems, there is often some confusion about whether or when OSHA standards apply. As clarification, DoD policy—according to DoDI 6055.1, *DoD Safety and Occupational Health (SOH) Program*—states that OSHA standards apply to non-military-unique operations and workplaces and, in certain circumstances, apply to military-unique systems if they are more stringent than military standards. In fact, military standards are usually more stringent because their objective is ensuring performance rather than merely avoiding injury as in OSHA standards.

NON-MILITARY UNIQUE SYSTEM COVERAGE BY OSHA

DoD Instruction 6055.1 (May 6, 1996) states that DoD Components shall comply with the standards promulgated by the OSHA in all non-military-unique DoD operations and workplaces (office, maintenance shops, and other non-frontline activities). This applies regardless of whether work is performed by military or civilian personnel. However, DoD components may develop and apply standards that are alternate or supplemental to such OSHA standards, and DoD standards may need to be more stringent than OSHA ones if the military situation warrants. DoD components shall apply OSHA and other non-DoD regulatory safety and health standards to military-unique equipment, systems, operations, or workplaces, in whole or in part, as far as practicable and when supported by good science. According to DoD Instruction 6055.1, if a

> DoD Component determines that compliance in a non-military unique work environment with an OSHA standard is not feasible, a proposed alternate standard shall be developed and submitted after consultation with other DoD Components and with affected employees or their representatives. *For example, OSHA health standards designed to protect personnel from eight-hour exposures to hazardous chemicals may not be applicable for 24 hr exposures, or for multiple exposures and various modes of entry into the body during military operations and deployment situations.* When military design, specifications, or deployment requirements render these standards unfeasible or inappropriate, or when no standard exists for such military application, DoD Components shall develop, publish, and follow special military standards, rules, or regulations prescribing SOH measures, dosimetry, and acceptable exposure levels. Acceptable exposure measures and limits shall be derived from use of the risk management process described elsewhere in this Instruction.

CONTINUING ISSUES FOR MILITARY STANDARDS

There are many fundamental differences between the HFE military standards and those specifications and standards that have been attacked as wasteful. There are at least seven issues distinguishing HFE military standards from other military standards and commercial standards:

- Obsolescence
- Specifying a solution
- Targeted systems and subsystems
- COTS/NDI (commercial off-the-shelf/non-developmental items)
- Performance standards
- Commercial standards
- Work breakdown structure

ISSUE OF OBSOLESCENCE

The industry's primary reason for attacking military specifications and standards was their use of obsolete technology and manufacturing processes. Such specifications and standards were probably valid at the time they were promulgated but became obsolete as technology evolved.

Some would argue that the HFE design standards are not up-to-date because the latest technologies are not covered. Those seeking a quick solution to high-technology human-system interfaces are often disappointed by not finding standard answers. *In this regard, it is important to understand that such standards are not intended to provide solutions; they are limits on design*. Moreover, if the HFE standards did attempt to provide design criteria or preferred practices for rapidly evolving technology issues, they would be open to the same criticism of the electronics and manufacturing standards. An HFE design standard should properly provide criteria for which there is common agreement. This means that the technology has settled down to the point where a consensus can be reached on a needed human engineering design provision. If there is no consensus on design limits or process issues, a standard is premature. So, when we say that the HFE military standards are current, we mean that all its provisions reflect current consensus.

Even older HFE standards are not obsolete because, although certain technologies have evolved rapidly, the human has not. Sensors have ever-increasing resolution and spectral range, but the human still has two eyes, with no evidence that visual perception is any better now than it was in the past. The speed of computers has progressed so much that an ordinary desktop PC now outperforms the early supercomputers. Computer memory has increased from kilobytes to gigabytes. Yet, the human brain cannot be said to have any greater memory or new capability. The only noticeable change—average human stature—that increased about 2% in the first half of the 20th century (attributed to better health and nutrition) appears to have leveled off.

Unlike the rapidly evolving digital electronics, the capabilities and limitations of humans exhibit negligible evolution and the design principles for the human-system interface are not obsolete. The HFE standards describe design limits to ensure that the system will be effectively, efficiently, safely, and cheaply operable and maintainable by its intended users (both men and women), irrespective of whether yesterday's or innovative technology is involved. The HFE standards community has always kept its standards and DIDs up-to-date and consistent with current DoD policy and needs. Many years ago, when DoD policy called for reducing the number of DIDs, the HFE DIDs were reduced from more than 30 to ten. When prior acquisition reform initiatives emphasized streamlining, the HFE community revised its process standard to bring it into compliance as the first such document to add comprehensive tailoring guides tied to each of the acquisition phases to avoid excessive requirements.

Issue of Specifying a Solution

MIL-STD-1472 does not specify any solutions; it provides time-tested design limits as requirements or guidelines. These represent performance standards in the sense that most of its criteria are human performance-driven. Failing to meet these minimum standards will cause performance to be degraded. By specifying performance-based design limits for routine elements of the human system interface, the designer (a) avoids repeating past mistakes, (b) devotes more effort to the new human systems issues (see Figure 42.1), and (c) is provided the flexibility to be innovative within relatively liberal design limits that reflect the consensus of the technical community. Some military-unique performance-driven design limits include weightlift maxima for the military population, label size and color for low light operations, and control guarding options to prevent inadvertent actuation under certain conditions. In addition to not specifying solutions, the military's Human Factors Engineering standards do not prescribe materials or manufacturing techniques. MIL-STD-1472 does not specify solutions.

Issue of Targeted Systems and Subsystems

Although the military's HFE standards deal with the broad spectrum of design issues, they contain unique human performance requirements peculiar to the military and the military operating environment that include worldwide operations and warfighting. On the other hand, most voluntary standards apply to the commercial marketplace and, justifiably, have different priorities—for example, aesthetics, use by a wider range population, use by an untrained population, sales appeal, and use in benign environments.

HFE military standards frequently address mission environments that are unique, or near unique, to the military or the battlefield. For example, design must accommodate operation and maintenance by military personnel wearing protective equipment and clothing, such as chemical and biological protective gear that retains body heat, reduces body mobility, and aggravates accessibility and operability of equipment. Commercial products very seldom have a need to deal with such issues.

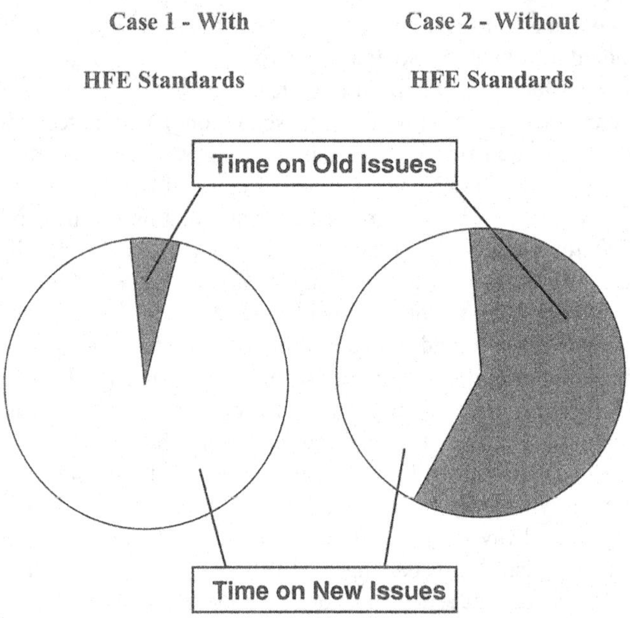

Case 1 - With   Case 2 - Without

HFE Standards   HFE Standards

FIGURE 42.1 Using HFE standards to design performance into the routine aspects of the system leaves more engineering labor to apply to the new design issues. Without HFE standards, every detail of every aspect of the human-system interface must be researched, designed, and tested. Much time is spent "reinventing the wheel." In most programs, the HFE budget is fixed. Using accepted military standards frees up labor for solving new issues.

APPLYING LABOR TO DESIGN HUMAN INTERFACE

ISSUES OF COTS/ND I (COMMERCIAL OFF-THE-SHELF/NONDEVELOPMENTAL ITEMS)

The market for consumer electronics has increased so rapidly that it is driving new technology. Seeing the advanced capability available to the public at small costs has made the military envious. Systems and equipment developed for the military are relatively more costly because of the smaller quantities purchased, the longer development period, and frequently required military-peculiar features such as ruggedness and operability in climatic extremes. A modem weapon system might take more than ten years to develop and field. As a result, computers embedded in new aircraft may use ten-year-old technology.

It might help clarify the proper role of non-military standard products if the acquisition system recognized that there are two distinct military functions that must be equipped: (a) the peacetime military plus DoD civilian personnel and (b) the combat military. The "peacetime military and DoD civilian personnel" is by far the largest group and is made up of the military members performing non-combat duties at their home base and the civilian workforce of the DoD (approximately one-third of the DoD is made up of civilians). The work performed by this category of DoD personnel is similar, in most respects, to the work performed by business and industry. In this role, it is both reasonable and necessary that these workers have the same commercial-grade equipment. A military office already has ordinary commercial furniture, carpet, telephone, desktop computer, fax machine, and copier.

When we talk about military systems and equipment, we are referring to systems and equipment to be used by the "combat military—those military men and women who are deployed somewhere in the world and performing or training for the traditional military mission, either fighting a war or keeping the peace. Consumer products are rarely suitable in a military combat environment. If you do not believe this, all you have to do is get out the manual and read the safety instructions and the warranty.

Here is a specific example: a new cordless phone intended for home and office use has an encryption feature that prevents its radio transmissions from being monitored. Such a phone would seemingly have utility for the military. It is state-of-the-art technology, mass-produced, and inexpensive. However, the instruction manual reveals five fatal flaws that would cause this product to fail miserably if used by the military in the field. First, the product may not be used near water or liquid of any kind. If the unit is exposed to rain or water, it must be unplugged immediately and returned to the manufacturer. This limitation may make it unattractive to the Army and Marines, or troops deployed in a tent in the rain. Second, the unit should not be operated in a hot environment, so it must be kept out of the sun and the desert. Third, one must not let anything rest on the power cord, nor should anyone walk on it. In other words, the power cord is very delicate. Fourth, this item may cause interference with other radio or television equipment. Fifth, the small, low-contrast labels cannot be read reliably in other than a well-lit area with good, unobstructed vision (i.e. incompatible with certain types of eye protection). It cannot be used at all with chemical defense gear on. When wearing gloves, one cannot use the keypad.

Clearly, this commercial product was designed for use in an air-conditioned home or office and will not function reliably in any other environment. For peacetime military and civilian use, this product would provide acceptable service. However, despite its economical price and ready availability, it would fail to perform in a combat military environment.

Traditionally, products for combat use are not designed for eye-appeal. A military jet has neither a simulated wood-grain instrument panel nor a carpeted floor, and yet it costs a fortune. With some obvious exceptions, such as dress uniforms, made-for-the-military emphasizes functionality, not appearance. On the other hand, the private consumer demands that products have a stylish appearance, with functionality being of secondary importance. The military human factors engineer cares about how well a product works, not how it looks. Consumer products must look attractive if they are to compete in the open marketplace. However, shiny trim on commercial products, while offering eye appeal, may reflect light and disclose the location to the enemy.

There have been occasions when the military has attempted to use consumer or commercial equipment in a war, usually with poor results. For example, in the Gulf War, to aid communication, many of the deploying troops took their office fax machines with them. In a very short time, the fine desert sand had found its way into every nook and cranny, and these marvels of modern technology literally ground to a halt. The hero of the day turned out to be the Air Force, who had acquired a few fax machines that had been hardened according to military standards. These machines continued to work, despite the sand, and were heavily used because they were shared with the other services whose COTS fax machines had quickly failed.

We tend to forget that equipment designed for use in air-conditioned offices will fail when operated in many other places in the world where our military must deploy. Taking commercial and consumer products designed for office or even factory use to a cold, hot, wet, or dusty region of the world can quickly convert high-tech equipment into expensive doorstops. Buying an inexpensive item that does not meet the need is never a bargain. DoD must consider both the lifecycle costs and the environment of use. This is the "military-unique" environment the policymakers were referring to when talking of exceptions to COTS.

Issues With Developing HFE Performance Standards

The Standards Process Action Team said: "Because of the uniqueness of military requirements, it is unlikely that the DoD will ever be able to rely completely on performance-based specifications." According to Lowell (1994), the DSIC had subsequently developed a common waiver process for military specifications and standards. There is a precedence where HFE and system safety engineering has been set aside in an exempt category, as illustrated by the following quote from a House of Representatives Report (1916):

Content of Performance Standards. A performance standard established for a device under the proposed legislation must provide reasonable assurance of a device's safe and effective performance. Although use of the term "performance standard" reflects a preference for standards which allow the fullest use of technological alternatives, the Committee does not intend the term to be construed as excluding design-related requirements, as it is when it is used in the engineering community. *Design-related requirements that are necessary to provide reasonable assurance of safe and effective performance or that improve device safety and effectiveness by reducing the likelihood of human error should be included in a performance standard.*

This quotation tells how even if performance standards are used, they must be supplemented with specific design-related requirements to assure safe and effective performance. This is precisely the essence of the military standards for HFE.

According to Lowell (1994), the DSIC approved the following definition of performance specifications:

A performance specification ... states requirements in terms of the required results with criteria for verifying compliance, but without stating the methods for achieving the required results. A performance specification defines the functional requirements for the item, the environment in which it must operate, and interface and interchangeability characteristics.

The military's HFE standards do not dictate materials or manufacturing techniques. Arguably, they are performance standards because they provide design criteria that (a) allow users to operate and maintain efficiency with minimal error and (b) because most of the criteria are human performance-based. For example, we know from studying human performance that as the keys of a keyboard get smaller and closer together, keying errors and time will increase exponentially. MIL-STD-1472 describes the size and separation of keys on a keyboard that humans can operate with minimal errors. For barehanded use, the minimum spacing for keys on a keyboard is 16.4 mm (0.675 inch) center to center, with 19 mm (0.75 inch) preferred. (Note the human performance basis for this provision.) The typical computer keyboard meets the preferred separation, so the standard does not limit access to computer keyboards and typewriters. This standard key separation is easy to understand, easy to apply, consistent with 50 years of competent research, and compliance can be economically verified in a few seconds with an ordinary ruler. It guarantees adequate performance on military equipment. In arriving at this standard, there was no opposition from the industry because the industry was interested in optimizing speed and accuracy.

To convert this key separation requirement to a true "performance standard," we must replace it with a statement of our performance goal. For example, a performance standard would not specify a key separation, but have words to the effect that "the key size and spacing must be such that keying errors should not exceed 'X'%, and the average time to correctly reach and operate push buttons should not exceed 'Y'-tenths of a second when operated by all members of target audience under all conditions of intended use." Unlike the previous standard, this is not straightforward in its application. What would a contractor have to do to comply with such a performance requirement? To start, the contractor must acquire data on the relationships described. As a minimum, this would involve a literature search by someone skilled in that technology. If relevant data were not found, the contractor may have to perform an expensive and time-consuming research study to gather the needed data.

After the product is delivered, verification by the government that the goal has been met may also require a very expensive and time-consuming test and evaluation. A real example illustrates the difficulty. One manufacturer makes a compact keyboard with keys 12 mm (0.48 inch) between centers. Designed for the less demanding consumer market, the manufacturer would also like to sell it to the military. After all, it is economical commercial off-the-shelf (COTS) equipment. It fails to meet the minimum separation in the old HFE standard, but what about the new performance standard? Reputable research has shown that the close key spacing causes many times more keying errors than does the preferred spacing. However, in addition to competent studies, the research literature also has studies with inadequate design or sample size that failed to find conclusive results. Because the literature contains studies with conflicting results, the contractor not only is free to choose one that supports the offered design but also has an economic incentive to do so.

Next, the government's program manager will be confronted by the company's self-serving arguments together with the military's human engineer pointing out the problems with the questionable research. The program manager has better things to do than adjudicate disputes over sample sizes in unfamiliar research studies. The contractor's arguments for low cost and saving of valuable control panel space will likely carry more weight than debatable research results. Sadly, none of these players will ever know if years later, the troops in the field had problems with this design.

In an alternate scenario, the contractor's strategy might be not to worry about research at all and take a chance on passing the government's final acceptance test. At the end of the program, when the product is delivered for test and evaluation, the key spacing problem will again surface as a concern and warrant a test study to be performed. Contentious issues with this study include such factors as hand size of the test subjects, number of subjects, the test procedure, and the definition of what the performance specification meant by "correctly reach and operate." So, in this case, the technical practitioners' comparing test measurements with design limits have been supplanted by lawyers and contract specialists' arguing about words. If a competent evaluation were performed, the results would show that the keyboard did not meet the spec. However, the odds are greatly against the study being performed. The program manager is now faced with an even more difficult problem because correcting the design at this late date will be prohibitively costly and further delay a program that is already behind schedule. With the program facing further delays and cost increases, the original requirement will come under fire. Why did the government specify a maximum of 1% errors, anyway? Isn't 2% good enough? The program manager may not want to reject the entire system based on a test that showed only 2% errors. What difference does an extra 1% make anyway?

Although performance standards for high-technology items will undoubtedly save money and increase opportunities for innovation where cutting-edge technology is involved, applying them to the human–system interface may cause the contractors unnecessary work, and may give the military error-prone designs and increase costs of test and evaluation. It is likely that, rather than heading off problematic designs early when they can be dealt with inexpensively, design issues will be postponed until late in the program when changes are difficult and prohibitively expensive. Experience has shown that industry and DoD can agree on reasonable HFE design standards, provided the decision is made outside the context of a specific program. Once the program begins, schedules, existing designs, and profit incentives tend to cloud the issues and make resolution expensive and contentious.

ISSUES WITH CONVERTING TO COMMERCIAL (NON-GOVERNMENT STANDARDS)

Standards reform gave priority to using voluntary standards and converting military specs and standards to NGSs, as applicable. Unfortunately, when military standards are converted to commercial standards, their scope is broadened to include non-military systems and equipment and non-military populations.[16] Because there are more non-military systems, expanding the scope dilutes the amount of time and effort that goes toward updating the standards.

Currently, there are no comprehensive, general HFE NGSs.[17] There are some point-designs, such as the video display terminals covered by ANSI/HFS 100, but the DoD is the only customer whose interest includes virtually all products. It is for this reason that the military has led and still leads the development of HFE design technology. The military has been a consistent user of HFE technology for the past 50 years because the consequences of failing to do so can be catastrophic. One of the reasons stated for emphasizing NGSs is to take advantage of the best consensus standards industry has to offer without burdening them. Industry supports the military's HFE standards because it has participated in their development (as voting participants) and kept them reasonable. Today, many non-defense companies use applicable provisions of MIL-STD-1472 voluntarily on their commercial products because it is recognized as the best available.

There are two fundamental reasons why the military should have its own HFE standards. First, the mission and weapons functions are unique to the military. The military should retain control of performance requirements for all equipment the troops take to the field in a military action. These requirements are almost always life-critical, with mission performance and system safety at stake. Certainly, these standards do not have to be applied to the everyday equipment used by military and DoD civilian personnel in the performance of non-combat duties. Second, the military needs an integrated HFE standard, not a large number of piecemeal standards. The mixing and matching from a set of hundreds or thousands of commercial standards not only is inefficient for HFE requirements but also will likely lead to omissions of important considerations. When the military considers commercial off-the-shelf equipment, it should always be tested to determine whether it is compatible with the military environment. When modified commercial off-the-shelf equipment is developed, it should be consistent with the military's human–system interface standards.

Reduced military budgets make it increasingly difficult for DoD HFE practitioners and researchers to participate in updates of commercial standards and guidelines. Reduced budgets drive the military to sacrifice all but strictly mission-essential activities and updated commercial standards and guidelines are not seen as mission-essential.

OBTAINING NON-GOVERNMENT STANDARDS

After October 1, 2000, the DODSSP ceased to provide DoD users NGSs. Instead, each organization will use government purchase cards to acquire NGSs as needed. The NSSN[18] website will refer all users to a designated source. Many commercial resellers provide ready access to both government and non-government standards through subscriptions or individual document sales. Commercial resellers often also add other value (e.g., improved indexing, document summaries, and full-text search capability), and some have extensive collections of historical documents. The Navy's surface ship community now uses a commercialized version of 1472 tailored for ships for the Navy and Coast Guard. ASTM FI 166-95a *Standard Practice for Human Engineering Design for Marine Systems, Equipment, and Facilities* is copyrighted and can be ordered for $60.00 per copy. MIL-STD-1472 is not copyrighted, and unlimited numbers can be reproduced or downloaded from the web free of charge.

Government Participation in Non-Government Standards Bodies (NGSB)

To reduce to a minimum the reliance by agencies on government-unique standards, the Office of Management and Budget has published Revised OMB Circular A-1[19] that directs government agencies to use voluntary consensus standards in lieu of government standards, except where inconsistent with law or otherwise impractical. It also provides guidance for agencies participating in voluntary consensus standards bodies and describes procedures for satisfying the reporting requirements. This policy encourages DoD employees to participate as "equal partners" with private-sector and other government agency employees on technical committees of NGSBs. Such participation ensures proper consideration of DoD requirements, enhances the technical knowledge of DoD personnel, and allows DoD to contribute the considerable technical capabilities of its employees in the development of "world-class" national standards.

Although government personnel are encouraged to participate in the development of NGSs, it is costly to do so. One should never forget that many voluntary standards are typically written to "the least common denominator" so that all the standard-writing participants' products are acceptable. This has both positive and negative implications.

Travel costs can be considerable. Moreover, some NGSBs charge government personnel a fee[20] to attend each standard committee meeting. Some government personnel sense a conflict of interest when paying a fee to donate the taxpayer's labor to develop NGSs that the NGSBs then sell back to the government, its contractors, and the public for a profit. These can be reasons why the number of

DoD participants on NGSB committees has dropped dramatically from over 2,200 DoD participants in 1994 to fewer than 500 in 1999.[21]

The Problems With Non-Government Standards (NGS)

When examining a list of NGSs in the human factors domain, the two most striking characteristics are the spotty coverage and the duplication in isolated popular areas. NGSs tend not to be comprehensive but focus on specific products, product lines, or product components. An example of overlapping standards is the group: agricultural equipment (with seven standards), earthmoving equipment (with ten standards), off-road equipment (three standards), and graders. The overlap among these is obvious. In addition to numerous standards on telecommunications, there are six human factors standards relating to telephones. With the repetition and overlap of NGSs, just selecting the one applicable for a military program would be a laborious task. It is also an expensive task because you first must buy all the NGS on your topic before you can determine which one, if any, is applicable to the design.

HUMAN ENGINEERING—PRINCIPLES AND PRACTICES (HEB1)

The G-45 (Human Factors) Committee of the Government Electronics and Information Technology Association (GEIA), with the support of the Human Factors Standardization SubTAG, prepared a Human Engineering—Principles and Practices Bulletin. The Bulletin and its annexes provide guidance to the application of human engineering principles and practices in the analysis, design, development, testing, fielding, support, and training for military and commercial systems, equipment, and facilities. As an industry (non-government) document, the use of this document is consistent with the DoD's acquisition reform and could be applied in DoD solicitations. The Electronic Industries Alliance (EIA) published this bulletin in June 2002.[22]

HEB1 resulted from the need to present to the Program Office a succinct human engineering management approach that would explain human engineering requirements for both government and industry systems, equipment, and facilities. HEB1 is a 21-page document based on Section 4 of MIL-HDBK-46855A (May 1999), an update of MIL-H-46855B, Rev 2. It has been edited to address both government and industry needs, and to include a list of, and links to, current Data Item Descriptions (DIDs) developed at the DoD and by the Federal Aviation Administration (FAA). The bulletin is a Human Engineering (HE) Best Practices document developed by both DoD and industrial HE practitioners. It also includes a list of acronyms used in the document and the terms they represent and a list of documents that give additional information.

With the acquisition policy of the DoD discouraging the use of military and government standards in favor of industry practices and standards, HEB1 may prove to be a valuable document. As an "engineering bulletin," HEB1 is not a standard, but a recommended practice. When the DoD's MIL-H-46855 was converted to MIL-HDBK-46855A in May 1999, access to the Data Item Descriptions (DIDs) was lost because they cannot be invoked by a handbook. As indicated in the following, the FAA currently uses some of the DIDs. By recommending and providing links to the DIDs listed here, HEB 1 returns them from limbo to a prominent place in the acquisition community.

- Human Engineering Simulation Concept [DI-HFAC-80742B and FAA-HF-005]
- Human Engineering Design Approach Document-Operator [DI-HFAC-80746A and FAA-HF-002]
- Human Engineering Design Approach Document—Maintainer [DI-HFAC-80747B and FAA-HF-003]
- Noise Measurement Report (NMR) [DI-HFAC-80938A]
- Critical Task Analysis Report [DI-HFAC-81399 and FAA-HF-004]
- Human Engineering Program Plan [FAA-HF-001]

THE DEFENSE STANDARDIZATION PROGRAM

The Department of Defense Instruction 4120.24 (June 18, 1998) Defense Standardization Program (DSP) establishes the DSP under the Defense Logistics Agency (DLA). It is DoD policy to promote standardization of materiel, facilities, and engineering practices to improve military operational readiness, reduce total ownership costs, and reduce acquisition cycle time. These objectives are accomplished by a single, integrated DSP and a uniform series of specifications, standards, and related documents. Specific implementation and guidance is found in the current issue of DoD 4120.24-M *Defense Standardization Program (DSP) Policies and Procedures (March 2000)*.[23]

RESOURCES

The DSP website (http://www.dsp.dla.mil/) provides ready access to current and recently obsolete[24] military specifications and standards. The DSP mission is to identify, influence, develop, manage, and provide access to standardization processes, products, and services for the warfighter, the acquisition community, and the logistics community to promote interoperability, reduce total ownership costs, and sustain readiness.

Search tools allow users to locate and view defense specifications, standards, handbooks, other documents listed in the DoD Index of Specifications and Standards (DODISS), and data item descriptions (DIDs). As of March 1, 2000, the ASSIST database is the official source for all active DIDs. Earlier versions of the most recently revised DIDs are also available.

NATIONAL STANDARDS SYSTEMS NETWORK (NSSN)[18]

The NSSN is a national resource for global standards that indexes documents of over 600 standards-developing organizations. There is no charge, nor do users need to register. NSSN is a service of the American National Standards Network (ANSI). Users perform searches by document number or by keywords within the document title or description. Once a document is located, the NSSN index describes where to obtain it. The 33 organizations with significant HFE standards or guidelines are shown in the "Index of Non-government Standards on Human Engineering Design Criteria and Program Requirements/Guidelines" at http://dtica.dtic.mil/hftag/product.html.

DEFENSE TECHNICAL INFORMATION CENTER (DTIC)

The DTIC lets users search the Public Scientific and Technical Information Network and retrieve copies of unclassified, unrestricted technical papers.

THE DoD SINGLE STOCK POINT (DODSSP)

All interested parties can request copies of defense specifications and standards, federal specifications and standards used by the DoD, and other DoD standardization documents from the DODSSP in Philadelphia, PA. The DODSSP maintains the official repository of all DoD standardization documents and publishes the DoD Index of Specifications and Standards (DoDISS) at http://assist.daps.dla.mil/quick search. Registered users can query the ASSIST database and download most document images as Adobe PDF files.

OTHER GOVERNMENT HFE STANDARDS

NASA STANDARDS

The principal human engineering standard used by NASA is NASA-STD-3000, Man–Systems Integration Standards (MSIS). This family of standards provides specific user information to ensure proper integration of the man–system interface requirements with those of other aerospace disciplines. These man–system interface requirements apply to launch, entry, on-orbit, and extraterrestrial space environments. This document is intended for use by design engineers, systems engineers, maintainability engineers, operations analysts, human factors specialists, and others engaged in the definition and development of manned space projects or programs. Concise design considerations, design requirements, and design examples are provided. Requirements specified are applicable to all U.S.-manned spaceflight programs.

In addition to the basic document, additional volumes of the MSIS are created and maintained, which specifically address the human factors and crew interface needs for that program. As specialized volumes of this type are updated and revised, the information gathered for them is also evaluated for possible inclusion in the basic MSIS volume. To date, there are three volumes planned, with four already published and released:

Vol. I, Man-Systems Integration Standards, first published in 1987 and last updated as Rev. B in June 1995 [http://msis.jsc.nasa.gov/]

Vol. II, Man-Systems Integration Standards—Appendices, first published in 1987 and last updated in 1995, at the same time as, and to the same revision letter as Vol. I [http://msis.jsc.nasa.gov/]

Vol. III, Man-Systems Integration Standards—Design Handbook (the data in this volume coincides with Rev. A, of Vol. I)

Vol. IV, Space Station Freedom Man-Systems Integration Standards, a subset of Vol. I, published in 1987 (Inactive)

FEDERAL AVIATION ADMINISTRATION (FAA) STANDARDS

The FAA HF-SFD-001 (June 2003) Human Factors Design Standard (HFDS) provides reference information to assist in the selection, analysis, design, development, and evaluation of new and modified FAA systems and equipment. This guide covers a broad range of human factors topics that pertain to automation, maintenance, human interfaces, workplace design, documentation, system security, safety, the environment, and anthropometry. This document also includes extensive human-computer interface guidance. The HFDG draws heavily from human factors information developed by the DoD, NASA, and DOE. This document (Wagner et al., 1996) is available to the public through the National Technical Information Service (NTIS), Springfield, VA 22161 and online at http://hf.tc.faa.gov/hfds.

CONCLUSIONS

A comparison of the HFE standards against the recommendations of the PAT reveals that they met most of the stated goals as-is because (a) they were jointly developed by government and industry, (b) they have been kept up-to-date for the issues they cover, and (c) the HFE design criteria are based on human performance. HFE standards do not require dual manufacturing processes because they are not manufacturing standards, do not define hardware, and are not obsolete. They save the government money by reducing the need for design studies, tests, and evaluation. They lower life-cycle costs by providing solutions early in design, not during T&E when repair is most costly. They are military-unique, embodying descriptions of the capabilities and limitations of military personnel (somewhat different from civilian populations) and their personal protective equipment (very different than civilian counterparts).

There are no non-government standard alternatives. Indeed, the superiority of the HFE military standards causes them to be used for civilian applications (dual use). The military and industry jointly developed and updated these standards. They are coordinated through industry groups, professional societies, and other standards organizations.

Standards should not be written for rapidly evolving technology, such as is now ongoing in the electronics technology area. If standards are to be efficient and effective, they should be based on the consensus of both the buyer and the seller. The HFE standards do not have the serious deficiencies that are addressed by the acquisition reform.

Although not the cause of the problems that drove standardization reform, HFE standards were flushed out with the rest. Although HFE standards are important to the HFE profession, they are probably too small to merit special consideration from the DoD management in these busy times.

Reforms and fine-tuning of acquisition are always needed. The HFE/HSI community must continue to keep up with the changing policies and comply with the new directives in a way that is efficient, effective, and, in the best interest of the military.

So far, the largest effect on HFE of de-emphasizing the use of military standards to date has been severe cuts in the size of HFE staff in defense industries (McDaniel, 1996). Part of these cuts result from the general decline in military spending, but the HFE staff have taken disproportionately large cuts. Anticipating less emphasis on HFE, some companies have significantly cut back on HFE staff. Government HFEs cannot take up the slack because their numbers have also been reduced. As a result, there are fewer HFE people in the industry to do the work, and fewer HFE people in the government to write the performance specs and test the finished products. This trend may have a chilling effect on the profession as a whole.

The prognosis for the future of military standards is favorable. On March 29, 2005, the Under Secretary of Defense for Acquisition, Technology, and Logistics signed Policy Memo 05-3 titled "Elimination of Waivers to Cite Military Specifications and Standards in Solicitations and Contracts." This memo allows program managers to use and cite MIL-STD-1472 and MIL-STD-1474 as contractually binding requirements for the first time in over ten years. This was not widely advertised but is known to those in the human factors community. The pendulum has swung the other way, and future acquisitions will benefit from it.

NOTES

1 Joe W. McDaniel is retired from Human Effectiveness Directorate, Air Force Research Lab, Wright-Patterson AFB, OH. The views presented are those of the authors and do not necessarily represent the views of DoD or its Components.

2 Gerald Chaikin died October 20, 2001. He is best known for his work with the military human factors standards, specifications, and handbooks, first as a civil servant, where for 20 years he chaired the Human Factors Standardization Steering Committee (HFSSC), and later as a contractor supporting the Lead Standardization Activity for Human Factors Standardization at U.S. Army Missile Command, Redstone Arsenal, AL.

3 If a military specification or standard is cited in a contract, it becomes a binding part of that contract. However, specifications and standards can be cited for guidance only, in which case they are not binding documents. Military handbooks are never binding.

4 The August 1998 issue of HFAC Highlights tabulates these standards, as well as other HFAC documents and (then) current key points-of-contact for the HFAC (Human Factors Standardization Area) Program, other U.S. Government human factors standards organizations, U.S. NGS Committee Chairs, and related information.

5 The commonly used phrase "triservice" began when the National Security Act of 1947 became law on July 26, 1947, it created the Department of the Air Force as a third service. Prior to 1947, the Air Force was the Army Air Forces (AAF) and prior to March 1942, the Army Air Corps.

6 On October 1, 1997, major components of the U.S. Army Missile Command and the U.S. Army Aviation and Troop Command (ATCOM) formed the U.S. Army Aviation and Missile Command (AMCOM) at Redstone Arsenal.

7 AFBM stands for Air Force Ballistic Missile.

8 This was later published as *Human Engineering Guide to Equipment Design*, (Morgan, Cook, Chapanis, & Lund, eds., McGraw-Hill Book Co., Inc., New York, 1963). Popularly called "the HEGED," it was widely used as a textbook.

9 Digital files from recent surveys are available from Human Systems Integration Information Analysis Center [HSIIAC].

10 MANPRINT is the Army's MANpower and PeRsonnel INTegration (MANPRINT) program.

11 This requirement was also in the November 5, 1959 MIL-STD-803.

12 See MIL-HDBK-46855A summary at the beginning of this section.

13 Department of Defense Human Factors Engineering Technical Advisory Group [http://dticam.dtic.mil/hftag/index,html].

14 The JACG is comprised of senior military and civilian representatives from the Army, Navy, Air Force, Marine Corps, Coast Guard, DLA, NASA, and FA A. The JACG's charter is to develop and continuously improve joint processes and procedures that will facilitate the design, development, and acquisition of aviation systems that are identical (to the maximum extent possible) or common, and that maximize interoperability.

15 Portions of JSSG-2000 are restricted to government employees and government contractors. Those currently restricted are JSSG-2005, JSSG-2006, JSSG-2008, and JSSG-2010. When these are updated, they may be available to the public. Qualified users can order JSSG-2000 by regular mail at ASC/ENOI; 2530 Loop Road West; Wright-Patterson AFB OH; 45433-7107 or email at Engineering, Standards@wpaf-b.af.mil. Those currently available to the public are JSSG-200b, JSSG-2001B, JSSG-2007A, and JSSG-2009.

16 Military populations have body height and weight limitations, and for all practical purposes, age restrictions that make their physical characteristics differ from the general civilian population.

17 Voluntary standards themselves are copyrighted products that must be purchased and cannot be reproduced. When commercial standards organizations considered converting MIL-STD-1472 to a voluntary standard, they planned to split the comprehensive document into many single-topic standards. Anybody in government and industry would have to purchase up to 16 separate documents to get the same material. In contrast, hard copies of military standards and handbooks are available for a modest page charge, may be reproduced without charge, and are instantly available in electronic format at http://www.dodssp.daps.mil/. Reinforcing the concern regarding the need to purchase or otherwise access an unreasonably large number of HFE NGS is identification of 364 HFE NGS under 36 topical categories by the DoD HFE TAG in September 2002. Compare having to search a large portion of these 364 HFE NGS versus using a single standard, MIL-STD-1472F with its 17 NGS cited as referenced documents (TS/I, 1997).

18 NSSN originally stood for National Standards Systems Network, but now prefers National Resource for Global Standards; http://www.nssn.org/index.html. The NSSN contains over 250,000 references to standards from more than 600 developers worldwide. These have been grouped into six categories: 200,000 Approved Industry Standards; 15,000 Approved International Standards; 46,000 Approved U.S. Government Standards; 10,000 Industry Standards Under Development; 3000 International Standards Under Development; and 4000 U.S. Government Standards Under Development

19 Revised OMB Circular A-119, Federal Participation in the Development and Use of Voluntary Consensus Standards and in Conformity Assessment Activities, February 10, 1998. This circular establishes policies on Federal use and development of voluntary consensus standards and on conformity assessment activities.

20 One of the major NGS organizations, for example, charges government personnel a $200 fee to attend each NGS development meeting.

21 According to Under Secretary of Defense for Acquisition and Technology J. S. Gansler, October 14, 1999.

22 See Global Engineering: http://global.ihs.com/.

23 DoD directives, instructions, regulations, manuals, etc. are available at http://www.dtic.mil/whs/directives/.

24 The version of military standards and handbooks placed on the original contract remain in force despite subsequent updates during the life of the program.

REFERENCES

Chaikin, G. (1978). Human engineering design criteria—The value of obsolete standards and guides. Proceedings of the *Human Factors Society—22nd Annual Meeting—1978*, 409–415.

Chaikin, G. (Ed.). (1998). *HFAC highlights August 1998*. Redstone Arsenal, AL: US Army Aviation and Missile Command.

House of Representatives Report. (1916). *Report No. 94-853, Medical Device Amendments of 1976* (p. 26). Washington, DC.

Jehan, H. I., Jr. (1994). MIL-SPECS and MIL-STDS no more? DoD changes prioritizing policy. In *Program Manager, July–August* (pp. 8–10). Ft Belvoir, VA: Defense Systems Management College Press.

Lowell, S. C. (November 1, 1994). *Effects of specs and standards reform on HFE.* Unpublished oral presentation to the Human Factors Standardization Steering Committee. Orlando, FL.

McDaniel, J. W. (1996). Demise of military standards may affect ergonomics. *International Journal of Industrial Ergonomics* (Vol. 18(5–6), pp. 339–348).

Perry, W. P. (1994). *Policy memorandum on military specifications and standards.* Washington, DC: Office of the Secretary of Defense.

Process Action Team (PAT). (1994). *Report of the process action team on military specifications and standards.* Washington, DC: Office of the Under Secretary of Defense for Acquisition Technology.

Technical Society/Industry Subgroup (TS/I). (1997). *Index of non-government standards on human engineering design criteria and program requirements/guidelines.* San Antonio, TX: DoD Human Factors Engineering Technical Advisory Group.

Wagner, D., Birt, J., Snyder, M., & Duncanson, J. (1996). *FAA human factors design guide (HFDG) for acquisition of commercial off-the-shelf subsystems, non-developmental items, and developmental systems.* Atlantic City, NJ: FAA William J. Hughes Technical Center.

Section IX

Sources of Human Factors and Ergonomics Standards

43 Sources and Bibliography of Selected Human Factors and Ergonomics Standards

Anna Szopa and Waldemar Karwowski

CONTENTS

INTRODUCTION

International Organization for Standardization (ISO) defines a standard as "a documented agreement containing technical specifications or other precise criteria, to be used consistently as rules, guidelines, or definitions of characteristics, to ensure that materials, products, processes, and services are fit for the purpose served by those making reference to the standard" (ISO, 2004). Over 50 years of research and practice in human factors and ergonomics discipline clearly demonstrated that consideration of workers as "human being" in designing work and production systems results in beneficial outcomes. The objective of this chapter is to identify selected human factors and ergonomics standards developed by the international, national, and local bodies and provide selected bibliography to assist researchers and practitioners in human factors and ergonomics domain.

The listing of the standards is reasonably current as of 2020. The standards that were selected contains the terms *human factors or ergonomics* in their titles. Those standards that did not contain these terms in their titles were not included. The bibliography section of the chapter contains a compilation of books and journal articles that reflect theoretical views and empirical research on existing human factors and ergonomics standards.

SOURCES OF SELECTED HF/E STANDARDS AND GUIDELINES

Ergonomics: General Guiding Principles

ISO / TC 159/SC 1 General Ergonomics Principles, International Organization for Standardization (http://www.iso.org/iso/en/CatalogueListPage.CatalogueList?COMMID=3906&scopelist=)

ISO 6385:2016 Ergonomic principles in the design of work systems

ISO 10075:1991 Ergonomic principles related to mental work-load—General terms and definitions

ISO 10075-2:1996 Ergonomic principles related to mental workload—Part 2: Design principles

ISO 10075-3:2004 Ergonomic principles related to mental workload—Part 3: Principles and requirements concerning methods for measuring and assessing mental workload

Anthropometry and Biomechanics

NASA RP-1024 Anthropometric Source Book (http://msis.jsc.nasa.gov/volume2/Appx_a_Bibli.htm)

Guidelines for Using Anthropometric Data in Product Design (ISBN 0-945289-23-5), Human Factors and Ergonomics Society, Santa Monica, CA (http://www.hfes.org/publications/anthropometryguide.html)

ANSI Bll.TR1 Technical Report: Ergonomic Guidelines for the Design, Installation, and Use of Machine Tools. (2016) (https://www.ansi.org/default)

TC 159/SC 3 Anthropometry and biomechanics, International Organization for Standardization

ISO 7250-1:2017 Basic human body measurements for technological design—Part 1: Body measurements definitions and landmarks

ISO 11226:2000/COR 1: 2006 Ergonomics—Evaluation of static working postures—Technical corrigendum 1

ISO 11228-1:2003 Ergonomics—Manual handling—Part 1: Lifting and carrying

ISO 14738:2002 Safety of machinery—Anthropometric requirements for the design of workstations at machinery

ISO 15534-1:2000 Ergonomic design for the safety of machinery—Part 1: Principles for determining the dimensions required for openings for whole-body access into machinery

ISO 15534-2:2000 Ergonomic design for the safety of machinery—Part 2: Principles for determining the dimensions required for access openings

ISO 15534-3:2000 Ergonomic design for the safety of machinery—Part 3: Anthropometric data

ISO 15535:2006 General requirements for establishing anthropometric databases

ISO/TS 20646-1:2004 Ergonomic procedures for the improvement of local muscular workloads—Part 1: Guidelines for reducing local muscular workloads

JIS Z8500 Basic Human Body Measurements for Technological Design. (2002). Japanese Industrial Standard (http://www.jsa.or.jp/default_english.asp)

DIN 33402-2 Ergonomics-Human Body Dimensions—Part 2: Values. (2005). Deutsches Institut für Normung

SAE J833 Human Physical Dimensions. (2003). Society of Automotive Engineers (http://www.sae.org/servlets/index)

ISO 7250-1:2017 Basic human body measurements for technological design—Part 1: Body measurements definitions and landmarks, European Committee for Standardization (CEN) (http://www.cenorm.be/cenorm/index.htm)

CLOTHING

ISO 13688 Protective Clothing—General Requirements. (2013). International Organization for Standardization (http://www.iso.ch/iso/en/prods-services/ISOstore/store.html)

ISO 9920 Ergonomics of the Thermal Environment—Estimation of the Thermal Insulation and Evaporative Resistance of a Clothing Ensemble. (2007). International Organization for Standardization (http://www.iso.ch/iso/en/prods-services/ISOstore/store.html)

COLLISION AVOIDANCE

SAE ARP 4153 Human Interface Criteria for Collision Avoidance Systems in Transport Aircraft. (2008). Society of Automotive Engineers (http://www.sae.org/servlets/index)

COMMUNICATION

SAE ARP 4791 Human Engineering Recommendations for Data Link Systems. (2008). Society of Automotive Engineers (http://www.sae.org/servlets/index)

ETSI ETR 070 The Multiple Index Approach (MIA) for the Evaluation of Pictograms. (1993). European Telecommunications Standardization Institute (http://www.etsi.org/, http://www.techstreet.com/)

TELECOMMUNICATION

ETSI ETR 029 Access to Telecommunications for People with Special Needs: Recommendations for Improving and Adapting Telecommunication Terminals and Services for People with Impairments. (1998). European Telecommunications Standardization Institute (http://www.etsi.org/, http://www.techstreet.com/)

ETSI ETR 068 European Standardization Situation of Telecommunications Facilities for People with Special Needs. (1998). European Telecommunications Standardization Institute (http://www.etsi.org/, http://www.techstreet.com/)

ETSI ETR 170 Generic User Control Procedures for Telecommunication Terminals and Services. (1995). European Telecommunications Standardization Institute (http://www.etsi.org/, http://www.techstreet.com/)

ETSI ETR 095 Guide for Usability Evaluations of Telecommunications Systems and Services. (1993). European Telecommunications Standardization Institute (http://www.etsi.org/, http://www. techstreet.com/)

ETSI ETR 160 Human Factors Aspects of Multimedia Telecommunications. (1995). European Telecommunications Standardization Institute (http://www.etsi.org/, http://www.tech-street.com/)

ETSI ETR 039 Human Factors Standards for Telecommunications Applications. (1992). European Telecommunications Standardization Institute (http://www.etsi.org/, http://www. techstreet.com/)

ETSI ETR 165 Recommendation for a Tactile Identifier on Machine Readable Cards for Telecommunication Terminals. (1995). European Telecommunications Standardization Institute (http://www.etsi.org/, http://www.techstreet.com/)

ETSI ETR 167 User Instructions for Public Telecommunications Services; Design Guidelines. (1995). European Telecommunications Standardization Institute (http://www.etsi.org/, http://www. techstreet.com/)

TELEPHONES

ETSI ETR 051 Human Usability Checklist for Telephones: Basic Requirements. (1992). European Telecommunications Standardization Institute (http://www.etsi.org/, http://www.techstreet.com/)

ETSI ETR 096 Human Factors Guidelines for the Design of Minimum Phone Based User Interface to Computer Services. (1993). European Telecommunications Standardization Institute (http://www.etsi.org/, http://www.techstreet.com/)

ETSI ETR 166 Evaluation of Telephones for People with Special Needs: An Evaluation Method. (1995). European Telecommunications Standardization Institute (http://www.etsi.org/, http://www.techstreet.com/)

ETSI ETR 187 Recommendation of Characteristics of Telephone Services Tones When Locally Generated in Telephony Terminals. (1995). European Telecommunications Standardization Institute (http://www.etsi.org/, http://www.techstreet.com/)

VIDEOPHONES

ETSI ETS 300 375 Pictograms for Point-to-Point Video Telephony. (1994). European Telecommunications Standardization Institute (http://www.etsi.org/, http://www.techstreet.com/)

ETSI ETR 175 User Procedures for Multipoint Video Telephony. (1995). European Telecommunications Standardization Institute (http://www.etsi.org/, http://www.techstreet.com/)

CONTROL ROOMS

ISO 11064-1 Ergonomic Design of Control Centres—Part 1: Principles for the Design of Control Centers. (2000). International Organization for Standardization (http://www.iso.ch/iso/en/prods-services/ISOstore/store.html)

ISO 11064-2 Ergonomic Design of Control Centers—Part 2: Principles for the Arrangement of Control Suites. (2000). International Organization for Standardization (http://www.iso.ch/iso/en/prods-services/ISOstore/store.html)

ISO 11064-3 Ergonomic Design of Control Centers—Part 3: Control Room Layout. (1999). International Organization for Standardization (http://www.iso.ch/iso/en/prods-services/ISOstore/store.html)

ISA RP 60.3 Human Engineering for Control Centers. (1985). Instrumentation, Systems, and Automation Society (http://www.isa.org)

BS 7517 Nuclear Power Plants—Control Rooms—Operator Controls. (1995). British Standard https://www.bsigroup.com/en-GB/

IEC 61227 Nuclear Power Plants—Control Rooms—Operator Controls. (2008). International Electrotechnical Commission (http://www.iec.ch/)

IEC 60964 Nuclear Power Plants—Control Rooms—Design. (2018). International Electrotechnical Commission (http://www.iec.ch/)

AIRCRAFT CONTROLS AND DISPLAYS

SAE ARP 4102 Flight Deck Panels, Controls, and Displays. (1988). Society of Automotive Engineers (http://www.sae.org/servlets/index)

SAE ARP 4102/7 Flight Deck Panels, Controls, and Displays, Part 7: Electronic Display Symbology for EADI/PFD. (2007). Society of Automotive Engineers (http://www.sae.org/servlets/index)

SAE ARP 4102/8A Flight Deck Panels, Controls, and Displays, Part 8: Flight Deck Head-Up Displays. (1998). Society of Automotive Engineers (http://www.sae.org/servlets/index) SAE

SAE ARP 4032B Human Engineering Considerations in the Application of Color to Electronic Aircraft Displays. (2013). Society of Automotive Engineers (http://www.sae.org/servlets/index)

CONTROL AND DISPLAY DESIGN

ISO 9355-1 Ergonomic Requirements for the Design of Displays and Control Actuators—Part 1: Human Interactions with Displays and Control Actuators. (1999). International Organization for Standardization (http://www.iso.ch/iso/en/prods-services/ISOstore/store.html) ISO 9355-2 Ergonomic Requirements for the Design of Signals and Control Actuators—Part 2: Displays. (1999). International Organization for Standardization (http://www.iso.ch/iso/en/prods-services/ISOstore/store.html)

SAE J107 Operator Controls and Displays on Motorcycles. (2019). Society of Automotive Engineers (http://www.sae.org/servlets/index)

SAE J680 Location and Operation of Air Brake Controls in Motor Trucks Cabs. (2015). Society of Automotive Engineers (http://www.sae.org/servlets/index)

Office Ergonomics

ANSI/HFES100 Human Factors Engineering of Computer Workstations (Draft standard for trial use). (2007). Human Factors and Ergonomics Society, Santa Monica, CA. [NATIONAL] (http://www.hfes.org/publications/HFES100.html)

ISO 9241 Usability Standards, International Organization for Standardization. Brief description reported in http://www.ergoweb.com/resources/reference/guidelines/iso9241.cfm

Part 1: General Introduction contains general information about the standard and provides an overview of each of the parts.

Part 2: Task Requirements discusses the enhancement of user interface efficiency and the well-being of users by applying practical ergonomic knowledge to the design of VDT work tasks.

Part 3: Display Requirements specifies requirements for visual displays and their images. Part 4: Keyboard Requirements specify the characteristics that determine the effectiveness in accepting keystrokes from a user.

Part 5: Workstation Requirements specify the design characteristics of workplaces in which VDTs are used.

Part 6: Environmental Requirements specifies characteristics of the working environment in which VDTs are used.

Part 7: Display requirements with reflections describe how to maintain usable and acceptable VDT image quality by evaluating the reflection properties of a screen and the image quality of the screen over a range of typical office lighting conditions.

Part 8: Requirements for displayed color state specifications for display color images, color measurement metrics, and visual perception tests.

Part 9: Requirements for non-keyboard-input devices specifies requirements for the design and usability of input devices other than keyboards.

Part 10: Dialogue Principles specifies a set of high-level dialogue design principles for command languages, direct manipulation, and form-based entries.

Part 11: Guidance on Usability explains the way in which the user, equipment, task, and environment should be described—as part of the total system—and how usability can be specified and evaluated.

Part 12: Presentation of Information specifies requirements for the coding and formatting of information on computer screens.

Part 13: User Guidance specifies requirements and attributes to be considered in the design and evaluation of the software user interfaces.

Part 14: Menu Dialogues provides conditional requirements and recommendations for menus in user-computer dialogues.

Part 15: Command Dialogues provides conditional recommendations for common languages.

Part 16: Direct Manipulation Dialogues provides guidance on the design of manipulation dialogues in which the user directly acts upon object or object representations (icons) to be manipulated.

Part 14: Menu Dialogues Part 15: Command Dialogues Part 16: Direct Manipulation Dialogues Part 17: Form-Filling Dialogues

BIFMA G1-2013 Ergonomics Guideline Furniture Used in Office Work Spaces Designed for Computer Use, The Business and Institutional Furniture Manufacturers Association [LOCAL/ ORGANIZATIONAL] (http://www.bifma.org/standards/index.html)

CSA Z412-00 Guideline on Office Ergonomics, The Canadian Standards Association (CSA). [NATIONAL] (http://www.csa-intl.org/onlinestore)

ACGIH 9331 Ergonomics in Computerized Offices. (2002). American Conference of Governmental Industrial Hygienists (http://www.acgih.org/home.htm)

HUMAN-SYSTEM INTERACTION

TC 159/SC 4 Ergonomics of human-system interaction, International Organization for Standardization (http://www.iso.org/iso/en/CatalogueListPage.CatalogueList?COMMID=3916&scopelist=)

ISO 1503:2008 Spatial Orientation and direction of movement—Ergonomic requirements.

ISO 9241-1:1997 Ergonomic requirements for office work with visual display terminals (VDTs)—Part 1: General introduction

ISO 9241-1:1997/Amd 1: 2001

ISO 9241-2:1992 Ergonomic requirements for office work with visual display terminals (VDTs)—Part 2: Guidance on task requirements

ISO/WD TR 9241-311 Ergonomics of human-system interaction—Part 311: Guidance of application of ISO 9241-307: LCD screens for workstation.

ISO 9241-304:2008 Ergonomics of human-system interaction—Part 304: User performance test methods for electronic visual displays.

ISO 9241-411:2012 Ergonomics of human-system interaction—Part 411: Evaluation methods for the design of physical input devices.

ISO 9241-5:1998 Ergonomic requirements for office work with visual display terminals (VDTs)—Part 5: Workstation layout and postural requirements

ISO 9241-6:1999 Ergonomic requirements for office work with visual display terminals (VDTs)—Part 6: Guidance on the work environment

ISO 9241-302:2008 Ergonomics of human-system interaction—Part 302: Technology for electronic visual display.

ISO 9241-303:2008 Ergonomics of human-system interaction—Part 303: Requirements for electronic visual display.

ISO 9241-305:2008 Ergonomics of human-system interaction—Part 305: Optical laboratory test methods for electronic visual display.

ISO 9241-307:2008 Ergonomics of human-system interaction—Part 307: Analysis and compliance test methods for electronic visual display.

ISO 9241-400:2007 Ergonomics of human-system interaction—Part 400: Principles and requirements for physical input devices.

ISO 9241-110:2020 Ergonomics of human-system interaction—Part 110: Interaction principles.

ISO 9241-11:2018 Usability: Definitions and concepts.

ISO 9241-125:2017 Ergonomics of human-system interaction—Part 125: Guidance on visual presentation of information.

ISO 9241-13:1998 Ergonomic requirements for office work with visual display terminals (VDTs)—Part 13: User guidance

ISO 9241-14:1997 Ergonomic requirements for office work with visual display terminals (VDTs)—Part 14: Menu dialogues

ISO 9241-15:1997 Ergonomic requirements for office work with visual display terminals (VDTs)—Part 15: Command dialogues

ISO 9241-16:1999 Ergonomic requirements for office work with visual display terminals (VDTs)—Part 16: Direct manipulation dialogues

ISO 9241-143:2012 Ergonomics of human-system interaction—Part 143: Forms.

ISO 9355-1:1999 Ergonomic requirements for the design of displays and control actuators—Part 1: Human interactions with displays and control actuators

ISO 9355-2:1999 Ergonomic requirements for the design of displays and control actuators—Part 2: Displays

ISO 11064-1:2000 Ergonomic design of control centers—Part 1: Principles for the design of control centers

ISO 11064-2:2000 Ergonomic design of control centers—Part 2: Principles for the arrangement of control suites

ISO 11064-3:1999 Ergonomic design of control centers—Part 3: Control room layout

ISO 11064-3:1999/Cor 1:2002

ISO 11064-4:2013 Ergonomic design of control centers—Part 4: Layout and dimensions of workstations

ISO 13407:1999 Human-centered design processes for interactive systems

ISO 14915-1:2002 Software ergonomics for multimedia user interfaces—Part 1: Design principles and framework

ISO 14915-2:2003 Software ergonomics for multimedia user interfaces—Part 2: Multimedia navigation and control

ISO 14915-3:2002 Software ergonomics for multimedia user interfaces—Part 3: Media selection and combination

ISO 9241-171:2008 Ergonomics of human-system interaction—Part 171: Guidance of software accessibility.

ISO/TR 16982:2002 Ergonomics of human-system interaction—Usability methods supporting human-centered design

ISO/PAS 18152:2010 Ergonomics of human-system interaction—Specification for the process assessment of human-system issues

ISO 9241-220: 2019 Ergonomics of human-system interaction—Part 220: Processes for enabling, executing, and assessing human-centered design within organizations.

PHYSICAL ENVIRONMENT

TC 159/SC 5 Ergonomics of the physical environment, International Organization for Standardization (http://www.iso.org/iso/en/CatalogueListPage.CatalogueList?COMMID=3916&scopelist=)

ISO 7243:2017 Ergonomics of the thermal environment—assessment of the heat stress using the WBGT (wet bulb globe temperature) index.

ISO 7726:1998 Ergonomics of the thermal environment—Instruments for measuring physical quantities

ISO 7730:2005 Ergonomics of the thermal environment—Analytical determination and interpretation of thermal comfort using calculation of the PMV and PPD indices and local thermal comfort criteria.

ISO 7731:2003 Ergonomics—Danger signals for public and work areas—Auditory danger signals

ISO 7933:2004 Ergonomics of the thermal environment—Analytical determination and interpretation of heat stress using calculation of the predicted heat strain

ISO 8996:2004 Ergonomics of the thermal environment—Determination of metabolic rate

ISO 9886:2004 Ergonomics—Evaluation of thermal strain by physiological measurements

ISO 9920:2007 Ergonomics of the thermal environment—Estimation of the thermal insulation and water vapor resistance of a clothing ensemble.

ISO 9921:2003 Ergonomics—Assessment of speech communication

ISO 10551:2019 Ergonomics of the physical environment—subjective judgment scales for assessing physical environments.

ISO 11079:2007 Evaluation of cold environments—Determination and interpretation of cold stress when using required clothing insulation (IREC) and local cooling effects.

ISO 11399:1995 Ergonomics of the thermal environment—Principles and application of relevant International Standards

ISO 11428:1996 Ergonomics—Visual danger signals—General requirements, design, and testing

ISO 11429:1996 Ergonomics—System of auditory and visual danger and information signals

ISO 12894:2001 Ergonomics of the thermal environment—Medical supervision of individuals exposed to extreme hot or cold environments

ISO 13731:2001 Ergonomics of the thermal environment—Vocabulary and symbols

ISO/TS 13732-2:2001 Ergonomics of the thermal environment—Methods for the assessment of human responses to contact with surfaces—Part 2: Human contact with surfaces at moderate temperature

ISO 15265:2004 Ergonomics of the thermal environment—Risk assessment strategy for the prevention of stress or discomfort in thermal working conditions

ISO/TR 19358:2002 Ergonomics—Construction and application of tests for speech technology

CEN/TC 122 Ergonomics, European Committee for Standardization (http://www.cenorm.be/cenorm/index.htm):

INFORMATION TECHNOLOGY/SOFTWARE ENGINEERING

ICS 35 Information technology. Office machines http://www.iso.org/iso/en/CatalogueListPage.CatalogueList?ICSl=35&scopelist

ISO/IEC 11581-1:2000 Information technology—User system interfaces and symbols—Icon symbols and functions Part 1: Icons—General

ISO/IEC 11581-2:2000 Information technology—User system interfaces and symbols—Icon symbols and functions Part 2: Object icons

ISO/IEC 11581-3:2000 Information technology—User system interfaces and symbols—Icon symbols and functions Part 3: Pointer icons

ISO/IEC 11581-5:2004 Information technology—User system interfaces and symbols—Icon symbols and functions Part 5: Tool icons

ISO/IEC 11581-6:1999 Information technology—User system interfaces and symbols—Icon symbols and functions Part 6: Action icons

NASA-STD-8719.13C Software Safety Standard. (2020). National Aeronautics and Space Administration (https://standards.nasa.gov)

JTC 1/SC 7 Software and system engineering (http://www.iso.org/iso/en/stdsdevelopment/tc/tclist/TechnicalCommitteeDetailPage.TechnicalCommitteeDetail?COMMID=40)

ISO/IEC 9126-1:2001 Software engineering—Product quality—Part 1: Quality model ISO/IEC ISO/IEC TR 9126-2:2003 Software engineering—Product quality—Part 2: External metrics

ISO/IEC TR 9126-3:2003 Software engineering—Product quality—Part 3: Internal metrics

ISO/IEC TR 9126-4:2004 Software engineering—Product quality—Part 4: Quality in use metrics

QUALITY AND ENVIRONMENTAL MANAGEMENT STANDARDS

ISO 9000. Quality management and related standard series: (http://www.iso.org/iso/en/iso9000-14000/iso9000/iso9000index.html)

ISO 9000:2015. Quality management systems—Fundamentals and vocabulary

ISO 9001:2015. Quality management systems—Requirements

ISO 9004:2018. Quality management—Quality of an organization—Guidance to achieve sustained success.

ISO 10005:2018. Quality management—Guidelines for quality plans

ISO 10006:2017. Quality management—Guidelines for quality management in projects

ISO 10007:2017. Quality management—Guidelines for configuration management

ISO/DIS 10012:2003. Measurement management systems—Requirements for measurements processes and measuring equipment

ISO/TR 10013:2001. Guidelines for developing quality manuals

ISO 10014:2006. Quality management-Guidelines for realizing financial and economic benefits

ISO 10015:2019. Quality management—Guidelines for competence management and people development

ISO/TS 16949:2009. Quality management systems—Particular requirements for the application of ISO 9001:2008 for automotive production and relevant service part organizations

ISO 14000:2015. Environment Management Systems—requirements with guidance for use

ISO 14040:2006. Environment Management—Life cycle assessment—Principals and framework

ISO 14062:2002. Environment Management—Integrating of environmental aspects into product design and development

ISO 14020:2000. Environmental labels and declarations—General principles

ISO 14063:2020. Environmental management-Environmental communication—Guidelines and examples

ISO 19011:2018. Guidelines for auditing management systems

OCCUPATIONAL SAFETY AND HEALTH STANDARDS

ILO-OSH 2001—Principles for occupational safety and health management system, International Labour Organization

AFMA Voluntary Ergonomics Guideline for the Furniture Manufacturing Industry, American Furniture Manufacturers Association (AFMA, 2003)

OSHA Voluntary Ergonomics Standards

1. Nursing Home Guideline (issued on March 13, 2003)
2. Draft Guideline for Poultry Processing (issued on June 3, 2003)
3. Guideline for the Retail Grocery Industry (issued on May 28, 2004)

AIHA ASC ZIO Occupational Health Safety Systems, American Industrial Hygiene Association (http://www.aiha.0rg/ANSIC0mmittees/html/z1Ocommittee.htm)

ASC Z-365 Management of Work-Related Musculoskeletal Disorders (MSD). (2002, final draft) (http://www.nsc.org/ehc/z365/finldrft.htm)

ACGIH 99-049 Ergonomics and Safety in Hand Tool Design. (1999). American Conference of Governmental Industrial Hygienists (http://www.acgih.org/home.htm)

MACHINERY SAFETY

CEN/TC 122 Ergonomics, European Committee for Standardization (http://www.cenorm.be/cenorm/index.htm):

EN 547-1:2009. Safety of machinery—Human body measurements—Part 1: Principles for determining the dimensions required for openings for whole body access into machinery

EN 547-2:2009. Safety of machinery—Human body measurements—Part 2: Principles for determining the dimensions required for access openings

EN 547-3:2009. Safety of machinery—Human body measurements—Part 3: Anthropometric data

ISO 13732-1:2006. Ergonomics of the thermal environment—Methods of the assessment of the human responses to contact with the surfaces—Part 1. Hot surfaces

ISO 13732-2:2001. Ergonomics of the thermal environment—Methods of the assessment of the human responses to contact with the surfaces—Part 2. Cold surfaces

ISO 13732-3:2005. Ergonomics of the thermal environment—Methods of the assessment of the human responses to contact with the surfaces—Part 3. Human contact with surfaces at the moderate temperature

EN 614-1:2009. Safety of machinery—Ergonomic design principles—Part 1: Terminology and general principles

EN 614-2:2008. Safety of machinery—Ergonomic design principles—Part 2: Interactions between the design of machinery and work tasks

EN 842:2009. Safety of machinery—Visual danger signals—General requirements, design and testing

EN 894-1:2009. Safety of machinery—Ergonomics requirements for the design of displays and control actuators—Part 1: General principles for human interactions with displays and control actuators

EN 894-2:2009. Safety of machinery—Ergonomics requirements for the design of displays and control actuators—Part 2: Displays

EN 894-3:2010. Safety of machinery—Ergonomics requirements for the design of displays and control actuators—Part 3: Control actuators

EN 981:2009. Safety of machinery—System of auditory and visual danger and information signals

EN 1005-1:2009. Safety of machinery—Human physical performance—Part 1: Terms and definitions

EN 1005-2:2009. Safety of machinery—Human physical performance—Part 2: Manual handling of machinery and component parts of machinery

EN 1005-3:2009. Safety of machinery—Human physical performance—Part 3: Recommended force limits for machinery operation

TRANSPORTATION

TC 22/SC 13 Ergonomics applicable to road vehicles (http://www.iso.org/iso/en/Catalogue ListPage.CatalogueList?COMMID=869&scopelist=CATALOGUE)

ISO 2575:2010. Road vehicles—Symbols for controls, indicators, and tell-tales

ISO 3409:1975. Passenger cars—Lateral spacing of foot controls

ISO 3958:1996. Passenger cars—Driver hand-control reach

ISO 4040:2009. Road vehicles—Location of hand controls, indicators, and tell-tales in motor vehicles

ISO 6549:1999. Road vehicles—Procedure for H- and R-point determination

ISO 12214:2018. Road vehicles—Direction-of-motion stereotypes for automotive hand controls

ISO 15005:2017. Road vehicles—Ergonomic aspects of transport information and control systems—Dialogue management principles and compliance procedures

ISO 15006:2011. Road vehicles—Ergonomic aspects of transport information and control systems—Specifications for in-vehicle auditory presentation

ISO 15007-1:2014. Road vehicles—Measurement of driver visual behavior with respect to transport information and control systems—Part 1: Definitions and parameters

ISO/TS 15007-2:2014. Road vehicles—Measurement of driver visual behavior with respect to transport information and control systems—Part 2: Equipment and procedures

ISO 15008:2017. Road vehicles—Ergonomic aspects of transport information and control systems—Specifications and test procedures for in-vehicle visual presentation

ISO/TS 16951:2004. Road vehicles—Ergonomic aspects of transport information and control systems (TICS)—Procedures for determining priority of on-board messages presented to drivers

ISO 17287:2003. Road vehicles—Ergonomic aspects of transport information and control systems—Procedure for assessing suitability for use while driving

TC 20/SC 14 Space systems and operations (http://www.iso.org/iso/en/CatalogueListPage.CatalogueList?COMMID=739&scopelist=CATALOGUE)

ISO 14620-1:2018. Space systems—Safety requirements—Part 1: System safety

ISO 14620-2:2019. Space systems—Safety requirements—Part 2: Launch site operations

ISO 17399:2003. Space systems—Man-systems integration

ISO 17666:2016. Space systems—Risk management

TC 23/SC 3 Tractor Safety and comfort of the operator (http://www.iso.org/iso/en/CatalogueListPage.CatalogueList?COMMID=938&scopelist=CATALOGUE)

ISO 3463:2006. Tractors for agriculture and forestry—Roll-over protective structures (ROPS)—Dynamic test method and acceptance conditions

ISO 3776-1:2006. Tractors and machinery for agriculture—Seat belt Part 1: Anchorage location requirements

ISO 3776-2:2013. Tractors and machinery for agriculture—Seat belt Part 2: Anchorage strength requirements

ISO 3776-3:2009. Tractors and machinery for agriculture—Seat belt Part 3: Requirements for assemblies

ISO 4254-1:2013. Agriculture machinery—Safety—Part 1: General

ISO 4254-5:2018. Agriculture machinery—Safety—Part 5: Power-driven soil-working equipment

ISO 4254-9:2018. Agriculture machinery—Safety—Part 9Seed drills

ISO 5700:2013. Tractors for agriculture and forestry—Roll-over protective structures—Static test method and acceptance conditions

ISO 12140:2013. Agricultural machinery—Agricultural trailers and trailed equipment—Drawbar jacks

ISO/TS 15077:2020. Tractors and self-propelled machinery for agriculture—Operator controls—Actuating forces, displacement, location, and method of operation

TC 23/SC 14 Tractor Operator controls, operator symbols and other displays, operator manuals (http://www.iso.org/iso/en/CatalogueListPage.CatalogueList?COMMID=991&scopelist=CATALOGUE)

ISO 3600:2015. Tractors, machinery for agriculture and forestry, powered lawn and garden equipment—Operator's manuals—Content and format

ISO 3767-1:2016. Tractors, machinery for agriculture and forestry, powered lawn and garden equipment—Symbols for operator controls and other displays—Part 1: Common symbols

ISO 3767-2:2016. Tractors, machinery for agriculture and forestry, powered lawn and garden equipment—Symbols for operator controls and other displays—Part 2: Symbols for agricultural tractors and machinery

ISO 3767-3:2016. Tractors, machinery for agriculture and forestry, powered lawn and garden equipment—Symbols for operator controls and other displays—Part 3: Symbols for powered lawn and garden equipment

ISO 3767-4:2016. Tractors, machinery for agriculture and forestry, powered lawn and garden equipment—Symbols for operator controls and other displays—Part 4: Symbols for forestry machinery

ISO 3767-5:2016. Tractors, machinery for agriculture and forestry, powered lawn and garden equipment—Symbols for operator controls and other displays—Part 5: Symbols for manual portable forestry machinery

ISO 11684:1995. Tractors, machinery for agriculture and forestry, powered lawn and garden equipment—Safety signs and hazard pictorials—General principles

TC 110/SC 2 Safety of powered industrial trucks (http://www.iso.org/iso/en/CatalogueListPage. CatalogueList?COMMID=3090&scopelist=CATALOGUE)

ISO 3287:1999. Powered industrial trucks—Symbols for operator controls and other displays

ISO 15870:2000. Powered industrial trucks—Safety signs and hazard pictorials—General principles

Manual Materials Handling

ISO 780:2015 Packaging—Distribution packaging—Graphical symbols for handling and storage of packages

ISO 11228-1:2003. Ergonomics—Manual handling—Part 1: Lifting and carrying

ISO 11228-2:2007. Ergonomics—Manual handling—Part 2: Pushing and pulling

ISO 11228-3:2007. Ergonomics—Manual Handling—Part 3: Handling of low loads at high frequency

ISO/TS 20646:2014. Ergonomic guidelines for the optimization of musculoskeletal workloads

NASA-STD-8719.9 Standard for Lifting Devices and Equipment. (2015). National Aeronautics and Space Administration (https://sma.nasa.gov/news/articles/newsitem/2015/09/22/streamlined-lifting-standard-reduces-redundancy-and-clarifies-processes)

Guidelines for Elderly/Disabled Users

ANSI A117.1 Accessible and Useable Buildings and Facilities. (2013). American National Standards Institute

BS 4467 Guide to Dimensions in Designing for Elderly People. (1991). British Standard (http://www.bsi-global.com/index.xalter)

Ground Vehicle Standards

SAE J2364: Navigation and Route Guidance Function Accessibility While Driving. (2015). Society of Automotive Engineers (https://www.sae.org)

SAE J2365: Calculation and measurements of the Time to Complete In-Vehicle Navigation and Route Guidance Tasks. (2016). Society of Automotive Engineers (https://www.sae.org)

SAE J1050 Describing and Measuring the Driver's Field of View. (2019). Society of Automotive Engineers (https://www.sae.org)

SAE J941 Motor Vehicle Driver's Eye Locations. (2010). Society of Automotive Engineers (https://www.sae.org)

ISO 7397-1 Verification of Driver's Direct Field of View—Part 1: Vehicle Positioning for Static Measurement. (1993). International Organization for Standardization

ISO 7397-2 Verification of Driver's Direct Field of View—Part 2: Test Method. (1993). International Organization for Standardization

FURNITURE STANDARDS

BS 3044 Guide to Ergonomics Principles in the Design and Selection of Office Furniture. (1991). British Standard (https://www.bsigroup.com/en-GB/)

HUMAN ERROR

ACGIH 9658 Human Error Reduction and Safety Management. (1996). American Conference of Governmental Industrial Hygienists (http://www.acgih.org/home.htm)

AICHE Gl5 Guidelines for Preventing Human Error in Process Safety. (1994). American Institute of Chemical Engineers (http://www.aiche.org/)

API 770A Manager's Guide to Reducing Human Errors: Improving Human Performance in the Process. (2001). American Petroleum Institute (http://api-ec.api.org/newsplashpage/index.cfm/)

GUIDELINES FOR MEDICAL DEVICES

AAMI HE48 Human Factors Engineering Guidelines and Preferred Practices for the Design of Medical Devices. (1993). Association for the Advancement of Medical Instrumentation (http://www.aami.org/)

AAMI HE74 Human Factors Engineering—Design of medical devices. (2009). Association for the Advancement of Medical Instrumentation (http://www.aami.org/)

GUIDELINES FOR HUMAN PERFORMANCE AND RELIABILITY

AIAA G-035 Guide to Human Performance Measurements. (2000). American Institute of Aeronautics and Astronautics (http://www.aiaa.org/)

ACGIH 9651 Evaluation of Human Work, A Practical Ergonomics Methodology. (1995). American Conference of Governmental Industrial Hygienists (http://www.acgih.org/home.htm)

IEEE 1082 Guide for incorporating human reliability analysis into probabilistic risk assessments for nuclear power generating stations and other nuclear facilities. (2017).

ROBOT DESIGN AND SAFETY

ANSI/RIA R15.02-1 Industrial Robots and Robot Systems—Hand-Held Robot Control Pendants—Human Engineering Design Criteria (1990). American National Standards Institute (http://www.ansi.org/, http://www.roboticsonline.com/store/)

ANSI/RIA R15.06 Industrial Robots and Robot Systems—Safety Requirements. (2012). American National Standards Institute (http://www.ansi.org/, http://www.roboticsonline.com/store/)

USER INTERFACE STANDARDS/GUIDELINES

IBM Web Design Guidelines (http://www-3.ibm.com/ibm/easy/eou_ext.nsf/publish/572)

IBM User Interface Architecture Administrative Guidelines (http://www-3.ibm.com/ibm/easy/eou_ext.nsf/publish/1392/$File/ibm_uia.pdf)

IBM OOBE Usability Guidelines (http://www-3.ibm.com/ibm/easy/eou_ext.nsf/publish/577)

GNOME Foundation Human Interface Guidelines. (2004). (http://developer.gnome.org/projects/gup/hig/2.0/)

ESD-TR-86-278 Guidelines for Designing User Interface Software. (1986). The MITRE Corporation, Bedford, MA (http://hcibib.org/sam/)

Apple Human Interface Guidelines (http://developer.apple.com/documentation/UserExperience/Conceptual/OSXHIGuidelines/)

GUI Standard by Human Factors International, Inc. (Local/Organizational) (http://www.humanfactors.com/downloads/GUIbooklet.asp)

Web Site Design Standards/Guidelines

36 CFR Part 1194 Information and communication technology (ICT)—Standards and guidelines (2017).

Research-Based Web Design and Usability Guidelines. (2003). The U.S. Department of Health and Human Services (HHS) and National Cancer Institute (http://usability.gov/pdfs/guidelines.html)

Web Style Guide, 4th edition (http://www.webstyleguide.com/index.html)

Telstra Online Standards (http://www.telstra.com.au/standards/standards/standards_all.cfm)

W3C User Agent Accessibility Guidelines. (2002). (http://www.w3.org/TR/UAAG10/) Internet Standard by Human Factors International, Inc. (Local/Organizational) (http://www.humanfactors.com/downloads/Intranetbooklet.asp)

U.S. Department of Defense Standards

DOD-HDBK-743A Anthropometry of U.S. Military Personnel. (1991). (http://assist.daps.dla.mil/docimages/0000/40/29/54083.pd0)

U.S. Department of Transportation Standards—Federal Aviation Administration

HF-STD-001 Human Factors Design Standard. (2003). Department of Transportation, Federal Aviation Administration (http://www.hf.faa.gov/docs/508/docs/wjhtc/hfds.zip) DOT-VNTSC-FAA-95-3 Human Factors in the Design and Evaluation of Air Traffic Control Systems. (1995). Department of Transportation, Federal Aviation Administration (http://www.hf.faa.gov/docs/volpehndk.zip)

FAA-HF-001A Human Factors Program Plan. (2009). Department of transportation, Federal Aviation Administration (http://www.hf.faa.gov/docs/did_001.htm)

FAA-HF-002A Human Engineering Design Approach Document—Operator. (2009). Department of Transportation, Federal Aviation Administration (http://www.hf.faa.gov/docs/did_002.htm)

FAA-HF-003A Human Engineering Design Approach Document—Maintainer. (2009). Department of Transportation, Federal Aviation Administration (http://www.hf.faa.gov/docs/did_003.htm)

FAA-HF-004A Critical Task Analysis Report (2009). Department of Transportation, Federal Aviation Administration (http://hfetag.dtic.mil/docs-hfs/faa-hf-004_critical_task_anal-ysis_report.doc)

FAA-HF-005A Human Engineering Simulation Concept. (2009). Department of Transportation, Federal Aviation Administration (http://hfetag.dtic.mil/docs-hfs/faa-hf-005_human-engineering_simulation.doc)

U.S. Department of Transportation Standards—Federal Highway Agency

FHWA-JPO-99-042 Preliminary Human Factors Guidelines for Traffic Management Centers. (1999). Department of Transportation, Federal Highway Agency (http://plan2op.fhwa.dot.gov/pdfs/pdf2/edl10303.pdf)

FHWA-RD-98-057 Human Factors Design Guidelines for Advanced Traveler Information Systems (ATIS) and Commercial Vehicle Operations (CVO). (1998). Department of Transportation, Federal Highway Agency (http://www.fhwa.dot.gov/tfhrc/safety/pubs/atis/index.html)

FHWA-RD-01-051 Guidelines and Recommendations to Accommodate Older Drivers and Pedestrians. (2001). Department of Transportation, Federal Highway Agency (http://www.tfhrc. gov/humanfac/01105/cover.htm)

FHWA-RD-01-103 Highway Design Handbook for Older Drivers and Pedestrians. (2001). Department of Transportation, Federal Highway Agency (http://www.tfhrc.gov/humanfac/01103/ coverfront.htm)

U.S. DEPARTMENT OF ENERGY STANDARD

DOE-HDBK-1140-2001 Human Factors/Ergonomics Handbook for the Design for Ease of Maintenance. (2017). Department of Energy (http://tis.eh.doe.gov/techstds/standard/hdbk1140/ hdbk11402001_part1.pdf)

MILITARY STANDARDS AND RELATED DOCUMENTATIONS

MIL-STD-882E Standard Practice for System Safety. (2012). (http://assist.daps.dla.mil/ docimages/0001/95/78/std882d.pd8)

MIL-STD-1472GCHG-1 Human Engineering. (2019). (http://assist.daps.dla.mil/docimages/ 0001/87/31/milstd14.pd1)

MIL-STD-1474E Noise Limits. (2015). (http://assist.daps.dla.mil/docimages/0000/31/59/1474d.pdl)

MIL-STD-1477C Symbols for Army Systems Displays. (2009). (http://assist.daps.dla.mil/ docimages/0000/42/03/69268.pd9)

MIL-STD-1787D Aircraft Display Symbology. (2018).

MIL-HDBK-759C Human Engineering Design Guidelines. (2012). (http://assist.daps.dla.mil/ docimages/0000/40/04/mh759c.pd8)

MIL-HDBK-767 Design Guidance for Interior Noise Reduction in Light-Armored Tracked Vehicles. (2004). (http://assist.daps.dla.mil/docimages/0000/13/24/767.pdl)

MIL-HDBK-1473A Color and Marking of Army Materiel. (2018). (http://assist.daps.dla.mil/ docimages/OOOO/85/40/hdbk1473.pd6)

MIL-HDBK-1908A Definitions of Human Factors Terms. (2004). (http://assist.daps.dla.mil/ docimages/0001/81/33/1908hdbk.pd9)

MIL-HDBK-46855A Human Engineering Requirements for Military Systems Equipment and Facilities (2011). https://quicksearch.dla.mil/qsSearch.aspx

NATIONAL AERONAUTICS AND SPACE ADMINISTRATION (NASA) MAN-SYSTEM INTEGRATION STANDARD

NASA-STD-3000B Man-Systems Integration Standards. (1995). National Aeronautics and Space Administration (http://msis.jsc.nasa.gov)

FEDERAL ACCESSIBILITY STANDARD

FED-STD-795 Uniform Federal Accessibility Standards. (1988). (http://assist.daps.dla.mil/ docimages/0000/46/05/53835.pd5)

SOURCES OF BIBLIOGRAPHY

Akoumianakis, D., Stephanidis, C. (1997). Supporting user-adapted interface design: The USE-IT system. *Interacting with Computers*, *9*, 73–104.

Albin, T. J. (2004). Board of Standards Review/Human Factor and Ergonomics Society 100—Human factors engineering of computer workstations—Draft Standard for Trial Use. In W. Karwowski (Ed.), 2005,

Handbook of human factors and ergonomics standards and guidelines. Hillsdale, NJ: Lawrence Erlbaum Publishers.

American Furniture Manufacturers Association. (2003). *Voluntary Ergonomics Guideline for the furniture manufacturing industry*. High Point, NC: AFMA.

Anshel, J. (1998). *Visual ergonomics in the workplace*. Taylor & Francis: London, UK.

Babakri, K. A., Bennett, R. A., and Franchetti, M. (2003). Critical factors for implementing ISO 14001 standard, in United States industrial companies. *Journal of Cleaner Production, 11*, 749–752.

Babakri, K. A., Bennett, R. A., Rao, S., & Franchetti, M. (2004). Recycling performance of firms before and after adoption of the ISO 14001 standard. *Journal of Cleaner Production, 12*, 633–637.

Baleani, M., Cristofolini, L., & Viceconi, M. (1999). Endurance testing of hip prostheses: A comparison between the load fixed in ISO 7206 standard and the physiological loads. *Clinical Biomechanics, 14*, 339–345.

Barre, F., & Lopez, J. (2001). On a 3D extension of the MOTIF method (ISO 12085). *International Journal of Machine Tools and Manufacture, 41*, 1873–1880.

Bastien, J. M. C., Scapin, D. L., & Leulier, C. (1999). The ergonomic criteria and the ISO/DIS 9241–10 dialogue principles: A pilot comparison in an evaluation task. *Interacting with Computers, 11*, 299–322.

Berlage, T. (1995). OSF/Motif as a user interface standard. *Computer Standards & Interface, 17*, 99–106.

Besuijen, K., & Spendelink, G. P. J. (1998). Standardizing visual display quality. *Displays, 19*, 67–76.

Button, K., Clarke, A., Palubinskas, G., Stough, R., & Thibault, M. (2004). Conforming with ICAO safety oversight standards. *Journal of Air Transport Management, 70*(4), 249–255.

Cakir, A., & Dzida, W. (1997). International ergonomic HCI standards. In M. Helander, T. K. Landauer, & P. Prabhu (Eds.). *Handbook of human-computer interaction* (pp. 407–420). Amsterdam, The Netherlands: Elsevier.

Calonius, O., & Saikko, V. (2002). Slide track analysis of eight contemporary hip simulator designs. *Journal of Biomechanics, 35*, 1439–1450.

Carson, B. E. Sr., Alper, M., Barrett, C. B., & Brink, K. (2004). ISO 9001:2000—A quality management system (QMS) to make your IVF center better. *Fertility and Sterility, 52*(Suppl. 2), S190–S191.

CEN. (2004). European Standardization Committee website. http://www.cenorm.be/cenorm/index.htm

Chapanis, A. (1996). *Human factors in systems engineering*. New York: Wiley.

Cho, D. S., Kim, J. H., Choi, T. M., Kim, B. H., & Manvell, D. (2004). Highway traffic noise prediction using method fully compliant with ISO 9613: Comparison with measurements. *Applied Acoustics, 65*, 883–892.

Chord-Auger, S., de Bouchony, E. T., Moll, M. C., Boudart, D., & Follea, G. (2014). [Satisfaction survey in general hospital personnel involved in blood transfusion: implementation of the ISO 9001: 2000 standard], *Transfusion Clinique et Biologique, 11*(3), 161–167

Department of Industrial Relation. (2004). California Department of Industrial Relation homepage, http://www.dir.ca.gov/

Dickinson, C. E. (1995). Proposed manual handling international and European Standards. *Applied Ergonomics, 26*(4), 265–270.

Dowlatshahi, S., & Urias, C. (2004). An empirical study of ISO certification in the maquiladora industry. *International Journal of Production Economics, 88*, 291–306.

Dui, J., de Vlaming, P. M., & Munnik, M. J. (1996). A review of ISO and CEN standards on ergonomics. *International Journal of Industrial Ergonomics, 77*(3), 291–297.

Dui, J., de Vries, H., Verschoof, S., Eveleens, W., & Feilzer, A. (2004). Combining economic and social goals in the design of production systems by using ergonomics standards. *Computers & Industrial Engineering, 47*(2–3), 207–222.

Dzida, W. (1995). Standards for user-interfaces. *Computer Standards & Interfaces, 17*(1), 89–97.

Dzida, W. (1997). International user-interface standardization. In A. B. Tucker (Ed.), *The computer science and engineering handbook* (pp. 1474–1493). Boca Raton, FL: CRC Press.

Earthy, J., Jones, B. S., & Bevan, N. (2001). The improvement of human-centred processes—Facing the challenge and reaping the benefit of ISO 13407. *International Journal of Human-Computer Studies, 55*, 553–585.

Eibl, M. (2005). International standards of interface design. In W. Karwowski (Ed.), 2005. *Handbook of human factors and ergonomics standards and guidelines*. Hillsdale, NJ: Lawrence Erlbaum Associates.

Emam, K. E., & Birk, A. (2000). Validating the ISO/IEC 15504 measures of software development process capability. *The Journal of Systems and Software, 51*, 119–149.

Emam, K. E., & Garro, I. (2000). Estimating the extent of standards use: The case of ISO/IEC 15504. *The Journal of Systems and Software, 53*, 137–143.

Emam, K. E., & Jung, H.-W. (2001). An empirical evaluation of the ISO/IEC 15504 assessment model. *The Journal of Systems and Software, 59*, 23–41.

Escanciano, C., Fernandez, E., & Vazquez, C. (2002). Linking the firm's technological status and ISO 9000 certification: Results of an empirical study. *Technovision, 22*, 509–515.

Ghisellini, A., & Thurston, D. L. (2005). Decision traps in ISO 14001 implementation process: Case study results from Illinois certified companies. *Journal of Cleaner Production, 13*, 763–777.

Gingele, J., Childe, S. J., & Miles, M. E. (2002). A modeling technique for re-engineering business processes controlled by ISO 9001. *Computers in Industry, 49*(3), 235–251.

Gordon, C. C., Churchill, T., Clauser, T.E., Bradtmiller, B., McConville, J.T., Tebbets, I., et al. (1989). *1988 Anthropometric survey of U.S. Army personnel: Summary statistics interim report* (Tech. Rep. NATICK/TR-89/027).

Grieco, A., Occhipinti, E., Colombini, D., & Molteni, G. (1997). Manual handling of loads: The point of view of experts involved in the application of EC Directives 90/269. *Ergonomics, 40*(10), 1035–1056.

Griefahn, B., & Brode, P. (1999). The significance of lateral whole-body vibrations related to separately and simultaneously applied vertical motions: A validation study of ISO 2631. *Applied Ergonomics, 30*, 505–513.

Griefahn, B. (2000). Limits of and possibilities to improve the IREQ cold stress model (ISO/TR 11079): A validation study in the field. *Applied Ergonomics, 31*, 423–431.

Harker, S. (1995). The development of ergonomics standards for software. *Applied Ergonomics, 26*(4), 275–279.

Hiyassat, M. A. S. (2000). Applying the ISO standards to a construction company: A case study. *International Journal of Project Management, 18*, 275–280.

Hoyle, D. (2001). *ISO 9000: Quality systems handbook*. Oxford: Butterworth-Heinemann.

Human Factors and Ergonomics Society. (2002). *Board of Standards Review/Human Factors and Ergonomics Society 100—Human factors engineering of computer workstations—Draft Standard for Trial Use*. Human Factors and Ergonomics Society, Santa Monica, CA.

ILO. (2004). International Labor Organization Web site, http://www.ilo.org/public/english/index.htm

ILO-OSH. (2001). *Guidelines on occupational safety and health management systems. ILO-OSH 2001*. Geneva, International Labour Office, http://www.ilo.org/public/english/protection/safework/managmnt/guide.htm

Ishitake, T., Miyazaki, Y., Noguchi, R., Ando, H., & Matoba, T. (2002). Evaluation of frequency weighting (ISO 2631-1) for acute effects of whole-body vibration on gastric motility. *Journal of Sound and Vibration, 253*(1), 31–36.

ISO (2004). International Standardization Organization Web site. http://www.iso.org/iso/en/ISOOnline.openerpage

Jabir, & Moore, J. W. (1998). A search for fundamental principles of software engineering, *Computer Standards & Interfaces, 19*, 155–160.

Jung, H.-W., & Hunter, R. (2001). The relationship between ISO/IEC 15504 process capability levels, ISO 9001 certification and organization size: An empirical study. *The Journal of Systems and Software, 59*, 43–55.

Kampmann, B., & Piekarski, C. (2000). The evaluation of workplaces subjected to heat stress: Can ISO 7933 (1989) adequately describe heat strain in industrial workplaces? *Applied Ergonomics, 31*, 59–71.

Karltun, J., Axelsson, J., & Eklund, J. (1998). Working conditions and effects of ISO 9000 in six furniture-making companies: Implementation and processes. *Applied Ergonomics, 29*(4), 225–232.

Karwowski, W. (Ed.). (in press). *Handbook of human factors and ergonomics standards and guidelines*. Hillsdale, NJ: Lawrence Erlbaum Associates.

Kenny, D. (2001). ISO and CEN documents on quality in medical laboratories. *Clinica Chimica Acta, 309*, 121–125.

Kosanke, K., & Nell, J. G. (1999). Standardization in ISO for enterprise engineering and integration. *Computers in Industry, 40*, 311–319.

Lindermeier, R. (1994). Quality assessment of software prototypes. *Reliability Engineering & System Safety, 43*(1), 87–94.

Lindfors, M. (1998). Accuracy and repeatability of the ISO 9241–7 test method. *Displays, 19*, 3–16.

Long, J. (1996). Specifying relations between research and the design of human-computer interactions. *International Journal of Human-Computer Studies, 44*, 875–920.

MacDonald, J. P. (2005). Strategic sustainable development using the ISO 14001 standard. *Journal of Cleaner Production, 13*, 631–643.

MATRIS. (2004). Directory of Design Support Methods (DSSM). http://dtica.dtic.mil/ddsm/

McDaniel, J. W. (1996). The demise of military standards may affect ergonomics. *International Journal of Industrial Ergonomics, 78*(5–6), 339–348.

Nachreiner, F. (1995). Standards for ergonomics principles relating to the design of work systems and to mental workload. *Applied Ergonomics, 26*(4), 259–263.

Nanthavanij, S. (2000). Developing national ergonomics standards for Thai industry. *International Journal of Industrial Ergonomics, 25*(6), 699–707.

Occupational Safety and Health Administration (OSHA). (2000). Ergonomics program rule (*Federal Register*) *65*(220).

Olesen, B. W., & Parsons, K. C. (2002). Introduction to thermal comfort standards and to the proposed new version of EN ISO 7730. *Energy and Buildings, 34*(6), 537–548.

Olesen, B. W. (1995). International standards and the ergonomics of the thermal environment. *Applied Ergonomics, 26*(4), 293–302.

OSHA. (2004). Occupational Safety and Health Administration Web site, http://www.osha-slc.gov

Parsons, K. (1995). Ergonomics and international standards. *Applied Ergonomics, 26*(4), 237–238.

Parsons, K. C. (1995). Ergonomics of the physical environment: International ergonomics standards concerning speech communication, danger signals, lighting, vibration and surface temperatures. *Applied Ergonomics, 26*(4), 281–292.

Parsons, K. C. (1995). Ergonomics and international standards: Introduction, brief review of standards for anthropometry and control room design and useful information. *Applied Ergonomics, 26*(4), 239–247.

Parsons, K. C. (2000). Environmental ergonomics: A review of principles, methods and models. *Applied Ergonomics, 31*(6), 581–594.

Parsons, K. C., Shackel, B., & Metz, B. (1995). Ergonomics and international standards: History, organizational structure and method of development. *Applied Ergonomics, 26*(4), 249–258.

Public Law 101–336. (1990). Americans with Disabilities Act. Public Law 336 of the 101st Congress, enacted July 26, 1990.

Ragothaman, S., & Korte, L. (1999). The ISO 9000 international quality registration: An empirical analysis of implications for business firms. *International Journal of Applied Quality Management, 2*(1), 59–73.

Raines, S. S. (2002). Implementing ISO 14001—An international survey assessing the benefits of certification. *Corporate Environment Strategy, 9*(4), 418–426.

Reed, R, Holdaway, K., Isensee, S., Buie, E., Fox, J., Williams, J., Lund, A. (1999). User interface guidelines and standards: Progress, issues, and prospects. *Interacting with Computers, 12*(2), 119–142.

Rosenthal, I., Ignatowski, A. J., & Kirchsteiger, C. K. (2002). A generic standard for the risk assessment process: Discussion on a proposal made by the program committee of the ER-JRC workshop on "Promotion of Technical Harmonization of Risk-base Decision Making." *Safety Science, 40*, 75–103.

Saito, S. Piccoli, B., Smith, M. J., Sotoyama, M., Sweitzer, G., Villanueva, M. B. G., et al. (2000). Ergonomic guidelines for using notebook personal computers. *Industrial Health, 38*, 421–434.

Seabrook, K. A. (2001). International Standards Update: Occupational Safety and Health Management Systems. In Proceedings of the *American Society of Safety Engineers' 2001 Professional Development Conference*, Anaheim, CA.

Sherehiy, B., Karwowski, W., & Rodrick, D. (2006). Human factors and ergonomics standards, In G. Salvendy (Ed.), *Handbook of human factors and ergonomics*, New York: Wiley.

Smith, W. J. (1996). *ISO and ANSI ergonomic standards for computer products*. Prentice Hall: Upper Saddle River, NJ.

Spivak, S. M., & Brenner, F. C. (2001). *Standardization essentials: Principles and practice*. New York: Dekker.

Staccini, P., Joubert, M., Quaranta, J.-F., & Fieschi, M. (2005). Mapping care processes within a hospital: From theory to a web-based proposal merging enterprise modelling and ISO normative principles. *International Journal of Medical Informatics, 74*, 335–344.

Stevenson, T. H., & Barnes, F. C. (May-June 2001). Fourteen years of ISO 9000: Impact, criticisms, costs, and benefits. *Business Horizons, 44*, 45–51.

Stewart, T. (1995). Ergonomics standards concerning human-system interaction: Visual displays, controls and environmental requirements. *Applied Ergonomics, 26*(4), 271–274.

Stuart-Buttle, C. (2006). Overview of international standards and guidelines. In W. Karwowski (Ed.), *Handbook of human factors and ergonomics standards and guidelines*. Hillsdale, NJ: Lawrence Erlbaum Publishers.

Umezu, N., Nakano, Y., Sakai, T., Yoshitake, R., Herlitschke, W., & Kubota, S. (1998). Specular and diffuse reflection measurement feasibility study of ISO 9241 Part 7. Method. *Displays, 19*, 17–25.

Vikari-Juntura, E. R. A. (1997). The scientific basis for making guidelines and standards to prevent work-related musculoskeletal disorders. *Ergonomics, 40*(10), 1097–1117.

Walker, A. J. (1998). Improving the quality of ISO 9001 audits in the field of software. *Information and Software Technology, 40*, 865–869.

Wegner, E. (1995). Quality of software packages: The forthcoming international standard. *Computer Standards & Interfaces, 17*, 115–120.

Welzel, D., & Hausen, H-L. (1995). A method for software evaluation: Contribution of the European project SCOPE to international standards. *Computer Standards & Interfaces, 17*, 121–129.

Wettig, J. (2002). New developments in standardization in the past 15 years—Product versus process related standards. *Safety Science, 40*(1–4), 51–56.

Zuo, L., & Nayfeh, S. A. (2003). Low order continuous-time filters for approximation of the ISO 2631-1 human vibration sensitivity weighting. *Journal of Sound and Vibration, 265*, 459–465.

Index

Printed in the United States
by Baker & Taylor Publisher Services